Medicinal Plant Responses to Stressful Conditions

Medicinal Plant Responses to Stressful Conditions discusses the effects of multiple biotic and abiotic stressors on medicinal plants. It features information on biochemical, molecular and physiological strategies used to mitigate or alleviate detrimental effects of biotic and abiotic stressors. The book contains chapters featuring medicinal plants of importance covering subjects including genomics, functional genomics, metabolomics, phenomics, proteomics and transcriptomics under biotic and abiotic stress of medicinal plants and their molecular responses. It suggests exogenous application of different types of stimulants to enhance medicinal plant production in such conditions.

Features:

- Details all aspects of biotic and abiotic stressors in various important medicinal plant species.
- Chapters cover evidence-based approaches in the diagnosis and management of medicinal plants under stressful conditions.
- Includes information on ways to mitigate effects from biotic stress (diseases and pests) or abiotic stress (high salinity, drought, temperature extremes, waterlogging, wind, high light intensity, UV radiation, heavy metals and mineral deficiencies).

A volume in the *Exploring Medicinal Plants* series, this book is an essential resource for plant scientists, botanists, environmental scientists and anyone with an interest in herbal medicine.

Exploring Medicinal Plants

Series Editor
Azamal Husen
Wolaita Sodo University, Ethiopia

Medicinal plants render a rich source of bioactive compounds used in drug formulation and development; they play a key role in traditional or indigenous health systems. As the demand for herbal medicines increases worldwide, supply is declining as most of the harvest is derived from naturally growing vegetation. Considering global interests and covering several important aspects associated with medicinal plants, the Exploring Medicinal Plants series comprises volumes valuable to academia, practitioners and researchers interested in medicinal plants. Topics provide information on a range of subjects including diversity, conservation, propagation, cultivation, physiology, molecular biology, growth response under extreme environment, handling, storage, bioactive compounds, secondary metabolites, extraction, therapeutics, mode of action and healthcare practices.

Led by Azamal Husen, PhD, this series is directed to a broad range of researchers and professionals consisting of topical books exploring information related to medicinal plants. It includes edited volumes, references and textbooks available for individual print and electronic purchases.

Traditional Herbal Therapy for the Human Immune System
Azamal Husen

Environmental Pollution and Medicinal Plants
Azamal Husen

Herbs, Shrubs and Trees of Potential Medicinal Benefits
Azamal Husen

Phytopharmaceuticals and Biotechnology of Herbal Plants
Sachidanand Singh, Rahul Datta, Parul Johri, and Mala Trivedi

Omics Studies of Medicinal Plants
Ahmad Altaf

Exploring Poisonous Plants: Medicinal Values, Toxicity Responses, and Therapeutic Uses
Azamal Husen

Plants as Medicine and Aromatics: Conservation, Ecology, and Pharmacognosy
Mohd Kafeel Ahmad Ansari, Bengu Turkyilmaz Unal, Munir Ozturk and Gary Owens

Sustainable Uses of Medicinal Plants
Learnmore Kambizi and Callistus Bvenura

Medicinal Plant Responses to Stressful Conditions
Arafat Abdel Hamed Abdel Latef

Medicinal Plant Responses to Stressful Conditions

Edited by
Arafat Abdel Hamed Abdel Latef

CRC Press
Taylor & Francis Group
Boca Raton London New York

CRC Press is an imprint of the
Taylor & Francis Group, an **informa** business

First edition published 2023
by CRC Press
6000 Broken Sound Parkway NW, Suite 300, Boca Raton, FL 33487-2742

and by CRC Press
4 Park Square, Milton Park, Abingdon, Oxon, OX14 4RN

CRC Press is an imprint of Taylor & Francis Group, LLC

© 2023 Taylor & Francis Group, LLC

ISBN: 9781032151960 (hbk)
ISBN: 9781032151984 (pbk)
ISBN: 9781003242963 (ebk)

DOI: 10.1201/9781003242963

Typeset in Times
by Newgen Publishing UK

Contents

Editor Bio

Professor Arafat Abdel Hamed Abdel Latef is a professor of plant physiology at the Department of Botany and Microbiology, Faculty of Science, South Valley University, Egypt. He received his BSc in 1998, M.Sc. in 2003 and Ph.D. in 2005 from South Valley University, Egypt. Prof. Arafat has published more than 80 research/review articles in high-quality international and local journals. He has edited one book for CRC Press, Taylor & Francis Group, LLC (published in July 2021) and one book for Springer-Verlag, Heidelberg (published in November 2022). He has also published 20 book chapters in international volumes including John Wiley and Sons Inc, Taylor & Francis Group, Academic Press (Elsevier) and Springer-Verlag, Heidelberg.

Prof. Arafat is the recipient of the ParOwn 1207 Post Doctor Fellowship 2007 granted by the Egyptian Ministry of Higher Education and Scientific Research, which was carried out in the Institute of Vegetables and Flowers, Chinese Academy of Agricultural Sciences, Beijing, China. He is serving as an editorial member for 15 reputed journals and a guest editor for four special issues in *Plants—Basel (2), Frontiers in Plant Science (1)* and *Phyton-International Journal of Experimental Botany (1)*. Furthermore, he is an expert reviewer for 120 international journals. Recently, he received the Top Peer Reviewer Award 2019 Powered by Publons.

Prof. Arafat has received several awards such as the Egypt State Award for Excellence 2021, South Valley University Award for Excellence 2019 and South Valley University Award for Scientific Publication (2010–2022). Prof. Arafat is actively engaged in studying the physio-biochemical and molecular responses of different plants under environmental stresses and their tolerance strategies under these stressors.

Contributors

Arafat Abdel Hamed Abdel Latef
Botany and Microbiology Department,
Faculty of Science, South Valley University,
Qena 83523, Egypt

Sawsan Abd-Ellatif
Bioprocess Development Department, Genetic
Engineering and Biotechnology Research
Institute, City of Scientific Research and
Technology Applications, Borg EL-Arab,
21934, Alexandria, Egypt

Elsayed S. Abdel Razik
Plant Protection and Biomolecular Diagnosis
Department, Arid Lands Cultivation
Research Institute, City of Scientific
Research and Technology Applications,
Borg EL-Arab, 21934, Alexandria, Egypt

Amir Abdullah Khan
Department of Plant Biology and Ecology,
Nankai University Tianjin 300071, China

Kaoutar Aboukhalid
National Institute of Agronomic Research
(INRA), 10 Bd. Mohamed VI, P.B 428,
Oujda 60000, Morocco

Faijuddin Ahammad
Department of Plant Pathology, Sam
Higginbottom University of Agriculture,
Technology and Sciences, Allahabad,
Uttar Pradesh, India

Hassan Ahmed Ibraheem Ahmed
Department of Botany, Faculty of Science,
Port Said University, Port Said,
42526, Egypt

Sk. Md. Ajaharuddin
Department of Agricultural Entomology,
Bidhan Chandra Krishi Viswavidyalaya
(BCKV), Mohanpur, Nadia, West Bengal,
India

Chaouki Al Faiz
National Institute of Agronomic Research,
CRRA-Rabat, PB 6570, 10101 Rabat,
Morocco

Haifa Abdulaziz Sakit AlHaithloul
Biology Department, College of Science,
Jouf University, Sakaka 2014, Kingdom of
Saudi Arabia

Wardah A. Alhoqail
Department of Biology, College of Education,
Majmaah University, 11952, Kingdom of
Saudi Arabia

Iftikhar Ali
Center for Plant Science and Biodiversity,
University of Swat, Charbagh, 19120
Pakistan

Mahatab Ali
Department of Genetics and Plant Breeding,
Bidhan Chandra Krishi Viswavidyalaya
(BCKV), Mohanpur, Nadia, West Bengal,
India

Ayshah Aysh ALrashidi
Department of Biology, Faculty of Science,
University of Hail, Hail, 81411, Saudi Arabia

Omar Mahmoud Al zoubi
Biology Department, Faculty of Science,
Taibah University, Yanbu El-Bahr,
Yanbu 46429, Kingdom of Saudi Arabia

Fatima Amanullah
Sustainable Development Study Centre,
Government College University,
Lahore-54000, Pakistan

Godwin Anywar
Department of Plant Sciences, Microbiology
and Biotechnology, Makerere University,
P.O. Box 7062, Kampala, Uganda

Mohammed Ater
Laboratory of Applied Botany,
   BioAgrodiversity Team, Faculty of Sciences
   of Tétouan, Abdelmalek Essaâdi University,
   Tétouan, Morocco

Rachid Azenzem
National Institute of Agronomic Research,
   INRA-CRRA, Marrakech-Safi, BP 533,
   Marrakech, Morocco

Mohamed Bakha
Unit of Plant Biotechnology and Sustainable
   Development of Natural Resources
   "B2DRN," Polydisciplinary Faculty of Beni
   Mellal, Sultan Moulay Slimane University,
   Mghila, PO Box 592, Beni Mellal 23000,
   Morocco

Ali Bandehagh
Department of Plant Breeding and
   Biotechnology, Faculty of Agriculture,
   University of Tabriz, Tabriz, Iran

Arghya Banerjee
Department of Plant Pathology, School of
   Agriculture and Allied Science, The Neotia
   University, Jhinga, South 24 Parganas,
   West Bengal, India

Soham Barik
Department of Agronomy, ChatrapatiSahu Ji
   Maharaj University, Kanpur, Uttar Pradesh,
   India

Mainak Barman
Department of Genetics and Plant Breeding,
   Bidhan Chandra KrishiViswavidyalaya
   (BCKV), Mohanpur, Nadia, West Bengal,
   India

Abdelhakim Bouyahya
Laboratory of Human Pathologies Biology,
   Department of Biology, Faculty of Sciences
   and Genomic Center of Human Pathologies,
   Mohammed V University, Rabat P.O. Box
   1014, Morocco

Soufian Chakkour
Laboratory of Applied Botany,
   BioAgrodiversity Team, Faculty of Sciences

of Tétouan, Abdelmalek Essaâdi University,
   Tétouan, Morocco

Liping Chen
Research Center of Intelligent Equipment,
   Beijing Academy of Agriculture and Forestry
   Sciences, Beijing, China

Mona F.A. Dawood
Botany and Microbiology Department, Faculty
   of Science, Assiut University (71615),
   Assiut, Egypt

Zahra Dehghanian
Department of Biotechnology, Faculty of
   Agriculture, Azarbaijan Shahid Madani
   University, Tabriz, Iran

Noureddine El Mtili
Laboratory of Biology and Health, Faculty of
   Sciences of Tétouan, Abdelmalek Essaâdi
   University, BP 2121, 93002 Tétouan,
   Morocco

Nasreddine El Omari
Laboratory of Histology, Embryology, and
   Cytogenetic, Faculty of Medicine and
   Pharmacy, Mohammed V University in
   Rabat, Morocco

Abdeltif El Ouahrani
Laboratory of Applied Botany,
   BioAgrodiversity Team, Faculty of Sciences
   of Tétouan, Abdelmalek Essaâdi University,
   Tétouan, Morocco

A. K. M. Golam Sarwar
Laboratory of Plant Systematics, Department
   of Crop Botany, Bangladesh Agricultural
   University, Mymensingh 2202, Bangladesh

Subrata Goswami
Department of Agricultural Entomology,
   Jawaharlal Nehru Krishi Vihwavidyalaya,
   Jabalpur, Madhya Pradesh, India

Khashayar Habibi
Department of Biotechnology, College
   of Agriculture, Isfahan University of
   Technology, Isfahan, Iran

Maksud Hasan Shah
Department of Agronomy, Bidhan Chandra
    Krishi Viswavidyalaya (BCKV), Mohanpur,
    Nadia, West Bengal, India

Hamed Hassanzadeh Khankahdani
Horticulture Crops Research Department,
    Hormozgan Agricultural and Natural
    Resources Research and Education Center,
    AREEO, Bandar Abbas, Iran

Peichen Hou
Research Center of Intelligent Equipment,
    Beijing Academy of Agriculture and Forestry
    Sciences, Beijing, China

Akbar Hossain
Bangladesh Wheat and Maize Research
    Institute, Dinajpur 5200, Bangladesh

Mhammad Houssni
Laboratory of Applied Botany,
    BioAgrodiversity Team, Faculty of Sciences
    of Tétouan, Abdelmalek Essaâdi University,
    Tétouan, Morocco

Amira A. Ibrahim
Botany and Microbiology Department,
    Faculty of Science, Arish University,
    Al-Arish 45511, Egypt

Saidul Islam
Nadia Krishi Vigyan Kendra, Bidhan Chandra
    Krishi Viswavidyalaya (BCKV), Gayeshpur,
    Nadia, West Bengal, India

Khalil Kadaoui
Laboratory of Applied Botany,
    BioAgrodiversity Team, Faculty of Sciences
    of Tétouan, Abdelmalek Essaâdi University,
    Tétouan, Morocco

Jalal Kassout
Laboratory of Applied Botany,
    BioAgrodiversity Team, Faculty of Sciences
    of Tétouan, Abdelmalek Essaâdi University,
    Tétouan, Morocco; National Institute of
    Agronomic Research, INRA-CRRA,
    Marrakech-Safi, BP 533, Marrakech,
    Morocco

Amrani Joutei Khalid
Laboratory of Microbial Biotechnology and
    Bioactive Molecules: Faculty of Science
    and Technology Fez, Sidi Mohamed Ben
    Abdellah University, B.P. 2202-Road of
    Imouzzer, Fez, Morocco

Sheharyar Khan
Department of Botany, University of Peshawar,
    Khyber Pakhtunkhwa, Pakistan

Waqas-ud-Din Khan
Sustainable Development Study Centre,
    Government College University,
    Lahore-54000, Pakistan

Abdelkarim Khiraoui
Ecology and Biodiversity Team, Faculty of
    Sciences and Technology, University of
    Sultan Moulay Slimane, BP 523,
    Beni-Mellal, Morocco

Santanu Kundu
Department of Agronomy, Professor
    JayashankarTelangana State Agricultural
    University, Hyderabad, Telangana, India

Aixue Li
Research Center of Intelligent Equipment,
    Beijing Academy of Agriculture and Forestry
    Sciences, Beijing, China

Bin Luo
Research Center of Intelligent Equipment,
    Beijing Academy of Agriculture and Forestry
    Sciences, Beijing, China

Suliman Mohammed Alghanem
Biology Department, Faculty of Science, Tabuk
    University, Tabuk 71491, Saudi Arabia

Karine Pedneault
Centre de Recherche et d'Innovation sur les
    Végétaux, Département de Phytologie,
    Université Laval, Québec, QC, Canada

Yuanying Peng
State Key Laboratory of Crop Gene Exploration
    and Utilization in Southwest China, Sichuan
    Agricultural University, Chengdu, China

Kalipada Pramanik
Assistant Professor, Department of
   Agronomy, PalliSikshaBhavana, Institute
   of Agriculture(Visva- Bharati), Sriniketan,
   Birbhum, West Bengal, India

Mei Qu
International Research Centre for
   Environmental Membrane Biology, Foshan
   University, Foshan, China

Samah Ramadan
Botany Department, Faculty of Science,
   Mansoura University, Mansoura 35516,
   Egypt

SK Sadikur Rahman
Assistant Technology Manager,
   Department of Agriculture,
   Howrah, West Bengal, India

Shoumik Saha
Department of Genetics and Plant Breeding,
   Bidhan Chandra Krishi Viswavidyalaya
   (BCKV),Mohanpur, Nadia, West Bengal,
   India

Abdelouahab Sahli
Laboratory of Applied Botany,
   BioAgrodiversity Team, Faculty of Sciences
   of Tétouan, Abdelmalek Essaâdi University,
   Tétouan, Morocco

Khaled F.M. Salem
Department of Plant Biotechnology, Genetic
   Engineering and Biotechnology Research
   Institute (GEBRI), University of Sadat City,
   Sadat City, Egypt
Current address: Department of Biology,
   College of Science and Humanitarian
   Studies, Shaqra University, Qwaieah,
   Saudi Arabia

Hoda H. Senousy
Botany and Microbiology Department, Faculty
   of Science, Cairo University, Giza 12613,
   Egypt

Fuchen Shi
Department of Plant Biology and
   Ecology, Nankai University Tianjin
   300071, China

Seyedeh-Somayyeh Shafiei-Masouleh
Department of Genetics and Breeding,
   Ornamental Plants Research Center (OPRC),
   Horticultural Sciences Research Institute
   (HSRI), Agricultural Research, Education
   and Extension Organization (AREEO),
   Mahallat, Iran

Sikandar Shah
Department of Botany, University of Peshawar,
   Khyber Pakhtunkhwa, Pakistan

Mona H. Soliman
Biology Department, Faculty of Science,
   Taibah University, Yanbu El-Bahr,
   Yanbu 46429, Kingdom of Saudi Arabia;
   Botany and Microbiology Department,
   Faculty of Science, Cairo University,
   Giza 12613, Egypt

Jamilu Edirisa Ssenku
Department of Plant Sciences, Microbiology
   and Biotechnology, Makerere University,
   P.O. Box 7062, Kampala, Uganda

Sulaiman
Department of Botany, University of Peshawar,
   Khyber Pakhtunkhwa, Pakistan

Jean-Frédéric Terral
ISEM, Université de Montpellier,
   Equipe DBA, CNRS, IRD, EPHE,
   Montpellier, France

Patience Tugume
Department of Plant Sciences, Microbiology
   and Biotechnology, Makerere University,
   P.O. Box 7062, Kampala, Uganda

Ping Yun
Tasmanian Institute of Agriculture, University
   of Tasmania, Tasmania 7001, Australia

Hazzoumi Zakaria
Laboratory of Plant and Microbial
  Biotechnology, Moroccan Foundation for
  Advanced Science, Innovation and Research,
  Rabat Design Center, Madinat Al Irfan,
  Rabat, Morocco

Es-sbihi Fatima Zohra
Laboratory of Microbial Biotechnology and
  Bioactive Molecules: Faculty of Science
  and Technology Fez, Sidi Mohamed Ben
  Abdellah University, B.P. 2202-Road of
  Imouzzer, Fez, Morocco

1 *Allium cepa* under Stressful Conditions

Mona F. A. Dawood[1*] *and Arafat Abdel Hamed Abdel Latef*[2]
[1]Botany and Microbiology Department, Faculty of Science,
Assiut University, Assiut, Egypt
[2]Botany and Microbiology Department, Faculty of Science,
South Valley University, Qena 83523, Egypt
*Corresponding author: MFAD; Mo_fa8@aun.edu.eg

CONTENTS

1.1 INTRODUCTION

Onion is one of the oldest vegetables, and has been reported in many ancient scriptures (Singh, 2008). By the Middle Ages, it had become one of the fundamentals in many cuisines in most parts of the world and therefore is always in demand throughout the year. Onion is ranked second after tomato (*Solanum lycopersicum*) (Rady et al. 2018), and has been widely used as a vegetable, spice, and medicine since time immemorial (Teshika et al. 2019). Green leaves and green and/or dry onion bulbs are rich sources of vitamins, minerals, carbohydrates, proteins, and antioxidants that play a central role in protecting humans from a number of diseases, such as chickenpox, influenza, cancer, diabetes, high blood pressure, and cardiovascular disorders (Teshika et al. 2019; Oboh et al. 2019; Kothari et al. 2020). The current worldwide onion production is estimated to be 78.31 million tons, with average productivity of 19.79 t/ha (FAO 2015). With the world's population rising at an exponential rate, market demand for onions is increasing. In addition to biotic stress, onions are highly vulnerable to abiotic stresses such as extreme temperature injuries, drought, and waterlogging (Ghodke et al. 2018). However, abiotic stresses have an adverse impact on the growth, development and yield of onion around the world (Ratnarajah and Gnanachelvam 2021). Thus, it is important to know the adverse impacts caused by the environment through the production of onions. Stress due to biotic and abiotic factors is among the major constraints in exploiting the yield potential of the onion crop.

1.2 CHEMICAL COMPOSITION AND MEDICINAL/NUTRITIVE USES OF *ALLIUM CEPA*

Onion (*Allium cepa* L.) belongs to the Family: Liliaceae. It is an easily digestible aromatic vegetable that is used throughout the world. Onion is one of the most consumed and grown vegetable crops in the world. The onion bulb, with its characteristic flavor, is the third most essential horticultural spice with a substantial commercial value (Teshika et al. 2019).

Various studies have explored its chemical analysis, flavor and discoloration precursors (Corzo-Martínez et al. 2007; Dong et al. 2010; Wiczkowski 2011; Kato et al. 2013). A wide range of phytocompounds, including phenolic acids, flavonoids (quercetin and kaempferol), anthocyanins and organosulfur compounds, have been identified in onion. These compounds have potential anti-inflammatory, anticholesterol, anticancer and antioxidant properties (Teshika et al. 2019). The onion bulb and skin contain various bioactive compounds, such as organosulfur compounds (OSCs), thiosulfinates, polyphenols, including flavonoids, and fructooligosaccharides (Sagar et al. 2022) and among them, flavonoids are the most effective bioactive compounds. Among the different classes of phytochemicals, phenolic compounds are important compounds due to their contribution to the biological properties of plants. A study by Prakash et al. (2007) was conducted on four varieties of *A. cepa* (red, violet, white and green) for their respective phenolic composition where ferulic acid, gallic acid, protocatechuic acid, quercetin and kaempferol were identified. Moreover, a number of flavonoids were also detected in different onion varieties: quercetin aglycon, quercetin-3,4′-diglucoside, quercetin-4′-monoglucoside, quercetin-3-monoglucoside (Zill-e et al. 2011), quercetin 3-glycosides, delphinidin 3,5-diglycosides (Zhang et al., 2016), quercetin 3,7,4′-triglucoside, quercetin 7,4′-diglucoside, quercetin 3,4′-diglucoside and isorhamnetin 3,4′-diglucoside (Pérez-Gregorio et al. 2010). Furthermore, various anthocyanin compounds have been identified in minute amounts from pigmented parts of red onion: cyanidin 3-O-(3″-O-β-glucopyranosyl-6″-O-malonyl-β-glucopyranoside)-4′-O-β-glucopyranoside, cyaniding 7-O-(3″-O-β-glucopyranosyl-6″-O-malonyl-β-glucopyranoside)-4′-O-β glucopyranoside, cyanidin 3,4′-di-O-β-glucopyranoside, cyanidin 4′-O-β-glucoside, peonidin 3-O-(6″-Omalonyl-β glucopyranoside)-5-O-β-glucopyranoside and peonidin 3-O-(6″-O-malonyl-β-glucopyranoside) (Pérez-Gregorio et al., 2010). Also, Fossen et al. (2003) reported four anthocyanins with 4-substituted aglycone, carboxypyranocyanidin, were identified from the red onion-methanolic extract. The structures of two of them were identified as 5-carboxypyranocyanidin 3-O-(6"-O-malonyl-β-glucopyranoside and 5-carboxypyranocyanidin 3-O-β-glucopyranoside. Moreover, Fredotović et al. (2017) reported the presence of peonidin 3′-glucoside petunidin 3′-glucoside acetate and malvidin 3′-glucoside. Also, ascorbic acid and fructooligosaccharides have been detected in onion (Sagar et al. 2022).

Ouyang et al. (2017) reported the 2,2-diphenyl-1-picrylhydrazyl (DPPH) radical scavenging activity, ferric reducing antioxidant power (FRAP) radical scavenging activity and hydroxyl radicle ($^\cdot$OH) radical scavenging activity of total polyphenols from onion. Dietary antioxidants play a crucial role in the suppression of oxidative stress, which may cause the initiation and progression of several diseases, including cancer, diabetes, inflammation and cardiovascular diseases (Razavi-Azarkhiavi et al., 2014). Ma et al. (2018) found that the polysaccharides extracted from *A. cepa* displayed strong antioxidant activity toward 2,2′-azinobis(3-ethylbenzothiazoline-6-sulfonic acid) (ABTS) radical cations, Fe^{2+} chelating and superoxide anion radical scavenging. The environment, cultivar type, agronomic practices, maturation stage and storage duration have significant effects on the bioactive compounds of onion (Galdón et al. 2009). Waste onion skin also contains a higher level of flavonoids than the edible part (Duan et al. 2015) due to the oxidation of quercetin flavonol into 3,4-hydroxybenzoic acid and 2,4,6- trihydroxyphenylglycosilic acid and concentrated in dry onion skin to protect the bulb from soil microbes (Takahama and Hirota 2000).

Allium cepa has been considered a potent antimicrobial agent against infectious diseases. Many bacteria, fungi and viruses have been found to be susceptible to different solvents extracts of *A. cepa* (Benmalek et al. 2013; ur Rahman et al. 2017). Sulfur compounds have proven to be the principal active antimicrobial agents that occur in onion (Rose et al., 2005). Many studies have reconsidered the effect of organosulfur-containing compounds on the growth of microorganisms (Liguori et al. 2017). An in vivo study by ur Rahman et al. (2017) showed that birds fed with onion at a rate of 2.5 g/kg of feed had a decrease in *E. coli* population and a significant increase in *Lactobacillus* spp.

1.3 RESPONSES OF *ALLIUM CEPA* TO VARIOUS STRESSFUL CONDITIONS

Because of the increasing tendency toward modern lifestyles and the use of land for building construction, agriculture is very much in a dangerous situation. In addition, extreme weather events such as floods, droughts, hurricanes and sea-level increases and salinity intrusion are key factors that influence the yield of crops such as onion. Furthermore, the consequences of biotic stress as fungal, bacterial, viral and insect invasion adversely affect the yield and quality of the bulb. As a result, the adaptation of agriculture to climate change is of crucial importance for nutritionally important crops like *Allium cepa*. *Allium cepa* is of significance because its inclusion in the diet helps improve the activity of the immune system to enable good health. Therefore, there is increasing momentum in studying the effects of biotic and abiotic stresses on immune-boosting food crops like onion and also in developing rescue strategies that will help policymakers to suggest better guidelines for food security. In the following sections, we will review the abiotic and biotic stressors on onion plants and mitigation strategies.

1.4 RESPONSES OF *ALLIUM CEPA* TO ABIOTIC STRESSORS AND SOME MITIGATION AGENTS

Abiotic stresses are now one of the major constraints to global crop development. A considerable portion of the population in developing countries where subsistence agriculture still exists is constantly threatened by abiotic stress factors and their interactions with biotic stress factors. As a result of climate change, the situation is likely to worsen. With the predicted increment in the global population and food demand, it will be important to enhance crop tolerance to abiotic stress factors to improve agricultural production and food security (Calanca et al. 2017). Drought, waterlogging, heat, cold and salinity stresses and their adverse effects on onion are discussed in the following subsections.

1.4.1 SALINITY

Salinity is one of the major abiotic stresses that significantly affects plant growth and yield (Sofy et al. 2021; Dawood 2022; Dawood et al. 2022; Ragaey et al. 2022). The continuous increase in salinity in arable land due to poor cultivation practices and climate change has devastating global effects, and it is estimated that about 50% of arable land will be lost by the middle of the 21st century (Islam et al. 2019). To date, about 1,125 million hectares of agricultural land have already been seriously affected by salinity, thus it is considered a serious threat to agriculture (Islam et al. 2019; Sanower-Hossain 2019). According to the United States Department of Agriculture (USDA), onions are more sensitive to salinity compared to other vegetables. Soil salinity affects the growth and photosynthetic metabolism of onions (Beinşan et al. 2015). Salt stress reduces plant growth and yield in several ways. Osmotic stress and ionic toxicity are two main effects of salt stress on crops. The osmotic pressure under salinity stress in a soil solution exceeds the osmotic pressure in plant cells due to the presence of more salt and thus limits the ability of plants to absorb water and minerals like K^+ and Ca^{2+} (Dawood 2022; Sheteiwy et al. 2022). The side effects caused by the primary effects of salinity stress are assimilate production, reduced cell expansion and membrane function, and reduced cytosolic metabolism (Sheteiwy et al. 2022). Onions are salt- and sulfate-sensitive (Ryang et al. 2009). Salinity has a significant effect on the number of bulbs per unit area, size and fresh weight of onion bulbs. It influences the bulbing and quality of harvested bulbs. Also, salinity at different growth stages of onion influences the fresh weight of bulbs at harvest (Sta-Baba et al. 2010). Camilia et al. (2013) showed that irrigating onions with saltwater reduced plant growth and biomass production when compared to tap water irrigation. Salinity also affects the flavor development and mineral content of onion bulbs.

Chang and Randle (2004) found that the application of different NaCl concentrations on onion cultivar "Granex 33" variety caused a decrement in bulb fresh weight of mature plants, and plants did not survive at 125 mM NaCl concentrations. A high NaCl content also reduces the sulfur content of bulbs, and enhances the bulb pungency, without affecting the soluble solid content of bulbs. The study by Pessoa et al. (2019) displayed that an increase of salt in the irrigation water led to a significant reduction in fresh and dry matter of the onion plants (bulbs and roots), as well as the fresh matter of the leaves, in addition to reducing the number and height of leaves and bulb diameter. There was a reduction in K and Mg contents in onion bulb dry matter concomitant with the disproportionate presence of Na^+ in bulb tissues. On the other hand, an increase in the Ca, Cl and Na levels was observed in the onion bulb dry matter, possibly through the supply of these elements in the irrigation waters that were prepared with NaCl and $CaCl_2$. This indicates the risk of the use of saline waters for the nutritional balance of onions and, consequently, for their productivity. Shoaib et al. (2018) observed that onion plants under salinity stress showed a reduction in osmotic potential, total chlorophyll content, membrane stability index and total protein content of the onion leaves, while total phenolics and sugars were increased. Salinity stress caused an increment of the activities of peroxidase and polyphenol oxidase, however, phenylalanine ammonia-lyase (PAL) and catalase (CAT) declined. Aghajanzadeh et al. (2019) reported that the exposure of onion to NaCl and Na_2SO_4 salinity resulted in decreased plant biomass production by 40 and 60% at 100 and 200 mM NaCl and by 35 and 55% at 50 at 100 mM Na_2SO_4, respectively. The exposure of plants to 200 mM NaCl and 100 mM Na_2SO_4 resulted in 35 and 50% increases in the root dry matter content, respectively. The dry matter content of the shoot was increased by 34 and 56% at 100 and 200 mM NaCl and 24 and 52% at 50 and 100 mM Na_2SO_4, respectively. Sulfate salinity resulted in an up to 3-fold and 1.5-fold increase in the total sulfur content of the root and shoot, respectively. This increase can mostly be ascribed to an accumulation of sulfate. Exposure of plants to 50 and 100 mM Na_2SO_4 resulted in a 2.5- and 3-fold increase in sulfate content of the root and 2- and 2.5-fold increase in its content of the shoot, respectively. The sulfate content of the root was 5-fold higher than that of the shoot, which may indicate that the translocation of sulfate from the root to the shoot of onion was restricted upon sulfate salinity. The chloride contents of both root and shoot were not affected by sulfate salinity. Also, both NaCl and Na_2SO_4 salinity resulted in a strongly reduced nitrate content of the shoot, whereas that of the root remained unaffected. Sodium salinity resulted in a strongly enhanced sodium content of both the root and shoot of onion. In the root, exposure to 200 mM NaCl and 100 mM Na_2SO_4 led to a slightly higher increase in sodium content. The exposure to both sodium salts strongly decreased the potassium and calcium content in shoot and root. Magnesium was similarly decreased by NaCl and Na_2SO_4 in shoot. Exposure to both sodium salts significantly decreased manganese content in root. Molybdenum content was decreased significantly in root and shoot by sulfate salt. Phosphorus, copper, iron and zinc contents remained unaffected by both salts in shoot and root. The alliin content of the root was only enhanced at 200 mM NaCl and 100 mM Na_2SO_4, and that of shoot was only significantly increased at 100 mM Na_2SO_4.

The application of H_2O_2 to onion plants potentially minimized the damaging impact of salinity stress in onion by increasing the photosynthetic efficiency (Semida, 2016). Semida et al. (2016) reported that external use of α-tocopherol improved photosynthetic performance (chlorophyll and carotenoid contents in plants), endogenous turgor status, enzymatic and non-enzymatic antioxidant activities/levels and decreased lipid peroxidation and H_2O_2 formation in salt-stressed onion plants. Mohamed and Aly (2008) applied α-tocopherol to mitigate the harmful effects of salt stress on onion plants and found enhanced levels of total phenolics and flavonoids in the plants.

1.4.2 DROUGHT

Onion is an irrigated crop, which consumes a large volume of irrigation water for its production. Poor water availability results in low productivity. High soil moisture is required for onions to produce a high yield (Kadayifci et al. 2005). Onions are considered a shallow-rooted crop, with most penetration up to 18 cm and very few roots extended up to 31 cm, thus it extracts very little water from depths beyond 60 cm. Potopova et al. (2018) stated that drought stress causes approximately 30% yield losses in onion. Previous studies have shown that drought stress significantly reduces onion bulb yield (Ghodke et al. 2018; Wakchaure et al. 2018). Srinivasa Rao et al. (2016) found that most of the soil water is absorbed from the top 30 cm, therefore, it is important to keep the soil moist to provide enough water for the plant. Srinivasa Rao (2016) also reported that when drought stress was applied for 1, 2, 3 and 4 weeks on onion cultivars, there was a significant decrease in bulb fresh and dry mass and bulb yield. Bulb dry matter was attenuated by 44.4–54.0% after 3 weeks of stress. Drought stress severely affects the plant height of onion plants, leading to stunted growth (Ghodke et al. 2018). Bekele and Tilahun (2007) reported that up to 23% reduction in water use efficiency and bulb weight was recorded for onion grown under deficit irrigation treatments. Different growth stages of onions that are subjected to soil-drought stress have a great impact on the yield and quality of bulbs. Zayton (2007) found that soil water stress applied at the later stage of onion growth produced a lower yield by impacting the weight and size of onion bulbs but the application of water-stressed plants at the initial stage has a lower effect on bulbs.

Piri and Naserin (2020) reported that reducing the amount of irrigation water can minimize the onion bulb yield and increase (IWUE). Physiological properties of bulbs are also affected by drought stress, in this regard, Wakchaure et al. (2018) reported that the growth rate, onion bulb yield, leaf area index and relative water content were limited by water stress under deficit irrigation. Lower RWC is the main factor accountable for the lowering of the transpiration rate, resulting in a higher canopy temperature at severe water deficits. Also, severe water stress adversely affected the physicochemical properties of the onion bulb in terms of total soluble sugar content, increased pungency values and total protein content of onion. Also, the same study denoted that the small-sized onion bulb obtained at higher water deficits contained higher total phenolic compounds compared to a larger onion bulb without stress. Drought stress negatively influenced the plant phenotype by decreasing plant height, photosynthetically active leaves and leaf area in onion genotypes. A reduction in the number of leaves and chlorophyll pigments, and a higher leaf senescence rate with a reduced leaf area were found in the tested genotypes under water-deficit stress compared with well-watered conditions. Increments in phenolic compounds and antioxidants were reported for tolerant onion cultivars (Gedam et al. 2021).

The continuous application of water in a limited amount increases the yield of onion. Thus, drip irrigation is one of the best ways to provide irrigation under drought-prone conditions. Drip irrigation at shorter intervals increased the bulb yield significantly (Bagali et al. 2012). A subsurface drip irrigation experiment conducted by Ensico et al. (2009) reported that a subsurface drip irrigation above 30 kPa at 20 cm depth yielded high and bigger onion bulb sizes. Piri and Naserin (2020) also observed that the combined effect of applying subsurface drip irrigation and supplying nitrogen mitigates the destructivity of deficit irrigation in terms of increased yield, yield components and IWUE of onion.

Also, the use of biostimulants had a profound effect on onion growth, yield and quality. Wakchaure et al. (2018) denoted that the application of plant bioregulators, viz., potassium nitrate, thiourea (TU), salicylic acid (SA), gibberellic acid (GA3) and sodium benzoate (SB) improved growth, helped to maintain higher RWC, enhanced water productivity, increased onion bulb yield of onion plants, enhanced the total soluble sugar, stimulated the total protein content of onion bulb, replenishment of K content, increased total phenolic compounds of onion and boosted pungency values. The KNO_3 showed maximum impact followed by TU, SA, GA3 and SB. Foliar sprays of PBRs helped to maintain a cooler canopy temperature, thus modulating canopy temperature and leaf water status. A recent study by Ibrahim et al. (2022), reported the enhancement of bulb attributes of onion by application of herbicides treatments, viz., fluazifop-p, oxyfluorfen and clethodium.

1.4.3 FLOODING

Waterlogging is one of the abiotic pressures that adversely impacts crops. It is caused by excessive unpredictable rainfall and poor soil drainage caused by compacted soils through the use of heavy agricultural machinery (Hirabayashi et al. 2013). Waterlogging affects plant growth, development and yield by reducing the oxygen supplied to the submerged tissues (Barickman et al. 2019). Onion is highly susceptible to flooding stress due to its shallow rooting nature. The extent of flooding impacts varies with season, soil property, variety, plant growth stage, excess water duration and intensity which affect the bulb yield and survival of plants. Prasanna and Rao (2014) found that flooding retarded the translocation of assimilates from source to sink, attenuating the bulb yield. Waterlogging at different growth stages has a significant impact on bulb production and yield. The response of different onion genotypes to waterlogging stress was studied by Dubey et al. (2020). The biochemical traits (leaf antioxidant, leaf flavonoids and phenol content) exhibited their maximum in genotype tolerant cultivar, thus producing more secondary metabolites for the defense strategy of the plant. Further, the yield attributes such as bulb yield percent were minimally decreased in tolerant cultivar, while the maximal reduction percentage was recorded for the sensitive cultivar. This may be due to the waterlogging inhibiting the translocation of assimilates from source organs to sink organs.

Ghodke et al. (2018) stated that waterlogging at early growth stages after transplanting and the bulb initiation stage reduces the bulb quality and marketable bulb size. Waterlogging at 1–10 days after transplanting (DAT) causes large bulbs followed by 10–20 DAT, but at 20–90 DAT it caused a sharp decrement of bulb weight. However, flooding at the bulb maturity stage (90–100 DAT and 100–110 DAT) has a less adverse impact on the bulb size. Gedam et al. (2022) observed that the tolerant genotypes exhibited higher plant survival and better recovery and bulb size, whereas sensitive genotypes exhibited higher plant mortality, poor recovery and small bulb size under waterlogging conditions. There has been found to be a considerable variation in the morphological, physiological and yield characteristics of tolerant cultivars compared to sensitive ones. Flooding-tolerant genotypes exhibited higher plant height, leaf number, leaf area, leaf length, chlorophyll content, membrane stability index, pyruvic acid, antioxidant content and bulb yield than sensitive genotypes under the same stress conditions. Yiu et al. (2009) found that the impact of flooding was reduced by the pretreatment of Welsh onion with spermidine or spermine by stimulating osmoticants and maintaining membrane stability, thus exogenous paclobutrazol protected Welsh onion from flooding stress.

1.4.4 HEAT STRESS

Increasing temperatures around the world, affecting not only plant growth but also crop productivity, are now a major concern (Sallam et al. 2018; Dawood et al. 2020). High temperature differentially affects plants based on the intensity, duration and rate of temperature rise. As the temperature increases above a certain threshold, the magnitude and extent of stress rise rapidly, resulting in complex adaptation influences that depend on the temperature and other environmental factors. Heat stress causes a reduction in the seed germination rate, decreased photosynthetic efficiency and performance, pollen viability, fertilization and grain/fruit formation can all be affected by excessively high temperatures during the reproductive stage (Dawood et al. 2020). Heat-related damage to reproductive tissues in plant varieties causes severe yield loss in agriculture worldwide (Dawood et al. 2020) that impacts onion bulb initiation and formation.

The prime temperatures for best seedling growth and plant growth before bulb initiation and bulb development are 20–25, 13–24, 15–21, and 20–25°C, respectively (Mathur et al. 2011). Bolting is favored by very low temperatures during the bulb production stage. The crop matures early during the winter season due to a sudden temperature rise, resulting in smaller bulbs. Temperature rise has been directly associated with a reduction in photosynthetic efficiency, and consequently, reduced crop yield (Mathur et al. 2011). Temperature also affects the bulbing response, and the degree to which it is influenced varies between varieties. Wickramasinghe et al. (2000) found that the lowest temperatures (17–22°C) induced production of large bulbs with thick necks. This could be due to structural alterations in the bulb at low temperatures. Photoperiod is also important in bulbing and plants should be exposed to a minimum photoperiod to produce the bulb. In the case of onion, temperature rather than photoperiod is responsible for bulbing. Some studies (Abdalla 1967; Robinson 1973; Currah 1985) reported that bulbing of onions is affected more by temperature than by day length. Lee and Suh (2009) observed that the bulb diameter, bulb index, pyruvic acid content and total sugar content were expressed better in the range 20–25°C, and the sweet flavor was best when onions were grown at 25°C. However, a higher temperature 30°C decreased the bulb weight, which was ascribed to heat stress.

Curing temperature and duration can affect bulb quality and yield losses, as they are closely related to the frequency of storage diseases (Schroeder et al. 2012; Bansal et al. 2015). Several studies have focused on the effects of different curing methods on post-harvest bulb quality during storage (Nega et al., 2015). Other studies have explored methodological improvements and modifications of the curing process by reducing the temperature and duration of the curing and drying periods, thereby conserving energy (Chope and Terry 2010). Eshel et al. (2014) developed a fast-curing method by applying a rapid and controlled heat treatment at 98% relative humidity (RH) and 30°C for 9 d, which improved onion bulb quality due to an increase in the number of outer dry skins, and their physical strength. Downes et al. (2009) found that different curing temperatures affect the biochemical composition of the skin, mainly with regard to flavonols and anthocyanins and their contribution to the color change of skins during curing.

1.4.5 COLD STRESS

Cold and freezing stresses are common environmental abiotic stresses affecting plant growth and development. Crops are affected by extremely low temperature periods in various regions worldwide (Ruelland et al. 2009; Wang et al. 2016), which hinders plant growth and therefore affects productivity (Tommasini et al. 2008). Cold injury can limit the production of onion bulbs. Cold stress in terms of chilling (20°C) and/or freezing (0°C) temperatures, harms onion growth and production, as well as limiting agricultural productivity. Cold stress inhibits plants from expressing their maximum genetic capacity by inhibiting metabolic processes directly and indirectly by cold-induced osmotic, oxidative and other stresses (Chinnusamy et al. 2007). Wickramasinghe et al. (2000) found that a low temperature of 17–22°C produces bigger bulbs with thick necks which may be due to changes in bulb structure. Cold stress can cause poor germination, stunted seedlings, chlorosis, reduced leaf expansion and wilting, and may lead to the death of tissue (necrosis). Cold stress also severely affects the reproductive development of plants. Cold stress induces severe membrane damage and this damage is primarily caused by the acute dehydration caused by freezing during cold stress (Yadav, 2010). As a protection mechanism against cold (0–15°C) and freezing (0°C) temperatures, plants undergo a series of physiological and biochemical modifications. Even though not all plants can withstand cold and freezing temperatures, many do so by a process known as "cold acclimation" (Guy 1990; Thomashow 1999). Su et al. (2007) found that Welsh onion cultivars cultivated in winter suffered stress at low temperatures. However, low-temperature application for short durations improved the bulb weight per plant (Khokhara et al. 2007).

1.4.6 HEAVY METAL STRESS

The increasing abundance of heavy metals in soil, water and foodstuffs is an alarming phenomenon around the world. Every year, the quality of sources (soil and water) viable for agriculture is reduced by contamination primarily owing to anthropogenic activities (Dawood and Azooz 2019, 2020). Geremias et al. (2010) found that copper at different levels (0, 0.1, 0.5, 1, 5 and 10 μg mL^{-1}) caused inhibition of root elongation with increasing effects at higher doses, with growth being reduced by almost 60% at 0.1 μg mL^{-1} and up to 95% at 10 μg mL^{-1}. The elongation of leaves was significantly lower only in specimens exposed at 0.5 μg mL^{-1}, but a total absence of newly formed tissues was observed at 10 μg mL^{-1}. Heavy metal (Co, Cu, Cr, Fe, Mn, Ni, Pb and Zn) uptake by onion was examined in a study by Amin et al. (2013). The concentration of the heavy metals in plants cultivated in wastewater-irrigated soil was higher than in those watered with well water-irrigated soil, and they ranked in the order Mn > Cu > Fe > Zn > Cr > Pb > Ni > Co. Moreover, Mn, Zn, Cr, Ni and Pb levels in some cases exceeded the WHO/FAO permissible limits. Onion samples irrigated with wastewater were found to be the most contaminated, with Mn in the edible parts up to 50 times higher than in those irrigated with non-contaminated water. Qin et al. (2015) studied the effect of ionic Cu (2.0 μM and 8.0 μM) on mitosis, the microtubule cytoskeleton, and DNA in the root tip cells of *Allium cepa* and found that Cu accumulated in roots causing root growth inhibition. In addition, chromosomal aberrations (viz., C-mitosis, chromosome bridges, chromosome stickiness and micronucleus) were observed, and the mitotic index decreased under Cu stress. Microtubules were one of the target sites of Cu toxicity in root tip meristematic cells, and Cu exposure substantially impaired microtubule arrangements. Also, the α-tubulin level reduced following 36 h of exposure to Cu compared to the control. Copper increased DNA damage and suppressed cell cycle progression. Blessings et al. (2018) found that cadmium caused inhibition of root and leaf elongation and the accumulation of Cd was induced in the bulbs at higher levels. In addition, the total phenolic content of the onion bulbs was decreased in response to Cd stress.

1.5 RESPONSES OF *ALLIUM CEPA* TO BIOTIC STRESSORS AND SOME MANAGEMENT STRATEGIES

Unlike most crops, onion has a very poor competitive ability with weeds due to its inherent characteristics, such as a shallow root system, narrow leaf and small leaf area index, and slow plant development (Sahoo et al. 2017). Weeds are one of the main problems in onion fields. Competition of weeds with onion fields is not only for growth factors but also as hosts of insects and fungal diseases, which lead to reducing the final yield of onion. The yield of onion cultivars can be reduced by 26–48% due to weed competition (Rai and Meena 2017). For controlling the effects of weeds, chemical weed control should be used as an alternative method to save the optimal yield. Chemical weed control such as oxyfluorfen, pendimethalin and metribuzin has been used to decrease and prevent weed populations and to increase onion yields (Ahuja and Sandhu 2003). Weeds' dry matter was significantly reduced due to the application of pendimethalin, metolachlor and oxyfluorfen either alone or in combination with hand weeding at 35 days after planting compared to weedy control of onion (Kolhe 2001).

Onion also suffers from many diseases from the preharvest to postharvest period. It has been found that about 35–40% of onion is lost due to damage caused by different diseases. A number of microorganisms are responsible for onion diseases, but among them, fungi are the main causal agent responsible for pre- and postharvest period losses of onion. Among the major leaf diseases, there are downy mildew and white tip, caused by *Peronospora destructor* and *Phytophthora porri*, respectively, causing pale green oval patches on leaves; *Botrytis* leaf blight (BLB), caused by *Botrytis squamosa* characterized by small, white spots with a light green halo; rust, caused by *Puccinia allii*, identifiable by rust-brown clusters of spores (pustules) over the leaf, surrounded by pale yellow tissues; *Stemphylium* and purple blotch, caused by *Stemphylium vesicarium* and *Alternaria porri*, respectively, resulting in leaf lesions with brownish purple rings, which can merge causing leaf death. Among the major insects in onion crops, there are *Thrips tabaci*, *Delia antiqua* and *Liriomyza cepae* (Sekara et al. 2017). The major diseases of onion have been represented in Table 1.1.

Several pathogens, however, were found to attack onions during the plant growing season in the field and when onions are stored. Various fungi have been isolated from onion seeds and reported as

TABLE 1.1
Fungal And Bacterial Diseases of Onion (*Allium Cepa*) and Their Causal Agents

Causal organism	Type of disease	Symptoms
Fungal diseases	***Botrytis* leaf blight**	*Botrytis* infection initially results in small, oval, white spots on the leaves. These lesions often are surrounded by a halo of green water-soaked tissue. Leaf tissue within the spots eventually collapses and becomes tan-colored. Numerous lesions on a single leaf result in dieback of the entire onion top, giving severely affected fields a "blasted" appearance.
	***Fusarium* basal plate rot**	Initially, yellowing of leaves and stunted growth of the plant are observed, which later on dry from the tip downwards.
		In the early stage of infection, the roots of the plants become pink in color and rotting takes place later. In the advanced stage, the bulb starts decaying from the lower ends and ultimately the whole plant dies.
	***Fusarium* damping-off**	Pre-emergence damping-off: This results in seed and seedling rot before they emerge out of the soil.
		Post-emergence damping-off: The pathogen attacks the collar region of seedlings on the surface of the soil. The collar portion rots and ultimately the seedlings collapse and die.
	Pink root rot	Reduced bulb size.
		Roots turn pink or maroon when infected.
		In severe cases, the roots may die and the plants become weakened or stunted, especially in drier areas of the field.
		Crops suffering from heat or drought stress due to poor soil is more prone to yield losses.
	Purple blotch	The symptoms occur on leaves and flower stalks as small, sunken, whitish flecks with purple-colored centers.
		The lesions may girdle leaves/stalks and cause their drooping. The infected plants fail to develop bulbs
	Rust	Symptoms begin as small flecks, which expand into slightly larger, oval- to diamond-shaped, reddish- to dull-orange pustules that develop on leaf blades. Lesions typically form first on older leaves and then spread to younger leaves. Reddish airborne spores (urediniospores) are produced copiously within the lesions. Later in the growing season, the lesions may appear dark because black survival spores (teliospores) develop within the pustules. Severely affected leaves can be completely covered with pustules, turn chlorotic (yellow) and then dry out and die. When infection is severe, bulb size and quality are reduced
	Onion smut	Lesions appear as dark brown streaks running up and down the leaves. The streaks initially appear as long blisters on the leaf surface. As the lesions mature, they turn brown and contain a mass of dark powdery spores that give the tops a sooty appearance. Diseased leaves may bend or twist abnormally and usually are shed prematurely. Smut-infected plants normally are stunted and produce bulbs highly prone to soft rot.

(*continued*)

TABLE 1.1 (Continued)
Fungal And Bacterial Diseases of Onion (*Allium Cepa*) and Their Causal Agents

Causal organism	Type of disease	Symptoms
	White rot (sclerotial rot)	The initial symptoms are yellowing and dieback of leaf tips. Later, the scales, stem plate and roots get destroyed.
		The bulbs become soft and water soaked.
		Later, white fluffy or cottony growth of mycelium with abundant black sclerotia resembling mustard grain develops on the infected bulbs.
	Pythium seed rot	*Pythium* root rot is characterized by seed rotting and pre-emergence damping off.
	Colletotrichum blight/ anthracnose/twister disease	The affected leaves shrivel and droop down.
	Stemphylium leaf blight	The symptoms appear as small yellowish to orange flecks or streaks in the middle of the leaves, which soon develop into elongated spindle-shaped spots surrounded by a pinkish margin.
		The disease on the inflorescent stalk causes severe damage to the seed crop.
	Black mold	Infection usually is through neck tissues as the foliage dies down at maturity.
		Infected bulbs are discolored black around the neck, and affected scales shrivel.
		Masses of powdery black spores develop as streaks along veins on and between outer dry scales.
		Infection may advance from the neck into the central fleshy scales.
	Downy mildew	Leaves turn to pale green.
		On leaves, cottony white mycelial growth develops and appears white. Gradually the leaves turn pale yellow to dark brown and dry up.
	Neck rot	Neck rot symptoms are marked by sunken, collapsed tissues around the neck of the onion bulb. Infected necks soon appear dried out, and a gray mold often occurs between the scales on the collapsed areas. Infection by the neck rot fungus often is followed by a watery soft rot of the bulb.
	Green mold	Infection usually is through neck tissues as the foliage dies down at maturity.
		Infected bulbs are discolored green around the neck, and affected scales shrivel.
		Masses of powdery green spores generally are arranged as streaks along veins on and between outer dry scales.
		Infection may advance from the neck into the central fleshy scales.
Bacterial diseases	Bacterial soft rot	Bacterial soft rot is mainly a problem on mature bulbs. Affected scales first appear water-soaked and pale yellow to light brown.
		As the soft rot progresses, invaded fleshy scales become soft and sticky with the interior of the bulb breaking down.
		A watery, foul-smelling thick liquid can be squeezed from the neck of diseased bulbs.
	Bacterial brown rot	Field symptoms often appear as one or two wilted leaves in the center of the leaf cluster. These leaves eventually turn pale yellow and die back from the tip, while older and younger leaves maintain a healthy green appearance.
		During the early stages of this disease, the bulbs may appear healthy except for a softening of the neck tissue. In a longitudinal section, one or more inner scales become watery or cooked.
		The disease progresses from the top of the infected scale to the base.

TABLE 1.1 (Continued)
Fungal And Bacterial Diseases of Onion (*Allium Cepa*) and Their Causal Agents

Causal organism	Type of disease	Symptoms
		Eventually, all the internal tissue will rot. Finally, the internal scales dry and the bulb shrivels.
		Squeezing the base of infected plants causes the rotted inner portion of the bulbs to slide out through the neck, hence the name slippery skin.
Viral diseases	Dwarf onion yellow dwarf virus	The first symptoms of onion yellow dwarf in young onions are yellow streaks at the bases of the first true leaves.
		All leaves developing after these initial symptoms show symptoms ranging from yellow streaks to complete yellowing of leaves.
		Leaves are sometimes crinkled and flattened and tend to fall over. Bulbs are undersized.
Pests diseases	Bulb mites	Bulbs infested with bulb mites may rot and fail to produce new growth, or new growth may be off-color, stunted and distorted.
	Red spider mite	Adults and nymphs feed primarily on the undersides of the leaves.
		The upper surface of the leaves becomes stippled with little dots that are the feeding punctures.
		The mites tend to feed in "pockets", often near the midrib and veins.
		Silk webbing produced by these mites is usually visible.
		The leaves eventually become bleached and discolored and may fall off.
	Eriophyid mite	Both adults and immature mites feed on the young leaves and between the layers in bulbs of onion.
		Their feeding causes stunting, twisting, curling and discoloration of foliage and scarification and drying of bulb tissue.
		This damage has been attributed to various viruses thought to be transmitted by the mites.
	Allium leaf miner	This is caused by leaf miners (*Phytomyza gymnostoma*).
		Adult females make repeated punctures in leaf tissue with their ovipositor, and both females and males feed on the plant exudates. These punctures may be the first sign of damage. Larvae mine leaves and move toward and into bulbs and leaf sheaths. Both the leaf punctures and mines serve as entry routes for bacterial and fungal pathogens. High rates of infestation have been reported: from 20 to 100 pupae per plant, and 100% of plants in fields.
	Onion maggots	Onion maggots are caused by root-maggot flies (*Delia antiqua*).
		Onion plants are most vulnerable during the seedling stage to onion maggots, and larval feeding may kill seedlings.
		Damage appears as wilted and yellowed foliage, followed by collapsed leaves. Leaves can become rotten, and plants may die.
	Iris yellow spot disease	Onion thrips (*Thrips tabaci*) transmit IYSV.
		Symptoms of the iris yellow spot virus are often seen as cream, elliptical spots on the leaves. The spots also appear on onion scapes or flower stalks of onions.
		As both infected leaves and scapes age, they can collapse at the site of the spots. The spots may be clear or less obvious. Although the spots may at first be insignificant, the disease has the potential of devastating the whole crop.

seed-borne pathogens, including *Aspergillus niger* and *Fusarium oxysporum* (Adongo et al. 2015), which can reduce seed viability and increase pre- and post-emergence damping-off diseases (Rathore and Patil 2019). Pathogens associated with onion seeds cause serious problems by decreasing the percentage of germinated seeds and impeding the growth of germinated seedlings leading to pre- and post-emergence damping-off. Abdelrhim et al. (2022) reported that infection with *Aspergillus niger* and *Fusarium oxysporum* reduced the germination percentage and vigor index of onion plants. The causal agents increased the hydrogen peroxide and differentially affected phenolic contents, salicylic acid and antioxidant system based on fungal severity. Shoaib et al. (2018) observed that onion plants infected with *Fusarium* showed a reduction in osmotic potential, total chlorophyll content, total sugar, membrane stability index and total protein content of the onions leaf, while total phenolics were increased. *Fusarium* infection did not change the activity of polyphenol oxidase (PPO), while it improved peroxidase (POD) and PAL and decreased CAT activity. Mixing salt stress with *Fusarium* infection intensified disease incidence, plant growth, and bulb and affected biochemical responses of onion plants.

Medina-Melchor et al. (2022) found that infection with *Stemphylium vesicarium* causing leaf blight disease begins in the foliar tissue where small yellow spots on the leaves later become elongated and light brown with purple margins. The leaves become yellow and dry, and this drying extends throughout the foliage until it collapses. In bulbs, infection with *S. vesicarium* generated an increase in the concentration of the amino acids leucine, valine, tyrosine and phenylalanine, acids pyruvic acid and fumaric acid and the nucleosides guanosine, adenosine and uridine, while infection with the pathogen caused a reduction in the concentrations of sucrose, 2-oxobutyrate and acetic, citric and formic acids. In leaves, the concentrations of lactic acid and pyruvic acid increased with *S. vesicarum* infection, while the concentrations of sucrose, myo-inositol, 4-aminobutirate, formic and succinic acids, ethanol and adenosine diminished with pathogen infection. In roots, the sugar concentration was not affected by *S. vesicarium* infection, but an increase in the concentration of the amino acids alanine, glutamine, glutamic acid, isoleucine, leucine, phenylalanine, tyrosine and valine was found; likewise, it increased the concentration of pyruvic acid, the nucleosides adenosine, cytidine, guanosine and uridine, and other metabolites, such as 2-hydroxybutyrate and choline.

Notably, the genetic and agronomic tools are applicable as a general strategy for attenuating parasite occurrences such as resistant cultivars, crop rotation, disposal of onion waste, appropriate planting time and density, correct management of water and nutrients, and removal of onion weeds (Brewster 2008). The chemical and biological strategies are referred to as associated with each parasite species or homogeneous group of species, instead.

Downy mildew (*Peronospora destructor*) and white tip (*Phytophthora porri*) can be chemically controlled by choosing among the following active ingredients: copper-based formulations, azoxystrobin, chlorothalonil, cymoxanil, mandipropamid, metalaxyl, dimethomorph, dodina, folpet, benthiavalicarb, iprovalicarb, mancozeb, metiram, pyraclostrobin, systemic acquired resistance (SAR) inducers and K phosphate (Romanazzi et al. 2012). Moreover, molecular diagnostic methods are important tools in detecting and exactly recognizing a pathogen in plant tissues with no symptoms (Mancini et al. 2012). The chemical control of *Botrytis* leaf blight (*Botrytis squamosa*) is achievable by using the following active ingredients: boscalid, chlorothalonil, dithiocarbamate, mancozeb, pyraclostrobin, fluopyram, metconazole, cyprodinil, fludioxonil, iprodione and vinclozolin. Moreover, the use of *Bacillus subtilis* and *Microsphaeropsis ochracea* represents an alternative or additional biological strategy (Carisse et al. 2015). As for rust (*Puccinia allii*), several active ingredients are available for chemical control: benzothiadiazole, propiconazole, tebuconazole, mancozeb, maneb, manzate, chlorothalonil, azoxystrobin, metalaxyl, myclobutanil, iprodione, sulfur and copper hydroxide (Koike et al. 2001). The control management of *Stemphylium* (*Stemphylium vesicarium*) and purple blotch (*Alternaria porri*) can be achieved both by chemical treatment with mefenoxam and copper or by the application of biological agents such as *Bacillus subtilis*, *Pseudomonas fluorescens* or *Saccharomyces cerevisiae*; the use of resistance inducers, i.e., Bion, K_2HPO_4 and salicylic acid, is also effective (Hussein et al. 2007). Among the soil-borne diseases, the most widespread in onions are white rot caused by *Sclerotium cepivorum*, which results in damage to roots and bulb base, pink root caused by *Pyrenochaeta terrestris*, which leads to roots and sometimes outer skins of bulb changing from pink to deep purple color before dying, and *Fusarium* basal rot caused by *Fusarium oxysporum* f. sp. *cepae*, which is identifiable by yellowing, twisted leaves, then dying, by roots turning dark brown, then rotting, with the entire plant wilting during the early stages of infection. The soil-borne diseases can be mainly controlled by performing seed disinfection, soil fumigation or solarization and biological control (Mancini and Romanazzi 2014). Moreover, real-time qPCR using primer pair P1 may be useful to differentiate between onion cultivars that are susceptible or resistant to *Fusarium* basal rot (Sasaki et al. 2015). *Thrips tabaci* causes damage by feeding on leaf tissue and by transmitting other parasites, including Iris yellow spot virus (Gent et al. 2006). The control strategies consist of: applications of predators of the genera *Aeolothrips*, *Toxomerus*, *Sphaerophoria*, *Orius*, *Coleomegilla* and *Hippodamia* (Fok et al. 2014); and chemical treatments with azadirachtin, spinosad, deltamethrin, methomyl, acephate and chlorpyrifos (Jensen 2005). Onion maggot (*Delia antiqua*) damage is caused by larvae that feed on developing epicotyls and roots of young onion plants, often resulting in seedling mortality. Among the available chemical ingredients, spinosad, clothianidin, imidacloprid, thiamethoxam and chlorpyrifos show effectiveness (Wilson et al. 2015). Onion leaf miner (*Liriomyza cepae*) burrows tunnels in the leaves and it can be chemically controlled by dimethoate, chlorpyrifos, phoxim and imidacloprid (Mešić et al. 2008). The main parasite nematodes in onion are *Ditylenchus dipsaci* and *Pratylenchus penetrans*, being the stem nematode and the root lesion nematode, respectively. These affections result in twisted and deformed leaves, root rot, soft and swelled bulbs, and stunted and rotting plants. They can be controlled by soil fumigation or solarization, and chemical treatments (Brewster 2008). Viruses affecting onion are transmitted by *Thrips tabaci* and the main ones are Iris yellow spot virus (IYSV), genus *Tospovirus*, and Onion yellow dwarf virus (OYDV), genus *Potyvirus*. They can be preventatively controlled by using meristem culture on symptomless plants and by preserving genetic material with virus resistance. Moreover, a clear and rapid diagnosis of these viruses is needed, as well as control of *Thrips* and imported materials (Gent et al. 2006).

1.6 CONCLUSION

Onion is an important vegetable crop used worldwide and its susceptibility to environmental conditions highly affects its yield and the nutritional status of the bulb. The richness of onion bulbs with various bioactive compounds allows its use as an antimicrobial potential therapeutic agent. Adverse conditions in terms of biotic or abiotic stresses lower the yield of onions, which results in various management techniques to alleviate their damaging impacts. Vital strategies should be further improved to discover relevant biostimulants with environmentally friendly feedback as ameliorating agents of the stress response. In addition, more studies should be focused on the anticancer properties of various extracts of onion bulbs or leaves.

REFERENCES

Abdalla AA (1967) Effect of temperature and photoperiod on bulbing of common onion (*Allium cepa* L.) under arid tropical conditions of the Sudan. Exp. Agric. 3:137–142.

Abdelrhim AS, Dawood MF, Galal AA (2022) Hydrogen peroxide-mixed compounds and/or microwave radiation as alternative control means against onion seed associated pathogens, *Aspergillus niger* and *Fusarium oxysporum*. J. Plant Pathol. 104(1), 49–63.

Adongo, B. A., Kwoseh, C. K. Moses, E. (2015). Storage rot fungi and seed-borne pathogens of onion. Journal of Science and Technology 35(2): 13–21.

Aghajanzadeh TA, Reich, M, Hawkesford MJ, Burow M (2019) Sulfur metabolism in Allium cepa is hardly affected by chloride and sulfate salinity. Archives of Agron. Soil. Sci. 65(7): 945–956.

Ahuja S, Sandhu KS (2003) Weed management through the use of herbicides in cabbage–onion relay cropping system. Annals Biol. 19: 27–30.

Amin N, Hussain A, Alamzeb S, Begum S (2013) Accumulation of heavy metals in edible parts of vegetables irrigated with waste water and their daily intake to adults and children, District Mardan, Pakistan. Food Chem. 136:1515–23.

Bagali AN, Agali HB, Patil MB, Guled R, Patil V (2012) Effect of scheduling of drip irrigation on growth, yield and water use efficiency of onion (*Allium cepa* L.). Karnataka J. Agric. Sci. 25: 116–119.

Bansal MK, Boyhan GE, MacLean DD (2015) Effects of postharvest curing, ozone, sulfur dioxide, or low oxygen/high carbon dioxide storage atmospheres on quality of short-day onions. Hort. Technol. 25: 639–644.

Barickman TC, Simpson CR, Sams CE (2019) Waterlogging Causes Early Modification in the Physiological Performance, Carotenoids, Chlorophylls, Proline, and Soluble Sugars of Cucumber Plants. Plants (Basel) 8: 160.

Beinşan C, Sumalan R, Vâtcă S (2015) Influence of Salt Stress on Quality of some Onion (*Allium cepa* L.) Local Landraces. Bulletin USAMV Series Agri. 72: 2.

Bekele S, Tilahun K (2007) Regulated deficit irrigation scheduling of onion in a semiarid region of Ethiopia. Agric. Water Manag. 89: 148–152.

Benmalek Y, Yahia OA, Belkebir A Fardeau M-L (2013) Anti-microbial and anti-oxidant activities of *Illicium verum*, *Crataegus oxyacantha* ssp monogyna and *Allium cepa* red and white varieties. Bioengineered 4: 244–248.

Blessings O, Joseph AA, Ifedayo OI (2018) Research Article Deleterious Effects of cadmium solutions on onion (*Allium cepa*) growth and the plant's potential as bioindicator of Cd Exposure. Res. J. Environ. Sci. 12 (3): 114–120.

Brewster JL (2008) Onions and other vegetable Alliums. Wallingford, CAB International.

Calanca PP (2017). Effects of Abiotic Stress in Crop Production. In: M. Ahmed, C. Stockle, (eds). Quantification of Climate Variability, Adaptation and Mitigation for Agricultural Sustainability. Springer, Cham. 2017.

Camilia Y, El-Dewiny M, Hussein M, Awad F (2013) Influence of mono potassium phosphate fertilizer on mitigate the negative effects of high saline irrigation water on onion crop. Middle East J. Agric. Res. 2: 152–158.

Carisse O, Tremblay D-M, Brodeur L, McDonald MR, McRoberts N (2015) Management of Botrytis leaf blight of onion. The Québec experience of 20 years of continual improvement. Plant Disease 95: 504–514.

Chang P, Randle WM (2004) Sodium chloride in nutrient solutions can affect onion growth and flavor development. Hort. Sci. 39: 1416–1420.

Chinnusamy V, Zhu J, Zhu JK (2004) Cold stress regulation of gene expression in plants. Trends Plant Sci. 12(10): 444–451.

Chope GA, Terry L (2010) Effect of curing at different temperatures on phytohormone and biochemical composition of onion 'Red Baron' during long-term postharvest storage. Acta Horticultur. 877: 699–705.

Corzo-Martínez M, Corzo N, Villamiel M (2007) Biological properties of onions and garlic. Trends Food Sci. Technol. 18: 609–625.

Currah L (1985) Review of three onion improvement schemes in the tropics. Trop. Agric. 62:131–136.

Dawood MF (2022) Melatonin: an elicitor of plant tolerance under prevailing environmental stresses. In: Emerging Plant Growth Regulators in Agriculture (pp. 245–286). Academic Press.

Dawood MF, Moursi YS, Amro A, Baenziger PS, Sallam A (2020) Investigation of heat-induced changes in the grain yield and grains metabolites, with molecular insights on the candidate genes in barley. Agronomy 10(11): 1730.

Dawood MFA, Azooz MA (2020) Insights into the oxidative status and antioxidative responses of germinating broccoli (*Brassica oleracea* var. *italica* L.) seeds in tungstate contaminated water. Chemosphere 261: 127585.

Dawood MFA, Azooz MA (2019) Concentration-dependent effects of tungstate on germination, growth, lignification-related enzymes, antioxidants, and reactive oxygen species in broccoli (*Brassica oleracea* var. *italica* L.). Environ. Sci. Pollut. Res. 26(36): 36441–36457.

Dong, Y., Wang, D., Li, M., Hu, X., and Zhao, G. (2010). One new pathway for *Allium* discoloration. Food Chemistry 119: 548–553

Downes K, Chope GA, Terry LA (2009) Effect of curing at different temperatures on biochemical composition of onion (*Allium cepa* L.) skin from three freshly cured and cold stored UK-grown onion cultivars. Postharvest Biol. Technol. 54: 80–86.

Duan Y, Jin DH, Kim HS, Seong JH, Lee YG, Kim DS, Jang SH (2015) Analysis of total phenol, flavonoid content and antioxidant activity of various extraction solvents extracts from onion (Allium cepa L.) peels. J. Korean Oil Chemical Soc. 32(3): 418–426.

Dubey S, Kuruwanshi VB, Ghodke PH, Mahajan V (2020) Biochemical and yield evaluation of onion (Allium cepa L.) genotypes under waterlogging condition. Int. J. Chem. Stu. 8: 2036–2040.

Enciso J., Wiedenfeld B, Jifon J, Nelson S (2009) Onion yield and quality response to two irrigation scheduling strategies. Sci. Horticultur. 120: 301–305.

Eshel D, Teper-Bamnolker P, Vinokur Y, Saad I, Zutahy Y, Rodov V (2014) Fast curing: a method to improve postharvest quality of onions in hot climate harvest. Postharvest Biol. Technol. 88: 34–39.

FAO (2015). Food and Agriculture Organization Statistical Pocketbook on world food and agriculture. ISBN 978-92-5-108802-9.

Fok EJ, Petersen JD, Nault BA (2014) Relationships between insect predator populations and their prey, Thrips tabaci, in onion fields grown in large-scale and small-scale cropping systems. Bio. Control 59: 739–748.

Fossen T, Slimestad R, Andersen ØM (2003) Anthocyanins with 4' -glucosidation from red onion, *Allium cepa*. Phytochemist 64: 1367–1374

Fredotović Ž, Šprung M, Soldo B, Ljubenkov I, Budić-Leto I, Bilušić T, Čikeš-Čulić V, and Puizina J (2017) Chemical composition and biological activity of Allium cepa L. and Allium cornutum (Clementi ex Visiani 1842) Methanolic extracts. Molecules 22: 448.

Galdón BR, Rodríguez CT, Rodríguez ER, Romero CD (2009). Fructans and major compounds in onion cultivars (*Allium cepa*). Journal of Food Composition and Analysis, 22(1), 25–32.

Gawande SJ, Hanjagi PS, Ramakrishnan RS and Singh M (2021) Screening of Onion (*Allium cepa* L.) Genotypes for drought tolerance using physiological and yield based indices through multivariate analysis. Front. Plant Sci. 12:600371.

Gedam, P. A., Thangasamy, A., Shirsat, D. V., Ghosh, S., Bhagat, K. P., Sogam, O. A., ... Singh, M. (2021). Screening of onion (Allium cepa L.) genotypes for drought tolerance using physiological and yield based indices through multivariate analysis. Front. Plant Sci. 12, 600371.

Gedam PA, Shirsat DV, Arunachalam T, Ghosh S, Gawande SJ, Mahajan V, Gupta AJ, Singh M (2022) Screening of onion (*Allium cepa* L.) genotypes for waterlogging tolerance. Front. Plant Sci. 12:727262.

Gedam PA, Thangasamy A, Shirsat DV, Ghosh S, Bhagat KP, Sogam OA, Gupta AJ, Mahajan V, Soumia PS, Salunkhe VN, Khade YP, Gent DH, du Toit LJ, Fichtner SF, Mohan SK, Pappu HR, Schwartz HF (2006) Iris yellow spot virus: an emerging threat to onion bulb and seed production. Plant Disease 90: 1468–1480.

Geremias R, Fattorini D, Fávere VT, Pedrosa RC (2010) Bioaccumulation and toxic effects of copper in common onion *Allium cepa* L. Chem. Ecol. 26(1): 19–26.

Ghodke PH, Shirsat DV, Thangasamy A, Mahajan V, Salunkhe VN, Khade Y, et al. (2018) Effect of water logging stress at specific growth stages in onion crop. Int. J. Curr. Microbiol. Applied Sci. 7:3438–48.

Guy CL (1990) Cold-acclimation and freezing stress tolerance—role of protein metabolism. Annu. Rev. Plant Physiol. Plant Mol. Biol. 41: 187–223.

Hirabayashi Y, Mahendran R, Koirala S, Konoshima L, Yamazaki D, Watanabe S, Kim H, Kanae S (2013) Global flood risk under climate change. Nat. Clim. Chang. 3: 816.

Hussein M, Hassan M, Allam ADA, Abo-elyousr KAM (2007) Management of Stemphylium blight of onion by using biological agents and resistance inducers. Egyptian J. Phytopathol. 35: 49–60.

Ibrahim HH, Abdalla AA, Salem WS (2022) Efficacy of irrigation intervals and chemical weed control on optimizing bulb yield and quality of onion (*Allium cepa* L.). Bragantia 81: e1722.

Islam F, Wang J, Farooq MA, Yang C, Jan M, Mwamba TM, et al. (2019) Rice responses and tolerance to salt stress," In Advances in Rice Research for Abiotic Stress Tolerance. eds. Hasanuzzaman M, Fujita M, Nahar K, Biswas J (Cambridge: Woodhead Publishing), 791–819.

Jensen L (2005) Insecticide trials for onion Thrips (*Thrips tabaci*) control–2004. Malheur Experiment Station Annual Report: 71–76.

Kadayifci A, Tuylu GI, Ucar Y, Cakmak B (2005) Crop water use of onion (*Allium cepa* L.) in Turkey. Agric. Water Manage. 72: 59–68.

Kato M, Kamoi T, Sasaki R, Sakurai N, Aoki K, Shibata D, Imai S (2013) Structures and reactions of compounds involved in pink discolouration of onion. Food Chem. 139: 885–892.

Khokhara KM, Hadleyb P, Pearson S (2007) Effect of cold temperature durations of onion sets in store on the incidence of bolting, bulbing and seed yield. Sci. Hortic. 112: 16–22.

Kidder D, Behrens R (1988) Plant Responses to Haloxyfop as influenced by water stress. Weed Sci. 36: 305–312.

Koike ST, Smith RF, Davis RM, Nunez JJ, Voss RE (2001) Characterization and control of garlic rust in California. Plant Disease 85: 585–591.

Kolhe SS (2001) Integrated weed management in onion (*Allium cepa* L.). Indian J Weed Sci. 33: 26–29.

Kothari D, Lee WD, Kim SK (2020) Allium flavonols: Health benefits, molecular targets, and bioavailability. Antioxidants 9: 888.

Lee E, Suh JK (2009) Effect of Temperature on the Growth, Pyruvic Acid and Sugar Contents in Onion Bulbs. Korean J. Hortic. Sci. Technol. 27: 554–559.

Liguori L, Califano R, Albanese D, Raimo F, Crescitelli A, Di Matteo M (2017) Chemical composition and antioxidant properties of five white onion (*Allium cepa* L.) landraces. J. Food Quality 2017: 1–9.

Ma Y-L, Zhu D-Y, Thakur K, Wang C-H, Wang H, Ren Y-F, Zhang J-G, Wei Z- J (2018) Antioxidant and anti-bacterial evaluation of polysaccharides sequentially extracted from onion (*Allium cepa* L.). Int. J. Biol. Macromolecul. 111: 92–101.

Mancini V, Murolo S, Romanazzi G (2012) Molecular diagnosis of *Peronospora destructor* in onion plants. Proceedings of the Workshop Updates on onion downy mildew and other phytosanitary problems of vegetable seed crops, March 29, 2012, Ancona, Italy: 94–96.

Mancini V, Romanazzi G (2014) Seed treatments to control seedborne fungal pathogens of vegetable crops. Pest Manage. Sci. 70: 860–868.

Mathur S, Allakhverdiev SI, Jajoo A (2011) Analysis of high temperature stress on the dynamics of antenna size and reducing side heterogeneity of photosystem II in wheat (*Triticum aestivum*). Biochim. Biophys. Acta 2011: 1807,

Medina-Melchor DL, Zapata-Sarmiento DH, Becerra-Martínez E, Rodríguez-Monroy M, Vallejo L, Sepúlveda-Jiménez G (2022) Changes in the metabolomic profiling of *Allium cepa* L.(onion) plants infected with *Stemphylium vesicarium*. European J. Plant Pathol. 162(3): 557–573.

Mešić A, Igrc Barčić J, Barčić J, Zvonar M, Filipović I (2008) Diptera pests control in onions. Fragmenta Phytomedica et Herbologica 30: 5–21.

Mohamed AA, Aly AA (2008) Alterations of some secondary metabolites and enzymes activity by using exogenous antioxidant compound in onion plants grown under seawater salt stress. Am.-Eur. J. Sci. Res. 3(2):139–146

Nega G, Mohammed A, Menamo T (2015) Effect of curing and top removal time on quality and shelf life of onions (*Allium cepa* L.). Global J. Sci. Front. Res.: D Agri. Vet. 15: 26–36.

Oboh G, Ademiluyi AO, Agunloye OM, Ademosun AO, Ogunsakin BG (2019) Inhibitory effect of garlic, purple onion, and white onion on key enzymes linked with type 2 diabetes and hypertension. J. Diet. Suppl. 16: 105–118.

Ouyang H, Hou K, Peng W, Liu Z, Deng H (2017) Antioxidant and xanthine oxidase inhibitory activities of totapolyphenols from onion. Saudi J. Biol. Sci. 25 (7): 1509–1513

Pérez-Gregorio RM, García-Falcón MS, Simal-Gándara J, Rodrigues AS, Almeida DP (2010) Identification and quantification of flavonoids in traditional cultivars of red and white onions at harvest. J. Food Composit. Anal. 23: 592–598.

Pessoa LGM, dos Santos Freire MBG, dos Santos RL, Freire FJ, dos Santos PR, Miranda MFA (2019) Saline water irrigation in semiarid region: II-effects on growth and nutritional status of onions. Australian J. Crop Sci. 13(7): 1177–1182.

Piri H, Naserin A (2020) Effect of different levels of water, applied nitrogen and irrigation methods on yield, yield components and IWUE of onion. Sci. Horticultur. 268: 109361.

Potopova V, Stepanek P, Farda A, Turkott L, Zahradnicek P, Soukup J (2016) Drought stress impact on vegetable crop yields in the Elbe River Lowland between 1961 and 2014. Cuadernos De Investigacion Geografica 42(1):127–43.

Prakash D, Singh BN, Upadhyay G (2007) Antioxidant and free radical scavenging activities of phenols from onion (*Allium cepa*). Food Chem. 102: 1389–1393.

Prasanna YL, Rao GR (2014) Effect of waterlogging on growth and seed yield in greengram genotypes. Int. J. Food Agri. Vet. Sci. 4: 124–128.

Qin R, Wang C, Chen D, Björn LO, Li S (2015) Copper-induced root growth inhibition of *Allium cepa* var. *agrogarum* L. involves disturbances in cell division and DNA damage. Environm. Toxicol. Chem. 34(5): 1045–1055.

Rady MO, Semida WM, Abd El-Mageed TA, Hemida KA, Rady MM (2018) Up-regulation of antioxidative defense systems by glycine betaine foliar application in onion plants confer tolerance to salinity stress. Sci. Hortic. 240: 614–622.

Ragaey MM, Sadak MS, Dawood MF, Mousa NH, Hanafy RS, Latef AAHA (2022) Role of signaling molecules sodium nitroprusside and arginine in alleviating salt-induced oxidative stress in wheat. Plants 11(14): 1786.

Rai, T, Meena M (2017) Impact of weed and fertilizer management on yield and quality parameters of onion (*Allium cepa* L.) Var. Pusa Red under Lucknow conditions. J. Pharmacognosy Phytochem. 6: 1934–1937.

Rathore, M. S., Patil, A. D. (2019). Isolation and identification of fungi causing damping-off disease in onion (*Alium cepa* L.) plants. Int. J. Curr. Microbiol. App. Sci. 8(8), 2277–2281.

Ratnarajah V, Gnanachelvam N (2021) Effect of abiotic stress on onion yield: a review. Adv. Technol. 2021: 147–160.

Razavi-Azarkhiavi K, Behravan J, Mosaffa F, Sehatbakhsh S, Shirani K, Karimi G (2014) Protective effects of aqueous and ethanol extracts of rosemary on H_2O_2-induced oxidative DNA damage in human lymphocytes by comet assay. J. Complement. Integr. Med. 11: 27–33.

Robinson JC (1973) Studies on the performance and growth of various short-day onion varieties (Allium cepa L.) in the Rhodesian Lowveld in relation to sowing.1. Growth analysis. Rhod. J. Agric. Res. 11: 51–68.

Romanazzi G, Mancini V, Murolo S, Feliziani E (2012) Evaluation of synthetic and natural fungicides for the protection of seed bearing onion from downy mildew. Proceedings of the workshop Updates on onion downy mildew and other phytosanitary problems of vegetable seed crops, March 29, Ancona, Italy, 2012: 76–79.

Rose P, Whiteman M, Moore PK, Zhu YZ (2005) Bioactive S-alk (en)yl cysteine sulfoxide metabolites in the genus *Allium*: the chemistry of potential therapeutic agents. Natural Prod. Reports 22: 351–368.

Ruelland E, Vaultier MN, Zachowski A, Hurry V (2009) Cold signalling and cold acclimation in plants. Adv. Bot. Res. 49: 35–150.

Ryang S, Woo S, Kwon S, Kim S, Lee SH, Kim K, Lee D (2009) Changes of net photosynthesis, antioxidant enzyme activities, and antioxidant contents of *Liriodendron tulipifera* under elevated ozone. Photosynthetica 47: 19–25.

Sagar NA, Pareek S, Benkeblia N, Xiao J (2022). Onion (*Allium cepa* L.) bioactives: Chemistry, pharmacotherapeutic functions, and industrial applications. Food Front. 3(3): 380–412.

Sahoo SK, Chakravorty S, Soren L, Mishraand C, Sahoo BB (2017) Effect of weed management on growth and yield of onion (*Allium cepa* L.). J. Crop Weed 13: 208–211.

Sallam A, Amro A, El-Akhdar A, Dawood MF, Kumamaru T, Stephen Baenziger P (2018) Genetic diversity and genetic variation in morpho-physiological traits to improve heat tolerance in Spring barley. Mol. Biol. Reports 45(6): 2441–2453.

Sanower-Hossain M (2019) Present scenario of global salt affected soils, its management and importance of salinity research. Int. Res. J. Biol. Sci. 1: 1–3.

Sasaki K, Nakahara K, Shigyo M, Tanaka S, Ito S (2015) Detection and quantification of onion isolates of *Fusarium oxysporum* f. sp. *cepae* in onion plant. J. General Plant Pathol. 81: 232–236.

Schroeder B, Humann J, Du Toit L (2012) Effects of postharvest onion curing parameters on the development of sour skin and slippery skin in storage. Plant Disease 96: 1548–1555.

Sekara A, Pokluda R, Del Vacchio L, Somma S, Caruso G (2017) Interactions among genotype, environment and agronomic practices on production and quality of storage onion (*Allium cepa* L.) – A review. Hort. Sci. (Prague) 44: 21–42.

Semida WM, Abd El-Mageed TA, Howladar SM, Rady MM (2016) Foliar-applied alpha-tocopherol enhances salt-tolerance in onion plants by improving antioxidant defence system. Aust. J. Crop Sci. 10(7):1030–1039

Semida WM (2016a) Hydrogen peroxide alleviates salt-stress in two onion (*Allium cepa* L.) cultivars. American-Eurasian J. Agric. Environ. Sci. 16: 294–301.

Sheteiwy M, Zaid Ulhassan, Qi W, Lu H, AbdElgawad H, Minkina T, Sushkova S, Rajput V D, El-Keblawy A, Josko I, Sulieman S, El-Esawi MA, El-Tarabily KA, AbuQamar S, Dawood MFA (2022) Association of jasmonic acid priming with multiple defense mechanisms in wheat. Front. Plant Sci. 2022:2614.

Shoaib A, Meraj S, Khan KA, Javaid MA (2018) Influence of salinity and *Fusarium oxysporum* as the stress factors on morpho-physiological and yield attributes in onion. Physiol. Mol. Biol. Plants 24(6): 1093–1101.

Singh U (2008) A history of ancient and early medieval India: from the Stone Age to the 12th century. Pearson Education India.

Sofy M, Mohamed H, Dawood M, Abu-Elsaoud A, Soliman M (2021) Integrated usage of Trichoderma harzianum and biochar to ameliorate salt stress on spinach plants. Archives Agronom. Soil Sci. 1–22.

Srinivasa Rao NK, Laxman RH, Bhatt RM (2010) Extent of impact of flooding and water stress on growth and yield of onion and tomato. In: Aggarwal PK (ed) Annual progress report of ICAR network project impact, adaptation and vulnerability of Indian Agri- culture to climate change 2009–10. 111–112.

Srinivasa Rao NK, Shivashankara KS, Laxman RH (2016) Abiotic Stress Physiology of Horticultural Crops. Abiotic stress physiology of horticultural crops, (Vol. 311). 1–368. India: Springer.

Sta-Baba R, Hachicha M, Mansour M, Nahdi H, Kheder MB (2010) Response of onion to salinity. Afr. J. Plant Sci. 4: 7–12.

Su H, Xu K, Liu W (2007) Cold tolerance and winter cultivation of Welsh onions. Acta Hortic. 760: 335–340.

Takahama U, Hirota S (2000) Deglucosidation of quercetin glucosides to the aglycone and formation of antifungal agents by peroxidase-dependent oxidation of quercetin on browning of onion scales. Plant Cell Physiol. 41(9): 1021–1029.

Teshika JD, Zakariyyah AM, Zaynab T, Zengin G, Rengasamy KR, Pandian SK, Fawzi MM (2019) Traditional and modern uses of onion bulb (*Allium cepa* L.): a systematic review. Critical Reviews Food Sci. Nutr. 59: S39-S70.

Thomashow MF (1999) Plant cold acclimation: freezing tolerance genes and regulatory mechanisms. Annu. Rev. Plant Physiol. Plant Mol. Biol 50: 571–99.

Tommasini L, Svensson JT, Rodriguez EM, Wahid A, Malatrasi M, Kato K, ... Close TJ (2008) Dehydrin gene expression provides an indicator of low temperature and drought stress: transcriptome-based analysis of barley (*Hordeum vulgare* L.). Funct. Integrat. Genomics 8(4): 387–405.

ur Rahman S, Khan S, Chand N, Sadique U, Khan RU (2017) In vivo effects of *Allium cepa* L. on the selected gut microflora and intestinal histomorphology in broiler. Acta Histochemica 119: 446–450.

Vásque, J, Alarcón JC, Jiménez SL, Jaramillo GI, Gómez-Betancur IC, Rey-Suárez JP, Jaramillo KM, Muñoz DC, Marín DM Romero JO (2015) Main plants used in traditional medicine for the treatment of snake bites n the regions of the department of Antioquia, Colombia. J. Ethnopharmacol. 170: 158–166.

Wakchaure GC, Minhas PS, Meena KK, Singh NP, Hegade PM, Sorty AM (2018) Growth, bulb yield, water productivity and quality of onion (*Allium cepa* L.) as affected by deficit irrigation regimes and exogenous application of plant bio-regulators. Agric. Water Manag. 199: 1–10.

Wang W, Chen Q, Hussain S, Mei J, Dong H, Peng S, Huang J, Cui K, Nie L (2016) Pre-sowing seed treatments in direct-seeded early rice: consequences for emergence, seedling growth and associated metabolic events under chilling stress. Sci. Rep. 6: 19637.

Wickramasinghe UL, Wright CJ, Currah L (2000) Bulbing responses of two cultivars of red tropical onions to photoperiod, light integral and temperature under controlled growth conditions. J. Hortic. Sci. Bio-Tech. 75: 304–311.

Wiczkowski W (2011) Garlic and onion: Production, biochemistry, and processing. Handbook of Vegetables and Vegetable Processing: 625–642.

Wilson RG, Orloff SB, Taylor AG (2015) Evaluation of insecticides and application methods to protect onions from onion maggot, Delia antiqua, and seedcorn maggot, Delia platura, damage. Crop Protection 67: 102–108.

Yadav SK (2010) Cold stress tolerance mechanisms in plants. A review. Agron. Sustain. Dev. 30(3): 515–527.

Yiu JC, Liu CW, Fang DY, Lai YS (2009) Waterlogging tolerance of Welsh onion (*Allium fistulosum* L.) enhanced by exogenous spermidine and spermine. Plant Physiol. Biochem. 47: 710–716.

Younes NA, Rahman MM, Wardany AA, Dawood MFA, Mostofa MG, Keya SS, Abdel Latef AAH, Tran L-SP (2021) Antioxidants and bioactive compounds in licorice root extract potentially contribute to improving growth, bulb quality and yield of onion (*Allium cepa*). Molecules 2021, 26: 2633.

Zayton AM (2007) Effect of soil-water stress on onion Yield. Misr. J. Ag. Eng. 24: 141–160.

Zhang S-l, Peng D, Xu Y-c, Lü S-w, WANG J-j (2016) Quantification and analysis of anthocyanin and flavonoids compositions, and antioxidant activities in onions with three different colors. Journal Integt. Agri. 15: 2175–2181.

Zill-e H, Vian MA, Fabiano-Tixier A-S, Elmaataoui M, Dangles O, Chemat F (2011) A remarkable influence of microwave extraction: Enhancement of antioxidant activity of extracted onion varieties. Food Chemist 127: 1472–1480.

2 Allium sativum and Stressful Conditions

Amira A. Ibrahim[1], Khaled F. M. Salem[2] and Samah Ramadan[3]*

[1]Botany and Microbiology Department, Faculty of Science,
Arish University, Al-Arish 45511, Egypt
[2]Department of Plant Biotechnology, Genetic Engineering and
Biotechnology Research Institute (GEBRI), University of Sadat City,
Sadat City, Egypt. Current address: Department of Biology,
College of Science and Humanitarian Studies, Shaqra University,
Qwaieah, Saudi Arabia
[3]Botany Department, Faculty of Science, Mansoura University,
Mansoura 35516, Egypt
*Corresponding Author: amiranasreldeen@sci.aru.edu.eg;
amiranasreldeen@yahoo.com

CONTENTS

2.1 INTRODUCTION

Garlic (*Allium sativum* L.) is the second most important annual medicinal bulb crop after onion (Adekpe et al. 2007). Garlic is an ancient horticultural crop belonging to the family Amaryllidaceae and its cultivation originated in central Asia. Its importance has increased in the Mediterranean region including Africa and Europe; China ranked first in the production of garlic with approximately 23 million tons and India ranked second with 3 million tons, while Bangladesh produced about 467,000 tons (Srivastava et al. 2021). The total global production of garlic is over 31 million tonnes annually, and Egypt produced about 319 thousand tons (FAOSTAT, 2022). *Allium sativum* is rich in protein, vitamins, fibers and minerals that give it its economic and pharmaceutical qualities and it is consumed widely in foods and drinks (Chakraborty and Majumder 2020). *Allium sativum* is considered an important ancient medicinal plant used by Pharaohs', as mentioned in Ebers Papyrus. Garlic contains alliin, allicin and volatile oils that can be used in disease treatments such as for cold, flu and cancer through the immune system (Williamson 2003; Gruenwald 2004; Wichtl 2004). Known for its culinary and therapeutic benefits, garlic (*Allium sativum* L.) is a crop that is asexually propagated and that has a high degree of morphological variety (Bradley et al. 1996) (Figure 2.1). For garlic to grow and develop, vernalization must be completed, followed by a higher temperature and an extended photoperiod (Wu et al. 2015). In Egypt, *Allium sativum* was used in meals and feeding of workers during the construction of the pyramids and it was used also as performance enhancers during competitive athletics in the Roman and Olympic games (Green and Polydoris 1993; Lawson 1998). The cloves and fresh leaves are used in cooking around the world and are consumed as a powder to enhance the meals' flavor, and they are also used in folk medicine (Sasi et al. 2021). It is difficult to create genetic variants or establish genetically improved cultivars using traditional breeding techniques. Although only a small number of genotypes of garlic flower in a particular place, these genotypes have demonstrated sterility and location specificity, which has constrained the creation of new genetically enhanced cultivars. Compared with other vegetable crops, very few genomic studies have been carried out on garlic in the current molecular breeding and genomic period.

Garlic's 15.9 GB genome size is slightly lower than that of onions (Arumuganathan and Earle 1991; Abdelrahman et al. 2017). Meristem culture, genetic modification and molecular breeding have nevertheless been implemented for the enhancement of garlic qualities; however, a more detailed study is needed to generate smart climate features in garlic. Garlic is subject to a variety of biotic and abiotic stressors, just like other crops. There is no documented evidence that the production of garlic is impacted by the changing climatic scenario. Reddy et al. (2000) assert that climatic changes will cause crop productivity to decline year after year, even in controlled environments. Garlic's yield component is sensitive to environmental factors (Panse et al. 2013). Therefore, it requires breeders' attention to create climate-smart garlic cultivars for improved climatic adaptation by employing available genetic resources and contemporary biotechnological methods to ensure production sustainability.

FIGURE 2.1   Cloves (a) and fresh leaves (b) of garlic (*Allium sativum* L).

2.1.1 ORIGIN AND TAXONOMY

Genus *Allium* is distributed from dry subtropic to boreal zones over the Holarctic region. *Allium sativum* growth is good on sunny, open, dry sites. Garlic is a medicinal herbaceous annual plant that is distinguished by bulbs composed of fibers, and it contains four to six segments with a spicy flavor. Garlic has a cylindrical stem of 50 cm height and 2–3 cm width. The flowers are replaced by bulbs consisting of six segments 4–6 mm long and seeds are rarely produced (Vanaclocha and Cañigueral 2019).

Allium sativum is a herbaceous perennial plant belonging to the class Monocotyledon plant, order Asparagales and family Amaryllidaceae. *Allium sativum* is one of the *Allium* genera containing garlic, onion, scallion, shallot, leek and chives.

Allium sativum was described by the Swedish botanist Linnaeus, with a basic chromosome number x = 8 with polyploidy in both series. The chromosomes of *A. sativum* were studied and described first by Khoshoo et al. (1960) and Battaglia (1963). Levan (1935) and Mensinkai, (1939) reported the diploid number of *A. sativum* as 2n = 16 and 2n = 18 in two varieties of garlic (Sharma and Bal 1959). Meanwhile, Bozzini and De Luca (1991) claimed that common garlic had a somatic number of 2n = 16 and a karyotypic formula of six metacentric, four submetacentric and six acrocentric chromosomes. Certain garlic plants discovered in Campania, Italy, were determined to be tetraploid, with 4n = 32. There were two sets of satellite chromosomes in a diploid garlic plant (2n = 16) (McCollum 1976). For the centromere's position, chromosomal length and number of chromosome satellites, karyological variants of garlic have been observed (Etoh 1983; Etoh 1985; Hong et al. 2000; Osman et al. 2007). The karyotype of *A. sativum* varied greatly among individual clones. Egyptian cultivars of garlic have a chromosome number 2n = 16 (Figure 2.2).

2.2 RESPONSES OF GARLIC TO VARIOUS STRESSFUL CONDITIONS

2.2.1 RESPONSES OF GARLIC TO SALINITY

Producing crops is significantly hampered by salinity, especially in arid and semi-arid environments (Flowers 2004; Munns and Tester 2008). By having toxic and osmotic effects, it inhibits plant development. The osmotic impact can decrease a plant's ability to store water, and the toxic consequences can cause early senescence and a decrease in the amount of photosynthetic leaf area in a plant to the point where it cannot support growth (Munns and Tester 2008; Nemati et al. 2011;

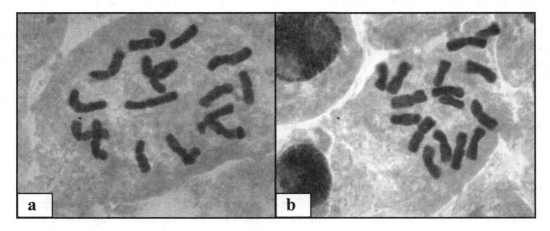

FIGURE 2.2 Chromosome number of Egyptian *Allium sativum* cultivars 2n = 16.

Source: Prepared by Amira A. Ibrahim.

Horie et al. 2012). A plant's capacity to thrive in saline environments is influenced by the amount of salt present, how well it can adapt to low water potential and the presence of high levels of sodium and chloride ions. Garlic, which is categorized as sensitive, is among the crop species that respond differently to salinity, ranging from tolerant to sensitive (Ayer 1985). Salinity makes cells drier and makes nutrients less available, which might limit growth (Campos et al. 2013).

Although garlic is typically slightly more salt-resistant than most plants, maintaining the lowest possible soil salinity levels is essential for maximizing production. Processing qualities such as yield and quality are reduced when salinity is higher than 100 mM.

Low soil water potential, nutritional imbalance, disrupting ionic homeostasis and particular ion impacts are all associated with the negative consequences of high salinity on plants. Additionally, excessive sodium (Na^+) ion concentrations in the cytosol exert direct harmful effects on cell membranes that result in electrolyte leakage and impair cytosolic metabolic processes (Zhu 2001). These consequences result in lowered biochemical and physiological processes. Reactive oxygen species (ROS) overproduction is also known as oxidative stress (Parvaiz and Satyawati 2008; Hasanuzzaman et al. 2013). Under salinity conditions, produced ROS harm cells by oxidizing lipids, proteins and nucleic acids (Foyer et al. 2000), which cause premature leaf senescence and loss of photosynthetic efficiency, which in turn causes reduced carbon uptake and, ultimately, lower crop production. Consequently, the ability of the system to survive under stress depends on the proper balance of ROS production and ROS detoxification by a variety of antioxidants, including phenyl-alanine ammonia-lyase (PAL).

Salinity is a significant stressor that has a global impact on agricultural productivity. There hasn't been a great deal of research on this topic in garlic. Knowing how much salt stress different garlic cultivars can withstand may help scientists identify and create salt-tolerant variants as the climate changes. Garlic, which is classed as sensitive, has different sensitivity to salinity than other agricultural species, ranging from tolerant to sensitive (Khademi et al. 2018). Nearly all agricultural locations can support the growth of garlic, although most of it is produced in regions with salt issues that are already present or that could arise. According to Francois (1994), garlic's threshold salinity was 3.9 dS m^{-1}, and at 7.4 dS m^{-1}, the yield was 50% lower. Exogenous protective agents, including osmoprotectants, phytohormones, polyamines, antioxidants and other trace elements, have recently been discovered to help reduce the harm caused by salt (Hasanuzzaman et al. 2013). Selenium is one of these defenses that demonstrated the capacity to enhance plant development and stress tolerance when growing in a high-salt (Se) environment. Both salt stress and selenium treatment had a sizable impact on total phenolic content. Garlic leaves' phenol content was dramatically decreased at salinity levels of 30, 60 and 90 mM NaCl. The administration of 4 and 8 mg Se L^{-1} produced 59% and 51% more phenols than the control treatment without Se, yielding the greatest concentration of phenols at the 90 mM NaCl salinity level. When aerobic or photo-synthetic metabolism is damaged by environmental stimuli, ROS are necessarily created and contribute to the defense against them (Ksouri et al. 2007). Comparing garlic plants under salinity stress to control plants without Se, the total amount of phenolic compounds in garlic leaves was dramatically reduced. According to Mangal et al. (1990), the genotype-dependent 50% yield loss in garlic occurs at between 5.60 and 7.80 dS m^{-1}. Additionally, they calculated that the average garlic yield decreased by 1.68% for every unit increase in soil salinity if the salinity was higher than 1.70 dS m^{-1}. According to Francois (1994), the yield declines by 14.3% for every unit rise in salinity above the tolerance level, or 3.9 dS m^{-1}. Garlic has a slightly greater salt tolerance threshold than most vegetable crops, but once the threshold is crossed, production quickly declines (Maas and Hoffman 1977).

Amorim et al. (2002) investigated how irrigation water affected the development and output of *A. sativum*. They discovered that during bulb development and the early growth stage, garlic plants can tolerate salt relatively well for up to 30 days. Additionally, they discovered that the last 30 days of the plant's life are the most sensitive time for bulb growth.

Siddiqui et al. (1996) studied how soil salinity affected the levels of soluble sugars, phenols, fatty acids, minerals and the rate of respiration of garlic cultivars. They found that while the initial levels of phenolic compounds and sulfur were relatively low in salt-tolerant cultivars, they increased under high salt concentrations; the opposite was true for salinity-sensitive cultivars. In saline conditions, there were no appreciable alterations in the fatty acids profile. There were no changes in the levels of N, Mn, Cu, Zn or Fe, whereas K and Ca levels fell and Na levels rose. Meanwhile, variations in respiration rate and the content of soluble saccharides were not connected to salt tolerance.

Being a crop that prefers potassium, applying K fertilizer can increase the intake of C and N while enhancing the transfer of assimilation products from the aerial leaves to the underground bulb in garlic (Shafeek et al. 2013). From 0 to 225 kg/ha, the yield of the garlic bulb and sprout rose with increasing KCl fertilizer applications. When compared to the control group, garlic biomass was efficiently boosted by KCl administration starting at the seedling stage, and the differences increased as the garlic grew. As a result, the final garlic output was significantly influenced by the early growth stage.

The photosynthetic materials were more effectively transferred from the leaves to the subterranean bulbs thanks to the K fertilizer. K fertilizer may additionally minimize sucrose hydrolysis and help with the synthesis of sucrose phosphate. The conversion of starch into soluble sugar could be enhanced by adding more K to the soil (Milford et al. 2000). Due to K's ability to accelerate the movement of nutrients from the stem and leaf to the underground bulb, the use of K fertilizer enhanced the ratio of N, P and K deposits in the bulb. In addition, the addition of K fertilizer increased the amount of soluble sugar in the soil, promoting the growth of the garlic bulb and ultimately increasing the bulb yield. The product's nutritional value was also improved by the addition of K fertilizer (Wang et al. 2022).

Salinity reduces the amount of chlorophyll and carotenoid; however, selenium supplementation helps these plants counteract the negative effects of salt stress on their chlorophyll and carotenoid levels. According to Santos (2004), increased chlorophyll hydrolysis or decreased chlorophyll production may be to blame for the drop in carotenoid levels in salinity-affected leaves (Blokhina et al. 2003; Bhattacharjee 2005). A common occurrence under salt stress is a decrease in photosynthetic pigments, which have been utilized as a sensitive indicator of the metabolic status of cells (Chutipaijit et al. 2011).

According to Amirjani (2011), salt stress reduced the amount of chlorophyll a and b in leaves. On the other hand, Se treatment raised the level of carotenoids in NaCl-containing media. Habibi (2013) and Wang (2011) looked at the effects of Se in plants under water shortage stress. As a result, there is a considerable decrease in K concentration and an increase in Na concentration in the leaves and roots of garlic, indicating that Se can assist maintain the water balance of plants under salinity stress or osmotic stress.

According to Saqib et al. (2004), the antagonistic interactions between Na and K at the absorption sites in the roots can cause the high concentration of Na in the root environment to reduce the absorption of K at the root level. In non-saline settings, the concentration of K in leaves increased as the Se concentration increased, whereas the concentration of K and Na in roots declined. The first instance of Se negatively affecting the K level in plants exposed to NaCl is the low K content in lettuce treated with Se, according to reports. The concentration of Na in the roots and leaves of garlic plants was dramatically lowered by the addition of Se (Astaneh et al. 2018).

This impact might be the result of complexes between Se and Na forming at the root level, which restricted the flow of apoplectic molecules upward by decreasing their transit through the root and to the leaves (Gong et al. 2006). Additionally, the negative impact of Na on K absorption is reduced when these substances are present in a saline medium. In addition, Liang et al. (2003) found that the addition of Se led to an increase in plasma membrane H-ATPase activity and a considerable decrease in Na absorption under salt stress. According to Kong et al. (2005), administering Se to salt-stressed sorrel plants led to K accumulation in the leaves and root shrinkage.

2.2.2 Responses of Garlic to Drought Stress

Drought stress has been proven to be a crucial limiting factor throughout the early stages of plant growth and development. It has an impact on both expansion and elongation growth (Amirjani 2011). Under water-limited conditions, higher plant fresh and dry weights are desirable traits. A common negative impact of water stress on crop plants is a decrease in the production of fresh and dry biomass (Farooq et al. 2009).

To withstand drought stress, plants exhibit a variety of physiological, biochemical and molecular responses. When there is a water shortage, leaves' stomata close and their photosynthesis slows down to reduce water loss (Cornic 2000). Abscisic acid (ABA), a hormone produced by plants and generated from a class of natural pigments called carotenoids found in plastids, plays a major role in mediating this process (Parry and Horgan 1991). By controlling stomatal closure, root growth and architecture, and the transcription of stress-responsive genes, ABA plays a crucial role in the ability of plants to withstand drought (Sah et al. 2016). Drought stress controls the expression of genes involved in ABA production and signal transduction (Yoshida et al. 2010; Huang et al. 2018).

The maximum output of garlic bulbs depends mostly on carefully planned irrigation because garlic is susceptible to water scarcity and drought stress (Abd-Elhady and Eldardiry 2016). According to Banon et al. (2006), drought is a major abiotic stressor that has increased primarily in arid and semi-arid regions like Egypt. It can directly affect crop growth by reducing cell elongation, cell turgor or cell volume, or by covering xylem and phloem vessels and obstructing any translocations (Lavisolo and Schuber 1998). The use of selenium (Se), which can improve growth and raise stress tolerance while decreasing the negative effects of abiotic stressors, is one of the preventive techniques (Astaneh et al. 2018).

The thin line separating selenium's roles as an environmental toxin and beneficial nutrition for humans and animals depend on the element's chemical makeup, concentration and other factors that affect the environment (Fan et al. 2002; Shardendu et al. 2003). Se has exhibited the capacity to control the water status of plants during drought stress, according to Kuznetsov et al. (2003). According to Kong et al. (2005), Se treatment preserved the ultrastructures of the chloroplast and mitochondria, boosting photosynthesis and improving salt tolerance in sorrel seedlings. Selenium is thought to have a beneficial effect on plants via regulating photosynthesis and antioxidant defense mechanisms. According to research by Hawrylak-Nowak (2009), exogenous selenium treatment boosted the growth rate, photosynthetic pigment and proline concentration of cucumber under salt-stress conditions. The plant's ability to tolerate salt has improved because of this application protected the cell membrane. The physiological processes in plants, such as seed germination, stomatal closure, ion uptake and transport, membrane permeability, photosynthesis and plant growth rate, are all regulated by silicon (Si), which is also described as a growth regulator (Aftab et al. 2010). When the plants were treated with Si under water stress conditions, Mustafa et al. (2017) and Bideshki and Arvin (2010) discovered that there were improvements in growth and yield parameters like fresh and dry weights of the total plant, bulb diameter and yield production.

Apart from a crop's genetic potential, growth and production are also influenced by the environmental factors present during its very stage-specific development. Water stress, whether it is an excess or a shortage, is one of the most difficult environmental factors for crop production, particularly for vegetable crops, which have a short lifespan and need a certain amount of moisture to grow and develop. The main problems restricting vegetable output due to the ongoing climatic changes are both an excess and a shortage of water. Being a shallow-rooted plant, garlic demonstrates a marked loss in anthocyanin, chlorophylls (a, b and total), carotenoids, growth metrics like the fresh weight of plant and root, bulb production, quality, and higher allicin content, as well as an increase in ion leakage, under dry circumstances (Bideshki et al. 2013; Diriba-Shiferaw 2017). The growth of plants and the development of bulbs are also harmed by excessive rain and waterlogging conditions (Diriba-Shiferaw 2017). When three *Allium* species were subjected to drought conditions in Romania, Csiszár et al. (2007) observed antioxidant enzyme activity. They discovered that, after one week, there were changes in the activities of glutathione-related enzymes glutathione reductase (GR), glutathione transferases (GST) and peroxidase (POD) in shoots linked to the relative water content of leaves. They also discovered that inducible antioxidants were very effective at protecting *Allium* ancient populations from drought. Badran (2015) performed a comparison investigation in Egypt using four commercial garlic varieties—Egaseed 2, Balady, Egaseed 1, and Sids 40—while the country was experiencing a drought. In terms of yield injury percent, superiority measure and relative performance, Egaseed 1 was determined to be the most tolerant variety, while Balady was discovered to be the most sensitive one. Additionally, he employed five ISSR primers and found 50.83% mean polymorphism, with just three primers (HB08, HB11 and 44B) exhibiting distinct bands. Any breeding effort could benefit from using the ISSR marker analysis to differentiate among garlic varieties.

Different cellular organelles produce activated ROS as waste products of regular metabolism, including superoxide ($O_2^{\cdot-}$), H_2O_2 and hydroxyl radicals (OH^-) (Scandalios 1993). AOS generation is said to increase under drought-stress conditions (Menconi et al. 1995; Borde et al. 2012), which could lead to damage from ROS-initiated damage. The antioxidant system is a term used to describe the various detoxification systems that plants have. Superoxide dismutase (SOD), POD and catalase (CAT) are among the enzymes that make up the antioxidant defense system. Under drought stress conditions, garlic plants that had been inoculated with the arbuscular mycorrhizal (AM) fungus *Glomus fasciculatum* exhibited better plant growth in terms of shoot length and biomass. By spreading extraradical hyphae in the soil, the mycorrhizal garlic plants were able to absorb more water when under drought stress, which aids in improving water usage efficiency. Due to the stress reaction to drought, the antioxidant defense response either increases or decreases. In garlic plants without AM inoculation, drought stress led to proline buildup in the shoots. However, AM garlic plants had a larger proline buildup in the roots (Wu et al. 2008; Borde et al. 2012). As proline accumulation helps to sustain high osmotic levels in plant cells experiencing a water deficit, proline is a marker of drought tolerance in drought-stressed plant tissue (Chaves et al. 2002). Similar outcomes in soybean roots during a drought are described. The buildup of shoot proline in drought-stressed garlic plants that are not mycorrhizal was shown to diminish after the 16th day of the drought period. This decline in proline accumulation in the shoots is a sign of drought damage (Ruiz-Lozano and Azcon 1997; Zhou et al. 2021). Abiotic stress primarily controls gene expression at the transcription level (Haak et al. 2017).

Pre-mRNAs are then transformed into mature mRNAs through RNA processing, which involves modifying RNA molecules and splicing out introns. Previous RNA-seq research has demonstrated alternative splicing under abiotic stress, proving that RNA-seq is a potent tool to examine gene expression regulation at the transcriptional and post-transcriptional levels (Filichkin et al. 2018; Zhu et al. 2018).

Additionally, it is important to remember that widespread expression of the eukaryotic genome results in the production of a large number of non-protein-coding RNAs (ncRNAs) in addition to protein-coding mRNAs (Dinger et al. 2009). Low levels of ncRNAs are found in plant cells, but mounting evidence points to their participation in abiotic stress responses, plant development and reproduction (du Toit. 2004; Wang et al. 2017). Totals of 57,561 protein-coding genes and 20,083 non-coding RNAs were found in garlic in an earlier study (Sun et al. 2020). Drought stress raises abscisic acid (ABA) levels, which are crucial for drought tolerance (Bray 1988).

The increased transcription of essential genes for carotenoid and ABA biosynthesis is primarily responsible for the increased buildup of ABA in response to drought. For instance, water stress was discovered to cause β-carotene hydroxylase (BCH) to be elevated in rice (Du et al. 2010). For the ABA biosynthesis brought on by dehydration, ABA deficient 4 (ABA4) expressions were required (North et al. 2007). ABA catabolism also affects the levels of ABA produced in response to drought stress. It was discovered that ABA 80-hydroxylase (CYP707A3) controls ABA threshold levels both before and after rehydration (Umezawa et al. 2006).

In garlic, BCH, beta-carotene hydroxylase (NCED), and ABA4 levels were dramatically increased in response to drought, whereas ABA 80-hydroxylase levels were significantly decreased. Garlic likely uses the same method to deal with drought stress, as shown by the greater accumulation of ABA in leaves as a result of the reverse control of ABA production and catabolism. These genes can be used as targets to increase the drought stress tolerance of garlic plants (Zhou et al. 2021). Protein denaturation and aggregation, which are harmful to plant cells, are frequently brought on by drought and other abiotic stressors (Reis et al. 2012).

Therefore, it is essential for plant survival under drought or other conditions to preserve protein structure and prevent the aggregation of misfolded ones. Plants employ a variety of tactics, one of which is the production of HSPs, which promote the refolding of denatured proteins (Wang et al. 2004). Significant abiotic factors like salt and drought represent a serious threat to agriculture. Cell dehydration and the generation of ROS are two common physiological effects they have (Forni et al. 2017).

Approximately 16.6% (867 out of 5215 genes) of the genes controlled by drought were similarly impacted by salinity stress. Comparatively, salinity harmed 5.8% (305 out of 5215 genes) of the genes controlled by drought. These genes may be the genetic basis of physiological responses to salinity and drought stress that are different and comparable. A novel method to breed garlic plants that are more tolerant to salinity and drought, or to increase drought or salinity tolerance without negatively impacting other stress defense mechanisms, should be revealed by the further functional investigation of these genes (Zhou et al. 2021).

In a climate change scenario, genetic diversity and the adaptive functional responses of crops to limiting climatic conditions are essential for sustained agricultural output. The study of functional features connected to stomata and phenotypic plasticity helps us understand how cultivars respond to various water supply circumstances. Using four different water regimens—fully irrigated (WW), water-stressed before (SW) or after bulbing stage (WS), and water-stressed (SS) throughout the entire growth period—we compared the biomass and functional properties of garlic cultivars. According to reports on other *Allium* crops (Mallor and Thomas 2008), garlic cultivars can alter resource allocation to the bulb in response to environmental variables. Several grain crops' harvest indices have improved recently, however, this characteristic is still poorly understood in some crops, including garlic (Mathew et al. 2011). Different antioxidant enzyme activity patterns could be used to distinguish among grown allium cultivars. After one week of water stress, changes in the glutathione-related enzymes (GR, GST) and POD activity in shoots were connected to the relative water content of leaves. Inducible antioxidant enzymes are thought to play a significant role in the drought stress response, according to studies. Ancient populations with higher antioxidant enzyme activity or populations that are highly inducible offer novel breeding options and opportunities to learn more about the mechanisms by which various allium plants tolerate drought (Csiszár et al. 2007).

2.2.3 Responses of Garlic to Temperature

One of the main abiotic factors that limit germination, plant development, metabolism and productivity globally is thermal stress. From seed germination to plant senescence, several biochemical events and enzyme activities that are extremely temperature-sensitive take place. The length and intensity of the temperature have an impact on how crop plants react to it. Crop breeders are currently particularly concerned about temperature stress to maintain crop output (Khar et al. 2020).

Allium's vegetative and reproductive growth and development are significantly influenced by environmental factors such as temperature, photoperiod, and others, according to the studies that have been published (Etoh and Simon 2002; Kamenetsky and Rabinowitch 2002; Kamenetsky et al.

2004a, 2004b). When garlic is in its active growing phase, the apical meristem changes from a vegetative to a reproductive condition (Kamenetsky and Rabinowitch 2001). Low temperatures encourage floral development, and a long photoperiod is necessary for the lengthening of the floral scape.

Long photoperiod and low temperatures are both necessary for flower scape elongation. According to Kamenetsky et al. (2004a, 2004b), a lengthy photoperiod and high temperature facilitated the transfer of reserves to the cloves and the deterioration of the growing inflorescence. In addition, it was found that altering the environment before and after planting might control the growth of flowers and restore fertility in genotypes of bolting garlic.

According to Wu et al. (2016), higher levels of endogenous phytohormones, particularly gibberellins (GA) and endogenous phytohormones, including MeJA (MeJA), are advantageous for garlic bolting and bulbing, which changed with different treatment combinations of photoperiod and temperature. Son et al. (2012) identified 15 upregulated and four downregulated cold-responsive genes in garlic in response to cold stress. When garlic is hibernating in the field, these cold-responsive (CR) genes can be modified to protect it against frost damage.

Garlic crop that is asexually propagated and has a high level of morphological variation is prized around the world for both its culinary and therapeutic applications. The completion of vernalization, followed by an increase in temperature and a long photoperiod, is essential for the growth and development of garlic (Wu et al. 2015). The inaccuracy of certain factors, the combination of temperature and photoperiod, and the bolting and bulbing mechanisms, on the other hand, have been significant obstacles to the control of garlic's developmental process as well as the planning of production seasons and systems. There are three types of garlic genotypes: non-bolting, semi-bolting and bolting (Kamenetsky et al. 2004a, 2004b). These genotypes vary greatly in terms of their ability to bolt, scape length and seed output (Mathew et al. 2010). Certain symmetries of temperature and photoperiod have a considerable impact on the reproductive processes in bolting accessions (Mathew et al. 2010). Initial flower stalk lengthening is brought on by lengthy photoperiod circumstances (Kamenetsky et al. 2004a; Mathew et al. 2010; Wu et al. 2016). Therefore, it has been suggested that the impact of temperature and photoperiod on scape and bulb growth should be taken into account in the context of the simultaneous but competitive development of storage (bulb) and reproductive organs in garlic (Mathew et al. 2010).

An extensive network of signalling channels is responsible for the regulation of bolting and bulbing, which consists of numerous complex procedures. According to studies by King et al. (2006) and Matthew et al. (2010), long photoperiods increase the amounts of endogenous gibberellins, which lead to improved flower bud differentiation. Gibberellic acid (GA), according to numerous research (Guo et al. 2004; Nyarko et al. 2007), could either completely or partially replace vernalization for various plants. Long-day or biennial plants with the induction of flowers had an increase in endogenous GA levels. The production of bulbs is thought to be inhibited by GA (Xu et al. 1998).

The key environmental influences on this crop's ontogeny are cold storage, growth temperature and photoperiod, although how it reacts to these influences depends on its phenological stage. In the absence of dormancy, temperature primarily regulates the emergence of the sprout, although growth temperature, photoperiod and ambient cold conditioning encourage bulb commencement. Low temperature is a significant environmental constraint that causes many structural, physiological and biochemical changes within plant cells, as well as altered gene expression patterns (Kosmala et al. 2009). While genomic and transcript-profiling studies have provided a wealth of information about the cold acclimation process, there is a growing recognition that the abundance of mRNA transcripts is not always a good predictor of cognate protein levels and that posttranslational regulation mechanisms must also play a role (Renaut et al. 2006). As a result, proteomics presents an alternative strategy to the traditional physiological method and molecular techniques, particularly for studies involving non-model plant species where minimal or no genome sequencing data are available.

The proteome study results showed that several proteins were freshly produced, accumulated or decreased due to the plant's response to the low-temperature environment. The differentially expressed proteins participate in a variety of metabolic processes, including those related to signaling, translation, host defense mechanisms, carbohydrate metabolism and amino acid metabolism (Renaut et al. 2006; Kosová et al. 2011). These proteins include both well-known stress-responsive proteins and some novel cold-responsive proteins.

Cellular growth, oxidative/antioxidative states, macromolecule transport, protein folding and transcription regulation activities are among the physiological processes that are impacted. The metabolic processes that are impacted include photosynthesis, photorespiration, energy production, carbohydrate and nucleotide metabolism, protein biosynthesis and quality control systems for proteins. These activities may combine to create a new state of cellular homeostasis that may be connected to the physiological and biochemical alterations seen in earlier research (Dufoo-Hurtado et al. 2013; Guevara-Figueroa et al. 2015); the low temperatures affect the protein profiles of garlic seed cloves.

Even though garlic is not perishable (Vázquez-Barrios et al. 2006), its quality typically degrades during storage for 6 months after harvest (Volk et al. 2004). Even when the bulbs eventually sprout, keeping the dried neck and outer skin on them will increase their shelf life. This indicates that the bulbs' hibernation period is over, their metabolism has accelerated, and the quality has changed (Tesfaye and Mengesha 2015). Garlic can be kept at room temperature (20–30 °C) for 1–2 months (Cantwell and Suslow 2004).

The loss of water, however, causes the bulbs to weaken over time and they become spongy and discolored. If the bulbs are stored between 5 and 18 °C, dormancy will soon end; the ideal storage temperature is between −1 and 0 °C. Depending on the variety, garlic bulbs respond differently to cold storage conditions. When kept at room temperature, the qualities (hardness and flavor) of garlic bulbs lasted for at least 2 months. When kept in cold storage (7.8–10.3 °C, 70% RH), garlic bulbs lost 4.38–5.57% of their weight, although losses when kept in room temperature were slightly higher (4.44–5.77%) (Bahnasawy and Dabee 2006).

There were more sprouting blubs present (24.68–64.80%) than when they were kept at ambient temperature (19.87–34.03%), when stored in a refrigerator. However, when kept in a refrigerator as opposed to at room temperature, there were fewer vacant bulbs (2.39–12.17%) (Bahnasawy and Dabee 2006). For preserving good edible quality and seed stock, it is crucial to identify the correct storage conditions for garlic bulbs. Some studies have been done on the effects of storage requirements on the quality of garlic bulbs (Amjad and Anjum 2002; Hughes et al. 2006; Sukkaew and Tira-Umphon 2012), however, there aren't many studies on the best ways to store a local variety like Lumbu Kuning.

At various storage temperatures, Purwanto et al. (2019) found that bulb weight losses varied. Those kept at 7 °C (25.08%) had the highest percentage, followed by those kept at room temperature (18.76%) and 0 °C (10.47%). The bulbs held at room temperature had the largest percentage of empty bulbs. The findings showed that low-temperature storage (7 °C; RH 50–70%) is an effective method for preserving garlic bulbs and could reduce the water content loss during prolonged storage. The percentage of empty bulbs for the garlic stored at 0 °C and room temperature was discovered.

The storage condition at room temperature produced more empty bulbs than at 0 °C. The study also revealed that garlic bulbs stored at 7 °C (RH 50–70%) had the highest sprouting percentage (25.16%). After being stored at 7 °C for 3 months, some garlic bulbs began to sprout, although none did so when they were stored at 0 °C or ambient temperature. As a result of the γ-glutamyl peptides being degraded by increased activity of a trans-peptidase induced after the breakdown of dormancy, the bulb sprouting is linked to an increase in isoalliin after storage at low temperature.

Physically, external conditions like a chilly temperature (7 °C) mostly determine when sprouting is initiated (Dufoo-Hurtado et al. 2015). Garlic bulbs are kept at the right temperature to preserve them for seed stock. Garlic bulbs from the cv. Lumbu Kuning suffered from poor quality during storage due to temperature circumstances. The lowest proportion of weight and empty bulbs were produced by storage at 0 °C (RH 50–70%) in complete darkness. Bulb sprouting rates were maximum when stored at 7 °C. According to one study, keeping garlic bulbs in storage at 0 °C is advised for long-term storage (Purwanto et al. 2019).

Bulbing only happened in some cultivars when two thresholds were met: a minimum thermal time of 600 degree days and a minimum photoperiod of 13.75 h. (Wu et al. 2016). To the best of our knowledge, however, little research has been done on the factors that are necessary for bolting and bulbing as well as the modifications that this process makes to the signal substance in *A. sativum* (Kamenetsky et al. 2004a, 2004b; Mathew et al. 2011).

2.2.4 Responses of Garlic to Heavy Metals

Heavy metals, particularly iron and copper, have been repeatedly demonstrated to be major soil contaminants in industrialized areas. To survive, plants must adapt their metabolism to these pollutants. One of these adaptations involves sulfur metabolism, which is known to play an important role in plant defense (Tchounwou et al. 2012). The pollution is typically produced by industrial, mining and milling processes, and radioactive waste. As contaminants, heavy metals have drawn increasing attention because of their potential harm to humans, animals and plants. Traditionally, heavy metals are described as elements with metallic characteristics, such as ductility, stability as cations, ligand specificity, conductivity, and as well as an atomic number over 20. One of the largest groups of harmful contaminants are heavy metals, which can interfere with the ecosystem's balance (Jaishankar et al. 2014).

A few investigations have been revealed to show their results, including the effects of Cd_2 concentration on garlic root, bulb and shoot accumulation. However, the plants transported only a small amount of Cd to their bulbs and shoots (Jiang et al. 2001). The differential uptake of Pb in roots and shoots may be due to the uptake of lead from the aqueous phase of the soil by the natural ion uptake mechanisms of the roots (Liu et al. 2009). The intercropping of garlic with other plants is helpful to absorb the heavy metal content from contaminated soils. Further, the interplanting application protects other companion plants by decreasing the heavy metal content of the rhizosphere (Ashrafi et al. 2015).

The ability of garlic to withstand biotic and abiotic environmental challenges, such as bacterial, viral and oxidative stresses, as well as its propensity to withstand some heavy metals, such as cadmium, is well documented (Zhang et al. 2005a). Given its impact on the microbiological activity of heavy metals in soil, garlic was chosen as an intercrop plant with hyper-accumulator properties (Ma et al. 2018). As far as interplanting is concerned, this area has caught the attention of soil scientists to treat soils with heavy metal contamination (Xiong et al. 2018).

Garlic has the ability to thrive while Cd levels are reduced with silicon treatment. To our knowledge, however, there is a dearth of data on garlic growth in response to heavy metal deposition in farming soils following organic amendments. The purpose of the current study was to determine whether any of the potential trace elements (PTEs) detected in some organic manures could contribute to pollution or be harmful to the general public's health after intake of vegetables (garlic) (Wang et al. 2016; Akhter et al. 2022).

A significant industrial agent and environmental pollutant that contributes significantly to plant disease is cadmium (Cd). This study assessed how Cd affected the physiological and molecular characteristics of garlic development and active its ingredient. The Cd concentrations used to irrigate the garlic cloves varied. The obtained data demonstrate that as the Cd concentration in the growth medium increased, the measured pigments and total soluble sugars dropped. Proline plays a critical function in plant osmoregulation, stabilizing the protein synthesis machinery, and acting as a powerful singlet oxygen quencher, all of which contribute to an increase in protein and proline concentration (Lalla et al. 2013).

Many other sulfuric compounds, including phytochelatin, thiosulfinate and sulfoxide, can be found in garlic. These chemicals share some metabolic processes with phytochelatins and use cysteine as a fundamental precursor (Lancaster and Shaw 1989; Block et al. 1992). As a result, a high capacity for heavy metal uptake and storage has been predicted. *A. sativum* was shown to hyperaccumulate cadmium (Jiang et al 2001; Soudek et al. 2009). The antioxidant enzymes SOD, POD and CAT were discovered to be stimulated in the leaves of garlic seedlings grown in the presence of cadmium by Zhang et al. (2005b). According to Liu et al. (2009), lead exposure caused oxidative stress in garlic.

The expression of the AsPCS1 and AsMT2a genes was significantly altered by stress (Zhang et al. 2005b). The new garlic gene, AsMT2b that responds to cadmium was also described by Zhang et al. (2006). Endoplasmic reticulum (ER) and dictyosomes are involved in the efficient defense system against Pb toxicity, which may be one reason explaining the decreased toxicity of Pb (Jiang and Liu, 2010). The creation of novel ER proteins involved in heavy metal tolerance was sped up after too many Pb ions were allowed to reach the cytoplasm. Metal-complexing proteins or polysaccharide components that assisted in the healing of damaged membranes and cell walls may be present in certain vesicles from the ER and dictyosomes (Jiang and Liu 2010).

The proteins that bind Pb by forming stable metal–PC complexes in the cytoplasm may have been conveyed by some vesicles. The content of free metal ions in the cytoplasm decreased in this manner. PC levels within cells can be adequate for Pb binding. Unquestionably, ER is crucial to the detoxification of Pb. Almost all harmful compounds that plants can be exposed to end up in the vacuole, and the vacuoles of root cells are the primary sites of metal sequestration (Clemens 2006). A detoxifying route for preventing cell harm and keeping the metal in certain vacuoles can be conceived of as cytoplasmic vacuolization and the increased level of electron-dense granules in vacuoles (Einicker-Lamas et al. 2002).

According to Sharma and Dubey (2005) the majority of Pb within a cell is confined to the vacuole in the form of complexes. In the leaf cells of several plants treated with Pb salt solutions, pinocytosis is seen. Pb particles can be released into the vacuole via pinocytotic vesicles (Wierzbicka and Antosiewicz 1993).

Many studies using electron energy loss (EEL) spectra provided unmistakable proof that Cd was localized as electron-dense granules accumulated in different cell areas during Cd stress (Liu and Kottke 2004). As a result, it is possible to imagine that these Cd-containing granules function as a Cd detoxifying mechanism. According to an earlier finding (Liu and Kottke 2004), the vesicles included phytochelatins (PCs), which form stable PC/Cd complexes that are encapsulated in plant vacuoles (Saxena et al. 1999).

2.3 RESPONSES OF GARLIC TO BIOTIC STRESS

Allium sativum is one of the 20 most important vegetables, and is the most significant bulbous crop in the world and one of the crops most sensitive to *Fusarium* spp. Garlic is primarily infected by *F. oxysporum* (f. sp. cepae) and *F. proliferatum*, but it is also susceptible to attacks from *F. acutatum*, *F. anthophilium*, *F. verticillioides*, *F. solani* and *F. acuminatum*. Typical symptoms of these infections include leaf wilting and discoloration (FW) or dry brown necrotic spots, and white mycelium. The immunological response of garlic to *Fusarium* infection is thought to be mediated by PR1–5 proteins, including chitinases and endo-1,3-glucanases, according to earlier research on FBR-resistant and -susceptible *A. sativum* cultivars (Filyushin et al. 2021; Anisimova et al. 2021; Filyushin et al. 2022).

White rot disease, one of the most severe global garlic illnesses, is brought on by the fungus *Sclerotium cepivorum* (Nabulsi et al. 2001). Al-Safadi et al. (2000) started mutation breeding of garlic in Syria to produce mutants immune to white rot using gamma radiation, and they were successful in producing such mutants. Furthermore, by the use of 13 random primers and random amplified polymorphic DNA (RAPD) analysis, Nabulsi et al. (2001) were able to determine the molecular diversity among eight mutants of garlic. Twelve primers showed polymorphism in the amplification products, while further highly resistant mutants had low correlation coefficients and were located fairly distant from the control. The pattern of bands produced by the highly resistant mutant and primer OPB-15 with sequence (GGAGGGTGTT) could be used as a genetic identifier for future garlic breeding programs.

Numerous *Penicillium* species are responsible for the blue mold disease in garlic, which has been linked to large yearly crop losses. Stunted, chlorotic plants with withered leaves and smaller bulbs are some of the symptoms (Valdez et al. 2006). Garlic accessions were tested in Argentina against *Penicillium hirsutum* by Cavagnaro et al. (2005 a, 2005b), who discovered considerable variations in the accessions. The most resistant cultivars were Castano and Morado, and it was shown that there was little link between allicin content and disease tolerance, suggesting that allicin is not the primary cause of resistance to *P. hirsutum*.

The fact that crop pests and diseases are changing in intensity, location and frequency is clear evidence of climate change (Lamichhane et al. 2015). Numerous diseases can affect garlic, including basal rot (*Fusarium culmorum*) (Mishra et al. 2014), white rot (*Sclerotium cepivorum*) (Zewde et al. 2007), downy mildew (*Peronospora destructor*) (Schwartz 2004), botrytis rot (*Botrytis porri*) (Wu et al. 2012) and *Penicillium* decay (*Penicillium hirsutum*). The majority of the primary garlic illnesses are soil-borne, thus conducting an accurate site assessment and rotating the plants annually are essential to maintaining a healthy garlic garden.

2.4 GENOMIC APPROACHES FOR CLIMATE-RESILIENT AND MOLECULAR BREEDING

2.4.1 TRANSCRIPTOMICS

RNA-seq is often used to study the transcriptomics process since it is inexpensive to sequence data and there are pipelines for interpreting readily available data. The study of transcription should facilitate the synchronized profiling of tissues at any stage of plant development or at any given moment, as well as the validation and discussion of any prospective transcripts that might be differentially expressed. Transcriptomes have been studied in the double haploid (DH) replication of garlic and onion (Baldwin et al. 2012; Sun et al. 2013). There are numerous references and studies for the use of RNA-seq in the investigation of organ growth, male sterility, marker discovery and response (Sun et al. 2012; Kamenetsky et al. 2015; Shemesh-Mayer et al. 2015; Zhu et al. 2017). Kim et al. (2015) attempted onion RNA-seq structural annotation using an integrated structural genomic annotation method. In order to build a draught reference transcript for onions, Sohn et al. (2016) used long-read sequencing technology. This suggests that future developments in sequencing chemistry will make it easier to sequence the entire genome of the *Allium* spp.

Many environmental factors, including low temperature, acetic acid complex and other amino compounds, can cause garlic to become discolored (Lukes 1986; Kubec and Velíšek 2007; Liang et al. 2014; Li et al. 2015). Despite this, the genome of garlic is huge (approximately 16 Gb), hence there is little genetic and genomic knowledge about this crop. Because of this, the precise mechanism underlying these environmental factors that turn garlic green is unknown. Recently, next-generation sequencing (NGS) has been used as a sequencing approach that is both affordable and effective (Van Dijk et al. 2014; Goodwin et al. 2016). Transcriptome analysis by the NGS method has several benefits, including being a quick, affordable and unrestricted method of genomic complexity that has been widely used as a fundamental method in many research fields, including gene identification, microsatellite marker evolution and the study of domesticated patterns of crops, particularly in the characterization of the profiling of gene expression (Wang et al. 2009; Qi et al. 2011; McGettigan 2013). A sizable number of expressed genes have been discovered as a result of prior research that generated the garlic transcript (Sun et al. 2012; Kamenetsky et al. 2015). These studies have widened our comprehension of the networks, structures and expression patterns of garlic traits and established a new field of study.

2.4.2 GENETIC DIVERSITY USING DNA MARKERS

Numerous DNA markers are frequently employed to assess genetic diversity due to their objectivity, consistency of results across laboratories and absence of environmental influence on their expression (Etoh and Simon 2002). Isozymes have been used to estimate the genetic diversity of garlic (Pooler and Simon 1993; Maass and Klaas 1995; Ipek et al. 2003), random amplified DNA (RAPD) markers (Choi et al. 2003; Ipek et al. 2003; Mario et al. 2008; Paredes et al. 2008; Abdoli et al. 2009) and inter-sample sequence repeats (Ipek et al. 2008 a &b). The evaluation of garlic diversity using SSR markers was initially reported by Ma et al. in 2009. Utilizing a large SSR collection, they were able to identify eight microsatellites to estimate genetic diversity. For investigations on DNA genetic variation, population structure analysis and core group evaluation, Zhao et al. (2011) employed the same eight SSRs. Jo et al. (2012) used seven markers from the same eight SSRs mentioned previously to rate the genetic diversity in 120 genotypes from five different countries. Using the enriched $(CT)_8$ and $(GT)_8$ collection, Cunha et al. (2012) published a recent collection of 16 microsatellites and found 10 polymorphic SSRs.

In contrast, Chen et al. (2013) used a comparable set of indicators and found eight polymorphic markers. Using 99 SSRs, Khar (2012) calculated the genetic diversity of garlic and described 18 polymorphic SSRs. Cunha et al. (2014) made use of 17 established SSR markers (Ma et al. 2009; Cunha et al. 2012). For Brazilian genotypes, Cunha et al. (2014) calculated the genetic diversity and population structure. After three decades of competition with imported garlic, Gomes Viana et al. (2020) recently utilized microsatellite markers to examine how patterns of genetic variation evolve in indigenous strains grown in family farming systems. Kıraç et al. (2022) studied the morphological and molecular characterization of garlic in Turkey.

2.4.3 MOLECULAR MARKER-ASSISTED BREEDING

Gene maps are useful tools for identifying genes and quantitative trait loci (QTLs), understanding the genetic basis of complex traits, using marker-assisted selection and cloning important genes. By enabling marker-assisted selection (MAS) and the identification of genes and QTLs that effectively influence desirable traits, the creation of a genetic map is anticipated to accelerate the evolution of garlic. Numerous genetic maps of garlic have been created. Zewdie et al. (2005) used 37 markers to create nine linkage groups to create the first genetic map of garlic. Additionally, Ipek (2005) used 321 markers separated in MP2 (13.9 per marker combination) and 360 markers segregated in MP1 (12.8 AFLP markers per marker combination) to illustrate a genetic linkage map with low density. From a collection of 53 garlic (*Allium* spp.) species, only two low-density genetic maps have been created thus far, and no QTLs have been found (Ipek et al. 2003).

However, utilizing a recently proposed transcription-weighted association research, Zhu et al. (2019) found 25, 2 and 30 single nucleotide polymorphisms (SNPs) in transcripts for bulb weight (BW), diameter (BD) and numerous garlic cloves (CN), respectively. QTL mapping to various characters can hasten garlic reproduction because it aids garlic breeders in (a) identifying some of the genes/QTLs that control the characteristic, (b) taking into consideration the influence of genes/ QTLs that control the characteristic, (c) locating the gene/QTLs and (d) examining the relationship among various genes/QTLs. All of these objectives support the hierarchy of numerous target genes in a single genotype and highlight the complexity of the examined germplasm (Abu-Ellail et al. 2021).

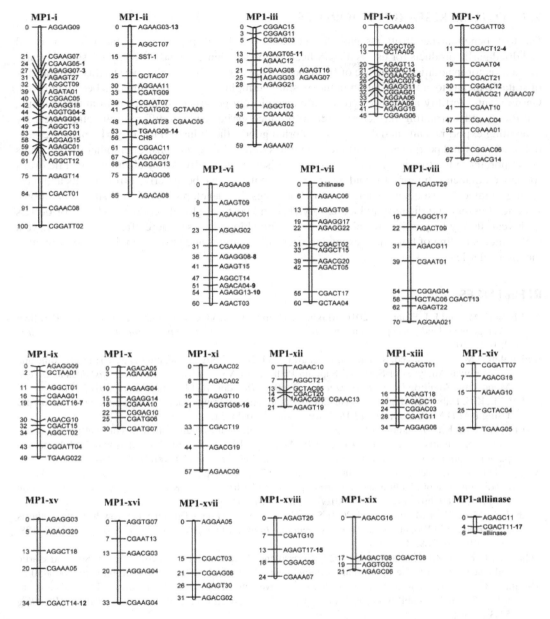

FIGURE 2.3 Linkage groupings in the genetic map of garlic constructed on the MP1 family. Map distances are in centiMorgans. Numbers in boldface are frequent markers segregating in both MP1 and MP2.

Source: Ipek et al. 2005.

Omics approaches offer potential resources for comprehending the biological functions of any genetic information, as well as displaying significant genetic regulatory networks, to boost yields (Keurentjes et al. 2008; Stinchcombe and Hoekstra 2008). Nearly all important crop improvement breeding programs use genomics techniques with traditional breeding to shorten the duration and examine elite germplasm (Salem 2004; Bevan and Waugh 2007) (Figure 2.3).

2.5 FUTURE RESEARCH PRIORITIES

Due to its beneficial health effects, garlic is a crucial horticultural crop. The manufacture of processed and dried garlic products for use as dietary health-food supplements and in food processing is a significant sector in addition to the use of fresh garlic. Due to garlic's reproductive sterility and the lack of suitable means to create variation in the existing germplasm, garlic breeding has been hampered. Garlic plants' fertility and flower growth are also closely governed by their environment, making it difficult and warranting further research to produce garlic seeds in different climatic zones. By generating export revenue and displacing imported goods, the industrial use of garlic plants' essential oil, flavors and perfumes support the economies of developing nations. Chemotherapeutic drugs are important in the treatment of human clinical diseases generally. Many chemotherapeutic drugs for cancer, diabetes, tumors, thyroid disorders and malaria have been derived from plants and have been generated in vast quantities using plant hormones that develop under abiotic stress conditions. Abiotic stressors significantly reduce crop yields and cost the global economy billions of dollars each year. It may be possible for plants to cope with abiotic stresses more effectively if we have a better understanding of how they respond to stressful situations and changes in their environment at the molecular level.

REFERENCES

Abd-Elhady, M., and E.I. Eldardiry. 2016. Maximize Crop Water Productivity of Garlic by Modified Fertilizer Management Under Drip Irrigation. *International Journal of Chemistry and Technology Research* 9: 144–150.

Abdelrahman M., M. El-Sayed, S. Sato, H. Hirakawa, et al. 2017. RNA-Sequencing-Based Transcriptome and Biochemical Analyses of Steroidal Saponin Pathway in A Complete Set of *Allium fistulosum-A. cepa* Monosomic Addition Lines. *PLoS One* 12 (8): e0181784. doi: 10.1371/journal.pone.0181784.

Abdoli, M., K.B. Habibi, K. Baghalian, et al. 2009. Classifcation of Iranian Garlic (*Allium sativum* L.) Ecotypes Using RAPD Marker. *Journal of Medicinal Plants* 8(5):45–51.

Abu-Ellail, F.F.B., K.F.M. Salem, M.M. Saleh, et al. 2021. Molecular Breeding Strategies of Beetroot (*Beta vulgaris* ssp. vulgaris var. conditiva Alefeld). In: Al-Khayri J.M., S.M. Jain, D.V. Johnson (eds) Advances in plant breeding strategies: Vegetable crops. Springer, Cham. doi.org/10.1007/978-3-030-66965-2.4

Adekpe, D.I., J.A.Y. Shebayan, U.F. Chiezey, et al. 2007. Yield responses of garlic (*Allium sativum* L.) to oxadiazon, date of planting and intra-row spacing under irrigation at Kadawa, Nigeria. *Crop Protection* 26, 1785–1789. https://doi.org/10.1016/j. cropro.2007.03.019.

Aftab, T., M.M.A. Khan, M. Idrees, et al. 2010. Salicylic Acid Acts as Potent Enhancer of Growth, Photosynthesis and Artemisinin Production in *Artemsis annua* L. *Journal of Crop Science Biotechnology* 13: 183–188. doi: 10.1007/s12892-010-0040-3

Akhter, P., P. Khan, Z.I. Hussain, et al. 2022. Assessment of Heavy Metal Accumulation in Soil and Garlic Influenced by Waste-Derived Organic Amendments. *Biology* 11: 850. doi:10.3390/biology11060850

Al-Safadi B., A.N. Mir, and M.I..E Arabi. 2000. Improvement of Garlic (*Allium sativum* L.) Resistance to White Rot and Storability Using Gamma Irradiation Induced Mutations. *Journal of Genetic Breeding* 54(3):175–182.

Amirjani, M.R. 2011. Effect of Salinity Stress on Growth, Sugar Content, Pigments and Enzyme Activity of Rice. *International Journal of Botany* 7(1): 73–81. doi: 10.3923/ijb.2011.73.81

Amjad, M., and M.A. Anjum. 2002. Evaluation of Physiological Quality of Onion Seed Stored for Different Periods. *International Journal of Agriculture and Biology* 4(3): 365–369. https://eurekamag.com/resea rch/003/759/003759890.php.

Amorim, J.R.d.A., P.D. Fernandes, H.R. Gheyi, et al. 2002. Effect of Irrigation Water Salinity and its Mode of Application on Garlic Growth and Production. *Pesquisa Agropecuária Brasileira* 37(2): 167–176. www. scielo.br/j/pab/a/x8WSmRy3kxzRTTmT8zBNsnG/abstract/?format=html&lang=en.

Anisimova, O.K., A.V. Shchennikova, E.Z. Kochieva, et al. 2021. Pathogenesis-Related Genes of PR1, PR2, PR4, and PR5 Families are Involved in the Response to Fusarium Infection in Garlic (*Allium sativum* L.). *International Journal of Molecular. Sciences* 22: 6688. doi:10.3390/ijms22136688

Arumuganathan K., and E.D. Earle. 1991. Nuclear DNA Content of Some Important Plant Species. *Plant molecular biology reporter* 9(3): 208–218. doi: 10.1007/BF02672069

Ashrafi, A., M. Zahedi, and M. Soleimani. 2015. Effect of Co-planted Purslane (*Portulaca oleracea* L.) on Cd Accumulation by Sunflower in Different Levels of Cd Contamination and Salinity: A pot Study. *International Journal of Phytoremediation*. 17: 853–860. doi: 10.1080/15226514.2014.981239

Astaneh, R.K., S. Bolandnazar, F.Z. Nahandi, et al. 2018. The Effects of Selenium on Some Physiological Traits and K, Na Concentration of Garlic (*Allium sativum* L.) Under NaCl Stress. *Information Processing in Agriculture* 5(1): 156–161. doi: 10.1016/j.inpa.2017.09.003

Atif, M.J., B. Amin, M.I. Ghani, et al. 2020. Variation in Morphological and Quality Parameters in Garlic (*Allium sativum* L.) Bulb Influenced by Different Photoperiod, Temperature, Sowing and Harvesting Time. *Plants* 9(2): 155. doi: 10.3390/plants9020155

Ayer, R. 1985. Water Quality for Agriculture. *FAO Irrigation and Drainage Paper, Rome* 29: 141.

Badran, A.E. 2015. Comparative Analysis of Some Garlic Varieties Under Drought Stress Conditions. *Journal of Agriculture Science* 7(10):271. doi: 10.5539/jas.v7n10p271

Bahnasawy, A.H., and S.A. Dabee. 2006. Technological Studies on Garlic Storage. Paper presented at the The 1st Conference of the Agricultural Chemistry and Enviroment Protection Society, Ain Shams university.

Baldwin, S., R. Revanna, S. Thomson, et al. 2012. A Toolkit for Bulk PCR-Based Marker Design from Next-Generation Sequence Data: Application for Development of a Framework Linkage Map in Bulb Onion (*Allium cepa* L.). *BMC Genomics* 13(1):637. doi: 10.1186/1471-2164-13-637

Banon, S.J., J. Ochoa, J.A. Franco, et al. 2006. Hardening of Oleander Seedlings by Deficit Irrigation and Low Air Humidity. *Environmental and Experimental Botany*. 56: 36–43.doi: 10.1016/j.envexpbot.2004.12.004

Battaglia, E. 1963. Mutazione Chromosomicae Cariotipe Fondamentale in *Allium sativum* L. *Caryologia* 16(1): 1–46. doi: 10.1080/00087114.1963.10796082

Bevan, M., and R. Waugh. 2007. Applying Plant Genomics to Crop Improvement. *Genome Biology* 8 (2): 302. doi: 10.1186/gb-2007-8-2-302

Bhattacharjee, S. 2005. Reactive Oxygen Species and Oxidative Burst: Roles in stress, Senescence and Signal Transducation in Plants. *Current Science* 1113–1121. www.jstor.org/stable/24110963.

Bideshki, A., M.J. Arvin, and M. Darini. 2013. Interactive Effects of Indole-3-Butyric Acid (IBA) and Salicylic Acid (SA) on Growth Parameters, Bulb Yield and Allicin Contents of Garlic (*Allium Sativum*) Under Drought Stress in Field. *International Journal of Agronomy and Plant Production* 4(2):271–279.

Bideshki, A., and M.J. Arvin. 2010. Effect of Salicylic Acid (SA) and Drought Stress on Growth, Bulb Yield and Allicin Content of Garlic (*Allium sativum*) in Field. *Plant Eco-physiology*. 2: 73–79.

Block, E., S. Naganathan, D. Putman, et al. 1992. *Allium* Chemistry: HPLC Analysis of Thiosulfinates from Onion, Garlic, Wild Garlic (Ramsoms), Leek, Scallion, Shallot, Elephant (Great-Headed) Garlic, Chive, and Chinese Chive. Uniquely high Allyl to Methyl Ratios in Some Garlic Samples. *Journal of Agricultural and Food Chemistry* 40(12): 2418–2430. doi: 10.1021/jf00024a017

Blokhina, O., E. Virolainen, and K.V. Fagerstedt. 2003. Antioxidants, Oxidative Damage and Oxygen Deprivation Stress: A Review. *Annals of Botany* 91(2): 179–194. doi: 10.1093/aob/mcf118

Borde M., M. Dudhane, and P. Jite. 2012. Growth, Water Use Efficiency and Antioxidant Defense Responses of Mycorrhizal and Non-Mycorrhizal *Allium sativum* L. Under Drought Stress Condition. *Annals of Plant Sciences* 1(1) 6–11.

Bozzini, A., and P. De Luca. 1991. Discovery of An Italian Fertile Tetraploid Line of Garlic. *Economic Botany* 45(3): 436–438. www.jstor.org/stable/4255377.

Bradley K.F., M.A. Rieger, and G.G. Collins. 1996. Classification of Australian Garlic Cultivars by DNA Fingerprinting. *Australian Journal of Experimental Agriculture* 36(5): 613-. doi:10.1071/EA9960613

Bray, E.A. 1988. Drought-and ABA-Induced Changes in Polypeptide and mRNA Accumulation in Tomato Leaves. *Plant Physiology* 88(4): 1210–1214. doi: 10.1104/pp.88.4.1210

Campos, M.R.S., K.R. Gómez, Y.M. Ordo, et al. 2013. Polyphenols, Ascorbic Acid and Carotenoids Contents and Antioxidant Properties of Habanero Pepper (*Capsicum chinense*) fruit. *Food and Nutrition Sciences* 4(8A): 47–54. doi: 10.4236/fns.2013.48A006

Cantwell, M., and T. Suslow. 2004. Garlic. Recommendations for Maintaining Postharvest Quality. http:/post harvest. ucdavis. edu/Produce/ProduceFacts.

Cavagnaro P.F., A. Camargo, R.J. Piccolo, et al. 2005a. Resistance to *Penicillium hirsutum* Dierckx in Garlic accessions. *European Journal of Plant Pathology* 112(2):195–199. doi: 10.1007/s10658-005-1750-6

Cavagnaro, P.F., D. Senalik, C.R. Galmarini, et al. 2005b. Correlation of Pungency, Thiosulfinates, Antiplatelet Activity and Total Soluble Solids in Two Garlic Families. *American Society for Horticulture Science* 40(4):1019. doi:10.21273/HORTSCI.40.4.1019E

Chakraborty D., Majumder A. 2020. Garlic (Lahsun)–an immunity booster against SARS-CoV-2. *Biotica Research Today* 2(8):755–757.

Chaves, M.M., J.S. Pereira, J. Maroco, et al. 2002, How plants Cope with Water Stress in the Field Photosynthesis and Growth. *Annalytical Botany* 89: 907–916. doi: 10.1093/aob/mcf105

Chen, S.X., J. Zhou, Q. Chen, et al. 2013. Analysis of the Genetic Diversity of Garlic (*Allium sativum* L.) Germplasm by SRAP. *Biochemical Systematics and Ecology* 50:139–146. doi:10.1016/j.bse.2013.03.004

Choi, H.S., K.T. Kim, Y.K. Ahn, et al. 2003. Analysis of Genetic Relationships in Garlic Germplasm and Fertile Garlic by RAPD. *Korean Society for Horticultura Science* 44(5): 595–600.

Chutipaijit, S., S. Cha-um, and K. Sompornpailin. 2011. High Contents of Proline and Anthocyanin Increase Protective Response to Salinity in 'Oryza sativa 'L. spp.'indica'. *Australian Journal of Crop Science* 5(10): 1191–1198. doi: 10.3316/informit.746208527906618

Clemens, S. 2006. Toxic metal Accumulation, Responses to Exposure and Mechanisms of Tolerance in Plants. *Biochimie* 88(11): 1707–1719. doi: 10.1016/j.biochi.2006.07.003

Cornic, G. 2000. Drought Stress Inhibits Photosynthesis by Decreasing Stomatal Aperture–Not by Affecting ATP Synthesis. *Trends in plant science* 5(5): 187–188. doi: 10.1016/S1360-1385(00)01625-3

Csiszár J., E. Lantos, I. Tari, et al. 2007. Antioxidant Enzyme Activities in Allium Species and their Cultivars Under Water Stress. *Plant Soil and Environment* 53(12):517–523. doi: 10.17221/2192-PSE

Cunha, C.P., E.S. Hoogerheide, M.I. Zucchi, et al. 2012. New Microsatellite Markers for Garlic, *Allium sativum* (Alliaceae). *American Journal of Botany* 99(1): e17–e19. doi: 10.3732/ajb.1100278

Cunha, C.P., F.V. Resende, M.I. Zucchi, et al. 2014. SSR-Based Genetic Diversity and Structure of Garlic Accessions from Brazil. *Genetica* 142:419–431. doi: 10.1007/s10709-014-9786-1

Di, C., J. Yuan, Y. Wu, et al. 2014. Characterization of Stress-Responsive Lnc RNAs in *A rabidopsis thaliana* by Integrating Expression, Epigenetic and Structural Features. *Plant Journal* 80(5): 848–861. doi: 10.1111/tpj.12679.

Dinger, M.E., P.P. Amaral, T.R. Mercer, et al, 2009. Pervasive Transcription of the Eukaryotic Genome: Functional Indices and Conceptual Implications. *Briefings in Functional Genomics and Proteomics* 8(6): 407–423. doi: 10.1093/bfgp/elp038.

Diriba-Shiferaw, G. 2017. Comparative Study of Different Compound Fertilizers on Garlic (*Allium sativum* L.) Productivity under Various Soils and Seasons. *Global Journal of Science Frontier Research* 17: 1–9.

du Toit, E.S., P.J. Robbertse and J.G. Niederwieser. 2004. Plant Carbohydrate Partitioning of Lachenalia cv. Ronina During Bulb Production. Scientia Horticulturae 102(4): 433–440. doi: https://doi.org/10.1016/j.scienta.2004.06.002

Du, H., N. Wang, F. Cui, et al. 2010. Characterization of the β-Carotene Hydroxylase Gene DSM2 Conferring Drought and Oxidative Stress Resistance by Increasing Xanthophylls and Abscisic Acid Synthesis in Rice. *Plant Physiology* 154(3): 1304–1318. doi:10.1104/pp.110.163741

Dufoo-Hurtado, M., D., Zavala-Gutieìrrez, K.G. Cao, et al. 2013. Low-Temperature Conditioning of "Seed" Cloves Enhances the Expression of Phenolic Metabolism Related Genes and Anthocyanin Content in 'Coreano' Garlic (*Allium sativum*) During Plant Development. *Journal of Agriculture and Food Chemistry* 61: 10439–10446. doi: 10.1021/jf403019t

Dufoo-Hurtado, M.D., J.Á. Huerta-Ocampo, A. Barrera-Pacheco, et al. 2015. Low Temperature Conditioning of Garlic (*Allium sativum* L.) "Seed" Cloves Induces Alterations in Sprouts Proteome. *Frontiers in Plant Science* 6. doi: 10.3389/fpls.2015.00332. doi: 10.3389/fpls.2015.00332

Einicker-Lamas, M., G.A. Mezian, T.B. Fernandes, et al. 2002. *Euglena gracilis* As A Model for the Study of Cu^{2+} and Zn^{2+} Toxicity and Accumulation in Eukaryotic Cells. *Environmental pollution* 120(3): 779–786. doi: 10.1016/S0269-7491(02)00170-7

Etoh, T. 1983. Germination of Seeds Obtained from A Clone of Garlic, *Allium sativum* L. *Proceedings of the Japan Academy*, Series B 59(4): 83–87. doi: 10.2183/pjab.59.83

Etoh, T. 1985. Studies on the Sterility in Garlic, *Allium sativum* L. *Memoirs of the Faculty of Agriculture Kagoshima University.* 21: 77–132.

Etoh, T. and P.W. Simon. 2002. Diversity, Fertility and Seed Production of Garlic. Allium Crop Science: Recent Advances. CABI International, Walling ford UK, pp 101–117. doi: 10.1079/9780851995106.0101

Fan, T.W.M., S.J. Teh, D.E. Hinton and R.M. Higashi, 2002. Selenium Bio-Transformations into Proteinaceous Forms by Food Web Organisms of Selenium-Laden Drainage Waters in California. *Aquatic Toxicology* 57:65–84. doi:10.1016/S0166-445X (01)00261-2

FAOSTAT, 2022. http://faostat.fao.org

Farooq, M., A. Wahid, N. Kobayashi, D. et al. 2009. Plant Drought Stress: Effects, Mechanisms and Management. In E. Lichtfouse, M. Navarrete, P. Debaeke, S. Véronique & C. Alberola (Eds.), Sustainable Agriculture (pp. 153–188). Dordrecht: Springer Netherlands.

Filichkin, S.A., M. Hamilton, P.D. Dharmawardhana, et al. 2018. Abiotic Stresses Modulate Landscape of Poplar Transcriptome Via Alternative Splicing, Differential Intron Retention, and Isoform Ratio Switching. *Frontiers in Plant Science* 9: 5. doi.org/10.3389/fpls.2018.00005

Filyushin, M.A., O.K Anisimova, E.Z. Kochieva, et al. 2021. Genome-Wide Identification and Expression of Chitinase Class I Genes in Garlic (*Allium sativum* L.) Cultivars Resistant and Susceptible to *Fusarium Proliferatum*. *Plants* 10: 720. doi: 10.3390/plants10040720

Filyushin, M.A., B.T. Shagdarova, A.V. Shchennikova, et al. 2022. Pretreatment with Chitosan Prevents Fusarium Infection and Induces the Expression of Chitinases and _-1,3-Glucanases in Garlic (*Allium sativum* L.). *Horticulturae* 8: 383. doi: 10.3390/horticulturae8050383

Flowers, T. 2004. Improving Crop Salt Tolerance. *Journal of Experimental Botany* 55(396): 307–319. doi: 10.1093/jxb/erh003

Forni, C., D. Duca, and B.R. Glick. 2017. Mechanisms of Plant Response to Salt and Drought Stress and their Alteration by Rhizobacteria. *Plant and Soil* 410(1): 335–356. doi: 10.1007/s11104-016-3007-x

Foyer, C.H., and G. Noctor. 2000. Oxygen Processing in Photosynthesis: Regulation and Signalling. *New Phytologist* 146(3): 359–388. doi: 10.1046/j.1469-8137.2000. 00667.x

Francois, L.E. 1994. Yield and Quality Response of Salt-Stressed Garlic. *Hortscience* 29(11), 1314–1317. doi: 10.21273/HORTSCI.29.11.1314

Gemma, J.N., R.E. Koske, E.M. Roberts, et al.1997. Mycorrhizal fungi improve drought resistance in creeping bentgrass. *Journal of Turfgrass Science* 73: 15–29.

Gomes Viana J.P., C.D.J. Pires, M.M. Bajay, et al. 2020. Do the importations of crop products affect the genetic diversity from landraces? A study case in garlic (*Allium sativum* L.). *Genetic Resources and Crop Evolution* 68: 1199–1211.

Gong, H., D. Randall, and T. Flowers. 2006. Silicon Deposition in the Root Reduces Sodium Uptake in Rice (*Oryza sativa* L.) Seedlings by Reducing Bypass Flow. *Plant, Cell and Environment* 29(10): 1970–1979. doi: 10.1111/j.1365-3040.2006.01572.x

Goodwin, S., J.D. McPherson, and W.R. McCombie. 2016. Coming of Age: Ten Years of Next-Generation Sequencing Technologies. *Nature Reviews Genetics* 17(6):333–351. doi: 10.1038/nrg.2016.49

Green, O.C., and N.G. Polydoris. 1993. Polydoris, Garlic, Cancer and Heart Disease: Review and Recommendations; GN Communications (Pub.) Limited:Chicago, IL, USA; Volume 3, pp. 21–41.

Gruenwald, J. 2004. The Medical Economics Team PDR Physicans Desk Reference Team. Eds. PDR for Herbal Medicines. 3rd Edn. Montvale, NJ: Thomson PDR.

Guevara-Figueroa, T., L. López-Hernández, M.G. Lopez, et al. 2015. Conditioning Garlic "Seed" Cloves at Low Temperature Modifies Plant Growth, Sugar, Fructan Content, and Sucrose Fructosyl Transferase (1-SST) Expression. *Scientia Horticulturae* 189, 150–158. doi: 10.1016/j.scienta.2015.03.030

Guo, D.P., G.A. Shah, G.W. Zeng, et al. 2004. The Interaction of Plant Growth Regulators and Vernalization on the Growth and Flowering of Cauliflower (*Brassica oleracea* var. botrytis). *Plant Growth Regulation* 43: 163–171. doi: 10.1023/B: GROW.0000040114.84122.11

Haak, D.C., T. Fukao, R. Grene, et al. 2017. Multilevel Regulation of Abiotic Stress Responses in Plants. *Frontiers in Plant Science* 8: 1564. doi: 10.3389/fpls.2017.01564

Habibi, G. 2013. Effect of Drought Stress and Selenium Spraying on Photosynthesis and Antioxidant Activity of Spring Barley. *Acta Agriculturae Slovenica* 101(1): 31:39. doi: 10.2478/acas-2013-0004

Hasanuzzaman, M., K. Nahar, and M. Fujita. 2013. Plant Response to Salt Stress and Role of Exogenous Protectants to Mitigate Salt-Induced Damages. Ecophysiology and Responses of Plants Under Salt Stress. Springer New York, 25–87. doi:10.1007/978-1-4614-4747-4_2

Hawrylak-Nowak, B. 2009. Beneficial Effects of Exogenous Selenium in Cucumber Seedlings Subjected to Salt Stress. *Biological Trace Element Research* 132: 259–69. doi: 10.1007/s12011-009-8402-1

Hong, C. J., H. Wattanabe, T. Etoh, et al. 2000. Morphological and Karyological Comparison of Garlic Clones Between the Center of Origin and Westernmost Area of Distribution. *Memoirs of the Faculty of Agriculture – Kagoshima University* 36: 1–10.

Horie, T., I. Karahara, and M. Katsuhara. 2012 Salinity Tolerance Mechanisms in Glycophytes: An Overview with the Central Focus on Rice Plants. Rice 5(1): 1–18. doi: 10.1186/1939-8433-5-11

Huang, Y., Y. Guo, Y. Liu, et al. 2018. 9-Cis-Epoxycarotenoid Dioxygenase 3 Regulates Plant Growth and Enhances Multi-Abiotic Stress Tolerance in Rice. *Frontiers in Plant Science* 9: 162. doi:10.3389/fpls.2018.00162.

Hughes, J., H.A. Collin, A. Tregova, et al. 2006. Effect of Low Storage Temperature on Some of the Flavour Precursors in Garlic (*Allium Sativum*). *Plant Foods for Human Nutrition* 61(2): 78–82. doi: 10.1007/s11130-006-0018-4.

Ipek, M., A. Ipek, and P.W. Simon. 2003. Comparison of AFLPs, RAPD Markers and Isozymes for Diversity Assessment of Garlic and Detection of Putative Duplicates in Germplasm Collections. *Journal of the American Society for Horticultural Science* 128(2): 246–52. doi: 10.21273/JASHS.128.2.0246.

Ipek, M., A. Ipek, and P.W. Simon. 2008a. Rapid Characterization of Garlic Clones with Locus-Specifc DNA Markers. *Turkish Journal of Agriculture and Forestry* 32(5): 357–362. https://journals.tubitak.gov.tr/agriculture/vol32/iss5/1/

Ipek, M., A. Ipek, and P.W. Simon. 2008b. Molecular Characterization of Kastamonu Garlic: An Economically Important Garlic Clone in Turkey. *Scientia horticulturae* 115(2): 203–208. https://pubag.nal.usda.gov/catalog/13287

Ipek, M., A. Ipek, S.G. Almquist, et al. 2005. Demonstration of Linkage and Development of the First Low-Density Genetic Map of Garlic, Based on AFLP Markers. *Theoretical and Applied Genetics* 110(2): 228–236. doi: 10.1007/s00122-004-1815-5

Jaishankar, M., T. Tseten, N. Anbalagan, et al. 2014. Toxicity, Mechanism and Health Effects of Some Heavy Metals. *Interdisciplinary Toxicology* 7, 60–72. doi: 10.2478/intox-2014-0009

Jiang, W., and D. Liu. 2010. Pb-Induced Cellular Defense System in The Root Meristematic Cells of *Allium sativum* L. *BMC Plant Biology* 10(1): 1–8. doi: 10.1186/1471-2229-10-40

Jiang, W., D. Liu, and W. Hou. 2001. Hyperaccumulation of Cadmium by Roots, Bulbs and Shoots of Garlic (*Allium sativum* L.). *Bioresource Technology* 76(1): 9–13. doi: 10.1016/S0960-8524(00)00086-9

Jo, M., I. Ham, K. Moe, et al. 2012. Classification of Genetic Variation in Garlic (*Allium sativum* L.) Using SSR Markers. *Australian Journal of Crop Science* 6(4):625–631 doi: 10.3316/informit.362158671579385

Kamenetsky, R., I. London, M. Shafir. et al. 2004a. Garlic (*Allium sativum* L.) and its Wild Relatives from Central Asia: Evaluation for Fertility. potential. *Acta Horticulturae* 637:83–91. doi: 10.17660/ActaHortic.2004.637.9

Kamenetsky, R., and D.H. Rabinowitch. 2001. Floral Development in Bolting Garlic. *Sexual Plant Reproduction* 13:23–241. doi: 10.1007/s004970000061

Kamenetsky, R., and H.D. Rabinowitch. 2002. Florogenesis. In: Rabinowitch HD, Currah L (eds) *Allium* Crop Sciences: Recent Advances. CAB International, Wallingford, UK, pp 31–57

Kamenetsky, R., A. Faigenboim, E.S. Mayer, et al. 2015. Integrated Transcriptome Catalogue and Organ-Specific Profiling of Gene Expression in Fertile Garlic (*Allium sativum* L.). *BMC Genomics* 16(1):12. doi: 10.1186/s12864-015-1212-2

Kamenetsky, R., I.L. Shafir, H. Zemah, et al. 2004b. Environmental Control of Garlic Growth and Florogenesis. *Journal of the American Society for Horticultural Science* 129(2): 144–151. doi: 10.21273/JASHS.129.2.0144

Keurentjes J.J., M. Koornneef, and D. Vreugdenhil .2008. Quantitative Genetics in the Age of Omics. *Current Opinion in Plant Biology* 11(2):123–128. doi: 10.1016/j.pbi.2008.01.006

Khademi A.R., S. Bolandnazar, F.Z. Nahandi, et al. 2018. Effect of Selenium Application on Phenylalanine Ammonia-Lyase (PAL) Activity, Phenol Leakage and Total Phenolic Content in Garlic (*Allium sativum* L.) Under NaCl Stress, *Information Processing in Agriculture* 5(3): 339–344.doi: 10.1016/j.inpa.2018.04.004

Khar, A. 2012. Cross Amplification of Onion Derived Microsatellites and Mining of Garlic EST Database for Assessment of Genetic Diversity in Garlic. *Acta Horticulture* 969:289–295. doi: 10.17660/ActaHortic.2012.969.37

Khar, A., S. Hirata, M. Abdelrahman, et al. 2020. Breeding and Genomic Approaches for Climate-Resilient Garlic. In: Kole, C. (eds) Genomic Designing of Climate-Smart Vegetable Crops. Springer, Cham. Doi:10.1007/978-3-319-97415-6_8

Khoshoo, T.N., C.K. Atal, and V.B. Sharma. 1960. Cytotaxonomical and Chemical Investigations on the North-West Indian Garlics. *Research Bulletin of the Panjab University of Science* 11: 37–47.

Kim, S., C.W. Kim, M. Park, et al. 2015. Identification of Candidate Genes Associated with Fertility Restoration of Cytoplasmic Male-Sterility in Onion (*Allium cepa* L.) Using a Combination of Bulked Segregant Analysis and RNA-seq. *Theoretical and Applied Genetics* 128(11): 2289–2299. doi: 10.1007/s00122-015-2584-z

King, R.W., T. Moritz, L.T. Evans, et al. 2006. Regulation of Flowering in the Long-Day Grass Lolium Temulentum by Gibberellins and the Flowering Locus T Gene. *Plant Physiology.* 141: 498–507. doi: 10.1104/pp.106.076760

Kıraç, H., Ş.A. Dalda, Ö.F. Coşkun, et al. 2022. Morphological and Molecular Characterization of Garlic (*Allium sativum* L.) Genotypes Sampled from Turkey. *Genetic Resources and Crop Evolution* 69(5): 1833–1841. doi: 10.1007/s10722-022-01343-4

Kong, L., M. Wang, and D. Bi. 2005. Selenium Modulates the Activities of Antioxidant Enzymes, Osmotic Homeostasis and Promotes the Growth of Sorrel Seedlings Under Salt Stress. *Plant Growth Regulation* 45(2): 155–163. doi: 10.1007/s10725-005-1893-7

Kosmala, K., A. Bocian, M. Rapacz, B. et al. 2009. Identification of Leaf Proteins Differentially Accumulated During Cold Acclimation Between Festuca Pratensis Plants with Distinct Levels of Frost Tolerance. *Journal of Experimental Botany* 60: 3595–3609. doi: 10.1093/jxb/erp205

Kosová, K., P. Vítámvás, I.T. Prášil, et al. 2011. Plant Proteome Changes under Abiotic Stress – Contribution of Proteomics Studies to Understanding Plant Stress Response. *Journal of Proteomics* 74, 1301–1322. doi: 10.1016/j.jprot.2011.02.006

Ksouri R., W. Megdiche, A. Debez, et al. 2007. Salinity Effects on Polyphenol Content and Antioxidant Activities in Leaves of the Halophyte Cakile maritima. *Plant Physiology and Biochemistry* 45(3): 244–249. doi: 10.1016/j.plaphy.2007.02.001

Kubec, R., and J. Velíšek. 2007. *Allium* Discoloration: The Color-Forming Potential of Individual Thiosulfinates and Amino Acids: Structural Requirements for the Color-Developing Precursors. *Journal of Agriculture and Food Chemistry* 55(9):3491–3497. doi: 10.1021/jf070040n

Kuznetsov, V.V., V.P. Kholodova, V.I.V. Kuznetsov, et al. 2003. Selenium Regulates the Water Status of Plants Exposed to Drought. *Doklady Biological Sciences* 390: 266–268. doi: 10.1023/a:1024426104894

Lalla F.D., B. Ahmed, A. Omar, et al. 2013. Mohieddine Chemical composition and biological activity of Allium Sativum essential oils against Callosobruchus maculatusEnviron. *Journal of Environmental Science, Toxicology and Food Technology* 3 (1): 30–36.

Lamichhane J.R., M. Barzman, K. Booij, et al. 2015. Robust Cropping Systems to Tackle Pests Under Climate Change. A Review. *Agronomy for Sustainable Development* 35(2):443–459. doi: 10.1007/s13593-014-0275-9

Lancaster, J.E., and M.L. Shaw. 1989. γ-Glutamyl Peptides in the Biosynthesis of S-alk (en) yl-L-Cysteine Sulphoxides (Flavour Precursors) in *Allium*. *Phytochemistry* 28(2): 455–460. doi: 10.1016/0031-9422(89)80031-7

Lavisolo, C. and A. Schuber, 1998. Effects of Water Stress on Vessel Size Xylem Hydraulic Conductivity in (*Vitis vinifera* L). *Journal of Experimental Botany*, 49: 693–700. doi:10.1093/JXB/49.321.693

Lawson, L.D. 1998. Garlic: A Review of its Medicinal Effects and Indicated Active Compounds. In Phytomedicines of Europe. *Chemistry and Biological Activity*; ACS Symposium Series 691; American Chemical Society: Washington, DC, USA; Volume 3, pp. 176–209.

Levan, A.K. 1935. Cytological Studies in Allium VI. The Chromosome Morphology of Some Diploid Species of *Allium*. *Hereditas* 20: 289–330.

Li, L., D. Wang, X. Li, et al. 2015. Elucidation of Colour Development and Microstructural Characteristics of *Allium sativum* Fumigated with Acetic Acid. *International Journal of Food Science and Technology* 50(5):1083–1088. doi: 10.1111/ijfs.12751

Liang, C.Y., J. Xiong, and J. Ye. 2014. Study on Storage Conditions of Greening of Garlic Purees. *Advanced Materials Research* 1052: 290–293. doi: 10.4028/www.scientific.net/AMR.1052.290

Liang, Y., Q. Chen, Q. Liu, et al. 2003. Exogenous Silicon (Si) Increases Antioxidant Enzyme Activity and Reduces Lipid Peroxidation in Roots of Salt-Stressed Barley (*Hordeum vulgare* L.). *Journal of plant physiology* 160(10): 1157–1164. doi: 10.1078/0176-1617-01065

Liu, D., and I. Kottke. 2004. Subcellular Localization of Copper in the Root Cells of *Allium sativum* by Electron Energy Loss Spectroscopy (EELS). *Bioresource Technology* 94(2): 153–158. doi: 10.1016/j.biortech.2003.12.003

Liu, D., J. Zou, Q. Meng, et al. 2009. Uptake and Accumulation and Oxidative Stress in Garlic (*Allium sativum* L.) Under Lead Phytotoxicity. *Ecotoxicology* 18(1): 134–143. doi: 10.1007/s10646-008-0266-1

Lukes, T.M. 1986. Factors Governing the Greening of Garlic Puree. *Journal of Food Science* 51(6):1577–1582. doi: 10.1111/j.1365-2621.1986.tb13869.x

Ma, J., E. Lei, M. Lei, et al. 2018. Remediation of Arsenic Contaminated Soil Using Malposed Intercropping of *Pteris vittata* L. and Maize. *Chemosphere* 194: 737–744.

Ma, K.H., J.G. Kwag, W. Zhao, et al. 2009. Isolation and Characteristics of Eight Novel Polymorphic Microsatellite Loci from the Genome of Garlic (*Allium sativum* L.). *Scientia Horticulturae* 122(3):355–361. doi: 10.1016/j.scienta.2009.06.010

Maas E.V., and G.J. Hoffman. 1977. Crop Salt Tolerance—Current Assessment. *Journal of the Irrigation and Drainage Division* 103:115–134.

Maass, H.I., and M. Klaas. 1995. Intraspecifc Diferentiation of Garlic (*Allium sativum* L.) By ISOZYME and RAPD MARKers. *Theoretical and Applied Genetics* 91:89–97. doi: 10.1007/BF00220863

Mallor, C., and B. Thomas . 2008. Resource Allocation and the Origin of Flavour Precursors in Onion Bulbs. *The Journal of Horticulture Science and Biotechnology* 83, 191–198. https://doi.org/10.1080/14620 316.2008.11512369.7

Mangal, J.L., R.K. Singh, A.C. Yadav, et al. 1990. Evaluation of Garlic Cultivars for Salinity Tolerance. Journal of Horticulture Science 65(6): 657–658. doi: 10.1080/00221589.1990.11516105

Mario, P.C., B.V. Viviana, and I.A. Maria 2008. Low Genetic Diversity Among Garlic Accessions Detected Using RAPD. *Chilean Journal of Agricultural Research* 68(1):3–12.

Mathew, D., Y. Forer, H.D. Rabinowitch, et al. 2011. Effect of Long Photoperiod on the Reproductive and Bulbing Processes in Garlic (*Allium sativum* L.) Genotypes. *Environmental and Experimental Botany* 71(2): 166–173. doi: https://doi.org/10.1016/j.envexpbot.2010.11.008

McCollum, G.D. 1976. Onion and Allies. In: Simmonds NW (ed.) Evolution of Crop Plants. Longman, London, pp 186–190.

McGettigan, P.A. 2013. Transcriptomics in the RNA-Seq Era. Current Opinion in Chemical Biology 17(1):4–11. doi: 10.1016/j.cbpa.2012.12.008

Menconi M., C.L. Sgherri, M.C. Pinzino, et al. 1995 Activated Oxygen Species Production and Detoxification in Wheat Plants Subjected to a Water Deficit Programme. *Journal of Experimental Botany*, 1995, 46:1123–1130.

Mensinkai, S.W. 1939. Cytogenetic Studies in Genus Allium. *Journal of Genetics* 39: 1–45.

Milford, G., M. Armstrong, P. Jarvis, et al. 2000. Effect of Potassium Fertilizer on the Yield, Quality and Potassium Offtake of Sugar Beet Crops Grown on Soils of Different Potassium Status. *The Journal of Agricultural Science* 135(1): 1–10. doi:10.1017/S0021859699007881.

Mishra, R.K., R.K. Jaiswal, D. Kumar, et al. 2014. Management of Major Diseases and Insect Pests of Onion and Garlic: A Comprehensive Review. *Journal of Plant Breeding and Crop Science* 6(11):160–170. doi: 10.5897/JPBCS2014.0467

Munns, R., and M. Tester. 2008. Mechanisms of Salinity Tolerance. *Annual Reveiw Plant Biology* 59: 651–681. doi: 10.1146/annurev.arplant.59.032607.092911

Mustafa, M.M.I., M.A. Wally, K.M. Refaie, et al. 2017. Effect of Different Irrigation Levels and Salicylic Acid Applications on Growth, Yield and Quality of Garlic (*Allium sativum* L.). *Journal of Biological Chemistry and Environmental Sciences* 12: 301–323.

Nabulsi I., B. Al-Safadi, N.M. Ali, et al. 2001. Evaluation of Some Garlic (*Allium Sativum* L.) Mutants Resistant to White Rot Disease by RAPD Analysis. *Annals of Applied Biology* 138(2): 197–202. doi: 10.1111/j.1744-7348.2001.tb00102.x

Nemati, I., F. Moradi, S. Gholizadeh, et al. 2011. The Effect of Salinity Stress on Ions and Soluble Sugars Distribution in Leaves, Leaf Sheaths and Roots of Rice (*Oryza sativa* L.) Seedlings. *Plant, Soil and Environment* 57(1): 26–33. doi: 10.17221/71/2010-PSE

North, H.M., A.D. Almeida, J.P. Boutin, et al. 2007. The Arabidopsis ABA-Deficient Mutant Aba4 Demonstrates that the Major Route for Stress Induced ABA Accumulation is Via Neoxanthin Isomers. *The Plant Journal* 50(5): 810–824. doi: 10.1111/j.1365-313X.2007.03094. x

Nyarko, G., P.G. Alderson, J. Craigon, et al. 2007. Induction of Flowering in Cabbage Plants by In Vitro Vernalization, Gibberellic Acid Treatment and Ratooning. *Ghana Journal of Horticulture* 6: 81–87.

Osman, S.A.M., A.T.M. Ata, and S.E.H.G. El-Hak. 2007. Morphological, Germination, Bolting and Cytogenetical Characteristics of Fourteen Promising Garlic Genotypes. Paper presented at the 8th African Crop Science Society Conference, El-Minia, Egypt, 27–31 October 2007.

Panse R.K., P.K. Jain, A. Gupta, et al. 2013. Morphological Variability and Character Association in Diverse Collection of Garlic Germplasm. *African Journal of Agricultural Research*, 8(23): 2861–2869. doi:10.5897/AJAR12.551

Paredes, C.M., V.V. Becerra, and A.M.I. González. 2008. Low Genetic Diversity Among Garlic (*Allium sativum* L.) Accessions Detected Using Random Amplifed Polymorphic DNA (RAPD). *Chilean Journal of Agricultural Research* 68:3–12. doi: 10.4067/S0718-58392008000100001

Parry, A.D., and R. Horgan. 1991. Carotenoids and Abscisic Acid (ABA) Biosynthesis in Higher Plants. *Physiologia Plantarum* 82(2): 320–326. doi: 10.1111/j.1399-3054. 1991.tb00100.x

Parvaiz A., S. Satyawati. 2008. Salt Stress and Phyto-Biochemical Responses of Plants-A review. *Plant, Soil Environment* 54(3): 89–99.

Pooler, M.R., and P.W. Simon. 1993. Characterization and Classifcation of Isozyme and Morphological Variation in a Diverse Collection of Garlic Clones. *Euphytica* 68:121–130. doi:10.1007/BF00024161

Purwanto, Y.A., N. Naibaho, S.Y. Pratama, et al. 2019. Effects of Temperature on The Quality of Garlic (*Allium sativum* L) cv. Lumbu Kuning During Storage. IOP Conference Series: *Earth and Environmental Science* 309(1): 012004. doi: 10.1088/1755-1315/309/1/012004

Qi, Y.X., Y.B. Liu, and W.H. Rong. 2011. RNA-Seq and its Applications: A New Technology for Transcriptomics. *Hereditas* 33(11):1191–202. doi: 10.3724/sp.j.1005.2011.01191

Reddy K.R., H.F. Hodges, and B.A. Kimball. 2000. Crop Ecosystem Responses to Global Climate Change: Cotton. In: Reddy KR, Hodges HF (eds) Climate change and global crop productivity. CABI, Wallingford, UK, pp 162–187.

Reis, S.P.d., A.M. Lima, and C.R.B. Souza. 2012. Recent Molecular Advances on Downstream Plant Responses to Abiotic Stress. *International Journal of Molecular Sciences* 13(7): 8628–8647. doi: 10.3390/ijms13078628

Renaut, J., J. F. Hausman, and M. E. Wisniewski. 2006. Proteomics and Low-Temperature Studies: Bridging the Gap Between Gene Expression and Metabolism. *Physiological Plantarum* 126: 97–109. doi: 10.1111/j.1399-3054.2006. 00617.x

Ruiz-Lozano J.M., and R. Azcon. 1997. Effect of Calcium Application on the Tolerance of Mycorrhizal Lettuce Plants to Polyethylene Glycol-Induced Water Stress. *Symbiosis* 23: 9–21.

Sah, S.K., K.R. Reddy, and J. Li. 2016. Abscisic Acid and Abiotic Stress Tolerance in Crop Plants. *Frontiers in Plant Science* 7: 571. doi: 10.3389/fpls.2016.00571

Salem, K.F.M. 2004. The Inheritance and Molecular Mapping of Genes for Post-Anthesis Drought Tolerance (PADT) in Wheat. Doctoral dissertation, Ph. D Dissertation, Martin Luther University, Halle-Wittenberg, Germany

Santos, C.V. 2004. Regulation of Chlorophyll Biosynthesis and Degradation by Salt Stress in Sunflower Leaves. *Scientia Horticulturae* 103(1): 93–99. doi: 10.1016/j.scienta.2004.04.009

Saqib, M., J. Akhtar, and R.H. Qureshi. 2004. Pot Study on Wheat Growth in Saline and Waterlogged Compacted Soil: I. Grain Yield and Yield Components. *Soil and Tillage Research* 77(2): 169–177. doi: 10.1016/j.still.2003.12.004

Sasi, M., S. Kumar, M. Kumar, et al. 2021. Garlic (*Allium sativum* L.) Bioactives and its Role in Alleviating Oral Pathologies. *Antioxidants* 10 (11): 1847. doi: 10.3390/antiox10111847

Saxena, P., S. Krishna, T. Raj, et al. 1999. Phytoremediation of Heavy Metal Contaminated and Polluted Soils. *Heavy metal stress in plants:* 305–329. doi: 10.1007/978-3-662-07745-0_14

Scandalios J.G.1993. Oxygen Stress and Superoxide Dismutases. *Plant Physiology.* 101: 7–12.

Schwartz, H. 2004. Botrytis, Downy Mildew and Purple Blotch of Onion. Colorado State University Cooperative Extension No. 2.941

Shafeek, M., M.H. Nagwa, S. Singer, et al. 2013. Effect of Potassium Fertilizer and Foliar Spraying with Etherel on Plant Development, Yield and Bulb Quality of Onion Plants (*Allium cepa* L). *Journal of Applied Sciences Research* 9(2): 1140–1146. www.aensiweb.com/old/jasr/jasr/2013/1140-1146.pdf.

Shardendu, S.N., S.F. Boulyga, and E. Stengel, 2003. Phytoremediation of Selenium by Two Helophyte Species in Subsurface Flow Constructed Wetland. *Chemosphere* 50: 967–73.

Sharma, A.K., and A.K. Bal. 1959 A Study of Spontaneous Fragmentation in Two Varieties of Allium sativum and Interpretation of their Karyotypes. Proc. 46th session of Indian Society Congress, Part III: 352, (cited after: Yuzbasioglu and Unal, 2004).

Sharma, P., and R.S. Dubey. 2005. Lead Toxicity in Plants. *Brazilian Journal of Plant Physiology* 17: 35–52. doi: 10.1590/S1677-04202005000100004

Shemesh-Mayer, E., T. Ben-Michael, N. Rotem, et al. 2015. Garlic (*Allium sativum* L.) Fertility: Transcriptome and Proteome Analyses Provide Insight into Flower and Pollen Development. *Frontiers in Plant Science* 6:271. doi: 10.3389/fpls.2015.00271

Siddiqui, S., K. Gupta, A.Yadav, et al. 1996. Soil Salinity Effect on Soluble Saccharides, Phenol, Fatty Acid and Mineral Contents, and Respiration Rate of Garlic Cultivars. *Biologia Plantarum* 38(4): 611–615. doi: 10.1007/BF02890618

Sohn, S.H., Y.K. Ahn, T.H. Lee, et al. 2016. Construction of a Draft Reference Transcripts of Onion (*Allium cepa*) Using Long-Read Sequencing. *Plant Biotechnology Reports* 10(6):383—390. doi: 10.1007/s11816-016-0409-4

Son J.H., K.C. Park, S. Lee, et al. 2012. Isolation of Cold-Responsive Genes from Garlic, *Allium sativum*. *Genes Genomics* 34:93–101. https://doi.org/10.1007/s13258- 011-0187-x

Soudek, P., J. Kotyza, I. Lenikusová, et al. 2009. Accumulation of Heavy Metals in Hydroponically Cultivated Garlic (*Allium sativum* L.), Onion (*Allium cepa* L.), Leek (*Allium porrum* L.) and Chive (*Allium schoenoprasum* L.). *Journal of Food, Agriculture and Environment* 7: 761–769.

Srivastava, S.C., U.C. Sharma, B.K Singh, et al. 2021. A Profile of Garlic Production in India: Facts, Trends and Opportunities. *International Journal of Agriculture Environment and Biotechnology* 5(4): 477–482. https://agris.fao.org/agrissearch/search.do?recordID=US202000307433

Stinchcombe, J.R., and H.E. Hoekstra. 2008. Combining Population Genomics and Quantitative Genetics: Finding the Genes Underlying Ecologically Important Traits. *Heredity* 100:158–170. www.nature.com/articles/6800937

Sukkaew, P., and A. Tira-Umphon. 2012. Effects of storage Conditions on Allicin Content in Garlic (*Allium sativum*). Paper presented at the VI International Symposium on Edible Alliaceae 969.

Sun, X., S. Zhou, F. Meng, et al. 2012. De Novo Assembly and Characterization of the Garlic (*Allium sativum*) Bud Transcriptome by Illumina Sequencing. *Plant Cell Reports* 31(10):1823–1828. doi: 10.1007/s00299-012-1295-z

Sun, X., S. Zhu, N. Li, et al. 2020. A Chromosome-Level Genome Assembly of Garlic (*Allium sativum*) Provides Insights into Genome Evolution and Allicin Biosynthesis. *Molecular Plant* 13(9): 1328–1339.

Sun, X.D., G.Q. Ma, B. Cheng, et al. 2013. Identification of Differentially Expressed Genes in Shoot Apex of Garlic (*Allium sativum* L.) Using Illumina Sequencing. *Journal of Plant Studies* 2(2):136–148. doi: 10.5539/jps.v2n2p136

Tchounwou, P. B., C.G. Yedjou, A. K. Patlolla, et al. 2012. Heavy metal toxicity and the environment. *Molecular, Clinical and Environmental Toxicology:* 133–164. doi: 10.1007/978-3-7643-8340-4_6

Tesfaye, A., and W. Mengesha. 2015. Traditional Uses, Phytochemistry and Pharmacological Properties of Garlic (*Allium Sativum*) and its Biological Active Compounds. *International journal of scientific research in science, engineering and technology* 1: 142–148. https://pubmed.ncbi.nlm.nih.gov/24379960/

Umezawa, T., M. Okamoto, T. Kushiro, et al. 2006. CYP707A3, a Major ABA 8′-Hydroxylase Involved in Dehydration and Rehydration Response in *Arabidopsis thaliana*. *The Plant Journal* 46(2), 171–182. doi: 10.1111/j.1365-313X.2006.02683. x

Valdez, J.G., M.A. Makuch, A.F. Ordovini, et al .2006. First Report of *Penicillium allii* as a Field Pathogen of Garlic (*Allium sativum*). Plant Pathol 55(4):583.

Van Dijk, E.L., H. Auger, Y. Jaszczyszyn, et al. 2014. Ten Years of Next-Generation Sequencing Technology. *Trend in Genetics* 30(9):418–426. doi: 10.1016/j.tig.2014.07.001

Vanaclocha B, S. Cañigueral Fitoterapia, Vademécum de prescripción. 5a ed. Barcelona: Elsevier; 2019.37p.

Vázquez-Barrios, M.E., G. López-Echevarría, E. Mercado-Silva, et al. 2006. Study and Prediction of Quality Changes in Garlic cv. Perla (*Allium sativum* L.) Stored at Different Temperatures. *Scientia Horticulturae* 108(2): 127–132. doi: https://doi.org/10.1016/j.scienta.2006.01.013

Volk, G.M., K.E. Rotindo, and W. Lyons. 2004. Low-Temperature Storage of Garlic for spring Planting. *HortScience,* 39(3), 571–573.

Wang, C. Q. 2011. Water-Stress Mitigation by Selenium in *Trifolium repens* L. *Journal of Plant Nutrition and Soil Science*, 174(2): 276–282. doi: 10.1002/jpln.200900011

Wang, D., Z. Qu, L. Yang, et al. 2017. Transposable Elements (TES) Contribute to Stress-Related Long Intergenic Noncoding RNAs in Plants. *The Plant Journal* 90(1) 133–146. doi: 10.1111/tpj.1348

Wang, M., Y. Ye, X. Chu, et al. 2022. Responses of Garlic Quality and Yields to Various Types and Rates of Potassium Fertilizer Applications. *HortScience* 57(1), 72–80. doi: 10.21273/HORTSCI15984-21

Wang, W., B. Vinocur, O. Shoseyov, et al. 2004. Role of plant Heat-Shock Proteins and Molecular Chaperones in the Abiotic Stress Response. *Trends in plant science* 9(5): 244–252. doi: 10.1016/j.tplants.2004.03.006

Wang, Y., Y. Hu, Y. Duan, et al. 2016. Silicon Reduces Long-Term Cadmium Toxicities in Potted Garlic Plants. *Acta Physiologiae Plantarum* 38(8): 1–9. doi:10.1007/s11738-016-2231-6

Wang, Z., M. Gerstein, and M. Snyder. 2009. RNA-Seq: A Revolutionary Tool for Transcriptomics. *Nature Review Genetics* 10(1):57–63. www.nature.com/articles/nrg2484.

Wichtl, M. 2004. Herbal Drugs and Phytopharmaceuticals, A Handbook for Practice on a Scientific Basis. 3rd Edn. Boca Raton, FL: CRC Press.

Wierzbicka, M., and D. Antosiewicz. 1993. How Lead Can Easily Enter the Food Chain—A Study of Plant Roots. *Science of the total environment*, 134: 423–429. doi: 10.1016/S0048-9697(05)80043-9

Williamson, E.M. 2003. Potter's Herbal Cyclopaedia. The Autoritative Reference Work on Plants with a Known Medicinal Use. Saffron Walden,Essex: London: CW Daniel Co.

Wu, C., M. Wang, Z. Cheng, et al. 2016. Response of Garlic (*Allium sativum* L.) Bolting and Bulbing to Temperature and Photoperiod Treatments. *Biology Open* 5(4):507–518. doi: 10.1242/bio.016444

Wu, C., M. Wang, Y. Dong, et al. 2015. Growth, Bolting and Yield of Garlic (*Allium sativum* L.) in Response to Clove Chilling Treatment. *Scientia Horticulturae* 194: 43–52. 10.1016/j.scienta.2015.07.018

Wu, Q.S., R.S. Xia, and Y.N. Zou. 2008. Improved Soil Structure and Citrus Growth After Inoculation with Three Arbuscular Mycorrhizal Fungi Under Drought Stress. *European Journal of Soil Biology* 44: 122–128. doi: 10.1016/j.ejsobi.2007.10.001

Wu, M., F. Jin, J. Zhang, et al. 2012. Characterization of a Novel Bipartite Doublestranded RNA Mycovirus Conferring Hypovirulence in the Pathogenic Fungus *Botrytis porri*. *Journal of Virology* 86: 6605–6619. doi: 10.1128/JVI.00292-12

Xiong, P.P., C. He, O. kokyo, et al. 2018. *Medicago sativa* L. Enhances the Phytoextraction of Cadmium and zinc by *Ricinus communis* L. on Contaminated Land In Situ. *Ecological Engineering* 116: 61–66. doi: 10.1016/j.ecoleng.2018.02.004

Xu, X., A.A.M. Van, E. Lammeren, et al. 1998. The Role of Gibberellin, Abscisic Acid, and Sucrose in the Regulation of Potato Tuber Formation In Vitro. *Plant Physiology* 117, 575–584. doi: 10.1104/pp.117.2.575

Yoshida, T., Y. Fujita, H. Sayama, et al. 2010. AREB1, AREB2, and ABF3 are Master Transcription Factors that Cooperatively Regulate ABRE-Dependent ABA Signaling Involved in Drought Stress Tolerance and Require ABA for Full Activation. *The Plant Journal* 61(4): 672–685. doi: 10.1111/j.1365-313X.2009.04092.x

Zewdie, Y., M.J. Havey, J.P. Prince, et al. 2005. The First Genetic Linkages Among Expressed Regions of the Garlic Genome. *Journal of the American Society for Horticultural Science* 130(4):569–574. doi: 10.21273/JASHS.130.4.569

Zewde, T., C. Fininsa, P. K. Sakhuja, et al. 2007. Association of white rot (*Sclerotium cepivorum*) of garlic with environmental factors and cultural practices in the North Shewa highlands of Ethiopia. *Crop Protection* 26 (10):1566–1573. http://dx.doi.org/10.1016/j.cropro.2007.01.007

Zhang, H., Y. Jiang, Z. He, et al. 2005a. Cadmium Accumulation and Oxidative Burst in Garlic (*Allium sativum*). *Journal of Plant Physiology* 162, 977–984. doi: 10.1016/j.jplph.2004.10.001

Zhang, H., W. Xu, W. Dai, Z.He, and M. Ma. 2006. Functional Characterization of Cadmium-Responsive Garlic Gene Asmt2b: A New Member of Metallothionein Family. *Chinese Science Bulletin* 51(4): 409–416. doi:10.1007/s11434-006-0409-9

Zhang, H., Y. Jiang, Z. He and M. Ma. 2005b. Cadmium Accumulation and Oxidative Burst in Garlic (*Allium sativum*). Journal of plant physiology 162(9): 977–984. doi: 10.1016/j.jplph.2004.10.001

Zhao, W.G., J.W. Chung, G.A. Lee, et al. 2011. Molecular Genetic Diversity and Population Structure of a Selected Core Set in Garlic and its Relatives Using Novel SSR Markers. *Plant Breed* 130:46–54. doi:10.1111/j.1439-0523.2010.01805.x

Zhou, X., J.A. Condori-Apfata, X. Liu, et al. 2021. Transcriptomic Changes Induced by Drought Stress in Hardneck Garlic during the Bolting/Bulbing Stage. *Agronomy* 11(2): 246. doi: 10.3390/agronomy11020246

Zhu, J.K. 2001. Plant Salt Tolerance. *Trends in Plant Science.* 6(2), 66–71. doi: 10.1016/s1360-1385(00)01838-0

Zhu, G., W. Li, F. Zhang, et al. 2018. RNA-seq Analysis Reveals Alternative Splicing Under Salt Stress in Cotton, *Gossypium davidsonii. BMC genomics* 19(1): 1–15. doi: 10.1186/s12864-018-4449-8

Zhu, S., S. Tang, Z. Tan, et al. 2017. Comparative Transcriptomics Provide Insight into the Morphogenesis and Evolution of Fistular Leaves in Allium. *BMC Genomics* 18(1):60. doi: 10.1186/s12864-016-3474-8

Zhu, S., X. Chen, X. Liu, et al. 2019. Transcriptome-Wide Association Study and Eqtl Analysis to Assess the Genetic Basis of Bulb-Yield Traits in Garlic (*Allium sativum*). *BMC Genomics* 20(1):657. doi:10.1186/s12864-019-6025-2

3 *Anethum graveolens* and Stressful Conditions

Jamilu Edirisa Ssenku, Patience Tugume*
and Godwin Anywar
Department of Plant Sciences, Microbiology and Biotechnology,
Makerere University, Kampala, Uganda
*Corresponding author: jssenku@gmail.com

CONTENTS

3.1 INTRODUCTION

Various biotic stresses, such as drought, heat, heavy metals, soil salinity, flooding and cold, are responsible for the reduction in the growth, development and productivity of crops worldwide (Gontia-Mishra et al., 2016). Abiotic stresses may trigger several physiological, biochemical and molecular responses that in turn affect various cellular processes in plants (Hasanuzzaman et al., 2013). The phytohormone abscisic acid (ABA) plays a central role in plant development and responses to abiotic stress as an important messenger that acts as the signaling mediator. ABA initiates a series of signal transduction pathways that regulate multiple defense mechanisms against various stress factors like drought, salt and cold. These include stomatal opening and the expression of genes imparting resistance to environmental stresses (Dar et al., 2017; Sah et al., 2016).

3.2   BOTANICAL DESCRIPTION OF *ANETHUM GRAVEOLENS* L.

Anethum graveolens L. (dill) is an important aromatic herb that belongs to the family Apiaceae and genus *Anethum* (Bulchandani and Shekhawat 2020; Najaran et al. 2016). The genus name *Anethum* is derived from the Greek word aneeson or aneeton, which means strong smelling (Jana and Shekhawat 2010), while the common name dill is derived from the Norse word dylla or dilla which probably means to soothe (Goodarzi et al. 2016; Singh and Panda 2005). *A. graveolens* is the sole species of the genus *Anethum* (Memariani et al. 2020; Pullaiah 2002). Depending on the growth environment, it can be an annual or perennial herb (Morcia et al. 2016), usually growing up to a height of 50–150 cm (Chahal et al. 2017). It has complete white to yellowish flowers that are arranged in compound terminal umbels of 2–9 cm diameter, which produce a dried ripe fruit commonly called schizocarps (Mirhosseini et al. 2014; Warrier et al. 1994) with numerous flat pungent and bitter seeds that are approximately 5 mm in length (Shelef 2003).

3.3   GLOBAL AND ECOLOGICAL DISTRIBUTION OF *A. GRAVEOLENS*

Dill is native to Algeria, Chad, Cyprus, the Gulf States, Iran, Lebanon, Syria, Libya, Morocco, Oman, Saudi Arabia and Tunisia (Plants of the World Online, 2021). *A. graveolens* is an annual herb with a long history of cultivation since ancient Egyptian times for its culinary and medicinal uses (Jana and Shekhawat, 2010). It is optimally cultivated in warm-to-hot summers with high sunshine levels and rich, well-drained soils (Morcia et al. 2016). In many countries in Europe, it has been cultivated since antiquity and used as a popular aromatic herb and spice (Ishikawa et al. 2002). It is also cultivated in the USA, Canada, India, Pakistan and Japan, and grows in the wild in Spain, Portugal and Italy (Shelef 2003). Throughout the Indian sub-continent, the Malaysian archipelago and Japan, it is cultivated for its foliage as a cold weather crop (Jana and Shekhawat, 2010). Dill has also been reported to escape from cultivation and naturalize in surrounding areas, often becoming adventive (Liogier et al. 2000; Randall 2012). Consequently, *A. graveolus* is listed in the Global Compendium of Weeds as an "agricultural weed, casual alien, cultivation escape, environmental weed, garden thug, naturalised and sleeper weed" (Randall 2012).

3.4   MEDICINAL APPLICATIONS OF *A. GRAVEOLENS*

Based on numerous historical reports, *A. graveolens* is among the oldest herbs known and used (Plinius 77 AD). According to Morcia et al. (2016), its use is not limited to food but also involves herbalist and pharmaceutical applications. There is a history of its use as an essential home remedy to treat infections related to the digestive system (Jana and Shekhawat 2010; Kaur and Arora 2010). In this regard, dill has been reported to be a galactagog and to have antihyperlipidemic, antihypercholesterolemic and antioxidant activities (Suresh et al. 2013). It has traditionally been used as a popular aromatic herb and spice for more than 5,000 years, as a remedy for indigestion and flatulence and as a milk secretion stimulant (Al-Snafi 2014; Stavri and Gibbons, 2005). It has also been used as an anticonvulsant, anti-emetic, anticramp (in children), wound healer and to increase appetite and strengthen the stomach (Al-Snafi 2014; Carrubba et al. 2008; Warrier 1993). *A. graveolens* is used as an ingredient in gripe water, and it is given to relieve colic pain in babies and flatulence in young children (Pulliah 2002). Fresh or dry leaves of *A. graveolens* are used in the traditional management of kidney disease in Ethiopia (Yineger et al. 2013).

3.5   PHARMACOLOGICAL EFFECTS OF BIOACTIVE COMPOUNDS IN *A. GRAVEOLENS*

Many *in vitro* and *in vivo* experimental studies have demonstrated the pharmacological effects of *A. graveolens* extracts. Dill is a powerful source of antioxidants and has antimicrobial and antispasmodic, anti-inflammatory and anticancer properties (Najaran et al., 2016; Bouyahya et al. 2021; Yu et al. 2017). Preclinical studies have suggested the anticancer, antigastric irritation, hypoglycemic and anti-inflammatory effects of dill Oshaghi et al. (2016). Dill also has antioxidant, anti-inflammatory and hypoglycemic properties that may explain its favorable effects against insulin resistance and dyslipidemia (Jana and Shekhawat 2010; Oshaghi et al. 2016). Dill extracts restore β-cell function and insulin secretion (Madani et al. 2005). The aqueous extracts of *A. graveolens* attenuated liver and kidney damage through neutralization of the free radicals in rats (Ramadan et al. 2013).

3.6 PHYTOCHEMICAL COMPOSITION OF *A. GRAVEOLENS*

Generally, plants contain phytochemicals that play a key role in promoting human health by reducing oxidative damage, modulating detoxifying enzymes, stimulating the immune system and showing chemopreventive actions (Giovannetti et al. 2013; Anywar 2021; Anywar et al. 2020a; 2020b; Olwenyi et al. 2021). Even though tremendous progress in synthetic chemistry has been achieved, plants still constitute an important source of pharmaceuticals and other compounds of economic importance (Zhang and Björn 2009). Various compounds have been isolated from the seeds, leaves and inflorescence of *A. graveolus* (Jana and Shekhawat 2010). The main components of the essential oil obtained from of dill are carvone, limonene, dihydrocarvone, carvacrol, p-cymen, α-phellandrene and dill apiole (Najaran et al. 2016; Santos et al. 2002). Other phytochemicals include α-pinene, β-pinene, sabinene, myrcene, α-terpinene, p-cymene, limonene, γ-terpinen, dill ether, dihydrocarvone, trans-dihydrocarvone and dill apiol (Davazdahemami and Allahdadi 2022). Dill fruit contains 3–4% essential oils (Ishikawa et al. 2002). α-phellandrene, dill ether and myristicin are the compounds which form the important odor of dill herb (Blank and Grosch 1991; Bonnländer and Winterhalter 2000). The composition and concentration of the different compounds vary depending on the source of the dill, as has been shown in different studies (Amanpour et al. 2017; Stavri and Gibbons 2005). The phytochemical constituents are also not evenly distributed across the organs of the *A. graveolens* plant (Tanko et al. 2005). Similarly, the phytochemical composition of the essential oils extracted from *A. graveolens* varies significantly across its organs (Davazdahemami and Allahdadi 2022). Dill seeds, leaves and roots contain tannins, terpenoids, cardiac glycosides and some flavonoids such as quercetin and isorhamnetin (Tchatchouang et al. 2017; Mahran et al. 1992).

3.7 BIOTIC AND ABIOTIC STRESS EFFECTS ON BIOACTIVE COMPOUND COMPOSITION IN *A. GRAVEOLENS*

Both biotic and abiotic stress signals are recognized as factors that regulate the biogenesis of secondary metabolites (Akula and Ravishankar 2011). Biotic and abiotic factors affect the secretion pathway, quantity and quality of bioactive constituents released by various secretory structures such as glands and trichomes (Sangwan et al. 2001). Plant secondary metabolites produced are important in the enhancement of the plant interaction with the environment for adaptation and defense against herbivores and pathogens. They also contribute to the specific odors, tastes and colors in plants (Ravishankar and Ramachandra Rao 2000). Furthermore, some of the secondary metabolites produced under abiotic and biotic stimuli have health-promoting properties and a pleasant scent or taste (Dutta and Neog 2016; Scagel and Lee 2012). Both biotic and abiotic factors can modulate the composition of secondary metabolites and essential oils (Egamberdieva and Teixeira da Silva 2015), and thus their bio-efficacy. Some of the biochemical modifications involve upregulated production of compounds with pharmacological value. Exposing plants to low doses of certain stress factors called eustressors or positive stressors can stimulate the production of positive responses in plants. Eustressors could be biological, such as biostimulants, or non-biological, such as physical or chemical compounds.

3.8 BIOTIC STRESS EFFECTS OF *A. GRAVEOLENS*

Plants as sessile organisms have evolved a defense system consisting of pre-existing and indu-cible responses to coping with different types of biotic and abiotic environmental stress factors (Cardenas-Manríquez et al. 2016; Mejía-Teniente et al. 2013). A range of biological stressors triggers inducible defense responses after being recognized by the plant, thus providing efficient resistance toward non-adapted pathogens (Wiesel et al. 2014). Biotic stresses, including pathogens and pests, account for up to 20% of crop losses (Goss et al. 2017). The aroma of dill is a deterrent to specific pests and their larvae (Osanloo et al. 2018) and it is protected against some biotic stress factors. Between the early seedling stage and flowering, *A. graveolens* when cultivated is often attacked by a large number of phytopathogenic fungi that inflict damage on specific parts. These include *Erysiphe anethi, Fusicladium depressum, Erysiphe heraclei, Cercosporidium punctum* and *Itersonilia perplexans,* and as the most serious insect, *Graphosoma lineatum* (Aćimović 2015). Wet weather at the flowering stage can cause a fungal attack of *Botrytis* sp. and *Alternaria* sp. which prevent fruit set (Marianne and Acevedo-Rodríguez 2015). Attacks by virus infections and insect pests (lice and caterpillars) are possible but are seldom very serious. Larvae of the chalcid fly *Systole albipensis* may infest fruits (Marianne and Acevedo-Rodríguez 2015). It should be noted that indu-cible biochemical responses in *A. graveolens* to cope with the aforementioned biotic stress factors have not been evaluated.

3.9 EFFECT OF ABIOTIC STRESS ON SECONDARY METABOLITE BIOGENESIS OF *A. GRAVEOLENS*

Plants are frequently exposed to a variety of stress conditions such as low/high temperature, salt, drought, flooding, heat, oxidative stress and heavy metal toxicity (Aishwath and Ratan 2016), which are harmful to them. However, such abiotic stress factors influence secondary metabolite produc-tion in medicinal plants (van Wyk and Prinsloo 2020). Eustress conditions, including decreased temperatures and supplementary UV light, have been pointed out by Vázquez-Hernández et al. (2019) as promising pre-harvest tools to improve the nutritional or sensorial quality of greenhouse crops by modulating plant metabolism (Kandel et al. 2020). A study by Haghshenas and Eskandari (2011) showed that drought stress reduces shoot and root dry weight, plant height and essential oil yield in *A. graveolens.* Similarly, Castro-Alves et al. (2021) showed that eustress such as low night temperature or supplementary UV-A- or UV-B-enriched exposure can improve the sensory qual-ities of greenhouse dill, associated with the accumulation of organic acids and free amino acids such as alanine, phenylalanine, glutamic acids, and valine. Eustress factors such as low temperature during the night or supplementary UV-A- or UV-B-enriched exposure can improve the sensory quality of greenhouse *A. graveolens,* through the enhanced accumulation of organic acids and free amino acids.

Soil salinity is an overwhelming environmental threat to global food production and agricultural sustainability (Abbasi et al. 2015). Soil salinity occurs via a complex series of steps by adversely affecting the physiological and biochemical pathways of plants (Nabati et al. 2011). The excess accumulation of Na^+ induces an efflux of cytosolic K^+ and Ca^{2+} through a cascade of events leading to the disruption of homeostasis, oxidative stress, retarded growth and eventually cell death (Ahanger and Agarwal 2017). High levels of salinity affect photosynthesis due to some stomatal restrictions (Kamran et al. 2020) and reduction in photosynthetic pigments in dill (Akhtar 2020). According to Ghassemi-Golezani et al. (2011), *A. graveolens* is slightly salt tolerant and its essential oil pro-duction is significantly enhanced by increasing salinity. Salinity stress only affected plant growth in the spring at the highest concentration of 8 dS/m^{-1} in Greece. However, the concentrations of chlorophyll and total phenolics in the leaves rose in the autumn with increasing salinity. In spring, increasing salinity caused fluctuations in the chlorophyll and vitamin C content of the leaves and a decrease in total phenolics. The concentrations of chlorophyll and antioxidants were higher in spring than in autumn at all levels of salinity. The essential oil content was also higher in the spring than in the autumn, irrespective of salinity. There was a relative decrease in dill ether within the herb oil as salinity increased. This was compensated for by an increase in α-phellandrene. In the flower oil, however, the increase in salinity caused a decrease in the relative concentrations of both α-phellandrene and dill ether, which was compensated for by an increase in carvone (Tsamaidi et al. 2017).

Salinity stress in *A. graveolens* caused oxidative stress and lead to an increase in malondialdehyde (MDA) concentration. It also caused a significant decrease in both the fresh and dry weights of the plant. Salinity in dill is more detrimental to the roots than shoots and causes lower root-to-shoot ratios. It also leads to a reduction in root anatomical characteristics such as epidermal thickness and vascular bundle (Akhtar 2020). However, exogenous application of silicon on roots of *A. graveolens* under saline conditions induced stress tolerance by regulating the generation of reactive oxygen species (ROS), reducing electrolytic leakage and MDA contents, and immobilizing and reducing the uptake of toxic ions (Kim et al. 2017; Shekari et al., 2017). Shekari et al. (2017) found that activities of catalase (CAT), ascorbate peroxidase (APX), superoxide dismutase (SOD) and guaiacol peroxidase (POD) were highly increased under silicon application with NaCl to *A. graveolens* plants. Silicon ameliorated the negative effects of salinity on plant dry matter and chlorophyll content, decreased Na^+ concentration and increased K^+ concentration in roots and shoots of dill plants. Thus, silicon under saline conditions could be used to maintain the productivity *A. graveolens*.

Plants exposed to heavy metal stress show differential responses in the synthesis and accumulation of pharmacologically active molecules (Nasim and Dhir 2010), ranging from negative effects (Murch et al. 2003; Pandey et al. 2007; Thangavel et al. 1999) to stimulatory effects that result in enhanced metabolite production (Aziz et al. 2007; Michalak 2006; Nasim and Dhir 2010; Zheng and Wu 2004). Heavy metals may alter some aspects of secondary metabolism in plants, thus affecting the concentration of different bioactive compounds (Verpoorte et al. 2002). They may also bioaccumulate in some plant tissues from contaminated soil (Zheljazkov and Warman 2003). In *A. graveolens* treated with different concentrations of Cd and Pb in the medium, the yields were not affected. However, high concentrations of Cu (60 and 150 mg/L) reduced both yields and plant height relative to the control, with the latter concentration suppressing growth and yields more severely than the latter. Cadmium increased proline production in *A. graveolens* compared with the control with a decreased electrolyte leakage, relative water content and all plant growth parameters (Aghaz et al. 2013).

Higher amounts of total phenols coincided with the highest levels of zinc that were detected in *A. graveolens* samples (Majdoub et al. 2017). This phenol accumulation may be a response to the oxidative damage induced by zinc in order to permit plant survival, such as has been previously reported for different plant species (Marichali et al. 2014; Morina et al. 2010). *A. graveolens* accumulates more zinc in the shoots than in the roots (Majdoub et al. 2017). Thus, more severe zinc stress effects could be realized in its aerial parts. The higher translocation of zinc to the aerial parts has been attributed to higher heavy metal ATPase 4 (HMA4) activity than that of heavy metal ATPase 3 (HMA3), because HMA3 protein mediates zinc accumulation in root vacuoles, whereas HMA4 is involved in the metal translocation from the root toward the shoot (Lin and Aarts 2012; Nouet et al. 2015). Much as the higher concentration of zinc may trigger the production of phytochemicals with pharmacological value, it is important to pay attention to the high accumulation of Zn in its aerial parts which makes its consumption dangerous for human health, due to the toxicity of zinc (Majdoub et al. 2017). Extracts of *A. graveolens* treated with zinc were found to have higher tyrosinase enzyme inhibitory activity values that were 1.3-fold higher than those of the *A. graveolens* nontreated extracts, suggesting that zinc treatment increased the ability of *A. graveolens* to inhibit tyrosinase enzyme. Tyrosinases catalyze the oxidation of mono- and di-phenolic compounds to corresponding quinones with the concomitant reduction of molecular oxygen to water (Nunes and

Vogel, 2018). The physiological role of tyrosinases has been exploited for the production of coloring and dyeing agents that have found application in food processing and the functionality of materials (Nunes and Vogel, 2018). Thus, heavy metal stress due to zinc in *A. graveolens* could limit its application in food processing.

The deficiency of some mineral nutrients increases levels of individual and total phenolic compounds in roots as well in leaves of *A. graveolens* (Wasli et al. 2018). Under iron deficiency, whereas the shoot and plant biomass production of *A. graveolens* was significantly decreased, that of roots was not affected. Iron deficiency also resulted in a significant reduction of chlorophyll and iron concentration. *A. graveolens* was also able to increase its shoot iron use efficiency when grown under iron deficiency conditions with a corresponding increase in total phenolic compounds in roots and leaves and enhancement of antioxidant activity (Wasli et al. 2018). Iron deficiency at the cellular level induces plant oxidative stress, mainly by impairing the electron transport chain functionality both in mitochondria and chloroplasts (Murgia et al. 2015). This is indicative of the ability of *A. graveolens* to confer sufficient protection against oxidative damage, and some level of tolerance to iron deficiency (Wasli et al. 2018).

Drought stress is one of the most important abiotic stresses that severely reduce crop production, especially in the era of rapidly changing climate (Jeandroz and Lamotte 2017). *A. graveolens* responded to drought stress by causing a significant increase in the levels of phenolic compounds in the roots and shoots. However, an increase in drought stress also caused a significant reduction in chlorophyll, carotenoid, soluble proteins, K^+, P and Ca^{+2} and shoot/root K^+ ratio (Setayesh-Mehr and Ganjeali 2013). The concentrations of chlorophyll, carotenoids, vitamin C and total phenolics within the leaves of *A. graveolens* were not affected by water stress (Tsamaidi et al. 2017). In an experimental study involving induction of drought stress with salicylic acid treatment by Ghassemi-Golezani and Solhi-Khajemarjan (2021), the essential oil production by *A. graveolens* organs increased with increasing water deficit up to moderate stress, but thereafter it decreased as water deficit became more severe. The highest essential oil yield of the vegetative parts and flowers was also produced in moderately stressed plants, but the greatest essential oil yield of seeds was recorded under mild water deficit.

3.10 PHARMACOLOGICAL METABOLITE PRODUCTION ENHANCEMENT IN *A. GRAVEOLENS*

3.10.1 PLANT GROWTH-PROMOTING RHIZOBACTERIA AND MYCORRHIZAL EFFECTS

Plants inoculated with beneficial microorganisms, such as plant growth-promoting rhizobacteria (PGPR), induce morphological and biochemical modifications resulting in increased tolerance to abiotic stresses defined as induced systemic tolerance (IST) (Vázquez-Hernández et al. 2019). Mycorrhization of *A. graveolus* enhanced both the quality and quantity of essential oils. Studies by Kapoor et al. (2002) on the effect of the mycorrhizars *Glomus macrocarpum* and *G. fasciculatum* on the concentration of the essential oil in tissues of *A. graveolus* reported increases of up to 90% over the respective controls. In the same study, the levels of limonene and carvone were enhanced in essential oil obtained from *Glomus macrocarpum*-inoculated dill plants. Arbuscular mycorrhizal (AM) fungi promote the yield and composition of secondary metabolites in aromatic and medicinal plants (Crişan et al. 2018). For instance, *G. macrocarpum* and *G. fasciculatum* increase the essential oil concentrations in fruits of *A. graveolens* which are important in the food and pharmaceutical industries (Gianinazzi et al. 2010). This has been attributed to phytochemical content that enhances nutrition by arbuscular mycorrhizal fungi and rhizosphere bacteria. AMF and PGPR confer heavy metal tolerance in plants (Upadhyay et al. 2015). They reduce the accumulation of heavy metals in contaminated sites by enhancing bioaccumulation and consequent sequestration of heavy metals in certain plants. Heavy metal-induced oxidative stress may alter the titer and nature of plant secondary metabolites.

3.10.2 PHYTOHORMONAL EFFECTS

Some of the harmful impacts of water shortage on crop performance may be alleviated by growth regulators such as salicylic acid (Ghassemi-Golezani and Solhi-Khajemarjan 2021). Treatment of *A. graveolens* seeds with gibberellic acid (GA) and salicylic acid (SA) significantly improved the emergence percentage of seedlings under moderate (8 dS/m) and severe (12 dS/m) salinities. The treatment also improved the chlorophyll a and b contents, while plants from seeds pretreated with gibberellic acid showed the highest plant weight (Nikpour-Rashidabad et al. 2016). Salicylic acid application increased the seed germination rate, plumule and radicle elongation, weight of the plant, and also vigor index in *A. graveolens* under Cd stress (Espanany et al. 2015). The anatomical characters of stem epidermal thickness and vascular bundle that are adversely affected by salinity can be restored through phytohormonal application. Doses of IAA and combined doses of GA and IAA showed enhancement in several anatomical parameters in salinity-stressed *A. graveolens*. Root anatomical characteristics indicated expansion of the cortical thickness (enhancing water storage) and vascular bundles (elevating water conduction) (Akhtar 2020).

Applications of moringa leaf extract (1:30) and seaweed (0.5 ml/l) on *A. graveolens* leaves improved vegetative growth, seed yield and essential oil characteristics (El-Gamal and Ahmed, 2016). This reduced plant stress due to delayed planting as a result of climatic change. In the same study, the use of 28-Homobrassinolid plant hormone improved the root and shoot dry weight, essential oil yield and percentage of essential oil in stressed plants. These results revealed that the application of 28-Homobrassinolid increases oil yield but reduces the damage caused by drought conditions. In a study by Bulchandani and Shekhawat (2020), exogenous application of salicylic acid was revealed to significantly enhance the MDA and proline content in *A. graveolens*. Furthermore, a lower SA concentration application of 0.1 mM resulted in high quantities of carvone (0.063%) in a cell suspension compared to 0.035% which was produced at a higher concentration of 0.75 mM. This effect of salicylic acid can be utilized *in vitro* to improve the production of carvone on a large scale for use in pharmaceutical industries. Dill contains higher levels of a monoterpene called carvone (Bulchandani and Shekhawat 2020), which is the most important constituent, synthesized in maximum proportions and responsible for imparting the plant with its most important medicinal properties including its potential antiviral activity (de Carvalho and da Fonseca 2006). Treatment of an *A. graveolus* cell suspension culture with 0.1 mM concentration of SA has been reported by Bulchandani and Shekhawat (2020) to elicit the highest amount of carvone (0.063%), whereas a higher concentration of 0.75 mM SA causes a reduction in the amount (0.035%) of carvone produced. An SA-elicited cell suspension culture offered an effective and favorable *in vitro* method to improve the production of carvone for potential use in pharmaceuticals.

3.10.3 AGROCHEMICAL APPLICATION EFFECTS

As the demand for medicinal plants continues to increase amidst high population growth and degradation of their natural habitats, some efforts have been diverted to the domestication of these plants. Such an effort implies the modification of different biotic and abiotic factors to trigger the production of secondary metabolites and improve productivity. Using agrochemicals in medicinal plant production is on the increase but results in high residual contamination and low rates of production of secondary metabolites (Pereira et al. 2019). These negative effects can be managed using humic substances and microorganisms. For instance, the application of *Azotobacter chroococcum* and *Azospirillum lipoferum* in *A. graveolens* increased seed and essential oil production. This was attributed to a higher uptake of nitrite, nitrate and phosphate (Cappellari et al. 2013), and an increased concentration of active principles (Damam et al. 2016; Rai et al. 2017).

3.10.4 Carbon Dioxide Enrichment

The functional food value of herbal plants is greatly related to their contents of valuable phytochemicals that are dependent on primary and secondary metabolic processes. Carbon dioxide enrichment improved the levels of soluble sugars, starch, organic acids, some EAAs, most USFAs, total phenolics, total flavonoids and vitamins A and E. Correspondingly, considerable improvements in the total antioxidant capacity, and antiprotozoal, antibacterial and anticancer activities were recorded for *A. graveolens* in response to CO_2 (Saleh et al., 2018).

3.10.5 Light Quality

Light quality affects plant growth, essential oil biogenesis and composition. According to Sangwan et al. (2001), far-red light induces internode growth, decreases leaf area and increases the plant growth rate, while red light also increases plant growth and elongates the internodes. The essential oil concentration was highest in plants grown under far-red light treatment. Red light treatment for 4 h also induced more oil production than control or blue light treatments. Compositionally, *A. graveolens* plants exposed to 4 hours of red and far-red light produced oil containing more phellandrene and less myristicin (Hälvä et al. 1992a, 1992b).

3.11 TRANSCRIPTOMICS AND PROTEOMICS OF *A. GRAVEOLENS*

There is little research that has been done on the transcriptomics of *A. graveolus*. Salicyclic acid inducement of stress in cell cultures of *Anethum graveolens* (Bulchandani and Shekhawat 2020) revealed an upregulation of enzymatic antioxidants (APX, SOD and CAT) activity with increasing concentrations of SA, whereas a reduction in guaiacol peroxidase (GPX) activity was recorded at the end of the growth phase.

3.12 CONCLUSION

Anethum graveolens is a widely used culinary and medicinal plant with a rich history of traditional use. It has been cultivated as a commercially important crop. Dill is affected by various stress factors, both abiotic and biotic. However, *A. graveolens* is slightly salt tolerant and its essential oil production significantly was increased with increasing salinity. Salinity stress only affected plant growth, whereas drought stress caused a significant increase in the levels of phenolic compounds in the roots and shoots. Dill is also affected by a variety of biotic factors, particularly fungal pathogens. The two most important secondary metabolites produced by dill are carvone and limonene, which are both monoterpenes, although it produces several other compounds in varying amounts such as α-phellandrene, dill ether and myristicin, which are responsible for its odor.

LIST OF ABBREVIATIONS

ABA   Abscisic acid
AM   Arbuscular mycorrhizal
AMF   Arbuscular mycorrhizal fungi
APX   Ascorbate peroxidase
CAT   Catalase
dS/m   deciSiemens per meter
GA   Gibberellic acid
GPX   Guaiacol peroxidase
HMA   4 Heavy metal ATPase
IAA   Indole acetic acid

IST   Induced systemic tolerance
MDA   Malondialdehyde
PGPR   Plant growth-promoting rhizobacteria
POD   Guaiacol peroxidase
POWO   Plants of the World Online
SOD   Superoxide dismutase
SA   Salicylic acid

REFERENCES

Abbasi, G. H., Akhtar, J., Ahmad, R., Jamil, M., Anwar-ul-Haq, M., Ali, S., and Ijaz, M. (2015). Potassium application mitigates salt stress differentially at different growth stages in tolerant and sensitive maize hybrids. Plant Growth Regulation, 76(1), 111–125. https://doi.org/10.1007/s10725-015-0050-1

Aćimović, M. (2015). Dill growing technology, pests and diseases. Biljni Lekar (Plant Doctor), 43(4), 353–359.

Aghaz, M., Bandehagh, A., Aghazade, E., Toorchi, M., and Ghassemi-Gholezani, K. (2013). Effects of cadmium stress on some growth and physiological characteristics in dill (*Anethum graveolens*) ecotypes. International Journal of Agriculture, 3(2), 409–413.

Ahanger, M. A., and Agarwal, R. M. (2017). Salinity stress induced alterations in antioxidant metabolism and nitrogen assimilation in wheat (*Triticum aestivum* L) as influenced by potassium supplementation. Plant Physiology and Biochemistry, 115, 449–460. https://doi.org/10.1016/j.plaphy.2017.04.017

Aishwath, O., and Ratan, L. (2016). Resilience of spices, medicinal and aromatic plants with climate change induced abiotic stresses. Annals of Plant and Soil Research, 18(2), 91–109.

Akhtar, M. (2020). Evaluation of Gibberellic acid and Indole acetic acid impacts on growth and physiological parameters of *Anethum graveolens* under salt stress http://itr.iub.edu.pk:8000/xmlui/handle/123456 789/1024

Akula, R., and Ravishankar, G. A. (2011). Influence of abiotic stress signals on secondary metabolites in plants. Plant signaling and behavior, 6(11), 1720–1731. https://doi.org/10.4161/psb.6.11.17613

Al-Snafi, A. E. (2014). The pharmacological importance of *Anethum graveolens*–A review. International Journal of Pharmacy and Pharmaceutical Sciences, 6(4), 11–13.

Amanpour, A., Kelebek, H., and Selli, S. (2017). Aroma constituents of shade-dried aerial parts of Iranian dill (*Anethum graveolens* L.) and savory (Satureja sahendica Bornm.) by solvent-assisted flavor evaporation technique. Journal of Food Measurement and Characterization, 11(3), 1430–1439. https://doi.org/ 10.1007/s11694-017-9522-5

Anywar, G. (2021). Traditional African medicinal plants for a strong immune system. In Traditional Herbal Therapy for the Human Immune System (First edit, p. 173). CRC Press. www.taylorfrancis.com/chapt ers/edit/10.1201/9781003137955-6/traditional-african-medicinal-plants-strong-immune-system-any war-godwin

Anywar, G., Kakudidi, E., Byamukama, R., Mukonzo, J., Schubert, A., and Oryem-Origa, H. (2020a). Data on medicinal plants used by herbalists for boosting immunity in people living with HIV/AIDS in Uganda. Data in Brief, 105097. https://doi.org/10.1016/j.dib.2019.105097

Anywar, G., Kakudidi, E., Byamukama, R., Mukonzo, J., Schubert, A., and Oryem-Origa, H. (2020b). Medicinal plants used by traditional medicine practitioners to boost the immune system in people living with HIV/AIDS in Uganda. European Journal of Integrative Medicine, 101011. https://doi.org/10.1016/ j.eujim.2019.101011

Aziz, E. E., Gad, N., and Badran, N. (2007). Effect of cobalt and nickel on plant growth, yield and flavonoids content of Hibiscus sabdariffa L. Australian Journal of Basic and Applied Sciences, 1(2), 73–78.

Blank, I., and Grosch, W. (1991). Evaluation of Potent Odorants in Dill Seed and Dill Herb (*Anethum graveolens* L.) by Aroma Extract Dilution Analysis. Journal of Food Science, 56(1), 63-67. https://doi.org/10.1111/ j.1365-2621.1991.tb07976.x

Bonnländer, B., and Winterhalter, P. (2000). 9-Hydroxypiperitone beta-D-glucopyranoside and other polar constituents from dill (*Anethum graveolens* L.) herb. J Agric Food Chem, 48(10), 4821–825. https://doi. org/10.1021/jf000439a

Bouyahya, A., Mechchate, H., Benali, T., Ghchime, R., Charfi, S., Balahbib, A., Burkov, P., Shariati, M. A., Lorenzo, J. M., and Omari, N. E. (2021). Health Benefits and Pharmacological Properties of Carvone. Biomolecules, 11(12), 1803. www.mdpi.com/2218-273X/11/12/1803

Bulchandani, N., and Shekhawat, G. S. (2020). Salicylic acid mediated up regulation of carvone biosynthesis during growth phase in cell suspension cultures of *Anethum graveolens*. 3 Biotech, 10(11), 482. https://doi.org/10.1007/s13205-020-02470-4

Cappellari, L. d. R., Santoro, M. V., Nievas, F., Giordano, W., and Banchio, E. (2013). Increase of secondary metabolite content in marigold by inoculation with plant growth-promoting rhizobacteria. Applied Soil Ecology, 70, 16–22. https://doi.org/10.1016/j.apsoil.2013.04.001

Cardenas-Manríquez, G., Vega-Muñoz, I., Villagómez-Aranda, A. L., León-Galvan, M. F., Cruz-Hernandez, A., Torres-Pacheco, I., Rangel-Cano, R. M., Rivera-Bustamante, R. F., and Guevara-Gonzalez, R. G. (2016). Proteomic and metabolomic profiles in transgenic tobacco (N. *tabacum xanthi* nc) to CchGLP from Capsicum chinense BG-3821 resistant to biotic and abiotic stresses. Environmental and Experimental Botany, 130, 33–41. https://doi.org/10.1016/j.envexpbot.2016.05.005

Carrubba, A., la Torre, R., Saiano, F., and Aiello, P. (2008). Sustainable production of fennel and dill by intercropping. Agronomy for Sustainable Development, 28(2), 247–256. https://doi.org/10.1051/agro:2007040

Castro-Alves, V., Kalbina, I., Nilsen, A., Aronsson, M., Rosenqvist, E., Jansen, M. A. K., Qian, M., Öström, Å., Hyötyläinen, T., and Strid, Å. (2021). Integration of non-target metabolomics and sensory analysis unravels vegetable plant metabolite signatures associated with sensory quality: A case study using dill (*Anethum graveolens*). Food Chemistry, 344, 128714. https://doi.org/10.1016/j.foodchem.2020.128714

Chahal, K., Monika, A. K., Bhardwaj, U., and Kaur, R. (2017). Chemistry and biological activities of *Anethum graveolens* L.(dill) essential oil: A review. Journal of Pharmacognosy and Phytochemistry, 6(2), 295–306.

Crişan, I., Vidican, R., and Stoian, V. (2018). Induced modifications on secondary metabolism of aromatic and medicinal plants: an endomycorrhizal approach. Hop Med Plants, 26, 15–29.

Damam, M., Kaloori, K., Gaddam, B., and Kausar, R. (2016). Plant growth promoting substances (phytohormones) produced by rhizobacterial strains isolated from the rhizosphere of medicinal plants. International Journal of Pharmaceutical Sciences Review and Research, 37(1), 130–136.

Dar, N. A., Amin, I., Wani, W., Wani, S. A., Shikari, A. B., Wani, S. H., and Masoodi, K. Z. (2017). Abscisic acid: A key regulator of abiotic stress tolerance in plants. Plant Gene, 11, 106–111.

Davazdahemami, S., and Allahdadi, M. (2022). Essential oil yield and composition of four annual plants (ajowan, dill, Moldavian balm and black cumin) under saline irrigation. Food Ther. Health Care, 4(5). https://doi.org/10.53388/FTHC20220124005

de Carvalho, C. C. C. R., and da Fonseca, M. M. R. (2006). Carvone: Why and how should one bother to produce this terpene. Food Chemistry, 95(3), 413–422. https://doi.org/10.1016/j.foodchem.2005.01.003

Dutta, S. C., and Neog, B. (2016). Accumulation of secondary metabolites in response to antioxidant activity of turmeric rhizomes co-inoculated with native arbuscular mycorrhizal fungi and plant growth promoting rhizobacteria. Scientia Horticulturae, 204, 179–184. https://doi.org/10.1016/j.scienta.2016.03.028

Egamberdieva, D., and Teixeira da Silva, J. A. (2015). Medicinal Plants and PGPR: A New Frontier for Phytochemicals. In D. Egamberdieva, S. Shrivastava, and A. Varma (Eds.), Plant-Growth-Promoting Rhizobacteria (PGPR) and Medicinal Plants (pp. 287–303). Springer International Publishing. https://doi.org/10.1007/978-3-319-13401-7_14

El-Gamal, S. M., and Ahmed, H. M. (2016). Response of dill (*Anethum graveloens* Linn.) to seaweed and moringa leaf extracts foliar application under different sowing dates. Alex. J. Agriculture. Sci, 61(5), 469–485.

Espanany, A., Fallah, S., and Tadayyon, A. (2015). The effect of halopriming and salicylic acid on the germination of fenugreek (*Trigonella foenum-graecum*) under different cadmium concentrations. Notulae Scientia Biologicae, 7(3), 322–329. https://doi.org/10.15835/nsb739563

Ghassemi-Golezani, K., and Solhi-Khajemarjan, R. (2021). Changes in growth and essential oil content of dill (*Anethum graveolens*) organs under drought stress in response to salicylic acid. Journal of Plant Physiology and Breeding, 11(1), 33–47. https://doi.org/10.22034/JPPB.2021.13717

Ghassemi-Golezani, K., Zehtab-Salmasi, S., and Dastborhan, S. (2011). Changes in essential oil content of dill (*Anethum graveolens*) organs under salinity stress. Journal of Medicinal Plants Research, 5(14), 3142–3145. https://doi.org/10.5897/JMPR.9000325

Gianinazzi, S., Gollotte, A., Binet, M.-N., van Tuinen, D., Redecker, D., and Wipf, D. (2010). Agroecology: the key role of arbuscular mycorrhizas in ecosystem services. Mycorrhiza, 20(8), 519–530. https://doi.org/10.1007/s00572-010-0333-3

Giovannetti, M., Avio, L., and Sbrana, C. (2013). Improvement of nutraceutical value of food by plant symbionts. In Ramawat K. and M. JM. (Eds.), Natural Products. Springer. https://doi.org/10.1007/978-3-642-22144-6_187

Gontia-Mishra, I., Sapre, S., Sharma, A., and Tiwari, S. (2016), Amelioration of drought tolerance in wheat by the interaction of plant growth-promoting rhizobacteria. Plant Biol J, 18: 992–1000. https://doi.org/10.1111/plb.12505

Goodarzi, M. T., Khodadadi, I., Tavilani, H., and Abbasi Oshaghi, E. (2016). The Role of *Anethum graveolens* L. (Dill) in the Management of Diabetes. Journal of Tropical Medicine, 2016, 1098916. https://doi.org/10.1155/2016/1098916

Goss, M. J., Carvalho, M., & Brito, I. (2017). *Chapter 1 - Challenges to Agriculture Systems* (M. J. Goss, M. Carvalho, & I. B. T.-F. D. of M. and S. A. Brito (eds.); pp. 1–14). Academic Press. https://doi.org/10.1016/B978-0-12-804244-1.00001-0

Haghshenas, J., and Eskandari, M. (2011). Growth parameters and essential oil percentage changes of dill (*Anethum graveolens*) as affected by drought stress and use of 28-homobrassinolide. Journal of Plant Ecophysiology, 3(9), 29–40.

Hälvä, S., Craker, L. E., Simon, J. E., and Charles, D. J. (1992a). Light Levels, Growth, and Essential Oil in Dill (*Anethum graveolens* L.). Journal of Herbs, Spices and Medicinal Plants, 1(1–2), 47–58. https://doi.org/10.1300/J044v01n01_06

Hälvä, S., Craker, L. E., Simon, J. E., and Charles, D. J. (1992b). Light Quality, Growth, and Essential Oil in Dill (*Anethum graveolens* L.). Journal of Herbs, Spices and Medicinal Plants, 1(1–2), 59–69. https://doi.org/10.1300/J044v01n01_07

Hasanuzzaman, M., Nahar, K., Alam, M. M., Roychowdhury, R., and Fujita, M. (2013). Physiological, biochemical, and molecular mechanisms of heat stress tolerance in plants. International Journal of Molecular Sciences, 14(5), 9643–9684.

Ishikawa, T., Kudo, M., and Kitajima, J. (2002). Water-Soluble Constituents of Dill. Chemical and Pharmaceutical Bulletin, 50(4), 501–507. https//doi.org/10.1248/cpb.50.501

Jana, S., and Shekhawat, G. S. (2010). *Anethum graveolens*: An Indian traditional medicinal herb and spice. Pharmacognosy reviews, 4(8), 179–184. https://doi.org/10.4103/0973-7847.70915

Jeandroz, S., & Lamotte, O. (2017). Editorial: Plant Responses to Biotic and Abiotic Stresses: Lessons from Cell Signaling. Frontiers in Plant Science, 8, 1772. https://doi.org/10.3389/fpls.2017.01772

Kamran, M., Parveen, A., Ahmar, S., Malik, Z., Hussain, S., Chattha, M. S., Saleem, M. H., Adil, M., Heidari, P., and Chen, J.-T. (2020). An Overview of Hazardous Impacts of Soil Salinity in Crops, Tolerance Mechanisms, and Amelioration through Selenium Supplementation. In International Journal of Molecular Sciences (Vol. 21, Issue 1). https://doi.org/10.3390/ijms21010148.

Kandel, D. R., Marconi, T. G., Badillo-Vargas, I. E., Enciso, J., Zapata, S. D., Lazcano, C. A., Crosby, K., and Avila, C. A. (2020). Yield and fruit quality of high-tunnel tomato cultivars produced during the off-season in South Texas. Scientia Horticulturae, 272, 109582.

Kapoor, R., Giri, B., and Mukerji, K. G. (2002). Glomus macrocarpum: a potential bioinoculant to improve essential oil quality and concentration in Dill (*Anethum graveolens* L.) and Carum (*Trachyspermum ammi* (Linn.) Sprague). World Journal of Microbiology and Biotechnology, 18(5), 459–463. https://doi.org/10.1023/A:1015522100497

Kaur, G. J., and Arora, D. S. (2010). Bioactive potential of *Anethum graveolens*, Foeniculum vulgare and Trachyspermum ammi belonging to the family Umbelliferae-Current status. Journal of Medicinal Plants Research, 4(2), 087–094. https://doi.org/10.5897/JMPR09.018

Kim, Y.-H., Khan, A. L., Waqas, M., and Lee, I.-J. (2017). Silicon Regulates Antioxidant Activities of Crop Plants under Abiotic-Induced Oxidative Stress: A Review. Frontiers in Plant Science, 8. https://doi.org/10.3389/fpls.2017.00510

Lin, Y.-F., and Aarts, M. G. M. (2012). The molecular mechanism of zinc and cadmium stress response in plants. Cellular and Molecular Life Sciences, 69(19), 3187–3206. https://doi.org/10.1007/s00018-012-1089-z

Liogier, A. H., Liogier, H. A., and Martorell, L. F. (2000). Flora of Puerto Rico and adjacent islands: a systematic synopsis. La Editorial, UPR.

Madani, H., Ahmady Mahmoodabady, N., and Vahdati, A. (2005). Effects of hydroalcoholic extracts of *Anethum graveolens* (Dill) on Plasma glucose and lipid levels diabets induced rats. Iranian Journal of Diabetes and Lipid Disorders, 5(2), 109–116. http://ijdld.tums.ac.ir/article-1-370-en.htmlhttp://ijdld.tums.ac.ir/article-1-370-en.pdf

Mahran, G. H., Kadry, H. A., Thabet, C. K., El-Olemy, M. M., Al-Azizi, M. M., Schiff, P. L., Wong, L. K., and Liv, N. (1992). GC/MS Analysis of Volatile Oil of Fruits of *Anethum graveolens*. International Journal of Pharmacognosy, 30(2), 139–144. https://doi.org/10.3109/13880209209053978

Majdoub, N., el-Guendouz, S., Rezgui, M., Carlier, J., Costa, C., Kaab, L. B. B., and Miguel, M. G. (2017). Growth, photosynthetic pigments, phenolic content and biological activities of Foeniculum vulgare Mill., *Anethum graveolens* L. and *Pimpinella anisum* L. (Apiaceae) in response to zinc. Industrial Crops and Products, 109, 627–636. https://doi.org/10.1016/j.indcrop.2017.09.012

Marianne, J. D., and Acevedo-Rodríguez, P. (2015, 2015). Invasive Species Compendium: Detailed Coverage of Invasive Species Threatening Livelihoods and the Environment Worldwide. Datasheet. *Anethum graveolens*. CAB International; London, UK. Retrieved 24 June 2022 from www.cabi.org/isc/datasheet/3472

Marichali, A., Dallali, S., Ouerghemmi, S., Sebei, H., and Hosni, K. (2014). Germination, morpho-physiological and biochemical responses of coriander (*Coriandrum sativum* L.) to zinc excess. Industrial Crops and Products, 55, 248–257. https://doi.org/10.1016/j.indcrop.2014.02.033

Mejía-Teniente, L., Durán-Flores, F. d. D., Chapa-Oliver, A. M., Torres-Pacheco, I., Cruz-Hernández, A., González-Chavira, M. M., Ocampo-Velázquez, R. V., and Guevara-González, R. G. (2013). Oxidative and Molecular Responses in Capsicum annuum L. after Hydrogen Peroxide, Salicylic Acid and Chitosan Foliar Applications. International Journal of Molecular Sciences, 14(5), 10178–10196. www.mdpi.com/1422-0067/14/5/10178

Memariani, Z., Gorji, N., Moeini, R., and Farzaei, M. H. (2020). Chapter Two – Traditional uses. In S. M. Nabavi, I. Suntar, D. Barreca, and H. Khan (Eds.), Phytonutrients in Food (pp. 23–66). Woodhead Publishing. https://doi.org/10.1016/B978-0-12-815354-3.00004-6

Michalak, A. (2006). Phenolic compounds and their antioxidant activity in plants growing under heavy metal stress. Polish journal of environmental studies, 15(4), 523–530.

Mirhosseini, M., Baradaran, A., and Rafieian-Kopaei, M. (2014). *Anethum graveolens* and hyperlipidemia: A randomized clinical trial. Journal of research in medical sciences: the official journal of Isfahan University of Medical Sciences, 19(8), 758–761. www.ncbi.nlm.nih.gov/pmc/articles/PMC4235097/

Morcia, C., Tumino, G., Ghizzoni, R., and Terzi, V. (2016). Chapter 35 – Carvone (Mentha spicata L.) Oils. In V. R. Preedy (Ed.), Essential Oils in Food Preservation, Flavor and Safety (pp. 309–316). Academic Press. https://doi.org/10.1016/B978-0-12-416641-7.00035-3

Morina, F., Jovanovic, L., Mojovic, M., Vidovic, M., Pankovic, D., and Veljovic Jovanovic, S. (2010). Zinc-induced oxidative stress in Verbascum thapsus is caused by an accumulation of reactive oxygen species and quinhydrone in the cell wall. Physiol Plant, 140(3), 209–224. https://doi.org/10.1111/j.1399-3054.2010.01399.x

Murch, S. J., Haq, K., Rupasinghe, H. V., and Saxena, P. K. (2003). Nickel contamination affects growth and secondary metabolite composition of St. John's wort (*Hypericum perforatum* L.). Environmental and Experimental Botany, 49(3), 251–257.

Murgia, I., Giacometti, S., Balestrazzi, A., Paparella, S., Pagliano, C., and Morandini, P. A. (2015). Analysis of the transgenerational iron deficiency stress memory in *Arabidopsis thaliana* plants. Frontiers in Plant Science, 6, 745.

Nabati, J., Kafi, M., Nezami, A., Moghaddam, P. R., Masoumi, A., and Mehrjerdi, M. Z. (2011). Effect of salinity on biomass production and activities of some key enzymatic antioxidants in kochia (*Kochia scoparia*). Pakistan Journal of Botany, 43(1), 539–548.

Najaran, Z. T., Hassanzadeh, M. K., Nasery, M., and Emami, S. A. (2016). Chapter 45 – Dill (*Anethum graveolens* L.) Oils. In V. R. Preedy (Ed.), Essential Oils in Food Preservation, Flavor and Safety (pp. 405–412). Academic Press. https://doi.org/10.1016/B978-0-12-416641-7.00045-6

Nasim, S. A., and Dhir, B. (2010). Heavy metals alter the potency of medicinal plants. Rev Environ Contam Toxicol, 203, 139–149. https://doi.org/10.1007/978-1-4419-1352-4_5

Nikpour-Rashidabad, N., Ghassemi-Golezani, K., Alizadeh-Salteh, S., and Valizadeh, M. (2016). Seed pre-treatment effect on seedling emergence, chlorophyll content and plant weight of dill under salt stress. Journal of Biodiversity and Environmental Sciences, 9, 158–164.

Nouet, C., Charlier, J.-B., Carnol, M., Bosman, B., Farnir, F., Motte, P., and Hanikenne, M. (2015). Functional analysis of the three HMA4 copies of the metal hyperaccumulator *Arabidopsis halleri*. Journal of Experimental Botany, 66(19), 5783–5795. https://doi.org/10.1093/jxb/erv280

Nunes, C. S., and Vogel, K. (2018). Chapter 20 – Tyrosinases—physiology, pathophysiology, and applications. In C. S. Nunes and V. Kumar (Eds.), Enzymes in Human and Animal Nutrition (pp. 403–412). Academic Press. https://doi.org/10.1016/B978-0-12-805419-2.00020-4

Olwenyi, O. A, Asingura, B., Naluyima,P., Anywar, GU., Nalunga, J., Nakabuye, M., Semwogerere, M., Bagaya, B., Cham, F., Tindikahwa, A., Kiweewa, F., Lichter, EZ., Podany, AT., Fletcher, CV., Byrareddy SN., Kibuuka H (2021). "*In-Vitro* Immunomodulatory Activity of *Azadirachta indica* A.Juss. Ethanol: Water Mixture against HIV Associated Chronic CD4+ T-Cell Activation/ Exhaustion." BMC Complementary Medicine and Therapies 21 (1): 114. https://doi.org/10.1186/s12906-021-03288-0

Osanloo, M., Sereshti, H., Sedaghat, M. M., and Amani, A. (2018). Nanoemulsion of Dill essential oil as a green and potent larvicide against Anopheles stephensi. Environmental Science and Pollution Research, 25(7), 6466–6473. https://doi.org/10.1007/s11356-017-0822-4

Oshaghi, E. A., Khodadadi, I., Mirzaei, F., Khazaei, M., Tavilani, H., and Goodarzi, M. T. (2017). Methanolic Extract of Dill Leaves Inhibits AGEs Formation and Shows Potential Hepatoprotective Effects in CCl_4 Induced Liver Toxicity in Rat. Journal of Pharmaceutics, 2017, 6081374. https://doi.org/10.1155/2017/6081374

Oshaghi, E. A., Khodadadi, I., Tavilani, H., and Goodarzi, M. T. (2016). Effect of dill tablet (*Anethum graveolens* L) on antioxidant status and biochemical factors on carbon tetrachloride-induced liver damage on rat. International journal of applied and basic medical research, 6(2), 111–114. https://doi.org/10.4103/2229-516X.179019

Pandey, S., Gupta, K., and Mukherjee, A. (2007). Impact of cadmium and lead on Catharanthus roseus-A phytoremediation study. Journal of Environmental Biology, 28(3), 655–662.

Pereira, M., Morais, L., Marques, E., Martins, A., Cavalcanti, V., Rodrigues, F., Gonçalves, W., Blank, A., Pasqual, M., and Dória, J. (2019). Humic substances and efficient microorganisms: elicitation of medicinal plants-a review. J Agric Sci, 11(7). https://doi.org/10.5539/jas.v11n7p268

Plants of the World Online (POWO) (2021). *Anethum graveolens* L. https://powo.science.kew.org/results?

Pullaiah, T. (2002). Medicinal Plants in Andhra Pradesh, India. Daya Books.

Rai, A., Kumar, S., Bauddh, K., Singh, N., and Singh, R. P. (2017). Improvement in growth and alkaloid content of Rauwolfia serpentina on application of organic matrix entrapped biofertilizers (*Azotobacter chroococcum, Azospirillum brasilense* and *Pseudomonas putida*). Journal of Plant Nutrition, 40(16), 2237–2247. https://doi.org/10.1080/01904167.2016.1222419

Ramadan, M. M., Abd-Algader, N. N., El-kamali, H. H., Ghanem, K. Z., and Farrag, A. R. H. (2013). Volatile compounds and antioxidant activity of the aromatic herb *Anethum graveolens*. Journal of the Arab Society for Medical Research, 8(2), 79–88. https://doi.org/10.4103/1687-4293.123791

Randall, R. (2012). A global compendium of weeds,(Ed. 2)[ed. by Randall, RP]. Perth, Australia: Department of Agriculture and Food Western Australia. 1124 pp. In.

Ravishankar, G., and Ramachandra Rao, S. (2000). Biotechnological Production of Phyto-Pharmaceuticals. Journal of Biochemistry Molecular Biology and Biophysics, 4, 73–102.

Sah, S. K., Reddy, K. R., & Li, J. (2016). Abscisic Acid and Abiotic Stress Tolerance in Crop Plants. *Frontiers in Plant Science, 7*. www.frontiersin.org/article/10.3389/fpls.2016.00571

Saleh, A., Selim, S., Al Jaouni, S., and Abdelgawad, H. (2018). CO 2 enrichment can enhance the nutritional and health benefits of parsley (*Petroselinum crispum* L.) and dill (*Anethum graveolens* L.). Food Chemistry, 269. https://doi.org/10.1016/j.foodchem.2018.07.046

Sangwan, N. S., Farooqi, A. H. A., Shabih, F., and Sangwan, R. S. (2001). Regulation of essential oil production in plants. Plant Growth Regulation, 34(1), 3–21. https://doi.org/10.1023/A:1013386921596

Santos, P. A. G., Figueiredo, A. C., Lourenço, P. M. L., Barroso, J. G., Pedro, L. G., Oliveira, M. M., Schripsema, J., Deans, S. G., and Scheffer, J. J. C. (2002). Hairy root cultures of *Anethum graveolens* (dill): establishment, growth, time-course study of their essential oil and its comparison with parent plant oils. Biotechnology Letters, 24(12), 1031–1036. https://doi.org/10.1023/A:1015653701265

Scagel, C. F., and Lee, J. (2012). Phenolic composition of basil plants is differentially altered by plant nutrient status and inoculation with mycorrhizal fungi. HortScience, 47(5), 660–671. https://doi.org/10.21273/HORTSCI.47.5.660

Setayesh-Mehr, Z., and Ganjeali, A. (2013). Effects of Drought Stress on Growth and Physiological Characteristics of Dill (*Anethum graveolens* L.). Journal Of Horticultural Science, 27(1), 27–35. https://doi.org/10.22067/jhorts4.v0i0.20782

Shekari, F., Abbasi, A., and Mustafavi, S. H. (2017). Effect of silicon and selenium on enzymatic changes and productivity of dill in saline condition. Journal of the Saudi Society of Agricultural Sciences, 16(4), 367–374. https://doi.org/10.1016/j.jssas.2015.11.006

Shelef, L. A. (2003). HERBS | Herbs of the Umbelliferae. In B. Caballero (Ed.), Encyclopedia of Food Sciences and Nutrition (Second Edition) (pp. 3090–3098). Academic Press. https://doi.org/10.1016/B0-12-227055-X/00594-0

Singh, M. P., and Panda, H. (2005). Medicinal herbs with their formulations. Vol 1. Daya Publishing House, Delhi, India.

Stavri, M., and Gibbons, S. (2005). The antimycobacterial constituents of dill (*Anethum graveolens*). Phytotherapy Research, 19(11), 938–941. https://doi.org/10.1002/ptr.1758

Suresh, S., Chung, J.-W., Sung, J.-S., Cho, G.-T., Park, J.-H., Yoon, M. S., Kim, C.-K., and Baek, H.-J. (2013). Analysis of genetic diversity and population structure of 135 dill (*Anethum graveolens* L.) accessions using RAPD markers. Genetic Resources and Crop Evolution, 60(3), 893–903. https://doi.org/10.1007/s10722-012-9886-7

Tanko, H., Carrier, D. J., Duan, L., and Clausen, E. (2005). Pre- and post-harvest processing of medicinal plants. Plant Genetic Resources, 3(2), 304–313. https://doi.org/10.1079/PGR200569

Tchatchouang, S., Beng, V. P., and Kuete, V. (2017). Chapter 11 – Antiemetic African Medicinal Spices and Vegetables. In V. Kuete (Ed.), Medicinal Spices and Vegetables from Africa (pp. 299–313). Academic Press. https://doi.org/10.1016/B978-0-12-809286-6.00011-X

Thangavel, P., Sulthana, A. S., and Subburam, V. (1999). Interactive effects of selenium and mercury on the restoration potential of leaves of the medicinal plant, Portulaca oleracea Linn. Science of The Total Environment, 243, 1–8.

Tsamaidi, D., Daferera, D., Karapanos, I. C., and Passam, H. C. (2017). The effect of water deficiency and salinity on the growth and quality of fresh dill (Anethum graveolens L.) during autumn and spring cultivation. International Journal of Plant Production, 11(1), 33–46.

Upadhyay, S., Koul, M., and Kapoor, R. (2015). Rhizosphere Microflora in Advocacy of Heavy Metal Tolerance in Plants. In D. Egamberdieva, S. Shrivastava, and A. Varma (Eds.), Plant-Growth-Promoting Rhizobacteria (PGPR) and Medicinal Plants (pp. 323–337). Springer International Publishing. https://doi.org/10.1007/978-3-319-13401-7_16

van Wyk, A. S., and Prinsloo, G. (2020). Health, safety and quality concerns of plant-based traditional medicines and herbal remedies. South African Journal of Botany, 133, 54–62. https://doi.org/10.1016/j.sajb.2020.06.031

Vázquez-Hernández, M. C., Parola-Contreras, I., Montoya-Gómez, L. M., Torres-Pacheco, I., Schwarz, D., and Guevara-González, R. G. (2019). Eustressors: Chemical and physical stress factors used to enhance vegetables production. Scientia Horticulturae, 250, 223–229. https://doi.org/10.1016/j.scienta.2019.02.053

Verpoorte, R., Contin, A., and Memelink, J. (2002). Biotechnology for the production of plant secondary metabolites. Phytochemistry Reviews, 1(1), 13–25.

Warrier, P., Nambiar, V., and Mankutty, C. (1994). Indian medicinal plants, orient lonman; Chennai, India.

Warrier, P. K. (1993). Indian medicinal plants: a compendium of 500 species (Vol. 5). Orient Blackswan.

Wasli, H., Jelali, N., Silva, A. M. S., Ksouri, R., and Cardoso, S. M. (2018). Variation of polyphenolic composition, antioxidants and physiological characteristics of dill (*Anethum graveolens* L.) as affected by bicarbonate-induced iron deficiency conditions. Industrial Crops and Products, 126, 466–476. https://doi.org/10.1016/j.indcrop.2018.10.007

Wiesel, L., Newton, A. C., Elliott, I., Booty, D., Gilroy, E. M., Birch, P. R. J., and Hein, I. (2014). Molecular effects of resistance elicitors from biological origin and their potential for crop protection. Frontiers in Plant Science, 5. https://doi.org/10.3389/fpls.2014.00655

Yineger, H., Kelbessa, E., Bekele, T., and Lulekal, E. (2013). Plants used in traditional management of human ailments at Bale Mountains National Park, Southeastern Ethiopia. Journal of Medicinal Plants Research, 2(6), 132–153.

Yu, L., Yan, J., and Sun, Z. (2017). D-limonene exhibits anti-inflammatory and antioxidant properties in an ulcerative colitis rat model via regulation of iNOS, COX-2, PGE2 and ERK signaling pathways. Molecular medicine reports, 15(4), 2339–2346. https://doi.org/10.3892/mmr.2017

Zhang, W. J., and Björn, L. O. (2009). The effect of ultraviolet radiation on the accumulation of medicinal compounds in plants. Fitoterapia, 80(4), 207–218. https://doi.org/10.1016/j.fitote.2009.02.006

Zheljazkov, V. D., and Warman, P. R. (2003). Application of high Cu compost to Swiss chard and basil. Science of the Total Environment, 302(1–3), 13–26.

Zheng, Z., and Wu, M. (2004). Cadmium treatment enhances the production of alkaloid secondary metabolites in *Catharanthus roseus*. Plant Science, 166(2), 507–514. https://doi.org/10.1016/j.plantsci.2003.10.022

4 *Avena sativa* under Drought Stress

Peichen Hou[1], Mei Qu[2,3], Ping Yun[3], Aixue Li[1*],*
Hassan Ahmed Ibraheem Ahmed[3,4], Yuanying Peng[5],*
Waqas-ud-Din Khan[6] and Liping Chen[1*]*
[1]Research Center of Intelligent Equipment, Beijing Academy of
Agriculture and Forestry Sciences, Beijing, China
[2]International Research Centre for Environmental Membrane Biology,
Foshan University, Foshan, China
[3]Tasmanian Institute of Agriculture, University of Tasmania,
Tasmania, 7005, Australia
[4]Department of Botany, Faculty of Science, Port Said University,
Port Said, 42526, Egypt
[5]State Key Laboratory of Crop Gene Exploration and Utilization in
Southwest China, Sichuan Agricultural University, Chengdu, China
[6]Sustainable Development Study Centre, Government College
University, Lahore, 54000, Pakistan
*Corresponding author: hassan.ahmed@utas.edu.au;
hassan.ahmed.sci@gmail.com; houpc@nercita.org.cn;
liax@nercita.org.cn; chenlp@nercita.org.cn; yy.peng@hotmail.com;
dr.waqasuddin@gcu.edu.pk

CONTENTS

DOI: 10.1201/9781003242963-4

4.1   INTRODUCTION

Plants in both arid and semi-arid regions of the world are threatened by water deficits (Anjum et al. 2011), and about 45% of fields suffer from frequent or continuous drought stress (Bot et al. 2000), leading to 50–70% crop yield reductions worldwide (Verma and Deepti 2016), threatening global food security. In recent years, the frequency of global drought has continued to increase, therefore, looking for genetic resources for drought-tolerant crops and cultivating drought-tolerant crops is an important trend in future agricultural production. Oat (*Avena sativa*) is an important grain and forage grass, functionally classified as a food rich in soluble fiber, β-glucan, lipid, protein and antioxidants. Since the oat-planted area has been decreasing, resulting in a decrease in yield, its price has risen in recent years (Isidro-Sánchez et al. 2020). Oat can be grown in various soil environments, especially with good adaptability to marginal soils (Suttie and Reynolds 2004). In areas with dry soil, oat cultivation is an important supplement to staple crops or forage production. Drought stress is an important factor limiting oat production, and so breeding excellent drought-tolerant oat cultivars is an important research topic for future oat production (Dita et al. 2006). Therefore, it is necessary to evaluate the physiological, biochemical and molecular mechanisms of oat, and analyze the complex biological processes of drought tolerance.

4.2   GROWTH RESPONSES

It is difficult for plants to avoid stress due to their sessile growth characteristics (Xiong and Yang 2003). In particular, drought stress is one of the common stresses in the life cycle of plants. Drought stress reduces the amount of water available around the rhizosphere of plants, and affects photosynthesis and respiration, inhibits plant growth and development, and ultimately reduces grain yield and quality (Garcia-Gomez et al. 2000; Anjum et al. 2011). Drought stress leads to adaptation and regulation of plant shoot phenotype, including changes in leaf morphology and structure (Poorter and Markesteijn 2008), regulation of the stomatal opening and closing (Qi et al. 2018), and regulation of the thickness and structure of the waxy layer (Xue et al. 2017). Moreover, plants can respond to drought stress by rapidly growing, increasing the root/shoot ratio, and using deep groundwater to ensure above-ground water supply (Wu and Cosgrove 2000; Hu and Xiong 2014), and changing root morphology or structures to respond to drought stress (Price et al. 2002). To date, there have been limited studies that have demonstrated the effects of drought on the growth and development of oat and the responses of oat to drought stress. Osmotic stress induced by PEG6000 affects oat seed germination (Mut et al. 2010). Some oat cultivars can tolerate mild drought stress, but the growth of leaves is inhibited, and drought stress has a significant effect on oat at its different developmental stages. However, the application of severe water deficiency at the 4–6 leaf stage, jointing stage and heading stage makes oat more sensitive, which is manifested by the reduction of leaf area, accelerated leaf senescence, limited carbon assimilation and inhibition of plant growth (Zhao et al. 2021). The addition of mannitol (0–12.5%) to root medium within 24 hours, with increasing mannitol concentration, the elongation of the youngest leaves (third and fourth leaves) decreased (Dastgheib et al. 1990), and mannitol solution in the 10–15% concentration range also significantly inhibited the shoot and root length of oat (Koka et al. 2015). However, it has been shown that the osmotic stress of PEG6000 at concentrations as low as 5% can significantly increase the length of roots and shoots of oat, as well as fresh/dry weight (Koka et al. 2015). Under drought stress, the relative water content (RWC) of two oat cultivars *CK1* and *F411* decreased significantly, resulting in a significant decrease in leaf water potential, inhibiting the growth and development of roots and shoots, and ultimately causing a significant decrease in biomass (Islam et al. 2011; Ghafoor et al. 2019). Drought stress inhibits seedling and root length, leading to a decrease in biomass, which has been reported successively (Shehzadi et al. 2019; Alyammahi and Gururani 2020). During severe water deficit, the biomass of oat decreased by 72%, nitrogen accumulation decreased by 72–80%, and the nitrogen absorption peak was earlier, reducing the daily nitrogen accumulation rate and shortening the life cycle of oat (Coelho et al. 2020).

Although studies show that oat can survive in arid environments, the inhibition of growth and development of oat under drought stress is also a common phenomenon. Drought stress resulted in 43–67% and 31–33% decreases in the yields of the two oat cultivars *Shadow* and *Bia*, respectively (Zhao et al. 2021). Drought stress adversely affected the growth, development and yield of oat, but it also highlighted that oat may reduce nutrient consumption under drought stress through "throttling strategies" such as slowing growth, accelerating senescence and dormancy(Chen et al. 2022). Therefore, oat passively evolved a self-regulating mechanism to adapt to or resist drought adversity to maintain basic life activities and complete the life cycle.

4.3 PHYTOHORMONE REGULATION AND SIGNAL TRANSDUCTION

Phytohormones regulate plant growth and development and participate in biotic and abiotic stress processes in the plant. After sensing drought stress signals, plants release phytohormones to activate and regulate various plant physiological and developmental processes, including stomatal closure, root growth stimulation and accumulation of osmotic regulators to alleviate drought stress damage (Ullah et al. 2018), so phytohormones play an important role in the regulation of plant drought tolerance mechanisms, and they cross-talk with each other to regulate plant growth and development under drought stress (Rowe et al. 2016). These hormones include abscisic acid (ABA), auxins (IAA), cytokines (CKs), salicylic acid (SA), jasmonic acid (JA), ethylene (ET), gibberellins (GAs) (Ullah et al. 2018) and melatonin (MT) (Chen et al. 2009; Gao et al. 2018).

ABA is a weak acid that promotes fruit abscission and was originally isolated from cotton fruit (Ohkuma et al. 1963), and it has been confirmed that ABA regulates plant growth and stress responses in the form of hormones (Zhu 2002). Osmotic stress promotes ABA synthesis, which activates relevant gene expression and physiological adaptive changes (Yamaguchi-Shinozaki and Shinozaki 2006). The response of the plant to osmotic stress can be regulated by both ABA-dependent and ABA-independent processes (Zhu, 2016). In the ABA-dependent process, the synthetic ABA induced by drought stress binds to the receptor protein PYR/PYL/RCAR, and further combines with PP2C to form an ABA-PYR/PYL/RCAR complex, resulting in the dissociation of SnRK2 from the PP2C-SnRK2 complex, followed by kinase phosphorylation, which ultimately induces SnRK2-dependent gene expression and various physiological and biochemical responses via signal transduction (Ma et al. 2009; Park et al. 2009; Miyakawa and Tanokura 2011). As the degree of drought increases, the ABA content in oat leaves increases, and the stomatal conductance decreases (Peltonen-Sainio and Mäkelä 1995). Some oat cultivars represented by *Flega* gradually closed stomata and reduced net photosynthesis (Sanchez-Martin et al. 2012). Therefore, ABA-dependent processes participate in the oat osmotic response to drought by regulating stomatal closure and photosynthetic rate.

Osmotic stress induced by sorbitol, mannitol or quebrachitol enhanced the polar transport of IAA in oat, resulting in a significant increase in the polar transport of IAA in the mesocotyl, and when the sorbitol concentration was 0.5M, the polar transport of IAA reached the maximum, but when the concentration of osmotic solvent increased, the polar transport of IAA decreased (Sheldrake 1979), which may be related to the cross-talk regulation of various phytohormones such as ABA, ET and CKs (Rowe et al. 2016).

MT promotes plant growth and regulates various biological processes, and so it has received increasing attention (Chen et al. 2009). In recent years, the mechanism of MT in regulating plant drought stress has been studied and reported (Meng et al. 2014; Fan et al. 2018; Li et al. 2018; Sharma and Zheng 2019). In oat, exogenous MT enhanced the activities of superoxide dismutase (SOD), peroxidase (POD), catalase (CAT) and ascorbate peroxidase (APX) in plant leaves and reduced H_2O_2 and O_2^- contents, and upregulated the expression levels of mitogen-activated protein kinases (MAPKs), *Asmap1* and *Aspk11* genes, and transcription factors *WRKY1*, *DREB2* and *MYB*, oat can resist damage due to oxidative stress by increasing the activities of antioxidant enzymes and scavenging ROS, upregulating the expression of transcription factor genes and promoting the trans-duction of drought stress signals (Gao et al. 2018).

In addition, studies have shown that the application of exogenous SA increased spermine levels, which modulate the response of oat to drought (Canales et al. 2019), and the inolenic acid (linoleic acid is a precursor of JA) content in the leaves of the drought-tolerant cultivar *Patones* under drought stress, and JA and jasmonoyl isoleucine (Ile-JA) contents increased, while the JA in the drought-sensitive cultivar *Flega* did not change, indicating that JA plays an important role in the response to drought stress and may be involved in oat drought tolerance signal transduction (Sanchez-Martin et al. 2018).

4.4 LEAF GAS EXCHANGE

Drought stress induces abscisic acid (ABA) accumulation, stomatal closure and gene expression changes within hours (Signorelli et al. 2015), and plants close stomata through ABA signaling to reduce water loss (Schroeder et al. 2001). When leaf photosynthesis happens, the stomatal guard cells expand by absorbing water, and the stomata open to absorb CO_2 and release O_2, but under drought stress, K^+ is pumped out of guard cells and the stomata are closed, preventing water loss in plants and thus protecting them from drought stress (Cochrane and Cochrane 2009). Stomatal closure is thought to be the main determinant of reduced photosynthesis under mild to moderate drought conditions (Medrano et al. 2002), and with the aggravation of soil drought, stomatal closure of all assessed oat cultivars represented by *Flega*, meanwhile, leaf net photosynthesis was reduced due to the ABA accumulation and signal transduction from roots to leaves under drought stress (Sanchez-Martin et al. 2012). Drought stress caused stomata closure, resulting in a decrease in stomatal conductance of the two genotypes of oat *Shadow* and *Bia*, leading to a decrease in the photosynthetic rate, intracellular CO_2 concentration and transpiration rate (Zhao et al. 2021), how-ever, stomatal closure can reduce water transpiration of oat leaves, which could be explained by a "throttling strategy" in response to drought stress (Chen et al. 2022). Osmotic stress induced by PEG6000 reduced stomatal conductance and inhibited the expression of *PsbA*, *PsbB*, *PsbC* and *PsbD* genes, demonstrating the inhibition of photosynthesis in oat at the molecular level (Alyammahi and Gururani 2020). The photosynthetic rate is influenced by stomatal conductance and non-stomatal regulation, and stomatal conductance reduction affects the ratio of photorespiration to dark respir-ation (Chastain et al. 2014). As the intensity and duration of drought stress are increased, the photo-synthetic rate is affected by non-stomata processes such as the reduction in rubisco activity, CO_2 availability in chloroplasts, and the efficiency of PSII photochemistry (Ullah et al. 2018).

4.5 WATER RELATIONS

Reduced water availability in the plant rhizosphere can trigger osmotic stress, interfere with cel-lular metabolism and affect plant growth and development. To alleviate the harmful effects of drought stress, regulating osmotic balance is very important to maintain cell membrane stability, and achieving high water use efficiency (WUE) at low water potential can maintain normal cellular physiological metabolism (Tripathy et al. 2000). The plant mainly relieves osmotic stress through the following ways, when the plant senses a drought signal, it rapidly absorbs and accumulates inorganic ions (K^+, Na^+, Cl^-) as osmotic regulators, which allows the plant to reduce the water potential and resist osmotic stress in a more cost-effective manner (Shabala and Newman 2000; Kaur and Asthir 2015; Franco-Navarro et al. 2016; Turner 2017; Munns et al. 2020; Hou et al. 2021) (Table 4.1). Under PEG6000 osmotic stress, exogenous potassium supplementation increased the content of free amino acids and free sugars to improve the osmotic regulation capacity of oat (Ahanger et al. 2015). Potassium responds most rapidly to drought stress and can use the "shortest path" in the early stage of drought conditions (Hou et al., 2021) to make up for the delay in the *de novo* synthesis of organic

TABLE 4.1

Osmotic Regulators and Osmotic Adjustment Related Genes of Oat (*Avena sativa*) under Drought Stress

Osmotic regulators and osmotic adjustment-related genes	Describe	Reference
Inorganic ions		
K^+	The application of exogenous K^+ can increase the content of free amino acids and sugars, and improve the ability of osmotic adjustment.	Ahanger et al. (2015)
Compatible solutes		
Proline	Improve the water-holding capacity and water-conducting property, increase the relative water content and adjust osmotic pressure.	Gong et al. (2010) Jones and Storey (1978)
Glycine betaine	Improve the water-holding capacity and water-conducting property, increase the relative water content and adjust osmotic pressure.	Gong et al. (2010) Jones and Storey (1978)
Glucose	Drought-tolerant varieties synthesize glucose to promote osmotic adjustment.	Sanchez-Martin et al. (2015)
Fructose	Drought-tolerant varieties synthesize fructose to promote osmotic adjustment.	Sanchez-Martin et al. (2015)
Osmotic adjustment related genes		
P5CS1 gene	The rate-limiting enzyme gene of the proline synthesis pathway.	Alyammahi and Gururani (2020)
PDH1 gene	Key enzyme genes of proline synthesis pathway.	Alyammahi and Gururani (2020)

osmotic regulators in a "rapid complement" manner, to ensure the continuity of the plant response to drought, in this condition K^+ is not only used for stomatal regulation (Cochrane and Cochrane 2009), but also can maintain abundant cation in the cytosol, which is very suitable for osmotic adjustment (OA) without toxicity to cells, and Na^+ and Cl^- are also rapidly absorbed and accumulated when the plant is under osmotic stress. They also are involved in OA in the plant within the concentration range that does not produce ion toxicity (Per et al. 2017; Hou et al. 2021), but in the halophyte *Sarcocornia quinqueflora*, Na^+ and Cl^- contribute 85% of OA, while organic solutes and K^+ account for 15% of OA (Ahmed et al. 2021b). The relationship between the field water capacity (FWC) and osmotic potentials (OPs) value of leaves and roots, soil water content (SWC) and plasma membrane H^+-ATPase activity of drought-tolerant oat cultivar *Dingyou6* and drought-sensitive cultivar *Bende* were inferred as follows. Before the significant reduction of RWC, root hairs can sense drought stress caused by surrounding water shortage, and the drought-tolerant cultivar *Dingyou6* can trigger a drought stress response earlier (Gong et al. 2010). Studies have shown that plasma membrane H^+-ATPase activity increases, which helps to hydrolyze ATP and pump H^+ out of the plasma membrane, triggering a rapid hyperpolarization of the plasma membrane potential (Oren-Shamir et al. 1990; Shabala and Newman 2000; Nieves-Cordones et al. 2014), facilitates enrichment of K^+ uptake via high- and low-affinity K^+ transport systems, H^+/K^+ symporters and K^+ permeable channels, *HAK/KUP* transporters, and inwardly rectifying K^+ channels *AKT* in a low-cost and convenient manner regulating cell osmotic pressure to maintain cell physiological function (Hou et al. 2021).

Another strategy is that plants in response to osmotic stress synthesize and accumulate compatible solutes (CS) (Table 4.1), which are highly soluble, low-molecular-weight compounds that can accumulate at high levels within cells without compromising cellular function, scavenging ROS, protecting membrane structure and stabilizing proteins. CS acts as an osmotic regulator to protect cells and organelles from dehydration damage, these CS include glucose, sucrose, polyols, amino acids, trehalose and quaternary ammonium compounds (Fang and Xiong 2015; Kaur and Asthir 2015; Blum 2017; Per et al. 2017; Shehzadi et al. 2019), however, the above CS needs to be synthesized *de novo* through substrates and complex enzymatic reaction steps, and consumes energy (Zhang et al. 2009; Rubio et al. 2020), and may regulate plant osmotic balance in a "long-lasting" manner. Different genotypes of oat have different tolerances to drought stress. The drought-tolerant cultivar *Dingyou6* has higher RWC in roots and leaves than the drought-sensitive cultivar *Bende*. *Dingyou6* plants have better water retention and water conductivity. Also, the root and leaf OPs of *Dingyou6* were higher than those of *Bende*, and the osmotic regulation ability of *Dingyou6* was stronger. This difference was attributed to the drought-tolerant cultivars having higher proline and glycine betaine (GB) synthesis ability (Gong et al. 2010). GB is a non-toxic osmotic regulator, which can bind to the hydrophilic and hydrophobic regions of biomacromolecules such as enzymes, and the accumulation of GB and proline in oat improves the drought resistance of the plant through OA (Jones and Storey 1978). However, there is evidence that proline does not play a major role in OA (Hou et al. 2021) and is more likely to be involved in ROS scavenging (Signorelli et al. 2015; Hou et al. 2021). In the oat cultivars *Flega* and *Patones*, glucose and fructose continued to increase within 12 days of drought stress, and the drought-sensitive cultivar *Flega* increased more. However, after 12 days, the glucose and fructose contents of *Flega* started to decrease, but in the drought-tolerant cultivar *Patones* the glucose and fructose contents were still increasing for 15 days, and the sucrose content in the two cultivars did not change significantly before 12 days of drought stress, after which the sucrose content began to decrease (Sanchez-Martin et al. 2015). It is obvious that the drought-tolerant cultivar can "long-term regulate" osmotic stress by continuously synthesizing glucose and fructose, and promote the growth and development of oat.

4.6 OXIDATIVE STRESS AND ANTIOXIDANTS

Drought stress also leads to increased ROS levels in plant cells, such as superoxide (O_2^-), hydrogen peroxide (H_2O_2), singlet oxygen ($^1O_2$) and hydroxyl radicals (OH·) (Munne-Bosch and Penuelas 2003; Apel and Hirt 2004; Cruz de Carvalho 2008; Dogra et al. 2019). When plants experience drought stress through the root system, stomata are closed via ABA signal transduction (Schroeder et al. 2001; Sanchez-Martin et al. 2012), thereby limiting the uptake and fixation of CO_2 by the chloroplast, and thus the light response electron flow exceeds that required for CO_2 assimilation, and this imbalance leads to ROS production (Cruz et al. 2005). In addition, NADPH oxidase (RBOH) is a source of ROS generation (Ozkur et al. 2009; Mittler et al. 2011; Wang et al. 2019). However, in the plant, ROS acts as a "double-edged sword" (Hou et al., 2021), and ROS in plants acts as a molecule in abiotic stress signals at normal levels (Apel and Hirt 2004; Ma et al. 2012; Baxter et al. 2014; Dogra et al. 2021). ROS controls ion homeostasis in the cytosol by regulating the activity of ROS-sensitive ion channels (Fu and Huang 2001; Shabala et al. 2016; Demidchik and Shabala 2018). ROS-induced K^+ and Ca^{2+} homeostasis under osmotic stress has important effects on *Sarcocornia quinqueflora* growth under optimal and non-optimal conditions (Ahmed et al. 2021a). However, excess ROS can damage plant cell membranes and chloroplasts, and induce peroxidation of cell membranes to produce the toxic metabolite malondialdehyde (MDA). MDA causes oxidative damage to key molecules such as proteins, lipids and deoxyribonucleic acid, and affects cellular biological functions (Foyer and Fletcher 2001; Munne-Bosch and Penuelas 2003; Farooq et al. 2009; Gill and Tuteja 2010; Xia et al. 2015). Drought stress caused lipid peroxidation in oat membranes and increased the MDA content, which may be the key cause of electrolyte leakage (Alyammahi and Gururani 2020).

Plants scavenge ROS through the synergistic action of enzymatic and non-enzymatic antioxidant mechanisms. Enzymatic-based antioxidant enzymes include SOD, CAT, glutathione reductase (GR) and APX. Non-enzymatic antioxidants include β-carotene, α-tocopherol, ascorbic acid, glutathione, anthocyanins and flavonoids (Halliwell and Lipids 1987; Asada 1992; Smirnoff 1993; Mittler and Blumwald 2010). In addition to providing energy and metabolic substrates for plants (Moustakas et al. 2011), carbohydrates can also act as non-enzymatic antioxidants (e.g. through the pentose phosphate pathway) to promote ROS scavenging (Proels and Huckelhoven 2014). These antioxidants dynamically scavenge ROS and regulate ROS balance in plants. The increase of ROS content in oat is often synchronized with the changes in antioxidant enzyme activities. Drought led to a significant increase in the H_2O_2 content of two oat cultivars, *F-411* and *CK-1*, and was accompanied by an increase in SOD, CAT and POD activities (Shehzadi et al. 2019). Under water deficiency, a significant increase in oat H_2O_2 content was also observed, which was also accompanied by increased activities of CAT and SOD, and an increase in MDA content was also observed (Ghafoor et al. 2019). As the degree of water shortage is increased, oat also shows a continuous increase in MDA content (Islam et al. 2011), which may be due to oxidative stress in oat, which may lead to different degrees of membrane ester peroxidation. At the same time, oat activates the antioxidant mechanism to regulate the content of ROS, by upregulating the expression levels of *SOD*, *CAT* and *APX* genes (Alyammahi and Gururani 2020), and increasing the activities of SOD, CAT, POD, APX or GR (Islam et al. 2011), thus alleviating the damaging effect of oxidative stress and avoiding excessive accumulation of ROS to damage the membrane system through the dynamic regulation of ROS (Hou et al. 2021). However, another mechanism of membrane lipid peroxidation that deserves attention is membrane lipid peroxidation caused by increased lipoxygenase activity (Morales and Munne-Bosch 2019). Under drought stress treatment, the MDA content of all the evaluated oat germplasms still increased gradually under control, and the drought-sensitive cultivar had a greater increase, but the MDA content of some drought-tolerant cultivars represented by *Patones* was significantly different from that of lipoxygenase (LOX). It has been found that there is a correlation between MDA content and lipoxygenase activity in drought-tolerant varieties, and most of the membrane peroxidation of polyunsaturated fatty acids may be mediated by enzymatic reactions rather than free radical chemistry (Sanchez-Martin et al. 2012; Morales and Munne-Bosch 2019). A further study found that drought induced an increase in the content of diacylglycerols (DAG) and jasmonate precursor linolenic acid in drought-tolerant cultivar *Patones*, and JA and Ile-JA was also increased, demonstrating that jasmonate and specific fatty acids in different lipid classes have important roles in response to drought stress in oat (Sanchez-Martin et al. 2018).

4.7 GENETIC RESPONSES AND KEY GENES EXPRESSED UNDER DROUGHT STRESS

The genus *Avena* (*Avena* L.) consists of several diploid, tetraploid and hexaploid species (Baum 1977; Ladizinsky 2012) (Figure 4.1), the diploid species all contain AA or CC genomes, and the tetraploid species contain AABB, AACC, CCCC or CCDD genomes, while hexaploid species have AACCDD genomes (Rajhathy and Thomas 1974), a larger oat genome change of 4.1–12.8 Gbp due to chromosomal fold changes and the type and number of DNA repeats (Yan et al. 2016). The oat genome is complex and its genomic studies are not very complete (Gutierrez-Gonzalez et al. 2013). In terms of molecular markers, based on restriction fragment length polymorphism (RFLP) technology, hexaploid oat cultivar (O'Donoughue et al. 1995) and hybrid diploid oat *A. strigosa* and *A. wiestii* (Kremer et al. 2001) were respectively performed molecular markers. More than 2,700 polymorphic markers were found after genetic diversity analysis of 60 oat cultivars worldwide based on DArT marker technology (Tinker et al. 2009), and further DArT was performed on the next generation of hexaploid common oat and tetraploid barley hybrids markers, and a genetic map was constructed (Oliver et al. 2011). Also using DArT marker technology, 1,205 oat genomes around the world were subjected to genome-wide scanning to explore the impact of oat population structure and linkage disequilibrium on genome-wide association analysis (Newell et al. 2011), after which 985 SNPs were included and 21 linkage groups of hexaploid oat were marked (Oliver et al. 2013). Another study constructed a high-density genetic map of the hexaploid oat genome based on SNPs and genotyping-by-sequencing (GBS) (Chaffin et al. 2016), based on the GBS study of 4,657 cultivated oat 164,741 tag-level (TL) genetic variants containing 241,224 SNPs were found, which facilitated oat genome analysis and genomic selection (Bekele et al. 2018). The above studies greatly increased the number of markers that can be used for population genomics studies of cultivated oat for drought stress-based breeding. At present, two complete diploid genomes of *A. atlantica* (AA genome) and *A. eriantha* (CC genome) have been sequenced, and their genomes were found to be 3.72 Gb and 4.17 Gb, respectively (Maughan et al. 2019), Since then, PepsiCo and Corteva Agriscience have released the first hexaploid oat OT3098 reference genome v1, which can be found on the Grain Genes website (https://wheat.pw.usda.gov/GG3/node/922). The release of the oat genome data also contributes to the study of the population genome, and the study of the whole genome information of oat will help to reveal the genetic diversity of oat, especially the location and exploration of the trait genes that adapt to the drought environment. Due to the increasing number of abnormal drought events in the world, as well as the excellent health benefits, grain and forage functions of oat, and the good adaptability of oat to arid environments, the oat genome data will surely be supplemented and improved in the future.

Although the research into the adaptive mechanism of oat to drought stress at the molecular level remains limited, which may be related to the delay in the study of the oat genome, the exploration and identification of agronomic trait genes related to drought stress in oat have also been reported a number of times. The osmotic stress response of plants is controlled by ABA-dependent and ABA-independent processes (Zhu 2016; Zhao et al. 2018), while ABA accumulation is controlled by expression of the NCED3 gene (Fujii et al. 2011). SnRK2 protein kinase plays a key role in the ABA signal transduction pathway. Drought stress-induced synthetic ABA binds to the receptor protein PYR/PYL/RCAR, and further combines with PP2C to form an ABA-PYR/PYL/RCAR complex, resulting in the dissociation of SnRK2 from the PP2C-SnRK2 complex, followed by kinase phosphorylation, through signal transduction, triggering SnRK2-dependent gene expression, as well as various physiological and biochemical responses (Ma et al. 2009; Park et al. 2009; Nishimura et al. 2010; Miyakawa and Tanokura 2011). The above mechanism has been further revised, and ABA-PYL-PP2C is responsible for the binding inhibition or release of SnRK2, while RAF mediates the self-activation of SnRK2.2/3/6 through phosphorylation to initiate SnRK2. The activation process of ABA rapidly amplifies the ABA signaling pathway, enabling the plant to initiate drought responses (Lin et al. 2021). AsSnRK2D, a member of the SnRK2 family from oat, has been isolated. It has been confirmed that the AsSnRK2D protein is mainly located in the nucleus, constitutively expressed, and has a positive response to dehydration stress, salinity, abscisic acid and low temperature. Heterologous expression of AsSnRK2D in tobacco resulted in improvements in primary root length and plant fresh weight under drought and salt stress, and transgenic tobacco lines exhibited lower electrolyte leakage, higher RWC, higher SOD and CAT compared to wild-type plants, as well as increased chlorophyll levels, decreased ROS accumulation and increased free proline content. Oat AsSnRK2D can improve the drought tolerance of transgenic tobacco (Xiang et al. 2020), and AsSnRK2D may be one of the candidate genes for drought tolerance in oat. MT treatment upregulated the expression levels of MAPK and TF genes in naked oat under drought stress, enhanced antioxidant enzyme activities, scavenged ROS oxidative stress and improved drought resistance, implying that mitogen-activated protein kinases (MAPKs) *Asmap1* and *Aspk11*, and the transcription factors *NAC*, *WRKY1*, *DREB2* and *MYB* may also serve as candidate genes for improving drought tolerance in oat (Chen et al. 2009). In oat, the proline synthesis rate-limiting enzymes and *P5CS1* and *PDH1* genes were significantly upregulated in PEG6000-induced drought stress, and the corresponding proline content was significantly increased, which may play an important role in regulating water stress or ROS scavenging (Alyammahi and Gururani 2020). In oat, genes related to adaptation to

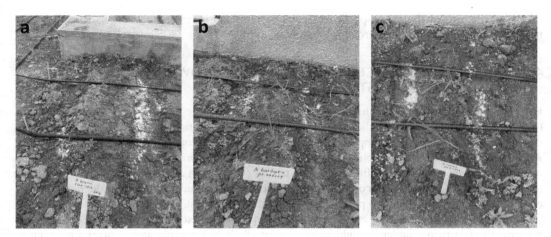

FIGURE 4.1 Different ploidy oats. (a) *A. brevis* is diploid, (b) *A. barbata* is tetraploid, (c) *A. sativa* is hexaploid. (Photo: Peichen Hou.)

drought stress still need to be continuously discovered and verified, which will be helpful for future oat gene-editing research or molecular design breeding research.

4.8 TRANSCRIPTOMIC AND METABOLOMIC ANALYSES RESPONSES UNDER DROUGHT STRESS CONDITIONS

The plant actively responds to drought stress at the physiological and molecular levels, initiating transcriptional and metabolic regulation (Cushman and Bohnert 2000; Claeys and Inze 2013). As an important technology in transcriptome research, RNA sequencing (RNA-seq) technology can provide a fast and efficient method for discovering the genes and regulatory networks expressed by the plant in response to abiotic stress (Shu et al. 2018). Especially in the absence or incomplete reference genome (Fan et al. 2013), high-throughput RNA-seq can analyze gene expression data for specific treatment conditions or specific developmental stages. Metabolomics is used to understand plant metabolic pathways and analyze the metabolic mechanisms of plant responses to stress (Saito and Matsuda 2010; Adamski and Suhre 2013), and metabolomics-based cultivar selection can improve crop yield and quality, as well as the production of functional foods with health benefits or drug efficiency. There have been many reports on the transcriptome and metabolome of oat (Wang et al. 2018; Yan et al. 2020; Jinqiu et al. 2021; Liu et al. 2021; Xu et al. 2021; Yan and Mao 2021; Zhang et al. 2021; Sun et al. 2022; Woo et al. 2022). Multi-omics studies have been conducted on environmental adaptability and agronomic traits of oat (Hu et al. 2021). The relationship between oat seed size and metabolite levels has also been reported (Brzozowski et al. 2021). Studies have shown that after treating oat seedlings with 300 mM PEG6000, the gene expression information at the transcriptional level of oat was obtained based on RNA-seq technology, and *acetyl-CoA carboxylation* (*ACCase*), which plays a key role in the initiation of fatty acid synthesis, was screened based on PPI network. The *ACCase* gene may improve the tolerance of oat to drought by regulating fatty acid saturation (Zhang et al. 2021). Recently, oat internal reference genes have been reported to facilitate the validation of transcriptome data, and it was confirmed that *ADPR* in leaves is suitable as an internal reference gene, and *GADPH* in roots is a suitable internal reference gene (Tajti et al. 2021). The release of the partial genome data of oat has promoted the transcriptomic and metabolomic research. In the future, there will be further research and supplements on the molecular and metabolic mechanisms of oat drought tolerance, including the identification of oat drought tolerance genes and the clarification of gene regulatory networks, which will help to increase understanding of the mechanism of the oat response to drought stress, and promote conventional breeding and molecular breeding to improve drought-tolerant oat cultivars.

4.9 STIMULANTS-INDUCED AMENDMENT OF GROWTH/PRODUCTION UNDER DROUGHT STRESS

Exogenous plant growth regulators play an important role in regulating plant stress resistance (Sharma and Zheng 2019). Studies have shown that the application of exogenous proline under drought stress can stabilize chloroplast structure and improve chlorophyll content (Hayat et al. 2012; Ashraf and Harris 2013). The application of exogenous proline improved the growth of leaves, stems and roots, and increased the proline content in tissues (Ghafoor et al. 2019), which may be due to the improvement of chlorophyll content caused by proline and reduced chloroplast degradation (Ali et al. 2007), possibly also due to proline stabilizing the chloroplast structure and activating chlorophyll synthase activity (Ghafoor et al. 2019). The application of exogenous potassium can improve the growth state of oat under drought stress. Under PEG6000 drought stress, potassium supplementation increased the number of oat tillers, increased the weight and length of roots, improved photosynthesis, restored the activity of nitrate reductase, and the increased activities of SOD, CAT and APX, which were used to scavenge ROS, and the content of free amino acids and free sugars increased to improve the osmotic regulation ability of oat (Ahanger et al. 2015). Under drought stress with PEG6000, the application of exogenous MT increased the antioxidant enzyme activity of naked oatmeal and reduced the excessive accumulation of ROS (Gao et al. 2018). MT treatment can increase the stomatal conductance of oat under drought stress, and regulate the expression of *PsbA*, *PsbB*, *PsbC* and *PsbD* genes to ensure the stability of PSII activity, improve leaf photosynthetic efficiency, increase the antioxidant capacity of oat by upregulating *SOD*, *CAT* and *APX* gene expression levels, protect the stability of cell membranes, reduce oxidative damage, upregulate the expression of *P5CS1* and *PDH1* genes, promote the accumulation of proline, and improve osmotic regulation or ROS scavenging ability (Alyammahi and Gururani 2020). Exogenous SA plays an important role in alleviating plant stress (Maruri-Lopez et al. 2019). SA can regulate drought tolerance in oat. Studies have shown that exogenous SA under drought stress increases spermine levels, thereby modulating the oat response to drought (Canales et al. 2019). A similar mechanism has also been found in rice, where drought stress inhibits the growth, development, and physiological and biochemical indicators of hydroponic and soil-cultivated rice, and accumulates H_2O_2. However, after exogenous SA treatment, rice can effectively alleviate oxidative damage by upregulating antioxidant enzyme activities. Increasing the proline level regulates the water status of the plant, and exogenous SA improves the growth and development of rice seedlings under drought stress (Sohag et al. 2020). GB is an important CS; a foliar application of 100 mM GB reduced the H_2O_2 content of both oat and ameliorated the negative effects of drought stress on roots, stems and leaf structures, but cultivar *CK-1* had superior anatomy compared to cultivar *F-411* (Shehzadi et al. 2019). GB plays an important role in regulating plant osmotic stress. For plants lacking GB synthesis ability, spraying exogenous GB can effectively improve their growth, development and yield (Cha-Um and Kirdmanee 2010; Shahbaz et al. 2011) which can increase wheat yields by 10%–50% in mild water-deficit situations (Shahbaz et al. 2011), but the effectiveness of GB may depend on internal and external factors such as crop developmental stage, dosage and stress conditions (Ashraf et al. 2008).

4.10 CONCLUSION

Although drought stress makes it difficult for oat roots to absorb water, inhibits growth and development, and reduces biomass and yield, oat can still survive in arid and semi-arid areas and provide grain for humans and forage for livestock, due to the unique drought response mechanism of oat (Figure 4.2). Oat has a unique stomatal regulation ability to reduce water loss, but photosynthesis is

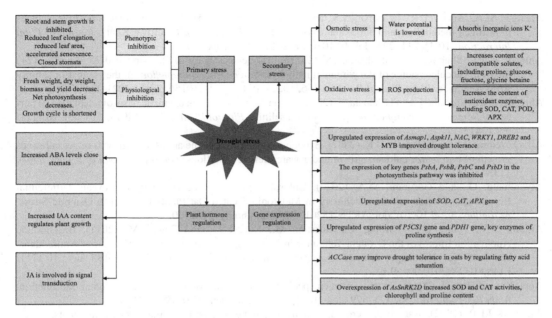

FIGURE 4.2 Mechanism of response and tolerance of oat (*Avena sativa*) to drought stress. ABA, abscisic acid; IAA, auxins; JA, jasmonic acid; ROS, reactive oxygen species; SOD, superoxide dismutase; CAT, catalase; POD, peroxidase; APX, ascorbate peroxidase.

inhibited, and carbon assimilation is reduced. Although ROS in oat is induced by drought, ROS are dynamically regulated by antioxidant enzymes that upregulate activity, and the osmotic balance of oat cells is reduced by inorganic and organic osmotic regulators, so that oat cells can survive under osmotic stress. Endogenous phytohormones and exogenous plant growth regulators can improve the physiological and molecular mechanisms involved in drought tolerance in oat. Some key genes related to drought tolerance in oat have been identified. However, studies on the adaptive mechanism of oat to drought stress at the molecular level remain limited, and oat comes second to staple crops (wheat, maize, rice, barley and sorghum) in agricultural production (Sanchez-Martin et al., 2012), although oat is a non-staple crop, which is related to the planter's pursuit of varied grain yields. This needs to be paid sufficient attention because of the infinite and disorderly development of water resources and the continuous deterioration of the natural environment, for the infinite future of human society, planting drought-tolerant and water-saving crops is an inevitable trend of agricultural production. It is more important than ever to represent the mechanisms of plant drought adaptation and carry out the breeding of crops with drought tolerance diversity.

ACKNOWLEDGMENTS

This study was supported by the Key Research and Development Project of Guangdong Province (2152324900016) and the Beijing Postdoctoral Research Foundation (2018-ZZ-052 and 2018-PC-04). There is no conflict of interest among the authors. We apologize for not citing all of the relevant references due to space limitations.

REFERENCES

Adamski, J., and Suhre, K. (2013). Metabolomics platforms for genome wide association studies-linking the genome to the metabolome. *Current Opinion in Biotechnology* 24(1), 39–47. doi: 10.1016/j.copbio.2012.10.003

Ahanger, M.A., Agarwal, R.M., Tomar, N.S., and Shrivastava, M. (2015). Potassium induces positive changes in nitrogen metabolism and antioxidant system of oat (*Avena sativa* L cultivar Kent). *Journal of Plant Interactions* 10(1), 211–223. doi: 10.1080/17429145.2015.1056260

Ahmed, H.A.I., Shabala, L., and Shabala, S. (2021a). Tissue-specificity of ROS-induced K(+) and Ca(2+) fluxes in succulent stems of the perennial halophyte *Sarcocornia quinqueflora* in the context of salinity stress tolerance. *Plant Physiol Biochem* 166, 1022–1031. doi: 10.1016/j.plaphy.2021.07.006

Ahmed, H.A.I., Shabala, L., and Shabala, S. (2021b). Understanding the mechanistic basis of adaptation of perennial *Sarcocornia quinqueflora* species to soil salinity. *Physiol Plant* 172(4), 1997–2010. doi: 10.1111/ppl.13413

Ali, Q., Ashraf, M., and Athar, H.U.R. (2007). Exogenously applied proline at different growth stages enhances growth of two maize cultivars grown under water deficit conditions. *Pakistan Journal of Botany* 39(4), 1133–1144.

Alyammahi, O., and Gururani, M.A. (2020). Chlorophyll-a Fluorescence Analysis Reveals Differential Response of Photosynthetic Machinery in Melatonin-Treated Oat Plants Exposed to Osmotic Stress. *Agronomy-Basel* 10(10). doi: ARTN 152010.3390/agronomy10101520

Anjum, S.A., Wang, L.C., Farooq, M., Hussain, M., Xue, L.L., and Zou, C.M. (2011). Brassinolide Application Improves the Drought Tolerance in Maize Through Modulation of Enzymatic Antioxidants and Leaf Gas Exchange. *Journal of Agronomy and Crop Science* 197(3), 177–185. doi: 10.1111/j.1439-037X.2010.00459.x

Apel, K., and Hirt, H. (2004). Reactive oxygen species: metabolism, oxidative stress, and signal transduction. *Annu Rev Plant Biol* 55, 373–399. doi: 10.1146/annurev.arplant.55.031903.141701

Asada, K.J.P.P. (1992). Ascorbate peroxidase–a hydrogen peroxide-scavenging enzyme in plants. 85(2), 235–241.

Ashraf, M., Athar, H.R., Harris, P.J.C., and Kwon, T.R. (2008). Some prospective strategies for improving crop salt tolerance. *Advances in Agronomy, Vol 97* 97, 45–110. doi: 10.1016/S0065-2113(07)00002-8

Ashraf, M., and Harris, P.J.C. (2013). Photosynthesis under stressful environments: An overview. *Photosynthetica* 51(2), 163–190. doi: 10.1007/s11099-013-0021-6

Baum, B.R. (1977). Oats: wild and cultivated. A monograph of the genus *Avena* L.(Poaceae). Ministry of Supply and Services (14).

Baxter, A., Mittler, R., and Suzuki, N. (2014). ROS as key players in plant stress signalling. *J Exp Bot* 65(5), 1229–1240. doi: 10.1093/jxb/ert375

Bekele, W.A., Wight, C.P., Chao, S., Howarth, C.J., and Tinker, N.A. (2018). Haplotype-based genotyping-by-sequencing in oat genome research. *Plant Biotechnol J* 16(8), 1452–1463. doi: 10.1111/pbi.12888

Blum, A. (2017). Osmotic adjustment is a prime drought stress adaptive engine in support of plant production. *Plant Cell Environ* 40(1), 4–10. doi: 10.1111/pce.12800

Bot, A., Nachtergaele, F., and Young, A. (2000). *Land resource potential and constraints at regional and country levels.* Food & Agriculture Org.

Brzozowski, L.J., Hu, H., Campbell, M.T., Broeckling, C.D., Caffe, M., Gutierrez, L., et al. (2021). Selection for seed size has uneven effects on specialized metabolite abundance in oat (*Avena sativa* L). *G3 (Bethesda).* doi: 10.1093/g3journal/jkab419

Canales, F.J., Montilla-Bascon, G., Rispail, N., and Prats, E. (2019). Salicylic acid regulates polyamine biosynthesis during drought responses in oat. *Plant Signal Behav* 14(10), e1651183. doi: 10.1080/15592324.2019.1651183

Cha-Um, S., and Kirdmanee, C. (2010). Effect of glycinebetaine on proline, water use, and photosynthetic efficiencies, and growth of rice seedlings under salt stress. *Turkish Journal of Agriculture and Forestry* 34(6), 517–527. doi: 10.3906/tar-0906-34

Chaffin, A.S., Huang, Y.F., Smith, S., Bekele, W.A., Babiker, E., Gnanesh, B.N., et al. (2016). A Consensus Map in Cultivated Hexaploid Oat Reveals Conserved Grass Synteny with Substantial Subgenome Rearrangement. *Plant Genome* 9(2). doi: 10.3835/plantgenome2015.10.0102

Chastain, D.R., Snider, J.L., Collins, G.D., Perry, C.D., Whitaker, J., and Byrd, S.A. (2014). Water deficit in field-grown Gossypium hirsutum primarily limits net photosynthesis by decreasing stomatal conductance, increasing photorespiration, and increasing the ratio of dark respiration to gross photosynthesis. *J Plant Physiol* 171(17), 1576–1585. doi: 10.1016/j.jplph.2014.07.014

Chen, Q., Qi, W.B., Reiter, R.J., Wei, W., and Wang, B.M. (2009). Exogenously applied melatonin stimulates root growth and raises endogenous indoleacetic acid in roots of etiolated seedlings of *Brassica juncea*. *J Plant Physiol* 166(3), 324–328. doi: 10.1016/j.jplph.2008.06.002

Chen, Q.C., Hu, T., Li, X.H., Song, C.P., Zhu, J.K., Chen, L.Q., et al. (2022). Phosphorylation of SWEET sucrose transporters regulates plant root:shoot ratio under drought. *Nature Plants* 8(1), 68-+. doi: 10.1038/s41477-021-01040-7

Claeys, H., and Inze, D. (2013). The Agony of Choice: How Plants Balance Growth and Survival under Water-Limiting Conditions. *Plant Physiology* 162(4), 1768–1779. doi: 10.1104/pp.113.220921

Cochrane, T.T., and Cochrane, T.A. (2009). The vital role of potassium in the osmotic mechanism of stomata aperture modulation and its link with potassium deficiency. *Plant Signal Behav* 4(3), 240–243. doi: 10.4161/psb.4.3.7955

Coelho, A.P., Faria, R.T.d., Leal, F.T., Barbosa, J.d.A., and Lemos, L.B.J.R.C. (2020). Biomass and nitrogen accumulation in white oat (*Avena sativa* L.) under water deficit 1. 67, 1–8.

Cruz de Carvalho, M.H. (2008). Drought stress and reactive oxygen species: Production, scavenging and signaling. *Plant Signal Behav* 3(3), 156–165. doi: 10.4161/psb.3.3.5536

Cruz, J.A., Avenson, T.J., Kanazawa, A., Takizawa, K., Edwards, G.E., and Kramer, D.M. (2005). Plasticity in light reactions of photosynthesis for energy production and photoprotection. *J Exp Bot* 56(411), 395–406. doi: 10.1093/jxb/eri022

Cushman, J.C., and Bohnert, H.J. (2000). Genomic approaches to plant stress tolerance. *Curr Opin Plant Biol* 3(2), 117–124. doi: 10.1016/s1369-5266(99)00052-7

Dastgheib, F., Andrews, M., Field, R., and Foreman, M.J.W.R. (1990). Effect of different levels of mannitol-induced water stress on the tolerance of cultivated oat (*Avena sativa* L.) to didofop-methyl. 30(3), 171–179.

Demidchik, V., and Shabala, S. (2018). Mechanisms of cytosolic calcium elevation in plants: the role of ion channels, calcium extrusion systems and NADPH oxidase-mediated 'ROS-Ca(2+) Hub'. *Funct Plant Biol* 45(2), 9–27. doi: 10.1071/FP16420

Dita, M.A., Rispail, N., Prats, E., Rubiales, D., and Singh, K.B.J.E. (2006). Biotechnology approaches to overcome biotic and abiotic stress constraints in legumes. 147(1), 1–24.

Dogra, V., Li, M., Singh, S., Li, M., and Kim, C. (2019). Oxidative post-translational modification of EXECUTER1 is required for singlet oxygen sensing in plastids. *Nat Commun* 10(1), 2834. doi: 10.1038/s41467-019-10760-6

Dogra, V., Singh, R.M., Li, M., Li, M., Singh, S., and Kim, C. (2021). EXECUTER2 modulates the EXECUTER1 signalosome through its singlet oxygen-dependent oxidation. *Mol Plant*. doi: 10.1016/j.molp.2021.12.016

Fan, H., Xiao, Y., Yang, Y., Xia, W., Mason, A.S., Xia, Z., et al. (2013). RNA-Seq analysis of Cocos nucifera: transcriptome sequencing and de novo assembly for subsequent functional genomics approaches. *PLoS One* 8(3), e59997. doi: 10.1371/journal.pone.0059997

Fan, J.B., Xie, Y., Zhang, Z.C., and Chen, L. (2018). Melatonin: A Multifunctional Factor in Plants. *International Journal of Molecular Sciences* 19(5). doi: ARTN 152810.3390/ijms19051528

Fang, Y., and Xiong, L. (2015). General mechanisms of drought response and their application in drought resistance improvement in plants. *Cell Mol Life Sci* 72(4), 673–689. doi: 10.1007/s00018-014-1767-0

Farooq, M., Wahid, A., Kobayashi, N., Fujita, D., and Basra, S.M.A. (2009). Plant drought stress: effects, mechanisms and management. *Agronomy for Sustainable Development* 29(1), 185–212. doi: 10.1051/agro:2008021

Foyer, C.H., and Fletcher, J.M. (2001). Plant antioxidants: colour me healthy. *Biologist (London)* 48(3), 115–120.

Franco-Navarro, J.D., Brumos, J., Rosales, M.A., Cubero-Font, P., Talon, M., and Colmenero-Flores, J.M. (2016). Chloride regulates leaf cell size and water relations in tobacco plants. *Journal of Experimental Botany* 67(3), 873–891. doi: 10.1093/jxb/erv502

Fu, J., and Huang, B. (2001). Involvement of antioxidants and lipid peroxidation in the adaptation of two cool-season grasses to localized drought stress. *Environ Exp Bot* 45(2), 105–114. doi: 10.1016/s0098-8472(00)00084-8

Fujii, H., Verslues, P.E., and Zhu, J.K. (2011). *Arabidopsis* decuple mutant reveals the importance of SnRK2 kinases in osmotic stress responses in vivo. *Proc Natl Acad Sci U S A* 108(4), 1717–1722. doi: 10.1073/pnas.1018367108

Gao, W., Zhang, Y., Feng, Z., Bai, Q., He, J., and Wang, Y. (2018). Effects of Melatonin on Antioxidant Capacity in Naked Oat Seedlings under Drought Stress. *Molecules* 23(7). doi: 10.3390/molecules23071580

Garcia-Gomez, B.I., Campos, F., Hernandez, M., and Covarrubias, A.A. (2000). Two bean cell wall proteins more abundant during water deficit are high in proline and interact with a plasma membrane protein. *Plant J* 22(4), 277–288. doi: 10.1046/j.1365-313x.2000.00739.x

Ghafoor, R., Akram, N.A., Rashid, M., Ashraf, M., Iqbal, M., and Lixin, Z. (2019). Exogenously applied proline induced changes in key anatomical features and physio-biochemical attributes in water stressed oat (*Avena sativa* L.) plants. *Physiol Mol Biol Plants* 25(5), 1121–1135. doi: 10.1007/s12298-019-00683-3

Gill, S.S., and Tuteja, N. (2010). Reactive oxygen species and antioxidant machinery in abiotic stress tolerance in crop plants. *Plant Physiol Biochem* 48(12), 909–930. doi: 10.1016/j.plaphy.2010.08.016

Gong, D.S., Xiong, Y.C., Ma, B.L., Wang, T.M., Ge, J.P., Qin, X.L., et al. (2010). Early activation of plasma membrane H$^+$-ATPase and its relation to drought adaptation in two contrasting oat (*Avena sativa* L.) genotypes. *Environmental and Experimental Botany* 69(1), 1–8. doi: 10.1016/j.envexpbot.2010.02.011

Gutierrez-Gonzalez, J.J., Tu, Z.J., and Garvin, D.F. (2013). Analysis and annotation of the hexaploid oat seed transcriptome. *BMC Genomics* 14, 471. doi: 10.1186/1471-2164-14-471

Halliwell, B.J.C., and lipids, P.o. (1987). Oxidative damage, lipid peroxidation and antioxidant protection in chloroplasts. 44(2-4), 327–340.

Hayat, S., Hayat, Q., Alyemeni, M.N., Wani, A.S., Pichtel, J., and Ahmad, A. (2012). Role of proline under changing environments: a review. *Plant Signal Behav* 7(11), 1456–1466. doi: 10.4161/psb.21949

Hou, P., Wang, F., Luo, B., Li, A., Wang, C., Shabala, L., et al. (2021). Antioxidant Enzymatic Activity and Osmotic Adjustment as Components of the Drought Tolerance Mechanism in *Carex duriuscula*. *Plants (Basel)* 10(3). doi: 10.3390/plants10030436

Hu, H., and Xiong, L. (2014). Genetic engineering and breeding of drought-resistant crops. *Annu Rev Plant Biol* 65, 715–741. doi: 10.1146/annurev-arplant-050213-040000

Hu, H.X., Campbell, M.T., Yeats, T.H., Zheng, X.Y., Runcie, D.E., Covarrubias-Pazaran, G., et al. (2021). Multi-omics prediction of oat agronomic and seed nutritional traits across environments and in distantly related populations. *Theoretical and Applied Genetics* 134(12), 4043–4054. doi: 10.1007/s00122-021-03946-4

Isidro-Sánchez, J., D'Arcy Cusack, K., Verheecke-Vaessen, C., Kahla, A., Bekele, W., Doohan, F., et al. (2020). Genome-wide association mapping of Fusarium langsethiae infection and mycotoxin accumulation in oat (*Avena sativa* L.). 13(2), e20023.

Islam, M.R., Xue, X., Mao, S., Ren, C., Eneji, A.E., and Hu, Y. (2011). Effects of water-saving superabsorbent polymer on antioxidant enzyme activities and lipid peroxidation in oat (*Avena sativa* L.) under drought stress. *J Sci Food Agric* 91(4), 680–686. doi: 10.1002/jsfa.4234

Jinqiu, Y., Bing, L., Tingting, S., Jinglei, H., Zelai, K., Lu, L., et al. (2021). Integrated Physiological and Transcriptomic Analyses Responses to Altitude Stress in Oat (*Avena sativa* L.). *Front Genet* 12, 638683. doi: 10.3389/fgene.2021.638683

Jones, R.W., and Storey, R.J.F.P.B. (1978). Salt stress and comparative physiology in the Gramineae. II. Glycinebetaine and proline accumulation in two salt-and water-stressed barley cultivars. 5(6), 817–829.

Kaur, G., and Asthir, B. (2015). Proline: a key player in plant abiotic stress tolerance. *Biologia Plantarum* 59(4), 609–619. doi: 10.1007/s10535-015-0549-3

Koka, J.A., Wani, A.H., Agarwal, R.M., Parveen, S., and Wani, F.A.J.E.A.R. (2015). Effect of Polyethylene glycol 6000, mannitol, sodium and potassium salts on the growth and biochemical characteristics of oat (*Avena sativa* L.). 3(1), 303–314.

Kremer, C.A., Lee, M., and Holland, J.B. (2001). A restriction fragment length polymorphism based linkage map of a diploid Avena recombinant inbred line population. *Genome* 44(2), 192–204.

Ladizinsky, G. (2012). *Studies in oat evolution: a man's life with Avena.* Springer Science & Business Media.

Li, J.J., Zeng, L., Cheng, Y., Lu, G.Y., Fu, G.P., Ma, H.Q., et al. (2018). Exogenous melatonin alleviates damage from drought stress in *Brassica napus* L. (rapeseed) seedlings. *Acta Physiologiae Plantarum* 40(3). doi: ARTN 4310.1007/s11738-017-2601-8

Lin, Z., Li, Y., Wang, Y., Liu, X., Ma, L., Zhang, Z., et al. (2021). Initiation and amplification of SnRK2 activation in abscisic acid signaling. *Nat Commun* 12(1), 2456. doi: 10.1038/s41467-021-22812-x

Liu, B., Zhang, D., Sun, M., Li, M., Ma, X., Jia, S., et al. (2021). PSII Activity Was Inhibited at Flowering Stage with Developing Black Bracts of Oat. *Int J Mol Sci* 22(10). doi: 10.3390/ijms22105258

Ma, L., Zhang, H., Sun, L., Jiao, Y., Zhang, G., Miao, C., et al. (2012). NADPH oxidase AtrbohD and AtrbohF function in ROS-dependent regulation of Na(+)/K(+)homeostasis in *Arabidopsis* under salt stress. *J Exp Bot* 63(1), 305–317. doi: 10.1093/jxb/err280

Ma, Y., Szostkiewicz, I., Korte, A., Moes, D., Yang, Y., Christmann, A., et al. (2009). Regulators of PP2C phosphatase activity function as abscisic acid sensors. *Science* 324(5930), 1064–1068. doi: 10.1126/science.1172408

Maruri-Lopez, I., Aviles-Baltazar, N.Y., Buchala, A., and Serrano, M. (2019). Intra and Extracellular Journey of the Phytohormone Salicylic Acid. *Frontiers in Plant Science* 10. doi: ARTN 423 10.3389/fpls.2019.00423

Maughan, P.J., Lee, R., Walstead, R., Vickerstaff, R.J., Fogarty, M.C., Brouwer, C.R., et al. (2019). Genomic insights from the first chromosome-scale assemblies of oat (*Avena spp.*) diploid species. *BMC Biol* 17(1), 92. doi: 10.1186/s12915-019-0712-y

Medrano, H., Escalona, J.M., Bota, J., Gulias, J., and Flexas, J. (2002). Regulation of photosynthesis of C_3 plants in response to progressive drought: stomatal conductance as a reference parameter. *Ann Bot* 89 Spec No, 895–905. doi: 10.1093/aob/mcf079

Meng, J.F., Xu, T.F., Wang, Z.Z., Fang, Y.L., Xi, Z.M., and Zhang, Z.W. (2014). The ameliorative effects of exogenous melatonin on grape cuttings under water-deficient stress: antioxidant metabolites, leaf anatomy, and chloroplast morphology. *J Pineal Res* 57(2), 200–212. doi: 10.1111/jpi.12159

Mittler, R., and Blumwald, E. (2010). Genetic engineering for modern agriculture: challenges and perspectives. *Annu Rev Plant Biol* 61, 443–462. doi: 10.1146/annurev-arplant-042809-112116

Mittler, R., Vanderauwera, S., Suzuki, N., Miller, G., Tognetti, V.B., Vandepoele, K., et al. (2011). ROS signaling: the new wave? *Trends Plant Sci* 16(6), 300–309. doi: 10.1016/j.tplants.2011.03.007

Miyakawa, T., and Tanokura, M. (2011). Regulatory mechanism of abscisic acid signaling. *Biophysics (Nagoya-shi)* 7, 123–128. doi: 10.2142/biophysics.7.123

Morales, M., and Munne-Bosch, S. (2019). Malondialdehyde: Facts and Artifacts. *Plant Physiol* 180(3), 1246–1250. doi: 10.1104/pp.19.00405

Moustakas, M., Sperdouli, I., Kouna, T., Antonopoulou, C.I., and Therios, I. (2011). Exogenous proline induces soluble sugar accumulation and alleviates drought stress effects on photosystem II functioning of *Arabidopsis thaliana* leaves. *Plant Growth Regulation* 65(2), 315–325. doi: 10.1007/s10725-011-9604-z

Munne-Bosch, S., and Penuelas, J. (2003). Photo- and antioxidative protection, and a role for salicylic acid during drought and recovery in field-grown Phillyrea angustifolia plants. *Planta* 217(5), 758–766. doi: 10.1007/s00425-003-1037-0

Munns, R., Day, D.A., Fricke, W., Watt, M., Arsova, B., Barkla, B.J., et al. (2020). Energy costs of salt tolerance in crop plants. *New Phytol* 225(3), 1072–1090. doi: 10.1111/nph.15864

Mut, Z., Akay, H., and Aydin, N. (2010). Effects of seed size and drought stress on germination and seedling growth of some oat genotypes (*Avena sativa* L.). *African Journal of Agricultural Research* 5(10), 1101–1107.

Newell, M.A., Cook, D., Tinker, N.A., and Jannink, J.L. (2011). Population structure and linkage disequilibrium in oat (*Avena sativa* L.): implications for genome-wide association studies. *Theor Appl Genet* 122(3), 623–632. doi: 10.1007/s00122-010-1474-7

Nieves-Cordones, M., Aleman, F., Martinez, V., and Rubio, F. (2014). K^+ uptake in plant roots. The systems involved, their regulation and parallels in other organisms. *J Plant Physiol* 171(9), 688–695. doi: 10.1016/j.jplph.2013.09.021

Nishimura, N., Sarkeshik, A., Nito, K., Park, S.Y., Wang, A., Carvalho, P.C., et al. (2010). PYR/PYL/RCAR family members are major in-vivo ABI1 protein phosphatase 2C-interacting proteins in *Arabidopsis*. *Plant Journal* 61(2), 290–299. doi: 10.1111/j.1365-313X.2009.04054.x

O'Donoughue, L.S., Sorrells, M.E., Tanksley, S.D., Autrique, E., Deynze, A.V., Kianian, S.F., et al. (1995). A molecular linkage map of cultivated oat. *Genome* 38(2), 368–380. doi: 10.1139/g95-048.

Ohkuma, K., Lyon, J.L., Addicott, F.T., and Smith, O.E. (1963). Abscisin II, an Abscission-Accelerating Substance from Young Cotton Fruit. *Science* 142(3599), 1592–1593. doi: 10.1126/science.142.3599.1592.

Oliver, R.E., Jellen, E.N., Ladizinsky, G., Korol, A.B., Kilian, A., Beard, J.L., et al. (2011). New Diversity Arrays Technology (DArT) markers for tetraploid oat (Avena magna Murphy et Terrell) provide the first complete oat linkage map and markers linked to domestication genes from hexaploid *A. sativa* L. *Theor Appl Genet* 123(7), 1159–1171. doi: 10.1007/s00122-011-1656-y.

Oliver, R.E., Tinker, N.A., Lazo, G.R., Chao, S., Jellen, E.N., Carson, M.L., et al. (2013). SNP discovery and chromosome anchoring provide the first physically-anchored hexaploid oat map and reveal synteny with model species. *PLoS One* 8(3), e58068. doi: 10.1371/journal.pone.0058068

Oren-Shamir, M., Pick, U., and Avron, M. (1990). Plasma membrane potential of the alga dunaliella, and its relation to osmoregulation. *Plant Physiol* 93(2), 403–408. doi: 10.1104/pp.93.2.403

Ozkur, O., Ozdemir, F., Bor, M., and Turkan, I. (2009). Physiochemical and antioxidant responses of the perennial xerophyte *Capparis ovata Desf.* to drought. *Environmental and Experimental Botany* 66(3), 487–492. doi: 10.1016/j.envexpbot.2009.04.003

Park, S.Y., Fung, P., Nishimura, N., Jensen, D.R., Fujii, H., Zhao, Y., et al. (2009). Abscisic acid inhibits type 2C protein phosphatases via the PYR/PYL family of START proteins. *Science* 324(5930), 1068–1071. doi: 10.1126/science.1173041

Peltonen-Sainio, P., and Mäkelä, P.J.A.A.S.B.-P.S.S. (1995). Comparison of physiological methods to assess drought tolerance in oats. 45(1), 32–38.

Per, T.S., Khan, N.A., Reddy, P.S., Masood, A., Hasanuzzaman, M., Khan, M.I.R., et al. (2017). Approaches in modulating proline metabolism in plants for salt and drought stress tolerance: Phytohormones, mineral nutrients and transgenics. *Plant Physiol Biochem* 115, 126–140. doi: 10.1016/j.plaphy.2017.03.018

Poorter, L., and Markesteijn, L. (2008). Seedling traits determine drought tolerance of tropical tree species. *Biotropica* 40(3), 321–331. doi: 10.1111/j.1744-7429.2007.00380.x

Price, A.H., Cairns, J.E., Horton, P., Jones, H.G., and Griffiths, H. (2002). Linking drought-resistance mechanisms to drought avoidance in upland rice using a QTL approach: progress and new opportunities to integrate stomatal and mesophyll responses. *Journal of Experimental Botany* 53(371), 989–1004. doi: DOI 10.1093/jexbot/53.371.989

Proels, R.K., and Huckelhoven, R. (2014). Cell-wall invertases, key enzymes in the modulation of plant metabolism during defence responses. *Mol Plant Pathol* 15(8), 858–864. doi: 10.1111/mpp.12139

Qi, J., Song, C.P., Wang, B., Zhou, J., Kangasjarvi, J., Zhu, J.K., et al. (2018). Reactive oxygen species signaling and stomatal movement in plant responses to drought stress and pathogen attack. *J Integr Plant Biol* 60(9), 805–826. doi: 10.1111/jipb.12654

Rajhathy, T., and Thomas, H. (1974). *Cytogenetics of oats (Avena L.).* Ottawa: Genetics Society of Canada.

Rowe, J.H., Topping, J.F., Liu, J., and Lindsey, K. (2016). Abscisic acid regulates root growth under osmotic stress conditions via an interacting hormonal network with cytokinin, ethylene and auxin. *New Phytol* 211(1), 225–239. doi: 10.1111/nph.13882

Rubio, F., Nieves-Cordones, M., Horie, T., and Shabala, S. (2020). Doing 'business as usual' comes with a cost: evaluating energy cost of maintaining plant intracellular K^+ homeostasis under saline conditions. *New Phytologist* 225(3), 1097–1104. doi: 10.1111/nph.15852

Saito, K., and Matsuda, F. (2010). Metabolomics for Functional Genomics, Systems Biology, and Biotechnology. *Annual Review of Plant Biology, Vol 61* 61, 463–489. doi: 10.1146/annurev. arplant.043008.092035

Sanchez-Martin, J., Canales, F.J., Tweed, J.K.S., Lee, M.R.F., Rubiales, D., Gomez-Cadenas, A., et al. (2018). Fatty Acid Profile Changes During Gradual Soil Water Depletion in Oats Suggests a Role for Jasmonates in Coping With Drought. *Front Plant Sci* 9, 1077. doi: 10.3389/fpls.2018.01077

Sanchez-Martin, J., Heald, J., Kingston-Smith, A., Winters, A., Rubiales, D., Sanz, M., et al. (2015). A metabolomic study in oats (*Avena sativa*) highlights a drought tolerance mechanism based upon salicylate signalling pathways and the modulation of carbon, antioxidant and photo-oxidative metabolism. *Plant Cell and Environment* 38(7), 1434–1452. doi: 10.1111/pce.12501

Sanchez-Martin, J., Mur, L.A.J., Rubiales, D., and Prats, E. (2012). Targeting sources of drought tolerance within an *Avena spp.* collection through multivariate approaches. *Planta* 236(5), 1529–1545. doi: 10.1007/s00425-012-1709-8

Schroeder, J.I., Kwak, J.M., and Allen, G.J.J.N. (2001). Guard cell abscisic acid signalling and engineering drought hardiness in plants. 410(6826), 327–330.

Shabala, S., Bose, J., Fuglsang, A.T., and Pottosin, I. (2016). On a quest for stress tolerance genes: membrane transporters in sensing and adapting to hostile soils. *J Exp Bot* 67(4), 1015–1031. doi: 10.1093/jxb/erv465

Shabala, S., and Newman, I.J.A.o.B. (2000). Salinity effects on the activity of plasma membrane H^+ and Ca^{2+} transporters in bean leaf mesophyll: masking role of the cell wall. 85(5), 681–686.

Shahbaz, M., Masood, Y., Perveen, S., and Ashraf, M. (2011). Is foliar-applied glycinebetaine effective in mitigating the adverse effects of drought stress on wheat (*Triticum aestivum* L.)? *Journal of Applied Botany and Food Quality* 84(2), 192–199.

Sharma, A., and Zheng, B. (2019). Melatonin Mediated Regulation of Drought Stress: Physiological and Molecular Aspects. *Plants (Basel)* 8(7). doi: 10.3390/plants8070190

Shehzadi, A., Akram, N.A., Ali, A., and Ashraf, M. (2019). Exogenously applied glycinebetaine induced alteration in some key physio-biochemical attributes and plant anatomical features in water stressed oat (Avena sativa L.) plants. *Journal of Arid Land* 11(2), 292–305. doi: 10.1007/s40333-019-0007-8

Sheldrake, A.R. (1979). Effects of osmotic stress on polar auxin transport in *Avena* mesocotyl sections. *Planta* 145(2), 113–117. doi: 10.1007/BF00388706

Shu, Y.J., Li, W., Zhao, J.Y., Liu, Y., and Guo, C.H. (2018). Transcriptome sequencing and expression profiling of genes involved in the response to abiotic stress in *Medicago ruthenica*. *Genetics and Molecular Biology* 41(3), 638–648. doi: 10.1590/1678-4685-Gmb-2017-0284

Signorelli, S., Dans, P.D., Coitino, E.L., Borsani, O., and Monza, J. (2015). Connecting Proline and gamma-Aminobutyric Acid in Stressed Plants through Non-Enzymatic Reactions. *Plos One* 10(3). doi: ARTN e011534910.1371/journal.pone.0115349

Smirnoff, N. (1993). The role of active oxygen in the response of plants to water deficit and desiccation. *New Phytol* 125(1), 27–58. doi: 10.1111/j.1469-8137.1993.tb03863.x

Sohag, A.A., Tahjib-Ul-Arif, M., Brestic, M., Afrin, S., Sakil, M.A., Hossain, M.T., et al. (2020). Exogenous salicylic acid and hydrogen peroxide attenuate drought stress in rice. *Plant Soil and Environment* 66(1), 7–13. doi: 10.17221/472/2019-Pse

Sun, M., Sun, S., Mao, C., Zhang, H., Ou, C., Jia, Z., et al. (2022). Dynamic Responses of Antioxidant and Glyoxalase Systems to Seed Aging Based on Full-Length Transcriptome in Oat (*Avena sativa* L.). *Antioxidants (Basel)* 11(2). doi: 10.3390/antiox11020395

Suttie, J.M., and Reynolds, S.G. (2004). *Fodder oats: a world overview*. Food & Agriculture Org.

Tajti, J., Pal, M., and Janda, T. (2021). Validation of Reference Genes for Studying Different Abiotic Stresses in Oat (*Avena sativa* L.) by RT-qPCR. *Plants (Basel)* 10(7). doi: 10.3390/plants10071272

Tinker, N.A., Kilian, A., Wight, C.P., Heller-Uszynska, K., Wenzl, P., Rines, H.W., et al. (2009). New DArT markers for oat provide enhanced map coverage and global germplasm characterization. *BMC Genomics* 10, 39. doi: 10.1186/1471-2164-10-39

Tripathy, J., Zhang, J., Robin, S., Nguyen, T.T., Nguyen, H.J.T., and Genetics, A. (2000). QTLs for cell-membrane stability mapped in rice (*Oryza sativa* L.) under drought stress. 100(8), 1197–1202.

Turner, N.C. (2017). Turgor maintenance by osmotic adjustment, an adaptive mechanism for coping with plant water deficits. *Plant Cell Environ* 40(1), 1–3. doi: 10.1111/pce.12839

Ullah, A., Manghwar, H., Shaban, M., Khan, A.H., Akbar, A., Ali, U., et al. (2018). Phytohormones enhanced drought tolerance in plants: a coping strategy. *Environmental Science and Pollution Research* 25(33), 33103–33118. doi: 10.1007/s11356-018-3364-5

Verma, A., and Deepti, S.J.A.P.A.R. (2016). Abiotic stress and crop improvement: current scenario. 4(4), 00149.

Wang, F.F., Chen, Z.H., Liu, X.H., Shabala, L., Yu, M., Zhou, M.X., et al. (2019). The loss of RBOHD function modulates root adaptive responses to combined hypoxia and salinity stress in *Arabidopsis*. *Environmental and Experimental Botany* 158, 125–135. doi: 10.1016/j.envexpbot.2018.11.020

Wang, Y., Lysoe, E., Armarego-Marriott, T., Erban, A., Paruch, L., van Eerde, A., et al. (2018). Transcriptome and metabolome analyses provide insights into root and root-released organic anion responses to phosphorus deficiency in oat. *J Exp Bot* 69(15), 3759–3771. doi: 10.1093/jxb/ery176

Woo, S.Y., Yang, J.Y., Lee, H., Ahn, H.J., Lee, Y.B., Do, S.H., et al. (2022). Changes in metabolites with harvest times of seedlings of various Korean oat (*Avena sativa* L.) cultivars and their neuraminidase inhibitory effects. *Food Chemistry* 373. doi: ARTN 13142910.1016/j.foodchem.2021.131429

Wu, Y., and Cosgrove, D.J. (2000). Adaptation of roots to low water potentials by changes in cell wall extensibility and cell wall proteins. *J Exp Bot* 51(350), 1543–1553. doi: 10.1093/jexbot/51.350.1543

Xia, F.S., Wang, X.G., Li, M.L., and Mao, P.S. (2015). Mitochondrial structural and antioxidant system responses to aging in oat (*Avena sativa* L.) seeds with different moisture contents. *Plant Physiology and Biochemistry* 94, 122–129. doi: 10.1016/j.plaphy.2015.06.002

Xiang, D.J., Man, L.L., Cao, S., Liu, P., Li, Z.G., and Wang, X.D. (2020). Ectopic expression of an oat SnRK2 gene, AsSnRK2D, enhances dehydration and salinity tolerance in tobacco by modulating the expression of stress-related genes. *Brazilian Journal of Botany* 43(3), 429–446. doi: 10.1007/s40415-020-00614-7

Xiong, L.Z., and Yang, Y.N. (2003). Disease resistance and abiotic stress tolerance in rice are inversely modulated by an abscisic acid-inducible mitogen-activated protein kinase. *Plant Cell* 15(3), 745–759. doi: 10.1105/tpc.008714

Xu, Z., Chen, X., Lu, X., Zhao, B., Yang, Y., and Liu, J. (2021). Integrative analysis of transcriptome and metabolome reveal mechanism of tolerance to salt stress in oat (*Avena sativa* L.). *Plant Physiol Biochem* 160, 315–328. doi: 10.1016/j.plaphy.2021.01.027

Xue, D., Zhang, X., Lu, X., Chen, G., and Chen, Z.H. (2017). Molecular and Evolutionary Mechanisms of Cuticular Wax for Plant Drought Tolerance. *Front Plant Sci* 8, 621. doi: 10.3389/fpls.2017.00621

Yamaguchi-Shinozaki, K., and Shinozaki, K. (2006). Transcriptional regulatory networks in cellular responses and tolerance to dehydration and cold stresses. *Annual Review of Plant Biology* 57, 781–803. doi: 10.1146/annurev.arplant.57.032905.105444

Yan, H., and Mao, P. (2021). Comparative Time-Course Physiological Responses and Proteomic Analysis of Melatonin Priming on Promoting Germination in Aged Oat (*Avena sativa* L.) Seeds. *Int J Mol Sci* 22(2). doi: 10.3390/ijms22020811

Yan, H., Martin, S.L., Bekele, W.A., Latta, R.G., Diederichsen, A., Peng, Y., et al. (2016). Genome size variation in the genus *Avena*. *Genome* 59(3), 209–220. doi: 10.1139/gen-2015-0132

Yan, H.F., Jia, S.G., and Mao, P.S. (2020). Melatonin Priming Alleviates Aging-Induced Germination Inhibition by Regulating beta-oxidation, Protein Translation, and Antioxidant Metabolism in Oat (*Avena sativa* L.) Seeds. *International Journal of Molecular Sciences* 21(5). doi: ARTN 189810.3390/ijms21051898

Zhang, D., Cheng, Y., Lu, Z., Wang, J., Ye, X., Zhang, X., et al. (2021). Global insights to drought stress perturbed genes in oat (*Avena sativa* L.) seedlings using RNA sequencing. *Plant Signal Behav* 16(2), 1845934. doi: 10.1080/15592324.2020.1845934

Zhang, Y., Primavesi, L.F., Jhurrea, D., Andralojc, P.J., Mitchell, R.A., Powers, S.J., et al. (2009). Inhibition of SNF1-related protein kinase1 activity and regulation of metabolic pathways by trehalose-6-phosphate. *Plant Physiol* 149(4), 1860–1871. doi: 10.1104/pp.108.133934

Zhao, B.P., Ma, B.L., Hu, Y.G., and Liu, J.H. (2021). Source-Sink Adjustment: A Mechanistic Understanding of the Timing and Severity of Drought Stress on Photosynthesis and Grain Yields of Two Contrasting Oat (*Avena sativa* L.) Genotypes. *Journal of Plant Growth Regulation* 40(1), 263–276. doi: 10.1007/s00344-020-10093-5

Zhao, Y., Zhang, Z.J., Gao, J.H., Wang, P.C., Hu, T., Wang, Z.G., et al. (2018). *Arabidopsis* Duodecuple Mutant of PYL ABA Receptors Reveals PYL Repression of ABA-Independent SnRK2 Activity. *Cell Reports* 23(11), 3340–3351. doi: 10.1016/j.celrep.2018.05.044

Zhu, J.K. (2002). Salt and drought stress signal transduction in plants. *Annual Review of Plant Biology* 53, 247–273. doi: 10.1146/annurev.arplant.53.091401.143329

Zhu, J.K. (2016). Abiotic Stress Signaling and Responses in Plants. *Cell* 167(2), 313–324. doi: 10.1016/j.cell.2016.08.029

5 Avena sativa under Toxic Elements

Waqas-ud-Din Khan[1], Fatima Amanullah[2],*
Hassan Ahmed Ibraheem Ahmed[3,4] and Peichen Hou[5*]*
[1]Sustainable Development Study Centre, Government College
University, Lahore, Pakistan
[2]Sustainable Development Study Centre, Government College
University, Lahore, Pakistan
[3]Department of Botany, Faculty of Science, Port Said University,
Port Said, 42526, Egypt
[4]Tasmanian Institute of Agriculture, University of Tasmania, Hobart,
Tasmania, 7005, Australia
[5]Intelligent Equipment Research Center, Beijing Academy of
Agriculture and Forestry Sciences, Beijing, China
*Corresponding author: dr.waqasuddin@gcu.edu.pk;
houpc@nercita.org.cn; hassan.ahmed@utas.edu.au;
hassan.ahmed.sci@gmail.com

CONTENTS

5.1 INTRODUCTION

Potentially toxic elements (PTE) result from various natural processes such as earthquakes and volcanic eruptions, and also anthropogenic activities such as urbanization, rapid industrial growth, conventional agricultural practices and improper sewage disposal (Figure 5.1; Palansooriya et al. 2020). Some common names of such PTEs include arsenic (As), cadmium (Cd), chromium (Cr), lead (Pb), nickel (Ni) and mercury (Hg) on the basis of their occurrence and impact on the environment. These toxic elements, when entering the food chain, can become a serious concern for the health of plants, animals and humans (Antoniadis et al. 2017). In recent years, an estimated 12.6 million people have died from > 100 diseases worldwide caused by unhealthy and poor environmental conditions, including contaminated soils (World Health Organization 2016). The situation is deteriorating in developing countries; for example, toxicity has affected nearly the lives of 35–77 million people in Bangladesh in the past decade, while in Pakistan, the lives of 60 million people are at risk due to the presence of As in groundwater (Shakoor et al. 2019; BBC news 2017). Similarly, a report by the World Health Organization (2018) stated that Pb poisoning remained a major reason for the deaths of young children in Nigeria and Senegal where they remained exposed to Pb-contaminated soil. The elevated level of Pb in the blood vessels remained a cause of mental retardation among >600,000 children on a yearly basis (O'Connor et al. 2018).

As human health is at stake with PTEs, this issue must be tackled on the first step of the food chain, that is, plants. Naturally, plants are bio-accumulators of heavy metals as they extract, accumulate and concentrate them from the soil and water (Ali et al. 2016; Ozturk et al. 2017); however, the accumulation potential and extent of tolerance may vary from species to species. Some common PTE toxicity symptoms in plants include stunted growth, chlorosis, root browning and, eventually, apoptosis (Öztürk et al. 2015). Arsenic toxicity inhibits crop growth and yield, reduces stomatal conductance and photosynthetic rate and causes an imbalance in nutrient uptake ability (Chandrakar et al. 2016). Primarily, Pb toxicity occurs in plant cytoplasm. Pb travels through the plasma membrane by nutrient carriers such as Ca. The damage to the plasma membrane causes chlorosis and necrosis in plants. Pb also disrupts the nutrient uptake ability of plants (Kumar and Kumari 2015).

Heavy metal toxicity in plants induces both oxidative stress and modulation (up- and downregulation) of various stress-related genes. The elevated concentration of PTE denatures the structure of important enzymes and other proteins and mimics their function in substitution reactions. Dysfunctional biomolecules disturb the integrity of membranes, resulting in the alteration of basic plant metabolic reactions such as photosynthesis, respiration and homeostasis (Hossain et al. 2012). Moreover, in response to heavy metal stress, plants synthesize reactive oxygen species (ROS) such as superoxide radical (O^{-2}), hydroxyl radical (OH) and hydrogen peroxide (H_2O_2). These highly reactive oxygen species lead to lipid peroxidation, especially of the cellular membranes, causing them to undergo cell membrane permeability, damaging biomolecules and also cleavage of DNA strands (Ahmad et al. 2012).

In response to oxidative stress, cereal plants such as wheat, rice and oat have also developed complex and schematic mechanisms to protect themselves from the detrimental effects of ROS. Plants produce several antioxidant enzymes including superoxide dismutase (SOD), ascorbate peroxidase (APX), monodehydroascorbate reductase (MDAR), dehydroascorbate reductase (DHAR) and catalase (CAT). Superoxide radical is the first ROS to be produced during the second leg of the electron transport chain, when under oxidative stress the electron is captured by oxygen to produce superoxide instead of using it to generate NADPH. Superoxide dismutase provides the first line of defense by converting the superoxide radical into H_2O_2. A high level of cadmium Cd (5–30 mg/kg) causes toxicity in the plant (Kabata-Pendias and Pendias 1992). It inhibits plant growth and alters its nutrient (Na, K and Ca) uptake mechanisms. It also disrupts the water balance and enzyme activity, leading to plant death (Das et al. 1997, Mohanpuria et al. 2007; Marchel et al. 2018).

Considerable work has been done in the last decade, analyzing the biochemical, physiological and molecular processes involved in the uptake and translocation of specific heavy metals from the soil by different plants (Shulman et al. 2017; Kabir et al. 2018; Kozak and Brygadyrenko 2018). The plant exhibits a different counter process when exposed to heavy metal toxicity (Figure 5.1). Various studies have concluded that one of the prominent models of heavy metal detoxification exhibited by plants initiates with the formation of complexes (heavy metals with small peptides, e.g., phytochelatins) and later their sequestration by vacuoles inside the cell (Shri et al. 2014; Anjum et al. 2015). In this chapter, we use *Avena sativa* L. as a model plant due to its high ecological and nutritious value. Due to its high protein and vital mineral content, oat is grown all over the globe as a distinctive cereal and fodder crop. It's also high in dietary fiber, particularly β-glucan, which has been shown to promote human health. Furthermore, because this plant is less lucrative than maize, soybean or wheat, it is frequently grown in locations where there are a variety of challenges, such as drought or excessive salinity. In addition, this species is vulnerable to variations in temperature (Wang et al. 2009).

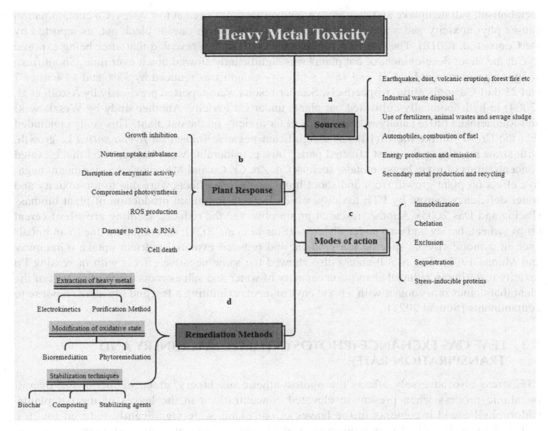

FIGURE 5.1 A flowchart representing (a) sources of heavy metals in the environment, (b) plant responses to elevated heavy metal concentrations, (c) mode of action in plants against stress and (d) remediation methods for heavy metals toxicity.

We discussed the effects of different PTEs applied on different aspects of *Avena sativa* L. growth and its physiology including photosynthetic machinery. This chapter provides recent information regarding the oxidative stress and genetic responses expressed by oat plants in response to heavy metal toxicity. Along with this information the chapter reviews the potential and future scenarios for the bio/phytoremediation of heavy metals from contaminated soil and water using *Avena sativa* L.

5.2 GROWTH RESPONSES (EFFECT OF STRESS RANGES APPLIED AND DURATIONS)

The accumulation of heavy metals such as Cd, Ni, Pb, Ag, Hg and Cr in both underground and aboveground parts of grain crops has been reported in a number of studies (Benedet et al. 2019; Marastoni et al. 2019; Souahi 2021; Souahi et al. 2021). The PTE toxic effect not only reduces the seed germination and seedling growth but also severely affects root development, plant dry weight, root and shoot length, total proteins and photosynthetic pigments in cereal crops (Rizvi et al. 2020). Similarly, the negative effects of PTE on *Avena sativa* growth, its metabolism and enzyme activity have been thoroughly studied by Tiecher et al. (2016). Shams et al. (2010) reported that Cr application reduced plant development and interferes with photosynthesis, glucose metabolism, sulfate uptake and numerous enzyme functions, even at low doses. Cu contamination causes phytotoxicity and a significant reduction in the growth rate of black oat, as reported by De Conti et al. (2018). The study of Tanhuanpää et al. (2007) revealed that after being exposed to Cd, the shoot development of oat plants was significantly slowed down over time. Shoot fresh weight was decreased by 32% and 48%, while dry weight was reduced by 33% and 43% after 7 and 21 d of Cd application, respectively. Similar results were reported previously by Astolfi et al. (2004) in hydroponically cultivated oat plants under Cd toxicity. Another study by Wyszkowski and Radziemska (2013) analyzed the effects of Cr toxicity on the oat plant. This study concluded that the Cr(VI) unlike the Cr(III) has a significant negative impact on *Avena sativa* L. growth, with straw remaining the most affected part. Thus, considerable work has reported that elevated concentrations of toxic heavy metals, such as Cd, Zn, Cr, Cu and Mn, showed a significant negative effect on plant growth (root and shoot biomass). These effects were due to ion toxicity and water deficiency caused by PTE toxicity, which leads to minimum production of plant biomass (Parida and Das 2005). Another research group observed the reduced seedling growth of cereal crops (wheat, barley and oat) under Pb stress (Souahi et al. 2021); this might be due to an imbalance in osmotic effects, specific ion gradient and reduced essential nutrient uptake (Greenway and Munns 1980). The root biomass also showed the same negative effects with increasing Pb toxicity resulting in reduced absorption capacity of water and salt as roots are the first part of the plant that comes into contact with a toxic environment, exhibiting a fast and sensitive response to contaminants (Souahi 2021).

5.3 LEAF GAS EXCHANGE (PHOTOSYNTHETIC MACHINERY AND TRANSPIRATION RATE)

PTE stress also adversely affects the photosynthetic machinery, green pigments and overall metabolic process when present in elevated concentrations in the leaves of the oat plants. Chlorophyll a and b contents in the leaves of oat plants were significantly reduced by 26% and 32% in artificially Cd-contaminated soil (Tanhuanpää et al. 2007); similarly, 27% and 13% reductions in chlorophyll a and b contents were reported in hydroponically cultivated oat plants (Astolfi et al. 2004), respectively, than the control. Several studies also reported a similar trend of a decrease in shoot chlorophyll contents, photosynthetic rate and transpiration rate with Cr(VI) stress (Oliveira, 2012; Kumar et al. 2019; Nikalje et al. 2019). The reported reduction in chlorophyll contents with Cd application might be related to disorders in chlorophyll synthesis or the photosynthetic equipment losing its structural integrity due to the loosening of thylakoid structures and the loss of chlorophyll pigment (Mosa et al. 2016; Shulman et al. 2017). Protein concentration was not altered in shoots of control and Cd-treated oat plants in both growing situations. Similar results were shown by other authors concluding that one of the primary effects of Pb toxicity on food (vegetables and cereal) crops is their significantly reduced photosynthetic content (Akinci et al. 2010; Li et al. 2014; Hou et al. 2017; Souahi et al. 2017). Other researchers suggested that the reduction in chlorophyll (Chl.)concentration was due to the suppression of specific enzymes necessary for chlorophyll synthesis and a decrease in uptake of minerals like Mg, which play a key role in the synthesis of chlorophyll pigments (Sirhindi et al. 2015; Wang et al. 2017).

On the contrary, the study of Souahi (2021) showed that Chl. A content was increased by 31.2%, 34.2% and 36% as compared to a control when exposed to increasing concentrations of Pb (0.15 g/L, 0.30 g/L and 0.60 g/L), respectively. A similar trend was followed for Chl. B and carotenoids. This might be attributed to the high tolerance of *Avena sativa* against Pb toxicity, however further experiments must be done to identify the underlying mechanisms responsible for this improved tolerance of oat against toxicity of Pb.

5.4 IONIC RELATIONS (TRANSMEMBRANE TRANSPORT, SEQUESTRATION AND HOMEOSTASIS)

The plasma membrane of plant root cells is the major membrane barrier between the cytoplasm and soil, as well as the root's first functional site of contact with any ion; hence, there can be a number of consequences for PTE toxicity. The interactions between root cells and Cd toxicity cause physiological alterations such as membrane damage and changes in enzyme activities (Prates, 2017). Calcium (Ca^{2+}) substitution at critical areas of cell membranes has been hypothesized as a possible reason for some of the Cd adverse effects. However, various evidences imply that the metal has directly affected the membrane components including protein and lipid (De Conti et al. 2018). The plasma membrane contains potential metal-sensitive enzymes (H^+ ATPase), responsible for delivering soluble compounds into cells, and its action is modulated in response to a range of environmental challenges, including lower root temperature and lower pH (Shulman et al. 2017).

Similarly, the effect of Cd on H^+ efflux from excised oat roots has been analyzed. The control roots had a significant H^+ efflux, which resulted in a pH drop of roughly 1.18 units at 180 minutes after K^+ injection. The roots, on the other hand, did not demonstrate any significant H^+ efflux after being pre-incubated for 30 minutes with 100 mol/L vanadates or 100 mol/L $CdSO_4$. Although plasmalemma H^+ATPase was implicated in the process of Cd poisoning, varied results were detected based on the test used by Petukhov et al. (2021). Similarly, they looked at H^+ efflux in oat roots that had been pre-incubated in deionized water, 100 mol/L vanadate or 100 mol/L $CdSO_4$, an action that is thought to be reliant on the plasma membrane H^+ATPase (Ben et al. 2020) (Figure 5.2).

Different studies have discovered that vanadate and $CdSO_4$ inhibited H^+ outflow, although control roots are competent to acidify the solution (Astolfi et al. 2003; Kotyk et al. 1996). Given that

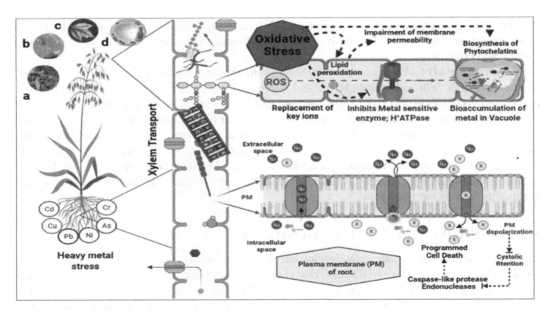

FIGURE 5.2 Potentially toxic elements induce (a) leaf roll, (b) spots, (c) leaf chlorosis and (d) metabolic and physiological alterations such as membrane permeability and inhibition of membrane-bound enzymes (H^+ ATPase). This leads to the reduced uptake of essential mineral nutrients required for growth. This figure was modified from Astolfi et al. (2003).

H⁺ATPase is thought to be involved in electrogenic H⁺ efflux, however, the inhibition of H⁺ATPase activity may explain these results. When Cd was introduced to the nutrient solution, H⁺ATPase function in the root plasma membrane was modulated, resulting in reduced ATP action and a comparable drop in H⁺ pump. This impact might be caused by a variety of metal processes, along with alterations in membrane lipid content and its permeability (Stuckey et al. 2021).

One of the key processes triggered by PTE toxicity may have been the alteration of sulfhydryl (–SH) groups of membrane proteins. Cadmium toxicity to *Avena sativa* L. causes changes in the lipid content of the membrane bilayer and, on the other side, it might have an impact on intrinsic-membrane protein function (Astolfi et al. 2004).

In vitro tests were also used to determine the reaction of plasmalemma H⁺ATPase due to the introduction of Cd to *Avena sativa* L. (Astolfi et al. 2003). The findings revealed that Cd inhibited ATP activity *in vitro* and that the limitation was dose-dependent. The tests were carried out using a reagent of 5 mmol/L Mg²⁺:ATP. Although all Cd was coupled to ATP, the Mg²⁺:ATP supply did not drop appreciably since the substrate concentration surpassed that of Cd. Moreover, when the enzyme function of *Avena sativa* was tested in the addition of the detergents, the inhibitory effect of Cd was still visible (Prates and Yu 2017). This can be viewed as a direct suppression of plasma membrane ATPase's phosphohydrolase action, rather than a change in the transmembrane electrical potential caused by the H⁺ pump. *In vitro*, Cd reduced H⁺-pump activity in the same way that it did for ATPase. The suppression trend was nearly the same in both situations, however, the rates of proton build-up in vesicles were inhibited to a greater extent than the hydrolytic action. This finding implies that PTE influences the membrane functioning of the plant. The changes in plasma membrane permeability might represent the plant cell's reaction to stress, resulting in changes to the membrane-bound H⁺ATPase enzyme that could aid the cell's adjustment to the new circumstances. In light of this, research shows that the plasmalemma, and especially the maintenance of plasma membrane permeability, may play a key role in metal tolerance (Marastoni et al. 2019).

Finally, the findings show that Cd inhibits ATP hydrolysis effectively. The reduced H⁺ proton accumulation as a result of ATP hydrolysis may be attributed to the Cd-induced changes in plasma membrane permeability.

An elevated level of Al³⁺ was reported in the roots and shoots of sensitive (UFRGS 930598), intermediate (UFRGS 280) and tolerant (UFRGS 17) genotypes of *Avena sativa* L. with high $Al_2(SO_4)_3$ concentrations (Pereira, et al. 2011). The accumulated concentration of Al³⁺ was significantly higher in roots than in shoots. Several studies have observed similar findings when dealing with other plant species' exposure to various Al³⁺ levels (Poschenrieder et al. 2008; Yamamoto et al. 2003). The presence of Al³⁺ in the root ball could reveal that the roots act as a partial protective layer against Al³⁺ ion to transfer to the shoots. When comparing the shoots of tolerant and sensitive seeds, the shoots of medium seedlings had a greater Al³⁺ content. These findings illustrate a genotypic variation in Al³⁺ partitioning among oat genotypes.

5.5 OXIDATIVE STRESS AND ANTIOXIDANTS

It is well established that plant cells develop defensive mechanisms against ROS, including H_2O_2, OH and O_2^-, when exposed to PTE stress (Dad et al. 2021; Sharma et al. 2016). Different antioxidant enzymes including SOD, POD and CAT were activated by the oat plant to normalize body functions (Manquián-Cerda et al. 2016). Plants usually respond against abiotic stresses through a series of high-regulatory reactions or a decrease in enzyme production and activity, and the plant response varies depending on the plant species and their stress-associated genes (Nahar et al. 2016). The four barley genotypes showed different levels of Cd stress tolerance in terms of plant growth, biomass, chlorophyll content, lipid peroxidation and antioxidant activities (Qi and Zhao 2020).

To control the amount of ROS production, equilibrium between the ROS scavenging enzymes (SOD, POD and CAT) is critical. APX and CAT have a differing affinity for H_2O_2, indicating that they relate to two separate groups of H_2O_2-scavenging enzymes, such as APX, that could be account-able for the detoxification of H_2O_2. Seedlings' precise control of ROS indicates that even modest amounts of H_2O_2 produced oxidative stress in the *Avena sativa* genotype (De Conti et al. 2018). This shows that the oat genotype (UFRGS 930598) is much more sensitive than the (UFRGS 17) resist-ance genotype, which is consistent with Sanchez-Chacon et al. (2000), who analyzed the variations between oat seedlings and found that the Al-sensitive genotype (UFGRS 930598) showed reduced root development and biomass.

Similarly, the exposure of plants to Cd showed a significant increase in GPX (an oxidative stress-controlling enzyme) activity (Fryzova et al. 2017). The findings reveal that Cd contributes to differences in GPX activity and that *in vivo* Cd intake is directly connected with the enzymatic O-acetylserine sulfhydrylase activity (De Conti et al. 2018). However, the release of active scaven-ging molecules differs with respect to plant genotype, its tissues and metal speciation (Emamverdian et al. 2015). The study of Kaur et al. (2013) revealed that when *Triticum aestivum* was exposed to Pb toxicity, in response to the H_2O_2 production the activity of CAT was enhanced, whereas the con-centration of POD was substantially reduced. The reason behind the reduced CAT activity could be attributed to the inhibited cell metabolism and H_2O_2 production, due to the high concentration of heavy metal toxicity.

The study of Astolfi et al. (2004) stated that two enzymes participating in C metabolism, phosphoenolpyruvate carboxylase (PEPCase) and ribulose-1,5-biphosphate carboxylase (RUBPC), were marginally affected by the Cd stress. In plants, the PEPCase activity was reduced by 11% and 18% after 7 and 21 days of Cd application, respectively. Following that, the impact of Cd was measured on nitrogen metabolism by measuring glutamine synthetase (GS) and NAD-dependent glutamate dehydrogenase (GDH) activities. In both soil and hydroponic growing conditions, GS exhibited no sensitivity to Cd treatment; while GDH revealed a distinct pattern that was influenced by growing conditions (Astolfi et al. 2004). In the soil medium, the concentration of GDH was rela-tively unchanged in plants by Cd exposure, while it was dramatically boosted in hydroponic cultured plants (30% and 26% after 7 and 21 days, respectively) (Guala et al. 2010).

5.6 GENETIC RESPONSES AND KEY GENES EXPRESSED IN RESPONSE TO HEAVY METAL TOXICITY

On a molecular scale, our understanding of the stress-adaptive response in oat remains limited, as the genomes of many crops are similar, such as rice and corn, and grains were clustered, which fur-ther aided inadequate investigations about these species. Recently, the completely sequenced and assembled oat genome became accessible (Yang et al. 2020).

Under various abiotic conditions (drought, salt, heavy metal, cold and heat), nine potential ref-erence genes (including ACT, TUB and CYP) were evaluated from the roots and leaves of two oat genotypes, and their phase stability was analyzed using four different statistical programs. The various systems' fundamental scores agreed on which genes were the least reactive for each treatment/tissue/genotype interaction, allowing them to be easily deleted. According to the multiple assessing software used in our investigation, glyceraldehyde-3-phosphate dehydrogenase (GAPDH) was shown to be stable during Cd stress (Tanhuanpää et al. 2007).

When plants are treated with Cd, GAPDH was one of the least consistent genes in gigantic reeds (*Arundo donax* L.), which is a Poaceae family member. In the leaves of both oat varieties studied, ACT1 was highly stable. Other members of the Actin gene family, such as ACT12 in switch-grass and ACT3 in soybean, demonstrated good translation stability following cadmium exposure. PGD (phosphogluconate dehydrogenase) was a steady reference gene in both oat genotypes' roots, whereas its transcription in soybean was dramatically altered by Cd stress. TUB transcription was unstable in the oat samples as well as in Cd-treated soybean plants (Tajti et al. 2021).

5.7 TRANSCRIPTOMIC AND METABOLOMIC ANALYSES RESPONSES UNDER PTE TOXICITY

Transcriptomic and metabolomic studies are used to study the adaptation behavior of plants when exposed to biotic and abiotic stresses. These studies allow us to identify the specific genes and their expression pattern during stress perception and response by the plant (Tchounwou et al. 2012). Phytochelatins (PCs) have been considered the prevalent mechanism of plant adaptation to heavy metals. They form complexes with heavy metals and deposit them in the vacuole. Unlike other PTEs, Cr is unable to induce PCs, and thus the detoxification mechanism for this metal is poorly understood (Shanker et al. 2009). The study of Chakrabarty et al. (2009) analyzed the genome-wide expression in rice (*Oryza sativa*) when exposed to As(V) toxicity. It showed that the expression of 10 genes (Os01g49710, Os10g34020, Os10g38350, Os10g38720, Os10g38189, Os10g38610, Os01g49720, Os09g20220, Os01g72150 and Os01g72140) coding for glutathione S-transferase (GSTs) are upregulated. These genes are involved in xenobiotic (As) metabolism. Similarly, another PTE stress-related gene glutaredoxin (Os02g40500) and four (Os12g38300, Os12g38290, Os12g38051 and Os12g38064) metallothioneins (MTs) are upregulated under As(V) exposure. Similarly, the genome expression of rice root was evaluated under Cr toxicity (Dubey et al. 2010). This study showed that 1138 genes were upregulated, and 1610 genes were downregulated by Cr (VI) stress. Among the downregulated genes, the most affected genes are energy metabolism (111 genes), carbohydrate metabolism (218), stilbene, coumarin and lignin biosynthesis (60), phenylalanine metabolism (53), cell growth and death (150), photosynthesis (26), lipid metabolism (103), biodegradation of xenobiotics (140), photosystem II (15), amino acid metabolism (103), cell cycle (46), etc. The most upregulated genes are related to amino acid metabolism (111), biosynthesis of secondary metabolites (122) and xenobiotics (127), membrane transport (53) and signal transduction (75). The expression of 18 GST coding genes (Os01g37750, Os09g20220, Os10g38350, Os10g38495, Os01g72150, Os10g38610, Os01g49710, Os01g49710, Os10g38600, Os10g38600, 001071656, Os10g38140, Os01g72160, Os10g38740, Os10g38501, Os10g38150, AF402802 and Os10g38489) involved in Cr metabolism was upregulated (Dubey et al. 2010).

Genetic factors affect the Cd translocation inside the plant body and within species; hence, the development of such cultivars that poorly accumulate Cd is a worthy future approach. The use of molecular markers linked to low Cd accumulation inside the plant body could be an alternative to phenotypic selection. In the study of Marastoni et al. (2019), such markers were derived by using bulked-segregant analysis in a F2 population from the cross between oat cultivars "Aslak" and "Salo" (the latter one is known to be a high Cd accumulator). Four markers associated with grain Cd concentration were found: two RAPDs (random amplified polymorphic DNAs), one REMAP (retrotransposon-microsatellite amplified polymorphism) and one SRAP (sequence-related amplified polymorphism). The first three were converted into more reproducible SCAR (sequence-characterized amplified region) markers. The four markers were assigned to one linkage group via QTL (quantitative trait locus) mapping, which presented a major gene for grain Cd concentration. Whether or not they have identified the correct candidate gene, the DNA markers can be used in future breeding programs to select low Cd accumulators in oat. Choosing any of the four markers increases the chance that even one of them will be polymorphic in the desired cross. The SRAP marker is polymorphic in a doubled haploid population that is being used to construct an oat linkage map at MTT Agrifood Research Finland. As a consequence, the gene for grain Cd accumulation will be localized on this map. The ultimate verification of the candidate gene will help in the transfer of the low Cd accumulation trait to other cereals and crops (Marastoni et al. 2019).

5.8 PHYTOREMEDIATION POTENTIAL OF *AVENA SATIVA* IN HEAVY-METAL-CONTAMINATED SOIL

Phytoremediation is an alternative green technique used to remove the hazardous pollutants and carcinogenic compounds present in the ecosystem that are adversely affecting the ecological balance and causing toxicity. In addition, comparatively plant-based remediation is cost-effective, rapid and environmentally sustainable. In this, different plants are used to extract toxic elements from the environment by using the plant's natural metabolic activity and depositing it in the plant shoot in relatively high concentrations (Khan and Doty, 2011). On the other hand, hyperaccumulators have less biomass, a slow growth rate and lack the commonly used cultivation and management practices (Sanz-Fernández et al. 2017). Therefore, various plant species have been tested for their potential to be used in phytoremediation based on their tolerance level to metal stress, ability to accumulate metal in their body, growth rate, biomass yield and post-harvest management protocols required. As a result, many vegetative and grain crops have been recommended as potential candidates to be used for the phytoremediation technique. This is due to their high biomass and progressive growth rates through which relatively high-moderate accumulation of toxic metals can be compensated (Takahashi et al. 2016). The studies of Mantry and Patra (2017) and Wang et al. (2017) showed that other grain crops such as rice (*Oryza sativa*) and sorghum (*Sorghum bicolor*) can be used for environmental restoration of soils contaminated with Cr, Cd and Sr, respectively. The results showed that these crops can sufficiently tolerate the increasing concentrations of heavy metals and can accumulate them in higher plant biomass, thus making them a potential candidate for phytoremediation.

In a field study, 26 different cultivars of three grain crops (wheat, oat and barley) were analyzed to evaluate their phytoremediation potential in strontium (Sr)-contaminated soil. Among all three crops, the oat variety (Neimengkeyi-1) exhibited the maximum potential for removing Sr from the soil (Qi et al. 2015). Another study showed that the translocation factor (TLF) of *Avena sativa* is greater than 1 (Qi and Zhao 2020). The oat variety (Neimengkeyi-1) showed a TLF ranging from 2.03 to 11.66 and BCF ranging from 2.76% to 1.49% when the Sr concentration was increased from 25 mg kg^{-1} to 1000 mg kg^{-1}, respectively. The possible mechanism behind the phytoremediation potential of *Avena sativa* lies in the mechanisms of phytochelatins (PCs). This has been considered the prevalent mechanism of plant adaptation to heavy metals. It forms complexes with heavy metals and deposits them in the vacuole (Shanker et al. 2009).

5.9 CONCLUSION

Avena sativa L. is grown all over the world as a distinctive cereal and fodder crop, due to its high protein and vital mineral content. Potentially toxic elements (PTEs) resulting from various natural and anthropogenic processes have shown a significant effect on oat plants, which not only modify their cellular processes but also affect the *Avena sativa* genome or gene expressions. However, oat plants have developed smart strategies to minimize the stress effects by accumulating heavy metals in their tissues and organs. These defensive mechanisms are operated at multiple levels: morphological, physiological, genetic and epigenetic modifications of the genome. The available literature shows that PTE exposure to *Avena sativa* significantly affects oat growth and development, metabolism and key membrane-bound enzymes activity, even at low doses. Heavy metal toxicity negatively affects oat genes involved in energy metabolism, carbohydrate metabolism, lignin biosynthesis, phenylalanine metabolism, cell growth and death, photosynthesis, lipid metabolism, biodegradation of xenobiotics, photosystem II, amino acid metabolism and cell cycle. The most upregulated genes are related to amino acid metabolism, biosynthesis of secondary metabolites and xenobiotics, membrane transport and signal transduction. The above notions may be of importance when recommending the proper practices for cultivating oat plants under saline conditions, with the ultimate aim of achieving the best nutritional value.

REFERENCES

Ahmad, P., Ozturk, M., & Gucel, S. (2012). Oxidative damage and antioxidants induced by heavy metal stress in two cultivars of mustard (*Brassica juncea* L.) plants. Fresenius Environ Bull, 21(10), 2953–2961.

Akinci, I. E., Akinci, S., & Yilmaz, K. (2010). Response of tomato (*Solanum lycopersicum* L.) to lead toxicity: Growth, element uptake, chlorophyll and water content. African Journal of Agricultural Research, 5(6), 416–423.

Ali, I., AL-Othman, Z. A., & Alwarthan, A. (2016). Molecular uptake of congo red dye from water on iron composite nano particles. Journal of Molecular Liquids, 224, 171–176.

Anjum, N. A., Hasanuzzaman, M., Hossain, M. A., Thangavel, P., Roychoudhury, A., Gill, S. S., ... & Ahmad, I. (2015). Jacks of metal/metalloid chelation trade in plants—an overview. Frontiers in Plant Science, 6, 192.

Antoniadis, V., Levizou, E., Shaheen, S. M., Ok, Y. S., Sebastian, A., Baum, C., ... & Rinklebe, J. (2017). Trace elements in the soil-plant interface: Phytoavailability, translocation, and phytoremediation–A review. Earth-Science Reviews, 171, 621–645.

Astolfi, S., Zuchi, S., & Passera, C. (2004). Effects of cadmium on the metabolic activity of *Avena sativa* plants grown in soil or hydroponic culture. Biologia plantarum, 48(3), 413–418.

Astolfi, S., Zuchi, S., Chiani, A., & Passera, C. (2003). In vivo and in vitro effects of cadmium on H+ ATPase activity of plasma membrane vesicles from oat (*Avena sativa* L.) roots. Journal of Plant Physiology, 160(4), 387–393.

Ben Chabchoubi, I., Mtibaa, S., Ksibi, M., & Hentati, O. (2020). Health risk assessment of heavy metals (Cu, Zn, and Mn) in wild oat grown in soils amended with sediment dredged from the Joumine Dam in Bizerte, Tunisia. Euro-Mediterranean Journal for Environmental Integration, 5(3), 1–14.

Benedet, L., De Conti, L., Lazzari, C. J. R., Júnior, V. M., Dick, D. P., Lourenzi, C. R., ... & Brunetto, G. (2019). Copper and zinc in rhizosphere soil and toxicity potential in white oats (Avena sativa) grown in soil with long-term pig manure application. Water, Air, & Soil Pollution, 230(8), 1–10.

Chakrabarty, D., Trivedi, P. K., Misra, P., Tiwari, M., Shri, M., Shukla, D., ... & Tuli, R. (2009). Comparative transcriptome analysis of arsenate and arsenite stresses in rice seedlings. Chemosphere, 74(5), 688–702.

Chandrakar, V., Naithani, S. C., and Keshavkant, S. (2016). Arsenic-induced metabolic disturbances and their mitigation mechanisms in crop plants: A review. Biologia, 71(4), 367–377.

Dad, H. A., Gu, T. W., Zhu, A. Q., Huang, L. Q., & Peng, L. H. (2021). Plant exosome-like nanovesicles: emerging therapeutics and drug delivery nanoplatforms. Molecular Therapy, 29(1), 13–31.

Das, P., Samantaray, S., & Rout, G. R. (1997). Studies on cadmium toxicity in plants: a review. Environmental Pollution, 98(1), 29–36.

De Conti, L., Ceretta, C. A., Tiecher, T. L., da Silva, L. O., Tassinari, A., Somavilla, L. M., ... & Brunetto, G. (2018). Growth and chemical changes in the rhizosphere of black oat (Avetrigoseosa) grown in soils contaminated with copper. Ecotoxicology and Environmental Safety, 163, 19–27.

Dubey, S., Misra, P., Dwivedi, S., Chatterjee, S., Bag, S. K., Mantri, S., ... & Tuli, R. (2010). Transcriptomic and metabolomic shifts in rice roots in response to Cr (VI) stress. BMC genomics, 11(1), 1–19.

Emamverdian, A., Ding, Y., Mokhberdoran, F., & Xie, Y. (2015). Heavy metal stress and some mechanisms of plant defense response. The Scientific World Journal, 2015.

Fryzova, R., Pohanka, M., Martinkova, P., Cihlarova, H., Brtnicky, M., Hladky, J., & Kynicky, J. (2017). Oxidative stress and heavy metals in plants. Reviews of environmental contamination and toxicology volume 245, 129–156.

Greenway, H., & Munns, R. (1980). Mechanisms of salt tolerance in nonhalophytes. Annual Review of Plant Physiology, 31(1), 149–190.

Guala, S. D., Vega, F. A., & Covelo, E. F. (2010). Heavy metal concentrations in plants and different harvestable parts: a soil–plant equilibrium model. Environmental pollution, 158(8), 2659–2663.

Gupta, A., Joia, J., Sood, A., Sood, R., Sidhu, C., & Kaur, G. (2016). Microbes as potential tool for remediation of heavy metals: a review. Journal of Microbial and Biochemical Technology, 8(4), 364–372.

Hossain, M. A., Piyatida, P., da Silva, J. A. T., & Fujita, M. (2012). Molecular mechanism of heavy metal toxicity and tolerance in plants: central role of glutathione in detoxification of reactive oxygen species and methylglyoxal and in heavy metal chelation. Journal of botany, 2012.

Hou, X. L., Cai, L., Han, H., Zhou, C., Wang, G., & Liu, A. (2017). Effect of Pb stress on the chlorophyll fluorescence characteristics and antioxidative enzyme activities of *Paspalum notatum*. Acta Agrestia Sinica, 26(3), 142–148.

Kabata-Pendias, A., & Pendias, H. (1992). Trace elements in soils and plants. CRC Press: Boca Raton, FL.

Kabir, M., Iqbal, M. Z., & Shafiq, M. (2018). The effects of lead and cadmium individually and in combinations on germination and seedling growth of Leucaena leucocephala (Lam.) de Wit. American Academic Scientific Research Journal for Engineering, Technology, and Sciences, 43(1), 33–43.

Kaur, G., Singh, H. P., Batish, D. R., & Kohli, R. K. (2013). Lead (Pb)-induced biochemical and ultrastructural changes in wheat (Triticum aestivum) roots. Protoplasma, 250(1), 53–62.

Khan, Z., & Doty, S. (2011). Endophyte-assisted phytoremediation. Current Topics in Plant Biology, 11, 97–105.

Kotyk, A., Kamínek, M., Pulkrábek, J., & Zahradníček, J. (1996). Effect ofin vivo andin vitro application of the cytokinin N6-(m-hydroxybenzyl) adenosine on respiration and membrane transport processes in sugar beet. Biologia plantarum, 38(3), 363–368.

Kozak, V. M., & Brygadyrenko, V. V. (2018). Impact of cadmium and lead on Megaphyllum kievense (Diplopoda, Julidae) in a laboratory experiment. Biosystems Diversity, 26(2), 128–131.

Kumar, H.G., & Kumari, P.J. (2015). Heavy metal lead influative toxicity and its assessment in phytoremediating plants—a review. Water, Air, & Soil Pollution, 226(10), 1–11.

Kumar, P., Tokas, J., & Singal, H. R. (2019). Amelioration of chromium VI toxicity in sorghum (Sorghum bicolor L.) using glycine betaine. Scientific reports, 9(1), 1–15.

Li, W., Xu, B., Song, Q., Liu, X., Xu, J., & Brookes, P. C. (2014). The identification of 'hotspots' of heavy metal pollution in soil–rice systems at a regional scale in eastern China. Science of the Total Environment, 472, 407–420.

Manquián-Cerda, K., Escudey, M., Zúñiga, G., Arancibia-Miranda, N., Molina, M., & Cruces, E. (2016). Effect of cadmium on phenolic compounds, antioxidant enzyme activity and oxidative stress in blueberry (*Vaccinium corymbosum* L.) plantlets grown in vitro. Ecotoxicology and Environmental Safety, 133, 316–326.

Mantry, P., & Patra, H. K. (2017). Combined effect of Cr+ 6 and chelating agents on growth and Cr bio-accumulation in flood susceptible variety of rice Oryza sativa (L.). Annals of Plant Science, 6, 1573–1578.

Marastoni, L., Tauber, P., Pii, Y., Valentinuzzi, F., Astolfi, S., Simoni, A., ... & Mimmo, T. (2019). The potential of two different Avena sativa L. cultivars to alleviate Cu toxicity. Ecotoxicology and environmental safety, 182, 109430.

Marchel, M., Kaniuczak, J., Hajduk, E., & Wlasniewski, S. (2018). Response of oat (Avena sativa) to the addition cadmium to soil inoculation with the genus trichoderma fungi. Journal of Elementology, 23(2).

McGrath, M. (2017, August 23). "Alarmingly high" levels of arsenic in Pakistan's ground water. BBC News. www.bbc.com/news/science-environment-41002005

Mohanpuria, P., Rana, N. K., & Yadav, S. K. (2007). Cadmium induced oxidative stress influence on glutathione metabolic genes of *Camellia sinensis* (L.) O. Kuntze. Environmental Toxicology: An International Journal, 22(4), 368–374.

Mosa, A., El-Banna, M. F., & Gao, B. (2016). Biochar filters reduced the toxic effects of nickel on tomato (*Lycopersicon esculentum* L.) grown in nutrient film technique hydroponic system. Chemosphere, 149, 254–262.

Nahar, K., Hasanuzzaman, M., Alam, M. M., Rahman, A., Suzuki, T., & Fujita, M. (2016). Polyamine and nitric oxide crosstalk: antagonistic effects on cadmium toxicity in mung bean plants through upregulating the metal detoxification, antioxidant defense and methylglyoxal detoxification systems. Ecotoxicology and Environmental Safety, 126, 245–255.

Nikalje, G. C., Saini, N., & Suprasanna, P. (2019). Halophytes and Heavy Metals: Interesting Partnerships. In Plant-Metal Interactions (pp. 99–118). Springer, Cham.

O'Connor, D., Hou, D., Ye, J., Zhang, Y., Ok, Y. S., Song, Y., ... & Tian, L. (2018). Lead-based paint remains a major public health concern: A critical review of global production, trade, use, exposure, health risk, and implications. Environment International, 121, 85–101.

Oliveira, H. (2012). Chromium as an environmental pollutant: insights on induced plant toxicity. Journal of Botany.

Ozturk, M., Altay, V., & Karahan, F. (2017). Studies on trace elements distributed in Glycyrrhiza taxa in Hatay-Turkey. International Journal of Plant and Environment, 3(02), 01–07.

Öztürk, M., Ashraf, M., Aksoy, A., Ahmad, M. S. A., & Hakeem, K. R. (Eds.). (2015). Plants, pollutants and remediation. Berlin/Heidelberg, Germany: Springer.

Parida, A. K., & Das, A. B. (2005). Salt tolerance and salinity effects on plants: a review. Ecotoxicology and Environmental Safety, 60(3), 324–349.

Pereira, L. B., de A Mazzanti, C. M., Cargnelutti, D., Rossato, L. V., Gonçalves, J. F., Calgaroto, N., ... & Schetinger, M. R. (2011). Differential responses of oat genotypes: oxidative stress provoked by aluminum. Biometals, 24(1), 73–83.

Petukhov, A. S., Kremleva, T. A., & Petukhova, G. A. (2021). Bioaccumulation of heavy metals by cultivated oat from industrially polluted soils of the Tyumen town. Agrochem Herald.

Poschenrieder, C., Gunsé, B., Corrales, I., & Barceló, J. (2008). A glance into aluminum toxicity and resistance in plants. Science of the Total Environment, 400(1–3), 356–368.

Pourrut B, Jean S, Silvestre J, Pinelli E (2011) Lead-induced DNA damage in Vicia faba root cells: potential involvement of oxidative stress. Mutation Research 726(2):123–128.

Prates, L. L., & Yu, P. (2017). Recent research on inherent molecular structure, physiochemical properties, and bio-functions of food and feed-type Avena sativa and processing-induced changes revealed with molecular micro spectroscopic techniques. Applied Spectroscopy Reviews, 52(10), 850–867.

Qi, L., & Zhao, W. (2020). Strontium uptake and antioxidant capacity comparisons of low accumulator and high accumulator oat (Avena sativa L.) genotypes. International Journal of Phytoremediation, 22(3), 227–235.

Qi, L., & Zhao, W. (2020). Strontium uptake and antioxidant capacity comparisons of low accumulator and high accumulator oat (Avena sativa L.) genotypes. International Journal of Phytoremediation, 22(3), 227–235.

Qi, L., Qin, X., Li, F. M., Siddique, K. H., Brandl, H., Xu, J., & Li, X. (2015). Uptake and distribution of stable strontium in 26 cultivars of three crop species: oats, wheat, and barley for their potential use in phytoremediation. International Journal of Phytoremediation, 17(3), 264–271.

Rizvi, A., Zaidi, A., Ameen, F., Ahmed, B., AlKahtani, M. D., & Khan, M. S. (2020). Heavy metal induced stress on wheat: phytotoxicity and microbiological management. RSC Advances, 10(63), 38379–38403.

Sanchez-Chacon, C. D., Federizzi, L. C., Milach, S. C. K., & Pacheco, M. T. (2000). Variabilidade genética e herança da tolerância à toxicidade do alumínio em aveia. Pesquisa Agropecuária Brasileira, 35, 1797–1808.

Sanz-Fernández, M., Rodríguez-Serrano, M., Sevilla-Perea, A., Pena, L., Mingorance, M. D., Sandalio, L. M., & Romero-Puertas, M. C. (2017). Screening Arabidopsis mutants in genes useful for phytoremediation. Journal of hazardous materials, 335, 143–151.

Shakoor, M. B., Niazi, N. K., Bibi, I., Shahid, M., Saqib, Z. A., Nawaz, M. F., ... & Rinklebe, J. (2019). Exploring the arsenic removal potential of various biosorbents from water. Environment International, 123, 567–579.

Shams, K. M., Tichy, G., Fischer, A., Sager, M., Peer, T., Bashar, A., & Filip, K. (2010). Aspects of phytoremediation for chromium contaminated sites using common plants Urtica dioica, Brassica napus and Zea mays. Plant and soil, 328(1), 175–189.

Shanker, A. K., Cervantes, C., Loza-Tavera, H., & Avudainayagam, S. (2005). Chromium toxicity in plants. Environment international, 31(5), 739–753.

Shanker, A. K., Djanaguiraman, M., & Venkateswarlu, B. (2009). Chromium interactions in plants: current status and future strategies. Metallomics, 1(5), 375–383.

Sharma, S. S., Dietz, K. J., & Mimura, T. (2016). Vacuolar compartmentalization as indispensable component of heavy metal detoxification in plants. Plant, Cell & Environment, 39(5), 1112–1126.

Shri, M., Dave, R., Diwedi, S., Shukla, D., Kesari, R., Tripathi, R. D., ... & Chakrabarty, D. (2014). Heterologous expression of Ceratophyllum demersum phytochelatin synthase, CdPCS1, in rice leads to lower arsenic accumulation in grain. Scientific reports, 4(1), 1–10.

Shulman, M. V., Pakhomov, O. Y., & Brygadyrenko, V. V. (2017). Effect of lead and cadmium ions upon the pupariation and morphological changes in Calliphora vicina (Diptera, Calliphoridae). Folia Oecologica, 44(1), 28–37.

Sirhindi, G., Mir, M. A., Sharma, P., Gill, S. S., Kaur, H., & Mushtaq, R. (2015). Modulatory role of jasmonic acid on photosynthetic pigments, antioxidants and stress markers of Glycine max L. under nickel stress. Physiology and Molecular Biology of Plants, 21(4), 559–565.

Souahi, H. (2021). Impact of lead on the amount of chlorophyll and carotenoids in the leaves of Triticum durum and T. aestivum, Hordeum vulgare and Avena sativa. Biosystems Diversity, 29(3), 207–210.

Souahi, H., Chebout, A., Akrout, K., Massaoud, N., & Gacem, R. (2021). Physiological responses to lead exposure in wheat, barley and oat. Environmental Challenges, 4, 100079.

Souahi, H., Gharbi, A., & Gassarellil, Z. (2017). Growth and physiological responses of cereals species under lead stress. International Journal of Biosciences, 11(1), 266–273.

Stuckey, J. W., Neaman, A., Verdejo, J., Navarro-Villarroel, C., Peñaloza, P., & Dovletyarova, E. A. (2021). Zinc Alleviates Copper Toxicity to Lettuce and Oat in Copper-Contaminated Soils. Journal of Soil Science and Plant Nutrition, 21(2), 1229–1235.

Tajti, J., Pál, M., & Janda, T. (2021). Validation of reference genes for studying different abiotic stresses in oat (*Avena sativa* L.) by RT-qPCR. Plants, 10(7), 1272.

Takahashi, R., Ito, M., Katou, K., Sato, K., Nakagawa, S., Tezuka, K., ... & Kawamoto, T. (2016). Breeding and characterization of the rice (Oryza sativa L.) line "Akita 110" for cadmium phytoremediation. Soil Science and Plant Nutrition, 62(4), 373–378.

Tanhuanpää, P., Kalendar, R., Schulman, A. H., & Kiviharju, E. (2007). A major gene for grain cadmium accumulation in oat (*Avena sativa* L.). Genome, 50(6), 588–594.

Tchounwou, P. B., Yedjou, C. G., Patlolla, A. K., & Sutton, D. J. (2012). Heavy metal toxicity and the environment in Molecular, clinical and environmental toxicology, vol 101 (ed. Luch, A.) 133–164.

Tiecher, T. L., Tiecher, T., Ceretta, C. A., Ferreira, P. A., Nicoloso, F. T., Soriani, H. H., ... & Brunetto, G. (2016). Physiological and nutritional status of black oat (*Avena strigosa* Schreb.) grown in soil with interaction of high doses of copper and zinc. Plant Physiology and Biochemistry, 106, 253–263.

Wang, X., Chen, C., & Wang, J. (2017). Phytoremediation of strontium contaminated soil by *Sorghum bicolor* (L.) Moench and soil microbial community-level physiological profiles (CLPPs). Environmental Science and Pollution Research, 24(8), 7668–7678.

Wang, Z., Gerstein, M., & Snyder, M. (2009). RNA-Seq: a revolutionary tool for transcriptomics. Nature reviews genetics, 10(1), 57–63.

World Health Organization. (2016). An estimated 12.6 million deaths each year are attributable to unhealthy environments. WHO. March, 15, 2016.

World Health Organization. 2018. Lead Poisoning and Health. www.who.int/en/news-room/ fact-sheets/detail/ lead-poisoning-and-health.

Wyszkowski, M., & Radziemska, M. (2013). Assessment of tri-and hexavalent chromium phytotoxicity on Oats (*Avena sativa* L.) biomass and content of nitrogen compounds. Water, Air, & Soil Pollution, 224(7), 1–14.

Yamamoto, Y., Kobayashi, Y., Devi, S. R., Rikiishi, S., & Matsumoto, H. (2003). Oxidative stress triggered by aluminum in plant roots. In Roots: the dynamic interface between plants and the earth (pp. 239–243). Springer, Dordrecht.

Yang, Z.; Wang, K.; Aziz, U.; Zhao, C.; Zhang, M. Evaluation of duplicated reference genes for quantitative real-time PCR analysis in genome unknown hexaploid oat (*Avena sativa* L.). Plant Methods 2020, 16, 138.

6 *Citrus lemon* and Stressful Conditions

Maksud Hasan Shah[1], Sk. Md. Ajaharuddin[2],*
SK Sadikur Rahman[3], Shoumik Saha[4], Subrata Goswami[5],
Saidul Islam[6], Soham Barik[7], Akbar Hossain[8] and
Arafat Abdel Hamed Abdel Latef[9]
[1]Department of Agronomy, Bidhan Chandra Krishi Viswavidyalaya
(BCKV), Mohanpur, Nadia, West Bengal, India
[2]Department of Agricultural Entomology, Bidhan Chandra Krishi
Viswavidyalaya (BCKV), Mohanpur, Nadia, West Bengal, India
[3]Assistant Technology Manager, Department of Agriculture, Howrah,
West Bengal, India
[4]Department of Genetics and Plant Breeding, Bidhan Chandra Krishi
Viswavidyalaya (BCKV),Mohanpur, Nadia, West Bengal, India
[5]Department of Agricultural Entomology, Jawaharlal Nehru Krishi
Vihwavidyalaya, Jabalpur, Madhya Pradesh, India
[6]Nadia Krishi Vigyan Kendra, Bidhan Chandra Krishi Viswavidyalaya
(BCKV), Gayeshpur, Nadia, West Bengal, India
[7]Department of Agronomy, ChatrapatiSahu Ji Maharaj University,
Kanpur, Uttar Pradesh, India
[8]Bangladesh Wheat and Maize Research Institute, Dinajpur 5200,
Bangladesh
[9]Botany and Microbiology Department, Faculty of Science,
South Valley University, Qena 83523, Egypt
*Corresponding author e-mail: maksudhasanshah@gmail.com

CONTENTS

DOI: 10.1201/9781003242963-6

6.1 INTRODUCTION

Citrus fruits are extensively cultivated around the world and are renowned for their delicious and refreshing juice, which has a number of health advantages (Okwu and Emenike 2006; Okwu 2008). They are rutaceaeous (fruit-bearing) trees that belong to the plant family Rutaceae. According to Okwu and Emenike (2006), the six kinds of citrus trees that produce citrus fruits are *Citrus sinensis* (sweet orange fruits tree), *Citrus reticulata* (tangerine fruits tree), *Citrus aurantifolia* (lime fruits tree), *Citrus limonum* (lemon fruits tree), *Citrus reticulata* (tangerine fruits tree) and *Citrus vitis* (the grape fruits tree). Citrus is grown on over 15 million hectares with production to the tune of about 157 million tonnes per year. The northern hemisphere contributes around half of the world's citrus area and production. The largest contributors to global citrus production are China (28%) and the Mediterranean areas (25%), followed by Brazil (13%). In the Mediterranean, Spain produces the most citrus (6 million tonnes), including exports of mandarins, limes, oranges and lemons. Brazil is the largest manufacturer of fresh sweet oranges as well as juice in the world. Mexico and India are two of the world's largest lime producers (Arata, Fabrizi et al. 2020). Pakistan has a small (1.6%) proportion of global citrus output, which includes major species such as mandarin, sweet oranges and minor species such as lemons, grapefruit and limes. Many biotic (citrus greening, CTV, sudden death, citrus canker, root weevils and *Phytophthora*) and abiotic (heat, water, salt and cold) stresses affect fruit crop production and yield in the global citrus industry (Mendonça et al. 2017).

6.2 ORIGIN AND DIVERSIFICATION

Citrus is the world's most popular tree fruit crop and one of the most diversified members of the Rutaceae family. Citrus classification has been a source of debate for a long time because of morphological diversity and intergeneric and interspecific sexual compatibility. However, genetic technologies have identified four species as the true parental species that have driven the enlargement of other species, namely mandarins, pummelos (*C. grandis*),citrons (*C. medica*) and wild cultivars of papeda (*C. micrantha*) (Garcia-Lor et al. 2013; Wu et al. 2018).Mandarin and citron are believed to have originated in northeast India and China, respectively, according to phylogenetic and genomic analyses. From these original citrus lines, random hybridization and natural mutation have resulted in the creation of limes, sweet oranges, sour oranges, lemons and hybrids (tangelos and tangors) (Oueslati et al. 2017).Numerous wild species, relatives, ancient and new varieties, and breeding lines are preserved in the vast citrus genetic resources collections of the United States, Japan, China, France, Spain and Brazil (Krueger and Navarro 2007).Citrus genetic resources are primarily preserved in India through the use of orchards or germplasm units at various academic and research institutions.

6.3 ABIOTIC STRESS

Climate change is causing weather changes such as severe droughts, cold waves, water logging due to excessive rainfall, to name a few. It may seem counterintuitive to believe that the greenhouse effect is responsible for cold waves in the face of climate change and global warming. There are numerous theories about its cause currently, but the fact remains that enhancement of extreme changes is occurring surrounded by seasons with regard to water regimes, such as long droughts followed by excessive rainfall or thermal imbalances, like winters with average temperatures 1–2°C above all-time highs, followed by extreme cold waves (Rosenzweig et al. 2001; Kodra et al. 2011; Cohen et al. 2013; Trouet et al. 2018). Therefore, new improvement initiatives are critical for developing new crops that are extra resistant to abiotic stressors. Plant breeding initiatives to combat abiotic stress are complicated because they involve a large number of genes. To carry them out, we must first have a thorough understanding of the crop in question, the effect of the studied stress on that crop and the mechanism of actions deployed by the plant to overcome the stress.

6.3.1 CITRUS CULTURE AND COLD STRESS

Globally, citrus is one of the most popular fruits, including lemons, limes, oranges and mandarins (mandarin, tangerine, clementine and satsuma are also part of the citrus industry). With the production of 17 and 10 million tonnes of oranges, respectively, Brazil and China were the world's largest orange producers in 2018. Grafting has been standard practice in agriculture for the propagation and development of various annual and arboreal plants. Farmers can use this technique to combine the shoot and root of two different plant species, to improve crop performance against a variety of issues, such as biotic and abiotic stressors during growth, low harvest yield, or just a shift in the harvest season to obtain better pricing for their harvest. Water and nutrient utilization and transport, hormone synthesis and transport, and large-scale movement of proteins, mRNAs and sRNAs are all affected by grafting (Warschefsky et al. 2016).When rootstock and variety are combined, the rootstock transmits unique traits that are beneficial to the variety. When grafted with a CTV-sensitive cultivar, *Poncirus trifoliata*, for example, is one of the most tristeza virus (CTV)-tolerant rootstocks, capable of imparting tolerance capacity (Castle 2010; Albrecht et al. 2012). To put it another way, rootstocks must be able to pass on a desirable feature to a variety, and these abilities will differ depending on the rootstock/variety combination (Forner-Giner et al. 2009; Martínez-Cuenca et al. 2019; Morales et al. 2021; Primo-Capella et al. 2021). The physiological causes of these graft-induced advantages are straightforward to understand, but it's also crucial to be able to explain them from a molecular angle and to provide a comprehensive picture of the long-distance communication between the rootstock and crop. Many transcripts can move between the rootstock and the variety, allowing for communication and long-distance regulation of many traits (Kim et al. 2001; Kudo and Harada 2007; Thieme et al. 2015). This has been demonstrated using techniques such as RNA sequencing. Understanding and addressing today's difficulties necessitates knowledge of genome transmissibility between both plants. Low temperature is one of the abiotic stresses that causes the majority of citrus losses. According to estimates from the AVA-Asaja (Valencian Association of Farmers), two consecutive frosts cost 142 million euros of losses in 2010. Citrus is a tropical and subtropical crop that can be damaged by cold weather (16–18⁰ Fahrenheit). The best-quality citrus fruits are cultivated in Spain (Forner-Giner et al. 2009), in temperatures ranging from 23 to 34^0 C (with a low of 13^0 C and a maximum of 39^0C, respectively). Temperatures that are outside of this range have the potential to be dangerous (Sakai and Larcher 2012). Citrus fruits are self-inductive crops, meaning they don't need any special climatic conditions to bloom, and they don't need lower temperatures to grow (Rebolledo Roa 2012). However, environmental factors such as a relatively low temperature regime in subtropical settings and water stress in tropical environments considerably speed up flowering and, as a result, production (Vu and Yelenosky 1993). Immature branches die at 12^0C, however certain citrus fruits can endure temperatures as low as 10^0C (Vu and Yelenosky 1991). The abundance of anthocyanins in both flavedo and pulp distinguishes varieties like Tarocco, Moro and Sanguinello, and these have the highest flavonoids content of any fruit or vegetable (Rapisarda and Giuffrida 1992). The content of these chemicals is influenced by a number of internal and external or environmental factors (light exposure, dietary balance, hormones, xenobiotics and temperature) that are both indigenous to the species (pigmented variety, maturation process, etc.) (de Pascual-Teresa and Sanchez-Ballesta 2008). Studying low-temperature stress with this type is particularly fascinating for these reasons. The cold sensitivity/tolerance of many citrus genotypes documented in Randall Driggers' group in Florida is shown in Table 6.1 (Oustric et al. 2017).

There have been few studies comparing grafted citrus under cold stress. Metabolomic and transcriptome studies have revealed that soon after exposure to cold stress certain mechanisms are activated in the plants which remain active in the long term. Abscisic acid (ABA) signaling may help Valencia sweet oranges grafted onto Carrizo citrus or *Citrus macrophylla* rootstocks to increase

TABLE 6.1
Citrus Types that are Sensitive/Tolerant to Cold Stress

Common Name	Scientific Name	Cold sensitive/tolerant
Large leaf Australian wild lime	*Microcitrus inodora*	Sensitive
Sunki mandarin	*C. sunki hort.* ex Tan.	Medium
Pineapple sweet orange	*C. sinensis* Osbeck	Tolerant
King tangor	*Citrus nobilis* Lour	Sensitive
Volkamer lemon hybrid	*C. volkameriana/C. limonia* Osbeck	Medium
Standard sour orange	*C. aurantium* L.	Tolerant
Lee mandarin	*C. reticulata* ("Clementine" x "Orlando")	Sensitive
Florida rough lemon	*C. jambhiri* Lush	Medium
Orange berry	*Glycosmis penthaphylla*	Tolerant
Australian round lime	*Microcitrus australis*	Sensitive
Alemow	*C. macrophylla* Wester	Medium
Swingle citrumelo	*C. paradisi* 'Duncan' (*P. trifoliata*)	Tolerant
Parson's special	*C. reticulata* Blanco	Sensitive
Citrus trifoliate	*Poncirus trifoliata*	Tolerant
Sun Chu Sha mandarin	*C. reticulata* Blanco	Medium

cold tolerance or sensitivity, according to hormone assessments and differential expression studies. This study sheds new light on the mechanisms by which rootstocks regulate grafted-on producing varieties' tolerance to abiotic stress (Primo-Capella et al. 2022).

6.3.2 Citrus's Molecular Reactions to Cold Stress

With *Citrus sinensis* (L.) Osbeck grafted onto *Poncirus trifoliata* rootstock, CAMTA genes are retained and interfere in the regulation of abiotic stresses like cold, drought and salinity, as well as to ABA and JA therapies (Rosenzweig, Iglesius et al. 2001). There was a three-fold induction of CsCAMTA1, CsCAMTA2, CsCAMTA3, CsCAMTA4, and to a lesser extent, CsCAMTA5 at 24 and 48 hours following cold stress treatment (Ouyang et al. 2019). MAPK has a conserved function, according to a study using cDNA amplified fragment length polymorphism (cDNA-AFLP) in *Poncirus trifoliata*. This team discovered a putative MAPK3 in *Poncirus trifoliata* and found that its expression was 17-fold higher at 4^0C for 17 hours compared to controlled expression (Meng et al. 2008). The CBF expression gene inducer, ICE1, is governed by a complex regulatory mechanism, as previously stated. Some citrus researchers are trying to discover what ICE1, a bHLH transcription factor, performs. Liu Jihong's team was able to identify a bHLB with a possible ICE1 function using *Poncirus trifoliate* (Huang et al. 2013; Huang et al. 2015; Geng et al. 2019). PtrbHLH overexpression in tobacco (*Nicotiana tabacum*) or lemon (*Citrus limon*) increased cold tolerance when subjected to chilling or freezing conditions but RNA interference (RNAi)-mediated downregulation of PtrbHLH in trifoliate orange resulted in significant cold sensitivity (Huang et al. 2013). In a second investigation, transgenic lines in lemon and tobacco that over-expressed putative ICE1 from *Poncirus trifoliate* showed higher antioxidant enzyme activity under cold circumstances, including superoxide dismutase and catalase. These data suggest that PtrICE1 promotes cold tolerance, possibly as a result of its interaction with the ADC gene, which regulates polyamine levels.

6.3.3 Physiological Response

Signaling causes plants to go through a number of physiological reactions. A drop in temperature that does not cause freezing harms plant tissues because membrane fluidity is lost as a result of an increase in the amount of unsaturated fatty acids that must adapt to innovative situations. Changes in the chlorophyll content, structure of chloroplast thylakoids, photosynthetic enzyme activity and electronic transport occur when temperatures drop. On the other hand, the link between physiological changes isn't always clear (Adam and Murthy 2014). These changes, along with stomatal closure, are assumed to be the primary cause of reduced photosynthesis in the winter (Driscoll et al. 1992). Ion leakage is caused by the inactivation of membrane-anchored channels and pumps (Verslues et al. 2006), which has an impact on every physiological function that takes place in the membrane, including organelle and cellular functions.

6.3.4 Citrus Membrane Alterations in Response to Cold

Previous studies suggest that citrus changes the shape of cell membranes and the concentration of lipids in response to cold stress. Many experiments carried out at low temperatures resulted in ion losses and membrane ruptures, jeopardizing the research (Nordby and Yelenosky 1985; Crifò et al. 2011; Yang et al. 2012; Abouzari et al. 2020; Jiang et al. 2021). George Yelenosky's group investigated the membrane alterations induced by cold stress in *Citrus sinensis*, *Poncirus trifoliata* and a hybrid which crosses (*Citrus paradisi* × *Poncirus trifoliata*) × *Citrus sinensis*. Phosphatidylcholine (89%), monogalactosyl diglyceride (79%), digalactosyl diglyceride (79%) and phosphatidyl glycerol (83%) all had higher rates of fatty acid breakdown (50%). In the cold-hardened hybrid, three kinds of triacylglycerols rich in linolenic acid increased and total leaf fatty acids rose by 12% throughout the freeze–thaw cycle (Nordby and Yelenosky 1985). This rise in highly unsaturated triacylglycerol in response to freezing–thaw stress suggested that triacylglycerol is involved in biomembrane fluidity (Nordby and Yelenosky 1985). Strong activation of genes involved in fatty acid production and phospholipid breakdown was seen in grapefruit (*Citrus paradisi*).The acyl-[acyl-carrier-protein] desaturase and the stearoyl-acyl carrier protein that may desaturase acyl-[acyl-carrier-protein] (Genbank CV706341 and CN182241CV706341CN182241, respectively) were specifically increased in the cold, whereas numerous fatty acid desaturases were downregulated. Blood orange and finger citron desaturase gene expression was similarly decreased (Crif et al. 2011; Yang et al. 2012).

6.3.5 Photosynthesis and Photo-Inhibition in Citrus under Cold Stress

Citrus clementina (Oustric, Morillon et al. 2017), Carrizo citrange (Oustric et al. 2017; Primo-Capella et al. 2022), *Citrus junos* (Jiang et al. 2021), Fortune mandarin and Ellendale tangor (Lourkisti et al. 2020), *Citrus deliciosa* and *Poncirus trifoliata* var. "Pomeroy" (Oustric et al. 2018), *Citrus medica* (fingered citron) and sweet orange (Primo-Capella et al. 2021) and satsuma mandarin (Barkataky et al. 2013) are just a few species of citrus where photosynthesis and photosystems are impaired. In all cases, photosynthesis (Pn) declines, stomatal closure (Gs) occurs, respiration declines, and internal CO_2 concentration (Ci) rises. Photo-inhibition happens as a result, and PSII's capacity to reduce Fv/Fm values is impaired.

6.3.6 Water Balance in Citrus under Cold Stress

The citrus water balance may be contradictory in the face of cold conditions. In extreme drought settings, when plants strive to avoid losing water through transpiration, low temperatures promote stomatal closure, reduced transpiration and a considerable fall in osmotic potential (Y). On warm sunny days, however, well-watered trees grown in commercial groves usually have the normal water potential Y stem and relative water content (RWC) values expected of actively growing plants (Barkataky al. 2013; Primo-Capella et al. 2021). In this circumstance, the transcellular water movement enabled by aquaporins is crucial. According to RNA-seq results obtained from Carrizo citrange and *Citrus macrophylla* grafted onto the *Valencia delta* seedless variety, the expression of putative aquaporins PIP1-2, PIP2-2 and PIP2-5 is higher in Carrizo citrange plants at low temperatures than in the low-temperature-sensitive rootstock *Citrus macrophylla* (Primo-Capella et al. 2021). This might be explained by improved transcellular water use and increased resistance to stress brought on by cold temperatures.

6.3.7 Osmoprotectors in Citrus under Cold Stress

Numerous studies in citrus have revealed that a conserved response to low-temperature stress is an increase in the content of the osmoprotectors. *Citrus junos* was utilized in a study that looked at sugar and starch production. The majority of differential abundance protein species (DAPS) involved in starch and sucrose metabolism were found using COG, KEGG and protein interaction network studies. Cold stress decreased the accumulation of proteins such as starch synthase (A0A067H6P7), 1,4-alpha-glucan-branching enzyme 1 (A0A2H5PA48) and glucose-1-phosphate adenylyltransferase (V4S7Z6), whereas polygalacturonase (A0A067H357) and mannitol dehydrogenase (A0A067H357) physiological data also demonstrated a significant increase in glucose, fructose and soluble carbohydrate levels (Jiang et al. 2021). In a related investigation, the PtrBAM1 gene was discovered in the *Poncirus trifoliata* rootstock and found to be overexpressed in tobacco plants. In comparison to citrus lemon plants, overexpression causes higher starch breakdown, maltose and soluble sugars (Peng et al. 2014). The PtrBAM1 promoter of the PtrCBF gene also has a recognition zone, according to bioinformatics research and yeast one-hybrid studies (Peng et al. 2014). Proline has been discovered to be very effective in the selection of resistant genotypes in citrus, in addition to providing cold tolerance (Rai and Penna 2013; Abouzari et al. 2020; Primo-Capella et al. 2021). Numerous investigations (Sanchez-Ballesta et al. 2000; Lo Piero et al. 2006; Crifò et al. 2011; Mohammadian et al. 2011; Perotti et al. 2015; Mohammadrezakhani et al. 2019; Sicilia et al. 2020), in addition to assessing anthocyanin and flavonoid contents, assessed gene expression of biosynthesis genes by comparing it across many citrus genotypes, have shown that the anthocyanin synthesis response to cold is well-known and preserved in citrus fruits. The anthocyanin production route in citrus has been studied by Angela Roberta Lo Piero's group.

6.3.8 Citrus Salinity Stress and Responsive Mechanisms

Citrus is grown in locations with minimal precipitation. In most cases, supplementary irrigation is required. Citrus is subjected to salinization regimes as a result of the use of low-quality irrigation water topped with the dry and hot conditions found across the citrus cultivation belt (Syvertsen et al. 2014). Plants are stressed by salinity when the Na^+ ion concentration in the soil solution surpasses 1500 ppm or 25 mM, which is the threshold quantity for the development of cultivated crops. Electrical conductivity (EC) above 3 dS m^{-1} and sodium adsorption ratio (SAR) above 9 in saturated soil extract are considered crucial for cultivation viability in citriculture. In addition, cultivating citrus is impossible if the chlorine content is more than 355 ppm (Simpson et al. 2014). Additionally, soil salinity above 2 dS m^{-1} has been shown to significantly affect citrus development and fruit output, with a fruit yield reduction of 13% noted for every 1 dS m^{-1} rise in salinity beyond 1.4 dS m^{-1}, the electrical conductivity threshold value for saturated soil extract (Murkute et al. 2005). Furthermore, ECs of 2.5–3.5 dS m^{-1} were described as the salinity threshold levels in the rhizosphere of orange plants cv. Valencia (Al-Yassin 2004).The hazardous threshold for salt stress syndromes in lemon trees cv. *Verna varies* depended on the rootstock utilized, however in sour orange, *Cleopatra mandarin* and macrophylla, the response threshold values were 1.53, 2.08 and 1.02 dS m^{-1}, respectively (Cerda et al. 1990).

6.3.8.1 Response to Salinity Stress

When exposed directly to salt, their physiology undergoes many modifications that are influenced by osmotic forces. Because of the gradient buildup of ions, these changes are abrupt and temporary, modifying plant water status and progressively causing toxic syndromes (Munns and Tester 2008). The presence of salt in the soil solution (García-Sánchez and Syvertsen 2006), hinders the growth of citrus leaves and roots by lowering the osmotic potential, which reduces the plant's water availability. There is a quick drop in the rate of leaf and root growth as an early reaction to the stress, but after activation of many physiological and biochemical systems, fast partial recovery in the case of leaf growth or total recovery of roots can occur (Acosta-Motos et al. 2017).Citrus plants, unlike halophytes, were thought to have insufficient capability to distribute salt ions into intercellular cell structures, resulting in poor osmotic alteration and dry stress (Ferguson and Grattan 2005). Recent research demonstrates that cell osmotic alteration is a vital response for plant life under salinity stress circumstances, even when detrimental ions are not adequately eliminated (Aragón and Rodríguez Navarro 2017). This suggests that in order to achieve proper osmotic adjustment, plants under salinity stress are more prone to increase root Na and Cl absorption and store these ions in plant sections, potentially resulting in long-term harm (Li et al. 2017).Citrus plants exposed to salt stress had a lower osmotic pressure than those exposed to drought stress, according to research using polyethylene glycol (PEG). This discovery leads to the conclusion that citrus plants utilize salt ions, which are mostly stored in the vacuole, to modify their osmotic balance and escape water stress (Ferguson and Grattan 2005).

6.3.8.2   Ion Toxicity Interplay under Salinity Stress

Under salty stress, hazardous contents of Na^+ and Cl^- ions build inside plant cells. Setting toxicity limits for these ions is difficult, since various aspects such as the kind of salts and the established rootstock/scion combination must be considered (Brito et al. 2014). In the root cell fluid, a concentration of 10–30 mM Na^+ has been deemed hazardous because it limits enzyme function, however, in the leaves, it can reach 100 mM in some situations (Tester and Davenport 2003). In terms of Cl^- ions, the quantities necessary to produce toxicity in citrus plant leaves begin at 0.7% dry matter (Ferguson and Grattan 2005). According to Al-Yassin (Al-Yassin 2004), the major source of toxicity in most plants is Na^+, which tends to collect in the woody roots and trunk, whereas Cl^- accumulates mostly in young shoots and leaves, creating necrotic lesions. Citrus plants exposed to mild salinity stress mostly show an osmotic-driven drop in fruit output, with no visible signs of severe toxicity owing to the buildup of Na^+ or Cl^- ions. Citrus trees build excessive quantities of Na^+ and Cl^- ions in the canopy during extreme salinity stress, reaching hazardous levels and substantially deregulating the photosynthetic system and tree development (García-Sánchez et al. 2002).The ability of the rootstock to remove ions from shoot tissues is one of the most critical components that require attention. It is important to note that Cl^- is not regarded as more metabolically harmful to citrus trees than Na^+. Citrus trees, like the majority of woody perennial plants, may store Na^+ in the woody root-sphere and basal stem sections of the plant and exclude it from the leaves by xylem retrieval (Li et al. 2017). As a result, the residual Cl^- ions in the saline solution become the most damaging and poisonous element (Munns and Tester 2008). As a result, the physiological framework that must be built in order to investigate citrus resistance to saline stress is connected to the plant's capacity to restrict Cl^- ion transit from the root to the scion, a mechanism closely controlled by the rootstock (Brumos et al. 2010). Swingle citrumelo rootstock's capacity to retain lower levels of Na^+ in leaves than rough lemon is due to the former's ability to sequestrate Na^+ in vacuoles of root tissue and immobilize it within the cell membrane (Gonzalez et al. 2012). Cl^- ions are implicated in the detrimental effects of leaf necrosis, growth halt and leaf abscission in citrus, according to scientific evidence (Li et al. 2020).Endogenous levels of phytohormones like ABA and 1-aminocycloprpane-1-carboxylic acid (ACC) induce leaf abscission, which increased gradually following the development of the salt stress factor (Gómez-Cadenas et al. 2003). Other compounds, such as polyamines, have also been postulated as signaling molecules during citrus plant adaptation to salt stress (Tanou et al. 2014).

6.3.8.3   Genetic Approaches to Improve Salinity Stress

The discovery of natural variants via direct selection or the mapping of quantitative trait loci (QTL) can both be used to enhance salt tolerance in citrus plants. Due to the limited effectiveness of direct selection in open fields, breeders have concentrated their efforts on identifying genes and gene products that may be incorporated to generate cultivars using genetic transformation and marker-assisted breeding (Singh et al. 2018). QTLs are sections of DNA associated with phenotypic variation of a quantitative characteristic in a population of animals (Myles and Wayne 2008).Such quantitative changes in plants might be the consequence of numerous genes and environmental cues working together. Crossing two parents that vary in one or more quantitative characteristics in order to find candidate genes underlying the desired trait is known as QTL analysis (Singh et al. 2017). In a hybrid citrus population [(*Cleopatra mandarin* (salt tolerant) trifoliate orange (salt sensitive)], a total of 98 QTLs putatively associated with salinity resistance were discovered. A cluster of QTLs affecting plant vigor and leaf boron content identified a chromosomal area in linkage group 3 as the most significant for improving salt resistance using Cleopatra mandarin as a donor (Raga et al. 2016).

6.3.9   CITRUS DROUGHT STRESS AND RESPONSIVE MECHANISMS

Among all the abiotic stresses on the physiology and productivity of plants, drought stress is considered to be one of the most detrimental globally. Drought affects normal development, disrupts water interactions and lowers the efficiency with which plants utilize water. Plants, on the other hand, exhibit a wide range of physiological and biochemical reactions at the cellular and organismal levels, making it a more complicated phenomenon. The photosynthetic rate decreases due to stomatal closure, membrane damage and disrupted activity of numerous enzymes, particularly those involved in ATP generation (Berthomieu et al. 2003).

Drought stress causes a considerable reduction in growth and cellular metabolism in citrus trees, as well as a drop in crop output and fruit quality (Pérez-Pérez et al. 2008). Physiological parameters such as stomatal conductance (Gs), net CO_2 assimilation (ACO_2) and leaf transpiration (E_{leaf}) when stressed by drought (Syvertsen and Garcia-Sanchez 2014). The genotype of the plant also plays an integral role in determining citrus' capacity to deal with the negative effects of drought resulting in the following drought resistance ranking: strong resistance: mandarins (*Citrus reticulata* spp.) > rangpur lime > rough lemon > sour orange >*Citrus macrophylla*; lemon > trifoliate orange > citrange hybrid >*Citrus chuana*; sweet orange >*Citrus verrucose*> grapefruit; medium resistance: lemon > trifoliate orange > citrange hybrid >*Citrus chuana*; poor tolerance: sweet orange >*Citrus verrucose*> grapefruit.

6.3.9.1 Drought Stress Resistance Mechanisms

Plants adapt to drought stress by changing their morphological, physiological, anatomical and biochemical characteristics. Drought triggers the plant's initial response to reduce stress, which prevents fluids or toxic ions from accumulating in sensitive leaf tissues. In the event of mild stress or short-term stress, avoidance strategies alone may be sufficient to preserve plant function (Verslues et al. 2006). The ability of the plant to retain tissue water potential (Y_w) and water content at the unstressed level by either limiting water loss or increasing water intake is called whole-plant water dehydration. Stomatal closure control is a basic short-term avoidance mechanism. In the long term, the most essential elements for agricultural plants may include increasing cuticular resistance, leaf rolling flexibility, lowering leaf biomass, raising the root/ shoot ratio by generating a deeper and thicker root system and controlling root water conductivity (Gowda et al. 2011; Hu and Xiong 2014). Osmotic moderation and cell wall hardening reactions are linked to cell dehydration avoidance mechanisms. As varied responses show that the accumulation of osmolytes might depend on the severity and duration of the stress and also on the genotype, osmotic adjustment through control of leaf osmolytes seems to be distinct from many other plants (Pérez-Pérez et al. 2009). With the exception of glycine betaine, osmoregulation is achieved by the production of suitable osmolytes, viz., proline and other betaines (García-Sánchez et al. 2007).Citrus is capable of changing the flexibility of the cell wall, which is one of the mechanisms for drought resistance. The pressure potential is decreased by increasing the flexibility of the cell wall, leading to reduction of the water potential while preserving the cell's turgor pressure due to shrinkage of the cell (Ruiz-Sánchez et al. 1997). The plant's reaction to drought is influenced by leaf age in citrus since older leaves contain traits allowing them to cope more efficiently with drought than younger ones (Ruiz-Sánchez et al. 1997). Controlling stomatal opening is the most important component for citrus tree survival when they are stressed by drought.

Antioxidant defenses are widely recognized as critical for abiotic stress tolerance. Another key mechanism of plant cell tolerance produced by water stress is the accumulation of defensive proteins like HSPs and hydrophilins. In vegetative tissues, HSPs and hydrophilins are present in trace quantities but high ABA-dependent transcriptional activation due to osmotic stress results in a considerable accumulation of these proteins (Colmenero-Flores et al. 1997; Reyes et al. 2005). Late embryogenesis abundant (LEA) proteins, for example, are hydrophilins that assist cells to survive protoplasmic water deprivation. Hydrophilins and HSP protection proteins have been found to have a role in the response of citrus plants to water stress in several studies (Podda et al. 2013; Pedrosa et al. 2015; Xiao et al. 2017) (Table 6.2).

TABLE 6.2
Citriculture and Salinity/Drought Stress Highlights

Salinity Stress	References
Citriculture and salinization limits	
1. Values of sodium adsorption ratio (SAR) above 9and electrical conductivity (EC) above 3 dS m^{-1} in saturated soil extract are characterized as crucial for the survival of the culture.	Simpson et al. 2014
2. The chlorine ion levels in citrus plant leaves that must be present to be toxic begin at 0.7% dry matter.	Ferguson and Grattan 2005
3. Citrus plants cannot be grown in areas where the chlorine content is greater than 355 ppm.	Simpson et al. 2014
Ion toxicity interplay	
1. Citrus trees accumulate enormous amounts of Na$^+$ and Cl$^-$ ions into the canopy when under extreme salinity stress, reaching hazardous levels and significantly deregulating the photosynthetic system and tree development.	García-Sánchez et al. 2002
2. It has been suggested that polyamines act as signaling molecules while citrus plants are adjusting to salt stress.	Tanou et al. 2014
3. Sequester Na$^+$ in the vacuoles of the root tissue, where it is then immobilized by the cell wall.	Gonzalez et al. 2012
4. When exposed to mild salinity stress, fruit production decreases mostly owing to osmotic forces; there are no observable symptoms of severe toxicity brought on by the buildup of Na$^+$ or Cl$^-$ ions.	García-Sánchez et al. 2002
5. Abscisic acid (ABA), a phytohormone, and 1-aminocyclopropane-1-carboxylic acid act as triggers for leaf abscission (ACC).	Gómez-Cadenas et al. 2003
6. a$^+$ is stored in the woody root-sphere and basal stem regions, and by xylem retrieval, it is kept out of the leaves.	Li et al. 2017
Response to salinity stress	
1. Inability to distribute salt ions into intercellular cell structures effectively.	Ferguson and Grattan 2005
2. To prevent a water shortage, modify salt ions' osmotic properties.	Ferguson and Grattan 2005
3. To accomplish adequate osmotic adjustment, encourage root Na$^+$ and Cl absorption and accumulate these ions to plant parts.	Li et al. 2017
Amelioration mechanisms	
1. Exogenous proline (5 mM) was used to dramatically reduce the detrimental effects of salinity stress in salt-sensitive orange cv. Late in Valencia.	Lima-Costa et al. 2010
2. The Na$^+$ and Cl exclusion mechanism is a heritable trait, leading to the development of citrus hybrids that effectively exclude salt ions and outperform their parent genotypes.	Forner-Giner et al. 2009
3. By preventing Cl$^-$ buildup in leaves, pretreatment with ABA lowers ethylene release and leaf abscission.	Li et al. 2020
4. Tetraploidy citrus seedlings exhibit stronger salt resistance than diploid genotypes and genetic ploidy level promotes the relative salt tolerance of rootstocks.	Ruiz et al.; 2016
Genetic approaches	
1. Quantitative trait loci (QTL) mapping or direct selection may be used to identify natural variations.	Raga 2016
Drought stress	
1. Diminish the physiological parameters such as net assimilation of CO_2 (ACO2), leaf transpiration (Eleaf) and stomatal conductance (gs).	Syvertsen and Garcia-Sanchez 2014

6.3.9.2 Agricultural and Irrigation Practices that Cope with Salinity and Drought in Citrus

Figure 6.1 shows citriculture farming approaches that could greatly reduce the harmful impacts of salinity or drought stress. The amount of water required by citrus trees varies depending on their age, size, citrus variety, climate and soil type. According to a study, mature citrus (orange) plants require around 4000–5000 m^3 of water per hectare per year (Carr 2012; Kourgialas et al. 2015). In citriculture, three primary irrigation methods are used: micro-sprinkler and sprinkler methods, which are suitable mostly for sandy soils, and drip irrigation, which is one of the best technological approaches owing to water economy. The irrigation schedule should be planned according to the soil moisture content, which may be assessed by augering soil samples or using tensiometers. For mature citrus trees, installing tensiometers in the midpoint between two emitters along the irrigation channel (0.20 m from the emitter, 1 m from the trunk) is advised. Tensiometers should be installed at a distance of 1 m from the sprinkler or 0.5 m in the case of a micro-sprinkler (Ziogas et al. 2021). Citrus has a rather shallow root system. Therefore, it's necessary to deliver irrigation water at the effective root zone (normally up to 30 cm soil depth) to avoid deep percolation of water. There is a key matric potential minimal threshold at that depth that must be maintained. Typical thresholds for sandy soils are 20 kPa and for clayey soils are 100 kPa (Kourgialas et al. 2019; Morianou et al. 2021). Although citrus is relatively resistant to saline water, salts can build up in the soil or on the plant, causing root dieback and leaf loss. On poorly drained clay or silt soils, salinity will always be a bigger issue than on permeable sandy or gravelly soils. If saline water is used for irrigation through the commonly used drip or micro-sprinkler system, it must be ensured that water does not come in contact with the leaves, which would otherwise lead to the burning of leaves. Also, salts can clog the emitter orifices. Frequent shallow irrigations may also lead to the deposition of salt on the surface of the soil (in the form of white crust) and rhizosphere. These accumulated salts can be removed through leaching by applying a huge amount of water once or twice a year (Simpson et al. 2014). Under salt stress and water deficiency situations, relief strategies have been evaluated to boost yield. To address this issue in a commercial orchard, it is first necessary to exercise careful control over irrigation

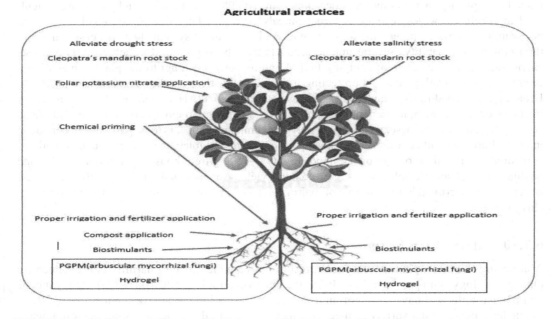

FIGURE 6.1 Mitigation of drought and salinity stress in plants through agricultural practices.

and fertilizer treatments. Other horticultural approaches that have improved the performance of citrus under drought and/or salinity stress conditions include the use of orange types as interstocks, the use of shade, hydrogels and treatments with persistent analogues of phytohormones, abscisic acid and polyamines (Gimeno et al. 2014). The impact of adding calcium or nitrates in the dietary solution, as well as therapies to reduce salt stress, has been studied. Plants reject Na^+ more efficiently and avoid its buildup in cells when there is an adequate supply of calcium (Ca^{2+}).It has been demonstrated several times that nitrate and other nitrogen-derived substances such as urea or ammonium have a favorable influence on citrus development (Álvarez-Aragón and Rodríguez-Navarro 2017). Furthermore, nitrate appears to have two distinct effects on citrus seedling performance in saline environments. First, nitrate supplementation was shown to increase photosynthesis and growth while decreasing leaf abscission. Second, the nitrogen-induced increase in leaf biomass has caused chloride dilution, which is a key element in salt damage (Khoshbakht et al. 2014). In citrus plants, reactive nitrogen species (RNS) such as nitric oxide (NO) and reactive oxygen species (ROS) such as hydrogen peroxide (H_2O_2) cause hardening toward salt and drought (Morianou et al. 2021). These chemical agents, like sodium hydrosulfide, sodium nitroprusside, polyamines and melatonin, have the potential to improve abiotic stress resistance in the field (Kostopoulou et al. 2015; Pedrosa et al. 2015; Ziogas et al. 2015). Furthermore, many farmers now utilize biostimulants as part of their farming practices. The usage of these chemicals has been shown to protect plants from abiotic stress factors and promote the overall development of plants (Van Oosten et al. 2017). These calming effects are achieved by the coordinated activity of plant metabolites like phytohormones, proline, carbohydrates, amino acids and other compounds whose synthesis is stimulated by the biostimulants (Ji et al. 2017). Recent scientific evidence suggests that the use of biostimulants enhance sroot growth or modulates the plant cell osmoregulatory system in citrus (Conesa et al. 2020).When watered with 50% restitution of evapotranspired water, orange trees (*Citrus sinensis* L.) had enhanced water relations and greater water usage efficiency (WUE) when sprayed with a commercial extract of *Ascophyllum nodosum* (Spann and Little 2011). Biostimulants are regarded as an agricultural strategy that may help to alleviate drought stress and boost WUE in citrus crops, particularly in drought-prone areas where citrus trees are economically significant but water resources are restricted owing to urbanization and ongoing climate change (Van Oosten et al. 2017). Compounds loaded with plant-growth-promoting microorganisms (PGPMs)are a unique and promising agricultural approach. The widespread use of agrochemicals has resulted in a significant loss of soil quality, necessitating the development and implementation of agricultural systems that will ensure agricultural output sustainability. Several unique products have been produced to achieve this purpose, encompassing the technique of applying PGPM and mycorrhizal fungi to the plant's rhizosphere, improving plant development and protecting it from abiotic stress conditions (Kumar et al. 2017). Drought stress conditions stimulated H^+-ATPase activity and PtAHA2 gene expression, resulting in enhanced root growth, nutrient uptake, and a lower soil pH microenvironment in one study (Zou et al. 2021). Apart from these, the hydrogel polymer component appears to be particularly beneficial in citriculture as a soil conditioner, enhancing crop growth and tolerance under drought situations. A medium or high dose of hydrogel composite (1000–1500 g/tree) boosts overall quantitative and qualitative (total soluble solids and sugars content) yield (Abobatta and Khalifa 2019). This might be related to the critical long-term role of hydrogels in enhancing water and nutrient availability for citrus crops.

6.3.10 HEAVY METALS STRESS

Abiotic stress inhibits plant development, resulting in a wide range of harmful consequences in plant physiology, with photosynthesis being the most vulnerable (Polle et al. 1993). Heavy metals that persist in the soil have an important role in plant stress induction (Azab and Hegazy 2020). Plants have developed a variety of mechanisms to cope with this type of stress, including heavy metals binding to cells and the synthesis of powerful ligands called phytochelation, sequestration in cells and intracellular compartments, efflux pumps, etc. The majority of heavy-metal-resistant plants inhibit heavy metal buildup in the tissues (Baker 1981).Toxicity symptoms revealed that Cu is more toxic to plants than Pb, and root growth is more vulnerable than shoot growth according to research. Cu generally does not affect the highest dark-adapted quantum yield of PSII, calculated as Fv/Fm (Ouzounidou et al. 1997), according to recent studies Cu stress experiments with ecologically realistic Cu concentrations (Ouzounidou 1993). Cu and Pb poisoning had a deleterious impact on shoot growth. The loss of K^+ from root cells caused a reduction in shoot length in maize seedlings following Pb poisoning (Ouzounidou, Moustakas et al. 1997). The same process is most likely to be operational also in the case of *Citrus aurantium* L. plants. Pb buildup in roots or unknown signaling from the roots to shoots due to Pb or Cu exposure are another two proposed pathways (Małkowski et al. 2002). Cu and Pb at 800 µM impede chlorophyll and carotenoid production and prevent these pigments from being incorporated into the photosynthetic mechanism. The measure of chlorophyll content has been recommended as a good *in vivo* biomarker of toxicity to heavy metal stress for estimating upper critical tissue concentrations. One study (Małkowski et al. 2002) discovered that at high heavy metal (Zn) concentrations, reducing Chla content while increasing total phenolic content in leaves of K. Cu stress caused dramatic changes in the chloroplasts of oregano plants (Heckathorn et al. 2004). As a result, chloroplasts in normal leaves were larger and more numerous, but chloroplasts in Cu-treated leaves showed extensive plastoglobuli and organelle dilatation, restricting the double membrane. The leaves of Cu + Pb-treated *Citrus aurantium* L. plants show a similar pattern, according to earlier reports. This decrease in chlorophyll and carotenoid content in *C. aurantium* L. plants exposed to Pb and Cu was viewed as a particular plant response to metal stress, resulting in Chl breakdown and photosynthesis suppression (Chen et al. 2019). Photosynthesis is one of the most vulnerable physiological pathways to Pb and Cu poisoning (, and the consequences are multi-faceted, reducing photosynthetic CO_2 fixation both *in vivo* and *in vitro*. PSII is known to be quite sensitive to a variety of different environmental conditions. For higher plants, Cu is an important micronutrient aiding in photosynthesis (Ahmad et al. 2008). Cu is a component of the major electron donor in plants' PSI as well as enzymes involved in superoxide radical elimination. There is a significant correlation between Pb treatment and a reduction in net photosynthesis in the plant, which is thought to be due to stomatal closure rather than a direct action of Pb in the photosynthesis process (Küpper et al. 2002). Cu and Pb poisoning adversely impact the photosynthetic process. Plants exposed to both metals showed a decrease in the rate of photosynthesis, which may be related to deformed chloroplast structure, limited chlorophyll and carotenoids photosynthesis, Calvin cycle suppression and CO_2 deficit owing to stomatal closure (Zhang et al. 2018). Furthermore, Pb hinders chlorophyll production by preventing plants from absorbing critical minerals like Mg or Fe (Moustakas et al. 1994). In *Citrus aurantium* L. plants, a high Pb concentration in the nutritional substrate generates a mineral nutrient imbalance. Pb also causes dissociation of the oxygen-evolving extrinsic polypeptide of PSII and displacement of Ca, Cl or Mn from the oxygen-evolving complex (. As a result, numerous variables influence photosynthetic activity (Sharma and Dubey 2005), including stomatal conductance, which decreased significantly as Cu + Pb concentrations rose. Cu poisoning, which causes a decrease in PSII effectiveness, is linked to chloroplast thylakoid membranes (Kosobrukhov et al. 2004). The reduction in the assimilation rate of *C. aurantium* L. plants treated with Pb and Cu was probably due to stomatal closure since Pb and Cu treated plants were accompanied by a lower Gs as well as transpiration rate, especially those growing at higher levels of Pb and Cu (800 µM). After combined Pb and Cu treatment at the highest level (800 µM), the negative effect was most significant in all photosynthetic metrics. *Citrus aurantium* L. plants exposed to Pb have a decrease in transpiration rate and water usage efficiency. Pb and Cu treatment slows development, resulting in a smaller leaf area, which is the primary organ for transpiration. The activation of oxidative stress in developing portions of *C. aurantium* L. due to increased formation of ROS appears to be one of Pb's phytotoxic effects (Malecka et al. 2009; Chaneva et al. 2010). During oxidative metabolism, a large variety of distinct ROS are formed, including superoxide ion ($O2^{\cdot-}$), hydrogen peroxide (H_2O_2) and the hydroxyl radical (OH^-). In earlier reports, the production of H_2O_2 and its distribution in leaf tissue aided in the elucidation of photoprotective mechanisms against Pb- and Cu-induced oxidative stress (Pandey et al. 2009). Plants often overproduce ROS due to heavy metal stress, which are extremely reactive and toxic and damage carbohydrates, proteins and lipids (Jubany-Marí et al. 2010). Pb causes the generation of ROS in plants, which is dependent on the severity of the stress. Pb causes lipid peroxidation and lowers fatty acid levels (Gill and Tuteja 2010). Furthermore, antioxidant content was shown to be higher in the leaves of *C. aurantium* L. plants treated with Pb and Cu. These techniques are important in protecting the photosynthetic mechanism from oxidative damage. Plants that generate large amounts of phenolic compounds in response to heavy metal stress have also been found to be suitable candidates for phytoremediation (Vidal et al. 2020). As a promising method for removing polluted cultivated areas, phytoremediation can assure the consumption of agricultural products like fruits, which are an essential component of the human diet.

6.4 CITRUS RESEARCH USING OMICS

Over the past two decades, extensive studies known as omics have been utilized to mimic plant studies, making a substantial contribution to plant science. Figure 6.2 depicts the "omics" conceptual framework. A genome is a complete haploid set of chromosomes in an organism, whereas a phenotype is an organism's outward appearance (phenotype). The whole set of RNAs and proteins derived from the genome are referred to as the transcriptome and proteome, respectively. Every biological sample such as a cell, organ, tissue or an entire organism, contains metabolites, hormones and ions, which are referred to as the metabolome, hormonome and ionome, respectively. Genomic, proteomic, metabolomic, transcriptomic, hormonomic, phenomic and ionomic analyses are all referred to as genomics, proteomics, metabolomics, transcriptomics, hormonomics, phenomics and ionomics, respectively. Horticultural crops have a limited amount of information available. In this chapter, we examine recent omics research on citrus trees.

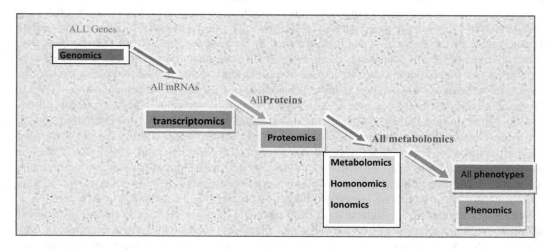

FIGURE 6.2   Omics studies framework.

6.4.1 GENOMICS

6.4.1.1 Citrus

Many modern genomic approaches, such as sequence determination of DNA, marker-assisted selection and the transfer from marker-assisted selection to genomic selection, aid in the rapid creation of varietals. It has also revolutionized citrus breeding, allowing researchers to better understand the link between genetic make-up and response to diverse abiotic and biotic stimuli (Hamblin et al. 2011). Butelli et al. (2012) developed a PCR assay for pulp anthocyanin concentration, Wang et al. (2017) linked polyembryony to AFLP markers, and dwarfism and fruit acidity to RAPD markers. QTLs have also been connected to other traits such as salt tolerance and worm resistance (Ben-Hayyim et al. 2007). The newly created SNP08 marker boosted the selection of resistant genotypes at the early growth stage. The ABD resistance gene is 0.4 cM away from this marker and has a function in preventing vulnerable types from being selected. New markers have been discovered on the other side of the gene, 0.7 cM from the ABSr locus. When these novel markers were combined with SNP08, the likelihood of selecting a resistant genotype increased by 0.0028%. This marker proved to be very helpful in determining whether genotypes are resistant or vulnerable, as well as in the analysis of resistant germplasm in order to assemble the ABS genes. As a result, it's a highly useful technique for finding vulnerable heterozygous cultivars that can be utilized as breeding parents, allowing for genetic variety modification in citrus and preventing susceptible homozygous genotypes. The SNP08 marker was utilized to assess about 40 mandarin genotypes (sensitive and resistant), and they were employed as breeding parents. The response to alternaria infections was found to be linked to the SNP08 marker in the end. SNP08 was recently employed in citrus breeding operations at CIRAD and IVIA to create ABS-resistant citrus genotypes. Since its inception, the SNP08 marker has been used to select about 2187 resistant hybrids from 4517 total hybrids resulting from 10 different parental pairings. This analysis was extremely useful in preventing the growth of over 2000 vulnerable lines that were deleted at an early stage of growth following selection, saving a significant amount of time, money, employees and resources.

6.4.2 TRANSCRIPTOMICS

Transcriptomics is the study of RNA (transcripts) in a cell, tissue, organ or organism. Several citrus transcriptomics studies, concentrating on citrus greening ([huanglongbing (HLB)] disease (Mafra et al. 2013; Martinelli et al. 2012; Martinelli et al. 2013), as well as a combination of proteomics studies, have been published (Mafra et al. 2013; Martinelli et al. 2012; Martinelli et al. 2013; (Fan et al. 2011; Zhong et al. 2015).A spontaneous late-ripening sweet orange mutant's transcriptome and proteome were analyzed (Wu et al. 2014b; Zeng et al. 2012; Zhang et al. 2014b). Several citrus ripening episodes or roles for abscisic acid (ABA), jasmonic acid (JA) and sucrose in citrus ripening have been hypothesized based on these omics investigations. Sun et al. (2012) and Zhang et al. (2011a) published transcriptomics of an early blooming tyrifoliate orange mutant. On an early-blooming tyrifoliate orange mutant, researchers presented multi-omics (transcriptomics and proteomics) data. Sun et al. (2012) used transcriptomics to study micro-RNAs (miRNAs), which are small non-coding RNAs (ncRNAs) that control posttranscriptional gene regulation and RNA silencing. The most common method for comprehensively analyzing ncRNA is RNAseq with NGS. Nishikawa et al. (2010) used microarray analysis to demonstrate that early-blooming transgenic tyrifoliate oranges expressing Flowering Locus T from citrus (CiFT) showed greater levels of expression of MADS-box transcription factors involved in floral organ assembly. When boron levels are low, corky split veins develop. Citrus boron deficiency has been reported by transcriptomics (Yang et al. 2013; Lu et al. 2014). The cytokinin signaling pathway was implicated in the corky split vein in the previous investigation. On the other hand, later research that was miRNA-focused suggested that miRNAs were involved in the body's reaction to boron deprivation.

(Voo et al. (2014) developed a method for extracting orange fruit peel essential oil gland cells for single-cell transcriptomics, which has lately become a hot issue in biology. Citrus transcriptomics has also been published on physiological fruit abscission, or "self-pruning," polyploidy, somatic hybridity, red fruit flesh reaction to ethylene in fruit and GA treatment on the bud (Zhang et al. 2014a; Goldberg et al. 2013).

6.4.3 Proteomics

The branch of science dealing with the study of proteins in an organism, tissue, organ or cell is known as proteomics. About 1500 proteins were identified and quantified by Katz et al. using a proteomic method in citrus fruit juice sac cells at three developmental stages. A complete sequence library of citrus genes, ESTs and proteins, dubbed i Citrus, has been constructed as a protein database for this proteome research. Citrus fruits undergo color change through development, going from green to yellow or orange on the peel. Chlorophyll present in the chloroplast gives the fruit its green color and develops into chromoplasts with the aging of the fruit and accumulates carotenoids. A total of 493 proteins, including those involved in the production and storage of carotenoid in the chromoplast, were found in sweet orange. Numerous citrus species exhibit "biennial bearing," also known as "alternate bearing," which refers to the occurrence of significant fruit load (on crop) and less yield (off-crop) in alternate years. In an attempt to unravel the mechanism of biennial bearing, they examined 2D-PAGE pictures of on-crop and off-crop proteins, and then used a proteomic technique to identify the differently expressed proteins. The findings revealed that the proteins involved in fundamental metabolic processes, such as glucose metabolism, were more prevalent in the buds and leaves of off-crop trees. The transcriptomes of buds from on-crop and off-crop trees were compared by Shalometal (2014). Despite the fact that their hormone testing revealed a decline in both abscisic acid and auxin levels in off-crop tree buds, the outcomes were visible. In off-crop trees, auxin polar transport is controlled by downregulating genes involved in ABA metabolism and upregulating genes involved in photosynthesis.

6.4.4 Metabolomics

Metabolomics deals with the study of metabolites in an organ, tissue, organism or cell. Lin et al. (2015) used metabolomics and transcriptomics to study ponkan fruit growth and post-harvest preservation of pumelo fruit (Sun et al. 2013). In an earlier study, sugar build up and organic acid break down were identified in fruit maturation according to alterations in metabolic enzymes. Enhancements in succinic acid, glutamine and gamma-amino butyric acid (GABA), as well as a decrease in oxo-glutaric acid, were observed in later research. Surprisingly, both studies showed that the GABA shunt is essential for the oxidation of fruits' organic acids. "Puffing," or the disintegration of albedo (the white component under the peel) and separation of pulp and flavedo, is one of the physiological problems of citrus fruit (outer colored part of the peel). Transcriptome and metabolome investigations were done to investigate the process of puffing (Ibanez et al. 2014). Multiomics analysis of puffed fruits revealed substantial modifications in primary metabolism, a decrease in citric acid content and downregulation of GA and cytokinin signaling. In triggered navel orange fruits, Ballester et al. (2013) looked at the albedo and flavedo metabolic profiles. The results indicated that *Penicillium* spp., the most prevalent citrus post-harvest infections, are induced to become resistant to biphenyl propanoids and their derivatives (Ballester et al. 2013). A comparison of the metabolomes of several citrus species and cultivars with different levels of sensitivity to the HLB disease revealed that leaves of the most susceptible sweet orange varieties contained higher levels of organic acids, amino acids and galactose (Cevallos-Cevallosetal. 2012). Fruits with symptoms and those without symptoms differ in their concentrations of histidine, phenylalanine, synephrine and limonin. According to Chin et al. (2014), metabolomics has the potential to create "metabolite-based biomarkers" of important characteristics in crops, such as resistance to HLB in citrus (Chin et al. 2014).

6.4.5 OTHER OMICS STUDIES

Metabolomics includes hormoneomics. Plant hormone concentrations are generally quite low as compared to other metabolites. As most phytohormones are not detectable using standard metabolome analysis, they must be analyzed using a specific MS detection technique. Hormonomics is not frequently employed in plant science or studies on fruit trees. A thorough examination of plant hormones in fruit trees has only been reported by Oikawa et al. (2015) using LC-triple QMS to determine fluctuations in 15 plant hormones in the fruit development of the European pear, including indole acetic acid, salicylic acid, ABA, GA1, GA4, JA, JA isoleucine conjugate, brassinolide (BL), isopentenyl adenine, castasterone (CS), dihydrozeatin and trans-zeatin. The hormonome data obtained can be found in the "Fruits Omics Database." The youngest fruit had the highest levels of all plant hormones, which thereafter sharply decreased. In the ripening stage, only the concentrations of ABA, CS and BL increased. Although an increase in ethylene and ABA levels is common during the ripening stage of fruit, a similar increase in BL or CS has almost never been detected. BRs, like BL and CS, may be associated with fruit ripening in European pear. Several processes in plants are regulated by plant hormones, including fruit formation. The study of hormones will provide light on the functions of phytohormones in the growth and development of fruit trees. Ionomics is the study of inorganic ions in an organism, organ, tissue or cell. Ionomics investigations are also uncommon in the realm of fruit tree study. A few fruit tree ionomics investigations have been published recently (Parent et al. 2013a, 2013b). An ionome study of European pear fruit was reported by Reuscher et al. (2014), and the data were released from the "Fruits Omics Database." Seventeen elements are required for plant growth, including C,H,N,O, Cl, Cu, Fe, K, Ca, Mg, Mn, Mo, P, S, B, Ni and Zn. Excess amounts of some elements, such as Na, Al and Cd, are detrimental to both plants and humans. As a result, ionomics is an important approach for monitoring both plant health and human food safety.

6.5 BIOTIC STRESS

Several biotic stresses harm citrus trees, including viral, bacterial and fungal diseases, as well as nematodes, phytoplasmas, spiroplasmas, viroids and some infections that can be transmitted through grafts. All of these pathogens can have a significant impact on a variety of biochemical and metabolic changes that are critical to citrus defense mechanisms (Cevallos-Cevallos et al. 2011,2012; Slisz et al. 2012; Malik et al. 2014; Killiny et al. 2016, 2017; Rosales et al. 2011; Nehela et al. 2018, 2019). Volatile organic compounds, amino acids, organic acids (Cevallos et al. 2011; Slisz et al. 2012; Killiny et al. 2017), fatty acids (Cevallos-Cevallos et al. 2012; Killiny et al. 2017), phytohormones (Rosales et al. 2011; Nehela et al. 2018), polyamines (PAs) (Malik et al. 2014; Nehela et al. 2019) and other secondary metabolites are among the metabolic alterations. It is currently unknown if these alterations represent a metabolic reaction of the plant, a response to aiding citrus pathogens or a combination of the two.

6.6 EFFECTS OF SOME BACTERIAL, VIRAL, ETC. DISEASES ON CITRUS PLANTS

6.6.1 CTV DISEASE OF CITRUS

One of the most significant commercial citrus viral diseases is citrus tristeza, which is brought on by the virus that causes it (CTV). CTV targets the phloem tissue of Rutaceae plants, primarily those of the genus *Citrus*. It belongs to the *Clostero* virus genus (Karasev et al. 1995), which has a highly diverse single-stranded positive-sense RNA genome. Plants infected with tristeza may be asymptomatic or show a variety of symptoms that fall into three categories: rapid dieback (QD), stem pitting (SP) and seedling yellowing (SY). The symptoms of tristeza disease include small, cupped-shaped chlorotic leaves, vein loss, decreased plant development and pits in woody areas. However, the virus' symptomatology changes according to the isolate's viral genotype, the environment and the rootstock/vine combination (Moreno et al. 2008). Plants that have been infected with tristeza disease may eventually succumb to it and die. Tristeza disease affects *Citrus reticulata* Blanco and sweet oranges (*Citrus aurantium* var. *sinensis* L.) planted on sour orange rootstocks (*Citrus aurantium* var. *aurantium* L.) and has destroyed millions of citrus plantations worldwide, particularly in South Africa (since 1910), Brazil and Argentina (since 1970) and the United States (since 1950). The interaction of several citrus species, cultivars and intergenic hybrids with CTV can lead to a range of physiological and biochemical reactions (Dawson et al. 2013). Although the specific mechanism causing viral pathogenesis is not known, 28% of the genes activated in host Mexican lime (*Citrus aurantifolia*) infected with CTV are associated with the stress response and defense (Gandia et al. 2007). According to recent studies, the CTV infection of sweet oranges enhances the activation of genes involved in the metabolism of cell walls, nutrient transport and plant defense (Cheng et al. 2016).

6.6.2 CITRUS CANKER DISEASE

A dangerous bacterial disease called Asian citrus canker (ACC) has an impact on citrus output, commerce and quarantine. The most prevalent and harmful kind of this illness is ACC, sometimes referred to as "Canker A." The disease is brought on by *Xanthomonas citri* (1915) (Gabriel et al. 1989) sub sp. *Citri* (Schaad et al. 2006) pathotype A (XccA). XccA has historically been referred to as *Xanthomonas citri*, *Xanthomonas axonopodis pv. citri* and *Xanthomonas campestris pv. citri* "strain" A. The main signs of ACC, a foliar and fruit spot disease, are cankers on susceptible citrus trees' fruit, leaves and young stems (. Although grapefruit is the citrus fruit that is most vulnerable to this disease, the majority of commercial citrus cultivars are moderately to severely sensitive. Severe disease progression can result in defoliation, dieback and fruit drop, making affected fruit less desirable or unmarketable (Graham et al. 1991). Tropical storms and hurricanes dramatically exacerbate the transmission of ACC. The disease is spread by rain splash in combination with wind (Gottwald et al. 2002).

6.6.3 CITRUS ROOT WEEVIL *DIAPREPES ABBREVIATUS* (L.)

The most crucial elements that can affect the health of citrus trees are soil water, pH, Fe, Mg, and Ca concentrations, and air/soil temperatures. All of these have been linked to the growth patterns of the *Diaprepes* weevil (Li et al. 2004a, 2007a). *Diaprepes* larvae, pupae and adults are soil-dwelling organisms, and the physical and chemical characteristics of the soil may have an impact on larval development and adult weevil density. Several greenhouse simulation experiments (Rogers et al. 2000; Li et al. 2003b, 2004b, 2006, 2007b) and a series of field studies in various citrus orchards in central and southern Florida all reached the same conclusion (Nigg et al. 2001,2003; McCoy et al. 2003; Li et al. 2003a,; 2004a, 2007a,; 2007c, 2007d). It has been demonstrated that citrus trees exposed to the *Phytophthora diaprepes* weevil complex are vulnerable (Graham et al. 2003). Symptoms of citrus tree death do not appear until the larvae have attached to the roots and caused significant damage (Graham et al. 2003). In Florida, there has been a lot of research on biotic stress caused by adult weevil leaf eating and *Diaprepes* root feeding by larvae (Nigg et al. 2001; Graham et al. 2003; Li et al. 2003a; Nigg et al. 2003; Li et al. 2006, 2007b). *Diaprepes* root weevil larvae were found to eat 20–80% of the roots of citrus seedlings within 6 weeks of infestation (Rogers et al., 2000). Both larval feeding and illness can kill trees because larvae do major damage to roots.

Long-term feeding by *Diaprepes* larvae can decrease the resilience of citrus roots to *Phytophthora* spp. infections (Graham et al. 2003). The occurrence, timing, development, dissemination and mobility of pests are all assumed to be significantly influenced by temperature (Viens et al. 2000; Li et al. 2007c, 2008). Insect outbreaks, changes in insect behavior and insecticide efficacy in pest control can be affected by warm temperatures (Viens et al. 2000; Amarasekare et al. 2004; Peacock et al. 2006). In Florida citrus orchards, high air and soil temperatures can result in higher densities of adult populations of *Diaprepes* root weevil (Li et al. 2007c).

6.6.4 Citrus Greening/Huanglongbing (HLB) Disease

HLB is one of the most devastating diseases affecting citrus trees worldwide. *Candidatus liberibacter* causes citrus greening disease, which is spread by insect vectors known as psyllids in a variety of climates (Spann et al. 2010; Berk et al. 2016). The health of citrus trees, fruit growth and the ripening and quality of citrus fruits and juice are all affected by citrus greening disease (Dala-Paula et al. 2019). Various management measures have been necessary to avert this potential hazard, including chemical and biological vector population control, developing plant resistance, antibiotics, nutrient augmentation and genetic manipulation for resistance. They are all used to treat graft wood (Ghosh et al., 2018). Because there is presently no cure for citrus greening disease, it is vital to remove afflicted trees as soon as possible to prevent the illness from spreading further. Symptomatic young trees, in particular, should be removed and replaced (Batool et al. 2007). Early and precise detection of the disease is critical for contaminated citrus-producing areas (Matos et al. 2013).

6.6.5 Control Measures of Some Citrus Diseases

Bacterial, fungal or viral diseases can occur in a region, spread rapidly, and have a significant economic impact on citrus production within a few years. There are a number of methods, including the use of chemical pesticides and other synthetic substances, to manage illnesses in crops, notably citrus (Prabha et al. 2017; Gimenes-Fernandes et al. 2000). However, it is impossible to overlook the detrimental impacts of chemicals, including issues with ecological balance, increasing pesticide resistance in illnesses, recurrence, pesticide residues in agricultural goods and environmental degradation (Pan et al. 2018). Therefore, resistance to chronic diseases is a major goal of all breeding efforts, and breeders are working hard to breed disease-resistant varieties. Traditional breeding methods have been successful in improving citrus plants and developing new cultivars. However, it is difficult to develop desired citrus features in the short term using standard breeding methods because of the large plant size and long juvenile period of this crop, as well as polyembryony, incompatibility and heterozygosity. Traditional breeding is also limited to fruit quality factors such as the number of seeds, ripening time and flesh color (Gong and Liu 2012). Certain transgenic research and genetic approaches have been created to lessen the many bacterial, viral, fungal and other illnesses that affect citrus, as described below.

6.6.6 Progress in Citrus Transgenic Research for Bacterial Disease Resistance

One of the most important methods to protect sensitive citrus varieties in a short time against canker or HLB caused by similar bacterial infections is genetic transformation. Thus, transgenic approaches that introduce exogenous genes (plant resistance genes, pathogenic genes, plant metabolism genes, positive key regulators of SAR genes, kinase genes and antimicrobial peptide genes) have been used to develop transgenic citrus plants resistant to cancer or HLB infections. All effective transformations that increase citrus plant's resistance to bacterial illness are addressed and listed in the section that follows.

6.6.7 Transgenic Research Related to Canker Resistance

It has been proposed that transgenic methods could increase citrus resistance to canker by increasing the overexpression of genes encoding disease-resistant proteins, antibacterial peptides, kinases, transcription factors and other exogenous proteins of plant or non-plant origin that increase natural plant defenses. Antimicrobial peptides are essential elements of innate immune defenses against microbial infections in a variety of animals (Zaslo et al. 1992; Boman et al. 2003). Attacin A, cecropin B, shiva A, D2A21, Stx IA and the dermaseptin gene are a few antimicrobial peptide genes that have been added to citrus and rootstocks to improve cancer resistance (He et al. 2011; Boscariol et al. 2006; Cardoso et al. 2009; Sala Junior et al. 2008; Kobayashi et al. 2017; Hao et al. 2017; Furman et al. 2013). Transgenic grape fruits and sweet oranges overexpressing the At NPR1 gene, a positive regulator of SAR or its equivalent CtNH1 from *Citrus maxima* were less susceptible to Xcc in two earlier studies (Zhang et al. 2010; Chen et al. 2013). The introduction of resistance genes (R-genes) is one method to boost plant disease resistance (Flor et al. 1971).

6.6.7.1 Genes Utilized to Genetically Modify Oranges to Provide Canker Resistance

R genes that are disease-resistant are frequently exploited in plant breeding. Using the rice Xa21 gene, susceptibility to Xcc was decreased in transgenic citrus (Omar et al. 2007,; 2018; Mendes et al. 2010; Li et al. 2014). ROS has been found to be a significant regulator of plant stress responses in a range of plant–pathogen interactions, including those involving bacteria, fungi and viruses (Ape et al. 2004). Reactive oxygen species (ROS) accumulation is regarded as the initial occurrence during plant–pathogen interaction and regulates and limits pathogen development. Through the activation of defense systems like kinases and signaling network components, a brief rise in ROS levels can enhance stress tolerance (Gechev et al.,2016). In citrus, overexpression of the pathogen-associated molecular pattern (PAMP) receptor NbFLS2 increases the production of ROS and activates PAMP-triggered immunity, and the incorporation and expression of NbFLS2 genes have been shown to increase resistance to cancer (Hao et al., 2016). CsMAPK1 is a mitogen-activated protein kinase gene that activates defense-related gene expression in the citrus cancer defense response.

6.6.8 Progress in Citrus Transgenic Research for Resistance to Viral and Fungal Diseases

New techniques and genetic engineering research have created previously unimaginable opportunities for the development of fungus- and virus-resistant plants outside of traditional breeding procedures. Additionally, the canker and HLB infections, several viral and fungal diseases of citrus, are also attracting attention. These diseases include citrus tristeza virus (CTV), citrus psorosis (CP), black spot disease, mal secco and gumminose, root rot, gray mold and citrus scab. Thanks to transgenic technology, resistance to several viral and fungal infections has also increased (Table 6.3). Most citrus cultivars rapidly decline due to CTV, one of the most prevalent citrus viruses, and some cultivars also have lower fruit production and quality (Bar-Joseph et al. 1989; Dominguez et al. 2000). Because of the agronomic characteristics of citrus trees, genetic modification appears to be the most practical way to produce CTV resistance because of the agronomic characteristics of citrus trees. The viability of creating transgenic plants with increased CTV resistance has been examined using a number of gene constructions and citrus genotypes (Dominguez et al. 2000). The pathogen-derived resistance (PDR) strategy has demonstrated efficacy in producing FTV resistance or tolerance material by inserting genes or specific regions of the FTV genome into citrus plants. Through genetic engineering, p25 and p23 transgenic Mexican limes were created. They demonstrated a notable delay in viral amplification and no CTV symptoms, indicating that they may be able to confer CTV resistance (Dominguez et al. 2002; Fagoaga et al. 2006). Two sweet

TABLE 6.3
The Genes that Provide Citrus Disease Resistance to Viruses and Fungi

Genes	Type	Sources	Species	Fungi and Virus Diseases	References
3DF1scFv	Monoclonal antibody	Hybridoma 3DF1 cell	*C. aurantifolia* (Christm.) Swing	Tristeza virus	Ballester et al. 2013
CTV-CP	CTV coat protein gene	Citrus Tristeza virus	*C. sinensis* cv. Valencia/Hamlin	Tristeza virus	Sun et al. 2013
p20/23/25	Silencing suppressor protein	Citrus Tristeza virus	*C. aurantifolia* (Christm.) Swing.; C. paradisi Macf. *C. sinensis* (L.) Osbeck	Tristeza virus	Oikawa et al. 2015
CTV-392/393	CTV-derived gene	Citrus Tristeza virus	*C. sinensis* cv. Itaborai	Tristeza virus	Ibáñez et al. 2014
CPsV-CP ihpCP	Coat protein gene/ siRNA	Citrus psorosis virus	*C. sinensis* (L.) Osbeck	Psorosis virus	Reuscher et al. 2014
Attacin E (attE)	Hyalophoracecropia	Antimicrobial peptide	*C. paradisi* Macf.	Citrus scab	Malik et al. 2014
chit42	Endochitinase	*Trichoderma harzianum*	*C. limon* (L.) Burm. f.	Mal secco and gray mold	Cevallos-Cevallos et al. 2012
CitMTSE1	Limonene synthase gene	*Citrus sinensis*	*C. sinensis* (L.) Osbeck	Black spot	Killiny et al. 2016
PR-5	Pathogenesis-related protein	*Solanum lycopersicum*	*C. sinensis* cv. Pineapple	Root rot and gummosis	Slisz et al. 2012

oranges were transformed by Muniz et al. using p25, an intron spliced hairpin p25, and a 559 nt long conserved 30-terminal region. They noticed that certain lines showed partial inhibition of viral replication (Muniz et al. 2012). Rarely, a rise in CTV resistance has been associated with the expression of the CTV coat protein gene. CTV replication was seen in protoplasts isolated from 10 sweet orange callus lines that had undergone genetic modification using the CTV-392/393 sequence from the CTV genome (Olivares-Fuster et al. 2003). Cervera et al. also provided a description of the first transgenic Mexican lime produced by ectopic production of variable single-chain fragment (scFv) recombinant antibodies. Additionally, they found that the majority of transgenic lines were either tolerant of or resistant to CTV injection (Cervera et al. 2010). RNA silencing or RNA interference (RNAi) is the primary strategy used by plants to fight viral infections (Waterhouse 2001). The basic method of RNAi to develop viral resistance is to generate viral sequences that, when expressed in the host, form a self-complementary hairpin RNA (hp-RNA) (Cheng et al. 2017). Nine lines of Mexican lime treated with sense, antisense and intro-hairpin variants of the CTV 549-nt-30 terminal were found to be CTV resistant (Lopez et al. 2010). Three transgenic Mexican lime lines that expressed the intron hairpin form of the three CTV silencing suppressor genes were totally resistant to CTV T36 infection in the lab (Soler et al. 2012). Some sour orange lines with a conservative p20 region with hairpin structure were also resistant or tolerable to extreme CTV stressors (Cheng et al. 2015, 2017). Researchers have made similar efforts to improve resistance to other citrus diseases. Several endo chitinase genes, including chit42 from *Trichoderma harzianum*, were also effectively transformed and expressed in lemon to enhance fungal tolerance. After injection with *Phomatra cheiphila* and *Botrytis cinerea*, the causal agents of mal secco and gray mold in citrus, transgenic lemons expressing chit42 showed significantly reduced lesion development (Gentile et al. 2007; Distefano et al. 2008). Transgenic sweet oranges and regenerates were found to produce a tomato PR-5 protein gene, and all transgenic lines showed notable resistance to the gumminose-and root-rot-causing *Phytophthora citrophthora* (Fagoaga et al. 2001).

6.6.9 OTHER GENETIC APPROACHES TO PROTECT CITRUS AGAINST VECTOR-BORNE BACTERIAL/VIRAL DISEASES

RNAi and CRISPR have recently shown promise in developing new control approaches for vector-borne viral and bacterial infections in citrus (Hunter and Sinisterra-Hunter 2018; Goulin et al. 2019). HLB diseases have been effectively controlled in bioassays using plant feeding systems (Andrade et al. 2016, 2017; Andrade et al. 2017; Ghosh et al. 2017, 2018; Kruse et al. 2017; Taning et al. 2016), bioassays with separate leaves (Ammar et al. 2013; Raiol-Junior et al. 2017), leaf discs, artificial feeding and sugar solutions (Killiny et al. 2014, 2017; Yu et al. 2017). ACP has been shown to be particularly sensitive to dsRNA, resulting in a significant reduction of the targeted transcript and enhanced mortality of psyllids. In addition, two studies have shown that the primary effectors PthA4/PthA4AT of TAL from Xcc's method of action may provide fresh perspectives on how to fight canker by interfering with the bacteria Xcc ANA-ASO/ASO. Recently, it was discovered that antisense oligonucleotides, such as 2'F-cancer, cause RNAi degradation to control diseases and pests in afflicted citrus trees. The CRISPR-Cas9 system was used to knock down the thioredoxin gene in ACP, and psyllids treated with CRISPR-Cas9 had longer developmental duration, shorter adult lifespan and lower fertility (Hunter et al. 2018; Zhang and Reed 2017). The CRISPR/Cas system was originally discovered in the CLas genus (Zheng et al. 2016) and has the potential to be used to develop effective techniques to combat HLB disease. RNAi-mediated protection techniques against *Homalodisca vitripennis* and *Aphis* (*Toxoptera*) *citricidus* (Kirkaldy) have all been shown to diminish the incidence of HLB, CTV and citrus variegated chlorosis (CVC) diseases (Rosa et al. 2010, 2012). As a result, RNAi and CRISPR could be used to develop novel vector and infection control strategies that benefit both growers and consumers while avoiding the need for crop genetic modification (Roeschlin et al. 2019).

6.7 CONCLUSIONS

Citrus fruits have anticancer, antiviral, antitumor, anti-inflammatory and capillary fragility properties, as well as the capability to suppress platelet aggregation. It has been established that the quantity of biochemically active chemicals a plant contains does not necessarily determine its health advantages. These bioactive substances (vitamins, phytochemicals, minerals and other nutrients) can serve as antioxidants, boost the immune system, trigger the liver's production of protective enzymes and shield genetic material from oxidative damage. Therefore, scientists from around the world working on citrus believe that it will be an important fruit for human consumption and industrial purposes. In the field of citrus stress breeding, many studies have already been conducted and progress has been made. Recent citrus genome sequencing data will provide a strong boost to molecular breeding and the study of genomics, proteomics, transcriptomics and metabolomics under biotic and abiotic stress in citrus. Future research efforts will play a critical role in improving citrus plants under various stress conditions. An integrated crop improvement program could lead to further progress.

ACKNOWLEDGMENT

All authors have declared that they have no conflict of interest. We apologize for not citing all of the relevant references due to space limitations.

REFERENCES

Abobatta, W.F., Khalifa, S.M. 2019. Influence of hydrogel composites soil conditioner on navel orange growth and productivity. J. Agric. Hortic. Res. 2: 1–6.

Abouzari, A., Solouki, M., Golein, B., Fakheri, B.A., Sabouri, A. 2020. The change trend in physiological traits of 110 citrus accessions in response to cold stress. Bangladesh J. Bot. 49: 375–385.

Abouzari, A., Solouki, M., Golein, B., Fakheri, B.A., Sabouri, A., Dadras, A.R. 2020. Screening of molecular markers associated to cold tolerance- related traits in Citrus. Sci. Hortic. 263: 109145.

Acosta-Motos, J.R., Ortuño, M.F., Bernal-Vicente, A., Diaz-Vivancos, P., Sanchez-Blanco, M.J., Hernandez, J.A. 2017. Plant responses to salt stress: Adaptive mechanisms. Agronomy. 7.

Adam, S. Murthy, S. 2014. Effect of cold stress on photosynthesis of plants and possible protection mechanisms. Approaches to plant stress and their management, Springer: 219–226.

Ahmad, M.S.A., Hussain, M., Ijaz, M., Alvi, A.K. 2008. Photosynthetic Performance of Two Mung Bean (Vigna radiata) Cultivars under Lead and Copper Stress. Int. J. Agri. Biol. 10: 167–172.

Albrecht, U., McCollum, G., Bowman, K.D. 2012. Influence of rootstock variety on Huanglongbing disease development in field-grown sweet orange (Citrus sinensis [L.] Osbeck) trees. Sci. Hortic. 138: 210–220.

Álvarez-Aragón, R., Rodríguez-Navarro, A.2017. Nitrate-dependent shoot sodium accumulation and osmotic functions of sodium in *Arabidopsis* under saline conditions. Plant J. 91: 208–219.

Al-Yassin, A., 2004. Influence of salinity on citrus: A review paper. J. Cent. Eur. Agric. 5: 263–272.

Amarasekare, K.G., and J.V. Edelson. 2004. Effect of temperature on efficacy of insecticides to differential grasshopper (Orthoptera: Acrididae). Journal of Economical Entomology 97 (1): 1595–1602.

Ammar, E.D., Walter, A.J., Hall, D.G.2013. New excised-leaf assay method to test inoculativity of Asian citrus psyllid (Hemiptera: Psyllidae) with Candidatus liberibacter asiaticus associated with citrus huanglongbing disease. J. Econ. Entomol. 106, 25–35.

Andrade, C., Hunter, W.B. 2016. RNA interference-natural gene-based technology for highly specific pest control (HiSPeC). RNA Interf. 391–409.

Andrade, E.C., Hunter, W.B. 2017.RNAi feeding bioassay: Development of a non-transgenic approach to control Asian citrus psyllid and other hemipterans. Entomol. Exp. Appl. 162, 389–396.

Apel, K., and H. Hirt. 2004. Reactive oxygen species: Metabolism, oxidative stress, and signal transduction. Annual Review of Plant Biology 55 (1): 373–399.

Arata, L., et al. 2020. A worldwide analysis of trend in crop yields and yield variability: Evidence from FAO data. Economic Modelling 90: 190–208.

Azab, E., Hegazy, A.K. 2020. Monitoring the efficiency of *Rhazya stricta* L. plants in phytoremediation of heavy Metal-contaminated soil. Plants 9: 1057.

Baker, A.J.M. 1981. Accumulators and excluders – strategies in the response of plants to heavy metals. J. Plant Nutr. 3: 643–654.

Ballester, A.R., M.T. Lafuente, R.C.H. de Vos, A.G. Bovy and L. González-Candelas. 2013. Citrus phenylpropanoids and defence against pathogens. Part I: metabolic profiling in elicited fruits. Food Chem. 136: 178–185.

Bar-Joseph, M., R. Marcus, and R.F. Lee. 1989. The continuous challenge of citrus tristeza virus control. Annual Review of Phytopathology 27 (1): 291–316.

Barkataky, S., Ebel, R.C., Morgan, K.T. and Dansereau, K.2013. Water relations of well-watered citrus exposed to cold-acclimating temperatures. Hort Science 48: 1309–1312.

Batool, A., Y. Iftikhar, S.M. Mughal, et al. 2007. Citrus greening disease—A major cause of citrus decline in the world—A Review. Horticultural Science 34 (1):159–166.

Ben-Hayyim, G., Moore, G.A. 2007. Recent advances in breeding citrus for drought and saline stress tolerance. Advances in Molecular Breeding toward Drought and Salt-Tolerant Crops 2007: 627–642.

Berk, Z. 2016. Diseases and pests. In Citrus Fruit Processing, 1st ed., Academic Press: London, UK. pp. 83–93.

Berthomieu, P., Conejero, G., Nublat, A., Brackenbury, W.J., Lambert, C., Savio, C., Uozumi, N., Oiki, S., Yamada, K. and Cellier, F., et al. 2003. Functional analysis of AtHKT1 in Arabidopsis shows that Na+ recirculation by the phloem is crucial for salt tolerance. EMBO J.22: 2004–2014.

Boman, H. 2003. Antibacterial peptides: Basic facts and emerging concepts. Journal of Internal Medicine 254 (1): 197–215.

Boscariol, R.L., M. Monteiro, E.K. Takahashi, et al. 2006. Attacin A gene from Tricloplusiani reduces susceptibility to *Xanthomonas axonopodisp* v. *citri* in transgenic Citrus sinensis 'Hamlin'. Journal of American
Society of Horticultural Science 131 (1): 530–536.

Brito, M.E.B., Arruda de Brito, K.S., Fernandes, P.D., Gheyi, H.R., Suassuna, J.F., Walter dos Santos S.F.,
Soares de Melo, A., & Xavier, D.A. (2014). Growth of ungrafted and grafted citrus rootstocks under
saline water irrigation. African Journal of Agricultural Research, 9 (50), 3600–3609.

Brumos, J., Talon, M., Bouhal, R. and Colmenero-Flores, J.M.2010. Cl-homeostasis in includer and excluder
citrus rootstocks: Transport mechanisms and identification of candidate genes. Plant Cell Environ.
33: 2012–2027.

Butelli, E., Licciardello, C., Zhang, Y., Liu, J., Mackay, S., Bailey, P., Reforgiato-Recupero, G., Martin, C.
2012. Retro-transposons control fruit-specific, cold-dependent accumulation of anthocyanins in blood
oranges. The Plant Cell 24 (3): 1242–1255.

Cardoso, S.C., J.M. Barbosa-Mendes, R.L. Boscariol-Camargo, et al. 2009. Transgenic sweet orange (*Citrus
sinensis* L. Osbeck) expressing the attacin A gene for resistance to *Xanthomonas citri* subsp. *citri*. Plant
Molecular Biology Reports 28 (1): 185–192.

Carr, M.K.V. 2012. The water relations and irrigation requirements of citrus (Citrus spp.): A review. Exp. Agric.
48: 347.

Castle, W.S.A. 2010. Career perspective on citrus rootstocks, their development, and commercialization. Hort
Science 45: 11–15.

Cerdá, A., Nieves, M. and Guillen, M.1990. Salt tolerance of lemon trees as affected by rootstock. Irrig.
Sci.11: 245–249.

Cervera, M., O. Esteban, M. Gil, et al. 2010. Transgenic expression in citrus of single-chain antibody fragments
specific to Citrus tristeza virus confers virus resistance. Transgenic Research 19 (1): 1001–1015.

Cevallos-Cevallos, J.M., R. García-Torres, E. Etxeberria, et al. 2011. GC-MS analysis of headspace and
liquid extracts for metabolomic differentiation of citrus Huanglongbing and zinc deficiency in leaves
of "Valencia" sweet orange from commercial groves. Phytochemical Analysis 22 (1):1–37.doi:10.1002/
pca.1271.

Cevallos-Cevallos, J.M., D.B. Futch, T. Shilts, et al. 2012. GC-MS metabolomic differentiation of selected
citrus varieties with different sensitivity to citrus huanglongbing. Plant Physiology and Biochemistry
53 (1): 69–76

Chaneva, G., Parvanova, P., Tzvetkova, N. and Uzunova, A.2010. Photosynthetic Response of maize plants
against cadmium and paraquat impact. Water Air Soil Pollut. 208: 287–293.

Chen, S., Wang, Q., Lu, H., Li, J., Yang, D., Liu, J. and Yan, C.2019. Phenolic metabolism and related heavy
metal tolerance mechanism in Kandelia Obovata under Cd and Zn stress. Ecotoxicol. Environ. Saf
169: 134–143.

Chen, X., J.Y. Barnaby, A. Sreedharan, et al. 2013. Over-expression of the citrus gene CtNH1 confers resistance
to bacterial canker disease. Physiological Molecular Plant Pathology 84 (1): 115–122.

Cheng, C., Y. Zhang, Y. Zhong, et al. 2016. Gene expression changes in leaves of Citrus sinensis (L.) Osbeck
infected by Citrus tristeza virus. Journal of Horticultural Science and Biotechnology 91 (1): 466–475.

Cheng, C., Y. Zhang, J. Yang, et al. 2017. Expression of hairpin RNA (hp RNA) targeting the three CTV-silencing
suppressor genes confers sweet orange with stem-pitting CTV tolerance. Journal of Horticultural Science
and Biotechnology 92 (1): 465–474.

Cevallos-Cevallos, J.M., Futch, D.B., Shilts, T., Folimonova, S.Y., Reyes-De-Corcuera, J.I. 2012. GC-MS
metabolomic differentiation of selected citrus varieties with different sensitivity to citrus Huanglongbing.
Plant Physiol. Biochem. 2012, 53:69–76 doi:10.1016/j.plaphy.2012.01.010.

Cheng, C.Z., Yang, J.W., Yan, H.B., Bei, X.J., Zhang, Y.Y., Lu, Z.M., Zhong, G.Y.. 2015. Expressing p20
hairpin RNA of Citrus tristeza virus confers Citrus aurantium with tolerance/resistance against stem
pitting and seedling yellow CTV strains. J. Integr. Agric, 14: 1767–1777.

Chin, E.L., D.O. Mishchuk, A.P. Breksa and C.M. Slupsky. 2014. Metabolite signature of Candidatus
Liberibacter asiaticus infection in two citrus varieties. J. Agric. Food Chem. 62: 6585–6591.

Cohen, J., Jones, J., Furtado, J., and Tziperman, E. 2013. Warm Arctic, Cold Continents: A Common Pattern
Related to Arctic Sea Ice Melt, Snow Advance, and Extreme Winter Weather. Oceanography 26: 150–160.

Colmenero-Flores, J.M., Campos, F., Garciarrubio, A. and Covarrubias, A.A.1997. Characterization of
Phaseolus vulgaris cDNA clones responsive to water deficit: Identification of a novel late embryogenesis abundant-like protein. Plant Mol. Biol. 35: 393–405.

Conesa, M.R., Espinosa, P.J., Pallarés, D. and Pérez-Pastor, A. 2020. Influence of plant biostimulant as technique to harden citrus nursery plants before transplanting to the field. Sustainability 12: 6190.

Crifò, T., Puglisi, I., Petrone, G., Recupero, G.R. and Lo Piero, A.R. 2011. Expression analysis in response to low temperature stress in blood oranges: Implication of the flavonoid biosynthetic pathway. Gene 476: 1–9.

Dala-Paula, B.M., A. Plotto, J. Bai, et al. 2019. Effect of Huanglongbing or Greening Disease on Orange Juice Quality, a Review. Frontiers in Plant Science 9 (1): 1976.

Dawson, W.O., S.M. Garnsey, S. Tatineni, et al. 2013. Citrus tristeza virus–host interaction. Frontiers in Microbiology 4 (1): 88.

De Pascual-Teresa, S., Sanchez-Ballesta, M.T. 2008. Anthocyanins: From plant to health. Phytochem. Rev. 7: 281–299.

Distefano, G., La Malfa, S., Vitale, A., Lorito, M., Deng, Z., Gentile, A.2008. Defence-related gene expression in transgenic lemon plants producing an antimicrobial Trichoderma harzianumendo chitinase during fungal infection. Transgenic Res. 17: 873–879.

Dominguez, A., A.H. de Mendoza, J. Guerri, et al. 2002. Pathogen-derived resistance to Citrus tristeza virus (CTV) in transgenic Mexican lime (Citrus aurantifolia (Christ.) Swing.) plants expressing its p25 coat protein gene. Molecular Breeding. 10 (1): 1–10.

Dominguez, A., J. Guerri, M. Cambra, et al. 2000. Efficient production of transgenic citrus plants expressing the coat protein gene of citrus tristeza virus. Plant Cell Reports 19 (1): 427–433.

Driscoll, P., et al. 1992. Sink-regulation of photosynthesis in relation to temperature in sunflower and rape. Journal of Experimental Botany 43 (2): 147–153.

Fagoaga, C., C. López, A.H. de Mendoza, et al. 2006. Post-transcriptional gene silencing of the p23 silencing suppressor of Citrus tristeza virus confers resistance to the virus in transgenic Mexican lime. Plant Molecular Biology 60 (1): 153–165.

Fagoaga, C., Rodrigo, I., Conejero, V., Hinarejos, C., Tuset, J.J., Arnau, J., Pina, J.A., Navarro, L., Peña, L. 2001. Increased tolerance to Phytophtora citrophthora in transgenic orange plants constitutively expressing a tomato pathogenesis related protein PR-5. Mol. Breed.7: 175–185.

Fan, J., C. Chen, Q. Yu, R.H. Brlansky, Z.G. Li and F.G. Gmitter. 2011. Comparative iTRAQ proteome and transcriptome analyses of sweet orange infected by 'Candidatus Liberibacter asiaticus'. Physiol. Plant. 143: 235–245.

Ferguson, L., Grattan, S.R. 2005. How salinity damages citrus: Osmotic effects and specific ion toxicities. Hort Technology 15: 95–99.

Flor, H.H. 1971. Current status of the gene-for-gene concept. Annual Review of Phytopathology 9 (1): 275–296.

Forner-Giner, M.A., Primo-Millo, E. and Forner, J.B. 2009. Performance of Forner-Alcaide 5 and Forner-Alcaide 13, hybrids of Cleopatra mandarin × Poncirus trifoliata, as salinity-tolerant citrus rootstocks. J. Am. Pomol. Soc. 63: 72–80.

Furman, N., K. Kobayashi, M.C. Zanek, et al. 2013. Transgenic sweet orange plants expressing a dermaseptin coding sequence show reduced symptoms of citrus canker disease. Journal of Biotechnology 167 (1): 412–419.

Gabriel, D.W., M.T. Kingsley, J.E. Hunter, et al. 1989. Reinstatement of Xanthomonas citri (ex Hasse) and X. phaseoli (ex Smith) to species and reclassification of all X. campestris pv. citri strains. International Journal of Systemic Bacteriology 39 (1): 14–22.

Gandía, M., A. Conesa, G. Ancillo, et al. 2007. Transcriptional response of Citrus aurantifolia to infection by Citrus tristeza virus. Virology 367 (1): 298–306.

Garcia-Lor, A., Curk, F., Snoussi-Trifa, H., Morillon, R., Ancillo, G., Luro, F., Navarro, L. and Ollitrault, P. A. 2013. Nuclear phylogenetic analysis: SNPs, indels and SSRs deliver new insights into the relationships in the 'true citrus fruit trees' group (Citrinae, Rutaceae) and the origin of cultivated species. Annals of Botany 111 (1): 1–9.

García-Sánchez, F., Martinez, V., Jifon, J., Syvertsen, J., Grosser, J. 2002. Salinity reduces growth, gas exchange, chlorophyll and nutrient concentrations in diploid sour orange and related allotetraploid somatic hybrids. J. Hortic. Sci. Biotechnol. 77: 379–386.

García-Sánchez, F., Syvertsen, J. 2006. Salinity tolerance of Cleopatra mandarin and Carrizo citrange citrus rootstock seedlings is affected by CO_2 enrichment during growth. J. Am. Soc. Hortic. Sci. 131: 24–31.

García-Sánchez, F., Syvertsen, J.P., Gimeno, V., Botía, P. and Perez-Perez, J.G. 2007. Responses to flooding and drought stress by two citrus rootstock seedlings with different water-use efficiency. Physiol. Plant. 130: 532–542.

Gechev, T.S., F. Van Breusegem, J.M. Stone, et al. 2016. Reactive oxygen species as signals that modulate plant stress responses and programmed cell death. Bioessays 28 (1): 1091–1101.

Geng, J., Wei, T., Wang, Y., Huang, X. and Liu, J.-H. 2019. Over expression of PtrbHLH, a basic helix-loop-helix transcription factor from Poncirus trifoliata confers enhanced cold tolerance in pummelo (Citrus grandis) by modulation of H2O2 level via regulating a CAT gene. Tree Physiol. 39: 2045–2054.

Gentile, A., Deng, Z., La Malfa, S., Distefano, G., Domina, F., Vitale, A., Polizzi, G., Lorito, M., Tribulato, E. 2007. Enhanced resistance to Phomatracheiphila and Botrytis cinerea in transgenic lemon plants expressing a Trichoderma harzianum chitinase gene. Plant Breed. 126: 146–151.

Ghosh, S.K.B., Hunter, W.B., Park, A.L., Gundersen-Rindal, D.E. 2017. Double strand RNA delivery system for plant-sap-feeding insects. PLoS ONE 12, e0171861.

Ghosh, D.K., Motghare, M., Gowda, S.2018. Citrus Greening: Overview of the Most Severe Disease of Citrus. Adv. Agric. Res. Tech. J. 2: 83–100.

Gill, S.S. and Tuteja, N. 2010. Reactive oxygen species and antioxidant machinery in abiotic stress tolerance in crop plants. Plant Physiology and Biochemistry, 48, 909–930. http://dx.doi.org/10.1016/j.pla phy.2010.08.016

Gimenes-Fernandes, N., J.C. Barbosa, A. Ayres, and C. Massari. 2000. Plantasdoentesnão detect a das nasinspeções difficult am a erradicação do cancrocítrico. Summa Phytopathologica 26 (1): 320–325.

Gimeno, V. , Díaz-López, L. , Simón-Grao, S. , Martínez, V. , Martínez-Nicolás, J.J. and García-Sánchez, F.2014. Foliar potassium nitrate application improves the tolerance of Citrus macrophylla L. seedlings to drought conditions. Plant Physiol. Biochem. 83: 308–315.

Goldberg-Moeller, R., L. Shalom, L. Shlizerman, S. Samuels, N. Zur, R. Ophir, E. Blumwald and A. Sadka. 2013. Effects of gibberellins treatment during flowering induction period on global gene expression and the transcription of flowering-control genes in citrus buds. Plant Sci. 198: 46–57.

Gómez-Cadenas, A., Iglesias, D., Arbona, V., Colmenero-Flores, J., Primo-Millo, E., Talon, M. 2003. Physiological and molecular responses of citrus to salinity. Recent Res. Dev. Plant Mol. Biol. 1: 281–298.

Gong, X.Q. and J.H. Liu. 2012. Genetic transformation and genes for resistance to abiotic and biotic stresses in Citrus and its related genera. Plant Cell, Tissue and Organ Culture 113 (1): 137–147.

Gonzalez, P., Syvertsen, J.P., Etxeberria, E. 2012. Sodium distribution in salt-stressed citrus rootstock seedlings. Hort Science 47: 1504–1511.

Gottwald, T.R., J.H. Graham, and T.S. Schubert. 2002. Citrus canker: the pathogen and its impact. Plant Health Progress. www.plantmanagementnetwork.org/ pub/php/review/citrus canker.

Goulin, E.H., Galdeano, D.M., Granato, L.M., Matsumura, E.E., Dalio, R.J.D., Machado, M.A. 2019. RNA interference and CRISPR: Promising approaches to better understand and control citrus pathogens. Microbiol. Res. 226: 1–9.

Gowda, V.R., Henry, A., Yamauchi, A., Shashidhar, H.E. and Serraj, R.2011. Root biology and genetic improvement for drought avoidance in rice. Field Crop. Res. 122: 1–13.

Graham, J.H., D.B. Bright, and C.W. McCoy. 2003 Phytophthora-Diaprepes complex: Phytophthora spp. relationship with citrus rootstocks. Plant Disease 87 (1): 85–90

Graham, J.H., and T.R. Gottwald. 1991. Research perspectives on eradication of citrus bacterial diseases in Florida. Plant Disease 75 (1): 1193–1200.

Hamblin, M.T., Buckler, E.S., Jannink, J.L. 2011. Population genetics of genomics-based crop improvement methods. Trends in Genetics 27 (3): 98–106.

Hao, G., E. Stover, and G. Gupta. 2016. Over expression of a modified plant thionin enhances disease Resistance to citrus canker and huanglongbing (HLB). Frontiers in Plant Science 7 (1): 1078.

Hao, G., S. Zhang, and E. Stover. 2017. Transgenic expression of antimicrobial peptide D2A21 confers resistance to diseases incited by Pseudomonas syringae pv. tabaci and Xanthomonas citri, but not Candidatus Liberibacter asiaticus. PLoS ONE 12 (1): e0186810.

He, Y., S. Chen, A. Peng, et al. 2011. Production and evaluation of transgenic sweet orange (Citrus sinensis Osbeck) containing bivalent antibacterial peptide genes (Shiva A and Cecropin B) via a novel Agrobacterium-mediated transformation of mature axillary buds. Scientia Horticulturae Amsterdam 128 (1): 99–107.

Heckathorn, S.A., Mueller, J.K., LaGuidice, S., Zhu, B., Barrett, T., Blair, B. and Dong, Y. 2004. Chloroplast small heat-shock proteins protect photosynthesis during heavy metal stress. Am. J. Bot. 91: 1312–1318.

Hu, H., Xiong, L. 2014. Genetic engineering and breeding of drought-resistant crops. Annu. Rev. Plant Biol. 65: 715–741.

Huang, X.-S., Wang, W., Zhang, Q. and Liu, J.-H. 2013. A Basic Helix-Loop-Helix Transcription Factor, PtrbHLH, of Poncirus trifoliate Confers Cold Tolerance and Modulates Peroxidase-Mediated Scavenging of Hydrogen Peroxide. Plant Physiol. 162: 1178.

Huang, X.-S., Zhang, Q., Zhu, D., Fu, X., Wang, M., Zhang, Q., Moriguchi, T. and Liu, J.-H. 2015. ICE1 of Poncirus trifoliata functions in cold tolerance by modulating polyamine levels through interacting with arginine decarboxylase. J. Exp. Bot. 66: 3259–3274.

Hunter, W.B., Sinisterra-Hunter, X.H. 2018. Emerging RNA Suppression Technologies to Protect Citrus Trees from Citrus Greening Disease Bacteria. Adv. Insect Physiol. 55: 163–197.

Ibáñez, A.M., F. Martinelli, R.L. Reagan, S.L. Uratsu, A. Vo, M.A. Tinoco, M.L. Phu, Y. Chen, D.M. Rocke and A.M. Dandekar. 2014. Transcriptome and metabolome analysis of citrus fruit to elucidate puffing disorder. Plant Sci. 217–218: 87–98.

Ji, S., Tong, L., Li, F., Lu, H., Li, S., Du, T. and Wu, Y. 2017. Effect of a new antitranspirant on the physiology and water use efficiency of soybean under different irrigation rates in an arid region. Front. Agric. Sci. Eng. 4: 155–164.

Jiang, J., et al. 2021. Physiological and TMT-labeled proteomic analyses reveal important roles of sugar and secondary metabolism in Citrus junos under cold stress. Journal of Proteomics 237: 104145.

Jiang, J., Hou, R., Yang, N., Li, L., Deng, J., Qin, G. and Ding, D. 2021. Physiological and TMT-labeled proteomic analyses reveal important roles of sugar and secondary metabolism in Citrus junos under cold stress. J. Proteom. 237: 104145.

Jubany-Marí, T., Munné-Bosch, S., Alegre, L. 2010. Redox regulation of water stress responses in field-grown plants. Role of hydrogen peroxide and ascorbate. Plant Physiol. Biochem. 48: 351–358.

Karasev, A.V., V.P. Boyko, S. Gowda, et al. 1995. Complete sequence of the Citrus tristeza virus RNA genome. Virology 208 (1): 511–520.

Khoshbakht, D., Ghorbani, A., Baninasab, B., Naseri, L.A., Mirzaei, M. 2014. Effects of supplementary potassium nitrate on growth and gas-exchange characteristics of salt-stressed citrus seedlings. Photosynthetica 52: 589–596.

Killiny, N., and F. Hijaz. 2016. Amino acids implicated in plant defense are higher in Candidatus Liberibacter asiaticus-tolerant citrus varieties. Plant Signaling & Behaviour 11: e1171449.

Killiny, N., and Y. Nehela. 2017. Metabolomic response to huanglongbing: role of carboxylic compounds in citrus sinensis response to 'Candidatus Liberibacter asiaticus' and its vector, Diaphorinacitri. Molecular Plant-Microbe Interactions. 30 (1):666–678. doi: 10.1094/MPMI-05-17-0106-R.

Killiny, N., Hajeri, S., Tiwari, S., Gowda, S., Stelinski, L.L. 2014. Double-stranded RNA uptake through topical application, mediates silencing of five CYP4 genes and suppresses insecticide resistance in *Diaphorina citri*. PLoS ONE 9, e110536.

Kim, M., Canio, W., Kessler, S., Sinha, N. 2001. Developmental Changes Due to Long-Distance Movement of a Homeobox Fusion Transcript in Tomato. Science. 293, 287.

Kodra, E., et al. (2011). Persisting cold extremes under 21st-century warming scenarios. Geophysical Research Letters. 38 (8).

Kostopoulou, Z., Therios, I., Roumeliotis, E., Kanellis, A.K., Molassiotis, A. 2015. Melatonin combined with ascorbic acid provides salt adaptation in Citrus aurantium L. seedlings. Plant Physiol. Biochem. 86, 155–165.

Kourgialas, N.N., Karatzas, G.P. 2015. A modeling approach for agricultural water management in citrus orchards: Cost-effective irrigation scheduling and agrochemical transport simulation. Environ. Monit. Assess. 187, 462.

Kourgialas, N.N., Koubouris, G.C., Dokou, Z. 2019. Optimal irrigation planning for addressing current or future water scarcity in Mediterranean tree crops. Sci. Total Environ. 654, 616–632.

Kobayashi, A.K., L.G.E. Vieira, J.C.B. Filho, et al. 2017. Enhanced resistance to citrus canker in transgenic sweet orange expressing the sarcotoxin IA gene. European Journal of Plant Pathology. 149 (1): 865–873.

Kodra, E., K. Steinhaeuser, A.R. Ganguly. 2011. Persisting cold extremes under 21st-century warming scenarios. Geophys. Res. Lett. 38: L08705.

Kosobrukhov, A., Knyazeva, I., Mudrik, V. 2004. Plantago major plants responses to increase of content of lead in soil: Growth and photosynthesis. Plant Growth Regul. 42, 145–151.

Krueger, R. R., Navarro, L. 2007. Citrus germplasm resources. Citrus Genetics, Breeding and Biotechnology. 45: 140.

Kruse, A., Fattah-Hosseini, S., Saha, S., Johnson, R., Warwick, E., Sturgeon, K., Mueller, L., MacCoss, M.J., Shatters Jr, R.G., Heck, M.C. 2017. Combining 'omics and microscopy to visualize interactions between the Asian citrus psyllid vector and the Huanglongbing pathogen Candidatus Liberibacter asiaticus in the insect gut. PLoS ONE 12, e0179531.

Kudo, H., Harada, T. 2007. A graft-transmissible RNA from tomato rootstock changes leaf morphology of potato scion. Hort Science 42 (2): 225–226.

Kumar, P., Rouphael, Y., Cardarelli, M., Colla, G.2017. Vegetable grafting as a tool to improve drought resistance and water use efficiency. Front. Plant Sci. 8: 1130.

Küpper, H., Šetlík, I., Spiller, M., Küpper, F.C., Prášil, O. 2002. Heavy metal-induced inhibition of photosynthesis: Targets of in vivo heavy metal chlorophyll formation. J. Phycol. 38: 429–441.

Li, H., S.H. Futch, R.J. Stuart, et al. 2007a. Association of soil iron with citrus tree decline and variability of soil pH, water, and magnesium and Diaprepes root weevil: two-site study. Environmental and Experimental Botany. 59 (1): 321–333.

Li, H., S.H. Futch, and J.P. Syvertsen. 2007c. Cross-correlation patterns of air and soil temperatures, rainfall and citrus Diaprepes abbreviatus (L.) root weevil. Pest Management Science. 63 (1): 1116–1123

Li, H., S.H. Futch, J.P. Syvertsen, and C.W. McCoy. 2007d. Time series forecast and soil characteristics-based simple and multivariate linear models for management of Diaprepes abbreviatus (L.) root weevil in citrus. Soil Biology and Biochemistry. 39 (1): 2436–2447.

Li, H., C.W. McCoy, and J.P. Syvertsen. 2007b. Controlling factors of environmental flooding, soil pH and Diaprepes abbreviatus (L.) root weevil feeding in citrus: larval survival and larval growth. Applied Soil Ecology. 35 (1): 553–565.

Li, H., W.A. Payne, G.J. Michels, and C.M. Rush. 2008. Reducing plant abiotic and biotic stress: Drought and attacks of green bugs, corn leaf aphids and virus disease in dryland sorghum. Environmental Experimental Botany. 63 (1): 305–316.

Li, H., J.P. Syvertsen, C.W. McCoy, and A. Schumann. 2003b. Soil redox potential and leaf stomatal conductance of two citrus rootstocks subjected to flooding and Diaprepes root weevil feeding. Proceedings of the Florida State Horticultural Society. 116 (1): 252–256.

Li, H., J.P. Syvertsen, C.W. McCoy, R.J. Stuart, and A. Schumann A. 2004b. Soil liming and flooding effects on Diaprepes root weevil survival and citrus seedling growth. Proceedings of the Florida State Horticultural Society. 117 (1): 139–143.

Li, H., J.P. Syvertsen, A. Schumann, C.W. McCoy, and R.J. Stuart. 2003a. Correlation of soil characteristics and Diaprepes root weevil in a flat woods citrus grove. Proceedings of the Florida State Horticultural Society. 116 (1): 242–248.

Li, H., J.P. Syvertsen, R.J. Stuart, C.W. McCoy, and A. Schumann. 2006. Water stress and root injury from simulated flooding and Diaprepes root weevil feeding in citrus. Soil Science. 171 (1): 138–151.

Li, H., J.P. Syvertsen, R.J. Stuart, C.W. McCoy, A. Schumann, and W.S. Castle. 2004a. Soil and Diaprepes root weevil spatial variability in a poorly drained citrus grove. Soil Science. 169 (1): 650–662.

Li, B., Tester, M., Gilliham, M. 2017. Chloride on the move. Trends Plant Sci. 22: 236–248.

Li, D.L., X. Xiao, and W.W. Guo. 2014. Production of Transgenic Anliucheng Sweet Orange (*Citrus sinensis* Osbeck) with Xa21 Gene for Potential Canker Resistance. Journal of Integrative Agriculture. 13 (1): 2370–2377.

Li, X., Li, S., Wang, J. and Lin, J.2020. Exogenous abscisic acid alleviates harmful effect of salt and alkali stresses on wheat seedlings. Int. J. Environ. Res. Public Health. 17.

Li, Z., Peng, D., Zhang, X., Peng, Y., Chen, M., Ma, X., Huang, L. and Yan, Y. 2017. Na+ induces the tolerance to water stress in white clover associated with osmotic adjustment and aquaporins-mediated water transport and balance in root and leaf. Environ. Exp. Bot. 144: 11–24.

Lin, Q., C. Wang, W. Dong, Q. Jiang, D. Wang, S. Li, M. Chen, C. Liu, C. Sun and K. Chen. 2015. Transcriptome and metabolome analyses of sugar and organic acid metabolism in Ponkan (*Citrus reticulata*) fruit during fruit maturation. Gene 554: 64–74.

Lima-Costa ME, Ferreira S, Duarte A, Ferreira AL (2010) Alleviation of salt stress using exogenous proline on a citrus cell line. Acta Hortic. 868: 109–112. doi: 10.17660/ActaHortic.2010.868.10

Lo Piero, A. R., et al. 2006. Gene isolation, analysis of expression, and in vitro synthesis of glutathione S-transferase from orange fruit [Citrus sinensis L. (Osbeck)]. Journal of Agricultural and Food Chemistry. 54 (24): 9227–9233.

Lopez, C., M. Cervera, C. Fagoaga, et al. 2010. Accumulation of transgene-derived siRNAs is not sufficient for RNAi-mediated protection against Citrus tristeza virus in transgenic Mexican lime. Molecular Plant Pathology. 11 (1): 33–41.

Lourkisti, R., et al. 2020. Triploid citrus genotypes have a better tolerance to natural chilling conditions of photosynthetic capacities and specific leaf volatile organic compounds. Frontiers in Plant Science. 11: 330.

Lu, Y.B., L.T. Yang, Y.P. Qi, Y. Li, Z. Li, Y.B. Chen, Z.R. Huang and L.S. Chen. 2014. Identification of boron-deficiency-responsive microRNAs in *Citrus sinensis* roots by Illumina sequencing. BMC Plant Biol. 14: 123.

Mafra, V., P.K. Martins, C.S. Francisco, M. Ribeiro-Alves, J. Freitas-Astúa and M.A. Machado. 2013. *Candidatus Liberibacter americanus* induces significant reprogramming of the transcriptome of the susceptible citrus genotype. BMC Genomics. 14: 247.

Malecka, A., Piechalak, A. and Tomaszewska, B. 2009. Reactive oxygen species production and anioxidative defense system in pea root tissues treated with lead ions the whole root level. Acta Physiol. Plant. 31: 1053–1063.

Malik, N.S.A., J.L. Perez, M. Kunta, J.M. Patt, and R.L. Mangan. 2014. Changes in free amino acids and polyamine levels in Satsuma leaves in response to Asian citrus psyllid infestation and water stress. Insect Science 6 (1):707–716.

Małkowski, E., Kita, A., Galas, W., Karcz, W., Kuperberg, J.M. 2002. Lead distribution in corn seedlings (Zea mays L.) and its effect on growth and the concentrations of potassium and calcium. Plant Growth Regul.37: 69–76.

Martinelli, F., R.L. Reagan, S.L. Uratsu, M.L. Phu, U. Albrecht, W. Zhao, C.E. Davis, K.D. Bowman and A.M. Dandekar. 2013. Gene regulatory networks elucidating huanglongbing disease mechanisms. PLoS ONE. 8: e74256.

Martinelli, F., S.L. Uratsu, U. Albrecht, R.L. Reagan, M.L. Phu, M. Britton, V. Buffalo, J. Fass, E. Leicht, W. Zhao, et al. 2012. Transcriptome profiling of citrus fruit response to huanglongbing disease. PLoS ONE. 7: e38039.

Martínez-Cuenca, M.-R., et al. 2019. Key role of boron compartmentalisation-related genes as the initial cell response to low B in citrus genotypes cultured in vitro. Horticulture, Environment, and Biotechnology. 60 (4): 519–530.

Matos, L.A., M.E. Hilf, X.A. Cayetano, A.O. Feliz, S.J. Harper, and S.Y. Folimonova. 2013. Dynamics of Citrus tristeza virus populations in the Dominican Republic. Plant Disease. 97 (1): 339–345.

McCoy, C.W., R.J. Stuart, and N.N. Nigg. 2003. Seasonal life stage abundance of Diaprepes abbreviatus (L.) in irrigated and non-irrigated citrus plantings in central Florida. Florida Entomology. 86 (1): 34–42

Mendes, B.M.J., S. Cardoso, R. Boscariol-Camargo, R. Cruz, F. Mourão Filho, and A. Bergamin Filho. 2010. Reduction in susceptibility to Xanthomonas axonopodis pv. citri in transgenic Citrus sinensis expressing the rice Xa21 gene. Plant Pathology 59 (1): 68–75.

Mendonca, L.B.P., Badel, J.L., Zambolim, L. 2017. Bacterial citrus diseases: Major threats and recent progress. Bacteriol Mycol Open Access 5 (4): 340–350.

Meng, S., et al. 2008. Gene expression analysis of cold treated versus cold acclimated Poncirus trifoliata. Euphytica. 164 (1): 209–219.

Miles, C., Wayne, M. 2008. Quantitative trait locus (QTL) analysis. Nat. Educ. 1: 208.

Miles, G.P., Stover, E., Ramadugu, C., Keremane, M.L., Lee, R.F. 2017. Apparent tolerance to Huanglongbing in Citrus and Citrus-related germplasm. Hort Science 52: 31–39.

Mohammadian, M.A. Mobrami, Z. and Sajedi, R.H. 2011. Bioactive compounds and antioxidant capacities in the flavedo tissue of two citrus cultivars under low temperature. Braz. J. Plant Physiol. 23: 203–208.

Mohammadrezakhani, S., Hajilou, J., Rezanejad, F., Zaare-Nahandi, F. 2019. Assessment of exogenous application of proline on antioxidant compounds in three Citrus species under low temperature stress. J. Plant Interact. 14:347–358.

Morales, J., et al. 2021. Rootstock effect on fruit quality, anthocyanins, sugars, hydroxycinnamic acids and flavanones content during the harvest of blood oranges 'Moro' and 'Tarocco Rosso' grown in Spain. Food Chemistry. 342: 128305.

Moreno, P., S. Ambros, M.R. Albiach-Marti, J. Guerri, and L. Peña. 2008. Citrus tristeza virus: a pathogen that changed the course of the citrus industry. Molecular Plant Pathology. 9 (1): 251–268.

Moustakas, M., Lanaras, T., Symeonidis, L., Karataglis, S.1994. Growth and some photosynthetic characteristics of field grown Avena sativa under copper and lead stress. Photosynthetica. 30, 389–396.

Muniz, F., A. De Souza, L.C.L. Stipp, et al. 2012. Genetic transformation of Citrus sinensis with Citrus tristeza virus (CTV) derived sequences and reaction of transgenic lines to CTV infection. Biologia Plantarum. 56 (1): 162–166.

Munns, R., and Tester, M. 2008. Mechanisms of salinity tolerance. Annu. Rev. Plant Biol. 59, 651–681.

Murkute, A.A., Sharma, S., Sing, S.K. 2005. Citrus in terms of soil and water salinity: A review. J. Sci. Ind. Res. 64: 393–402.

Nehela, Y., F. Hijaz, A.A. Elzaawely, H.M. El-Zahaby, and N. Killiny. 2018. Citrus phytohormonal response to Candidatus Liberibacter asiaticus and its vector Diaphorina citri. Physiology and Molecular Plant Pathology. 102 (1):24–35.

Nehela, Y., and N. Killiny. 2019. 'Candidatus Liberibacter asiaticus' and its vector, Diaphorina citri, augment the tricarboxylic acid cycle of their host via the γ-amino butyric acid shunt and polyamines pathway. Molecular Plant–Microbe Interactions. 32 (1): 413–427. doi: 10.1094/MPMI-09-18-0238-R.

Nigg, H.N., S.E. Simpson, L.E. Ramos, T. Tomerlin, J.M. Harrison, and N. Cuyler. 2001. Distribution and movement of adult Diaprepes abbreviatus (Coleoptera: Curculionidae) in a Florida citrus grove. Florida Entomology. 84 (1): 641–651.

Nigg, H.N., S.E. Simpson, R.J. Stuart, L.W. Duncan, C.W. McCoy, and F.G. Gmitter Jr. 2003. Abundance of Diaprepes abbreviatus (L.) (Coleoptera: Curculionidae) neonates falling to the soil under tree canopies in Florida citrus groves. Horticultural Entomology. 96 (1): 835–843.

Nishikawa, F., T. Endo, T. Shimada, H. Fujii, T. Shimizu, Y. Kobayashi, T. Araki and M. Omura. 2010. Transcriptional changes in CiFT introduced transgenic trifoliate orange (Poncirus trifoliata L. Raf.). Tree Physiol. 30: 431–439.

Nordby, H. and Yelenosky, G. 1985. Change in citrus leaf lipids during freeze-thaw stress. Phytochemistry. 24 (8): 1675–1679.

Oikawa, A., T. Otsuka, R. Nakabayashi, Y. Jikumaru, K. Isuzugawa, H. Murayama, K. Saito, and K. Shiratake. 2015. Metabolic profiling of developing pear fruits reveals dynamic variation in primary and secondary metabolites, including plant hormones. PLoS ONE. 10: e0131408.

Okwu, D.E., and Emenike, I.N. 2006. Evaluation of the phytonutrients and vitamins contents of citrus fruits. International Journal Molecular Medicine and Advance Sciences. 2 (1):1–6.

Okwu, D.E. 2008. Citrus fruits: A rich source of phytochemicals and their roles in human health. International Journal of Chemical Science. 6 (2): 451–471

Olivares-Fuster, O., G. Fleming, M. Albiach-Marti, S. Gowda, Dawson. J., 2003. Citrus tristeza virus (CTV) resistance in transgenic citrus based on virus challenge of protoplasts. In Vitro Cellular & Developmental Biology- Plant. 39 (1): 567–572.

Omar, A., W.Y. Song, and J. Grosser. 2007. Introduction of Xa21, a Xanthomonas-resistance gene from rice, into 'Hamlin' sweet orange [Citrus sinensis (L.) Osbeck] using protoplast-GFP co-transformation or single plasmid transformation. Journal of Horticultural Science and Biotechnology. 82 (1): 914–923.

Omar, A.A., M.M. Murata, H.A. El-Shamy, J.H. Graham, and J.W. Grosser. 2018. Enhanced resistance to citrus canker in transgenic mandarin expressing Xa21 from rice. Transgenic Research. 27 (1): 179–191.

Oueslati, A., Salhi-Hannachi, A., Luro, F., Vignes, H., Mournet, P., and Ollitrault, P. 2017. Genotyping by sequencing reveals the interspecific C. maxima/C. reticulata admixture along the genomes of modern citrus varieties of mandarins, tangors, tangelos, orangelos and grapefruits. PLoS One. 12 (10): e0185618.

Oustric, J., et al. 2017. Tetraploid Carrizo citrange rootstock (Citrus sinensis Osb.× Poncirus trifoliata L. Raf.) enhances natural chilling stress tolerance of common clementine (Citrus clementina Hort. ex Tan). Journal of Plant Physiology. 214: 108–115.

Oustric, J., et al. 2018. Somatic hybridization between diploid Poncirus and Citrus improves natural chilling and light stress tolerances compared with equivalent doubled-diploid genotypes. Trees. 32 (3): 883–895.

Ouyang, Z., et al. 2019. Differential expressions of citrus CAMTAs during fruit development and responses to abiotic stresses. Biol. Plant. 63: 354–364.

Ouzounidou, G. 1993. Changes in variable chlorophyll fluorescence as a result of Cu treatment: Dose-response relations in Silene and Thlaspi. Photosynthetica. 29: 455–462.

Ouzounidou, G., Moustakas, M., Strasser, R.J. 1997. Sites of action of copper in the photosynthetic apparatus of maize leaves: Kinetic analysis of chlorophyll fluorescence, oxygen evolution, absorption changes and thermal dissipation as monitored by photoacoustic signals. Funct. Plant Biol. 24, 81–90.

Pan, X.L., F.S. Dong, X.H. Wu, J. Xu, X.G. Liu, and Y.Q. Zheng. 2018. Progress of the discovery, application, and control technologies of chemical pesticides in China. Journal of Integrative Agriculture. 18 (1): 840–853.

Pandey, N., Patkak, G.C., Pandey, D.K., Pandey, R.2009. Heavy metals, Co, Ni, Cu, Zn and Cd, produce oxidative dam-age and evoke differential antioxidant responses in spinach. Braz. J. Plant Physiol. 21: 103–111.

Parent, S.É., L.E. Parent, D.E. Rozane, and W. Natale. 2013b. Plant ionome diagnosis using sound balances: case study with mango (Mangifera Indica). Front. Plant Sci. 4: 449.

Parent, S.É., L.E. Parent, J.J. Egozcue, D.E. Rozane, A. Hernandes, L. Lapointe, V. Hébert-Gentile, K. Naess, S. Marchand, J. Lafond, et al. 2013a. The plant ionome revisited by the nutrient balance concept. Front. Plant Sci. 4: 39.

Peacock, L., S. Worner, and R. Sedcole. 2006. Climate variables and their role in site discrimination of invasive insect species distributions. Environmental Entomology. 35 (1): 958–963

Pedrosa, A.M., Martins, C.d.P.S., Gonçalves, L.P., Costa, M.G.C. 2015. Late Embryogenesis Abundant (LEA) constitutes a large and diverse family of proteins involved in development and abiotic stress responses in sweet orange (Citrus sinensis L. Osb.).PLoS ONE. 10, e0145785.

Peng, T., et al. 2014. P tr BAM 1, a β-amylase-coding gene of Poncirus trifoliata, is a CBF regulon member with function in cold tolerance by modulating soluble sugar levels. Plant, Cell & Environment. 37 (12): 2754–2767.

Pérez-Pérez, J.G., Robles, J.M., Tovar, J.C., Botía, P.2009. Response to drought and salt stress of lemon 'Fino 49' under field conditions: Water relations, osmotic adjustment and gas exchange. Sci. Hortic. 122: 83–90.

Pérez-Pérez, J.G., Romero, P., Navarro, J.M., Botía, P. 2008. Response of sweet orange cv 'Lane late' to deficit irrigation in two rootstocks. I: Water relations, leaf gas exchange and vegetative growth. Irrig. Sci. 26: 415–425.

Perotti, V. E., et al. 2015. Proteomic and metabolomic profiling of Valencia orange fruit after natural frost exposure. Physiologia Plantarum. 153 (3): 337–354.

Podda, A., Checcucci, G., Mouhaya, W., Centeno, D., Rofidal, V., del Carratore, R., Luro, F., Morillon, R., Ollitrault, P., Maserti, B.E. 2013. Salt-stress induced changes in the leaf proteome of diploid and tetraploid mandarins with contrasting Na⁺ and Cl⁻ accumulation behaviour. J. Plant Physiol. 170: 952–959.

Polle, A., Pfirrmann, T., Chakrabarti, S., Rennenberg, H. 1993. The effects of enhanced ozone and enhanced carbon dioxide concentrations on biomass, pigments and antioxidative enzymes in spruce needles (Picea abies L.). Plant Cell Environ. 16: 311–316.

Prabha, R., D.P. Singh, and M.K. Verma. 2017. Microbial Interactions and Perspectives for Bioremediation of Pesticides in the Soils. In Plant-Microbe Interactions in Agro-Ecological Perspectives, Singh, D.P., Singh, H.B., Prabha, R., Eds., Springer: Singapore, 2017. 2: 649–671.

Primo-Capella, A., et al. 2022. Comparative transcriptomic analyses of citrus cold-resistant vs. sensitive rootstocks might suggest a relevant role of ABA signaling in triggering cold scion adaption. BMC Plant Biology 22 (1): 1–26.

Primo-Capella, A., Forner-Giner, M.A. 2021. Selección de Patrones de Cítricos a Bajas Temperaturas, Instituto Valenciano de Investigaciones Agrarias: Moncada, Valencia, Spain.

Primo-Capella, A., Forner-Giner, M.A., Martínez-Cuenca, M.R., Terol, J. 2021. Comparative transcriptomic analyses of Citrus cold resistant vs. sensitive rootstocks suggest a crucial role of ABA signaling in triggering cold scion adaption. BMC Plant Biol. under review, submitted on 23 April 2021.

Primo-Capella, A., Martínez-Cuenca, M.-R., Gil-Muñoz, F., Forner-Giner, M.A. 2021. Physiological characterization and proline route genes quantification under long-term cold stress in Carrizo citrange. Sci. Hortic. 276: 109744.

Raga, V. 2016. Genetic analysis of salt tolerance in a progeny derived from the citrus rootstocks Cleopatra mandarin and trifoliate orange. Tree Genet. Genomes. 12: 34.

Rai, A.N., Penna, S. 2013. Molecular evolution of plant P5CS gene involved in proline biosynthesis. Mol. Biol. Rep. 40: 6429–6435.

Raiol-Junior, L.L., Baia, A.D., Luiz, F.Q., Fassini, C.G., Marques, V.V., Lopes, S.A. 2017. Improvement in the excised citrus leaf assay to investigate inoculation of 'Candidatus Liberibacter asiaticus' by the Asian citrus psyllid Diaphorina citri. Plant Dis. 101: 409–413.

Rapisarda, P. and Giuffrida, A. 1992. Anthocyanins level in Italian blood oranges. Proc. Int. Soc. Citriculture.

Rebolledo Roa, A. 2012. Fisiología de la floración y fructificación enloscítricos, Corporación Universitaria Lasallista.

Reyes, J.L., Rodrigo, M.J., Colmenero-Flores, J.M., Gil, J.V., Garay-Arroyo, A., Campos, F., Salamini, F., Bartels, D., Covarrubias, A.A. 2005. Hydrophilins from distant organisms can protect enzymatic activities from water limitation effects in vitro. Plant Cell Environ. 28, 709–718.

Reyes, C.A., De Francesco, A., Peña, E.J., Costa, N., Plata, M.I., Sendin, L., Castagnaro, A.P., García, M.L. 2011. Resistance to Citrus psorosis virus in transgenic sweet orange plants is triggered by coat protein–RNA silencing. J. Biotechnol. 151: 151–158.

Reuscher, S., K. Isuzugawa, M. Kawachi, A. Oikawa, and K. Shiratake. 2014. Comprehensive elemental analysis of fruit flesh from European pear 'La France' and its giant fruit bud mutant indicates specific roles for B and Ca in fruit development. Sci. Hortic. 176: 255–260.

Roeschlin, R.A., Uviedo, F., García, L., Molina, M.C., Favaro, M.A., Chiesa, M.A., Tasselli, S., Franco-Zorrilla, J.M., Forment, J., Gadea, J., et al. 2019. PthA4AT, a 7.5-repeats transcription activator-like (TAL) effector from Xanthomonas citri ssp. citri, triggers citrus canker resistance. Mol. Plant Pathol. 20: 1394–1407.

Rogers, S., J.H. Graham, and C.W. McCoy. 2000. Larval growth of Diaprepes abbreviatus (L.) and resulting injury to three citrus varieties in two soil types. Journal of Economical Entomology. 93 (1): 380–387.

Rosa, C., Kamita, S.G., Dequine, H., Wuriyanghan, H., Lindbo, J.A., Falk, B.W. 2010. RNAi effects on actin mRNAs in Homalodisca vitripennis cells. J. RNAi Gene Silenc. 6, 361.

Rosa, C., Kamita, S.G.,, Falk, B.W. 2012. RNA interference is induced in the glassy winged sharpshooter Homalodisca vitripennis by actin dsRNA. Pest Manag. Sci. 68: 995–1002.

Rosales, R., and J.K. Burns. 2011. Phytohormones changes and carbohydrate status in sweet orange fruit from huanglongbing-infected trees. Journal of Plant Growth & Regulation. 30 (1):312–321.

Rosenzweig, C., A. Iglesias, X.B. Yang, P.R. Epstein, and E. Chivian. (2001). Climate change and extreme weather events: Implications for food production, plant diseases, and pests. Glob. Change Human Health. 2: 90–104.

Ruiz-Sánchez, M.C., Domingo, R., Savé, R., Biel, C., Torrecillas, A. 1997. Effects of water stress and rewatering on leaf water relations of lemon plants. Biol. Plant. 39: 623.

Ruiz-Sánchez, M.C., Domingo, R., Savé, R., Biel, C., Torrecillas, A. 2016. Effects of water stress and rewatering on leaf water relations of lemon plants. Biol. Plant. 39, 623.

Sakai, A. and Larcher, W. 2012. Frost survival of plants: responses and adaptation to freezing stress. Springer Science & Business Media.

Sala Junior, V., V.R. Celloto, L.G.E. Vieira, J.E. Gonçalves, R.A.C. Gonçalves, and A.J.B. de Oliveira. 2008. Floral nectar chemical composition of floral nectar in conventional and transgenic sweet orange, Citrus sinensis (L.) Osbeck, expressing an antibacterial peptide. Plant Systematics and Evolution. 275 (1): 1–7.

Sanchez-Ballesta, M. T., et al. 2000. Involvement of phenylalanine ammonia-lyase in the response of Fortune mandarin fruits to cold temperature. Physiologia Plantarum. 108 (4): 382–389.

Schaad, N.W., E. Postnikova, G. Lacy, et al. 2006. Emended classification of xanthomonad pathogens on citrus. Papers in Plant Pathology. Systematic and Applied Microbiology. 29 (1): 690–695.

Shalom, L., S. Samuels, N. Zur, L. Shlizerman, A. Doron-Faigenboim, E. Blumwald and A. Sadka. 2014. Fruit load induces changes in global gene expression and in abscisic acid (ABA) and indole acetic acid (IAA) homeostasis in citrus buds. J. Exp. Bot. 65: 3029–3044.

Sharma, P., and Dubey, R.S. 2005. Lead toxicity in plants. Braz. J. Plant Physiol. 17: 1677.

Sicilia, A., Scialò, E., Puglisi, I., Lo Piero, A.R. 2020. Anthocyanin biosynthesis and DNA methylation dynamics in sweet orange fruit [Citrus sinensis L. (Osbeck)] under cold stress. J. Agric. Food Chem. 68: 7024–7031.

Simpson, C.R., Nelson, S.D., Melgar, J.C., Jifon, J., King, S.R., Schuster, G., Volder, A. 2014. Growth response of grafted and ungrafted citrus trees to saline irrigation. Sci. Hortic. 169: 199–205.

Singh, A., Sharma, S., Singh, B. 2017. Effect of germination time and temperature on the functionality and protein solubility of sorghum flour. J. Cereal Sci. 76: 131–139.

Slisz, A.M., A.P. Breksa, D.O. Mishchuk, G. McCollum, and C.M. Slupsky. 2012. Metabolomic analysis of citrus infection by "Candidatus Liberibacter" reveals insight into pathogenicity. Journal of Proteome Research. 11 (1): 4223–4230. doi: 10.1021/pr300350x

Soler, N., Plomer, M., Fagoaga, C., Moreno, P., Navarro, L., Flores, R., Peña, L. 2012. Transformation of Mexican lime with an intron-hairpin construct expressing untranslatable versions of the genes coding for the three silencing suppressors of Citrus tristeza virus confers complete resistance to the virus. Plant Biotechnol. J. 10: 597–608.

Spann, T.M., R.A. Atwood, J.D. Yates, M.E. Rogers, and R.H. Brlansky. 2010. Dooryard Citrus Production: Citrus Greening Disease, EDIS HS1131, University of Florida, Institute of Food and Agricultural Sciences: Gainesville, FL, USA, 2010.

Spann, T.M., and Little, H.A. 2011. Applications of a commercial extract of the brown seaweed Ascophyllum nodosum increases drought tolerance in container-grown 'Hamlin' sweet orange nursery trees. Hort Science. 46: 577–582.

Sun, L.M., X.Y. Ai, W.Y. Li, W.W. Guo, X.X. Deng, C.G. Hu, and J.Z. Zhang. 2012. Identification and comparative profiling of miRNAs in an early flowering mutant of trifoliate orange and its wild type by genome-wide deep sequencing. PLoS ONE 7: e43760.

Sun, X., A. Zhu, S. Liu, L. Sheng, Q. Ma, L. Zhang, E.M.E. Nishawy, Y. Zeng, J. Xu, Z. Ma, et al. 2013. Integration of metabolomics and sub cellular organelle expression microarray to increase understanding the organic acid changes in post-harvest citrus fruit. J. Integr. Plant Biol. 55: 1038–1053.

Syvertsen, J.P., Garcia-Sanchez, F.2014. Multiple abiotic stresses occurring with salinity stress in citrus. Environ. Exp. Bot.103: 128–137.

Taning, C.N., Andrade, E.C., Hunter, W.B., Christiaens, O., and Smagghe, G. 2016. Asian Citrus Psyllid RNAi Pathway-RNAi evidence. Sci. Rep. 6: 38082.

Tanou, G., Ziogas, V., Belghazi, M., Christou, A., Filippou, P., Job, D., Fotopoulos, V. and Molassiotis, A. 2014. Polyamines reprogram oxidative and nitrosative status and the proteome of citrus plants exposed to salinity stress. Plant Cell Environ. 37: 864–885.

Tester, M., Davenport, R.2003. Na+ tolerance and Na+ transport in higher plants. Ann. Bot. 91: 503–527.

Thieme, C.J., Rojas-Triana, M., Stecyk, E., Schudoma, C., Zhang, W., Yang, L., Miñambres, M., Walther, D., Schulze, W.X., Paz-Ares, J., et al. 2015. Endogenous Arabidopsis messenger RNAs transported to distant tissues. Nat. Plants. 1: 15025.

Trouet, V., et al. 2018. Recent enhanced high-summer North Atlantic Jet variability emerges from three-century context. Nature Communications 9 (1): 1–9.

Van Oosten, M.J., Pepe, O., De Pascale, S., Silletti, S., and Maggio, A. 2017. The role of biostimulants and bioeffectors as alleviators of abiotic stress in crop plants. Chem. Biol. Technol. Agric. 4: 5.

Verslues, P. E., et al. 2006. Methods and concepts in quantifying resistance to drought, salt and freezing, abiotic stresses that affect plant water status. The Plant Journal. 45 (4): 523–539.

Vidal, C., Ruiz, A., Ortiz, J., Larama, G., Perez, R., Santander, C., Ferreira, P.A.A., and Cornejo, P. 2020. Antioxidant responses of phenolic compounds and immobilization of copper in Imperata cylindrica, a plant with potential use for bioremediation of Cu contaminated environments. Plants 9: 1397.

Viens, N., and Bosch, J. 2000. Weather-dependent pollinator activity in an apple orchard, with special reference to Osmia cornuta and Apis mellifera (Hymenoptera: Megachilidae and Apismellifer). Environmental Entomology. 29 (1): 413–420.

Voo, S.S. and Lange, B.M. 2014. Sample preparation for single cell transcriptomics: essential oil glands in Citrus fruit peel as an example. Methods Mol. Biol. 1153: 203–212.

Vu, J. C. and Yelenosky, G.1991. Photosynthetic responses of citrus trees to soil flooding. Physiologia Plantarum. 81 (1): 7–14.

Vu, J.C.V., and Yelenosky, G. 1993. Photosynthesis and freeze tolerance comparisons of the newly released "amber sweet" hybrid with "valencia" orange. Environ. Exp. Bot. 33: 391–395.

Vu, J.C.V., and Yelenosky, G. 1991. Photosynthetic responses of citrus trees to soil flooding. Physiol. Plant. 81: 7–14.

Wang, X., Xu, Y., Zhang, S., Cao, L., Huang, Y., Cheng, J., Wu, G., Tian, S., Chen, C., Liu, Y., Yu, H. 2017. Genomic analyses of primitive, wild and cultivated citrus provide insights into asexual reproduction. Nature Genetics. 49 (5): 765–772.

Warschefsky, E. J., et al. 2016. Rootstocks: diversity, domestication, and impacts on shoot phenotypes. Trends in Plant Science. 21 (5): 418–437.

Waterhouse, P.M., M.B. Wang, and T. Lough. 2001. Gene silencing as an adaptive defence against viruses. Nature. 411 (1): 834.

Wu, G.A., Terol, J., Ibanez, V., Lopez Garcia, A., Perez Roman, E., Borredá, C., Domingo, C., Tadeo, F.R., Carbonell-Caballero, J., Alonso, R., Curk, F., Du, D., Ollitrault, P., Roose, M.L., Dopazo, J., Gmitter, F.G., Rokhsar, D.S., and Talon, M. 2018. Genomics of the origin and evolution of Citrus. Nature. 554 (1): 311–316.

Wu, J., Z. Xu, Y. Zhang, L. Chai, H. Yi, and X. Deng. 2014b. An integrative analysis of the transcriptome and proteome of the pulp of a spontaneous late-ripening sweet orange mutant and its wild type improves our understanding of fruit ripening in citrus. J. Exp. Bot. 65: 1651–1671.

Xiao, J.P., Zhang, L.L., Zhang, H.Q., and Miao, L.X. 2017. Identification of genes involved in the responses of tangor (C. reticulata × C. sinensis) to drought stress. Bio. Med. Res. Int.

Yang, C.Q., Y.Z. Liu, J.C. An, S. Li, L.F. Jin, G.F. Zhou, Q.J. Wei, H.Q. Yan, N.N. Wang, L.N. Fu, et al. 2013. Digital gene expression analysis of corky split vein caused by boron deficiency in 'Newhall' navel orange (Citrus sinensis Osbeck) for selecting differentially expressed genes related to vascular hypertrophy. PLoS ONE. 8: e65737.

Yang, L., et al. 2012. Differences in cold tolerance and expression of two fatty acid desaturase genes in the leaves between fingered citron and its dwarf mutant. Trees. 26 (4): 1193–1201.

Yang, L., Ye, J., Guo, W.-D., Wang, C.-C., and Hu, H.-T. 2012. Differences in cold tolerance and expression of two fatty acid desaturase genes in the leaves between fingered citron and its dwarf mutant. Trees. 26: 1193–1201.

Yu, X., Gowda, S., and Killiny, N. 2017. Double-stranded RNA delivery through soaking mediates silencing of the muscle protein 20 and increases mortality to the Asian citrus psyllid, Diaphorinacitri. Pest Manag. Sci. 73: 1846–1853.

Zaslo, M.1992. Antibiotic peptides as mediators of innate immunity. Current Opinion in Immunology. 4 (1): 3–7.

Zhang, A., Cortés, V., Phelps, B., Van Ryswyk, H., and Srebotnjak, T.2018. Experimental analysis of soil and mandarin orange plants treated with heavy metals found in oilfield-produced waste water. Sustainability. 10: 1493.

Zhang, J.Z., K. Zhao, X.Y. Ai, and C.G. Hu. 2014a. Involvements of PCD and changes in gene expression profile during self-pruning of spring shoots in sweet orange (Citrus sinensis). BMC Genomics. 15: 892.

Zhang, L., and Reed, R.D. A Practical Guide to CRISPR/Cas9 Genome Editing in Lepidoptera. In Diversity and Evolution of Butterfly Wing Patterns, Sekimura, T., Nijhout, H., Eds., Springer: Singapore, 2017. 155–172.

Zhang, X., M.I. Francis, W.O. Dawson, et al. 2010. Over-expression of the Arabidopsis NPR1 gene in citrus increases resistance to citrus canker. European Journal of Plant Pathology. 128 (1): 91–100.

Zhang, Y.J., X.J. Wang, J.X. Wu, S.Y. Chen, H. Chen, L.J. Chai, and H.L. Yi. 2014b. Comparative transcriptome analyses between a spontaneous late-ripening sweet orange mutant and its wild type suggest the functions of ABA, sucrose and JA during citrus fruit ripening. PLoS ONE. 9: e116056

Zheng, Z., Bao, M., Wu, F., Chen, J., and Deng, X. 2016. Predominance of single prophase carrying a CRISPR/cas system in "Candidatus Liberibacter asiaticus" strains in southern China. PLoS ONE 11: e0146422.

Zeng, J., C. Gao, G. Deng, B. Jiang, G. Yi, X. Peng, Y. Zhong, B. Zhou and K. Liu .2012. Transcriptome analysis of fruit development of a citrus late-ripening mutant by microarray. Sci. Hortic. 134: 32–39.

Zhang, J.Z., K. Zhao, X.Y. Ai and C.G. Hu. 2011a. Involvements of PCD and changes in gene expression profile during self-pruning of spring shoots in sweet orange (*Citrus sinensis*). BMC Genomics. 15: 892.

Zhong, Y., C.Z. Cheng, N.H. Jiang, B. Jiang, Y.Y. Zhang, B. Wu, M.L. Hu, J.W. Zeng, H.X. Yan, G.J. Yi, et al. 2015. Comparative transcriptome and iTRAQ proteome analyses of citrus root responses to Candidatus Liberibacter asiaticus infection. PLoS ONE. 10: e0126973.

Ziogas, V., Tanou, G., Belghazi, M., Filippou, P., Fotopoulos, V., Grigorios, D., and Molassiotis, A. 2015. Roles of sodium hydrosulfide and sodium nitroprusside as priming molecules during drought acclimation in citrus plants. Plant Mol. Biol. 89: 433–450.

Ziogas, V., Tanou, G., Belghazi, M., Filippou, P., Fotopoulos, V., Grigorios, D., Molassiotis, A. 2021. Roles of sodium hydrosulfide and sodium nitroprusside as priming molecules during drought acclimation in citrus plants. Plant Mol. Biol. 89, 433–450.

Zou, Y., Zhang, F., Srivastava, A.K., Wu, Q., and Ku˘ca, K. 2021. Arbuscular mycorrhizal fungi regulate polyamine homeostasis in roots of trifoliate orange for improved adaptation to soil moisture deficit stress. Front. Plant Sci. 11.

7 *Crocus sativus* and Stressful Conditions

Seyedeh-Somayyeh Shafiei-Masouleh[1]* and Hamed Hassanzadeh Khankahdani[2]
[1]Department of Genetics and Breeding, Ornamental Plants Research Center (OPRC), Horticultural Sciences Research Institute (HSRI), Agricultural Research, Education and Extension Organization (AREEO), Mahallat, Iran
[2]Horticulture Crops Research Department, Hormozgan Agricultural and Natural Resources Research and Education Center, AREEO, Bandar Abbas, Iran
*Corresponding author: shafiee.masouleh@areeo.ac.ir; shafyii@gmail.com

CONTENTS

7.1 INTRODUCTION

Crocus sativus is known as saffron crocus or autumn crocus (family Iridaceae). It is widely known as a spice. It is a valuable spice produced from the red stigmas of the plant. These plants resemble *Colchicum* genus (family Colchicaceae), but saffron has three stamens and three styles (Bowles 1952; Askari-Khorasgani and Pessarakli 2019; Cardone et al. 2020), and *Colchicum* has six stamens and one style. The genera are propagated by corms (Bowles 1952). Other applications of saffron include in the food, drug, fabric and cosmetic industries. Recently, the medicinal properties of saffron have been of interest in both traditional and modern medicines, due to its particular cytotoxic, liver detoxifying, analgesic, antitumor and mood-elevating properties, which contribute to making saffron the most expensive spice worldwide (Gómez-Gómez et al. 2012; Siracusa et al. 2010; Anastasaki et al. 2010).

DOI: 10.1201/9781003242963-7

The main regions of saffron cultivation are Iran (as the main producer country), India, Turkey, Italy, Spain, Morocco, Switzerland, Greece and Central Asia, and more recently Australia, Argentine, New Zealand, and Mexico (Jalali-Heravi et al. 2010). Optimal climatic conditions for the species are a rainy autumn, mild winter and warm summer (Amirnia et al. 2014).

Askari-Khorasgani and Pessarakli (2019) described the edaphoclimatic conditions that affect the production of daughter corms and the yield of stigma, including soil properties, precipitation, geographical coordinates and especially temperature. They also affirmed that agricultural management such as fertilization, irrigation, planting space and date, plant depth, mother corm size as well as harvest and postharvest activities can affect the corm and stigma performances.

There are some problems in saffron cultivation areas such as lack of corm rot-resistant cultivars, continuous cultivation, poor agronomic practices, mono-culturing, frequent droughts, long cropping cycle, lack of irrigation facilities, diseases and pests, and nutrient depletion, which can all affect the saffron yield (Gupta and Razdan 2012).

One of the plant mechanisms against stressful conditions is the production of secondary metabolites, which protect plants from stress while making adaptations as well as benefits for human health (Albergaria et al. 2020). Plants suffer from different continuous threats from the environment, including biotic (fungi, viruses, insects, nematodes) and abiotic stresses (drought, salinity, temperature, UV radiation). These threats are recognized by plants as signals through receptors, and they respond to these signals, called elicitors, by activation of defense mechanisms. One of these mechanisms is the synthesis of secondary metabolites, including volatile oils, flavonoids, alkaloids, glycosides, tannins, resins, etc., which also are important as vital sources of food additives (flavors) and pharmaceuticals. Therefore, stressful conditions could be a beneficial source for improving the secondary metabolites of medicine plants (Thakur et al. 2019). We reviewed the effects of stressful conditions, including biotic and abiotic stress, on saffron plants as a medicinal plant as well as its other applications such as spice, cosmetic, dye, and so on, and how stress conditions change the medicinal qualification and economic value of saffron beneficially or detrimentally.

Another plant mechanism against stressful conditions is the synthesis of osmolytes, including soluble carbohydrates, amino acids such as proline, and antioxidant enzymes such as catalase, peroxidase, superoxide dismutase, and so on, which scavenge the oxidative stress-produced reactive oxygen species (ROS). Therefore, understanding how saffron plants respond to common stressful conditions helps to decide on suitable agricultural management for increasing profitability economically and pharmaceutically.

7.2   BIOTIC STRESS

Plants face various living organisms, including fungi, viruses, bacteria, insects and weeds, which are destructive to plant growth and development. Among them, fungi are the greatest enemy of plants compared to the other biotic stress, with a lot of species that act parasitically against plants. Furthermore, viruses are more dangerous for plants because they function on all parts of plants and can infect other plants rapidly. Moreover, weeds are considered another threat to crops because they limit crop growth and development directly, or indirectly by reducing the nutrient availability for crops (Saddique et al. 2018).

Saffron corm rot is a disease caused by various soil-borne fungi such as *Fusarium* spp., *Macrophomina phaseolina*, *Sclerotium rolfsii* and *Pencillium corymbiferum*. The outbreak of the disease is affected by various factors such as poor soil aeration, injured corms, soil temperature up to 32°C, and so on (Gupta and Razdan 2012; Gupta et al. 2021). Virus infections in saffron plants are rarely reported, with a new potyvirus called saffron latent virus (SaLV) being found in the saffron fields of Iran (Parizad et al. 2019). Weeds are wild plants that are another biotic stress that may affect saffron plants in different areas of saffron cultivation. Saffron plants are perennial crops that are affected by annual, biennial and perennial weeds (Mzabri et al. 2021).

7.3 ABIOTIC STRESS

Saffron plant growth and development, as well as saffron spice yield, composition and quality, are affected by the geographical origin, abiotic stresses, agronomical practices and processing conditions (such as stigma separation, drying and storage). The relative percentages of particular secondary metabolites, including crocin, picrocrocin and safranal, determine the organoleptic features of the spice and its quality based on the International Organization for Standardization (Caser et al. 2019b).

The cultivation regions of saffron impose some abiotic stress on plants, including salinity, water scarcity and/or drought stress, waterlogging, freezing and/or chilling stress, and even heat stress in hot summers, as well as heavy metal pollution, which are common stresses for saffron plants. Farrokhi et al. (2021) studied the effects of climatic and environmental characteristics of six saffron cultivation regions of Iran on the yield and quality of saffron. They found the regions that have low maximum summer temperatures to be the best regions for saffron production. They reported that Razavi Khorasan province has the best saffron in terms of contents of safranal, picrocrocin and crocin, and stigma yield, while North Khorasan province was superior in terms of biochemical characteristics. Even as commonly mentioned stresses, another environmental stress that may affect saffron plants is ultraviolet-B (UV-B) radiation. It leads to generating oxygen free radicals, which disrupt the balance of metabolism in cells (Rikabad et al. 2019).

7.3.1 SALINITY

Salt stress as one of the environmental limitations whose high levels causes hyper-ionic and hyper-osmotic stresses due to limitations of water absorption and nutrition deprivation as well as excess uptake of Na^+ and Cl^-. These all affect plant growth and productivity, and finally, cause plant death. Plants cope with these conditions through morphological and biochemical mechanisms. Moslemi et al. (2021) found that saffron plants under salinity showed increased Na^+ accumulation and decreased uptake of K^+, and the production of crocin, remarkably, was stopped. They found that key genes of apocarotenoid biosynthesis pathways (CsCCD2, CsCCD4, CsUGT2, CsCHY-β, and CsLCYB) were affected by salinity levels. Avarseji et al. (2013) showed that salinity can cause a decrease in leaf water content and an increase in electrolyte leakage, however, the use of 50% extra concentration of potassium fertilizers in the soil and rhizosphere can improve the detrimental effects of salinity up to 9.4 dS m^{-1}. Salinity causes a decrease in chlorophyll contents due to a reduction of the photosynthesis level and Na^+ emission-induced chlorophyll degradation (Liu et al. 2011).

As is known, no cultivars of saffron plants were introduced or released up to now, and there are different ecotypes of saffron, which are cultivated, therefore, ecotypes may have various responses to environmental changes compared to cultivars. Based on two reports (Torabi Pashai et al. 2016; Rahiman et al. 2018), saffron ecotypes have different antioxidant activities and different pharmaceutical activities, as well as differentiated tolerances to various abiotic stresses such as salinity. Therefore, to cope with salinity and achieve greater growth and development of saffron plants, farmers need to care about the management related to the agricultural practices, including fertilizers, bio-stimulants, planting date and space, and irrigation. Decision-making to have the best agricultural practices needs a greater understanding of plant responses and mechanisms. In other sections of this chapter, strategies and mechanisms for this are described.

7.3.2 WATER STRESS AND WATERLOGGING STRESS

Saffron corms are planted in the late summer to early fall, and so heavy rain in the fall may cause waterlogging stress on plants at the beginning stages of growth cycles. Waterlogging-stressed plants have been observed to have reduced root growth and development due to hypoxia (oxygen deficiency) and ethylene accumulation (Sorooshzadeh et al. 2012). Keyhani et al. (2003) reported that waterlogging effects on *Crocus sativus* depend on the development stage of plants, and hypoxia/ anoxia, as a signal, increases either alcohol dehydrogenase or NAD-dependent lactate dehydrogenase in plants.

Drought stress as well as waterlogging is an abiotic agent that affects saffron morphologically and physiologically. In a report by Sorooshzadeh et al. (2012), plants under drought stress and exogenous abscisic acid (ABA) had more root length and improved protein content of corms. They observed that antioxidant enzymes, including superoxide dismutase (SOD) and peroxidase (POX), increased in roots, corms and leaves, respectively. As is well known, antioxidant enzymes play a role in diminishing reactive oxygen species and protecting plants against these detrimental factors in stressful conditions. It has been stated that one of the roles of ABA as a plant hormone is the induction of gene expression and protecting plants from stress, especially water deficit. Pazoki et al. (2013) could not find any interaction between the planting method and drought stress, but each of them separately affected saffron yield. They found the flat plot method to be the best for the highest saffron yield.

Drought stress-reduced plant growth may be due to various changes in physiological and biochemical pathways such as photosynthesis, ion uptake and translocation, respiration, carbohydrate and other assimilates metabolism, and growth promoters (Farooq et al. 2009).

7.3.3 Freezing Stress

Iran is the most important country for the production of saffron worldwide, and its cultivation area features include cool to cold winters and warm summers with very little rainfall, therefore the plants are exposed to cold and winter frost, which limits saffron products. However, different ecotypes of saffron respond to freezing stress differently and have different tolerances (Nezami et al. 2016). Nezami et al. (2016) studied different ecotypes of Iranian saffron under freezing stress based on electrolyte leakage percent in leaves and corms, and a death percent of 50% of plants. They found the ecotype Torbat-e-Heydariyeh to be more tolerant, and corms were more tolerant than leaves against freezing stress. The yield of saffron is directly related to the flower number, and the flower number of saffron, similarly to many bulbous plants, depends on temperature as a critical factor for flower formation and development. Therefore, temperature stresses affect saffron yield. Saffron flower initiation occurs best at 23–27°C, and the ambient temperature for flower primordium formation and flower bud formation have special importance rather than the formation of leaf buds (Chen et al. 2021).

As mentioned above, saffron plants are mostly cultivated in regions with cold and dry winters, which induces freezing stress and causes detrimental effects on corm growth due to increasing electrolyte leakage. Koocheki and Seyyedi (2019) found that 4-year-old mother corms had greater contents of N, P and K in their leaves, roots and corms compared to younger corms. They observed higher electrolyte leakage in plants (in above- and underground organs) as well as increasing the proline content under decreasing temperature. They also reported that greater planting depth of corms reduced freezing stress effects. They stated that photosynthesis pigments were reduced under freezing conditions, however, the recovery of pigments was easier under −20°C compared to −10°C. Therefore, they recommended the use of older corms and greater depth of corm cultivation to alleviate freezing stress.

Kargar et al. (2020), based on cluster analysis, showed that ecotypes of saffron have different responses to chilling stress, and they recommended ecotypes of Torbat-e-Heydariyeh and Ferdows for cultivation in cold regions due to their higher biological functions and yields under chilling stress.

7.3.4 Nutrient Toxicity and Heavy Metal Stresses

In plants, nutrient toxicity may cause some detrimental and stressful effects. One of these nutrients that may affect saffron plants is copper toxicity, which alters lipid composition and the content of protein, and may decrease biomass production and root elongation as well as causing a reduction of plant growth, DNA mutations and blocked photosynthetic electron transport. This toxicity has a key mechanism which is the production of reactive oxygen species (ROS), which causes oxidative stress (Keyhani et al. 2006), and so, to protect against it, plants have some strategies and mechanisms.

In addition to nutrient toxicity, heavy metal pollution is another cultivation factor found in regions of saffron growth. Rostami et al. (2015) performed an interesting study of soil pollution with heavy metals, including nickel nitrate, silver nitrate, zinc nitrate, copper carbonate, lead nitrate and manganese sulfate, before sowing saffron corms. They found that photosynthetic pigments of leaves, except carotenoids, were affected by pollution as well as effects on proline and soluble carbohydrates, and not reducing carbohydrates. An interesting finding was the increased content of crocin under heavy metal pollution, except for nickel nitrate. Therefore, higher contents of apocarotenoids may be an alarm to be a heavy metal pollution, and it needs more care.

Chromium (Cr) is not considered a heavy metal as it is a transition metal, and is not an essential element for plants, although it has some effects on plant physiology and stressful impacts on plant growth and development. Rao et al. (2016) reported nine proteins that were downregulated under Cr stress such as cell division cycle protein 48 homolog, ATP synthase subunit alpha, a large chain of ribulose bisphosphate carboxylase, proteasome subunit alpha type, and six that were upregulated, including sucrose synthase 2, eukaryotic initiation factor 4A, alpha-1,4-glucan-protein synthase, 1-aminocyclopropane-1-carboxylate oxidase, and isoflavone reductase homolog IRL.

Cadmium is a carcinogenic heavy metal currently used in industry that also affects saffron growth and yield as well as having adverse health implications for humans. Hadizadeh and Keyhani (2006) reported that cadmium affects the rooting of corms and increases catalase activity, which controls cadmium toxicity.

As regards food and medicinal usage of saffron, heavy metal pollution and stress more than other stress are important to address. Therefore, agricultural management of stress must be focused on diminishing the metals in plant tissues more than focusing on plant growth and development. Ebrahimi et al. (2021) showed that the investigated organic fertilizers (humic acid, organic mineral emulsion and cow manure) had effects on stigma yield in the second year, with the best being cow manure. They also reported that major metal contaminants (arsenic, lead and mercury) were the lowest found in organic fertilizers.

7.4 STRATEGIES TO ALLEVIATE STRESS

7.4.1 PLANT STRATEGIES

One of the mechanisms in plants to address ROS-induced stress is the enzyme activities of plants such as peroxidases. For example, among various peroxidase isoenzymes with different biological roles, lignin peroxidases have a role in lignin degradation as well as in strengthening of the cell wall, and ascorbate peroxidases are involved in the defense against oxidative stress (Ward et al. 2001; Shigeoka et al. 2002). Keyhani et al. (2006) assayed the activities of ascorbate peroxidase and lignin peroxidase in saffron corms under copper toxicity stress. They observed the specific function of each enzyme and stimulation of root elongation but growth of roots was aborted when the element reaches more than critical concentrations. They showed the activity of ascorbate peroxidase based on H_2O_2-mediated oxidation of ascorbate and observed a decrease in its activity, and showed lignin peroxidase activity based on H_2O_2-mediated oxidation of ferulic acid and H_2O_2-mediated oxidation of 2,2'-azinobis (3-ethylbenzothiazoline-6-sulfonic acid) (ABTS), and observed various enzyme activities based on substrate monitoring.

The promotion of osmolytes content, including pigments, soluble carbohydrates and amino acids, is another plant mechanism against stress. The taste, color and aroma of saffron stigma are due to rare apocarotenoids, including crocin, picrocrocin and safranal, respectively. These compounds give analgesic and sedative properties to herbal medicine. Furthermore, they have roles in the reduction of oxidative damage and plant resistance to biotic stresses besides proteins, carbohydrates, flavonoid pigments, vitamins, etc. All of these metabolites are involved in chemical signaling and plant growth, development and environmental acclimation (Baba et al. 2015). Carotenoids are known as components with a photoprotection role and also a role in membrane protection against oxidative stress. Salinity stress can cause oxidative stress, and an increase in carotenoid was reported under salinity conditions (Verma and Misra 2005). Under salinity conditions, boosting total sugar content is a stress tolerance mechanism (Rajaei et al. 2009). Babaei et al. (2021) found greater total sugar content in saffron plants subjected to salinity, even more than in nitric oxide (NO)' and salicylic acid (SA)-treated plants. Iqbal et al. (2014) revealed that NO plays a role in proline metabolism regulation through promoting the nitrogen content.

Mzabri et al. (2016) investigated salinity impacts on 4-year-old saffron plants under NaCl concentrations. They reported some morpho-biochemical responses of plants under salt concentrations. They found a decrease in numbers, lengths and areas of leaves, daughter corms, chlorophyll content and quantum yield of PSII, and an increase in the contents of proline, soluble sugars and total phenols. They stated that the accumulation of organic solutes protects plants against salinity by maintaining the relative water content and malondialdehyde (MDA). They did not observe any detrimental effects of salinity on stigma yields. They also observed some positive effects of NaCl on growth parameters in Morocco.

In another study, Mzabri et al. (2017) investigated drought stress (100%, 60% and 40% irrigation) on saffron plants in Morocco. They found decreases in chlorophyll content and PSII quantum yield, and increases in proline content, soluble carbohydrates and total phenols. They stated that increasing the organic solutes maintains the relative water content and malondialdehyde, which are the plant-adaptative responses against stress. They observed a decrease in stigma yield under drought conditions. Therefore, despite salinity, drought stress can affect saffron plants economically.

Esf et al. (2009) showed that light conditions under low temperature in saffron lead to more activities of catalase, glutathione reductase and ascorbate peroxidase compared to dark conditions. However, superoxide dismutase activity was reported inversely. They stated that the increased activities of the enzymes maintained the cell membranes under the detrimental effects of light and low temperature.

Therefore, it can be said that plant strategies against stresses include morphology changes, synthesis of secondary metabolites, synthesis of organic compounds and antioxidant enzymes.

7.4.2 Agricultural Management

Phenological stages of saffron plants include reproductive, vegetative and dormancy. Ecological conditions and agricultural management complicatedly and directly affect flower induction of saffron. Saffron is a geophyte plant and the main environmental factors are both seasonal and daily thermoperiodism. Storage of corms at high temperatures (23–27°C) is needed for flower induction, then moderately low temperature (17°C) is necessary for flower emergence. In Mediterranean regions, these physiological factors occur in early spring to mid-summer (for flower induction), and early to late fall (for flower emergence). The time of flower initiation of saffron is mostly related to corm size. After flowering, vegetative growth during the winter will occur at the base of shoots. Finally, leaf length is increased and then starts to senescence, and the dormancy of corms begins (Caser et al. 2019b).

Corm geographical origin (ecotype), as well as agronomic practices (such as fertilization and the use of bio-stimulants), can affect saffron yield in terms of stigma yield and quality. Agronomic practices can affect water and nutrient retention capacity (Cardone et al. 2020). Saffron is largely produced in arid and semi-arid climates that are subjected to low to moderate salinity stress.

Therefore, agricultural management and other preparations are crucial to increase the quality and quantity yields of saffron (Torghaban and Ahmadi 2011). Dastranj and Sepaskhah (2019) simultaneously investigated the effects of salinity, drought stress and planting method (basin and in-furrow planting) on the growth and yield of saffron. They found that the plants are more sensitive to drought stress than salinity. They recommended different levels of salinity and irrigation based on the planting method. They showed that 100% irrigation in all levels of salinity, including 0.45 (well water) to 3 dS m^{-1} (saline water) in basin planting produces desirable results, however, in in-furrow planting, higher levels of irrigation are needed to remove higher levels of salinity. Askari-Khorasgani and Pessarakli (2019) stated that a plant depth of less than 10 cm affects saffron plant growth and corm production, exposing it to abiotic stresses such as salinity and drought stress in less deep soil was named as one of the reasons. Therefore, plants have the best growth and corm production in the first year when they are planted at 10 cm soil depth, but in the next year the daughter corms are formed on the mother corms and so their depth will be less and this affects stigma and corm yields, and replanting of corms is necessary.

Saffron flowers should emerge before the leaves, which is important because of the ease of harvesting and the decrease in labor costs. Unsuitable irrigation time, such as early irrigation or precipitation after planting the corms, may cause leaf emergence before flowers. Therefore, mild water stress can be effective for suitable growth of saffron depending on timing, as well as increasing crocin and picrocrocin contents in stigma, and phosphorus content in corms (Askari-Khorasgani and Pessarakli 2019). Gobadi et al. (2015) indicated that larger corms and an earlier planting date (early summer) resulted in a greater yield of stigma compared to smaller corms and later planting (early fall). The latter resulted in a higher proline content in plants, which indicates the plant response to stress conditions, especially water deficit and salinity, as an indicator of plant resistance. Koocheki et al. (2014) showed that supplying 100% of the saffron water requirement is needed to promote flower characteristics and yield in the second year more than in the first year. Therefore, drought stress is a key factor in the second year compared to the first. They reported that small cormlets have less phosphorous than larger corms, and then, during cultivation, these cormlets need more nutrients with phosphorous fertilizers. Renau-Morata et al. (2012) stated that the development of saffron daughter corms occurs after the maximum sizes of roots and leaves of mother corms are achieved and depends on photosynthesis. They found that cultivation in a greenhouse causes small daughter corms compared to those grown in the field due to low irradiance even after transferring to the field. They also reported water stress as a limiting factor of the photosynthesis rate. They expressed the role of mother corms in supplying the replacement corm biomass as only 10%, and so supplying water is important.

Saffron cultivation in Iran is annual, but in some countries such as Spain (3–4 years), Italy (4–5 years), India and Greece (6–8 years) it is performed on a multi-year cycle. The advantage of annual cultivation is the effective control of biotic stress, especially diseases, as well as more accurate corm selection. Annual cultivation is recommended because corm multiplication and its size may remarkably decrease over the third year, which affects the flower yield (Caser et al. 2019a). Rasool et al. (2021) isolated 13 bacterial strains (plant growth-promoting rhizobacteria; PGPR) from rhizospheric soil of in-flowering saffron corms and investigated the effects of these strains *in vitro* for solubilization of phosphate, production of indole acetic acid, siderophore, hydrocyanic acid and ammonia production, and antagonism by a dual culture test against *Sclerotium rolfsii* and *Fusarium oxysporum*. They found that some strains produced IAA, some caused phosphate solubilization, siderophore and HCN, all of them caused ammonia production, and some of them had higher activities of protease/lipase/amylase/cellulase and/or chitinase. They reported three isolates, AIS-3, AIS-8 and AIS-10, that had the most plant growth properties, and effectively slowed the growth of *Sclerotium rolfsii* and *Fusarium oxysporum*. Integrated disease management, including the use of fungicides and bio-controls, can help farmers to increase the yield (Gupta and Razdan 2012).

7.5   MULTI-OMICS APPROACHES TO COPE WITH STRESS

Crocus sativus L. is a triploid plant and it is a sterile plant that is only propagated by corms. Therefore, the classic methods of improvement are not available for this crop, and genetic engineering may be an effective way to improve saffron. Then, an inventory of expressed genes in a specific tissue and developmental stage by partial sequencing of cDNA called expressed sequence tags (ESTs) can be established for a future program of saffron improvement. Gómez-Gómez and co-workers (2003) made libraries of ESTs sequenced in two stages of corms, including storage accumulation, and corm growth and dormancy. They found ESTs similar to antimicrobial peptides, pathogenesis-related proteins (chitinases, thaumatin and polygalacturonase inhibitor), other proteins involved in virus resistance (RNAases and a protein related to post-translational gene silencing), other proteins related to abiotic stress (late embryogenesis abundant protein and other stress-induced proteins) and oxidative stress (catalase and dehydroascorbate reductase).

Apocarotenoids are multi-functional compounds in plants, for example, they have various roles as hormones, pigments and volatiles. These compounds are formed through catalyzing by carotenoid cleavage dioxygenase (CCD). In saffron, two CCD4 genes of the CCD4 family (the largest family of plant CCDs) are expressed in the stigma tissue, and their roles are the production of volatiles to attract pollinators. One of these genes is CsCCD4c, which is special to the stigma tissue in saffron and other *Crocus* species. Upregulation of CsCCD4c is affected by different environmental stresses, including wounding, heat, cold and osmotic stress. Abscisic acid (ABA) is one of the biologically active apocarotenoids which acts as a positive regulator in abiotic stress responses (Rubio-Moraga et al. 2014). Furthermore, the synthesis of apocarotenoids such as β-ionone, β-cyclocitral and geranylacetone was reported under low temperatures or ultraviolet (UV) stress (Lamikanra et al. 2002).

Chen et al. (2021) investigated the protein profiles of saffron plants under cold stress using iTRAQ-based proteomics analysis, and identified 5624 proteins, which included 201 DAPs (differentially abundant protein species). They reported that the key functions of the upregulated DAPs included the glutathione metabolic process, sucrose metabolic process, gene silencing by RNA and lipid transport. Furthermore, the key functions of downregulated DAPs were reported as starch biosynthetic processes and response pathways to several oxidative stress. They profiled three flower-related proteins, CseIF4a, CsFLK and CsHUA1, and eight key genes in flowering and non-flowering buds, CsFLK, CseIF4A, CsHUA1 and CsGSTU7 (floral induction- and floral organ development-related genes), CsSUS1 and CsSUS2 (sucrose synthase activity-related genes), and CsGBSS1 and CsPU1 (starch synthase activity-related genes). They found that the sucrose content of non-flowering buds remarkably decreased under cold stress without any changes in starch content. In normal and no-stress conditions, the sucrose content increases, while the starch content decreases during floral development. Therefore, a change in starch biosynthesis and sucrose metabolism could alleviate the detrimental effects of cold stress.

Rao et al. (2017) carried out the proteomics on a pot experiment of saffron plant responses to Cd toxicity. They found that upregulation of 15 proteins occurs in saffron leaves under cadmium stress, including proteins related to metabolism and signal transduction. Furthermore, they reported downregulation of 11 proteins, including ribulose bisphosphate carboxylase/oxygenase (Rubisco), ferredoxin-NADP reductase, a 70-kDa heat shock-related protein, and three protein synthesis-associated proteins.

MicroRNAs (miRNA) in plants are endogenous small non-coding RNAs that play roles in many biological processes of growth, development and stress response by post-transcriptional silencing of the main target genes (Barchi and Acquadro, 2019). Biotic and abiotic stresses and climate change are the most important challenges that threaten the saffron industry. It has been reported that using an EST library from mature saffron stigmas, miRNAs and their targets in *Crocus sativus*'s stigmas and their relation with the genes of biosynthetic pathways carotenoid/apocarotenoid were discovered. Two putative microRNAs, miR414 and miR837-5p, and their corresponding stem looped precursors were identified in *Crocus sativus* for the first time. Based on co-expression analyses, miR414 and miR837-5p play key roles in the metabolic pathways of saffron plants involved in the carotenoid/apocarotenoid biosynthetic pathway. Generally, the predicted targets for respective miRNAs were reported to have roles in the regulation of plant growth, mRNA export, protein synthesis, senescence, disease resistance, stress responses and post-translational modifications (Pandita 2021). The miRNAs are important regulators of gene expression. In saffron, CsCCD and CsUGT genes that are involved in the apocarotenoids biosynthetic pathway are targeted by miRNAs of csa-miR156g and csa-miR156b-3p, and then it reveals a unique post-transcriptional regulation dynamic in saffron. The identified miRNAs and their targets could explain many biological roles of miRNAs in saffron to control the apocarotenoid biosynthetic pathway in this valuable plant (Taheri-Dehkordi et al. 2021).

7.6 EXOGENOUS STIMULANTS

Stimulants are any natural compounds or microorganisms that can affect plant growth and development through the efficiency of nutrition, tolerance to abiotic stress, crop quality and finally sustainable agriculture. In Table 7.1, some of these stimulants and their effects on saffron plants in stressed and non-stressed conditions are presented. Feizi et al. (2021) stated that stimulants as foliar applications increased root density and caused larger corms in size and weight, with greater exclusion of Na^+ and higher accumulation of K^+ being the main tolerance mechanisms of saffron that produce a favorable economic performance under salinity conditions.

Soil microorganisms, i.e., arbuscular mycorrhizal fungi (AMF) and plant growth-promoting bacteria (PGPB), could affect saffron productivity as a bio-fertilizer. The members of fungi phylum *Glomeromycota* form arbuscular mycorrhizae that have the most common symbiotic association with plants. These fungi improve mineral nutrition and plant growth, and influence the whole plant physiology by changing the hormonal profile, the production of secondary metabolites and plant flowering (D'Agostino et al. 2009). Rhizobacteria are natural soil bacteria that promote plant growth and establish a beneficial relationship with their hosts. These communities exist in the rhizosphere and plant tissues and stimulate plant growth through direct and/or indirect mechanisms (Díez-Méndez and Rivas 2017). Arbuscular mycorrhizal fungi form a symbiosis with various plants and play a role in exchanging carbohydrates and minerals. These bio-stimulants can improve tolerance to abiotic and biotic stresses and increase crop quality by promoting the yield of terpenes, essential oils, polyphenols and antioxidant activity. Therefore, they are of interest in the production of several medicinal and aromatic plants (Caser et al. 2019a). Lone et al. (2016) showed that colonization of arbuscular mycorrhizal fungi in saffron plants has promising potential for the enhancement of metabolites like sugars, protein, phenolics and minerals in corm and nitrogen-assimilating enzymes in plants.

Besides soil microorganisms that are used as stimulants, some synthetic and chemical compound effects have been investigated as bio-stimulants to alleviate the stress effects on saffron. For instance, Babaei et al. (2021) examined nitric oxide and salicylic acid for the treatment of corms to alleviate salinity effects on saffron plants. They reported that nitric oxide could beneficially reduce stress effects through an increase in accumulation of compatible solutes, activities of antioxidant enzymes and the synthesis of secondary metabolites. They stated that salicylic acid does not affect growth under salinity conditions. Ehsanfar et al. (2018) examined the interactions of three sizes of corms (3–5, 5–7 and 7–10 g) and five concentrations of polyamines (zero in control, zero in distilled water, spermidine 0.5 and 1, putrescine 0.5 and 1) as the treatments on saffron plants. Based on the vegetative and biochemical traits of saffron, they recommended putrescine and corms at a size of more than 7 g for achieving the best results. Babaei et al. (2021) showed that high levels of salinity in *C. sativus* decreased the leaf length. They reported more growth of roots under nitric oxide treatment. That said,

TABLE 7.1
Saffron Plant Responses to Exogenous Stimulants

Exogenous stimulants	Biotic/abiotic stress	Plant responses	References
Arbuscular mycorrhizal fungi: *Rhizophagus intraradices* (Ri) *R. intraradices* + *Funneliformis mosseae* (Ri+Fm)	–	More sizes of replacement corms in both inoculates More contents of picrocrocin, safranal and crocin by Ri compared to the control and Ri+Fm, respectively. More isoquercitrin (flavonols): Ri+Fm More ellagic acid (benzoic acids): Ri+Fm More crocin I and crocin I (carotenoids): Ri	Caser et al. (2019a)
Gamma aminobutyric acid (GABA), salicylic acid (SA) and vermicompost extract (VCE)	Salt stress	Foliar application (especially VCE); removed harmful impacts of salinity, improved low soil osmotic potential- induced nutrient deficiency. Enhancement of aerial K^+/Na^+ ratio under salinity Promotion of antioxidant defense system and alleviating the detrimental impacts of salinity (especially with GABA and SA)	Feizi et al. (2021)
Nanosilver	Salinity	Saline and non-saline irrigation with or without nanosilver (80 and 100 mg/L): removing detrimental effects of salinity and promoting the antioxidant system only with 80 mg/L, and harmful effects of 100 mg/L nanosilver	Rezvani and Sorooshzadeh (2014)
Bacillus subtilis	–	*Bacillus subtilis* (rhizosphere bacteria) improved agronomic parameters, including length and number of leaves, shoot and root biomasses, corm weight, flowers number and stigmas length through increasing the solubility and availability of nutrients	Prisa (2020)
Nutritional management [humic acid, mycorrhiza (*Glomus intraradices*)], superabsorbent	Drought stress (irrigation every 2, 3 and 4 weeks)	Increasing the tunic of corms under drought. Increased vegetative and saffron yield by nutritional management and superabsorbent under drought	Aghhavani Shajari et al. (2018)
Silicon as soluble form, orthosilicic acid $(SiOH)_4$	Salinity (NaCl)	Increased number and weight of corms, maintained chlorophyll content and decreased proline content in shoots and underground parts of plants The valorizing capacity of saffron cultivation regions for the species adaptation	Fahimi et al. (2017)
Superabsorbent polymers	Drought stress	Growth criteria of saffron improved through increasing water-holding capacity in the root zone by application of superabsorbent polymers and increased size of daughter corms	Fallahi et al. (2016)

TABLE 7.1 (Continued)
Saffron Plant Responses to Exogenous Stimulants

Exogenous stimulants	Biotic/abiotic stress	Plant responses	References
Spermidine, bio-fertilizer, spermidine × bio-fertilizer	-	Both the spermidine and the bio-fertilizer in their high concentrations enhanced plant growth and yield parameters, including plant height, number of leaves, fresh and dry weight of herb, number of flowers and flowers (fresh weight), yield (stigmas fresh and dry weight, stigmas yield, number of corms, corms weight and corms yield) and some active constituents (crocin, picrocrocin and safranal) of the saffron plant. Bio-fertilizer was more effective in increasing safranal content to crocin and picrocrocin	Ihsan et al. (2014)
Salicylic acid, humic acid and jasmonic acid	Salinity	The use of growth-promoting hormones (salicylic acid) and organic acids (humic acid) under environmental stress conditions can improve morphological and vegetative characteristics, including chlorophyll a and b, and total chlorophyll, the number of flowers, flowering stem and leaves, petiole length, fresh and dry weights of leaves, stigma weight and fresh weight of flowers in saffron	Moradizadeh et al. (2020)
Abscisic acid (ABA)	Drought stress	Decreasing the leaf relative water content (RWC), length and number of leaves Increasing the root Improving the protein content in corms, leaves, and roots, the maximum protein content of corms Increasing the activities of superoxide dismutase (SOD) and peroxidase (POX) in roots, leaves, and corms	Maleki et al. (2011)
Salicylic acid (SA)	Salinity (NaCl)	Foliar application; increasing the contents of hydrogen peroxide, MDA and proline increased in all accessions (Natanz, Deyhook, Ghaenat) under salinity stress Reducing the content of hydrogen peroxide, malondialdehyde and proline in stressed-accession Natanz by SA Activities of antioxidant enzymes, peroxidase, catalase, ascorbate peroxidase, polyphenol oxidase and superoxide dismutase in three accessions: in Natanz reduced the activities, in Ghaenat, SA reduced the activities of enzymes under maximum salinity. Ghaenat is the most resistant accession	Torabi (2017)

(continued)

TABLE 7.1 (Continued)
Saffron Plant Responses to Exogenous Stimulants

Exogenous stimulants	Biotic/abiotic stress	Plant responses	References
Calcium nanoparticles (CaNP) and putrescine polyamine	Without stress	Corm treatment; there is the functional potential of CaNP and putrescine combination to increase growth and phytochemical properties, but CaNP treatment alone is more effective on crocin, picrocrocin and safranal content than the combined effect of CaNP and putrescine	Badihi et al. (2021)
Titanium dioxide nanoparticles (TiO_2 NPs)	Ultraviolet-B (UV-B) radiation	Daily foliar application before daily stress; the loss of dissolved sugars and the promoted content of total anthocyanins and MDA in leaves Increase in radical scavenging activity (DPPH) in saffron stigmas as well as the content of UV-B absorbents (total phenolics and total flavonoids) The use of TiO_2 NPs promotes the nutritive value due to the increased antioxidant activity of saffron stigmas.	Rikabad et al. (2019)
Beeswax waste biochar (a novel organic amendment)	Cadmium (Cd) phytotoxicity	Reduction of Cd pollution through soil alkalinization, immobilizing Cd in soil, and reducing Cd partitioning in the aboveground tissues Improving the reduced biomass Alleviating the oxidative damage-induced Cd stress	Moradi et al. (2019)
Chemical nitrogen fertilizer plant growth-promoting rhizobacteria including *Pseudomonas* and *Bacillus* as bio-fertilizer	Without stress	Simultaneous application of *Pseudomonas* and *Bacillus* inoculation increased most of the qualitative (safranal and crocin, and picrocrocin) and quantitative traits of saffron due to increasing the nutrients uptake.	Heydari et al. (2014)
Vermicompost and mycorrhiza	–	Increase in stigma yield as well as plant growth and development, including leaf area, flower number and photosynthetic pigments in both experimental years	Jami et al. (2020)
Bio-fertilizer and nano-fertilizer of Fe	Salinity	Increase in safranal under highest salinity and the application of bio-fertilizer and nano-fertilizer Increase in crocin under highest salinity and without bio-fertilizer and nano-fertilizer, and versus with bio-fertilizer and without nano-fertilizer Nano-fertilizer was not effective in the lowest salinity	Salariyan et al. (2022)

TABLE 7.1 (Continued)
Saffron Plant Responses to Exogenous Stimulants

Exogenous stimulants	Biotic/abiotic stress	Plant responses	References
1. Seed treatment with biocontrol agents viz., *Tricoderma viride*, *Tricoderma herzanium* and *Pseudomonas florescence* 2. Bio-fertilizers, viz., *Azatobacter*, *Azospirillium* and vermicompost 3. Soil amendment; Mustard cake, dal weed and neem cake	Corm rot (*Fusarium monliforme* var. *intermedium*, *Fusarium oxysporum* and *F. gladioliis*)	Combination of *Tricoderma herzanium*, *Azospirillum* spp. and Mustard cake increased corm yield, number of daughter corms/mother corm, fresh flower weight and saffron yield.	Sameer et al. (2012)
Bio-fertilizers (Nitroxin) and superabsorbent polymer (Stockosorb)	–	In two planting methods (streaking and cluster): In the first year: cluster method, application of bio-fertilizer and superabsorbent for more flower, more dry and fresh weights of stigma In the second year: cluster method and superabsorbent without bio-fertilizer were superior	Heidari et al. (2018)
Arbuscular mycorrhizal fungi (AMF) and plant growth-promoting bacteria (PGPB) as bio-fertilizers	–	Bigger corms in the first year, and more flower and yield of stigma and stigma's ingredients, especially crocin	D'Agostino et al. (2009)
Putrescine, spermidine and spermine	Aluminum (Al) stress	Alleviation of stress effects by putrescine was more than for spermidine and spermine. Polyamines could lower Al content in the root tips, and then decrease lipid peroxidation and oxidative stress.	Chen et al. (2008)
Benzyl aminopurine hormone treatment, application of bio-fertilizers (phosphate Barvar2 and mycorrhiza), and maternal corm weight	–	Combination of bigger mother corms, hormone and bio-fertilizers increased yield of cormlets, dry weight of flowers, and the element content of corms, dry weight of saffron stigma.	Bekhradiyaninasab et al. (2020)
Curtobacterium herbarum Cs10, isolated from *Crocus seronitus* subsp. *clusii*, as a bioinoculant	–	Multifunctional ability; production of siderophores, solubilize phosphate, plant growth hormones like IAA *Curtobacterium herbarum* Cs10 improves the number of flowers and remarkably promotes the length of the saffron stigma and saffron yield.	Díez-Méndez and Rivas (2017)

salicylic acid could affect the growth of other plants such as *Vicia faba*, *Brassica junca* (Nazar et al. 2011), *Medicago sativa* and *V. radiata* (Khan et al. 2012), despite its insignificant effects on saffron growth. Tajik et al. (2019) pretreated saffron corms with salicylic acid and planted the corms in greenhouse conditions. They found more total phenolic and flavonoid contents as well as the activity and gene expression of the PAL enzyme in the leaves of treated plants. Babaei et al. (2021) reported that salinity stress in saffron decreases the anthocyanin content of leaves, and it does not increase when the stress level is doubled and remains steady. They observed that SA could not make any changes compared to treated plants under salinity conditions. However, they observed an increase in anthocyanin with NO at every salinity level without any differences. Based on Walia et al. (2005), in their study in rice, the effects of salinity stress depend on genotype, and when the genotype is more sensitive it has a greater anthocyanin content under salinity conditions. The most common strategy of plants to cope with salinity is the accumulation of osmolytes (compatible soluble), with the key osmolyte being proline (Nazar et al. 2011; Khan et al. 2012). The proline content is increased under salinity conditions due to the expression of genes that encode enzymes of proline synthesis, and the repression of genes that encode enzymes of proline oxidation (Amini and Ehsanpour 2005). Babaei et al. (2021) showed that NO-treated saffron plants under salinity conditions have a higher proline content. They observed that saffron plants treated with NO had more total soluble protein content under salinity conditions compared to NO-treated and SA-treated plants with and without salinity. They referred to Zheng et al. (2010), who stated that the role of NO is balancing between carbon and nitrogen metabolism and then enhancing the total protein content of a course in salinity conditions. They claimed that NO can induce endopeptidase and carboxypeptidase activities under salinity conditions. Babaei and co-workers (2021) managed chlorophyll contents of saffron leaves under salinity conditions with NO and SA. They interpreted some roles of NO in increasing photosynthetic pigments under salinity conditions while relying upon others, including the induction of ATP synthesis, two respiratory electron transport pathways in mitochondria, and then modulation of ROS by promoting the antioxidant defense system (Siddiqui et al. 2009; Fan et al. 2007). It is interpreted that when plants decrease their chlorophyll content under salinity conditions that this is a type of tolerance response to salinity (Akram and Ashraf 2011), and when plants under treatments of NO and SA have a lower reduction of chlorophyll it means that they are more tolerant of salinity (Babaie et al. 2021).

7.7 CONCLUSIONS AND PROSPECTS

Saffron is a valuable plant worldwide because of its pharmaceutical features as well as its usage as a spice being its main application, and also the production of industrial food dyes. As a valuable plant with the name of "red gold," the management of agricultural practices to ensure the high yield and quality of plants in both the food and medicine industries needs more attention. This plant is cultivated in some special regions with certain climates that face some biotic and abiotic stresses, which could change the economic yield, food and pharmaceutical quality as well as plant growth and development. Understanding plant reactions to various stresses and the importance of these responses to stress related to plant quality and quantity will help farmers to ensure profitability.

Interest in hydroponic cultivation of saffron has been increasing recently. This could be interesting in regards to increasing the products, production engineering in terms of organic production, or the targeted production of secondary metabolites of saffron with pharmaceutical features and more active ingredients.

With regards to prospects, *in vitro* propagation of saffron corms and the use of new molecular technologies, including new genetic markers such as genotyping-by-assay and genotyping-by-sequencing are proposed for breeding of saffron, to have trustworthy and economic cultivation of high-quality and high-yield saffron cultivars rather than ecotypes.

REFERENCES

Aghhavani Shajari, M., P. Rezvani Moghaddam, R. Ghorbani, and A. Koocheki. 2018. Increasing saffron (*Crocus sativus* L.) corm size through the mycorrhizal inoculation, humic acid application and irrigation managements. *Journal of Plant Nutrition* 41(8): 1047–1064.

Akram, N.A., and M. Ashraf. 2011. Pattern of accumulation of inorganic elements in sunflower (*Helianthus annuus* L.) plants subjected to salt stress and exogenous application of 5-aminolevulinic acid. Pakistan Journal of Botany 43(1): 521–530.

Albergaria, E.T., A.F.M. Oliveira, and U.P. Albuquerque. 2020. The effect of water deficit stress on the composition of phenolic compounds in medicinal plants. *South African Journal of Botany* 131: 12–17.

Amini, F., and A.A. Ehsanpour. 2005. Soluble proteins, proline, carbohydrates and Na$^+$/K$^+$ changes in two tomato (*Lycopersicon esculentum* Mill.) cultivars under in vitro salt stress. *American Journal of Biochemistry and Biotechnology* 1(4): 204–208.

Amirnia, R., M. Bayat, and M. Tajbakhsh. 2014. Effects of nano fertilizer application and maternal corm weight on flowering at some saffron (*Crocus sativus* L.) ecotypes. *Turkish Journal of Field Crops* 19(2): 158–168.

Anastasaki, E., C. Kanakis, C. Pappas, et al. 2010. Differentiation of saffron from four countries by mid-infrared spectroscopy and multivariate analysis. *European Food Research and Technology* 230(4): 571–577.

Askari-Khorasgani, O., and M. Pessarakli. 2019. Shifting saffron (*Crocus sativus* L.) culture from traditional farmland to controlled environment (greenhouse) condition to avoid the negative impact of climate changes and increase its productivity. *Journal of Plant Nutrition* 42(19): 2642–2665.

Avarseji, Z., M. Kafi, M. Sabet Teimouri, and K. Orooji. 2013. Investigation of salinity stress and potassium levels on morphophysiological characteristics of saffron. *Journal of Plant Nutrition* 36(2): 299–310.

Baba, S. A., A.H. Malik, Z.A. Wani, et al. 2015. Phytochemical analysis and antioxidant activity of different tissue types of *Crocus sativus* and oxidative stress alleviating potential of saffron extract in plants, bacteria, and yeast. *South African Journal of Botany* 99: 80–87.

Babaei, S., V. Niknam, and M. Behmanesh. 2021. Comparative effects of nitric oxide and salicylic acid on salinity tolerance in saffron (*Crocus sativus*). *Plant Biosystems-An International Journal Dealing with all Aspects of Plant Biology* 155(1): 73–82.

Badihi, L., M. Gerami, D. Akbarinodeh, M. Shokrzadeh, and M. Ramezani. 2021. Physio-chemical responses of exogenous calcium nanoparticle and putrescine polyamine in Saffron (*Crocus sativus* L.). *Physiology and Molecular Biology of Plants* 27(1): 119–133.

Barchi, L., and A. Acquadro. 2019. miRNome. In *The Globe Artichoke Genome* (pp. 195–203). Springer, Cham.

Bekhradiyaninasab, A., H. Balouchi, M. Movahhedi Dehnavi, and A. Sorooshzadeh, 2020. Effect of benzyl aminopurine, phosphate solubilizing bio-fertilizers and maternal corm weight on the qualitative indices of saffron (*Crocus sativus* L.) flowers and cormlets in Yasouj region. *Journal of Saffron Research* 8(1): 99–113.

Bowles, E.A. 1952. Handbook of *Crocus* and *Colchicum* for gardeners. *Handbook of Crocus and Colchicum for gardeners.*, (Edn 2).

Cardone, L., D. Castronuovo, M. Perniola, N. Cicco, and V. Candido. 2020. Saffron (*Crocus sativus* L.), the king of spices: An overview. *Scientia Horticulturae*, 272: 109560.

Caser, M., S., Demasi, Í.M.M. Victorino, et al. 2019a. Arbuscular mycorrhizal fungi modulate the crop performance and metabolic profile of saffron in soilless cultivation. *Agronomy* 9(5): 232.

Caser, M., Í.M.M., Victorino, S., Demasi, et al. 2019b. Saffron cultivation in marginal Alpine environments: How AMF inoculation modulates yield and bioactive compounds. *Agronomy* 9(1): 12.

Chen, J., G., Zhou, Y., Dong, et al. 2021. Screening of key proteins affecting floral initiation of saffron under cold stress using iTRAQ-based proteomics. *Frontiers in plant science* 12: 708.

Chen, W., C. Xu, B. Zhao, X. Wang, and Y. Wang. 2008. Improved Al tolerance of saffron (*Crocus sativus* L.) by exogenous polyamines. *Acta Physiologiae Plantarum* 30(1): 121–127.

D'Agostino, G., E. Gamalero, V. Gianotti, et al. 2009, May. Use of arbuscular mycorrhizal fungi and beneficial soil bacteria to improve yield and quality of saffron (*Crocus sativus* L.). In *III International Symposium on Saffron: Forthcoming Challenges in Cultivation, Research and Economics 850* (pp. 159–164).

Dastranj, M., and A.R. Sepaskhah, 2019. Saffron response to irrigation regime, salinity and planting method. *Scientia Horticulturae* 251: 215–224.

Díez-Méndez, A., and R. Rivas. 2017. Improvement of saffron production using *Curtobacterium herbarum* as a bioinoculant under greenhouse conditions. *AIMS microbiology* 3(3): 354.

Ebrahimi, M., M., Pouyan, T., Shahi, et al. 2021. Effects of organic fertilisers and mother corm weight on yield, apocarotenoid concentration and accumulation of metal contaminants in saffron (*Crocus sativus* L.). *Biological Agriculture & Horticulture* 1–21.

Ehsanfar, S., A. Sorooshzadeh, S.AM. Modarres-Sanavy, and M. Ghorbani Javid. 2018. Effect of corm size and corm soaking in polyamines on yield and vegetative and qualitative traits of saffron. *Journal of Crops Improvement* 20(2): 467–485.

Esf, E., S.S. Alavi-Kia, A. Bahmani, and M.A. Aazami. 2009. The effect of light on ROS-scavenging systems and lipid peroxidation under cold conditions in saffron (*Crocus sativus* L.). *African Journal of Agricultural Research* 4(4): 378–382.

Fahimi, J., Z., Bouzoubaa, F., Achemchem, N., Saffaj, and R. Mamouni. 2017. Effect of silicon on improving salinity tolerance of Taliouine *Crocus sativus* L. *Acta horticulturae* (1184): 219–227.

Fallahi, H.R., G. Zamani, M. Mehrabani, M. Aghhavani-Shajari, and A. Samadzadeh, 2016. Influence of superabsorbent polymer rates on growth of saffron replacement corms. *Journal of Crop Science and Biotechnology* 19(1): 77–84.

Fan, H., S. Guo, Y. Jiao, R. Zhang, and J. Li. 2007. Effects of exogenous nitric oxide on growth, active oxygen species metabolism, and photosynthetic characteristics in cucumber seedlings under NaCl stress. *Frontiers of Agriculture in China* 1(3): 308–314.

Farooq, M., A. Wahid, N. Kobayashi, D. Fujita, and S.M. Basra. 2009. Plant drought stress: Effects, mechanisms and management. *Agronomy for sustainable development* 29(1): 185–212.

Farrokhi, H., A. Asgharzadeh, and M.K. Samadi. 2021. Yield and qualitative and biochemical characteristics of saffron (*Crocus sativus* L.) cultivated in different soil, water, and climate conditions. *Italian Journal of Agrometeorology* (2): 43–55.

Feizi, H., R. Moradi, N. Pourghasemian, and H. Sahabi, 2021. Assessing saffron response to salinity stress and alleviating potential of gamma amino butyric acid, salicylic acid and vermicompost extract on salt damage. *South African Journal of Botany* 141: 330–343.

Ghobadi, F., M. Ghorbani Javid, and A. Sorooshzadeh, 2015. Effects of planting date and corm size on flower yield and physiological traits of saffron (*Crocus sativus* L.) under Varamin plain climatic conditions. *Saffron agronomy and technology* 2(4): 265–276.

Gómez, L.G., A. Rubio, J. Escribano, et al. 2003, October. Development and gene expression in saffron corms. In *I International Symposium on Saffron Biology and Biotechnology 650* (pp. 141–153).

Gómez-Gómez, L., A. Trapero-Mozos, M.D. Gómez,, A. Rubio-Moraga, and O. Ahrazem. 2012. Identification and possible role of a MYB transcription factor from saffron (*Crocus sativus*). *Journal of plant physiology* 169(5): 509–515.

Gupta, V., and V.K. Razdan. 2012, October. Evaluation of integrated disease management technologies for corm rot of saffron in Kishtwar district of J&K. In *IV International Symposium on Saffron Biology and Technology 1200* (pp. 115–120).

Gupta, V., A. Sharma, P.K. Rai, et al. 2021. Corm rot of saffron: epidemiology and management. *Agronomy*, *11*(2): 339.

Hadizadeh, M., and E. Keyhani. 2006, October. Toxic Effect of Cadmium and Catalase Activity in the Corms of *Crocus sativus* L. In *II International Symposium on Saffron Biology and Technology 739* (pp. 443–449).

Heidari, S., K. Azizi, and A. Ismaili. 2018. Effects of Nitroxin bio-fertilizer, superabsorbent polymer and planting method on yield of flower and corm of saffron (*Crocus sativus* L.) in rainfed-farming condition of Khorramabad, Iran. *Saffron agronomy and technology* 6(4): 461–472.

Heydari, Z., H. Besharati, and S. Maleki Farahani. 2014. Effect of some chemical fertilizer and biofertilizer on quantitative and qualitative characteristics of Saffron. *Saffron agronomy and technology* 2(3): 177–189.

Ihsan, S.A., M.H. Al-Mohammad, and S.N. Al-Thamir. 2014. The influence of spermidine and biofertilizer application on the growth, yield and some active constituents of saffron plant (*Crocus sativus* L.). *Journal of Biology, Agriculture and Healthcare* 4(24): 131–135.

Iqbal, N., S., Umar, N.A. Khan, and M.I.R. Khan. 2014. A new perspective of phytohormones in salinity tolerance: regulation of proline metabolism. *Environmental and Experimental Botany* 100: 34–42.

Jalali-Heravi, M., H. Parastar, and H. Ebrahimi-Najafabadi. 2010. Self-modeling curve resolution techniques applied to comparative analysis of volatile components of Iranian saffron from different regions. *Analytica chimica acta* 662(2): 143–154.

Jami, N., A. Rahimi, M. Naghizadeh, and E. Sedaghati, 2020. Investigating the use of different levels of Mycorrhiza and Vermicompost on quantitative and qualitative yield of saffron (*Crocus sativus* L.). *Scientia Horticulturae* 262: 109027.

Kargar, S.M.A., F. Amiriyan, and A. Mostafaei. 2020. The Investigation of Genetic Diversity of Saffron Ecotypes Traits under Chilling Stress. *Journal of Saffron Research* 8(2): 191–206.

Keyhani, E., J. Keyhani, M. Hadizadeh, L. Ghamsari, and F. Attar. 2003, October. Cultivation techniques, morphology and enzymatic properties of *Crocus sativus* L. In *I International Symposium on Saffron Biology and Biotechnology 650* (pp. 227–246).

Keyhani, J., E. Keyhani, and L. Arzi. 2006, October. Alterations in lignin peroxidase and ascorbate peroxidase activities in *Crocus sativus* L. corms exposed to copper. In *II International Symposium on Saffron Biology and Technology 739* (pp. 427–434).

Khan, M.I.R., N., Iqbal, A., Masood, and N.A. Khan. 2012. Variation in salt tolerance of wheat cultivars: role of glycinebetaine and ethylene. *Pedosphere* 22(6): 746–754.

Koocheki, A., and S.M. Seyyedi. 2019. Mother corm origin and planting depth affect physiological responses in saffron (*Crocus sativus* L.) under controlled freezing conditions. *Industrial Crops and Products* 138: 111468.

Koocheki, A., S.M. Seyyedi, and M.J. Eyni, 2014. Irrigation levels and dense planting affect flower yield and phosphorus concentration of saffron corms under semi-arid region of Mashhad, Northeast Iran. *Scientia Horticulturae* 180; 147–155.

Lamikanra, O., O.A. Richard, and A. Parker. 2002. Ultraviolet induced stress response in fresh cut cantaloupe. *Phytochemistry* 60(1): 27–32.

Liu, C., Y. Liu, K. Guo, et al. 2011. Effect of drought on pigments, osmotic adjustment and antioxidant enzymes in six woody plant species in karst habitats of southwestern China. *Environmental and Experimental Botany* 71(2): 174–183.

Lone, R., R. Shuab, and K.K. Koul, 2016. AMF association and their effect on metabolite mobilization, mineral nutrition and nitrogen assimilating enzymes in saffron (*Crocus sativus*) plant. *Journal of Plant Nutrition* 39(13): 1852–1862.

Maleki, M., H. Ebrahimzade, M. Gholami, and V. Niknam. 2011. The effect of drought stress and exogenous abscisic acid on growth, protein content and antioxidative enzyme activity in saffron (*Crocus sativus* L.). *African Journal of Biotechnology* 10(45): 9068–9075.

Moradi, R., N. Pourghasemian, and M. Naghizadeh. 2019. Effect of beeswax waste biochar on growth, physiology and cadmium uptake in saffron. *Journal of Cleaner Production* 229: 1251–1261.

Moradizadeh, S., H.A. Asadi-Gharneh, and M.R. Naderi Darbaghshahi. 2020. How does immersion of saffron corm in some hormones and humic acid affect the morphological characteristics of plant under salinity stress. *Journal of Crop Nutrition Science* 6(2): 26–43.

Moslemi, F.S., A. Vaziri, G. Sharifi, and J. Gharechahi. 2021. The effect of salt stress on the production of apocarotenoids and the expression of genes related to their biosynthesis in saffron. *Molecular Biology Reports* 48(2): 1707–1715.

Mzabri, I., M. Legsayer, F. Aliyat, et al. 2017. Effect of drought stress on the growth and development of saffron (*Crocus Sativus* L) in Eastern Morocco. *Atlas Journal of Biology* 364–370.

Mzabri, I., M. Legsayer, F. Aliyat, et al. 2016, November. Effect of salt stress on the growth and development of saffron (*Crocus sativus* L.) in eastern Morocco. In *V International Symposium on Saffron Biology and Technology: Advances in Biology, Technologies, Uses and Market 1184* (pp. 55–62).

Mzabri, I., M. Rimani, K. Charif, N. Kouddane, and A. Berrichi. 2021. Study of the effect of mulching materials on weed control in saffron cultivation in Eastern Morocco. *The Scientific World Journal*.

Nazar, R., N. Iqbal, S. Syeed, and N.A. Khan. 2011. Salicylic acid alleviates decreases in photosynthesis under salt stress by enhancing nitrogen and sulfur assimilation and antioxidant metabolism differentially in two mungbean cultivars. *Journal of plant physiology* 168(8): 807–815.

Nezami, A., S. Rezvan Beidokhti, and S. Sanjani. 2016. Study of saffron (*Crosus sativus* L.) reaction to freezing stress under controlled condition. *Environmental Stresses in Crop Sciences* 9(1): 75–86.

Pandita, D. 2021. Saffron (*Crocus sativus* L.): Phytochemistry, therapeutic significance and omics-based biology. In *Medicinal and Aromatic Plants* (pp. 325–396). Academic Press.

Parizad, S., A. Dizadji, M.K. Habibi, et al. 2019. The effects of geographical origin and virus infection on the saffron (*Crocus sativus* L.) quality. *Food chemistry* 295: 387–394.

Pazoki, A., M. Kariminejad, and A.F. Targhi. 2013. Effect of planting patterns on yield and some agronomical traits in saffron (*Crocus sativus* L.) under different irrigation intervals in Shahr-e-Rey Region. *International Journal of farming and Allied Sciences*, 2(S2): 1363–8.

Prisa, D. 2020. Improving quality of *Crocus sativus* through the use of *Bacillus Subtilis,*". *International Journal of Scientific Research in Multidisciplinary Studies* 6(2): 9–15.

Rahiman, N., M. Akaberi, A. Sahebkar, S.A. Emami, and Z. Tayarani-Najaran. 2018. Protective effects of saffron and its active components against oxidative stress and apoptosis in endothelial cells. *Microvascular research* 118: 82–89.

Rajaei, S.M., V. Niknam, S.M. Seyedi, H. Ebrahimzadeh, and K. Razavi. 2009. Contractile roots are the most sensitive organ in *Crocus sativus* to salt stress. *Biologia Plantarum* 53(3): 523–529.

Rao, J., W. Lv, and J. Yang. 2017. Proteomic analysis of saffron (*Crocus sativus* L.) grown under conditions of cadmium toxicity. *Bioscience Journal* 33(3): 713–720.

Rao, J., W. Lyu, and F. Cao. 2016. Effect of chromium stress on protein profiles in saffron. *Journal of Zhejiang University (Science Edition)* 43(6): 751–755.

Rasool, A., M.I. Mir, M. Zulfajri, et al. 2021. Plant growth promoting and antifungal asset of indigenous rhizobacteria secluded from saffron (*Crocus sativus* L.) rhizosphere. *Microbial Pathogenesis* 150: 104734.

Renau-Morata, B., S.G. Nebauer, M. Sánchez, and R.V. Molina, 2012. Effect of corm size, water stress and cultivation conditions on photosynthesis and biomass partitioning during the vegetative growth of saffron (*Crocus sativus* L.). *Industrial Crops and Products* 39: 40–46.

Rezvani, N., and A. Sorooshzadeh.2014. Effect of nano-silver on root and bud growth of saffron in flooding stress condition. *Saffron Agronomy and Technology* 2(1): 91–104.

Rikabad, M.M., L. Pourakbar, S.S. Moghaddam, and J. Popović-Djordjević. 2019. Agrobiological, chemical and antioxidant properties of saffron (*Crocus sativus* L.) exposed to TiO_2 nanoparticles and ultraviolet-B stress. *Industrial Crops and Products* 137: 137–143.

Rostami, M., R. Karamian, and Z. Joulaei. 2015. Effect of different heavy metals on physiological traits of saffron (*Crocus sativus* L.). *Saffron Agronomy and Technology* 3(2): 83–96.

Rubio-Moraga, A., J.L. Rambla, A. Fernández-de-Carmen, et al. 2014. New target carotenoids for CCD4 enzymes are revealed with the characterization of a novel stress-induced carotenoid cleavage dioxygenase gene from *Crocus sativus*. *Plant molecular biology* 86(4): 555–569.

Saddique, M., M. Kamran, and M. Shahbaz, 2018. Differential responses of plants to biotic stress and the role of metabolites. In *Plant metabolites and regulation under environmental stress* (pp. 69–87). Academic Press.

Salariyan, A., S. Mahmoodi, M. Behdani, and H. Kaveh. 2022. Effects of bio fertilizer and nanoparticles of Fe on quantitative and qualitative properties of saffron (*Crocus sativus* L.) under salinity stress. *Saffron agronomy and technology*.

Sameer, S.S., S. Bashir, F.A. Nehvi, et al. 2012, October. Effect of biofertilizers, biological control agents and soil amendments on the control of saffron corm rot (*Crocus sativus* L.). In *IV International Symposium on Saffron Biology and Technology 1200* (pp. 121–124).

Shigeoka, S., T. Ishikawa, M. Tamoi, et al. 2002. Regulation and function of ascorbate peroxidase isoenzymes. *Journal of experimental botany* 53(372): 1305–1319.

Siddiqui, M.H., F. Mohammad, and M.N. Khan. 2009. Morphological and physio-biochemical characterization of *Brassica juncea* L. Czern. & Coss. genotypes under salt stress. *Journal of Plant Interactions* 4(1): 67–80.

Siracusa, L., F. Gresta, G. Avola, G.M. Lombardo, and G. Ruberto. 2010. Influence of corm provenance and environmental condition on yield and apocarotenoid profiles in saffron (*Crocus sativus* L.). *Journal of Food Composition and Analysis* 23(5): 394–400.

Sorooshzadeh, A., S. Hazrati, H. Oraki, M. Govahi, and A. Ramazani. 2012. Foliar application of nanosilver influence growth of saffron under flooding stress. *Brno, Czech Republic, EU* 10: 23–25.

Taheri-Dehkordi, A., R. Naderi, F. Martinelli, and S.A. Salami. 2021. Computational screening of miRNAs and their targets in saffron (*Crocus sativus* L.) by transcriptome mining. *Planta* 254(6): 1–22.

Tajik, S., F. Zarinkamar, B.M. Soltani, and M. Nazari. 2019. Induction of phenolic and flavonoid compounds in leaves of saffron (*Crocus sativus* L.) by salicylic acid. *Scientia Horticulturae*, 257: 108751.

Thakur, M., S. Bhattacharya, P.K. Khosla, and S. Puri. 2019. Improving production of plant secondary metabolites through biotic and abiotic elicitation. *Journal of Applied Research on Medicinal and Aromatic Plants* 12: 1–12.

Torabi, P.S., V. Niknam, H. Ebrahimzadeh, and G.A. Sharifi. 2016. Comparative study of biochemical responses of different saffron (*Crocus sativus*) accessions to salt stress and alleviative effects of salicylic acid.

Torabi, S. 2017. Comparative study of biochemical responses of different saffron (*Crocus sativus*) accessions to salt stress and alleviative effects of salicylic acid. *Journal of Plant Research (Iranian Journal of Biology)* 29(4): 728–740.

Torbaghan, M.E., and M.M. Ahmadi. 2011. The effect of salt stress on flower yield and growth parameters of saffron (*Crocus sativus* L.) in greenhouse condition. *International Research Journal of Agricultural Science and Soil Science* 1(10): 421–427.

Verma, S., and S.N. Mishra. 2005. Putrescine alleviation of growth in salt stressed *Brassica juncea* by inducing antioxidative defense system. *Journal of plant physiology* 162(6): 669–677.

Walia, H., C. Wilson, P. Condamine, et al. 2005. Comparative transcriptional profiling of two contrasting rice genotypes under salinity stress during the vegetative growth stage. *Plant physiology* 139(2): 822–835.

Ward, G., Y. Hadar, I. Bilkis, L. Konstantinovsky, and C.G. Dosoretz. 2001. Initial steps of ferulic acid polymerization by lignin peroxidase. *Journal of Biological Chemistry* 276(22): 18734–18741.

Zheng, C., D. Jiang, T. Dai, Q. Jing, and W. Cao. 2010. Effects nitroprusside, a nitric oxide donor, on carbon and nitrogen metabolism and the activity of the antioxidation system in wheat seedlings under salt stress. *Acta Ecol Sinica* 30: 1174–1183.

8 *Cuminum cyminum* and Stressful Conditions

Patience Tugume*, Jamilu Edirisa Ssenku and Godwin Anywar
Department of Plant Sciences, Microbiology and Biotechnology,
Makerere University, Kampala, Uganda
*Corresponding author: patiebeys@gmail.com

CONTENTS

8.1 INTRODUCTION

Cuminum cyminum L., commonly known as cumin, is an annual herbaceous flowering plant belonging to the family Apiaceae. The plant is known by different vernacular names such as Jeeru (Gujarati), Zeera (Hindi/Punjabi), Jeerkam (Malayalam) and Amla (Urdu) (Rai et al. 2012). The seeds have a strong and distinctive aroma leading to their widespread use as a culinary spice and condiment in many cultures across the world (Sheikholeslami et al. 2020). Cumin is a highly valuable crop due to its medicinal and nutritional properties as well as its revenue-generating ability (Kumar et al. 2015).

Biotic and abiotic stress factors are known to trigger defense mechanisms in the plant which result in the production of secondary metabolites that are of nutritional and pharmaceutical importance. On the contrary, some biotic and abiotic stresses may result in reduced crop yield, which necessitates the use of biotechnological techniques to try to overcome them (Munaweera et al. 2022; Piasecka et al. 2019).

8.2   GLOBAL AND ECOLOGICAL DISTRIBUTION OF
CUMINUM CYMINUM L.

Cumin is native to Egypt, the East Mediterranean and South Asia, but is currently grown worldwide because of its aromatic seeds. The word cumin is derived from the Latin word cuminum, which originated from the Greek word "kyminon." Cumin seeds are mainly grown in India, Pakistan, Egypt, Iraq, Morocco, Turkey, Syria, Sicily, Yugoslavia, Bulgaria, Malta, Sudan, Cyperus, Czechoslovakia, China, Indonesia Uzbekistan, Tajikistan, Mexico and Chile (Azeez 2008; Sebastian 2006).

Cumin requires a moderately cool and dry climate for its growth in a temperature range of 25–30°C. *Cuminum cyminum* grows in a variety of soils from sandy topsoils with a pH from 6.8 to 8.3 that comprise a lot of organic matter (Dave et al. 2021). Highly acidic and alkaline soils reduce crop yield unless the pH is reduced to 7.5 (Weiss 2002). The plant effectively develops in cool and dry atmospheres but a shady climate does not favour its blossoming and fruiting. In the tropics, the plant grows more successfully at high elevations compared to low ones. The optimum temperature for the plant is in the range of 17–26°C but it can tolerate 9–30°C (Dave et al. 2021). Cumin is highly sensitive to rain, which reduces crop quality during harvesting. During flowering and early seed formation, frost causes immense damage. Such damage could be mitigated by spraying with sulfuric acid (0.1%), irrigating the crop prior to frost incidence, setting up windbreaks against cool waves or creating an early morning smoke cover (Weiss 2002).

8.3   BOTANICAL DESCRIPTION OF *CUMINUM CYMINUM* L.

Cumin is a monotypic species with low genetic and phenotypic variability. Its synonyms include: *Cuminia cyminum* J.F.Gmel., *Cuminum aegyptiacum* Mérat ex DC., *Cuminum hispanicum* Mérat ex DC., *Cuminum odorum* Salisb., *Cuminum officinale* Garsault, and *Cuminum sativum* J.Sm. *Cuminum cyminum* grows up to 10–50 cm high with a slender branched stem. The leaves are about 5 cm long, bluish-green alternate, simple or compound with a sheathing leaf base, sparsely hairy and bear thread-like leaflets. Flowers are monoecious and borne in umbrella-like clusters of three to five (Koohsari et al. 2020). The petals are oblong, white or red in color with an intense long indented tip. The bracts are long and simple. Stamens are antisepalous with long filaments and dithecous anthers. The ovary is inferior and develops into a distinct fruit called a cremocarp (Prashar et al. 2014). The schizocarp has yellowish-green or brown-colored fruits 6 mm in length and 1.5 mm in breadth and is crowned with awl-shaped calyx tips. The fruit invariably breaks at maturity into two one-seeded bits, with a ribbed wall that has a number of longitudinal oil canals. These canals produce the characteristic odour and flavour that reflect the value of the fruit (Madhuri et al. 2014). Seeds are elongated or oblong-shaped, yellow-brown in colour and possess nine protuberances (Belal et al. 2017).

8.4   REPRODUCTIVE PHYSIOLOGY OF *CUMINUM CYMINUM* L.

Cumin is propagated by seeds. Since the crop is frost-intolerant, farmers in cold climates grow seeds indoors in biodegradable pots for about 4–8 weeks after which seedlings are transplanted outdoors when conditions become favorable. An alternative to biodegradable pots is homemade blocks of compressed soil, which eliminates the need for plastic pots and reduces the threat of transplant shock. In warmer climates, however, seeds are directly sown outdoors. Alternatively, cumin can be grown in soilless potting mixes such as Pro-Mix, Sunshine Mix, perlite, vermiculite, rockwool, coco peat and Oasis Rootcubes. Sowing depth is ¼ inch at a spacing of 4–8 inches in rows arranged 18 inches apart. In order to realize maximum yields, the optimum seed rate for cumin is 12–15 kg/hectare (Sharma et al. 1999). Seeds may be sown either by the broadcasting method or in lines sown at 30 cm row-to-row spacing. After broadcasting, the seeds should be covered lightly by soil with help of iron-toothed rakes (Bhati et al., 1984). Care should be taken to ensure that the seeds do not go deep inside the soil while being covered. Better and rapid germination is ensured by soaking seeds in water for 8 hours and then surface drying under shade before sowing. Maurya (1986) suggested seed pretreatment with potassium nitrate at 100 mg/L, followed by ammonium nitrate and urea, both at 1000 mg/L.

Cumin is a cross-pollinated crop wherein plenty of floral visitors attend the flowers. Its inflorescence is a compound umbel and its tiny flowers are either white or pink (Chittora and Tiwari 2013). Its small and slender flowers render artificial pollination difficult and varieties are developed by sib mating in enclosed chambers (Al Faiz 2006).

8.5 PHYTOCHEMICAL COMPOSITION OF *CUMINUM CYMINUM* L.

Cumin seeds contain a variety of nutrients including high amounts of monosaturated fats, proteins, fiber and vitamins B and E. Cumin seeds yield 2.3–4.8% volatile oil that is yellow amber but darkens on aging. The major volatile components of cumin are cuminaldehyde, cymene and terpenoids (Bettaieb et al. 2011). The distinctive aroma of cumin is due to cuminaldehyde and cuminic alcohol. In roasted cumin, aromatic compounds comprise substituted pyrazines, 2-ethoxy-3-isopropylpyrazine, 2-methoxy-3-sec-butylpyrazine, and 2-methoxy-3-methylpyrazine, γ-terpinene, safranal, p-cymene and β-pinene (Li and Jiang 2004). The distinctive odour of cumin is attributed to the aldehydes present in the seeds, cuminaldehyde, p-menth-3-en-7-al and p-menth-1,3-dien-7-al (Agrawal 2001).

The cuminaldehyde content varies depending on whether the oil is from fresh or ground seeds, or the origin of the seeds. For instance, grinding results in the loss of up to 50% of the volatile oil. Turkish cumin seed oil was reported to have 19.2% (Baser et al. 1992), while Egyptian oil had 25.01% of cuminaldehyde (Shaath and Azzo 1993). Monoterpene hydrocarbons are another major component of the oil while sesquiterpenes are minor constituents (Ahmad and Saeidnia 2011). Limonene, eugenol and β-pinenes have been found in cumin oil (Johri 2011; Dorman and Deans 2000).

8.6 MEDICINAL USES AND APPLICATIONS OF *CUMINUM CYMINUM* L.

Cumin seeds have been used in ethnomedicine since ancient times in different systems of medicine such as Ayurveda, Unani and Siddha (Shivakumar et al. 2010; Gruenwald et al. 2007). Due to its numerous medicinal properties, cumin has been used as an ingredient in many herbal home remedies for both humans and animals (Rai et al. 2012). The fruits of *C. cyminum* were used in Unani medicine to manage ulcers, boils, cough, styes, inflammation and corneal opacities (Agarwal et al. 2017; Tahir et al. 2016). Cumin is used in Iran, Indonesia, India and other Asian countries to stimulate milk production, relieve flatulence, pain, as an antispasmodic agent (Tabarsa et al., 2020), to treat bloody and chronic diarrhea, headache, kidney and bladder stones, leprosy and eye disease, stomach pain, dyspepsia, indigestion, hoarseness, toothache, hypertension, weight loss, jaundice (Al-Snafi 2016; Siow and Gan 2016; Taghizadeh et al. 2016), spitting up of blood, gonorrhoea, snake bites and scorpion stings (Prajapati et al. 2003). In Ayurvedic medicine, cumin seeds are used as a carminative, emmenogogic, antispasmodic, astringent, anti-epileptic, appetizer and diuretic agent (Tahir et al. 2016; Raikwar and Maurya 2015). In veterinary medicine *C. cyminum* is used to manage mastitis colic, dyspepsia (Rai et al. 2012) and as a digestant in cattle (Raikwar and Maurya 2015).

Cumin seed is a rich reservouir of bioactive compounds that are widely used for various healthcare purposes (Belal Ahmed and Ali 2017; Goodarzi et al. 2020; Kang et al. 2019). *Cuminum cyminum* possesses numerous biological activities such as antibacterial, antifungal, antioxidant, antidiabetic, hypolipidemic, immunomodulatory, anti-osteoporotic, anti-inflammatory, anti-asthmatic, antitussive, anti-infertility, anticancer, antistress, antipyretic, hypotensive, antiplatelet coagulation, hepatoprotective and analgesic properties (Agarwal et al. 2017; Al-Snafi 2016). These activities validate the widespread use of cumin in different traditional medical systems. Petroselinic acid from *C. cyminum* is applied in cosmetic formulations as a moisturizing and anti-aging agent and as a skin-irritation reducing agent in α-hydroxy acid-containing compositions (Delbeke et al. 2016; Hayes 2004).

Through enhanced production of cuminaldehyde, drought stress could enhance the therapeutic value of *C. cyminum* for the treatment of antibacterial and antifungal effects (Hajlaoui et al., 2010), and as an antimalarial (Zheljazkov et al. 2015), and it inhibited the growth of aflatoxin-producing fungi such as *Aspergillus flavus* (Kedia et al. 2014). Cuminaldehyde protected pancreatic *β*-cells against cytotoxicity induced by streptozotocin (Patil et al. 2013) and anti-inflammatory effects (Ebada 2017; Wei et al. 2015), and had protective effects against neurodegenerative diseases, particularly Parkinson's disease (Ebada 2017), and also anthelminthic properties mediated through physical damage to the anterior and posterior ends, intestinal, ovarian and esophageal regions of the worms (Goel et al. 2020). However, drought was also found to greatly decrease the percentage of α-thujene (Rebey et al. 2012), a compound that is rich in essential oils that has shown antimicrobial and antioxidant activities (Gupta et al. 2017).

8.7 EFFECT OF STRESS ON THE PHYTOCHEMICAL COMPOSITION OF *CUMINUM CYMINUM* L.

8.7.1 COMMON BIOTIC AND ABIOTIC STRESSORS OF CUMIN

Plants are faced with a wide range of environmental stresses that limit their productivity. Such stressors could either be abiotic or biotic. Abiotic stress is the negative impact of non-living factors on living organisms in a specific environment. Abiotic stresses are the most harmful to the growth and productivity of crops worldwide. They include drought, salinity, heat, cold, chilling, freezing, nutrient, high light intensity and anaerobic stresses (Wang et al. 2003; Chaves and Oliveira 2004; Agarwal and Grover 2006), which seriously threaten agricultural productivity.

Biotic stresses are negative impacts caused by living organisms. They include pathogens like bacteria, fungi, viruses, nematodes, herbivores, insects, nematodes, diseases or weeds that damage crops (Atkinson and Urwin 2012). The habitat range of pests and pathogens can be influenced by increasing temperatures (Luck et al. 2011; Madgwick et al. 2011). Many abiotic stress factors weaken the plant defense mechanisms, making them susceptible to pathogen infection (Amtmann et al. 2008; Goel et al. 2020; Mittler and Blumwald 2010; Atkinson and Urwin 2012). Biotic and abiotic stresses cause special defense reactions in plant organs resulting in the production of secondary metabolites (Guo et al. 2011). These compounds are instrumental in the therapeutic properties of medicinal plants (Pessarakli et al. 2015). For instance, ultraviolet radiation is an important cause of abiotic stress that affects the production of secondary metabolites in medicinal plants (DiCosmo and Misawa 1985).

8.7.2 BIOTIC STRESS IN CUMIN

Treatment of medicinal plants with suboptimal and excess quantities of the environmental factors and infestation with biotic stressors trigger off special defense reactions in plant organs, which, after a series of reactions, produce secondary metabolites (Ghasemi et al. 2019) that may be of medicinal value. Due to the widespread use of cumin for medicinal, nutritional and commercial purposes, its production is very pertinent. Despite the importance of cumin, its production is limited by several stress factors. The most serious biotic stress factor that affects *C. cyminum* is attack by fungal diseases, especially *Fusarium oxysporum* f. sp. *cumini* and *Alternaria bursnsii* (Agarwal 1996). Extracts of *Xanthium strumarium*, *Amaranthus retroflexus* and *Chenopodium album* significantly inhibited the germination of cumin seeds after application (Shams et al. 2014).

8.7.3 ABIOTIC STRESS IN CUMIN

8.7.3.1 Cold Stress

Cumin is susceptible to the effects of chilling and frost injury (Kumar et al. 2015). Frost reduces seed production, causing a serious loss in yield. Cold stress affects the water relations of a plant at the cellular and whole-plant levels causing damage and adaptation reactions (Beck et al. 2007). Low-temperature stress causes serious damage at the physiological, cellular and molecular levels. For instance, under cold conditions, the green leaves of cumin became purple, whereas long exposure to cold conditions leads to the death of the plant (Agrawal 1993). Plant height, dry matter and leaf area decreased significantly under lower temperatures (Nezami et al. 2012).

8.7.3.2 Salinity Stress

Increasing salt concentration delayed germination, and decreased the seed emergence rate and biomass in cumin. However, salt stress caused an increase in the content of amino acids in cumin except for asparagine, and led to a decrease in total sugar, flavonoid and phenolic contents (Pandey et al. 2015). Hydrogen peroxide (H_2O_2) and proline contents increased with increasing NaCl concentration to more than double at 100 mM salinity stress compared to control plants. The malondialdehyde (MDA) content increased by 1.5- and 2.2-fold under 50 and 100 mM NaCl treatments, respectively. Salinity stress in cumin led to the synthesis of metabolites methylhexacosane and dimethylpentacosane, which are important in nutrition and as a sex pheromone, respectively (Pandey et al. 2015). It also resulted in the synthesis of docetaxel with anticancer activity, and megalomicin with antiparasitic, antiviral and antibacterial properties. Moderate salinity (50 mM NaCl) resulted in the production of phospholipids, 2,2-hydroxy-6-ketononatrienedioate and/or 3-(2-carboxyethenyl)-cis,cismuconate, which have antioxidant activity. Under high salinity stress two natural flavonoids, kaempferol and quercetin, and O-methylated anthocyanidin peonidin, a primary pigment, were synthesized.

Cumin growth, specifically seed yield, was significantly reduced by increasing the concentration of NaCl. The treatment also decreased the composition of petroselinic and linoleic acids but increased the proportion of palmitic acid (Rebey et al. 2017). In the same study, the production and quality of essential oil in *C. cyminum* seeds were enhanced, but γ-terpinene/1-phenyl-1,2 ethanediol was modified to γ-terpinene/β-pinene. Salinity improved the amount of individual phenolic compounds. Additionally, highly stressed cumin plants exhibited high antioxidant activities (Rebey et al. 2017). These results imply that cumin seeds produced under saline conditions may be a source of essential oil and antioxidant compounds, to support the utilization of this plant in the food industry.

8.7.3.3 Water Stress

Cumin plants treated with moderate water deficit improved the number of umbels per plant, the number of umbellets per umbel and the seed yield. However, these decreased under severe water deficit (Rebey et al. 2012). Water deficit enhanced the palmitic acid percentage and affected the oil quality. The essential oil yield increased under moderate water deficit but decreased under severe water deficit. Total phenolic contents were higher in moderate and severe water deficit treated seeds (Rebey et al. 2012). Results suggest that water deficit treatment may regulate the production of bioactive compounds in cumin seeds, influencing their nutritional and industrial values.

Drought limits crop expansion and is becoming increasingly important due to global climate change (Chaves et al. 2004). Besides the quantitative and qualitative effects of drought on the essential oils, the total phenolic contents of *C. cyminum* seeds have also been reported to be higher in drought-stressed than non-stressed seeds (Rebey et al. 2012). These phenolic compounds are primarily responsible for the potent antioxidative, digestive stimulative, hypolipidemic, antibacterial, anti-inflammatory, antiviral and anticancer properties of spices and herbs Jiang et al. 2019;

Viuda-Martos et al. 2010). Thus, drought stress may regulate the production of bioactive compounds in cumin seeds, influencing their nutritional and industrial values.

Analysis of the fatty acid composition by Rebey et al. (2012) under normal growth conditions indicated that petroselinic acid was the major fatty acid (55.9%) followed by palmitic (23.82%) and linoleic (12.40%) acids. The composition of the fatty acid has been demonstrated to increase under mild drought conditions but decrease under severe drought stress. Besides the quantitative modifications, drought conditions also modify the essential oil chemotype of *C. cyminum* from - terpinene/phenyl-1,2 ethanediol in the control seeds to -terpinene/cuminaldehyde in stressed seeds. Terpinene has been reported to scavenge radicals to protect erythrocytes, DNA and methyl linoleate against oxidation (Li and Liu 2009).

8.7.3.4   Ultraviolet Radiation Stress

Plants react to biotic and abiotic stresses by producing secondary metabolites (Guo et al. 2011) that are important for plants as therapeutic agents. Ultraviolet radiation is one important factor that in many cases stimulates the production of secondary metabolites (Zhang and Björn 2009). In a study by Ghasemi et al. (2019), the application of UV-B stress to cumin significantly increased the expression of geranyl diphosphate synthase (GPPs), hydroxymethylglutaryl-CoA reductase, deoxy xylose phosphate synthase (DXS), deoxyribonino heptulosinate 7-phosphate synthase (DAHP) and phenylalanine ammonia-lyase (PAL) genes which are associated with different secondary metabolites. Ultraviolet radiation affected the expression of key genes involved in the biosynthesis of secondary metabolites in *C. cyminun* causing a significant increase in total levels of terpenoids, phenols, flavonoids, anthocyanins, alkaloids, beta-carotene and lycopene (Ghasemi et al. 2018). The increase in beta-carotene and lycopene is associated with their vital role in protecting plant cells against highlight stress and thus eliminating excess energy that is absorbed (Edreva 2005).

8.7.3.5   Foliar Application of Minerals and Acids

A study by Baghizadeh and Shahbazi (2013) showed that Zn+Fe foliar application significantly affected the number of umbels per plant, number of umbellets per umbel and seed weight in *C. cyminum*. The same treatment significantly affected ($P>0.05$) the relative water content, MDA, H_2O_2 accumulation, proline and protein contents. The same study revealed that treatment of cumin with Zn+Fe further increased the activity of superoxide dismutase (SOD), catalase (CAT) and peroxidase. These results show the effectiveness of zinc and iron foliar application on cumin resistance in dry farming. Concentrations of 0.01 and 0.1 mM salicylic acid significantly promoted plant height, and number of branches and umbels per plant in cumin. A concentration of 0.1 mM salicylic acid significantly increased fruit and essential oil yields (Rahimi et al. 2013).

8.8   CONCLUSION

Cumin is an important multipurpose commercial crop. It is widely used as a food, spice and medicine. Cumin seeds contain various nutrients, including high amounts of monosaturated fats, proteins, fiber, vitamins B and E, as well as volatile oils. The major volatile components of cumin essential are cuminaldehyde, cymene and terpenoids. Cumin is affected by both abiotic and biotic factors in various ways but appears to be most susceptible to cold stress and fungal diseases. Whereas the different stress factors generally caused a decrease in yield, different factors such as salt stress caused an increase in the number of amino acids in cumin. The essential oil yield also increased under moderate water deficit but decreased under severe water deficit. The secondary metabolites in cumin are responsible for their therapeutic effects and respond in various ways to the different intensities and exposures to various abiotic and biotic stress factors.

LIST OF ABBREVIATIONS

H_2O_2	Hydrogen peroxide
NaCl	Sodium chloride
UV-B	Ultraviolet B
GPPs	Geranyl diphosphate synthase
DXS	Deoxy xylose phosphate synthase
PAL	Phenylalanine ammonia-lyase
MDA	Malondialdehyde content
SOD	Superoxide dismutase
CAT	Catalase

REFERENCES

Agarwal, S. (1996). Volatile oil constituents and wilt resistance in cumin (*Cuminum cyminum* L.). Curr Sci 7:1177–1178.

Agarwal, S., and Grover, A. (2006). Molecular biology, biotechnology and genomics of flooding-associated low O2 stress response in plants. *Critical Reviews in Plant Sciences*, *25*(1), 1–21.

Agarwal, U., Pathak, D. P., Kapoor, G., Bhutani, R., Roper, R., Gupta, V., and Kant, R. (2017). Review on *Cuminum cyminum*—nature's magical seeds. *Journal of Chemical and Pharmaceutical Research*, *9*(9), 180–187.

Agrawal, P. K. (1993). Handbook of seed testing. National Seeds Corporation Limited. New Delhi, India.

Agrawal, S. (2001) Seed spices – an introduction. In: Agrawal, S., Sastry, E.V.D. and Sharma, R.K. (eds) Seed Spices – Production, Quality, Export. Pointer Publishers, Jaipur, India, pp. 1–18.

Ahmad Reza Gohari and Soodabeh Saeidnia. Phytochemistry of *Cuminum cyminum* seeds and its Standards from Field to Market. Pharmacognosy Journal, 2011; *3*(25):1–5c.

Al Faiz, C. (2006). Biological Diversity, Cultural and Economic Value of Medicinal, Herbal and Aromatic Plants in Morocco. Annual Report 2005–2006, Institut National de la Recherche Agronomique, Rabat, Maroc, 51p.

Al-Snafi, A. E. (2016). The pharmacological activities of *Cuminum cyminum*-A review. *IOSR Journal of Pharmacy*, *6*(6), 46–65.

Amtmann, A., Troufflard, S., and Armengaud, P. (2008). The effect of potassium nutrition on pest and disease resistance in plants. *Physiologia plantarum*, *133*(4), 682–691.

Atkinson, N. J., and Urwin, P. E. (2012). The interaction of plant biotic and abiotic stresses: from genes to the field. *Journal of experimental botany*, *63*(10), 3523–3543.

Azeez, S. (2008) Cumin. In: Parthasarathy, V.A., Champakam, B. and Zachariah, T.J., eds.: Chemistry of Spices. CABI International, Wallingford, UK.

Baghizadeh, A., and Shahbazi, M. (2013). Effect of Zn and Fe foliar application on yield, yield components and some physiological traits of cumin (*Cuminum cyminum*) in dry farming. *International Journal of Agronomy and Plant Production*, *4*(12), 3231–3237.

Baser, K. H. C., Özek, T., and Tümen, G. (1992). Essential oils of Thymus cariensis and Thymus haussknechtii, two endemic species in Turkey. *Journal of essential oil research*, *4*(6), 659–661.

Beck EH, Fettig S, Knake C, Hartig K, Bhattarai T (2007) Specific and unspecific responses of plants to cold and drought stress. J Biosci. 32: 501–510.

Belal, A. A., Ahmed, F. B., and Ali, L. I. (2017). Antibacterial activity of *Cuminum cyminum* L. oil on six types of bacteria. *American Journal of BioScience*, *5*(4), 70–73.

Belal, A. A., Ahmed, F. B., and Ali, L. I. (2017). Antibacterial activity of *Cuminum cyminum* L. oil on six types of bacteria. *American Journal of BioScience*, *5*(4), 70–73.

Bettaieb, I., Bourgou, S., Sriti, J., Msaada, K., Limam, F., Marzouk, B. (2011). Essential oils and fatty acids composition of Tunisian and Indian cumin (*Cuminum cyminum* L.) seeds: a comparative study. Journal of the Science of Food and Agriculture, *91*: 2100–2107.

Bhati, DS, MP Jain, RK Sharma, BN Batt and G.R Choudhary 1984. Indian Cocoa, Arecanut and Spices Journal, 8:1:89.

Chaves, M. M., and Oliveira, M. M. (2004). Mechanisms underlying plant resilience to water deficits: prospects for water-saving agriculture. *Journal of experimental botany*, *55*(407), 2365–2384.

Chittora M, Tiwari K. Biology and biotechnology of cumin. International J Bioassays. 2013; 2(7):1066–1068.

Dave, K., Mehta, K., and Patel, R. (2021). Cumin (*Cuminum cyminum* L.): The Flavor of India (Cultivation, Nutrifacts, Pharmacological effect, Disease control and Economical value).

Deepak, Arora, D.K., Saran P.L.s and Lal, G. (2008). Evaluation of cumin varieties against blight and wilt diseases with time of sowing. Annals of Plant Protection Sciences, *16*(2): 441–43.

Delbeke, E. I. P., Everaert, J., Uitterhaegen, E., Verweire, S., Verlee, A., Talou, T., Soetaert, W., Van Bogaert, I. N. A., and Stevens, C. V. (2016). Petroselinic acid purification and its use for the fermentation of new sophorolipids. AMB Express, *6*(1), 28. https://doi.org/10.1186/s13568-016-0199-7

DiCosmo, F., and Misawa, M. (1985). Eliciting secondary metabolism in plant cell cultures. *Trends in Biotechnology*, *3*(12), 318–322.

Dilshad, S. R., Rehman, N. U., Ahmad, N., and Iqbal, A. (2010). Documentation of ethnoveterinary practices for mastitis in dairy animals in Pakistan. *Pakistan Veterinary Journal*, *30*(3), 167–171.

Dorman HJD, Deans SG. Antimicrobial agents from plants: Antibacterial activity of plant volatile oils. J Appl Microbiol, 2000; *88*:308–16.

Ebada, M. E. (2017). Cuminaldehyde: A potential drug candidate. J Pharmacol Clin Res, *2*(2), 1–4. https://doi.org/10.19080/JPCR.2017.02.555585

Edreva, A. (2005). Generation and scavenging of reactive oxygen species in chloroplasts: a submolecular approach. Agriculture, Ecosystems and Environment, *106*(2), 119–133. https://doi.org/10.1016/j.agee.2004.10.022

Ghasemi, S., Kumleh, H. H., and Kordrostami, M. (2019). Changes in the expression of some genes involved in the biosynthesis of secondary metabolites in Cuminum cyminum L. under UV stress. *Protoplasma*, *256*(1), 279–290.

Goel, V., Singla, L. D., and Choudhury, D. (2020). Cuminaldehyde induces oxidative stress-mediated physical damage and death of *Haemonchus contortus*. Biomedicine and Pharmacotherapy, *130*, 110411. https://doi.org/10.1016/j.biopha.2020.110411

Goodarzi, S., Tabatabaei, M. J., Mohammad Jafari, R., Shemirani, F., Tavakoli, S., Mofasseri, M., and Tofighi, Z. (2020). *Cuminum cyminum* fruits as source of luteolin-7-O-glucoside, potent cytotoxic flavonoid against breast cancer cell lines. *Natural product research*, *34*(11), 1602–1606.

Gruenwald, J., Brendler, T., and Jaenicke, C. (2007). *PDR for herbal medicines*. Thomson, Reuters.

Guo, R., Yu, F., Gao, Z., An, H., Cao, X., and Guo, X. (2011). GhWRKY3, a novel cotton (*Gossypium hirsutum* L.) WRKY gene, is involved in diverse stress responses. *Molecular biology reports*, *38*(1), 49–58.

Gupta, M., Rout, P. K., Misra, L. N., Gupta, P., Singh, N., Darokar, M. P., Saikia, D., Singh, S. C., and Bhakuni, R. S. (2017). Chemical composition and bioactivity of Boswellia serrata Roxb. Essential oil in relation to geographical variation. *Plant Biosystems – An International Journal Dealing with all Aspects of Plant Biology*, *151*(4), 623–629. https://doi.org/10.1080/11263504.2016.1187681

Hajlaoui, H., Mighri, H., Noumi, E., Snoussi, M., Trabelsi, N., Ksouri, R., and Bakhrouf, A. (2010). Chemical composition and biological activities of Tunisian *Cuminum cyminum* L. essential oil: A high effectiveness against *Vibrio* spp. Strains. *Food and Chemical Toxicology*, *48*(8), 2186–2192. https://doi.org/10.1016/j.fct.2010.05.044

Hayes, D. G. (2004). Enzyme-Catalyzed modification of oilseed materials to produce eco-friendly products [https://doi.org/10.1007/s11746-004-1024-2]. *Journal of the American Oil Chemists' Society*, *81*(12), 1077–1103. https://doi.org/10.1007/s11746-004-1024-2

Jiang, L., Belwal, T., Huang, H., Ge, Z., Limwachiranon, J., Zhao, Y., … and Luo, Z. (2019). Extraction and characterization of phenolic compounds from bamboo shoot shell under optimized ultrasonic-assisted conditions: a potential source of nutraceutical compounds. *Food and Bioprocess Technology*, 12(10), 1741–1755.

Johri RK. *Cuminum cyminum* and Carum carvi. An update. Phcog Rev, 2011; *5*:63–72.

Kang, N., Yuan, R., Huang, L., Liu, Z., Huang, D., Huang, L., ... and Yang, S. (2019). Atypical nitrogen-containing flavonoid in the fruits of cumin (*Cuminum cyminum* L.) with anti-inflammatory activity. *Journal of agricultural and food chemistry*, *67*(30), 8339–8347.

Kedia, A., Prakash, B., Mishra, P. K., and Dubey, N. K. (2014). Antifungal and antiaflatoxigenic properties of *Cuminum cyminum* (L.) seed essential oil and its efficacy as a preservative in stored commodities. *International Journal of Food Microbiology*, 168–169, 1–7. www.sciencedirect.com/science/article/abs/pii/S0168160513004741?via%3Dihub. https://doi.org/10.1016/j.ijfoodmicro.2013.10.008

Koohsari, S., Sheikholeslami, M. A., Parvardeh, S., Ghafghazi, S., Samadi, S., Poul, Y. K., ... and Amiri, S. (2020). Antinociceptive and antineuropathic effects of cuminaldehyde, the major constituent of *Cuminum cyminum* seeds: Possible mechanisms of action. *Journal of ethnopharmacology*, *255*, 112786.

Kumar, S., Saxena, S. N., Mistry, J. G., Fougat, R. S., Solanki, R. K., and Sharma, R. (2015). Understanding *Cuminum cyminum*: An important seed spice crop of arid and semi-arid regions. Int. J. Seed Spices, *5*(2), 1–19.

Li, R., Jiang, Z. (2004). Chemical composition of the essential oil of *Cuminum cyminum* L. from China. *Flavour and Fragrance Journal*, *19*: 311–313.

Li, G.-X., and Liu, Z.-Q. (2009). Unusual antioxidant behavior of α- and γ-terpinene in protecting methyl linoleate, DNA, and erythrocyte. *Journal of Agricultural and Food Chemistry*, *57*(9), 3943–3948. https://doi.org/10.1021/jf803358g

Luck, J., Spackman, M., Freeman, A., Tre. bicki, P., Griffiths, W., Finlay, K., and Chakraborty, S. (2011). Climate change and diseases of food crops. *Plant Pathology*, *60*(1), 113–121.

Madhuri, P., Jakhar, M. L., and Malik, C. P. (2014). A review on biotechnology, genetic diversity in cumin (*Cuminum cyminum*). *International Journal of Life science and Pharma Research*, *4*(4).

Madgwick, J. W., West, J. S., White, R. P., Semenov, M. A., Townsend, J. A., Turner, J. A., and Fitt, B. D. (2011). Impacts of climate change on wheat anthesis and fusarium ear blight in the UK. *European Journal of Plant Pathology*, *130*(1), 117–131.

Maurya, K. R. (1986). Effect of IBA on germination, yield, carotene and ascorbic acid content of carrot. *Indian Journal of Horticulture*, *43*(1&2), 118–120.

Mittler R, Blumwald E. 2010. Genetic engineering for modern agriculture: challenges and perspectives. *Annual Review of Plant Biology 61*: 443–462.

Munaweera, T. I. K., Jayawardana, N. U., Rajaratnam, R., and Dissanayake, N. (2022). Modern plant biotechnology as a strategy in addressing climate change and attaining food security. *Agriculture & Food Security*, *11*(1), 26. https://doi.org/10.1186/s40066-022-00369-2

Nezami, A., Sanjani, S., Ziaee, M., Soleimani, M. R., Nassiri-Mahallati, M., and Bannayan, M. (2012). Evaluation of freezing tolerance of cumin (*Cuminum cyminum* l.) Under controlled conditions. *Agricultura*, *81*.

Pandey, S., Patel, M. K., Mishra, A., and Jha, B. (2015). Physio-biochemical composition and untargeted metabolomics of cumin (*Cuminum cyminum* L.) make it promising functional food and help in mitigating salinity stress. *PLoS One*, *10*(12), e0144469.

Parashar, M., Jakhar, M.L., and Malik, C.P. (2014). A review on biotechnology, genetic diversity in Cumin (*Cuminum cyminum*), Int. *J. of Life Science and Pharma Research*, *4*(4): 17–34.

Patil, S. B., Takalikar, S. S., Joglekar, M. M., Haldavnekar, V. S., and Arvindekar, A. U. (2013). Insulinotropic and β-cell protective action of cuminaldehyde, cuminol and an inhibitor isolated from *Cuminum cyminum* in streptozotocin-induced diabetic rats. British Journal of Nutrition, 110(8), 1434–1443. https://doi.org/10.1017/S0007114513000627

Pessarakli, M., Haghighi, M., and Sheibanirad, A. (2015). Plant responses under environmental stress conditions. *Adv. Plants Agric. Res*, *2*(6), 73.

Piasecka, A., Kachlicki, P., and Stobiecki, M. (2019). Analytical Methods for Detection of Plant Metabolomes Changes in Response to Biotic and Abiotic Stresses. *International Journal of Molecular Sciences*, *20*(2), 379. https://doi.org/10.3390/ijms20020379

Prajapati, N. D., Purohit, S. S., Sharma, A. K., and Kumar, T. (2003). A handbook of medicinal plants: A complete source book. In *A handbook of medicinal plants: a complete source book* (pp. 554–554).

Rai, N., Yadav, S., Verma, A. K., Tiwari, L., and Sharma, R. K. (2012). A monographic profile on quality specifications for a herbal drug and spice of commerce-*Cuminum cyminum* L. *International Journal of Advanced Herbal Science and Technology*, *1*(1), 1–12.

Raikwar, A., and Maurya, P. (2015). Ethnoveterinary medicine: in present perspective. *Int. J. Agric. Sc. and Vet. Res*, *3*(1), 44–49.

Rahimi, A. R., Rokhzadi, A., Amini, S., and Karami, E. (2013). Effect of salicylic acid and methyl jasmonate on growth and secondary metabolites in *Cuminum cyminum* L. *Journal of Biodiversity and Environmental sciences*, *3*(12), 140–149.

Rebey, I. B., Bourgou, S., Rahali, F. Z., Msaada, K., Ksouri, R., and Marzouk, B. (2017). Relation between salt tolerance and biochemical changes in cumin (*Cuminum cyminum* L.) seeds. *journal of food and drug analysis*, *25*(2), 391–402.

Rebey, I. B., Jabri-Karoui, I., Hamrouni-Sellami, I., Bourgou, S., Limam, F., and Marzouk, B. (2012). Effect of drought on the biochemical composition and antioxidant activities of cumin (*Cuminum cyminum* L.) seeds. *Industrial Crops and Products*, *36*(1), 238–245.

Rebey, I. B., Jabri-Karoui, I., Hamrouni-Sellami, I., Bourgou, S., Limam, F., and Marzouk, B. (2012). Effect of drought on the biochemical composition and antioxidant activities of cumin (*Cuminum cyminum* L.) seeds. *Industrial Crops and Products, 36*(1), 238–245. https://doi.org/10.1016/j.indcrop.2011.09.013

Shaath, N. A., and Azzo, N. R. (1993). Essential oils of Egypt. *Developments in food science.*

Sebastian, P. (2006). Ayurvedic medicine: the principles of traditional practice. London: Churchill Livingstone Elsevier. Pp 399.

Shams, M., Esfahan, E. Z., Ramezani, M., and Ghandkanlou, M. (2014). *Cuminum cyminum.*

Sharma, R.K., Bhati, D.S., and Sharma, M.M. (1999). Cumin. In: Bose, T.K., Mishra, S.K., Farooqi, A.A. and Sardhu, M.K. (eds) 1: 740–745.

Sheikholeslami, M. A., Parvardeh, S., Ghafghazi, S., Samadi, S., Poul, Y. K., Pouriran, R., & Amiri, S. (2020). Antinociceptive and antineuropathic effects of cuminaldehyde, the major constituent of Cuminum cyminum seeds: Possible mechanisms of action. *Journal of Ethnopharmacology*, *255*, 112786. https://doi.org/10.1016/j.jep.2020.112786

Shivakumar, S. I., Shahapurkar, A. A., Kalmath, K. V., and Shivakumar, B. (2010). Antiinflammatory activity of fruits of *Cuminum cyminum* Linn. *Der Pharmacia Lettre*, *2*(1), 22–24.

Siow, H. L., and Gan, C. Y. (2016). Extraction, identification, and structure–activity relationship of antioxidative and α-amylase inhibitory peptides from cumin seeds (*Cuminum cyminum*). *Journal of Functional Foods*, *22*, 1–12.

Tabarsa, M., You, S., Yelithao, K., Palanisamy, S., Prabhu, N. M., and Nan, M. (2020). Isolation, structural elucidation and immuno-stimulatory properties of polysaccharides from *Cuminum cyminum*. *Carbohydrate polymers*, *230*, 115636.

Taghizadeh, M., Memarzadeh, M. R., Abedi, F., Sharifi, N., Karamali, F., Kashan, Z. F., and Asemi, Z. (2016). The effect of Cumin cyminum L. Plus lime administration on weight loss and metabolic status in overweight subjects: A randomized double-blind placebo-controlled clinical trial. *Iranian Red Crescent Medical Journal*, *18*(8).

Tahir, H. U., Sarfraz, R. A., Ashraf, A., and Adil, S. (2016). Chemical composition and antidiabetic activity of essential oils obtained from two spices (*Syzygium aromaticum* and *Cuminum cyminum*). *International journal of food properties*, *19*(10), 2156–2164.

Taghizadeh, M., Ostad, S. N., Asemi, Z., Mahboubi, M., Hejazi, S., Sharafati-Chaleshtori, R., ... and Sharifi, N. (2017). Sub-chronic oral toxicity of *Cuminum cyminum* L.'s essential oil in female Wistar rats. *Regulatory Toxicology and Pharmacology*, *88*, 138–143.

Viuda-Martos, M., Ruiz Navajas, Y., Sánchez Zapata, E., Fernández-López, J., and Pérez-Álvarez, J. A. (2010). Antioxidant activity of essential oils of five spice plants widely used in a Mediterranean diet. *Flavour and Fragrance Journal, 25*(1), 13–19.

Wang, W., Vinocur, B., and Altman, A. (2003). Plant responses to drought, salinity and extreme temperatures: towards genetic engineering for stress tolerance. *Planta*, *218*(1), 1–14.

Wei, J., Zhang, X., Bi, Y., Miao, R., Zhang, Z., and Su, H. (2015). Anti-Inflammatory Effects of Cumin Essential Oil by Blocking JNK, ERK, and NF-κB Signaling Pathways in LPS-Stimulated RAW 264.7 Cells. Evid Based Complement Alternat Med, 2015, 474509. https://doi.org/10.1155/2015/474509

Weiss E.A. (2002). Spice Crops. CABI International, Wallingford, UK., 299p. [Family wise description of spice crops belonging to cruciferae, lauraceae, leguminosae, myristicaceae, myrtaceae umbelliferae and zingiberaceae.]

Zhang, W. J., and Björn, L. O. (2009). The effect of ultraviolet radiation on the accumulation of medicinal compounds in plants. *Fitoterapia, 80*(4), 207–218. https://doi.org/10.1016/j.fitote.2009.02.006

Zheljazkov, V. D., Gawde, A., Cantrell, C. L., Astatkie, T., and Schlegel, V. (2015). Distillation time as tool for improved antimalarial activity and differential oil composition of cumin seed oil. *PloS one, 10*(12), e0144120. https://doi.org/10.1371/journal.pone.0144120

9 Medicinal Plant *Hibiscus sabdariffa* L. and Its Responses to Various Stresses

A. K. M. Golam Sarwar

Laboratory of Plant Systematics, Department of Crop Botany, Bangladesh Agricultural University, Mymensingh 2202, Bangladesh

*Corresponding author: drsarwar@bau.edu.edu

CONTENTS

9.1 INTRODUCTION

"LET THY FOOD BE THY MEDICINE AND MEDICINE BE THY FOOD"—HIPPOCRATES (**400 BC**)

Medicinal plants could be defined as a group of plants that possess drug constituents and therapeutic agents, and that are used for medicinal purposes and/or precursors for the synthesis of useful drugs. Over 50% of prescription drugs are derived from chemicals first identified in plants (Evans 2009; Sarwar 2020). The World Health Organization has estimated that 80% of people worldwide rely on herbal medicines for some aspect of their primary healthcare needs and around 21,000 plant species have the potential for use as medicinal plants (www.who.int/). More than 1,200 plant species have been used for medicinal purposes in Bangladesh (Uddin and Lee 2020). By offering a direct source of medicines, healthcare remedies and treatments, medicinal plants play an important role in achieving the SDGs (#3 To ensure healthy lives and promote well-being for all at all ages). The cultivation of medicinal plants, as an adaptive crop, could be considered an alternative in stress-prone areas due to their low resource requirements and the implementation of organic and inorganic amendments.

DOI: 10.1201/9781003242963-9

TABLE 9.1
Botanical Classification of *H. sabdariffa*

Kingdom: Plantae
  Phylum: Spermatophyta
    Subphylum: Angiospermae
      Class: Dicotyledonae
        Subclass: Dilleniidae
          Order: Malvales
            Family: Malvaceae
              Subfamily: Malvoideae
                Tribe: Hibisceae
                  Genus: *Hibiscus* L.
                    Species: *H. sabdariffa* L.

The underutilized medicinal plant *Hibiscus sabdariffa* L., commonly known as roselle or red sorelle, belonging to the Mallow family Malvaceae is an annual plant with a basic chromosomal number $x = 18$ ($4n = 72$) (Table 9.1; Wilson and Menzel 1964). Its native range is India (Cobley 1957), another is from western tropical Africa to Sudan, with tropical Africa having extreme diversity (Wilson and Menzel 1964), and it is now widely naturalized in tropical and subtropical regions of the world, particularly in India and Southeast Asia, including China and Thailand, being the major global suppliers, and throughout the Old World tropics. It is known by different synonyms and vernacular names based on its geographical origins, including chukur, chukair, mestapat, lalmesta, kharapata, patwa, roselle, razelle, sorrel, red sorrel, Guinea sorrel, Jamaican sorrel, Indian sorrel, sour-sour, Asam susar, Queensland jelly plant, karkade´, pusa hemp, rohzelu, laalambaar, sabdriqa, jelly okra, lemon bush and Florida cranberry (Lim 2014; Mahunu 2021). The *H. sabdariffa* plant was first described in 1576 (Cobley 1957); it has been cultivated throughout India and parts of Asia for centuries, is grown throughout Africa and has been in cultivation in America since the eighteenth century and in Australia since the nineteenth century.

The *H. sabdariffa* plant is highly regarded from the standpoint of nutritional, culinary and medicinal values as well as economic value (Da-Costa-Rocha et al. 2014). The red calyces of *H. sabdariffa* were used in ancient healing practices by African traditional medical practitioners and ultimately incorporated into traditional Chinese medicine (Tackett 2017). Among uses, strong fiber is obtained from the stem (called roselle hemp) which offers attractive possibilities as a jute substitute crop and which is used for various household purposes including making sackcloth, carpet, twine and cord (Luvonga 2012). The salt-resistant trait in *H. sabdariffa* fiber makes it a perfect material for cordage production, packaging sacks, assorted paper material, upholstery, and fabric shoes and bags production (Mahunu 2021). Browning processes, either enzymatic or non-enzymatic, deteriorate the postharvest nutritional and sensory properties of fruits and vegetables, and food safety as well. The floral extract of *H. sabdariffa* could be used as a (plant-based) edible coating on fruits and vegetables for post-harvest quality assurance; as it possesses natural anti-browning elements (Yang et al. 2019).

9.2 BOTANY OF *H. SABDARIFFA*

Hibiscus sabdariffa is an annual, that is erect, bushy, herbaceous, deep-rooted sub-shrub up to 5.0 m tall, with smooth or nearly smooth, cylindrical, variously colored (some varieties have bright red coloration in stems, whilst some are green, often with white coloration in other parts of the plant); leaves are alternate, glabrous, 7.5–12.5 cm long, green with reddish veins, and long or short petioles,

20–50 mm long, with one green gland present in the mid-vein on the undersurface, stipules 6–8 mm long, filiform, usually palmately divided into three or five lobes with toothed margins, possessing a definite pulvinus at the junction of blade and petiole and with small linear stipules at the base of the petiole; flowers are showy, solitary and axillary, on very short peduncles, regular, and the epicalyx consists of up to 10 small thickened pointed bracts, 0.5–2 cm long, fused at the base with the calyx tube; the five sepals are large, 15–30 mm long (enlarging in fruit to 5.5 cm), accrescent, rounded and pointed, tending to become fleshy, hairy along the margins and grooved on each side of the midrib, without prominent glands, both the epicalyx and calyx are bright red, green or almost white in color depending on the variety; the five pale yellow or pale pink petals are spreading and showy, obovate, up to 5 cm × 3.5 cm, widely opening at maturity but with their tips crumpled together in the bud, usually having a deep red or purple spot at the throat; the staminal column gives rise to numerous stamens collected into five groups, up to 2 cm long, and surrounds the style with its five-lobed stigma; pollen grains are spiny; the ovary is superior, 0.3–0.4 cm long, having five locules each with several ovules in axile placentation and giving rise to a pointed ovate capsule covered with fine hairs, and closely invested with the persistent calyx and epicalyx, which often become very fleshy during fruit development; the fruit is an ovoid capsule up to 2.5 cm long, almost glabrous to appressed-pubescent, enclosed by the calyx, turning brown and splitting open when mature and dry, and is many-seeded; seeds are reniform, up to 7 mm long, dark brown, and surrounded by miniature, stout and stellate hairs; seedlings have epigeal germination; cotyledons are rounded, up to 2.5 cm × 3 cm and leafy.

There are two botanical varieties recognized: *H. sabdariffa* var. *sabdariffa*, a bushy subshrub with a red or green stem and red or pale yellow inflated edible calyces, which are important for food uses; and *H. sabdariffa* var. *altissima* Wester, a tall, vigorous, unbranched plant, with fibrous, spiny, inedible calyces, which is important for the high-quality fiber (Purseglove 1968). In addition, four races are recognized in *H. sabdariffa* var. *sabdariffa*: *bhagalpuriensi* with green, red-streaked, inedible calyces; *intermedius* with yellow-green edible calyces that also yields fiber; *albus* with green or yellow-green edible calyces that also yields fiber; *ruber* with red or variegated edible calyces (Morton 1987).

The green varieties of the plant seem to produce more fiber than the red types. The fibers are produced in layers just under the bark of the stem, outside the phloem, and consist of aggregates of small sclerenchyma cells, 1.2–3.3 mm long, with tapering ends that overlap and cohere together giving strands of fiber up to six or seven feet in length, that are fine and silky in appearance. The best quality fiber is obtained after 3–4 months of vegetative growth when the plants are entering the reproductive phase.

Hibiscus sabdariffa is a hermaphroditic and insect-pollinated shrub, usually propagated by seed but readily growing from cuttings. A very small amount (0.02%) of outcrossing is found in *H. sabdariffa* (Islam et al. 2021). The species could hybridize with *H. cannabinus* (Orwa et al. 2009). The plants exhibit marked photoperiodism, not flowering at short days of 13.5 hours, but flowering at 11 hours. It tolerates a warm and humid tropical climate (12) 18–30 (38)°C and grows best where the rainfall is 45–50 cm spread over 3–4 months, but is susceptible to damage from frost and fog, from sea level to 1250 m in altitude (Orwa et al. 2009). The seeds contain about 17% of oil, and is similar in properties to cottonseed oil (Cobley 1957). The calyx, stems and leaves are acidic and closely resemble the cranberry (*Vaccinium* spp.) in flavor. It is extensively cultivated in tropical Africa, Asia, Australia and Central America (Schippers 2000).

9.3 MEDICINAL/NUTRITIVE USES OF *H. SABDARIFFA*

Hibiscus sabdariffa was primarily grown on a small scale in native gardens as a beverage crop and also as a pot herb (Cobley 1957; Purseglove 1968). Ethno-botanical uses including as food, medicine, fodder and feed, industrial raw materials, etc. have been known from time immemorial and

TABLE 9.2
Food Value per 100 g of Edible Portion (Morton 1987)

Component	Calyces (Fresh)	Leaves (Fresh)	Seeds
Moisture	9.2 g	86.2 g	12.9 g
Protein	1.145 g	1.7–3.2 g	3.29 g
Fat	2.61 g	1.1 g	16.8 g
Fiber	12.0 g	10 g	–
Ash	6.90 g	1 g	–
Calcium	1,263 mg	180 mg	–
Phosphorus	273.2 mg	40 mg	–
Iron	8.98 mg	5.4 mg	–
Carotene	0.029 mg	–	–
Thiamine	0.117 mg	–	–
Riboflavin	0.277 mg	–	–
Niacin	3.765 mg	–	–
Ascorbic acid	6.7 mg	–	–
Malic acid	–	1.25 g	–
Cellulose	–	–	16.8 g
Pentosans	–	–	15.8 g
Starch	–	–	11.1 g

are reviewed in many publications (Da-Costa-Rocha et al. 2014; Lim 2014). In terms of nutritional value, it is a good source of essential nutrients. The calyces are rich in vitamin C and antioxidants, and also minerals. The fleshy calyces of *H. sabdariffa* have been used in various countries in Africa and the Caribbean as a food or food ingredient in herbal teas, jellies, sauces, chutneys or pickles, syrups, beverages, ice cream, chocolates, puddings, cakes and wines. The approximate composition of *H. sabdariffa* is presented in Table 9.2.

The fleshy and vivid red calyces have high concentrations of L-ascorbic, arachidic, citric, stearic and malic acids, in addition to pectins, phytosterols (e.g. β-sitosterol and ergosterol) and polyphenols (Guardiola and Mach 2014); and possess inherent natural and pharmacological properties, *viz.* are antihypertensive, antidiabetic, antiobesity, antioxidant, hepatoprotective, anti-hyperammonemia, anticancer, hematological activity, antispasmodic, aflatoxin production inhibition, antityrosinase and cosmeceutical, nephroprotective, diuretic, anticholesterol, anti-anemic, and many more (Lim 2014; Riaz and Chopra 2018). In traditional medicine, various parts of the *H. sabdariffa* plant have also been used for the prevention of diseases such as cardiovascular diseases and hypertension (Izquierdo-Vega et al. 2020). A diagrammatic presentation of the therapeutic and/or pharmacological properties of *H. sabdariffa* extracts is shown in Figure 9.1.

The amount of anthocyanins, a subgroup of the phenolic compounds, varies with extraction temperature and pH. Although the anthocyanins exhibited a certain degree of heat resistance and good color stability in an acidic environment, these degraded very quickly and exhibited significant changes in color in a low-acid environment (Wu et al. 2018). Nguyen et al. (2022) recently reported that yeast-hulls possess a good ability to protect anthocyanin against the influences of temperature, light and moisture compared to freeze-dried anthocyanin-rich extracts, and even better than maltodextrin under high humidity conditions. Different parts of *H. sabdariffa* plants, *viz.* calyces, leaves and seeds, are consumed in different preparations, for example, soup, stew, sauces, drinks, vegetables and salad, animal feed and fodder, oil for cooking, soap making, cosmetics, etc., however, most of the previous works on *H. sabdariffa* commonly targeted calyces (Da-Costa-Rocha et al. 2014; Salami and Afolayan 2021). The chemical and nutritional composition can be varied with the growth stages, and strongly depends on geographic origin (Da-Costa-Rocha et al. 2014; Lim 2014;

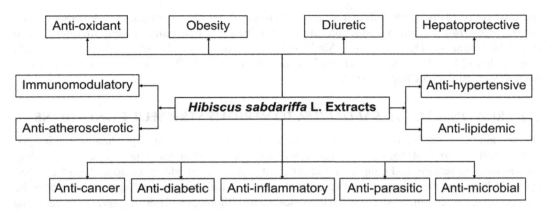

FIGURE 9.1 Therapeutic and/or pharmacological properties of *H. sabdariffa* extracts. (Redrawn from Izquierdo-Vega et al. 2020).

Islam et al. 2021). The nutritional and biochemical compositions of *H. sabdariffa* cultivars and their medicinal importance have been discussed in detail in Salami and Afolayan (2021).

The top 10 deadliest diseases, include ischemic heart disease, stroke, chronic obstructive pulmonary disease, lower respiratory infections, neonatal conditions, trachea, bronchus and lung cancers, Alzheimer's disease and other forms of dementia, diarrheal diseases, diabetes and kidney diseases, accounted for 55% of the 55.4 million deaths worldwide in 2019 (www.who.int/en/news-room/fact-sheets/detail/the-top-10-causes-of-death). *Hibiscus sabdariffa* plant parts are administered in many cases as a protective and/or curative agent for these diseases including diarrhea (Sarkar et al. 2012), Alzheimer's disease (El-Shiekh et al. 2020), diabetes type 2 (Banwo et al. 2022), obesity (Melchert et al. 2016; Herranz-López et al. 2019; Diantini et al. 2021; El-Hashash et al. 2022), hypertension (Mozaffari-Khosravi et al. 2009; Guardiola and Mach 2014; Zheoat et al. 2019; Bourqui et al. 2021), lipid disorder (Guardiola and Mach 2014; Zhang et al. 2020), antiviral activity (Sunday et al. 2010; Joshi et al. 2015; Baatartsogt et al. 2016; D'Souza et al. 2016; Hassan et al. 2017; Takeda et al. 2020; Parga-Lozano and Guerrero 2021; Sarwar 2021; Ramadhani et al. 2022), neutrophil elastase-driven diseases, including chronic obstructive pulmonary diseases, asthma and cystic fibrosis (Loo et al. 2016), and others. *Hibiscus sabdariffa* extracts also possess immunosuppressive capabilities and might be beneficial in the treatment or management of pathological conditions associated with a hyperactive immune system (Umeoguaju et al. 2021). Furthermore, these calyces are consumed as jam, syrup and hot or cold drinks, and also as natural food colorants (Hashemi and Shahani 2019).

The therapeutic properties of *H. sabdariffa* have been accredited to the bioactive compounds, mainly phenolic acids, volatiles, flavonoids, anthocyanins and organic acids (citric, hydroxy-citric, hibiscus, tartaric, malic, oxalic, and ascorbic) (Da-Costa-Rocha et al. 2014). However, in the case of therapeutic potential studies, the red cultivars have attracted additional attention. The phytochemical screening of (only) green *H. sabdariffa* extract has identified 323 compounds, including 5-hydroxymethyl-2-furaldehyde, 7-hydroxycoumarine, betaine, curcumin, nicotinamide and jasmonic acid (Fitriaturosidah et al. 2022). Phenolic compounds also play important functions in protecting plants from biotic and abiotic stressors (Dramane et al. 2019). The pharmacological properties of organic acids found in *H. sabdariffa* have also been reviewed by Izquierdo-Vega et al. (2020). In recent times, El-Naeem et al. (2022) highlighted the prospect of developing new anthocyanin drugs for the treatment of gout and other diseases related to xanthine oxidase increased activity such as hypertension.

However, the natural product might not always be safe for medicated individuals or vulnerable groups, for example, children, pregnant and lactating women, the elderly, etc., due to the possibility of the herbs interacting with synthetic drugs. For instance, the aqueous extract of *H. sabdariffa* caused a reduction in the elimination of acetaminophen, diclofenac and hydrochlorothiazide (Kolawole and Maduenyi 2004; Fakeye et al. 2009). The *H. sabdariffa* extract should be administered with care, as higher doses (>1000 mg kg^{-1} day^{-1}) might have a toxic effect (Riaz and Chopra 2018).

9.4 RESPONSES OF *H. SABDARIFFA* TO VARIOUS STRESSFUL CONDITIONS

Stress conditions—deficit or excess—caused yield losses and/or diminished product quality in agriculture and forestry. Global climate change will cause major changes in precipitation, including the amount and pattern, and temperature in different parts of the world (www.ipcc.ch/sr15/chapter/spm/); it increases the frequency and severity of extreme weather events and aggravates further the adverse effects of combined stresses and increases the resulting losses. These events may increase the susceptibility of plants, facilitate the spread, reproduction and development of pathogens and pests, and weaken or destroy their natural enemies and competitors (Teshome et al. 2020). Furthermore, anthropogenic activities and climate change impacts on biodiversity and ecosystems are affecting the ecology of pathogens and populations of disease vectors such as mosquitoes, further increasing the risk of the emergence and spread of infectious diseases in animals, plants and humans. A global survey of the major food crops indicated that pests—pathogens, insects and weeds—cause average yield losses ranging from 17.2% in potatoes up to 30.0% in rice (Savary et al. 2019). Similarly, the major abiotic stresses such as temperature extremes, drought, as well as the deficiency and toxicity of plant nutrients cause up to 51–82% annual losses of crop yield in the world (Oshunsanya et al. 2019). The yield loss (%) in *H. sabdariffa* due to different stressful conditions varies from 10% to 90%, and is presented in Table 9.3.

The formation of reactive oxygen species (ROS, consisting of superoxide radicals, hydrogen peroxide, hydroxyl anions and single oxygen atoms) and their subsequent detoxification is a common response of plants to both biotic and abiotic stresses. The ROS and phytohormone signaling pathways have been considered to be two of the main "converging points" between responses to biotic and abiotic stresses in plants (Zhang and Sonnewald 2017). The physiological adaptive responses such as biosynthesis of secondary metabolites, concentration, mode of action, etc. depend on the plant species, genotype, physiology, developmental stage and environmental factors during growth, coping with the stress and defensive stimuli (Isah 2019). The stress response sometimes involves reversible salicylic and jasmonic acid production, ethylene and ROS production, ion fluxes, phosphorylation, promoter elements and transcription factors. The application of exogenous nitric oxide, a small but important redox signaling molecule, plays an important role in alleviating the negative effects of stresses and generally improves the antioxidant activity in most plant species (Nabi et al. 2019). In addition, S-nitrosylation and tyrosine nitration are two NO-mediated posttranslational

---

TABLE 9.3
Yield Reduction (%) in *H. sabdariffa* Crops under Different Stress Conditions

Stress condition	Yield reduction	References
Drought	30–80%	Narejo et al. 2016; Dramane et al. 2019; Besharati et al. 2022
Waterlogging	66%	Changdee et al. 2010
Salinity	10–90%	Hashemi and Shahani 2019; Abou-Sreea et al. 2022
Heavy metal	75%	Saleem et al. 2020
Nematode	48%	Adegbite et al. 2008
Insect	85%	Simon et al. 2018
Diseases	40%	Amusa 2004

modifications, and all these factors are important in protecting plants from diverse stresses varying with the species. The application of salicylic acid, an oxidative plant growth regulator, also plays a significant role in plant protection systems against biotic and abiotic stresses in different crops (Ahmad et al. 2021). The effects and some of the remedial measures of specific stressors on *H. sabdariffa* plants are described below.

9.4.1 Responses of *H. sabdariffa* to Abiotic Stressors

9.4.1.1 Responses of *H. sabdariffa* to Drought

Plant water status plays an important role in economic yield by affecting different physiological properties such as leaf turgor potential, plant growth, stomatal conductance, transpiration rate, photosynthesis and respiration. Water (deficiency) stress is likely to become more frequent and intense as a result of global climate change, which may severely impact on the growth and yield of agricultural, horticultural and medicinal plants, especially in arid and semi-arid parts of the world (Qiao et al. 2020). Drought is considered one of the major limiting factors affecting the growth and productivity of crop plants.

The unpredictable nature of drought severity depends on several factors, for instance, rainfall amount and distribution, evaporative demands and moisture-storing ability of soils. It severely affects the morphological and physiological activities of plants and hampers the seed germination, root proliferation, biomass accumulation and final yield of field crops. The drought and osmotic stress-induced damage in plants are mainly due to water shortage or the imbalance of plant water absorption and loss. The accumulation of osmolytes and soluble sugars are among the most common responses of plants to osmotic stress resulting from abiotic stresses such as drought.

The *H. sabdariffa* plant maintained the turgidity of its cells despite a decrease in soil water content below the field capacity, by an increase in osmotic potential through the accumulation of osmolytes in the cytoplasm of cells, to resist water stresses (Dramane et al. 2019). Stunted growth of the leaves, stems and roots, and ameliorating the synthesis of proline and phenolic compounds are the adaptive strategies of plants to avoid and/or resist water deficit. Both the phenolic compounds and proline production gradually increased in *H. sabdariffa* leaves with the decline in water availability. Phenolic compounds, induced commonly in the presence of stressors (Ruiz-Garcia et al. 2012), protect plants from biotic and abiotic stressors. Proline acts as a solute for osmotic adjustment and serves as a reservoir of nitrogen and carbon compounds for subsequent use in plant growth (Oukara et al. 2017). Besharati et al. (2022) reported that ion leakage and total phenolic content in *H. sabdariffa* were significantly increased upon drought stress, and the DPPH (1,1-diphenyl-2-picrylhydrazyl) scavenging activity was increased under severe drought stress compared to mild and no-stress conditions. The combined application of hydrogel and compost can significantly improve plant physiological functions, leading to increase growth, and yield components under drought stress conditions.

At the 65% moisture irrigation regime, the total anthocyanin content in calyx was higher relative to the control (Hinojosa-Gómez et al. 2020). Among the cultivars of *H. sabdariffa*, UAN16-2 showed the greatest increases in the contents of cyanidin, delphinidin 3-O-glucoside, cyanidin 3-O-glucoside and cyanidin 3-O-sambubioside. The content of cyanidin 3-O-sambubioside showed the greatest increase, increasing by 55% relative to the control level. The contents of these compounds are correlated with color attributes such as luminosity. The market value also varied on its visual quality, size, anthocyanin content and chemical characteristics such as acidity and aroma. Moreover, water stress under the 33% moisture condition during plant development led to decreased anthocyanin contents in all of the *H. sabdariffa* cultivars (Hinojosa-Gómez et al. 2020). An increase in total anthocyanin content, maltase activity, fructose, glucose and malic acid in the dry planting seasons was reported by Ifie et al. (2018); in contrast, the citric acid content was lower. The bitterness in *H. sabdariffa* extracts could be reduced by exposing the plant to controlled water stress since malic acid enhances sucrose perception while citric and quinic acids mask the perception of sugars (Ifie et al. 2018).

The root–shoot ratio was significantly increased in stressed *H. sabdariffa* plants (Narejo et al. 2016). The reduction in leaf area was considered to be an avoidance mechanism to minimize the evaporative surface area. The number of branches per plant and chlorophyll content were also reduced, though not significantly. The drought tolerance mechanism in *H. sabdariffa* might be by increasing the root–shoot ratio and stability of chlorophyll content (Narejo et al. 2016). For *H. sabdariffa*, drought stresses were found to be most detrimental at the flowering stage compared to the vegetative or fruit development stages (Nahed and Moursi 2016). Irrespective of water stress, the foliar application of Fe, Zn and Mn, individually and in combination, significantly improved all the yield descriptors, *viz.* fresh and dry weights of calyces, seeds yield/plant, and calyces, fibers and seeds yield/fed, total anthocyanin, total soluble solids (TSS), (titratable) acidity, pH, TSS acidity ratio, and water relation descriptors compared with the control plants in two consecutive seasons. Skipping irrigation at the vegetative growth stage offsets saving nearly 9% of applied water (Nahed and Moursi 2016).

The re-watering of *H. sabdariffa* plants, grown under drought stress, at 100% water saturation level resulted in complete recovery of the photosynthetic rate within 48 hours; however, re-watering at the 25% level showed only 30% recovery. Photosynthetic pigments, including chlorophyll a, b and carotenoid, and total phenolic compound concentration of the calyx were not significantly influenced by the drought, although the anthocyanin concentration of calyx was significantly higher under the drought treatment (Evans and Al-Hamdani 2015). However, Ali and Hassan (2017) observed that water stress significantly decreased the chlorophyll content and membrane stability index (MSI) in stressed plants. Silicon (Si) application had ameliorative effects on *H. sabdariffa* growth, yield, physiological attributes and antioxidant enzymes under water stress and improved the growth even with non-stressed plants. The application of Si increased the amount of N, P, K, Mg, TSS, pH value, anthocyanin and chlorophyll contents, MSI, total soluble sugars, proline content, catalase (CAT) and superoxide dismutase (SOD), while only Ca and peroxidase (POD) contents were decreased due to Si treatment in stressed or non-stressed plants. Shala and Mahmoud (2018) reported that the exogenous (foliar) application of glycine betaine (GB), a naturally occurring amino acid derivative, alleviates the adverse effects of drought stress in *H. sabdariffa* plants. It acts as an osmolyte or a catabolic source of methyl groups to accelerate numerous biochemical processes, protects the plant cells against osmotic inactivation and enhances the water retention in cells. The GB at 25 and 50 mM concentrations enhanced vegetative growth and yield characters, TSS, acidity and total anthocyanin. Foliar spray of 25 mM GB with 65% of available soil moisture depletion recorded the highest branch number, fruit number, fruit fresh weight, calyces fresh and dry weights, seed yield per plant and calyces yield fed⁻¹, photosynthetic pigments, TSS, acidity % and total anthocyanin content while it reduced pH value, which could be a feasible technique for *H. sabdariffa* production that increases the yield and quality in water deficit conditions.

The application of arbuscular-mycorrhizal fungi (AMFs) and plant growth-promoting rhizobacteria (species of the genera *Azotobacter*, *Azospirillum* and *Pseudomonas*) increased 1000-seed weight and seed yield of *H. sabdariffa* plant under drought stress. The mycelium of AMFs plays an important role in enhancing fungus–host plant water relations through absorbing water from the very fine pores of the soil, and absorption of nutrients (in available forms) by improving soil chemical properties (Sembok et al. 2015). The application of AMFs improves the water stress tolerance in *H. sabdariffa* plants through increased activity of enzymatic antioxidants and higher production of non-enzymatic antioxidant compounds, as well as photosynthetic pigments in symbiotic association with AMFs, which can alleviate ROS damage resulting in increased growth and yield parameters (Sanayei et al. 2021). Thus, soil microorganisms showed their ability to alleviate water stress by regulating different morphological and physiological mechanisms of the *H. sabdariffa* plant. Along with AMF inoculation, the humic acid application could also be used as a reliable approach for *H. sabdariffa* cultivation in arid and semi-arid areas (Fallahi et al. 2016). The application of humic acid also stimulates the functions of mycorrhizal developments. Plant–mycorrhizal symbiosis adopts several mechanisms, including maintenance of leaf water potential and turgor pressure, increase in expression levels of drought resistance genes, regulation of abscisic acid production and enhancement of plant recovery, etc., to protect plants and/or avoid the negative impacts under drought condition.

These characters, *viz.* reduction in plant height, leaf osmotic potential, photosynthetic rate, carbon dioxide assimilation rate (i.e., stomatal conductance), transpiration rate, leaf relative water content and chlorophyll content and increment of proline content, could be used for evaluation, screening and manipulation of *H. sabdariffa* genotypes for improvement of drought tolerance through conventional breeding or drought-responsive gene isolation (Mohamed et al. 2015).

9.4.1.2 Responses of *H. sabdariffa* to Waterlogging

Waterlogging (WL), which has emerged as a serious environmental problem in world agriculture (Keya et al. 2022), is caused by heavy rainfall events, improper irrigation and inadequate drainage facilities, uneven fields and plow pan/soil compaction. It induced a reduction in soil oxygen availability with hypoxic or anoxic conditions. The growth and development of many crops are susceptible to this stressful environment, leading to a reduction in crop yields. The intricate effects of WL stress vary not only with plant species but also with plant age, temperature, soil condition, humidity, sunlight, evaporation, stress duration, water level and other factors. In hypoxic or anoxic conditions, anaerobic microorganisms trigger the accumulation of lethal metabolites (including H_2S, N_2, Mn^{2+}, Fe^{2+}, lactic acid, ethanol and aldehydes) and ROS, and affect stress hormone (e.g., abscisic acids and ethylene) balance in roots. On the contrary, physiological activities, i.e., respiration, photosynthesis, nutritional traits, plant growth and survival of plants are altered by WL (Sharma et al. 2021). Plants develop different morphological (*viz.* new adventitious roots, aerenchyma and a barrier to radial oxygen loss in the roots), physiological (e.g., alteration of photosynthesis, stomatal conductance and gas exchange, etc.) and biochemical (such as increased fermentative enzyme content, energy crisis and increased glycolysis supply, etc.) tolerance or resistance mechanisms to cope with WL stress.

The WL regimes had a significant effect on plant height, stem diameter, leaf area, biomass production, root growth (taproot length and lateral root number), aerenchyma tissue formation and fiber yield of *H. sabdariffa* plants (Changdee et al. 2009). However, the WL stress at later stages of growth (e.g., around 100 days after sowing or later) did not display a significant reduction in growth and yield as compared to the control (Changdee et al. 2010). The length and dry weight of adventitious roots increased depending on the duration of WL, and play an important role in supplying oxygen for the biological activities of underground roots under prolonged flooding. The WL treatment induced the onset of Casparian band formation and lignification of cell walls that occurred in the root exodermis of all three species, *viz. H. sabdariffa*, *H. cannabinus* and *C. olitorius*, but was not stable. The endodermal Casparian band formation was slightly closer to root tips in *H. sabdariffa* in comparison with the control (Changdee et al. 2008). Casparian bands and suberized cell walls may be a barrier against toxic materials and prevent oxygen loss from aerenchyma.

The production of flowers and leaves of *H. sabdariffa* were influenced by production systems (*viz.* green roof, field row and high tunnel) and genotypes in urban environments. Most minerals were influenced by the production systems, moreover, genotypes differed in nutrient content regardless of the production system (Richardson and Arlotta 2021). The application of exogenous ascorbic acid might be a strategy to mitigate WL stress through lowering malondialdehyde (MDA), CAT, SOD, polyphenol oxidase (PPO), POD and ascorbate peroxidase (APX) concentrations, and reduced glutathione (GSH) and proline accumulation in waterlogged plants (Ullah et al. 2017). Ascorbic acid-treated plants had significantly lower abscisic acid (ABA) concentration and an increase in jasmonic acid (JA). Keya et al. (2022) concluded that the exogenous application of reduced glutathione (GSH) effectively modulated multiple physiological and biochemical mechanisms to enhance the WL tolerance of sesame plants. GSH supplementation to WL plants also increased the transcript levels of numerous antioxidant defense-related genes, such as *SiCAT1, SiCATX1, SiAPX3, SiPOD40, SiPOD44, SiGPX4, SiPHGPX1, SiGSTU23* and *SiGSTU25* in shoots, corresponding to enhanced activities of the respective antioxidant enzymes. The recent advancements in WL tolerance/adaptation mechanisms in plants have been reviewed by Pan et al. (2021).

9.4.1.3   Responses of *H. sabdariffa* to Salinity

Salinity stress, a complex abiotic phenomenon, causes osmotic stress, specific ion impacts and deprivation of nutritive materials, which impact many physiological and biochemical pathways in plant development. *Hibiscus sabdariffa* plants are moderately tolerant to salinity up to 10 dS m^{-1} of water salinity; the highly saline irrigation water reduced the germination and vegetative growth and induced morphological, physiological and biochemical changes (Liu et al. 2021). Salinity stress mostly induced a marked increase in soluble and total carbohydrates, soluble and total proteins contents, and a marked accumulation in total free amino acids and proline contents, while a significant reduction in anthocyanin content was observed in *H. sabdariffa* cultivars (Abdel Latef et al. 2009). Hashemi and Shahani (2019) reported that salt stress significantly decreased plant height, number of leaves, number of flowers and shoot fresh weight except for the fresh, and dry weight of flowers, the total phenolic content also showed a significant reduction except for 40 mM NaCl. The total anthocyanin content showed a significant increase in 40 mM NaCl, which indicates the anthocyanin content of *H. sabdariffa* might be enhanced in response to low saline concentration, without any change in flower yield.

Kadamanda and Natarajan (2017) reported that the leaf area, stomatal index and stomatal frequency of *H. sabdariffa* also significantly differed among and/or between landraces, salts and salinity levels. The increased stomatal density and decreased stomatal area may be critical for stomatal regulation in salt-prone environments. High stomatal regulation depended largely on stomatal density, area and degree of encryption under salinity, which is of great ecophysiological significance for plants growing under osmotic stresses. Trivellini et al. (2014) reported an increase in electrolyte leakage and abscisic acid content, and a reduction in flower fresh weight, anthocyanin content and chlorophyll fluorescence parameters. Galal (2017) and Gadwal and Naik (2014) showed that NaCl affected the germination rate, vegetative growth of seedlings, length of seedlings, fresh weight, dry matter of seedlings, seed viability index, chloroplast pigments as well as chlorophyll and carotenoid biosynthesis.

Azooz (2009) reported that 1000 ppm aqueous solution of riboflavin (vitamin B$_2$) might act as an effective antioxidant and enhance the salinity stress resistance of *H. sabdariffa* seedlings by regulating osmotic and ionic balance. The exogenous application of benzyladenine, one of the naturally occurring cytokinins, appeared to nullify the adverse effect of salinity, either partially or completely, on the growth and development of *H. sabdariffa* (Abdel Latef et al. 2009). The accumulation and distribution of sugar in different plant parts could be a valid trait to identify tolerance to saline and osmotic stresses and to explain the sink–source relationships in plants grown under stress. Ascorbic acid, citric acid and thiamin application enhanced both the growth and production of *H. sabdariffa* plants grown under saline conditions (up to 75 mM NaCl) by increasing the leaf relative water content, membrane stability index, total soluble phenols, flavonoids and proline, and reduced glutathione concentrations in leaves and anthocyanins in dry calyces (Abdellatif and Ibrahim 2018). The negative impacts of salts in saline water irrigated plants were lowered and the growth of plants was enhanced by the individual and/or combined addition of biochar and compost by increasing the essential nutrients released from the added organic materials, microbes and enzymes activity, and improving the physiochemical properties of soil (Liu et al. 2021). Although the single addition of compost or biochar demonstrated a notable improvement in soil and *H. sabdariffa* quality, the mixing of these two amendments showed a distinct superiority. The foliar spray of *Aloe saponaria* extract (Ae) and/or potassium silicate (KSi) has recently been reported to be effective in reducing the stressful salinity impacts on the development, yield and features of *H. sabdariffa* plants (Abou-Sreea et al. 2022). They described that Ae (1%) and KSi (60 g L^{-1}) enhanced the growth, physiological and nutritional characteristics of *H. sabdariffa* plants under saline conditions (up to 16 dS/m) either individually or combined.

Both WL and salinity stress individually and in combination result in oxidative stress. To cope with oxygen deficiency, plants develop numerous anatomical and morphological adaptations such as aerenchyma formation, formation of adventitious roots, etc. (Duhan et al. 2019). The prevention of ROS production and/or the efficient ROS scavenging ability by a strong induction of the anti-oxidant defense system (enzymatic and non-enzymatic antioxidants) are important physiological factors for plant tolerance under combined WL and salt stress (WL/NaCl) conditions (Duhan et al. 2019). Plants also require to maintain a positive balance between promoting Na$^+$ efflux and limiting K$^+$ efflux in the cytoplasm to tolerate combined stresses. There are two main possible pathways for K$^+$ efflux out of the root cells, including non-selective cation channels and guard cell outward-rectifying potassium that are activated by ROS under combined WL/NaCl conditions. The efficient antioxidant metabolism is important in maintaining ionic homeostasis for plant tolerance to combined WL/NaCl conditions.

A novel nuclear-localized R2R3-type RMYB gene was isolated from *H. sabdariffa* roots under salt stress which is involved in the stress tolerance mechanism (Mohamed et al. 2017). Overexpression of RMYB was observed in transgenic plants under abiotic stresses, which further suggests its regulatory role in response to stressful conditions. The RMYB transcription factor overexpressing in transgenic plants may be used as a potential agent for the breeding and development of stress-tolerant or -resistant crop cultivars (Mohamed et al. 2017).

9.4.1.4 Responses of *H. sabdariffa* to Heavy Metals

Heavy metals (HMs) are a group of metals and metalloids that have relatively high density, an atomic number greater than 20 and atomic density above 5 g cm^{-3} and are toxic even at ppb levels (Duruibe et al. 2007). The uptake of toxic (heavy) metals, *viz.* lead (Pb), chromium (Cr), arsenic (As), zinc (Zn), cadmium (Cd), copper (Cu), mercury (Hg) and nickel (Ni), from soil depends upon several factors such as the concentration of HMs, pH and the plant species/characteristics among others (United States Environmental Protection Agency 2007). The toxic effects of HMs on plants depend on a variety of factors including the type and concentration of HMs and plant species, plant growth stage and exposure time. Although, Cu is an essential micronutrient for plants, toxic concentrations of Cu in the soil can cause nutrient imbalance by binding with organic matter, clay minerals and hydrated oxides of iron (Fe), aluminum (Al) and manganese (Mn), which ultimately affects the plant productivity. Tissue Cu/Pb concentrations increased linearly as metal doses increased in unamended soils and the transferability of Cu to *H. sabdariffa* was higher than Pb, indicating the greater phyto-availability of the former (Wuana and Mbasugh 2013). Tissue metal concentrations and soil-to-plant transfer factors portrayed *H. sabdariffa* as a promising phytoextractor of Cu and Pb. Although bio-mass yields decreased as the metal dose increased, the plants were typically greenish with linear growth profiles at all metal doses, indicating some level of tolerance in *H. sabdariffa* plants.

Mutari (2019) reported that the uptake (transfer of HMs from the soil to leaves) capability and bio-accumulation were greatest for Zn followed by Cu, then Pb and Cd in both *H. sabdariffa* and jute mallow. Strongly acidic soils increase Zn, Cu and Cd uptake and increase their phytotoxicity and the continuous consumption of the leaves of the crops makes the HMs bioavailable to the human body. *Hibiscus sabdariffa* could be used as a promising plant for the phytoremediation of toxic metals in contaminated soils due to its metal accumulation ability and relatively high biomass yield (Ondo et al. 2012). The application of Zn and Cu (at the concentration of 300 mg L^{-1}) caused, on average, 15% reductions in the size of the plant canopy (leaf area index) and stem diameter compared to plants without application (Apáez-Barrios et al. 2018). The effect of Cu and Zn application on calyx yield depended upon cultivars; in turn, these elements increased the amount of anthocyanins, the percentage of titratable acidity increased and decreased the concentration of ascorbic acid. The increased titratable acidity is important since the acidity has an antibacterial effect and contributes to the absorption of metal ions in the human body (Apáez-Barrios et al. 2018). However, two sprayings of Cu, at the concentration of 150 mg L^{-1}, increased the anthocyanin content and percentage of titratable acidity and decreased the ascorbic acid concentration in *Hibiscus* calyces without affecting the yield (Barrios et al. 2017).

Lead, one of the major anthropogenic pollutants, has been accumulated in different aquatic and terrestrial ecosystems since the Industrial Revolution. Low solubility and strong binding capacity with soil colloids aggravate the toxicity situation more. *Hibiscus sabdariffa* plants exposed to Pb exhibited a substantial decline in growth, yield, pigment content, anthocyanin, total carbohydrates, nutrient content and catalase enzymes, and an increase in lead concentration, proline content and APX activity in plant. The exogenous applications of benzyl adenine (100 mg L^{-1}) mitigated Pb toxicity in *H. sabdariffa* plants exposed to Pb stress, which probably includes not only the regulation of the chloroplast and antioxidant system but also the reduction of Pb uptake and the improvement of mineral nutrient absorption (Matter 2016). Benzyladenine enhances the uptake of elementals N, P and K, thus leading to a corresponding increase in chlorophyll which could be an indicator of the stress-induced balance of endogenous hormones (Marschner 1995). On the other hand, *H. sabdariffa* red calyx beverages could be used as both curative and protective agents against Pb-associated disorders (El-Hashash et al. 2022) and Cd-induced toxicity (Orororo et al. 2018) in rat models. Exposure to Cd significantly reduced GSH levels and glutathione-s-transferase activity, SOD and CAT, but increased lipid peroxidation in the plasma and tissues of rats. In contrast, *H. sabdariffa* anthocyanins improved tissue SOD, CAT and GST activities with a concomitant increase in GSH levels and a reduction in tissue lipid peroxidation. Rats administered *H. sabdariffa* calyx extract in addition to $CdSO_4$ were less susceptible to the toxic effects of Cd. Zinc antagonizes the uptake of Cd, i.e. Zn deficiency increases the incidence of Cd toxicity (Muhammad and Charles 2008).

Cadmium is extremely toxic to plants, animals and humans; it is non-biodegradable, extremely persistent and cannot be substantially removed from plants. Zeolite, a Si-rich material, increases the resistance of plants to Cd stress by the accumulation of mono- and polysilicic acids in the leaves, increases the activities of the antioxidant enzymes SOD, APX and guaiacol peroxidase (GPX) and decreases the contents of thiobarbituric acid reactive substances (TBARs) (Balakhnina et al. 2015). The stimulation of the antioxidant system of plants and the formation of complexes between toxic metal ions and Si-compounds are the most important mechanisms to explain the positive influence of active silicon on plants under HM toxicity. Biochar and foliar application of ascorbic acid, individually and/or collectively, were found to be useful to amend for Cd stress alleviation in barley plants (Yaseen et al. 2021).

Shuhaimi et al. (2019) reported that the best *H. sabdariffa* plant growth and development could be obtained from the application of food waste compost to the soil at a 1:1 ratio. The HM uptake was insignificant for most of the metals studied, except for Ni, Mn and Pb; this indicates that the *H. sabdariffa* plants showed low tolerance to HM mobility to its tissues in different types of treatments (Shuhaimi et al. 2019). The growth performance of *H. sabdariffa* did not show a significant difference due to the application of Fe_3O_4 nanoparticles, which appeared to be experimental condition dependent. Admixture of the residue of flower and sepal of *H. sabdariffa* (after extraction) and black tea residue can remove and decrease the amount of Cd and Ni significantly from contaminated wastewater, although the rate of adsorption of cobalt (Co) was relatively lower (Ziarati et al. 2018). Plants act both as "accumulators" (biodegrade or bio transform the contaminants into inert forms in their tissues) and "excluders" (restrict contaminant uptake into their biomass). The removal of metal ions, metal biosorption, by inactive, non-living biomass might be due to highly attractive forces present between them, especially the presence of certain functional groups, such as amine, carboxyl, hydroxyl, phosphate, sulfhydryl, etc., on the cell wall of the biomass. The bio-adsorption processes could be influenced by temperature, pH, biomass concentration and the type of metal contaminants (ions) (Ziarati et al. 2018).

Arsenic, a known human carcinogen and teratogen, is ubiquitously present in the soil environment as As^{3+} and As^{5+} (Akter et al. 2005). Contamination of drinking water with As is a serious environmental problem worldwide, especially in the Bengal basin region. The seed germination, plant height, shoot and root biomass of *H. sabdariffa* plants were reduced in As-contaminated soil compared to uncontaminated soil. In contrast, the concentration of As in plant tissue was higher due to the uptake of As increasing with increasing levels of As in soil (Uddin et al. 2016). The lower concentration of As in postharvest soils indicated the removal of As from soils; *H. sabdariffa* could be used as "hyperaccumulator plants" for the phytoremediation of metal pollutants from contaminated soil (Uddin et al. 2016). The HM accumulation capability of plants could be enhanced by soil additives (vermiculite, sulfur and compost) and foliar spraying (humic acid and gibberellins) (Shehata et al. 2019). HM toxicity-tolerant or -resistant plants activate a complex signaling network of a defense response that includes accumulation of (i) antioxidants for detoxification of excess ROS, (ii) secondary metabolites for sequestration of HMs to vacuoles and (iii) compatible solutes for the osmoregulation along with regulation of metal transporters, among others (Riyazuddin et al. 2022). The considerable antioxidant and a substantial amount of flavonoids and polyphenols present in *H. sabdariffa* could contribute to its antigenotoxic properties and inhibit the DNA damage induced by sodium arsenite in a dose-dependent manner (Ghosh et al. 2015). The crude aqueous extract of *H. sabdariffa* calyces could be used as a dietary supplement to prevent the clastogenic effects of arsenic exposure through drinking water.

Furthermore, some of these HMs, *viz.* Fe, Cu, Zn, Mn and Ni, are members of essential plant (micro-)nutrients. Their deficiency (quantity below the required amount or absence) retards plant growth by inhibiting the activity of crucial enzymes associated with the detoxification of oxidants and regulation of important physiological processes like photosynthesis, respiration, nitrogen assimilation, sugar metabolism, etc. These plants, therefore, experience inherent oxidative stress and are more susceptible to multiple abiotic stresses due to inefficient defense machinery (Banerjee and Roychoudhury 2022). Banerjee and Roychoudhury (2022) have elaborately described the molecular and physiological disconcertion due to micronutrient deficiency and explicate the underlying biochemical, metabolic and signaling cross-talks operative at the cellular and systemic levels.

9.4.2 Responses of *H. sabdariffa* to Biotic Stressors

Biotic stresses—diseases and pests—are significant threats to the yield and quality of all agricultural products including *H. sabdariffa*. Generally, the incidence of pests and diseases and their levels of invasion are influenced by various factors, including temperature, (air) relative humidity, soil moisture, rainfall and plant growth stage. Pathogens and pests in presence of a stressful environment become more detrimental to plants. Under favorable conditions, minor pests could become potential aggressive threats in the future. Temperature is one of the most important drivers of pest outbreaks, although the effects of drought on defoliator pest outbreaks are not consistent in the literature. Moreover, different phenologies (stages of plant development) of *H. sabdariffa* crops are susceptible to and/or attacked by different diseases and pests which also cause significant economic damage (Sarkar and Gawagde 2016).

9.4.2.1 Responses of *H. sabdariffa* to Insect Pests

The Agroforestry Database listed the following insects as pests of *H. sabdariffa*: *Anomis erosa*, *Chaetocnema* spp., *Cosmophila erosa*, *Dysdercus cingulatus*, *D. poecilus*, *Drosicha townsendi*, *Nistora gemella*, *Phenacoccus hirsutus*, *Pseudococcus filamentosus* and *Tectocoris diophthalmus*

(Orwa et al. 2009). Only four insect pest species, *viz.* spiral borer (*Agrilus acutus*), kenaf/flea beetle (*Nisotra orbiculata*), mealybug (*Pseudococcus virgatus*) and Mexican bean weevil (*Zabrotes subfasciatus*) have been reported in Bangladesh (Anon. 2019). According to Mahunu et al. (2021), *H. sabdariffa* is affected by a variety of pests including stem borer, spiral borer, mealybug (*Maconellicoccus hirsutus*), cotton mealybug (*Phenacoccus solenopsis*), flea beetle (*N. orbiculata*), abutilon moth, cotton bollworm, cotton stainer, cutworm, leaf miners, and leafhoppers. Sixteen species belonging to seven orders of insect pests on *H. sabdariffa* were reported by Abdel-Moniem et al. (2011); of these species, piercing-sucking pests *Empoasca* spp., *Spilostethus longulus*, *Nezara viridula* and *Oxycarenus hyalinipennis*, and the predators *Polistes* sp., *Coccinella undecimpunctata* and *Scymnus syriacus*, were the predominant species.

In Egypt, a total of 22 insect taxa were recorded (Abdel-Moniem and El-Wahab 2006). The number and dominance percentage of phytophagous insect species were higher (14; 82.44%) than those of the predators (6; 17.56%). The most dominant pests were piercing-sucking insects *Empoasca* spp., *M. hirsutus*, *Aphis gossypii*, *Bemisia tabaci*, *O. hyalinipennis* and *Earias insulana*. In contrast, the most dominant predators were *Polistes* sp., *Orius* spp., *C. undecimpunctata* and *Scymnus interruptus* (Abdel-Moniem and El-Wahab 2006). Tuo et al. (2019) identified insects belonging to 49 families and eight orders from the northern part of Côte D'Ivoire; 27 types of pesticides, by five pesticide application methods, were used by farmers to control these insect pests. In the Nigerian context, the dominant insect pests of *H. sabdariffa* were *Monolepta thomsoni*, *Nisotra sjostedti*, *Dysdercus volkeri* and *O. hyalinipennis* (Simon et al. 2018). Infestation and damage by the insects varied between the types, green *vs.* red. Insecticide spray improved both the quantity and quality of products. The red *H. sabdariffa* cultivation was more profitable compared to the green, the late crop to the early, and protection at both vegetative and reproductive stages was more cost-effective than other spray regimes. The cultivation of resistant varieties, cultural techniques, physical barriers, semi-chemical-based technologies and, as the last option, the use of selective chemicals which conserve beneficial insects in fields, might be some of the probable ecofriendly and economically feasible options to control the key field pests.

The introduction of biomolecules and transgenic technologies, *viz.* Bt-Cry toxins, lectins, proteinase inhibitors, α-amylase inhibitors and chitinase, have revolutionized the area of insect control; and these are toxic to the insects of orders Lepidoptera, Hymenoptera, Diptera and Coleoptera, and also to nematodes (Upadhyay and Singh 2016). Although *H. sabdariffa* commonly showed resistance to root-knot nematodes (*Meloidogyne* spp.), it did not to free-living nematodes (*Heterodera* spp.). Upadhyay and Singh (2016) described the insect control strategies by the insecticidal proteins and through insect-resistant transgenic crop development and advocated for the exploration of RNAi-mediated insect control by targeting high-expressing and/or important vital genes as an effective approach.

9.4.2.2 Responses of *H. sabdariffa* to Diseases

Various diseases occur in *H. sabdariffa* plants, including root rot and wilt diseases (*Fusarium oxysporum*, *F. solani* and *Macrophomina phaseolina*), gray mold (*Botrytis cinerea*), irregular leaf spot (*Cercospora malayensis*), anthracnose (*Colletotrichum gloeosporioides*), Sclerotinia rot (*S. sclerotiorum*), *Phytophthora* infection (*P. nicotianae*), foot and stem rot (*Phytophthora parasitica* var. *sabdariffae*), leaf blight (*Phylosticta hibiscini*), bacterial wilt (*Ralstonia solanacearum*), wrinkled leaves and phyllody disorder (16SrI phytoplasma), yellow vein mosaic, and diseases caused by nematodes and insect pests (Sarkar and Gawagde 2016; Tzean 2019). A total of 22 diseases of *H. sabdariffa* caused by fungi, viruses or mycoplasma, and also nematodes were reported from Bangladesh (Anon. 2019); of these diseases, stem rot (*M. phaseolina*, *Botryodiplodia theobromae*, *Fusarium* spp.), leaf yellowing (virus/mycoplasma) and seedling blight (*Rhizoctonia solani*, *M. phaseolina*, *Pythium* sp., *Phytopthora parasitica*, *F. solani*) are among the most devastating diseases in the field. The occurrences, symptoms, etiology, epidemiology and management strategies of some diseases have been described in Sarkar and Gawagde (2016). Mahunu et al. (2021) recently documented the effect of biotic stressors on *H. sabdariffa* yield and quality and their control measures.

Fusarium solani K1 (FsK1) can cause wilting and a rotted pith in *H. sabdariffa* (Wang et al. 2021). The *H. sabdariffa* wilt-infected soils also contain abundant populations of Xanthomonadaceae, Microbacteriaceae, Enterobacteriaceae, Flavobacteriaceae, Rubritaleaceae and Ascobolaceae. A confrontation assay of bacteria isolated from healthy rhizosphere soil identified *Bacillus velezensis* SOI-3374, a strain isolated from a healthy *H. sabdariffa* rhizosphere, has the best antagonistic effects toward FsK1; and formed a distinct inhibition zone between the two microorganisms. *Bacillus velezensis* SOI-3374 could be recommended as a potential biocontrol agent against *H. sabdariffa* wilt disease (Wang et al. 2021).

Organic soil amendment, foliar spraying by commercial natural products and their interactions were found to be effective in reducing wilt disease incidence and the severity of insect pest infestation as well as increasing microbial total counts of *H. sabdariffa* rhizosphere. Furthermore, these treatments improve soil's physical and chemical properties to encourage nutrient uptake by plants, enhanced plant growth and development, and significantly increased the yield of dry sepals, and total chlorophyll and anthocyanin content compared to untreated ones (Hashem et al. 2017). The application of microbiota i.e., a mixture of *Pseudomonas fluorescens* (PSR-11), *B. subtilis* (BSR-8), *Pleurotus ostreatus* and mycorrhizeen® individually or in combination reduced *Fusarium* wilt incidence and increased the percentage of surviving plants (Gaafar et al. 2021). A substantial increase in phenol, photosynthetic pigments and dehydrogenase contents, as well as root colonization per cent, was reported due to the application of these bio-agents. The conglomeration of the bacterial mixture of *P. fluorescens* and *B. subtilis*, *P. ostreatus* and mycorrhizeen® was more effective in controlling wilt disease than using any of them individually. Moreover, the mixture of all microbiota gave the best results for the growth and yield parameters of the *H. sabdariffa* plant. Gaafar et al. (2021) also reported that *B. subtilis* and *P. fluorescens* produce lytic enzymes such as β-1-3 glucanase, β-1-4 glucanase, chitinase and protease, in addition to their production of siderophore and HCN.

The combination of two stresses (abiotic–abiotic or abiotic–biotic) does not always lead to a negative impact on plants. Some stress combinations negate the effect of each other, leading to a net neutral or positive impact on plants (Pandey et al. 2017). Various biotic stress combinations and their impacts on plants have been discussed by Lamichhane and Venturi (2015), and drought and pathogen stress combinations as a model to understand the impact of abiotic and biotic stress combinations on plants (Pandey et al. 2017). Root system architecture, leaf pubescence, leaf water potential regulation, presence of cuticular wax and canopy temperature have been identified as some of the potential traits for screening genotypes for tolerance to combined drought and pathogen infection. In recent molecular studies, some potential (gene) candidates, *viz.* methionine homeostasis gene, methionine gamma-lyase (AtMGL), rapid alkalinization factor-like 8 (AtRALFL8) and azelaic acid-induced 1 (AZI1), were identified for improvement of plant tolerance to combined stresses (Atkinson et al. 2013).

9.5 FUTURE RESEARCH PRIORITIES

Multiple uses of *H. sabdariffa* appear to be excellent and promising sources, including functional foods, food colorants, food condiments and feed supplements. Useful food products from *H. sabdariffa* may provide health benefits to consumers due to their substantial contribution to phytochemicals which make this ancient underutilized crop, a food/feed for the future.

- Genetic improvement of *H. sabdariffa* resistance or tolerance to biotic and abiotic stresses through conventional breeding techniques and/or genetic engineering.

- Development of climatic hazard indicators for *H. sabdariffa* L. crop (Montiel-Gonzâlez et al. 2021) as a new tool that helps in planning and decision-making in the face of the growing occurrence of extreme climate events.
- The cellular, biological and epigenetic mechanisms of the specific effects of *H. sabdariffa* are to be elucidated.
- Isolation, characterization and use of the plant growth-promoting endophytic fungi for improving stress tolerance ability in plants.
- Prevention strategies such as strict quarantine have been useful in minimizing the introduction of exotic pathogens and pests.
- Integrative efforts from crop modeling experts, agronomists, field pathologists, breeders, physiologists and molecular biologists may lead to a holistic approach to the development of combined stress-tolerant crops (i.e., ideotype) that can perform well under field conditions.
- Necessity for molecular studies for identification, isolation and characterization of stress-responsive genes for different stress conditions. Their transfer and altering expression of different stress-responsive genes from other sources might be an effective way to develop (i.e., engineer) tolerant or resistant plants.
- Next-generation metabolic engineering, phyto-metabolic editing, using miRNA-mediated CRISPR–Cas9 (Sabzehzaria et al. 2020) in *H. sabdariffa* plants for higher secondary metabolites production.
- Producing transgenic plants, with stacked genes associated with overexpression of antioxidant enzymes, has an enhanced effect on stress tolerance.
- Further combined research on the exact mechanism of action of different medicinal properties, control of doses, active components, bioavailability and other critical variables, and value addition are required to make the cultivation of *H. sabdariffa* more popular and profitable to local farmers.

ACKNOWLEDGMENTS

The author thank, Professor Dr Md. Habibur Rahman Pramanik, and anonymous reviewer(s) for their valuable comments, constructive criticism and/or modification suggestions; and also Dr Subrata Kuri for reading the final draft and language improvement suggestions. The author apologizes for not citing all of the relevant references due to space limitations.

REFERENCES

Abdel Latef, A. A., M. A. K. Shaddad, A. M. Ismail, and M. F. A. Alhmad. 2009. Benzyladenine can alleviate saline injury of two roselle (*Hibiscus sabdariffa*) cultivars via equilibration of cytosolutes including anthocyanins. *International Journal of Agriculture and Biology* 11: 151–157.

Abdellatif, Y. M. R. and M. T. S. Ibrahim. 2018. Non-enzymatic anti-oxidants potential in enhancing *Hibiscus sabdariffa* L. tolerance to oxidative stress. *International Journal of Botany* 14: 43–58. doi: 10.3923/ijb.2018.43.58

Abdel-Moniem, A. S. H. and T. E. A. El-Wahab. 2006. Insect pests and predators inhabiting roselle plants, *Hibiscus sabdariffa* L., a medicinal plant in Egypt. *Archives of Phytopathology and Plant Protection* 39(1): 25–32. doi: 10.1080/03235400500103869

Abdel-Moniem, A. S. H., T. E. A. El-Wahab, and N. A. Farag. 2011. Prevailing insects in Roselle plants, *Hibiscus sabdariffa* L., and their efficiency on pollination. *Archives of Phytopathology and Plant Protection* 44(3): 242–252. doi: 10.1080/03235400903024662

Abou-Sreea, A. I. B., M. H. H. Roby, H. A. A. Mahdy, et al. 2022. Improvement of selected morphological, physiological, and biochemical parameters of roselle (*Hibiscus sabdariffa* L.) grown under different salinity levels using Potassium silicate and *Aloe saponaria* extract. *Plants* 11: 497. doi: 10.3390/plants11040497

Adegbite, A. A., G. O. Agbaje, and J. Abidoye. 2008. Assessment of yield loss of roselle (*Hibiscus sabdariffa* L.) due to root-knot nematode, *Meloidogyne incognita* under field conditions. *Journal of Plant Protection Research* 48(3): 267–273. doi: 10.2478/v10045-008-0035-4

Ahmad, A., Z. Aslam, M. Naz, S. Hussain, T. Javed, S. Aslam, et al. 2021. Exogenous salicylic acid-induced drought stress tolerance in wheat (*Triticum aestivum* L.) grown under hydroponic culture. *PLoS ONE* 16(12): e0260556. doi: 10.1371/journal.pone.0260556

Akter, K. F., G. Owens, D. E. Davey, and Naidu R. 2005. Arsenic speciation and toxicity in biological systems. In *Reviews of Environmental Contamination and Toxicology*. eds Ware G. W., L. A. Albert, D. G. Crosby, et al., 184: 97–149. doi: 10.1007/0-387-27565-7_3

Ali, E. F. and F. A. S. Hassan. 2017. Water stress alleviation of roselle plant by silicon treatment through some physiological and biochemical responses. *Annual Research and Review in Biology* 21(3): 1–17. doi: 10.9734/ARRB/2017/37670

Amusa, N. A. 2004. Foliar blight of roselle and its effect on yield in tropical forest region of southwestern Nigeria. *Mycopathologia* 157: 333–338. doi: 10.1023/B:MYCO.0000024175.84908.86

Anonymous 2019. *Final Report on Developing Pest List of Plants and Plant Products in Bangladesh*. Dhaka: Plant Quarantine Wing, Department of Agricultural Extension. http://dae.portal.gov.bd/site/publi cations/6f54b7ef-028d-4e75-b804-a9b043eab213/-

Apáez-Barrios, P., M. E. Pedraza-Santos, M. N. Rodríguez-Mendoza, Y. A. Raya-Montaño, and D. Jaén-Contreras. 2018. Yield and anthocyanin concentration in *Hibiscus sabdariffa* L. with foliar application of micronutrients. *Revista Chapingo Serie Horticultura* 24(2): 107–120. doi: 10.5154/r.rchsh.2017.06.020

Atkinson N. J., C. J. Lilley, and Urwin PE. 2013. Identification of genes involved in the response to simultaneous biotic and abiotic stress. *Plant Physiology* 162: 2028–2041. doi: 10.1104/pp.113.222372

Azooz, M. M. 2009. Foliar application with riboflavin (vitamin B$_2$) enhancing the resistance of *Hibiscus sabdariffa* L. (deep red sepals variety) to salinity stress. *Journal of Biological Sciences* 9: 109–118. doi: 10.3923/jbs.2009.109.118

Baatartsogt, T., V. N. Bui, D. Q. Trinh, et al. 2016. High antiviral effects of hibiscus tea extract on the H5 subtypes of low and highly pathogenic avian influenza viruses. *Journal of Veterinary Medical Science* 78(9): 1405–1411. doi:10.1292/jvms.16-0124

Balakhnina, T. I., P. Bulak, V. V. Matichenkov, A. A. Kosobryukhov, and T. M. Włodarczyk. 2015. The influence of Si-rich mineral zeolite on the growth processes and adaptive potential of barley plants under cadmium stress. *Plant Growth Regulators* 75: 557–565. doi: 10.1007/s10725-014-0021-y

Banerjee, A., and A. Roychoudhury. 2022. Dissecting the phytohormonal, genomic and proteomic regulation of micronutrient deficiency during abiotic stresses in plants. *Biologia* doi: 10.1007/s11756-022-01099-3

Banwo, K., A. Sanni, D. Sarkar, O. Ale, and K. Shetty. 2022. Phenolics-linked antioxidant and anti-hyperglycemic properties of edible roselle (*Hibiscus sabdariffa* Linn.) calyces targeting type 2 diabetes nutraceutical benefits *in vitro*. *Frontiers in Sustainable Food Systems* 6: 660831. doi: 10.3389/fsufs.2022.660831

Barrios, P. A., M. C. R. Granados, M. E. P. Santos, and Y. A. R. Montaño. 2017. Effect of foliar copper application on yield and anthocyanin concentration in *Hibiscus sabdariffa* calyces. *Revista de la Facultad de Ciencias Agrarias* 50(2): 65–75.

Besharati, J., M. Shirmardi, H. Meftahizadeh, M. D. Ardakani, and M. Ghorbanpour. 2022. Changes in growth and quality performance of Roselle (*Hibiscus sabdariffa* L.) in response to soil amendments with hydrogel and compost under drought stress. *South African Journal of Botany* 145: 334–347. doi: 10.1016/j.sajb.2021.03.018

Bourqui, A., E. A. B. Niang, B. Graz, et al. 2021. Hypertension treatment with *Combretum micranthum* or *Hibiscus sabdariffa*, as decoction or tablet: a randomized clinical trial. *Journal of Human Hypertension* 35: 800–808. doi: 10.1038/s41371-020-00415-1

Changdee, T., A. Polthanee, C. Akkasaeng, and S. Morita. 2009. Effect of different waterlogging regimes on growth, some yield and roots development parameters in three fiber crops (*Hibiscus cannabinus* L., *Hibiscus sabdarifa* L. and *Corchorus olitorius* L.). *Asian Journal of Plant Sciences* 8: 515–525. doi: 10.3923/ajps.2009.515.525

Changdee T., A. Polthanee, C. Akkasaeng, and S. Morita. 2010. Effect of waterlogging at different growth stages on growth and yield in *Hibiscus cannabinus* (Kenaf), *H. sabdariffa* (Roselle) and *Corchorus olitorius* (Jute). *Khon Kaen Agriculture Journal* 38: 349–360.

Changdee, T., S. Morita, J. Abe, K. Ito, R. Tajima, and A. Polthanee. 2008. Root anatomical responses to waterlogging at seedling stage of three cordage fiber crops. *Plant Production Science* 11(2): 232–237. doi: 10.1626/pps.11.232

Cobley, L. S. 1957. *An Introduction to the Botany of Tropical Crops*. 71–73. London: Longmans, Green and Co.

D'Souza, D. H., L. Dice, and P. M. Davidson. 2016. Aqueous extracts of *Hibiscus sabdariffa* calyces to control Aichi virus. *Food and Environmental Virology* 8: 112–119. doi: 10.1007/s12560-016-9229-5

Da-Costa-Rocha, I., B. Bonnlaender, H. Sievers, I. Pischel, and M. Heinrich. 2014. *Hibiscus sabdariffa* L. – A phytochemical and pharmacological review. *Food Chemistry* 165: 424–443. doi: 10.1016/j.foodchem.2014.05.002

Diantini, A., S. Rahmat, A. Alpiani, S. A. Sumiwi, L. Lubis, and J. Levita. 2021. Effect of the roselle (*Hibiscus sabdariffa* L.) calyces drink on the physiological parameters of healthy adult subjects. *Biomedical Reports* 15: 89. doi: 10.3892/br.2021.1465

Dramane, K., K. O. K. Samuel, T. Seydou, N.- A. Laurent, N.- A. S. Rachelle, and K. T. Hilaire. 2019. Impact of water deficit on morpho-physiological parameters of young roselle plants (*Hibiscus sabdariffa* var *sabdariffa*). *International Journal of Advance Agricultural Research* 7: 128–137. doi: 10.33500/ijaar.2019.07.010

Duhan, S., A. Kumari, M. Lal, and S. Sheokand. 2019. Oxidative stress and antioxidant defense under combined waterlogging and salinity stresses. In *Reactive Oxygen, Nitrogen and Sulfur Species in Plants: Production, Metabolism, Signaling and Defense Mechanisms*. eds Hasanuzzaman, M., V. Fotopoulos, K. Nahar, and M. Fujita. London: John Wiley and Sons Ltd. doi: 10.1002/9781119468677.ch5

Duruibe, J. O., M. O. C. Ogwuegbu, and J. N. Egwurugwu. 2007. Heavy metal pollution and human biotoxic effects. *International Journal of Physical Sciences* 2(5): 112–118.

El-Hashash, S. A., M. El-Sakhawy, E. E. El-Nahass, M. A. Abdelaziz, W. K. Abdelbasset, and M. M. Elwan. 2022. Prevention of hepatorenal insufficiency associated with lead exposure by *Hibiscus sabdariffa* L. beverages using in vivo assay. *BioMed Research International* 2022: 7990129. doi: 10.1155/2022/7990129

El-Naeem, A., Abdalla, S., Ahmed, I., and Alhassan, G. 2022. Phytochemicals and in silico investigations of Sudanese roselle. *South African Journal of Science* 118(1/2). doi: 10.17159/sajs.2022/10383

El-Shiekha, R. A., R. M. Ashoura, E. A. Abd El-Haleimb, K. A. Ahmed, and E. Abdel-Sattara. 2020. *Hibiscus sabdariffa* L.: A potent natural neuroprotective agent for the prevention of streptozotocin-induced Alzheimer's disease in mice. *Biomedicine and Pharmacotherapy* 128: 110303. doi: 10.1016/j.biopha.2020.110303

Evans, D., and S. Al-Hamdani. 2015. Selected physiological responses of roselle (*Hibiscus sabdariffa*) to drought stress. *Journal of Experimental Biology and Agricultural Sciences* 3: 500–507. doi: 10.18006/2015.3(6).500.507

Evans, W. C. 2009. *Trease and Evans' Pharmacognosy*. 16th ed. London: WB Saunders.

Fakeye, T.O., A. Pal, D.U. Bawankule, N.P. Yadav, and S.P.S. Khanuja, 2009. Toxic effects of oral administration of extracts of dried calyx of *Hibiscus sabdariffa* Linn. (Malvaceae). *Phytotherapy Research* 23(3): 412–416. doi: 10.1002/ptr.2644

Fallahi, H.-R., M. Ghorbany, A. Samadzadeh, M. Aghhavani-Shajari, and A. H. Asadian. 2016. Influence of arbuscular mycorrhizal inoculation and humic acid application on growth and yield of Roselle (*Hibiscus sabdariffa* L.) and its mycorrhizal colonization index under deficit irrigation. *International Journal of Horticultural Science and Technology* 3(2): 113–128. doi: 10.22059/IJHST.2016.62912

Fitriaturosidah, I., J. Kusnadi, E. Nurnasari, Nurindah, and B. Hariyono. 2022. Phytochemical screening and chemical compound of green roselle (*Hibiscus sabdariffa* L.) and potential antibacterial activities. *IOP Conference Series: Earth and Environmental Science* 974: 012118. doi:10.1088/1755-1315/974/1/012118

Gaafar, D. E. M., Z. A. M. Baka, M. I. Abou-Dobara, H. S. Shehata, and H. M. A. El-Tapey. 2021. Effect of some microorganisms on controlling *Fusarium* wilt of roselle (*Hibiscus sabdariffa* L.). *Egyptian Journal of Phytopathology* 49(1): 98–113. doi: 10.21608/ejp.2021.61270.1022

Gadwal, R., and G. Naik. 2014. A comparative study on the effect of salt stress on seed germination and early seedling growth of two *Hibiscus* species. *IOSR Journal of Agriculture and Veterinary Science* 7(3): 90–96. doi: 10.9790/2380-07319096

Galal, A. 2017. Physico-chemical changes in karkade (*Hibiscus sabdariffa* L.) seedlings responding to salt stress. *Acta Biologica Hungarica* 68(1): 73–87. doi: 10.1556/018.68.2017.1.7

Ghosh, I., S. Poddar, and A. Mukherjee. 2015. Evaluation of the protective effect of *Hibiscus sabdariffa* L. calyx (Malvaceae) extract on arsenic induced genotoxicity in mice and analysis of its antioxidant properties. *Biology and Medicine* (Aligarh) 7: 218. doi: 10.4172/0974-8369.1000218

Guardiola, S. and N. Mach. 2014. Therapeutic potential of *Hibiscus sabdariffa*: A review of the scientific evidence. *Endocrinology Nutrition* 61(5): 274–295. doi: 10.1016/j.endonu.2013.10.012

Hashem, H. A. E. A., A. E. A. EL-Hadidy, and E. A. Ali. 2017. Impact of some safe agricultural treatments on insect pests, vascular wilt disease management and roselle (*Hibiscus sabdariffa* L.) productivity under Siwa Oasis conditions. *International Journal of Environment* 6(4): 139–162.

Hashemi, A. and A. Shahani. 2019. Effects of salt stress on the morphological characteristics, total phenol and total anthocyanin contents of Roselle (*Hibiscus sabdariffa* L.). *Plant Physiology Report* 24(2): 210–214. doi: 10.1007/s40502-019-00446-y

Hassan, S. T. S., E. Švajdlenka, and K. Berchová-Bímová. 2017. *Hibiscus sabdariffa* L. and its bioactive constituents exhibit antiviral activity against HSV-2 and anti-enzymatic properties against urease by an ESI-MS based assay. *Molecules* 22(5): 722. doi: 10.3390/molecules22050722

Herranz-López, M., M. Olivares-Vicente, M. Boix-Castejón, N. Caturla, E. Roche, and V. Micol. 2019. Differential effects of a combination of *Hibiscus sabdariffa* and *Lippia citriodora* polyphenols in overweight/obese subjects: A randomized controlled trial. *Scientific Reports* 9: 2999. doi: 10.1038/s41598-019-39159-5

Hinojosa-Gómez, J., C. S. Martín-Hernández, J. B. Heredia, J. León-Félix, T. Osuna-Enciso, and M. D. Muy-Rangel. 2020. Anthocyanin induction by drought stress in the calyx of roselle cultivars. *Molecules* 25: 1555. doi:10.3390/molecules25071555

Ifie, I., B. E. Ifie, D. O. Ibitoye, L. J. Marshall, and G. Williamson. 2018. Seasonal variation in *Hibiscus sabdariffa* (Roselle) calyx phytochemical profile, soluble solids and α-glucosidase inhibition. *Food Chemistry* 261: 164–168. doi: 10.1016/j.foodchem.2018.04.052

Isah, T. 2019. Stress and defense responses in plant secondary metabolites production. *Biological Resesarch* 52: 39. doi: 10.1186/s40659-019-0246-3

Islam, A. K. M. A., M. B. Osman, M. B. Mohamad, and A. K. M. M. Islam. 2021. Vegetable Mesta (*Hibiscus sabdariffa* L. var. *sabdariffa*): A Potential Industrial Crop for Southeast Asia. In *Roselle: Production, Processing, Products and Biocomposites*. eds S. M. Sapuan, R. Nadlene, A.M. Radzi, and R.A. Ilyas, 25–42. London: Academic Press. doi: 10.1016/B978-0-323-85213-5.00016-0

Izquierdo-Vega, J. A., D. A. Arteaga-Badillo, M. Sánchez-Gutiérrez, et al. 2020. Organic acids from roselle (*Hibiscus sabdariffa* L.) – a brief review of its pharmacological effects. *Biomedicines* 8: 100. doi:10.3390/biomedicines8050100

Joshi, S. S., L. Dice, and D. H. D'Souza. 2015. Aqueous extracts of *Hibiscus sabdariffa* calyces decrease hepatitis A virus and human norovirus surrogate titers. *Food and Environmental Virology* 7: 366–373. doi: 10.1007/s1256 0-015-9209-1

Kadamanda, R. and S. R. Natarajan. 2017. Salinity effects on leaf on roselle landraces (*Hibiscus sabdariffa* L.). *International Journal of Pure Applied Bioscience* 5(6): 158–165. doi: 10.18782/2320-7051.5289

Keya, S. S., M. G. Mostofa, M. M. Rahman, et al. 2022. Effects of glutathione on waterlogging-induced damage in sesame crop. *Industrial Crops and Products* 185: 115092. doi: 10.1016/j.indcrop.2022.115092

Kolawole, J.A. and A. Maduenyi. 2004. Effect of zobo drink (*Hibiscus sabdariffa* water extract) on the pharmacokinetics of acetaminophen in human volunteers. *European Journal of Drug Metabolism and Pharamakokinetics* 29(1): 25–29. doi: 10.1007/BF03190570

Lamichhane, J. R. and V. Venturi. 2015. Synergisms between microbial pathogens in plant disease complexes: a growing trend. *Frontiers in Plant Science* 6: 385. doi: 10.3389/fpls.2015.00385

Lim, T. K. 2014. *Hibiscus sabdariffa*. In: *Edible Medicinal and Non Medicinal Plants*. Vol. 8: Flowers. pp. 324–370. London: Springer. doi: 10.1007/978-94-017-8748-2_23

Liu, D., Z. Ding, E. F. Ali, A. M. S. Kheir, M. A. Eissa, and O. H. M. Ibrahim. 2021. Biochar and compost enhance soil quality and growth of roselle (*Hibiscus sabdariffa* L.) under saline conditions. *Scientific Reports* 11: 8739. doi: 10.1038/s41598-021-88293-6

Loo, S., A. Kam, T. Xiao, G. K. T. Nguyen, C.- F. Liu, and J. P. Tam. 2016. Identification and characterization of roseltide, a knottin-type neutrophil elastase inhibitor derived from *Hibiscus sabdariffa*. *Scientific Reports* 6: 39401. doi: 10.1038/srep39401

Luvonga, A. W. 2012. *Nutritional & Phytochemical Composition, Functional Properties of Roselle (Hibiscus sabdariffa) and Sensory Evaluation of Some Beverages made from Roselle Calyces.* MS in Food Science and Technology, Jomo Kenyatta University of Agriculture and Technology, Kenya.

Mahunu, G. K., M. T. Apaliyaa, and M. Osei-Kwarteng. 2021. Effect of pests and diseases on *Hibiscus sabdariffa* quality. In *Roselle (Hibiscus sabdariffa): Chemistry, Production, Products, and Utilization.* eds A. A. Mariod, H. E. Tahir, and G. M. Mahunu, 33–46. London: Academic Press. doi: 10.1016/B978-0-12-822100-6.00004-5

Mahunu, G. M. 2021. Breeding, genetic diversity, and safe production of Hibiscus sabdariffa under climate change. In *Roselle (Hibiscus sabdariffa): Chemistry, Production, Products, and Utilization.* eds A. A. Mariod, H. E. Tahir, and G. M. Mahunu, 1–14. London: Academic Press. doi: 10.1016/B978-0-12-822100-6.00005-7

Marschner, H. 1995. *Mineral Nutrition of Higher Plants.* 2nd ed. 559–579. New York: Academic Press.

Matter, F. M. A. 2016. Benzyladenine alleviates the lead toxicity in roselle (*Hibiscus sabdariffa* L.) plants. *Middle East Journal of Agriculture Research* 5(2): 144–151.

Melchert, A., A. C. Rosa, V. Genari, M. S. G., et al. 2016. Effect of *Hibiscus sabdariffa* supplementation on renal function and lipidic profile in obese rats. *Asian Journal of Animal and Veterinary Advances* 11: 693–700. doi: 10.3923/ajava.2016.693.700

Mohamed, B. B., B. Aftab, M. B. Sarwar, et al. 2017. Identification and characterization of the diverse stress-responsive R2R3-RMYB transcription factor from *Hibiscus sabdariffa* L. *International Journal of Genomics* 2017: 2763259. doi: 10.1155/2017/2763259

Mohamed, B. B., M. B. Sarwar, S. Hasan, B. Rashid, B. Aftab, and T. Hussain. 2015. Tolerance of roselle (*Hibiscus sabdariffa* L.) genotypes to drought stress at vegetative stage. *Advances in Life Sciences* 2(2): 74–82.

Montiel-Gonzâlez, C., C. Montiel, A. Ortega, A. Pacheco, and F. Bautista. 2021. Development and validation of climatic hazard indicators for roselle (*Hibiscus sabdariffa* L.) crop in dryland agriculture. *Ecological Indicators* 121: 107140. doi: 10.1016/j.ecolind.2020.107140

Morton, J. 1987. Roselle. In *Fruits of Warm Climates.* ed J. F. Morton, 281–286. Florida: Flair Books. https://hort.purdue.edu/newcrop/morton/roselle.html

Mozaffari-Khosravi, H., B.-A. Jalali-Khanabadi, M. Afkhami-Ardekani, F. Fatehi, and M. Noori-Shadkam. 2009. The effects of sour tea (*Hibiscus sabdariffa*) on hypertension in patients with type II diabetes. *Journal of Human Hypertension* 23: 48–54. doi: 10.1038/jhh.2008.100

Muhammad, Y. Y. and U. U. Charles. 2008. Effect of calyx extract of *Hibiscus sabdariffa* against cadmium-induced liver damage. *Bayero Journal of Pure and Applied Sciences* 1(1): 80–82.

Mutari, A. 2019. Yield, nutritional quality and heavy metals toxicity in roselle (*Hibiscus sabdarrifa* L.) and jute mallow (*Corchorus olitorius* L.) crops as affected by two commercial composts and chicken manure. PhD diss., University for Development Studies, Ghana.

Nabi, R. B. S., R. Tayade, A. Hussain, et al. 2019. Nitric oxide regulates plant responses to drought, salinity, and heavy metal stress. *Environmental and Experimental Botany* 161: 120–133. doi: 10.1016/j.envexpbot.2019.02.003

Nahed, M. R. and E. A. Moursi. 2016. Effect of skipping one irrigation during different growth stages and foliar application of micronutrients on roselle (*Hibiscus sabdariffa* L.) plants and some water relations in heavy clay soils. *Journal of Agricultural Research Kafr El-Sheikh University* 42: 1–24. doi: 10.21608/JSAS.2016.2768

Narejo, M.-N., P. E. M. Wahab, S. A. Hassan, and C. R. C. M. Zain. 2016. Effects of drought stress on growth and physiological characteristics of roselle (*Hibiscus sabdariffa* L.). *Journal of Tropical Plant Physiology* 8: 44–51.

Nguyen, T.-T., A. Voilley, T. T. T. Tran, and Y. Waché. 2022. Microencapsulation of *Hibiscus sabdariffa* L. calyx anthocyanins with yeast hulls. *Plant Foods for Human Nutrition* 77: 83–89. doi: 10.1007/s11130-022-00947-6

Ondo, J. A., P. Prudent, R. M. Biyogo, J. Rabier, F. Eba, and M. Domeizel. 2012. Translocation of metals in two leafy vegetables grown in urban gardens of Ntoum, Gabon. *African Journal of Agricultural Research* 7: 5621–5627. doi: 10.5897/AJAR12.1613

Orororo, O. C., S. O. Asagba, N. J. Tonukari, O. J. Okandeji, and J. J. Mbanugo. 2018. Effects of *Hibiscus sabdariffa* L. anthocyanins on cadmium-induced oxidative stress in Wistar rats. *Journal of Applied Sciences and Environmental Management* 22(4): 465–470. doi: 10.4314/jasem.v22i4.4

Orwa, C., A. Mutua, R. Kindt, R. Jamnadass, and S. Anthony. 2009. Agroforestry Database: a tree reference and selection guide version 4.0 www.worldagroforestry.org/sites/treedbs/treedatabases.asp

Oshunsanya, S. O., N. J. Nwosu, and Y. Li. 2019. Abiotic stress in agricultural crops under climatic conditions. In *Sustainable agriculture, forest and environmental management.* eds M. K. Jhariya, A. Banerjee, and R. S. Meena, 71–100. Singapore: Springer.

Oukara, F. Z., C. Chaouia, and F. Z. Benrebiha. 2017. Contribution a l'étude de l'effet du stress hydrique sur le comportement morphologique et physiologique des plantules du pistachier de l'atlas *Pistacia atlantica* Desf. *Agrobiology* 7(1): 225–232.

Pan, J., R. Sharif, X. Xu, and X. Chen. 2021. Mechanisms of waterlogging tolerance in plants: research progress and prospects. *Frontiers in Plant Science* 11: 627331. doi: 10.3389/fpls.2020.627331

Pandey, P., V. Irulappan, M. V. Bagavathiannan, and M. Senthil-Kumar. 2017. Impact of combined abiotic and biotic stresses on plant growth and avenues for crop improvement by exploiting physio-morphological traits. *Frontiers in Plant Science* 8: 537. doi: 10.3389/fpls.2017.00537

Parga-Lozano, C. H. and N. E. S. Guerrero. 2021. Combination of two promising methodologies for possible treatment against COVID-19. *Biomedical Jounal of Science and Technology Research* 35(5): 28000–28004. doi: 10.26717/BJSTR.2021.35.005761

Purseglove, J. W. 1968. *Tropical Crops – Dicotyledons.* London: Longman.

Qiao, Y., J. Ren, L. Yin, et al. 2020. Exogenous melatonin alleviates PEG-induced short-term water deficiency in maize by increasing hydraulic conductance. *BMC Plant Biology* 20: 218. doi: 10.1186/s12870-020-02432-1

Ramadhani, N. F., A. P. Nugraha, D. Rahmadani, et al. 2022. Anthocyanin, tartaric acid, ascorbic acid of roselle flower (*Hibiscus sabdariffa* L.) for immunomodulatory adjuvant therapy in oral manifestation coronavirus disease-19: An immunoinformatic approach. *Journal of Pharmacology and Pharmacognosy Research* 10(3): 418–428.

Riaz, G, and R. Chopra. 2018. A review on phytochemistry and therapeutic uses of *Hibiscus sabdariffa* L. *Biomedicine and Pharmacotherapy* 102: 575–586. doi: 10.1016/j.biopha.2018.03.023

Richardson, M. L. and C. G. Arlotta. 2021. Differential yield and nutrients of *Hibiscus sabdariffa* L. genotypes when grown in urban production systems. *Scientia Horticulturae* 288: 110349. doi: 10.1016/j.scienta.2021.110349

Riyazuddin, R., N. Nisha, B. Ejaz, et al. 2022. Comprehensive review on the heavy metal toxicity and sequestration in plants. *Biomolecules* 12: 43. doi: 10.3390/biom12010043

Ruiz-Garcia, Y., I. Romero-Cascales, R. Gil-Muñoz, J. I. Fernandez-Fernandez, J. M. Lopez-Roca, and E. Gómez-Plaza. 2012. Improving grape phenolic content and wine chromatic characteristics through the use of two different elicitors: Methyl Jasmonate versus Benzothiadiazole. *Journal of Agricultural and Food Chemistry* 60: 1283–1290. doi: 10.1021/jf204028d

Sabzehzaria, M., M. Zeinalib, and M. R. Naghavia. 2020. CRISPR-based metabolic editing: Next-generation metabolic engineering in plants. *Gene* 759: 144993. doi: 10.1016/j.gene.2020.144993

Salami, S. O. and A. J. Afolayan. 2021. Evaluation of nutritional and elemental compositions of green and red cultivars of roselle: *Hibiscus sabdariffa* L. *Scientific Reports* 11:1030. doi: 10.1038/s41598-020-80433-8

Saleem, M. H., S. Fahad, M. Rehman, et al. 2020. Morpho-physiological traits, biochemical response and phytoextraction potential of short-term copper stress on kenaf (*Hibiscus cannabinus* L.) seedlings. *PeerJ* 8:e8321 doi: 0.7717/peerj.8321

Sanayei, S., M. Barmaki, A. Ebadi, and M. Torabi-Giglou. 2021. Amelioration of water deficiency stress in roselle (*Hibiscus sabdariffa*) by arbuscular mycorrhizal fungi and plant growth-promoting rhizobacteria. *Notulae Botanicae Horti Agrobotanici Cluj-Napoca* 49: 11987. doi: 10.15835/nbha49211987

Sarkar, M. R., S. M. M. Hossen, M. S. I. Howlader, M. A. Rahman, and A. Dey. 2012. Anti-diarrheal, Analgesic and Anti-microbial activities of the plant Lalmesta (*Hibiscus sabdariffa*): A review. *International Journal of Pharmaceutical and Life Sciences* 1(3): 1–11. doi: 10.3329/ijpls.v1i3.12978

Sarkar, S. K. and S. P. Gawande. 2016. Diseases of Jute and allied fibre crops and their management. *Journal of Mycopathological Research* 54(3): 321–337.

Sarwar, A. K. M. Golam. 2020. Medicinal and aromatic plant genetic resources of Bangladesh and their conservation at the Botanical Garden, Bangladesh Agricultural University. *International Journal of Minor Fruits Medicinal and Aromatic Plants* 6: 13–19.

Sarwar, A. K. M. Golam. 2021. Ethnomedicinal plant genetic resources of Bangladesh and COVID-19. *International Conference on Science and Technology for Celebrating the Birth Centenary of Bangabandhu (ICSTB-2021)*, March 11-13 2021, pp. 328. Dhaka, Bangladesh.

Savary, S., L. Willocquet, S. J. Pethybridge, P. Esker, N. McRoberts, and A. Nelson. 2019. The global burden of pathogens and pests on major food crops. *Nature Ecology and Evolution* 3: 430. doi: 10.1038/s41559-018-0793-y

Schippers, R. R. 2000. *African Indigenous vegetables: An Overview of Cultivated Species*. National Resource Institute, UK.

Sembok, W. W., N. Abu-Kassim, Y. Hamzah, and Z. A. Rahman. 2015. Effect of mycorrhizal inoculation on growth and quality of roselle (*Hibiscus sabdariffa* L.) grown in soilless culture system. *Malaysian Applied Biology* 44(1): 57–62.

Shala, A. Y. and M. A. Mahmoud. 2018. Influence of Glycinebetaine on water stress tolerance of *Hibiscus sabdariffa* L. plant. *Journal of Plant Production, Mansoura University* 9(12): 981–988. doi: 10.21608/JPP.2018.36615

Sharma, S., J. Sharma, V. Soni, H. M. Kalaji, and N. I. Elsheery. 2021. Waterlogging tolerance: A review on regulative morpho-physiological homeostasis of crop plants. *Journal of Water and Land Development* 49(IV–VI): 16–28. doi: 10.24425/jwld.2021.137092.

Shehata, S. M., R. K. Badawy, and Y. I. E. Aboulsoud. 2019. Phytoremediation of some heavy metals in contaminated soil. *Bulletin of the National Research Centre* 43: 189. doi: 10.1186/s42269-019-0214-7

Shuhaimi, S. A.-D. N., D. Kanakaraju, and H. Nori. 2019. Growth performance of roselle (*Hibiscus sabdariffa*) under application of food waste compost and Fe_3O_4 nanoparticle treatment. *International Journal of Recycling of Organic Waste in Agriculture* 8(Suppl 1): S299–S309. doi: 10.1007/s40093-019-00302-x

Simon, L. D., E. O. Ogunwolu, and E. Okoroafor. 2018. Impact of insect infestation on plant damage and yield of roselle (*Hibiscus sabdariffa* L.) in Benue State, Nigeria. *International Journal of Plant and Soil Science* 26(1): 1–10. doi: 10.9734/IJPSS/2018/45446

Sunday, O. A., A. B. Munir, O. O. Akeeb, B. A. Adesanya, and S. O. Badaru. 2010. Antiviral effect of *Hibiscus sabdariffa* and *Celosia argentea* on measles virus. *African Journal of Microbiology Research* 4(4): 293–296.

Tackett, L. 2017. Tracing Ancient Healing Practices through the *Hibiscus*. https://scholarworks.uni.edu/mcna irsymposium/2017/all/3/ (accessed May 2, 2022)

Takeda, Y., Y. Okuyama, H. Nakano, Y. Yaoita, K. Machida, H. Ogawa, and K. Imai. 2020. Antiviral activities of *Hibiscus sabdariffa* L. tea extract against human influenza A virus rely largely on acidic pH but partially on a low-pH-independent mechanism. *Food and Environmental Virology* 12: 9–19. doi: 10.1007/s12560-019-09408-x

Teshome, D. T., G. E. Zharare, and S. Naidoo. 2020. The threat of the combined effect of abiotic and biotic stress factors in forestry under changing climate. *Frontiers in Plant Science* 11: 601009. doi: 103389/fpls.2020.601009

Tomani, J. C. D., V. Kagisha, A. T. Tchinda, et al. 2020. The inhibition of NLRP3 Inflammasome and IL-6 production by *Hibiscus noldeae* Baker f. derived constituents provides a link to its anti-inflammatory therapeutic potentials. *Molecules* 25: 4693. doi: 10.3390/molecules25204693

Trivellini, A., B. N. Gordillo, F. J. Rodrı́guez-Pulido, et al. 2014. Effect of salt stress in the regulation of anthocyanins and color of *Hibiscus* flowers by digital image analysis. *Journal of Agricultural and Food Chemistry* 62(29): 6966–6974. doi: 10.1021/jf502444u

Tuo, Y., M. Kone, M. L. Yapo, K. Kone, and K. H. Koua. 2019. Insect pests of guinea sorrel (*Hibiscus sabdariffa* L., 1753) and farmers' control methods in the district of Korhogo, Northern Côte D'ivoire. *International Journal of Advance Research* 7(4): 748–754. doi: 10.21474/IJAR01/8881

Tzean, S. S. 2019. *List of Plant Diseases in Taiwan*. 5th ed., Taiwan: Taiwan Phytopathological Society, pp. 149–150.

Uddin, M. N., M. Wahid-Uz-Zaman, M. M. Rahman, and J.-E. Kim. 2016. Phytoremediation potential of Kenaf (*Hibiscus cannabinus* L.), Mesta (*Hibiscus sabdariffa* L.), and Jute (*Corchorus capsularis* L.) in Arsenic-contaminated soil. *Korean Journal of Environment and Agriculture* 35(2): 111–120. doi: 10.5338/KJEA.2016.35.2.15

Uddin, M.S., and S.W. Lee. 2020. MPB 3.1: A useful medicinal plants database of Bangladesh. *Journal of Advancement in Medical and Life Sciences* 8: 02. doi: 10.5281/zenodo.3950619

Ullah, I., M. Waqas, M. A. Khan, I.-J. Lee, and W.-C. Kim. 2017. Exogenous ascorbic acid mitigates flood stress damages of *Vigna angularis*. *Applied Biological Chemistry* 60(6): 603–614. doi: 10.1007/s13765-017-0316-6

Umeoguaju, F. U., B. C. Ephraim-Emmanuel, J. O. Uba, G. E. Bekibele, N. Chigozie, and O. E. Orisakwe. 2021. Immunomodulatory and mechanistic considerations of *Hibiscus sabdariffa* (HS) in dysfunctional immune responses: a systematic review. *Frontiers in Immunology* 12: 550670. doi: 10.3389/fimmu.2021.550670

United States Environmental Protection Agency. 2007. Framework for Metal Risk Assessment. EPA 120/R-07/001. Office of the Science Advisor Risk Assessment Forum, Washington DC 20460.

Upadhyay, S. K. and S. P. Singh. 2016. Molecules and Methods for the Control of Biotic Stress Especially the Insect Pests — Present Scenario and Future Perspective. In *Abiotic and Biotic Stress in Plants - Recent Advances and Future Perspectives*. eds A. K. Shanker, and C. Shanker, 339–362. London: IntechOpen. doi: 10.5772/62034

Wang, C.-W., Y.-H. Yu, C.-Y. Wu, et al. 2021. Detection of pathogenic and beneficial microbes for roselle wilt disease. *Frontiers in Microbiology* 12:756100. doi: 10.3389/fmicb.2021.756100

Wilson. F. and M. Y. Menzel. 1964. Kenaf (*Hibiscus cannabinus*), Roselle (*Hibiscus sabdariffa*). *Economic Botany* 18(1): 80–91.

Wu, H.-Y., K.- M. Yang, and P.- Y. Chiang. 2018. Roselle anthocyanins: antioxidant properties and stability to heat and pH. *Molecules* 23: 1357. doi: 10.3390/molecules23061357

Wuana, R. A. and P. A. Mbasugh. 2013. Response of roselle (*Hibiscus sabdariffa*) to heavy metals contamination in soils with different organic fertilisations. *Chemistry and Ecology* 29(5): 437–447. doi: 10.1080/02757540.2013.770479

Yang, Z. K., X. B. Zou, Z. H. Li, X. W. Huang, X. D. Zhai, W. Zhang, J. Y. Shi, and H. E. Tahir. 2019. Improved postharvest quality of cold stored blueberry by edible coating based on composite gum arabic/roselle extract. *Food and Bioprocess Technology* 12: 1537–1547. doi: 10.1007/s11947-019-02312-z

Yaseen, S., S. F. Amjad, N. Mansoora, et al. 2021. Supplemental effects of biochar and foliar application of ascorbic acid on physio-biochemical attributes of barley (*Hordeum vulgare* L.) under Cadmium-contaminated soil. *Sustainability* 13: 9128. doi: 10.3390/su13169128

Zhang, B., R. Yue, Y. Wang, et al. 2020. Effect of *Hibiscus sabdariffa* (Roselle) supplementation in regulating blood lipids among patients with metabolic syndrome and related disorders: A systematic review and meta-analysis. *Phytotherapy Research* 34(5): 1083–1095. doi: 10.1002/ptr.6592

Zhang, H. and U. Sonnewald. 2017. Differences and commonalities of plant responses to single and combined stresses. *Plant Journal* 90: 839–855. doi: 10.1111/tpj.13557

Zheoat, A. M., A. I. Gray, J. O. Igoli, V. A. Ferro and R. M. Drummond. 2019. Hibiscus acid from *Hibiscus sabdariffa* (Malvaceae) has a vasorelaxant effect on the rat aorta. *Fitoterapia* 134: 5–13. doi: 10.1016/j.fitote.2019.01.012

Ziarati P, S. Namvar and B. Sawicka. 2018. Heavy metals bio-adsorption by *Hibiscus sabdariffa* L. from contaminated water. *Technogenic and Ecological Safety* 4(2): 22–32. doi: 10.5281/zenodo.1244568

10 *Mentha piperita* and Stressful Conditions

Ayshah Aysh ALrashidi[1], Haifa Abdulaziz Sakit AlHaithloul[2], Omar Mahmoud Al zoubi[3] and Mona H. Soliman[3,4]*
[1]Department of Biology, Faculty of Science, University of Hail, Hail, 81411, Saudi Arabia
[2]Biology Department, College of Science, Jouf University, Sakaka, Kingdom of Saudi Arabia
[3]Biology Department, Faculty of Science, Taibah University, Yanbu El-Bahr, Yanbu 46429, Kingdom of Saudi Arabia
[4]Botany and Microbiology Department, Faculty of Science, Cairo University, Giza 12613, Egypt
*Corresponding author: hmona@sci.cu.edu.eg

CONTENTS

10.1   INTRODUCTION

The medicinal and aromatic plants (MAPs) are recognized as great source of wealth and one of the most important plant species used traditionally in a wide range of economic activities by mankind in one way or other. Currently, 80% of the world's population uses traditional MAPs and their derived products (especially plant extracts and essential oils) to meet their various health needs. In addition to their importance in health areas, these plants play a significant role in the natural environment. MAPs are botanicals that are used in various medications to treat illnesses and various ailments/disorders. Worldwide, MAPs have played a critical role in the social–cultural, financial and environmental factors of diverse individuals as they are used in the hygiene, care and diet, as well as for their distinct fragrant-containing compounds, and in the treatment of illnesses/injuries. MAPs are present ubiquitously in aquatic and terrestrial environments across the world. However, due to the increasing global population and disturbances caused by urbanization and industrialization, the habitats of many important wild-harvested MAPs are at risk (Marshall 2011). Nonetheless, these plants contain distinct types of primary and secondary metabolites (SMs) that gave characteristics such as fragrance, flavors and also known insecticidal and therapeutic properties (Facchini et al. 2012). It has been found that in the entire plant kingdom there are 200,000–1,000,000 plant metabolites (Dixon 2003; Afendi et al. 2011). The SMs play an important role in various metabolic processes, and help plants to interact with the physical environment to block ultraviolet radiation, inhibit enzymes, and increase and/or decrease the antioxidant activity and development of pigments (Bourgaud et al. 2001; Kennedy and Wightman 2011). In plants, the maximum amount of energy is utilized for

DOI: 10.1201/9781003242963-10

biosynthesis of the SMs than primary metabolites (Gershenzon 1994). In plants, the principal SMs are alkaloids, terpenoids and phenolic compounds (Ahmed et al. 2017; Kant et al. 2015). These are not directly engaged in regulating primary growth and developmental processes, *viz.* respiration, photosynthesis, nutrient uptake and assimilation dynamics, reproduction and synthesis of proteins (Akula and Ravishankar 2011; Pusztahelyi et al. 2015). The SMs help plants to survive and modify physiological responses under different ecological pressures and invasions of different pathogens (Aftab 2019; Zaid et al. 2021). At present, a large number of MAPs show radical scavenging activity. Different types of MAPs are used for the extraction of essential oils (EOs). EOs have commercial applications in various pharmaceutical industries in making detergents, soaps, cosmetics, insect repellent and perfumes, and also in food-processing industries. In addition, MAPs are also used in enhanced clean-up/phytoremediation techniques (Bello et al. 2004; Gupta et al. 2013). Currently, however, due to overexploitation and the influence of various abiotic pressures, the growth and yield of various MAPs have decreased. *Mentha piperita* L. (peppermint) is an important MAP that belongs to the Lamiaceae that has more than 4000 species in 200 genera. Peppermint is a perennial herb and is obtained by natural hybridization between spearmint and watermint. The EO of peppermint possesses significant antimicrobial and antiviral capacity and also contains antioxidant, antitumor and anti-allergic activities (Iscan et al. 2002; Loolaie et al. 2017). The main SM of the EO of peppermint is menthol (Mahmoud and Croteau 2003). Both EO and menthol find applications in cosmetics industries as a result of their antimicrobial and cooling properties.

The growth and development of crop plants including MAPs under a complex and ever-changing environment are negatively affected by a range of ecological pressures. According to an estimate by the Food and Agricultural Organization of the United Nations (2007), only 3.5% of the total area of the world is not affected by any stress (www.fao.org/docrep/010/a1075e/a1075e00.html). Bray et al. (2000) proposed that various abiotic stress factors result in a 51–82% reduction in annual agricultural crop production. Abiotic stresses like heavy metal/metalloid (HMs), temperature extremes, salt and drought, and biotic ones like pests, pathogens and bacterial and fungal stress decrease the growth and yield of various crop plants, including MAPs (Zaid et al. 2021). The abiotic ones are non-living, whereas biotic stress comprises animate entities (Meena et al. 2017; He et al. 2018). In natural habitats, plants are sessile and under the influence of both biotic and abiotic stressors simultaneously, which act in combination to alter myriads of physiological, morphological and physiological processes to a great extent. The responses of various MAPs to combined stresses (biotic and abiotic) differ significantly from that to a single stress factor. In MAPs, a change in their normal processes induces cascades of complex alterations in various physiological and biochemical processes. Such changes alter their optimum metabolism to cause a principal reduction in the overall growth and development of MAPs. In peppermint, which is also an important MAP, various kinds of abiotic stresses decrease the overall growth and productivity in terms of quality and quantity. Nevertheless, the application of various phytohormones to various MAPs, including peppermint, under stressful and optimal conditions showed promising results in ameliorating various growth, photosynthesis and quality traits. In the following sections, we discuss the impacts of various abiotic stress factors and their alleviating effects by the application of various exogenous elicitors in peppermint plants.

10.2 *MENTHA PIPERITA* L. UNDER DROUGHT STRESS

Drought stress affects the growth and development of crop plants by causing marked alterations in myriads of morpho-physiological and biochemical activities that result in reduced productivity of crop plants. Currently, drought stress poses a risk of devastating negative impacts on the overall agricultural production (Fahad et al. 2017; Hassan et al. 2018; Pérez-Borroto et al. 2021; Ali et al. 2021). In today's era, due to global warming, limited water resources have affected agricultural production due to drought stress (Zolin and Rodrigues 2015; Mirajkar et al. 2019). As a result, drought stress is increasingly more frequent than other stresses due to unpredictable rainfall and marked changes in climate patterns (Whitmore 2000). In addition, due to various anthropogenic emissions, the rapid increase in atmospheric temperature taking place has accelerated crop exposure to drought-induced conditions (Jat et al. 2016; Fahad et al. 2017; Hussain et al. 2018; Nasim et al. 2018; Raza et al. 2019). Drought stress is unpredictable, as its persistence is dependent upon many critical factors which include the amount and distribution of rainfall, evapotranspiration and the ability of soil to retain moisture (Jaleel et al. 2009; Farooq et al. 2009, 2017). Drought stress also imposes severe restrictions on MAPs, including peppermint plants. Khorasaninejad et al. (2011) studied the effect of drought stress on growth traits, EO yield and constituents of peppermint under various levels of water deficit stress conditions (100%, 85%, 70%, 60% and 45% field capacity). The results revealed that drought stress caused a significant reduction in the growth traits and EO yield as well as EO percentage. An experiment was conducted to investigate the effects of plant growth promoting rhizobacteria (PGPR), namely *Pseudomonas fluorescens* (WCS417r) and *Bacillus amyloliquefaciens* (GB03) on *Mentha piperita* grown under moderate and severe drought stress conditions (Chiappero et al. 2019). It was observed that an increment in the level of drought stress resulted in reductions in plant fresh weight, leaf number and leaf area. Nonetheless, under drought stress, PGPR caused higher total phenolic content and enzymatic activities (peroxidase, superoxide dismutase), but reduced the proline content and membrane lipid peroxidation. Alhaithloul et al. (2020) studied the changes in ecophysiology, osmolyte and secondary metabolite contents of *Mentha piperita* under combined drought and heat stress conditions. Drought stress (50%) was imposed by the cessation of watering for 2 weeks in tested plants. The results revealed that plant height and weight (both fresh and dry weight) were significantly decreased by drought stress. Drought stress increased the accumulation of osmolytes (proline, sugars, glycine betaine, inositol and mannitol) but decreased total phenol, flavonoid and saponin. The secondary metabolites (tannins, terpenoids and alkaloids) were found to be increased under drought stress conditions. Rahimi et al. (2017), under drought stress conditions (75%, 50% and 25% field capacity [FC]), studied the changes in the expression of key genes involved in the biosynthesis of menthol and menthofuran, *viz.* limonene synthase (*lS*), limon-3-hydroxylase (*l3oh*), trans-isopiperitenol dehydrogenase (*ipd*), isopiperitenone reductase (*ipr*), pulegone reductase (*pr*), menthol dehydrogenase (*mdeh*) and menthofuran synthase (*mfs*) in peppermint plants using qRT-PCR. The results showed that under drought stress conditions, the expression levels of *ipd*, *ipr* and *mfs* were increased, whereas the gene expression levels of *pr* and *mdeh* were decreased. By further studying the gas chromatography–mass spectrometry (GC–MS) results, the EO components (menthol, menthofuran and plugene), were analyzed and found to have a positive correlation with the expression levels of genes. Imposition of drought stress also decreased morphological traits and caused a reduction in menthol percentages but increased the pulegone and menthofuran contents. A recent 2-year field experiment by Jahani et al. (2021) in *Mentha piperita* L. calculated the influence of foliar zinc (0, 2 and 4 mg L^{-1}) and salicylic acid (SA; 0 and 1 mM) on total chlorophyll, phenolic, flavonoid contents, EO percent and total dry matter and oil composition under drought (moderate and severe) stress conditions. It was found that severe drought stress decreased the chlorophyll content and EO percentage, but moderate drought stress increased the EO percentage. The total phenolic and flavonoid contents were increased with increasing drought stress levels. However, the application of Zn increased the total phenolic compound content. The Zn and SA treatments resulted in higher EO percentage and yield. The menthol and menthone concentrations were found to be increased significantly in response to moderate drought stress but were reduced under severe drought stress. Figueroa-Pérez et al. (2014) studied the metabolite profile and antioxidant capacity of *Mentha piperita* L. plants grown under drought stress (65%, 35%, 24% and 12%). The results revealed that 35%, 24% and 12% decreased the fresh and dry weight of plants but increased the total phenolic and flavonoid contents as well as antioxidant capacity and improved the inhibition of the activity of pancreatic lipase and α-amylase. In related research, to study the effects of drought stress and jasmonic acid (JA) application on the quality and quantity of peppermint plants, a factorial experiment in a completely randomized design was conducted (Kheiry et al.

2017). The drought stress was imposed at three levels (50%, 75% and 100%) and also three concentrations of JA (0, 50 and 100 mg L^{-1}). The results showed that 100 mg L^{-1} JA provided the highest fresh and dry weight, height, total chlorophyll and relative leaf water content in peppermint plants. Further, the highest peroxidase (POX) activity was observed at 50% drought stress and with 100 mg L^{-1} JA application. The treatments (75% drought stress and 50 mg L^{-1} JA) showed the highest EO content. The secondary metabolites in EO like menthol, menthone and 1,8-cineole were obtained in the highest amounts by 100 mg L^{-1} JA application, whereas the highest amount of cyclohexanol was obtained in plants grown under 50 mg L^{-1} JA supply. Nevertheless, JA application decreased the amount of methyl acetate at the 100 mg L^{-1} treatment. Keshavarz et al. (2020) applied organic and urea fertilizer to study changes in phenolic compounds, antioxidant activity, yield and yield components of *Mentha piperita* L. under drought (75%, 60% and 45%) stress in a randomized complete block design with three replicates. The resulting data showed that imposition of drought stress caused a reduction in chlorophyll content, plant height, leaf area index, shoot dry weight and EO yield but flavonoids, carotenoids and total phenols amount and EO percentage first increased and then showed a reduction under increasing levels of drought stress. Under drought stress conditions, Parsa et al. (2021) conducted a greenhouse-based experiment and applied chemical and biological fertilizers to study changes in the morphological and physiological traits and activity of antioxidant enzymes of peppermint. Three irrigation regimes, 100%, 75% and 50% FC, were applied to ascertain drought stress. It was found that the highest proline amount was obtained at 100% FC and fertilizer treatments. The biofertilizer application decreased the antioxidant enzyme activity as compared to other chemical fertilizer supplies. García-Caparrós et al. (2019) imposed drought stress to study biomass, EO content and leaf nutrient concentration in six Lamiaceae species including peppermint. The *Mentha piperita* L. were subjected to two drought stress treatments (100% and 70%). The peppermint plants showed a significant decrease in fresh weight on imposition of drought stress conditions. Salehi Sardoei (2020) in a classical study applied drought stress and SA to study the response of growth traits, photosynthetic pigments and EO yield in *Mentha piperita* L. in a factorial randomized complete block design experiment. The drought stress consisted of four levels (90%, 75%, 45% and 20%) and SA at two levels (0 and 60 ppm). The results showed that drought stress alone had a significant effect on the plant traits studied. The combined effect of drought stress and SA was also significant in the case of the length of lateral branches, fresh and dry weight, chlorophyll a, b and total chlorophyll, and EO percentage as well as on yield-related traits. The highest values of EO percentage and yield were obtained under 90% drought stress and SA application. In a recent 2-year field experiment, Keshavarz Mirzamohammadi et al. (2021a) studied the effect of drought stress and fertilizer treatments on rosmarinic acid accumulation, total phenolic content and antioxidant potential in peppermint plants. The drought stress had three treatments: 25%, 40% and 55%. The fertilizer application had six treatments: control; 140 kg/ha urea; 105 kg/ha urea + 3.3 ton/h vermicompost; 70 kg/ha urea + 6.6 ton/h vermicompost; 35 kg/ h urea + 10 ton/h vermicompost; and 13.5 ton/h vermicompost. The results showed that the maximum EO content was observed under drought stress conditions. The highest 2,2-diphenyl-1-picrylhydrazyl (DPPH) and superoxide dismutase (SOD) activity were noticed under 140 kg/ha^{-1} urea fertilization. The plants treated with vermicompost and under drought stress showed reduced antioxidant features, total phenol and rosmarinic acid content. Saedi et al. (2020) calculated the potentiality of nano-potassium fertilizer application under drought stress on the morpho-physiological traits of peppermint plants. The drought consisted of 100%, 80% and 60% and nano-K application had the following foliar applications: control, and spray treatments (2/1000), (4/1000) and (6/1000). According to the results, the increase in the application of nano-K fertilizer application was accompanied by an increase in the plant height, number of tillers per plant, fresh and dry weight, yield and percent of EO, chlorophyll a, chlorophyll b, water content, proline and total soluble sugars. As far as EO is concerned, its value increased with increasing drought stress regimes. In yet another experiment, Keshavarz-Mirzamohammadi et al. (2021b) illustrated the relationship between agronomic traits, soil properties and EO profile of peppermint plants under fertilizer treatment applications and drought stress regimes. It was found that the maximum leaf area index and dry matter weight were in control plants, while the highest EO content was seen under mild drought stress conditions. The results further revealed that the dry matter weight and content of EO increased in a dose (fertilizer)-dependent manner. The menthol content in the plants was found to be increased in response to integrated drought stress and fertilizer dose. Ghanbari and Ariafar (2013), in a factorial randomized complete block design experiment with three replications, applied drought stress (70%, 50% and 30%) and zeolite (0, 1.5, 2 and 2.5 g/1 kg soil) to study the growth traits and oil yield of peppermint. The results showed that all the growth parameters and EO yield were affected by the application of both drought stress and zeolite supply. All the growth traits were found to be significantly decreased except oil percentage. In order to study the long-term impacts of water stress in *Mentha piperita* L. the modulation in the antioxidant activity was assessed (Rahimi et al. 2018). The water deficit stress consisted of normal irrigation as control, 0.75, 0.5 and 0.25 field capacity (FC). It was found that the tested morphological traits, total phenol content and flavonoid contents of peppermint were decreased under stress conditions. However, the onset of water deficit (0.25 FC) caused a significant increase in hydrogen peroxide (H_2O_2) and malondialdehyde (MDA) contents. Nonetheless, the activity of polyphenol oxidase (PPO) and SOD were maximally increased by the imposition of 0.5 and 0.25 FC treatments. In the case of EO percentage, the maximum values were observed under 0.25 FC treatments.

10.3 *MENTHA PIPERITA* L. UNDER HEAVY METAL/METALLOID STRESS

Heavy metals (HMs) are those elements whose density is five times more than the density of water (Järup 2003; Wani et al. 2018; Rehman et al. 2021). HM shows characteristic properties such as conductance of current and heat and possesses luster. Some HMs, in very low quantities, are required by humans and plants to perform normal functions, however their excess quantity has toxic impacts. HMs are potent pollutants in current times, as concerns about their addition are accelerating at an alarming rate on a daily basis (Nagajyoti et al. 2010; Jaishankar et al. 2014; Zaid et al. 2020a,b; Hasanuzzaman et al. 2021). There are various sources of HMs. Anthropogenic sources include insecticides, pesticides and municipal sewage expelled from industries and mining, whereas natural ones include weathering of rocks and erosion of soil by water and wind (Bradl 2005; Morais et al. 2012; Alloway 2013). In plants, the ions of HMs are known to enter either through root hairs by water absorption or through stomata (Salt et al. 1995). Once inside plant tissues, these ions induce ionic, oxidative and osmotic stress (Zaid et al. 2019; Zaid et al. 2020c; Akhtar et al. 2021; Sarraf et al. 2022), thereby causing marked alterations in all vital plant physiological and metabolic activities. The buildup of ions of HMs has been reported to restrict water (Haroun et al. 2003), mobilization as well as uptake of mineral elements (Zaid et al. 2020c; Arduini et al. 1998), chlorophyll formation and photosynthetic potential (Khan et al. 2008; Masood et al. 2016; Nazir et al. 2021), respiratory system (Lösch and Köhl 1999; Mesa-Marín et al. 2018), gene expression (Qian et al., 2010), increase the production of various of reactive oxygen species (ROS) (Gratão et al. 2015; Naveed et al. 2021; Ahmad et al. 2021) and alter the enzymatic actions and equilibrium of various plant growth regulators (Bücker-Neto et al. 2017; Pal et al. 2018; Emamverdian et al. 2020; Sharma et al. 2020). In addition, HMs restrict various cellular and molecular functions by causing the oxidation of proteins, lipids and thylakoid and mitochondrial membranes. In a pot experiment conducted at the Department of Botany, Aligarh Muslim University, Aligarh, India, Ahmad et al. (2018) applied Cd at three different levels (30, 60 and 120 mg/kg) to the soil. The peppermint plants were then sprayed with (10^{-4} M) salicylic acid (SA). The results revealed that Cd-grown plants had remarkably significant growth, photosynthesis and carbon- and nitrate-assimilating enzymes inhibition as well as active yield and active constituents decrement in the EO of tested plants. However, Cd caused a marked increase in the oxidative stress traits.

Application of SA nullified the effects of Cd-induced toxicity by increasing photosynthesis and minimization of oxidative stress biomarkers by scavenging the production of ROS. Furthermore, the SA application to Cd-stressed plants alleviated the decrease in the concentration of EO and menthol content as evident from the gas chromatograms. Khoramivafa et al. (2012) studied the potential of microbial associations to the uptake of Cd (50 mg/kg) in *Mentha piperita* L. It was found that applications of microbes resulted in a significant additive effect on the uptake of Cd in roots. The authors showed that after distillation all Cd remained in the residues and Cd was not present in EO. Peyvandi et al. (2016) unraveled the effect of $CdCl_2$ (0, 100, 500 and 1000 μM) on the growth and composition of EO of *Mentha piperita* L. The results showed that Cd-grown plants had minimum stem length and fresh and dry weight of leaves. The content of Cd in the rhizomes and leaves was increased in a dose-dependent manner and it was found that leaves accumulated more Cd than the rhizomes. However, there were no significant differences in the EO contents between treatments. Azimychetabi et al. (2021), in a recent classical study in *Mentha piperita* L. plants, determined the effects of Cd stress in the EO composition and the underlying gene expression involved in their biosynthesis. It was found that the antioxidant activities, malondialdehyde, hydrogen peroxide and proline contents were increased under Cd stress. The expression of key genes showed that the menthol content dropped, and menthofuran and pulegone concentrations were increased under Cd stress conditions. As the dose of Cd increased in the soil, there was a reduction in the expression of *MR* (menthone reductase) and *PR* (pulegone reductase) genes. However, the expression of *MFS* (menthofuran synthase) was increased. Amirmoradi et al. (2012), in a greenhouse experiment, applied Cd (10, 20, 40, 60, 80 and 100 ppm) and Pb (100, 300, 600, 900, 1200 and 1500 ppm) to study quantitative and EO changes in *Mentha piperita* L. The results demonstrated that both Cd and Pb with their increasing dose decreased the fresh and dry weights, main stem height, leaf area per plant, leaf number, number of nodes per stem and EO. It was concluded that plants can tolerate a medium range of Cd and Pb stress and the EO remained unaffected by these doses. Akoumianaki-Ioannidou et al. (2015) grew *Mentha piperita* L. plants in pots under glasshouse conditions and supplied Cd and Zn at four concentrations (0, 1, 5, 10 mg/kg each). It was observed that the concentration of Cd increased significantly at each Cd level with increasing Zn additions in the leaves of peppermint. In leaves, the Cd concentration increased under increasing Zn supply but the concentration of Zn was not affected by Cd supplementation. In a hydroponic experiment performed in controlled glass containers with continuous aeration, Khair et al. (2020) applied citric acid (CA; 0, 5 mM) to enhance the phytoextraction of Ni (0, 100, 250, 500 μM) and alleviate Ni-induced changes in *Mentha piperita* L. The application of Ni significantly decreased the plant agronomic traits and photosynthesis but enhanced the antioxidant enzyme activities and production of ROS. Nevertheless, the follow-up treatment with CA under Ni stress significantly improved the plant morpho-physiological and biochemical parameters. The application of CA to Ni-stressed plants enhanced the Ni concentrations in roots, stems and leaves. Candan and Tarhan (2012) studied the alterations in the activities of the antioxidative enzyme, lipid peroxidation levels, and chlorophyll and carotenoid contents in peppermint plants under deficiency and/or excess Cu conditions. Under Cu deficiency, the values of chlorophyll and carotenoid contents and the ratio of chlorophylls to carotenoids were significantly low, but the greatest decrease was observed under excess Cu stress conditions. The activities of SOD, catalase (CAT), ascorbate-dependent peroxidase and guaiacol-dependent peroxidases in the absence of Cu were higher than in control plants. Dinu et al. (2021) showed the influence of As, Cd, Ni and Pb (25 mg/kg As, 5 mg/kg Cd, 150 mg/kg Ni and 100 mg/kg Pb) in peppermint. It was observed that the As, Cd, Ni and Pb applications did not influence the translocation of micro- and macronutrients from the root to the aerial parts of the plant. Akhtar et al. (2009) calculated the effect of Zn (1 ppm, 1.5 ppm, 2 ppm, 2.5 ppm, 3 ppm and 3.5 ppm as zinc chloride) as foliar sprays on the growth, yield and EO content of *Mentha piperita* L. It was found that the maximum increase in leaves/hector was observed under 3 ppm Zn followed by 1 ppm treatments. In the case of EO, a maximum increase (28.20%) was noticed with 3 ppm Zn followed by 6.70% of foliar treatment with 2.5 ppm. Hassani et al. (2015) devised a study to compare the conventional application of chemical fertilizers and nano-fertilizer of Zn, Fe and K on the quantitative yield of peppermint in a completely randomized block design with three replications. It was observed that the highest plant height, the number of branches, leaves of branches and nodes, and wet weight of leaves, dry weight of leaves, wet weight of stems, dry weight of stems, wet weight of plant and dry weight of plants were increased. Mohammadi et al. (2016) applied four (0, 0.25, 0.5 and 0.75 g/L) doses of nano-ferric oxide and three (0, 25 and 50 kg/ha) doses of zinc sulfate to study changes in chlorophyll, anthocyanin, flavonoid and leaf mineral elements of *Mentha piperita* L. The results demonstrated that both fertilizers induce significant effects on leaf flavonoid and total leaf chlorophyll; however, leaf anthocyanin remained insignificant. The interactive effects of these two fertilizers were also significant on all studied peppermint traits, except for leaf anthocyanin content. The nano-ferric oxide doses (0.25 and 0.75 g/L) showed the highest and lowest rates of leaf iron content. Conclusively, the applications of zinc sulfate (25 kg/ha) and nano-ferric oxide (0.75 g/L) were proved to be best. In a recent study, Nemati Lafmejani et al. (2021) applied chelate and nano-chelate of Zn micronutrient (0, 0.5, 1 and 1.5 g/L) to examine the morpho-physiological traits and EO compounds of peppermint plants in a randomized complete block design with three replications. The results revealed that both nano- and chelated Zn micronutrients significantly increased all morpho-physiological traits with higher values in Zn-chelate (1.5 g/L) and nano-zinc (1 g/L) doses. It was found that the application of nano-zinc (1.5 g/L) increased menthol, menthone and menthofuran contents by 28.00%, 61.00% and 237.00% as compared to the control. The foliar spraying of nano-Zn (1 g/L) increased the dry matter yield and EO content.

10.4 *MENTHA PIPERITA* L. UNDER SALT STRESS

Soil salinity, a threatening challenge to farming soils, is a principal abiotic stress factor hampering crop growth and productivity across the globe, particularly in irrigated regions of arid and semiarid zones (Zörb et al. 2019). According to an estimate on daily basis about 2000–4000 ha of land under irrigation is affected by salinity across the globe, thereby making it unfit for agriculture purposes (Qadir et al. 2014; Zaman et al. 2016). Contamination of agricultural land by salt ions is a continuing natural process, which is further accelerated by poor irrigation and/or low-cost drainage systems (Jayakannan et al. 2015). On a global scale, salt toxicity is extensively increasing daily, with approximately 6% of the total global land and approximately 20% of the cultivated lands being affected by salt stress, thereby causing up to 50% economic failure of the major food crops across the globe (Munns and Tester 2008; Gupta and Huang 2014). Among salts, sodium chloride (NaCl) is the most soluble and causes salt stress by affecting many morpho-physiological and biochemical activities in crop plants, starting from germination of seedlings to vegetative and reproductive stage and finally senescence (Ashraf and Harris 2004; Ashraf 2009; Shahbaz and Ashraf 2013; Negrão et al. 2017). In plants, the adverse effects of salinity stress on plant growth and development are ascribed to osmotic, ionic and oxidative stress because of the high absorption and accumulation of Na^+ and Cl^- ions (Khan et al. 2010; Fariduddin et al. 2019; Rasheed et al. 2020; Kumar et al. 2022). Moreover, the characteristic impact of salt toxicity in plants is the excessive production of ROS, which alters the optimum cellular and physiological metabolism through oxidation of biomolecules, enzyme denaturation, DNA and RNA damage and increased lipid peroxidation and reduction in chlorophyll synthesis and disrupted the integrity of biological membranes (Ding et al. 2010; Zhang et al. 2016; Ahanger et al. 2017; Fariduddin et al. 2019; Liu et al. 2021; Choudhary et al. 2022). The growth, development and productivity of MAPs are directly and indirectly impacted by salt stress. Furthermore, the contents of secondary metabolites (SMs) are also modulated by salt stress, thereby affecting the quality and quantity of SM production. The MAPs' response to salt stress is multifaceted as they adopt diverse physiological and biochemical adaptive mechanisms to counteract stress.

To reduce salt-induced damage, MAPs check the overproduction of ROS, hasten the activity of antioxidants such as SOD, CAT and POX, and synthesize several metabolites and compatible osmolytes like total soluble sugars, proline and also non-enzymatic antioxidants. Çoban and Baydar (2016) applied brassinosteroids (BRs; 0, 0.5, 1.5 and 2.5 mg/L) to study the effects on selected physical and biochemical properties and accumulation of SMs in *Mentha piperita* L. under salt (0, 100 and 150 mM) stress. The results showed that fresh and dry weights of aerial parts and leaves were decreased in a dose-dependent manner under salinity. Increasing levels of salinity reduced EO content but increased lipid peroxidation, proline, antioxidant enzyme activities and total phenolic contents. It was also found that 150 mM NaCl toxicity resulted in the death of peppermint plants. The application of brassinosteroids prohibited the death of the peppermint plant. BR application improved the activities of antioxidant enzymes and EO content simultaneously, while decreasing the membrane permeability and lipid peroxidation. Khanam and Mohammad (2018) devised a pot experiment at the Department of Botany, Aligarh Muslim University, Aligarh, India, to study the ameliorative role of selected plant growth regulators (PGRs) under salt stress in *Mentha piperita* L. The peppermint plants were exposed to four graded levels of salt toxicity (0, 50, 100 or 150 mM NaCl) and PGRs (gibberellic acid, salicylic acid or triacontanol) were sprayed, each at 10^{-6} M. The graded levels of salt stress decreased growth, photosynthesis potential, carbonic anhydrase (CA) activity, contents of menthol, EO, N, P and K, and also peltate glandular trichome (PGT) density, but augmented the activities of CAT, POX and SOD and also proline content. Of the three PGRs tested, the foliar spray of SA improved all studied parameters under both salt and salt-free conditions. Cappellari and Banchio (2020) applied *Bacillus amyloliquefaciens GB03* to ameliorate the effects of salt (0, 75 or 100 mM) stress in *Mentha piperita* L. It was found that the microbial volatile organic compounds (mVOCs) emitted by *Bacillus amyloliquefaciens* GB03 in plants ameliorated the stress under different levels of NaCl. The increasing salt stress resulted in a decrease in plant growth. Plants treated with salt and exposed to rhizobacteria showed a significant increment in morphological characteristics and higher total chlorophyll content. The endogenous levels of PGRs like jasmonic acid (JA), SA and abscisic acid (ABA) were increased under salt-stressed treated plants. The rhizobacterial inoculation in salt-stressed plants did not show any change in JA level but the amount of SA increased remarkably. The levels of ABA in contrast decreased. In a recent study, Alavi et al. (2021) showed that the enhanced nutrient uptake in salt-stressed (0, 40, 80 and 120 mM) *Mentha piperita* L. was through the use of magnetically treated water (0, 100, 200 and 300 mT). The magnetically treated water was also found to significantly enhance the Fe and Zn concentration under salinity stress. Li et al. (2016), investigating NaCl stress in peppermint, explained the involvement of mitogen-activated protein kinase (MAPK)-mediated regulation of growth and EO composition. The results showed that plants showed normal growth, and higher yield and EO content under salt stress than wild-type (WT) plants. By using techniques such as gas chromatography–mass spectrometry (GC-MS) and qPCR, salt (150 mM) stress showed no change in EO composition, the transcriptional level of enzymes related to the metabolism of EO or the activity of pulegone reductase (*Pr*). Furthermore, MAPK undergoes a time-dependent activation and was associated with modulation of the metabolism of EO in the treatment of salt stress. Askary et al. (2017) applied different (0, 10, 20 and 30 μM) levels of iron nanoparticles (Fe_2O_3 NPs) on *Mentha piperita* L. under salinity (0, 50, 100 and 150 mM) stress. The application of Fe_2O_3 NPs increased leaf fresh and dry weight, and P, K, Fe, Zn and Ca contents under salinity stress. The 30 μM dose of Fe_2O_3 NPs was found to be more significant. The proline and lipid peroxidation decreased significantly by applying Fe_2O_3 NPs under salt stress. The highest activities of CAT, SOD and GPOX were observed in 150 mM of NaCl-treated plants but Fe_2O_3 NPs decreased the activities. Cappellari et al. (2020) calculated the tendency of VOC to increase the biosynthesis of SMs and improvement in the antioxidant status in *Mentha piperita* L. grown under salt (0, 75 and 100 mM) stress. The increasing salt doses decreased EO but these effects were alleviated in VOCs emitting rhizobacteria. The total phenolic compounds (TPCs) were increased under salt-stressed plants, and membrane lipid peroxidation showed a reduction under rhizobacterial application. Khalvandi and Gholami (2017) showed the symbiotic effect of *Piriformospora indica* under salt (0, 3, 6, 9 dS/m) stress on the quantity and quality of EO and physiological parameters of peppermint plants in a completely randomized design experiment with three replications. The results demonstrated that increasing salt doses significantly decreased root colonization, leaf EO content, chlorophyll a, chlorophyll b, carotenoids, plant dry matter and relative water content (RWC), but triggered a marked increase in EL, menthol, menthone and methyl acetate and soluble sugars. Fungal application improved total dry weight and photosynthetic pigments. The application of fungi reduced the negative effects of salinity stress by increasing the EO content, stability of cell membrane and RWC in tested plants. Vatankhah et al. (2017) applied methyl jasmonate (MeJA) under salt stress to study the changes in some physiological and phytochemical characteristics of *Mentha piperita* L. Plants were initially exposed to 1.86 (control), 5, 75 and 10 dS/m of the NaCl concentrations and after that foliar MeJA was applied at 0, 60 and 120 μM doses. The results revealed that the application of salt stress decreased the fresh and dry weights, K, Ca, Mg, K^+/Na^+ ratio and phenolic compounds, while increasing the Na content. The EO yield initially increased with the increasing levels of salt stress but decreased under the highest salt levels. The foliar application of MeJA enhanced the mineral contents, EO yield and phenolics, while it reduced the Na content under salt-stressed conditions. In an older study, Karray-Bouraoui et al., (2009) observed the salt stress effects on the yield and composition of EO and morphology and density of trichomes in *Mentha pulegium* L. The application of salt stress enhanced EO yield and modulated the biosynthesis of menthone and increased pulegone. By examining the anatomy, the study showed three types of trichomes to be present: (i) non-glandular, multicellular, simple hairs; (ii) small, capitate glandular; and (iii) peltate glandular. Plants grown under salt stress showed significant modifications in the distribution and size of trichomes. Li et al. (2015) elucidated cellular events for studying the molecular mechanisms of tolerance in peppermint under different (100, 150, 200 or 250 mM) concentrations of salt toxicity. The plants grown under 200 mM stress treatments showed increased cell death with chromatin condensation and caspase-3-like activation and an increase in ROS in the mitochondria and chloroplasts. The activity of mitochondria and photosynthesis were also decreased under 200 mM NaCl stress, while the activities of SOD, ascorbate peroxidase (APX), glutathione reductase (GR) and dehydroascorbate reductase (DHAR), as well as the contents of ascorbate and glutathione, changed in a concentration-dependent manner under salt stress. Alavi et al. (2020) studied the responses of *Mentha piperita* L. in terms of growth and biochemical properties under magnetized saline water in a split plot design experiment with three replications. The main magnetic fields were control (M1), 100 mT (M2), 200 mT (M3) and 300 mT (M4), and the salt stress included control (S1), 4 dS/m (S2), 8 dS/m (S3) and 12 dS/m (S4). The results showed that different magnetic fields significantly increased plant growth with increasing levels of salinity. At 8 dS/m and 100 and 200 mT, the menthol concentration was found to be increased. The results concluded the potentiality of magnetized salty water to alleviate the stress of salinity on peppermint plants. Ghorbani et al. (2018) studied the effect of salinity stress (0, 2, 4 and 6 dS/m) of peppermint plants on some morpho-physiological traits and the quantity and quality of EO. The results indicated that the levels of salinity stress had significant effects on all the studied traits. The lowest amounts of RWC, leaf area, root dry weight, shoot dry weight, leaf dry weight, chlorophyll a, chlorophyll b, total chlorophyll, carotenoid and EO (0.13%) were under 6 ds/m. Under increasing salt stress, the concentrations of most of the EO components such as menthol, D-limonen, α-pinene, sabinene, menthofuran and 1,8-cineole were reduced but the menthone content was increased. Aziz et al. (2008) studied the effect of salt stress (0, 1.5, 3.0 or 4.5 g/L NaCl for 74 days) on growth and EO production in peppermint along with pennyroyal and apple mint plants. Salt stress reduced the growth and EO in all three species including peppermint. From the above established experimental reports, it is evident that the growth and development of MAPs are modulated by the application of salt stress conditions.

10.5 CONCLUSIONS AND FUTURE OUTLOOK

It is evident from the above-collected literature that MAPs under drought, HM and salt stress, experienced alterations in growth, photosynthesis and yield and quality characteristics. These stresses also affect the production of SMs to a large extent. The abiotic stress factors modulate the gene expressions that encode the diverse proteins responsible for optimum growth and metabolism of various MAPs. The application of various signaling agents discussed in this chapter mediates intimate cross-talk with other signaling agents under stress conditions in diverse MAPs that imparts stress resistance. Therefore, the supplementation of these agents in MAPs under various abiotic pressures could increase the quality and quantity of MAPs.

REFERENCES

Afendi, F. M., Okada, T., Yamazaki, M., Hirai-Morita, A., Nakamura, Y., Nakamura, K., Saito, K. (2011). KNApSAcK family databases: integrated metabolite–plant species databases for multifaceted plant research. *Plant Cell Physiol*, *53*, e1–e1.

Aftab, T. (2019). A review of medicinal and aromatic plants and their secondary metabolites status under abiotic stress. *Journal of Medicinal Plants*, *7*, 99–106.

Ahanger, M. A., Tomar, N. S., Tittal, M., Argal, S., & Agarwal, R. M. (2017). Plant growth under water/salt stress: ROS production; antioxidants and significance of added potassium under such conditions. *Physiology and Molecular Biology of Plants*, *23*, 731–744.

Ahmad, B., Jaleel, H., Sadiq, Y., A Khan, M. M., & Shabbir, A. (2018). Response of exogenous salicylic acid on cadmium induced photosynthetic damage, antioxidant metabolism and essential oil production in peppermint. *Plant Growth Regulation*, *86*, 273–286.

Ahmad, P., Raja, V., Ashraf, M., Wijaya, L., Bajguz, A., & Alyemeni, M. N. (2021). Jasmonic acid (JA) and gibberellic acid (GA₃) mitigated Cd-toxicity in chickpea plants through restricted Cd uptake and oxidative stress management. *Scientific Reports*, *11*, 1–17.

Ahmed, E., Arshad, M., Khan, M. Z., Amjad, M. S., Sadaf, H. M., Riaz, I., Ahmad, N. (2017). Secondary metabolites and their multidimensional prospective in plant life. *Journal of Pharmacognosy and Phytochemistry*, *6*, 205–214.

Akhtar, N., Khan, S., Rehman, S. U., Rehman, Z. U., Khatoon, A., Rha, E. S., & Jamil, M. (2021). Synergistic effects of zinc oxide nanoparticles and bacteria reduce heavy metals toxicity in rice (*Oryza sativa* L.) plant. *Toxics*, *9*, 113.

Akhtar, N., Sarker, M. A. M., Akhter, H., & Nada, M. K. (2009). Effect of planting time and micronutrient as zinc chloride on the growth, yield and oil content of *Mentha piperita*. *Bangladesh Journal of Scientific and Industrial Research*, *44*, 125–130.

Akoumianaki-Ioannidou, A., Kalliopi, P., Pantelis, B., & Moustakas, N. (2015). The effects of Cd and Zn interactions on the concentration of Cd and Zn in sweet bush basil (*Ocimum basilicum* L.) and peppermint (*Mentha piperita* L.). *Fresenius Environ Bull*, *24*, 77–83.

Akula, R., & Ravishankar, G. A. (2011). Influence of abiotic stress signals on secondary metabolites in plants. *Plant signaling & behavior*, *6*, 1720–1731.

Alavi, S. A., Ghehsareh, A. M., Soleymani, A., & Panahpour, E. (2021). Enhanced nutrient uptake in salt-stressed *Mentha piperita* using magnetically treated water. *Protoplasma*, *258*, 403–414.

Alavi, S. A., Ghehsareh, A. M., Soleymani, A., Panahpour, E., & Mozafari, M. (2020). Peppermint (*Mentha piperita* L.) growth and biochemical properties affected by magnetized saline water. *Ecotoxicology and Environmental Safety*, *201*, 110775.

Alhaithloul, H. A., Soliman, M. H., Ameta, K. L., El-Esawi, M. A., & Elkelish, A. (2020). Changes in ecophysiology, osmolytes, and secondary metabolites of the medicinal plants of *Mentha piperita* and *Catharanthus roseus* subjected to drought and heat stress. *Biomolecules*, *10*(1), 43.

Ali, E. F., El-Shehawi, A. M., Ibrahim, O. H. M., Abdul-Hafeez, E. Y., Moussa, M. M., & Hassan, F. A. S. (2021). A vital role of chitosan nanoparticles in improvisation the drought stress tolerance in *Catharanthus roseus* (L.) through biochemical and gene expression modulation. *Plant Physiology and Biochemistry*, *161*, 166–175.

Alloway, B. J. (2013). Sources of heavy metals and metalloids in soils. In *Heavy metals in soils* (pp. 11–50). Springer, Dordrecht.

Amirmoradi, S., Moghaddam, P. R., Koocheki, A., Danesh, S., & Fotovat, A. (2012). Effect of cadmium and lead on quantitative and essential oil traits of peppermint (*Mentha piperita* L.). *Notulae Scientia Biologicae, 4*, 101–109.

Arduini, L., Godbold, D. L., Onnis, A., & Stefani, A. (1998). Heavy metals influence mineral nutrition of tree seedlings. *Chemosphere, 36*, 739–744.

Asgher, M., Khan, N. A., Khan, M. I. R., Fatma, M., & Masood, A. (2014). Ethylene production is associated with alleviation of cadmium-induced oxidative stress by sulfur in mustard types differing in ethylene sensitivity. *Ecotoxicology and environmental safety, 106*, 54–61.

Ashraf, M, & Harris, P.J C. (2004). Potential biochemical indicators of salinity tolerance in plants. *Plant science, 166*, 3–16.

Ashraf, M. (2009). Biotechnological approach of improving plant salt tolerance using antioxidants as markers. *Biotechnology advances, 27*, 84–93.

Askary, M., Talebi, S. M., Amini, F., & Bangan, A. D. B. (2017). Effects of iron nanoparticles on *Mentha piperita* L. under salinity stress. *Biologija, 63*(1).

Azimychetabi, Z., Nodehi, M. S., Moghadam, T. K., & Motesharezadeh, B. (2021). Cadmium stress alters the essential oil composition and the expression of genes involved in their synthesis in peppermint (*Mentha piperita* L.). *Industrial Crops and Products, 168*, 113602.

Aziz, E. E., Al-Amier, H., & Craker, L. E. (2008). Influence of salt stress on growth and essential oil production in peppermint, pennyroyal, and apple mint. *Journal of herbs, spices & medicinal plants, 14*, 77–87.

Bello, M. O., Ibrahim, A. O., Ogunwande, I. A., & Olawore, N. O. (2004). Heavy trace metals and macronutrients status in herbal plants of Nigeria. *Food Chemistry, 85*, 67–71.

Bourgaud, F., Gravot, A., Milesi, S., & Gontier, E. (2001). Production of plant secondary metabolites: a historical perspective. *Plant science, 161*, 839–851.

Bradl, H. B. (2005). Sources and origins of heavy metals. In *Interface science and technology* (Vol. 6, pp. 1–27). Elsevier.

Bray EA, Bailey-Serres J, Weretilnyk E (2000) Responses to abiotic stresses. In: Buchanan B, Gruissem W, Jones R (eds) Biochemistry and molecular biology of plants. American Society of Plant Physiologists, Rockville, pp 5477–5486.

Bücker-Neto, L., Paiva, A. L. S., Machado, R. D., Arenhart, R. A., & Margis-Pinheiro, M. (2017). Interactions between plant hormones and heavy metals responses. *Genetics and molecular biology, 40*, 373–386.

Candan, N., & Tarhan, L. (2012). Alterations of the antioxidative enzyme activities, lipid peroxidation levels, chlorophyll and carotenoid contents along the peppermint (*Mentha piperita* L.) leaves exposed to copper deficiency and excess stress conditions. *Journal of Applied Botany and Food Quality. 83*, 103–109.

Cappellari, L. D. R., & Banchio, E. (2020). Microbial volatile organic compounds produced by *Bacillus amyloliquefaciens* GB03 ameliorate the effects of salt stress in *Mentha piperita* principally through acetoin emission. *Journal of Plant Growth Regulation, 39*, 764–775.

Cappellari, L. D. R., Chiappero, J., Palermo, T. B., Giordano, W., & Banchio, E. (2020). Volatile organic compounds from rhizobacteria increase the biosynthesis of secondary metabolites and improve the antioxidant status in *Mentha piperita* L. grown under salt stress. *Agronomy, 10* (8), 1094.

Chiappero, J., del Rosario Cappellari, L., Alderete, L. G. S., Palermo, T. B., & Banchio, E. (2019). Plant growth promoting rhizobacteria improve the antioxidant status in *Mentha piperita* grown under drought stress leading to an enhancement of plant growth and total phenolic content. *Industrial Crops and Products, 139*, 111553.

Choudhary, S., Wani, K.I., Naeem, M., Khan M.M.A., Aftab, T (2022). Cellular Responses, Osmotic Adjustments, and Role of Osmolytes in Providing Salt Stress Resilience in Higher Plants: Polyamines and Nitric Oxide Crosstalk. *J Plant Growth Regul.* https://doi.org/10.1007/s00344-022-10584-7.

Çoban, Ö., & Baydar, N. G. (2016). Brassinosteroid effects on some physical and biochemical properties and secondary metabolite accumulation in peppermint (*Mentha piperita* L.) under salt stress. *Industrial Crops and Products, 86*, 251–258.

Ding, M., Hou, P., Shen, X., Wang, M., Deng, S., Sun, J., & Chen, S. (2010). Salt-induced expression of genes related to Na+/K+ and ROS homeostasis in leaves of salt-resistant and salt-sensitive poplar species. *Plant molecular biology, 73*, 251–269.

Dinu, C., Vasile, G., Tenea, A. G., Stoica, C., Gheorghe, S., & Serban, E. A. (2021). The influence of toxic metals As, Cd, Ni and Pb on nutrients accumulation in *Mentha piperita*. *Romanian Journal of Ecology & Environmental Chemistry*, 3(2). https://doi.org/10.21698/rjeec.2021.217.

Dixon, R. A. (2003). Phytochemistry meets genome analysis, and beyond. *Phytochem*, *62*, 815–816.

Emamverdian, A., Ding, Y., Mokhberdoran, F., & Ahmad, Z. (2020). Mechanisms of selected plant hormones under heavy metal stress. *Polish J Environ Stud*, *30*, 497–507.

Facchini, P.J., Bohlmann, J., Covello, P.S., De Luca, V., Mahadevan, R., Page, J.E., Ro, D.K., Sensen, C.W., Storms, R. and Martin, V.J. (2012) Synthetic biosystems for the production of high-value plant metabolites. Trend Biotechnol, 30, 127–131.

Fahad, S., Bajwa, A. A., Nazir, U., Anjum, S. A., Farooq, A., Zohaib, A., & Huang, J. (2017). Crop production under drought and heat stress: plant responses and management options. *Frontiers in plant science*, 1147.

Fariduddin, Q., Zaid, A., & Mohammad, F. (2019). Plant growth regulators and salt stress: mechanism of tolerance trade-off. In *Salt stress, microbes, and plant interactions: causes and solution* (pp. 91–111). Springer, Singapore.

Farooq, M., Gogoi, N., Barthakur, S., Baroowa, B., Bharadwaj, N., Alghamdi, S. S., & Siddique, K. H. (2017). Drought stress in grain legumes during reproduction and grain filling. *Journal of Agronomy and Crop Science*, *203*, 81–102.

Farooq, M., Wahid, A., Kobayashi, N. S. M. A., Fujita, D. B. S. M. A., & Basra, S. M. A. (2009). Plant drought stress: effects, mechanisms and management. In *Sustainable agriculture* (pp. 153–188). Springer, Dordrecht.

Figueroa-Pérez, M. G., Rocha-Guzmán, N. E., Perez-Ramirez, I. F., Mercado-Silva, E., & Reynoso-Camacho, R. (2014). Metabolite profile, antioxidant capacity, and inhibition of digestive enzymes in infusions of peppermint (*Mentha piperita*) grown under drought stress. *Journal of agricultural and food chemistry*, *62*, 12027–12033.

García-Caparrós, P., Romero, M. J., Llanderal, A., Cermeño, P., Lao, M. T., & Segura, M. L. (2019). Effects of drought stress on biomass, essential oil content, nutritional parameters, and costs of production in six Lamiaceae species. *Water*, *11*(3), 573.

Gershenzon, J. (1994). Metabolic costs of terpenoid accumulation in higher plants. *Journal of chemical ecology*, *20*, 1281–1328.

Ghanbari, M., & Ariafar, S. (2013). The effect of water deficit and zeolite application on Growth Traits and Oil Yield of Medicinal Peppermint (*Mentha piperita* L.). *Int. J. Med. Arom. Plants*, *3*, 33–39.

Ghorbani, M., Movahedi, Z., Kheiri, A., & Rostami, M. (2018). Effect of salinity stress on some morpho-physiological traits and quantity and quality of essential oils in Peppermint (*Mentha piperita* L.). *Environmental Stresses in Crop Sciences*, *11*, 413–420.

Gratão, P. L., Monteiro, C. C., Tezotto, T., Carvalho, R. F., Alves, L. R., Peters, L. P., & Azevedo, R. A. (2015). Cadmium stress antioxidant responses and root-to-shoot communication in grafted tomato plants. *Biometals*, *28*, 803–816.

Gupta B, Huang B (2014) Mechanism of salinity tolerance in plants: physiological, biochemical, and molecular characterization. Int J Genom. https://doi.org/10.1155/ 2014/701596.

Gupta, A. K., Verma, S. K., Khan, K., & Verma, R. K. (2013). Phytoremediation using aromatic plants: a sustainable approach for remediation of heavy metals polluted sites. *Environmental Science & Technology*, 47, 10115–10116.

Haroun, S. A., Aldesuquy, H. S., Abo-Hamed, S. A., & El-Said, A. A. (2003). Kinetin-induced modification in growth criteria, ion contents and water relations of sorghum plants treated with cadmium chloride. *Acta Botanica Hungarica*, *45*, 113–126.

Hasanuzzaman, M., Garcia, P., Zulfiqar, F., Parvin, K., Ahmed, N., Fujita, M., & Nahar, K. (2021). Selenium supplementation and crop plant tolerance to metal/metalloid toxicity. *Frontiers in Plant Science*, 2957.

Hassan, F. A. S., Ali, E. F., & Alamer, K. H. (2018). Exogenous application of polyamines alleviates water stress-induced oxidative stress of *Rosa damascena* Miller var. trigintipetala Dieck. *South African Journal of Botany*, *116*, 96–102.

Hassani, A., Tajali, A. A., Mazinani, S. M. H., & Hassani, M. (2015). Studying the conventional chemical fertilizers and nano-fertilizer of iron, zinc and potassium on quantitative yield of the medicinal plant of peppermint (*Mentha piperita* L.) in Khuzestan. *International Journal of Agriculture Innovations and Research*, *3*, 1078–1082.

He, M., He, C. Q., & Ding, N. Z. (2018). Abiotic stresses: general defenses of land plants and chances for engineering multistress tolerance. *Frontiers in plant science*, *9*, 1771.

Hussain, H. A., Hussain, S., Khaliq, A., Ashraf, U., Anjum, S. A., Men, S., & Wang, L. (2018). Chilling and drought stresses in crop plants: implications, cross talk, and potential management opportunities. *Frontiers in plant science*, *9*, 393.

İşcan, G., Kirïmer, N., Kürkcüoğlu, M., Başer, H. C., & Demirci, F. (2002). Antimicrobial screening of *Mentha piperita* essential oils. *Journal of agricultural and food chemistry*, *50*, 3943–3946.

Jahani, F., Tohidi-Moghadam, H. R., Larijani, H. R., Ghooshchi, F., & Oveysi, M. (2021). Influence of zinc and salicylic acid foliar application on total chlorophyll, phenolic components, yield and essential oil composition of peppermint (*Mentha piperita* L.) under drought stress condition. *Arabian Journal of Geosciences*, *14*, 1–12.

Jaishankar, M., Mathew, B. B., Shah, M. S., & Gowda, K. R. S. (2014). Biosorption of few heavy metal ions using agricultural wastes. *Journal of Environment Pollution and Human Health*, *2*, 1–6.

Jaleel, C. A., Manivannan, P., Wahid, A., Farooq, M., Al-Juburi, H. J., Somasundaram, R., Panneerselvam, R. (2009). Drought stress in plants: a review on morphological characteristics and pigments composition. *Int. J. Agric. Biol*, *11*, 100–105.

Järup, L. (2003). Hazards of heavy metal contamination. *British medical bulletin*, *68*, 167–182.

Jat, M. L., Dagar, J. C., Sapkota, T. B., Govaerts, B., Ridaura, S. L., Saharawat, Y. S., Stirling, C. (2016). Climate change and agriculture: adaptation strategies and mitigation opportunities for food security in South Asia and Latin America. In *Advances in agronomy* 137, 127–235. Academic Press.

Jayakannan M, Bose J, Babourina O, Rengel Z, Shabala S (2015) Salicylic acid in plant salinity stress signalling and tolerance. Plant Growth Regul. 76, 25–40.

Kant, M. R., Jonckheere, W., Knegt, B., Lemos, F., Liu, J., Schimmel, B. C. J., Alba, J. M. (2015). Mechanisms and ecological consequences of plant defence induction and suppression in herbivore communities. *Annals of botany*, *115*, 1015–1051.

Karray-Bouraoui, N., Rabhi, M., Neffati, M., Baldan, B., Ranieri, A., Marzouk, B., Smaoui, A. (2009). Salt effect on yield and composition of shoot essential oil and trichome morphology and density on leaves of *Mentha pulegium*. *Industrial Crops and Products*, *30*, 338–343.

Kennedy, D. O., & Wightman, E. L. (2011). Herbal extracts and phytochemicals: plant secondary metabolites and the enhancement of human brain function. *Advances in Nutrition*, *2*, 32–50.

Keshavarz-Mirzamohammadi, H., Modarres-Sanavy, S. A. M., Sefidkon, F., Mokhtassi-Bidgoli, A., & Mirjalili, M. H. (2021a). Irrigation and fertilizer treatments affecting rosmarinic acid accumulation, total phenolic content, antioxidant potential and correlation between them in peppermint (*Mentha piperita* L.). *Irrigation Science*, *39*, 671–683.

Keshavarz, H., Modarres, S. S., Sefidkon, F., & MokhtassI, B. A. (2020). Effect of Organic Fertilizers and Urea Fertilizer on Phenolic Compounds, Antioxidant Activity, Yield and Yield Components of Peppermint (*Mentha piperita* L.) under Drought Stress. Iranian Journal of Field Crops Research. 17, 661–672.

Keshavarz-Mirzamohammadi, H., Tohidi-Moghadam, H. R., & Hosseini, S. J. (2021b). Is there any relationship between agronomic traits, soil properties and essential oil profile of peppermint (*Mentha piperita* L.) treated by fertiliser treatments and irrigation regimes? *Annals of Applied Biology*, *179*, 331–344.

Khair, K. U., Farid, M., Ashraf, U., Zubair, M., Rizwan, M., Farid, S., Ali, S. (2020). Citric acid enhanced phytoextraction of nickel (Ni) and alleviate *Mentha piperita* (L.) from Ni-induced physiological and biochemical damages. *Environmental Science and Pollution Research*, *27*, 27010–27022.

Khalvandi, M., & Gholami, A. (2017). *Piriformospora indica* symbiotic effect on the quantity and quality of essential oils and some physiological parameters of peppermint (*Mentha piperita*) under salt stress. *Journal of Plant Process and Function*, *6*, 169–184.

Khan, M. N., Siddiqui, M. H., Mohammad, F., Naeem, M., & Khan, M. M. A. (2010). Calcium chloride and gibberellic acid protect linseed (*Linum usitatissimum* L.) from NaCl stress by inducing antioxidative defence system and osmoprotectant accumulation. *Acta Physiol Plant*, *32*, 121–132.

Khan, N. A., Singh, S., Anjum, N. A., & Nazar, R. (2008). Cadmium effects on carbonic anhydrase, photosynthesis, dry mass and antioxidative enzymes in wheat (*Triticum aestivum*) under low and sufficient zinc. *Journal of Plant Interactions*, *3*, 31–37.

Khanam, D., & Mohammad, F. (2018). Plant growth regulators ameliorate the ill effect of salt stress through improved growth, photosynthesis, antioxidant system, yield and quality attributes in *Mentha piperita* L. *Acta Physiologiae Plantarum*, *40*, 1–13.

Kheiry, A., Tori, H., & Mortazavi, N. (2017). Effects of drought stress and jasmonic acid elicitors on morphological and phytochemical characteristics of peppermint (*Mentha piperita* L.). *Iranian Journal of Medicinal and Aromatic Plants Research, 33*, 268–280.

Khoramivafa, M., Shokri, K., Sayyadian, K., & Rejali, F. (2012). Contribution of microbial associations to the cadmium uptake by peppermint (*Mentha piperita*). *Annals of Biological Research, 3*, 2325–2329.

Khorasaninejad, S., Mousavi, A., Soltanloo, H., Hemmati, K., & Khalighi, A. (2011). The effect of drought stress on growth parameters, essential oil yield and constituent of Peppermint (*Mentha piperita* L.). *Journal of Medicinal Plants Research, 5*, 5360–5365.

Kumar, S., Ahanger, M. A., Alshaya, H., Jan, B. L., & Yerramilli, V. (2022). Salicylic acid mitigates salt induced toxicity through the modifications of biochemical attributes and some key antioxidants in *Capsicum annuum. Saudi Journal of Biological Sciences.* https://doi.org/10.1016/j.sjbs.2022.01.028.

Li, Z., Wang, W., Li, G., Guo, K., Harvey, P., Chen, Q., & Yang, H. (2016). MAPK-mediated regulation of growth and essential oil composition in a salt-tolerant peppermint (*Mentha piperita* L.) under NaCl stress. *Protoplasma, 253*, 1541–1556.

Li, Z., Yang, H., Wu, X., Guo, K., & Li, J. (2015). Some aspects of salinity responses in peppermint (*Mentha× piperita* L.) to NaCl treatment. *Protoplasma, 252*, 885–899.

Liu, J., Fu, C., Li, G., Khan, M. N., & Wu, H. (2021). ROS homeostasis and plant salt tolerance: Plant nanobiotechnology updates. *Sustainability, 13*(6), 3552.

Loolaie, M., Moasefi, N., Rasouli, H., & Adibi, H. (2017). Peppermint and its functionality: A review. *Arch Clin Microbiol, 8*(4), 54.

Lösch, R., & Köhl, K. I. (1999). Plant respiration under the influence of heavy metals. In *Heavy metal stress in plants* (pp. 139–156). Springer, Berlin, Heidelberg.

Mahmoud, S. S., & Croteau, R. B. (2003). Menthofuran regulates essential oil biosynthesis in peppermint by controlling a downstream monoterpene reductase. *Proceedings of the National Academy of Sciences, 100*, 14481–14486.

Marshall E (2011) Health and wealth from medicinal aromatic plants. Food and Agriculture Organization, Rome

Masood, A., Khan, M. I. R., Fatma, M., Asgher, M., Per, T. S., & Khan, N. A. (2016). Involvement of ethylene in gibberellic acid-induced sulfur assimilation, photosynthetic responses, and alleviation of cadmium stress in mustard. *Plant Physiology and Biochemistry, 104*, 1–10.

Meena, K. K., Sorty, A. M., Bitla, U. M., Choudhary, K., Gupta, P., Pareek, A., Minhas, P. S. (2017). Abiotic stress responses and microbe-mediated mitigation in plants: the omics strategies. *Frontiers in plant science, 8*, 172.

Mesa-Marín, J., Del-Saz, N. F., Rodríguez-Llorente, I. D., Redondo-Gómez, S., Pajuelo, E., Ribas-Carbó, M., & Mateos-Naranjo, E. (2018). PGPR reduce root respiration and oxidative stress enhancing *Spartina maritima* root growth and heavy metal rhizoaccumulation. *Frontiers in plant science, 9*, 1500.

Mirajkar, S. J., Dalvi, S. G., Ramteke, S. D., & Suprasanna, P. (2019). Foliar application of gamma radiation processed chitosan triggered distinctive biological responses in sugarcane under water deficit stress conditions. *International journal of biological macromolecules, 139*, 1212–1223.

Mohammadi, M., Majnoun Hosseini, N., & Dashtaki, M. (2016). Effects of nano-ferric oxide and zinc sulfate on chlorophyll, anthocyanin, flavonoid and leaf mineral elements of peppermint (*Mentha piperita* L.) at Karaj climatic conditions. *Iranian Journal of Medicinal and Aromatic Plants Research, 32*, 770–783.

Morais, S., Costa, F. G., & Pereira, M. D. L. (2012). Heavy metals and human health. *Environmental health–emerging issues and practice, 10*, 227–245.

Munns R, Tester M (2008) Mechanisms of salinity tolerance. Ann Rev Plant Biol. 59, 651–681.

Nagajyoti, P. C., Lee, K. D., & Sreekanth, T. V. M. (2010). Heavy metals, occurrence and toxicity for plants: a review. *Environmental chemistry letters, 8*, 199–216.

Nasim W, Amin A, Fahad S, Awais M, Khan N, Mubeen M, Wahid A, Rehman MH, Ihsan MZ, Ahmad S, Hussain S (2018) Future risk assessment by estimating historical heat wave trends with projected heat accumulation using SimCLIM climate model in Pakistan. Atmos Res 205, 118–133.

Naveed, M., Tanvir, B., Xiukang, W., Brtnicky, M., Ditta, A., Kucerik, J., Mustafa, A. (2021). Co-composted Biochar Enhances Growth, Physiological, and Phytostabilization Efficiency of *Brassica napus* and Reduces Associated Health Risks Under Chromium Stress. *Frontiers in plant science, 12*, 775785–775785.

Nazir, F., Fariduddin, Q., Hussain, A., & Khan, T. A. (2021). Brassinosteroid and hydrogen peroxide improve photosynthetic machinery, stomatal movement, root morphology and cell viability and reduce Cu-triggered oxidative burst in tomato. *Ecotoxicology and Environmental Safety, 207*, 111081.

Negrão, S., Schmöckel, S. M., & Tester, M. (2017). Evaluating physiological responses of plants to salinity stress. *Annals of Botany*, 119, 1–11.

Nemati Lafmejani, Z., Jafari, A. A., Moradi, P., & Ladan Moghadam, A. (2021). Application of Chelate and Nano-Chelate Zinc Micronutrient Onmorpho-physiological Traits and Essential Oil Compounds of Peppermint (*Mentha piperita* L.). *Journal of Medicinal plants and By-product*, *10*, 21–28.

Pál, M., Janda, T., & Szalai, G. (2018). Interactions between plant hormones and thiol-related heavy metal chelators. *Plant Growth Regulation*, *85*, 173–185.

Parsa, M., Kamaei, R., & Yousefi, B. (2021). Effect of chemical and biological fertilizers on the physiological characteristics and activity of some antioxidant enzymes of peppermint (*Mentha piperita*) under drought stress conditions. *Journal of Plant Research (Iranian Journal of Biology)*. 2021.

Pérez-Borroto, L. S., Toum, L., Castagnaro, A. P., González-Olmedo, J. L., Coll-Manchado, F., Pardo, E. M., & Coll-García, Y. (2021). Brassinosteroid and brassinosteroid-mimic differentially modulate *Arabidopsis thaliana* fitness under drought. *Plant Growth Regulation*, *95*, 33–47.

Peyvandi, M., Aboie Mehrizi, Z., & Ebrahimzadeh, M. (2016). The effect of cadmium on growth and composition of essential oils of *Mentha piperita* L. *Iranian Journal of Plant Physiology*, *6*, 1715–1720.

Pusztahelyi, T., Holb, I. J., & Pócsi, I. (2015). Secondary metabolites in fungus-plant interactions. *Frontiers in plant science*, *6*, 573.

Qadir M, Quille'rou E, Nangia V, Murtaza G, Singh M, Thomas RJ, Drechsel P, Noble AD (2014) Economics of salt-induced land degradation and restoration. Nat Res Forum 38, 282–295.

Qian, H., Li, J., Pan, X., Jiang, H., Sun, L., & Fu, Z. (2010). Photoperiod and temperature influence cadmium's effects on photosynthesis-related gene transcription in *Chlorella vulgaris*. *Ecotoxicology and Environmental Safety*, *73*, 1202–1206.

Rahimi, Y., Taleei, A., & Ranjbar, M. (2017). Changes in the expression of key genes involved in the biosynthesis of menthol and menthofuran in *Mentha piperita* L. under drought stress. *Acta Physiologiae Plantarum*, *39*, 1–9.

Rahimi, Y., Taleei, A., & Ranjbar, M. (2018). Long-term water deficit modulates antioxidant capacity of peppermint (*Mentha piperita* L.). *Scientia Horticulturae*, *237*, 36–43.

Rasheed F, Anjum NA, Masood A, Sofo A, Khan NA (2020) The key roles of salicylic acid and sulfur in plant salinity stress tolerance. J. Plant Growth Regul. https://doi.org/10.1007/s00344-020-10257-3.

Raza, A., Razzaq, A., Mehmood, S. S., Zou, X., Zhang, X., Lv, Y., & Xu, J. (2019). Impact of climate change on crops adaptation and strategies to tackle its outcome: A review. *Plants*, *8*(2), 34.

Rehman, A. U., Nazir, S., Irshad, R., Tahir, K., ur Rehman, K., Islam, R. U., & Wahab, Z. (2021). Toxicity of heavy metals in plants and animals and their uptake by magnetic iron oxide nanoparticles. *Journal of Molecular Liquids*, *321*, 114455.

Saedi, F., Sirousmehr, A., & Javadi, T. (2020). Effect of nano-potassium fertilizer on some morpho-physiological characters of peppermint (*Mentha piperita* L.) under drought stress. *Journal of Plant Research (Iranian Journal of Biology*, *33*, 35–45.

Salehi Sardoei, A. (2020). Effect of drought stress and salicylic acid on some of growth traits, photosynthetic pigments and yield essential oil of peppermint (*Mentha piperita* L.). *New Finding in Agriculture*. *11*, 125–137.

Salt, D. E., Blaylock, M., Kumar, N. P., Dushenkov, V., Ensley, B. D., Chet, I., & Raskin, I. (1995). Phytoremediation: a novel strategy for the removal of toxic metals from the environment using plants. *Bio/technology*, *13*, 468–474.

Sarraf, M., Vishwakarma, K., Kumar, V., Arif, N., Das, S., Johnson, R., Hasanuzzaman, M. (2022). Metal/Metalloid-Based Nanomaterials for Plant Abiotic Stress Tolerance: An Overview of the Mechanisms. *Plants*, *11*(3), 316.

Shahbaz, M., & Ashraf, M. (2013). Improving salinity tolerance in cereals. *Critical reviews in plant sciences*, *32*, 237–249.

Sharma, A., Sidhu, G. P. S., Araniti, F., Bali, A. S., Shahzad, B., Tripathi, D. K., Landi, M. (2020). The role of salicylic acid in plants exposed to heavy metals. *Molecules*, *25*(3), 540.

Vatankhah, E., Kalantari, B., & Andalibi, B. (2017). Effects of methyl jasmonate and salt stress on physiological and phytochemical characteristics of peppermint (*Mentha piperita* L.). *Iranian Journal of Medicinal and Aromatic Plants Research*, *33*, 449–465.

Wani, W., Masoodi, K. Z., Zaid, A., Wani, S. H., Shah, F., Meena, V. S., Mosa, K. A. (2018). Engineering plants for heavy metal stress tolerance. *Rendiconti Lincei. Scienze Fisiche e Naturali*, *29*, 709–723.

Whitmore, T. C. (2000). The case of tropical rain forests. The sustainable development of forests: aspirations and the reality. *Naturzale-Cuadernos de Ciencias Naturales*, (15), 13–15.

Zaid, A., Ahmad, B., & Wani, S. H. (2021). Medicinal and aromatic plants under abiotic stress: a crosstalk on phytohormones' perspective. *Plant Growth Regulators*, 115–132.

Zaid, A., Bhat, J. A., & Wani, S. H. (2020a). Influence of metalloids and their toxicity impact on photosynthetic parameters of plants. *Metalloids in plants: advances and future prospects*, 113–124.

Zaid, A., Mohammad, F., & Fariduddin, Q. (2020b). Plant growth regulators improve growth, photosynthesis, mineral nutrient and antioxidant system under cadmium stress in menthol mint (*Mentha arvensis* L.). *Physiology and Molecular Biology of Plants*, 26, 25–39.

Zaid, A., Ahmad, B., Jaleel, H., Wani, S. H., & Hasanuzzaman, M. (2020c). A critical review on iron toxicity and tolerance in plants: role of exogenous phytoprotectants. *Plant Micronutrients*, 83–99.

Zaid, A., Mohammad, F., Wani, S. H., & Siddique, K. M. (2019). Salicylic acid enhances nickel stress tolerance by up-regulating antioxidant defense and glyoxalase systems in mustard plants. *Ecotoxicology and environmental safety*, 180, 575–587.

Zaman M, Shahid SA, Pharis RP (2016) Salinity a serious threat to food security – where do we stand? Soils Newsletter 39, 9–10.

Zhang, M., Smith, J. A. C., Harberd, N. P., & Jiang, C. (2016). The regulatory roles of ethylene and reactive oxygen species (ROS) in plant salt stress responses. *Plant molecular biology*, 91, 651–659.

Zolin, C.A., Rodrigues, R.d.A.R., (2015). Impact of Climate Change on Water Resources in Agriculture. CRC Press, Taylor & Francis Group, 232.

Zörb C, Geilfus CM, Dietz KJ (2019). Salinity and crop yield. Plant Biol 21, 31–38.

11 *Moringa oleifera* under Stressful Conditions

Saidul Islam[1], Mahatab Ali[2], Arghya Banerjee[3], Maksud Hasan Shah[4], SK Sadikur Rahman[5], Faijuddin Ahammad[6] and Akbar Hossain[7]*

[1]Nadia Krishi Vigyan Kendra, Bidhan Chandra Krishi Viswavidyalaya (BCKV), Gayeshpur, Nadia, West Bengal, India
[2]Department of Genetics and Plant Breeding, Bidhan Chandra Krishi Viswavidyalaya (BCKV), Mohanpur, Nadia, West Bengal, India
[3]Department of Plant Pathology, School of Agriculture and Allied Science, The Neotia University, Jhinga, South 24 Parganas, West Bengal, India
[4]Department of Agronomy, Bidhan Chandra Krishi Viswavidyalaya (BCKV), Mohanpur, Nadia, West Bengal, India
[5]Assistant Technology Manager, Department of Agriculture, Howrah, Government of West Bengal, India
[6]Department of Plant Pathology, Sam Higginbottom University of Agriculture, Technology and Sciences, Allahabad, Uttar Pradesh, India
[7]Department of Agronomy, Bangladesh Wheat and Maize Research Institute, Dinajpur 5200, Bangladesh
*Corresponding author: islamsaidulr23@gmail.com(Saidul Islam)

CONTENTS

DOI: 10.1201/9781003242963-11

11.1 INTRODUCTION

Moringa oleifera Lam (syn. *M. ptreygosperma* Gaertn.) is a well-known, extensively spread and naturalized species of the Moringaceaemono generic family (Nadkarni 1976; Ramachandran et al. 1980). *Moringa oleifera* Lam (Moringaceae) is a valuable plant found in many tropical and sub-tropical areas (Anwar et al. 2007). *Moringa*, drumstick tree (from the tall, thin, triangular seed-pods), horseradish tree (from the flavor of the roots, which is similar to horseradish), and ben oil tree or benzolive tree (from the oil extracted from the seeds) are among its common names (Anwar et al. 2007).

The fragrant, hermaphroditic blooms are surrounded by five unequally veined, whitish-yellow petals. The flowers are around 1.0–1.5 cm long and 2.0 cm in width. They bloom in spreading or drooping flower cluster centers with a length of 10–25 cm on narrow, hairy stalks. The fruits are hung from a three-sided brown capsule that is 20–45 cm in diameter and contains dark brown, globular seeds with a diameter of 1 cm. Wind and water disseminate the seeds, which have three pale papery wings. It is frequently pruned down to 1–2 m during cultivation and left to sprout so that the pods and leaves are always within reach (Parrotta 1993).

Moringa leaves have antioxidant qualities (Lalas and Tsaknis 2002; Sreelatha and Padma 2009), and the seeds have antibacterial characteristics (Olsen 1987; Madsen et al. 1987). External factors that adversely affect plant growth, development or output are referred to as stress in plants (Verma et al. 2013). Abiotic and biotic are the two main types of stress that plants face. Living organisms, such as viruses, bacteria, fungus, nematodes, insects, arachnids and weeds, induce biotic stress in plants. Drought (water stress), over-watering (water logging), severe temperatures (cold, frost, and heat), salt and mineral toxicity all have deleterious effects on crops' and other plants' growth, yield and seed quality (Gull et al. 2019).

11.2 ORIGIN AND DISTRIBUTION OF *MORINGA*

There are 13 species of *Moringa* in the single genus *Moringa* of the Moringa family, i.e., *M. arborea*, native to Kenya, *M. rivae* native to Kenya and Ethiopia, *M. borziana*, *M. peregrina* from Arabia and Horn of Africa, and *M. pygmaea* native to Somalia and Kenya, *M. longituba* native to Kenya, Ethiopia and Somalia, *M. stenopetala* native to Kenya and Ethiopia, *M. ruspoliana* native to Ethiopia, *M. ovalifolia* native to Namibia and Angola *M. drouhardii* and *M. hildebrandi* from Madagascar, and *M. concanensis* and *Moringa oleifera* (Paliwal and Sharma, 2011) from the sub-Himalayan region of northern India, with *Moringa oleifera* being the most widely used and studied to date.

This fast-growing soft wood tree native to the Himalayan foothills (northern India, Pakistan, and Nepal) can reach a height of 12 m (Roloff et al. 2009; Sharma et al. 2011). Farmers and academics in the past were drawn to it because of its many applications and possibilities. *Moringa oleifera*, according to Ayurvedic traditional medicine, can prevent 300 diseases, and its leaves have been used for both preventive and therapeutic purposes (Ganguly 2013). Furthermore, according to a study conducted in the Virudhunagar district of Tamil Nadu, India, *Moringa* is one of the species used by traditional Siddha healers (Mutheeswaran et al. 2011). Even though the species never became popular among the Greeks and Romans, they were aware of its medical virtues (Fahey 2005). Ancient Egyptians utilized *Moringa oleifera* oil for its beauty value and skin preparation (Mahmood et al. 2010). *Moringa oleifera* was grown and consumed in its native areas when a few scholars started to investigate its potential use in water refinement, and it was only later that the nutritional and medicinal properties were "discovered," and the species spread throughout almost all tropical countries. The first international conference on *Moringa oleifera* was conducted in Tanzania in 2001, and since then, the number of congresses and studies has grown, distributing information about *Moringa oleifera*'s remarkable capabilities. This species is now known as the "miracle tree," "natural gift," and "mother's best friend." *Moringa oleifera* can grow in any tropical or subtropical country with unique environmental characteristics, such as a dry to moist tropical or subtropical climate with annual precipitation ranging from 760 to 2500 mm (less than 800 mm irrigation is required) and temperatures ranging from 18 to 28°C. It thrives in a variety of soil types but prefers thick clay and wet soils with a pH of 4.5–8, at altitudes up to 2000 m (Nouman et al. 2014; Palada 1996). "Though considered a non-indigenous species, *Moringa oleifera* has found wide acceptance among various ethnic Nigerians, who have exploited different uses," according to a study on local uses and geographical distribution of *Moringa oleifera* (Popoola and Obembe 2013), which covered the major agro-ecological region of Nigeria (e.g., food, medicine, fodder, etc.).

11.3 PROSPECTS OF *MORINGA* TO ADAPT TO BIOTIC STRESS

Insects, weeds and diseases are among the major biotic stresses causing considerable reductions in crop productivity. Although synthetic pesticides such as fungicides, insecticides and herbicides are often very effective in controlling biotic stresses with significant increment in productivity, many concerns such as potential harms to human health and livestock, elevated environmental pollution, and occasional toxicity and resistance among genotypes have compelled researchers to explore natural environment-friendly approaches (Demoz and Korsten 2006; Macedo et al. 2007; Farooq et al. 2011, Hussain et al. 2013).

11.3.1 DISEASES OF *MORINGA*

Moringa plants are well known as resistant to the most common plant pathogenic pests and diseases, and numerous researchers have found that these plants rarely suffer from serious diseases in the various parts of the world in which they are found. However, under certain conditions, the plants can suffer from numerous pests and diseases. Several harmful fungi have recently been reported from pods in the Kingdom of Saudi Arabia (Mridha and Al-Barakah 2015). A total of six different fungus species belonging to five genera have been identified: *Aspergillus niger, Aspergillus flavus, Alternaria alternata, Fusarium oxysporum, Macrophomina phaseolina* and *Rhizopus stolonifera* (Mridha and Barakah 2017). Patricio and Palada (2017) noticed only stem rot disease during an adaptation and

TABLE 11.1
***Moringa* Diseases and Plant Parts Infected**

Sl. No.	Diseases	Causal Organism	Reported/ Distribution	Infected part	References
1	Root rot	*Diplodia* sp.	Egypt	Roots	(Palada 2003; Orwa et al. 2009)
2	Fruit rot	*Cochliobolushawaiiensis* Alcorn	India	Pods	(Kshirsagar and D'Souza 1989; Ramachandran et al. 1980)
3	Dieback	*Fusarium semitectum* Berk	China	Stems and roots	(Guizh 2008)
4	Fusarium wilt	*Fusarium oxysporium* f. sp. *Moringae*	Egypt	Seedlings	(Pande 1998)
5	Twig canker	*Fusarium pallidoroseum* (Cooke) Sacc	India	Stems	(Mandokhot et al. 1994)
6	Brown leaf spot	*Cercosporamoringicola*	Egypt	Leaves	(Saint 2010)
7	Alternaria leaf spot	*Alternaria solani* Sorauer	Egypt	Leaves	(Kshirsagar and D'Souza 1989; Ramachandran et al. 1980)
8	Powdery mildew	*Leveillulataurica* (Lev) Arn	Egypt	Leaves and pods	(Kumar et al. 2013)
9	Damping off	*Rhizoctonia solani* Kuehn	Egypt	Seedlings	(Abu-Taleb 2011)
10	Anthracnose	*Colletotrichum chlorophyti*	China	Leaves	(Cai et al. 2016)

horticulture trial with a large number of *Moringa* accessions in the Philippines. Carbungco et al. (2017) isolated, described and identified fungal endophytes associated with *M. oleifera* leaves from the Philippines. They found a total of 24 fungal species, with *Fusarium*, *Xylaria*, *Pestalotiopsis*, *Aspergillus*, *Nigrospora*, *Stachybotrys*, *Rhizoctonia* and *Macrophomina* among the fungi detected. Some major and minor diseases of *Moringa* are discussed in Table 11.1.

11.3.2 PESTS OF *MORINGA*

M. oleifera growing on agricultural farms of the King Saud University in Saudi Arabia showed widespread incidence and significant defoliation of the crop. In India, some major and minor pests on *Moringa* have been recorded. The pests' distribution and status, as well as the cause of damage and management, were all recorded. Pod fly (*Gitonadistigma*), bud worm (*Noordamoringae*), a major pest in South India, leaf caterpillar (*Noordablitealis*), a serious pest of drumstick trees, particularly in South India hairy caterpillars [*Eupterotemollifera*, *Pericalliaricini*, *Metanastriahyrtaca* and *Streblote (Taragama) siva*], bark borer (*Indarbelatetraonis*) and long horn beetles (*Batocerarubus*) are all present throughout the Indian subcontinent (Mridha and Barakah, 2017). Aphids (*Aphis gossypii*) are polyphagous minor pests of *Moringa*. The twigs provide crucial sap to the nymphs and adults. Most reproduction is parthenogenic, and population growth is quick. Scale insects (*Ceroplastodescajani*), bud midges (*Stictodiplosismoringae*), and leaf-eating weevils (*Myllocerus* spp.) are other minor pests. Moringa pests have also been mentioned on another website (TNAU Agrictech, 2016). Bud worms (*Noordamoringae*), pod flies (*Gitonadistigma*), leaf caterpillars (*Noordablitealis*), bark caterpillars (*Indarbelatetraonis*) and hairy caterpillars (*Eupterotemollifera*) are a few examples. During an adaptability and horticultural trial with a large number of *Moringa* accessions in the Philippines, Patricio and Palada (2017) observed red mites (*Tetranyctrusurticia*), defoliators, leaf-footed bug (*Leptoglossus phyllospus*), and whiteflies (*Bermisia* sp.) in the field and noted that the insects caused little damage to the plants. Kant et al. (2017) conducted an insect pest study in different regions of *M. oleifera*-producing fields in Samoa and found only modest floral damage. They also described a plant hopper and stem borer causing minor damage to the plants.

11.4 ABIOTIC STRESS ADAPTATION THROUGH *MORINGA* PLANTS

Abiotic stresses such as drought, temperature extremes, salinity and nutrient imbalances unfavorably affect plant growth and progress through a significant decrease in crop productivity (Farooq et al. 2009; Hamdia and Shaddad 2010; Hussain et al. 2013).

11.4.1 *MORINGA OLEIFERA* EXPOSED TO WATER STRESS

M. oleifera can modify its primary and secondary metabolism in response to increased water stress. *M. oleifera* has an effective isohydric behavior. Water stress causes a rapid and forceful stomatal closure, triggered by the accumulation of abscisic acid (ABA), resulting in photosynthesis suppression and detrimental consequences on biomass production. In stressed leaves, however, photochemistry was intact and maximal fluorescence and saturating photosynthesis were unaffected. Isoprene production in *M. oleifera* increased three-fold as stress was progressed, according to the researchers. Higher isoprene production may assist leaves to cope with water stress by acting as an antioxidant or membrane stabilizer, and it may also suggest activation of the MEP (methylerythritol 4-phosphate) pathway, which further protects photosynthesis under water stress (Brunetti et al. 2018).

Increased carbon flow through the MEP pathway also stimulates the synthesis of isoprene and non-volatile isoprenoids such as carotenoids and ABA in drought-stressed leaves (Marino et al. 2017; Tattini and Velikova 2014). Drought-stressed photosynthesis is known to be protected by carotenoids (Beckett et al. 2012; Tattini et al. 2015). Carotenoids have photo-protective functions in the chloroplasts, including quenching of triple-state chlorophyll, thermal dissipation of excess energy through de-epoxidation of xanthophylls (non-photochemical quenching, NPQ) (Brunetti et al. 2015) and an antioxidant role of zeaxanthin (Zea) in chloroplasts by strengthening thylakoid membranes underneath heat stress (Dall'Osto et al. 2010; Esteban et al. 2015; Havaux and Dall'Osto 2007). Notably, Zea biosynthesis during hydroxylation of carotene (car) may improve drought resistance (Davison and Hunter 2002; Du et al. 2010), possibly because Zea interacts with light-harvesting complex-b (LHCb), reducing O_2 production and maintaining NPQ in high light conditions (Dall'Osto et al. 2010; Johnson et al. 2007). In turn, β-car (together with isoprene; see Velikova and Edreva 2004) is a chemical quencher of O_2 (Ramel et al. 2012). Isoprene and foliar ABA have been linked in several studies (Barta, 2006; Marino et al. 2017; Tattini and Velikova 2014). Brodribb (2013) Coupel-Ledru et al. (2017) and McAdam (2014) found that abscisic acid plays a key role in the regulation of stomatal motions in plants capable of sustaining leaf water potential and relative water content under drought stress circumstances (isohydric behavior). The modification of secondary metabolism in drought-stricken plants may be primarily aimed at reducing excess ROS by boosting the production of antioxidant metabolites such as isoprenoids and phenylpropanoids (Loreto 2015; Nakabayashi and Yonekura-Sakakibara 2014; Tattini et al. 2015).

11.4.2 *MORINGA OLEIFERA* EXPOSED TO SALINE CONDITIONS

The *Moringa* tree is salt tolerant and has low nutrient requirements (Abay et al. 2016; Baker et al. 2014; Nadeem et al. 2015). According to Baker et al. (2014), a negative relationship was shown between salt stress degree and plant growth characteristics, i.e., plant height, green leaves area and dry weight of root, stem, leaves and shoot which decreased as the salt concentration increased in diluted seawater. Root and stem dry weights were more affected by salinity than any other growth parameters (Baker et al. 2014). A study conducted by Yap et al. (2021) on salinity stress was imposed on *S. marianum* plants by watering with different doses of saline. The effect of salinity on *S. marianum* was studied as the abiotic stresses that are threatening plant growth and production globally and being a stress aspect, it could be manipulated to improve the production of secondary metabolites, mainly silymarin. Plant growth and yield in *S. marianum* plants cultivated without salt stress (control salinity) increased (Yap et al. 2021). Some of the inhibitory effects of soil salinity on plant growth and productivity include reductions in water availability to plant roots, which negatively affect plant tissue water status, and disturbances in metabolic processes, which lead to decreases in meristematic activity and cell enlargement, as well as an increase in respiration rate due to higher energy requirements (Kaydan and Okut 2007; Abdul Qados 2015). Salinity stress also induces an excess of ROS in plan tissues. To avoid oxidative injury and sustain metabolic functioning under stress, a balance between the formation and breakdown of ROS is essential. The quantity of ROS in implant tissues is regulated by an antioxidant system that includes antioxidant enzymes as well as nonenzymatic low-molecular-weight antioxidant compounds such as proline, ascorbic acid and carotenoids (Schutzendubel and Polle 2002; Rady and Mohamed 2015). The osmotic effect caused by salt stress causes an increase in growth inhibitors (i.e. abscisic acid), a decrease in growth promoters (i.e. indole-3-acetic acid gibberellins) and a disturbance of the water balance of saline-stressed plants, which could explain the reduction in plant growth and productivity under adverse soil salinity conditions. These inhibitory effects of salinity result in stomatal closure, ionic imbalance, photosynthesis reduction, ionic homeostasis disturbance, harmful ion accumulation and growth suppression (Rady 2011; Rady et al. 2013).

11.5 PHYSIOLOGICAL AND BIOCHEMICAL ACTIVITIES OF *MORINGA*

Salicylic acid (SA) was first discovered as a prominent component in extracts from the *Salix* (willow) tree, whose bark has been used for medicinal purposes since ancient times (Hayat et al.

2010). It is a plant growth regulator (PGR) that is activated by a variety of biotic and abiotic stressors. It's been discovered to be an endogenous regulatory signal that helps plants defend themselves against infections. It's also a natural signal molecule for defense system activation (Anosheh et al. 2012; Ashraf et al. 2010; Raskin 1992). Plant stress tolerance has been demonstrated to improve when SA is used. Exogenous SA treatment on plants has been shown to protect plants from salinity in numerous studies (Mutlu et al. 2009; Senaratna et al. 2000). Plants exposed to abiotic stressors experience increased oxidative stress, which results in the creation of highly hazardous reactive oxygen species (ROS), which damage proteins, lipids, carbohydrates and DNA (Apel and Hirt 2004). The herbal plant regulation system is raised or stimulated by plant growth retailers, improving growth and improvement methods from seed germination to senescence. having a wealthy supply of plant growth regulators such as zeatin and gibberellins, proteins, nutrients which include B, C, D and K, minerals which include Ca and K, and more than 40 herbal antioxidants which include ascorbic acid, flavonoids, phenolics and carotenoids (Makkar and Becker 1996; Anwar et al. 2005; Mahmood et al. 2010; Luqman et al. 2012; Hussain et al. 2013), *Moringa* extract can modulate a vegetation increase and productiveness despite abiotic stresses. The *Moringa* leaves are high in total polyphenols (TPP) (260 mg 100 g^{-1}), quercetin (100 mg 100 g^{-1}), kaempferol (34 mg 100 g^{-1}) and β-carotene (34 mg 100 g^{-1}), and have a high total antioxidant capacity (TAC) (260 mg 100 g^{-1}) (Lako et al. 2007). *Moringa* leaves, roots and seeds all have strong antibacterial and antifungal properties, making them a potential biopesticide for controlling a variety of diseases (Anwar et al. 2007). *Trichophyton rubrum*, *Trichophyton mentagrophytes*, *Epidermophyton xoccosum*, *Aspergillus tamarii*, *Rhizopus solani*, *Mucor mucedo*, *Aspergillus niger* and *Microsporum canis* demonstrated high *in vitro* antifungal activity against several harmful fungal species (Chuang et al. 2007; Jamil et al. 2007). Stem rot and damping-off are important diseases occurring in more than 500 plant species in the tropics and subtropics caused by the soilborne pathogenic fungus *Sclerotium rolfsii* (Punja 1985; Adandonon et al. 2003). These antibacterial properties are due to the presence of crystalline alkaloids, fatty acids, proteins, peptides, glycosides and niazirins in *Moringa* leaves (Adandonon et al. 2003; Jamil et al. 2007; Hussain et al. 2013). Higher phenolic levels in *Moringa* leaves contributed to maize's increased performance under temperature stress; because plant phenolics are advantageous during the oxidative burst, soluble phenolics also contribute to strong antioxidant activity (Randhir et al. 2004). However, higher antioxidant content in *Moringa* chloroplasts, such as ascorbic acid and glutathione, is critical for plant resistance to oxidative stress (Noctor and Foyer 1998; Gill and Tuteja 2010). To avoid oxidative injury and sustain metabolic functioning under stress, a balance between the formation and breakdown of ROS is essential. In plant tissues, the level of ROS is controlled by an antioxidant system that consists of antioxidant enzymes and non-enzymatic low-molecular-weight antioxidant molecules, including proline and ascorbic acid. In the current study, the application of cytokinin-rich *Moringa* leaf extract (MLE) with SA, which induces cytokinin biosynthesis, increased the sink capacity by supplying photo-assimilates from stay-green leaves (Thomas and Howarth 2000), and re-translocation of stem reserves as a result of the application of cytokinin-rich MLE with SA, which induces cytokinin biosynthesis. The presence of zeatin-like cytokinin in MLE inhibits early leaf senescence and increases the leaf area available for photosynthetic activity. Exogenously applied cytokinin (found in MLE) can delay this process (Tetley and Thimann 1974; Gepstein and Glick 2013), possibly by activating cytokinin-dependent isopentenyl transferase (ipt) biosynthesis and thus increasing chlorophyll concentrations (Tetley and Thimann 1974; Mukherjee 2010; Gepstein and Glick 2013). SA, in combination with MLE, may help to mitigate the negative effects of salt stress by increasing antioxidant levels. SA is a component of a highly complicated signal transduction network, and its mechanism of action varies depending on the system. SA may influence the activity of specific enzymes directly or indirectly (Horváth et al. 2007). Exogenous treatments of MLE and/or SA dramatically enhance soluble sugars in salt-stressed bean plants (Rady and Mohamed 2015b). They play a role in osmotic adjustment (Hayashi et al. 1997) and can affect the expression of genes involved in metabolic activities, storage functions and defense directly or indirectly (Hebers and Sonnewald 1998). The oxidative damage caused by salt stress has been linked to an imbalance in ROS generation and changes in antioxidant activity (Hernández et al. 1993; Rady and Mohamed 2015b). Plants have developed many antioxidant systems to avoid the damage caused by oxidative stress; among non-enzymatic ones, the accumulation of proline is one of the most common changes induced by salinity or drought, though there is debate about whether its accumulation is a stress resistance mechanism or simply an indicator of the presence of stress (Thakur and Sharma 2005). Secondary metabolite accumulation is thought to be highly reliant on growth circumstances (Lommen et al. 2008). Polyphenols and proline have a variety of bioactivities, including protecting cells from stress, metal chelation (Sharma and Dietz 2009) and acting as ROS scavengers (Liang et al. 2013). Plants exposed to drought stress produce increased amounts of secondary metabolites, antioxidants and proline, according to a wide range of investigations. The most recent study examined the distribution of proline accumulation, the presence of polyphenolic compounds, antioxidant activity and proline in one type of *M. oleifera* plant element under varied water regimes. Results from these studies indicated that drought pressure impacts the synthesis and awareness of the osmolyte proline and polyphenolic compounds in *M. oleifera* plants. The research proved that *M. oleifera* responds to water scarcity by increasing biosynthesis and accumulating phenolic, condensed tannin, and proline contents in all plant parts, but primarily in leaves (Chitiyo et al., 2021). Therefore, *M. oleifera* represents a promising species that is able to minimize the damaging consequences of drought pressure and that might improve the soil quality in arid regions (Boumenjel et al. 2021). Other adaptive tendencies associated with secondary metabolism ought to play a determinant position within the method of plant acclimation to harsh environments, which have not been explored in *M. oleifera*. Secondary metabolites generated from the methylerythritol 4-phosphate (MEP) and phenylpropanoid pathways have been shown to play a vital role in the adaptation of 'mesic' species to low water availability (Tattini et al. 2015; Velikova et al. 2016; Zandalinas et al. 2018). Isoprene emission, for example, is more prevalent in hygrophytes than in xerophytes (Loreto et al. 2014), and isoprene is thought to help fast-growing species cope with drought stress (Loreto and Fineschi 2015; for reviews, see also: Sharkey et al. 2008 and Fini et al. 2017). Isoprene protects thylakoid membrane integrity (Velikova et al. 2011) and scavenges reactive oxygen species (ROS), particularly singlet oxygen (O_2) (Velikova et al. 2004; Vickers et al. 2009a; Zeinali et al., 2016), which can be generated at significant rates in drought-stricken leaves (Velikova et al. 2004). The advantages of isoprene biosynthesis on chloroplast membrane-related strategies can also additionally enhance the usage of radiant electricity for carbon fixation below suitable conditions (Pollastri et al. 2014; Vanzo et al. 2015), as a consequence decreasing the hazard of photo-oxidative damage (Vickers et al. 2009b).

11.6 OMICS STUDY OF *MORINGA*

The term omics refers to any branch of biology that ends in omics, such as genomics, transcriptomics, proteomics and metabolomics. The suffix '-ome' is used to refer to the study of objects in such domains as the genome, proteome, transcriptome, metabolome, etc. Genomic science is the study of the structure, function, evolution and mapping of genomes, with the goal of characterization and quantification of genes, which direct the production of proteins with the help of enzymes and messenger molecules. A cell, tissue or organism's transcriptome is the collection of all messenger RNA molecules in that cell, tissue or organism. In addition to the molecular identities, it indicates the amount or concentration of each RNA molecule. The term proteome refers to the sum of all of the proteins in a cell, tissue or organism. The study of proteins in relation to their biochemical properties and practical functions, as well as how their amounts, alteration, and systems change during growth and in response to internal and external stimuli, is known as proteomics. The collection of all metabolites in an organic cell, tissue, organ or organism is represented by the metabolome, which may be the waste product of mobile procedures. Metabolomics is the technology that uses chemical techniques to research metabolites (Boja et al. 2014; Dettmer and Hammock 2004; Vailati-Riboni et al. 2017).

11.6.1 BIOTIC STRESS AND OMICS STUDY

When an elicitor is prologued, it triggers a massive reprogramming of plant gene expression and defense responses, which could serve as a natural source for drug research and insertional mutagenesis. Using the differential display RT-PCR approach, a differential expression study of a medicinal plant *M. oleifera* for bioactive genes was performed during the seedling stage (Jabeen et al. 2014). Jabeen et al. 2014 used differential display PCR (DDRT-PCR) to isolate genes from *M. oleifera* after fungal stress by *F. solani*, which is a very helpful approach for identifying differentially expressed genes by the comparative display of arbitrarily amplified complementary DNA (cDNA) subsets (Jamil et al. 2007). The antibacterial activity of *M. oleifera* extracts mostly targeted bacterial species. From the plant that displayed an astonishing and broad-spectrum antibacterial activity, a 14.4 kDa peptide was isolated. A 14.4 kDa peptide becomes purified from the plant that exhibited an amazing and broad-spectrum antimicrobial pastime. It is possible to infect the seed with a few fungus in order to produce such antibacterial proteins and other proteins related to plant defense (Falak and Jamil 2013; Kwaik and Pederson 1996; Ragno et al. 1998). In response to fungal attack, plants activate a variety of defense mechanisms, and in order to understand the molecular aspects of these defense mechanisms, a thorough analysis of changes in the expression of these resistance genes must be conducted. For the screening of modifications in expression profiles of everyday and fungal-caused mRNAs, DDRT-PCR has been observed to be successful (Jabeen et al. 2014). The early oxidative burst that is induced by utilizing pathogens that are incompatible with the plant is crucial in the formation of resistance in plants (Mandal et al. 2014; Wojtaszek 1997). cDNA amplicons F7 and R8 correspond to bacterial-caused peroxidase genes. Peroxidases (POD) that are regarded to be activated in reaction to bacterial or pathogen assaults are a collection of heme-containing glycosylated proteins. In host–pathogen interactions (Chittoor et al. 1999) numerous roles of plant PODs have been found, such as xanthomonad resistance (Choi and Hwang 2015; Do et al. 2003; Hilaire et al. 2001; Jabeen et al. 2014).

11.6.2 ABIOTIC STRESS AND OMICS STUDY

Abiotic stresses, particularly drought stress, cause significant production losses, posing a serious threat to future food security. Understanding the molecular response of plants to abiotic stress is critical for minimizing the effects of climate change (Shyamli et al. 2021). Shyamli et al. (2021) reported whole-genome sequencing (WGS) of *M. oleifera*, assembling roughly 90% of the genome of *M. oleifera* var. Bhagya into 915 contigs with an N50 value of 4.7 Mb and predicting 32,062 potential protein-coding genes. On the SMRT platform, the genome of *M. oleifera* var. Bhagya generated 23.02 Gb of total reads and 90.9 Gb of short reads, while on the Illumina sequencing platform, the genome generated 90.9 Gb of short reads. This produced cumulative insurance of more than 300× for the reason that the said genome length of *Moringa* is 315 Mb (Tian et al. 2015). The smooth reads had been assembled into 915 contigs with MaSuRCa (Zimin et al. 2013) and yielded a genome meeting with an N50 value of 4.7 Mb, the longest contig duration of 13.8 Mb, and representing approximately 281 Mb (~90% of the general genome) of the *Moringa* genome. There are earlier reviews of WGS and the meeting of *M. oleifera*. Tian et al. (2015) claimed 33,332 contigs with an N50 value of 1.14 Mb and a complete number of 19,465 protein-coding genes, while Chang et al. (2019) reduced the wide variety of assembled contigs to 22,329, albeit with an N50 value of 0.9 Mb and anticipated 18,451 protein-coding genes. Simple series repeats (SSRs) are among the maximum giant molecular markers in plant genomes and are preferred for estimating genetic variety and molecular breeding (Pan et al. 2020). A total of 92,163 SSRs have been recognized in 594 of the 915 contigs of the assembled genome of *M. oleifera*. The most common repeats were tetranucleotide repeats, followed by dinucleotide and trinucleotide repeats. In plants, genic SSRs are particularly useful for assessing functional diversity and marker-assisted selection (MAS) (Li et al. 2021). Drought had a more pronounced effect on leaves than it did on roots in general. A homology-based search of the UniProt/Swiss-Prot database identified several differentially expressed genes as transcription factors/regulators. Zn finger-containing transcription factors, ethylene-responsive transcription factors, WRKY, bHLH and several secondary metabolite transporters and genes were among them. Many heat shock proteins (HSPs) and heat shock factors (HSFs) were also shown to be differently expressed in response to drought stress by Shyamli et al. (2021). Enrichment evaluation discovered the over-illustration of genes associated with "biosynthesis of secondary metabolites" and "reaction to stimuli." This provided Shyamli et al. (2021) with a concept of the regulatory mechanisms regulating drought pressure reaction in *M. oleifera* (Shyamli et al. 2021). Heat shock transcription elements have emerged as essential regulators of the reaction to abiotic pressure in plants, and even though *Moringa* is specifically valued for its medicinal properties, the fact that it could resist drought situations makes this plant an amazing supply of genomic sources for plant improvement. Researchers have found 21 HSFs in the *M. oleifera* genome (MolHSF1–MolHSF21), with lengths ranging from 110 to 1,530 amino acids. NLS, NES or both were projected to be present in all MolHSFs (Shyamli et al. 2021). In the promoter regions of MolHSF genes, Shyamli et al. (2021) discovered a number of patterns related to the abiotic stress response in plants. The existence of such motifs suggests that MolHSFs have a function in the plant's abiotic stress response. A qPCR study revealed that the HSF genes MolHSF-2 and MolHSF-19 were significantly upregulated in ODC3 leaf tissues in response to drought stress, with a fivefold improvement (Shyamli et al. 2021). The position of WRKY and zinc finger-containing transcription elements in abiotic strain reactions has been thoroughly documented in numerous flowers (Khan et al. 2018; Yoon et al. 2020). Both of those are classically huge TF households comprising many members, which makes their position in plant improvement variously and now no longer restricted to strain reaction (Li et al. 2020; Lyu et al. 2020; Wani et al. 2021). More recently, the point of interest has shifted to smaller households of transcription elements that have an extra direct position in abiotic strain reactions in flowers (Shyamli et al. 2021). Heat shock transcription elements in flowers adjust the expression of HSPs and thereby mediate the flowers' reaction to abiotic strain (Wang et al. 2017). Despite what their name suggests, further to the heat shock reaction, HSFs also are recognized to adjust the reaction to different stresses including cold, salinity and drought (Li et al., 2020). They are a small family of TFs, with 22 and 25 members in *Arabidopsis* and rice, respectively (Guo et al. 2008), 38 HSFs in soybean (Li et al. 2014), 6 and 17 HSFs in two species of wild peanuts, respectively (Wang et al. 2017), 33 HSFs in radish (Tang et al., 2019), and 32 HSFs in lettuce. Researchers found 21 putative HSFs in the *M. oleifera* genome and classified them into three groups based on the structure of the HR-A/B domain, which is responsible for HSFs' protein interaction activity (Scharf et al. 2012). MiRNAs are well known for controlling a variety of plant functions in response to biotic and abiotic stressors. Many studies have shown that abiotic stress can cause plants to produce a large number of miRNAs (Gupta et al. 2014; Shriram et al. 2016). MicroRNAs (miRNAs) are non-coding RNAs that regulate gene expression in eukaryotes and are tiny (21–24 nucleotides). MiRNAs are transcribed as primordial miRNAs (pri-miRNAs) and as precursor RNAs by RNA polymerase-II inside the nucleus (Bartel 2009). A vast number of stress-related miRNAs have been identified using high-throughput sequencing and computational techniques in recent years. High-throughput sequencing analysis indicated cold stress-induced overexpression of 31 miRNAs in tea [*Camellia sinesis* (L.) Kuntze] plants in a study by Zhang et al. (2014). Similarly, Cao et al. 2014 studied the chilling strain reaction in wild tomato (*Solanum habrochaites*) and discovered that 192 miRNAs confirmed elevated expression. Maximum miRNAs are involved in callus' amazing heterochronic gene law of improvement, especially under strain settings. In a work conducted by Pirrò et al. (2019), high-throughput sequencing technology was used to identify 431 conserved and 392 novel microRNA families, as well as nine novel short RNA libraries, from leaf and bloodless strain-handled callus.

While microRNA393 was found to be most abundantly expressed in seeds, microRNA159 was found to be the most abundant conserved microRNA in leaves and calluses. Most anticipated microRNA target genes were transcriptional components involved in plant growth, reproduction and abiotic and biotic strain reactions (Pirrò et al. 2019).

11.7 THE POSITIVE ROLE OF EXOGENOUS APPLICATION OF DIFFERENT TYPES OF STIMULANTS

11.7.1 ROLE OF SALICYLIC ACID FOR PROTECTION IN BIOTIC AND ABIOTIC CONDITIONS

Many studies have shown that abiotic stress can cause plants to produce a large number of miRNAs (Gupta et al. 2014). Regardless of salinity treatments, all growth characteristics improved with Si+ SA addition, followed by a single silicate application, and finally the control. Baker et al. (2014) and Shamsul and Aqil (2007) indicated that SA is a phenol that is ubiquitous in plants and has a significant impact on plant growth and development, photosynthesis, transpiration, ion uptake and transport, and also induces specific changes in leaf anatomy and chloroplast structure. Salicylic acid (SA) and *Moringa oleifera* leaf extract (MLE) improved the growth characteristics of salt-stressed common bean plants (i.e. shoot length, number and area of leaves per plant, and plant dry weight) (Rady and Mohamed, 2015a). Plant growth and development are sometimes aided by foliar applications of growth regulators, nutrients and antioxidants, but these are expensive procedures for commercial farmers (Fuglie 1999; Foidl et al. 2001; Hussain et al. 2013). Exogenous SA treatments combined with MLE significantly improved the negative effects of salt stress on plant characteristics. This is because SA promotes growth in CO_2 assimilation and photosynthetic rate and also promotes growth in mineral uptake by employing salt-damaged plants (Fariduddin et al. 2003; Szepesi et al., 2005; Fariduddin et al. 2018; Yusuf et al. 2012, 2013). In addition, the growth in increased parameters of salt-affected vegetation in reaction to SA is probably attributed to the protecting position of SA on membranes that could increase the tolerance of vegetation to salt pressure (Aftab et al. 2010).

11.7.2 ROLE OF *MORINGA* LEAF EXTRACT FOR PROTECTION IN BIOTIC AND ABIOTIC CONDITIONS

MLE, which is high in zeatin, ascorbates, carotenoids, phenols, antioxidants and vital plant nutrients, can influence plant growth and is frequently used as an exogenous plant growth enhancer (Fuglie 1999; Foidl et al. 2001; Hussain et al. 2013). Canola (*Brassica napus*) yield was improved by 35% over the control after three foliar applications of 2% moringa + 2% brassica water extract (Iqbal 2020; Hussain et al. 2013). The MLE spray used only at heading gave 6.84%, 3.17%, 6.80% and 3.51% more than the control 1000 grain weight, biological yield, grain yield, and harvest index, respectively (Figure 11.1) (Yasmeen et al. 2012). By foliar spraying of MLE in late-seeded wheat, Yasmeen et al. (2012) found significant improvements in grain weight, biological yield, grain yield and harvest index. In another study, Yasmeen ret al. (2012) found that MLE spraying improved tomato growth and yield significantly. MLE promotes young plant growth, generates solid stems, improves tolerance to biotic and abiotic challenges, extends lifetime, increases the weight of roots, stems and leaves, as well as providing more and larger fruits, and often increases the yield by 20–35% (Fuglie 2001; Foidl et al. 2001). Similarly, Younis et al. (2018) found that using *Moringa* increased crop growth and yield due to the presence of zeatin, a cytokinin-like plant growth hormone. Foliar utility of MLE improved the productiveness of numerous arable plants including soybean

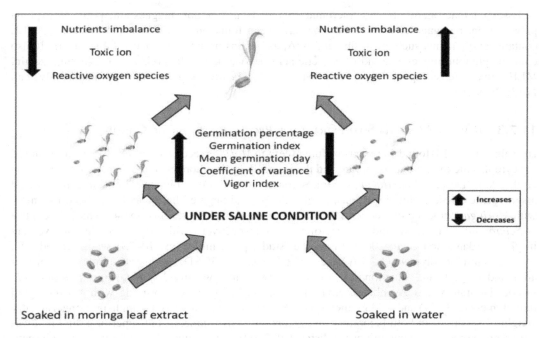

FIGURE 11.1 Hydropriming effect of MLE (*Moringa* leaf extract) and water on wheat seeds in a saline environment (Ahmed et al. 2021).

(*Glycine max* L.), sugarcane (*Saccharum officinarum*), corn, sorghum, black bean (*P. vulgaris*), coffee (*Coffee arabica* L.), bell pepper (*Capsicum annuum* L.), and onion (*Allium cepa* L.) starting from 6.57–47.88% recently, however exclusive cultivars behave differently. One cultivar of black bean (Dor163) exhibited a 16.04% increase, whilst a different cultivar (Steli 150) exhibited a 5.34% decrease in grain yield (Foidl et al. 2001). *Moringa* has also been shown to boost crop growth and yield, according to Younis et al. (2018). MLE, which is high in growth regulators, vital plant elements and antioxidants, can boost the development and productivity of a variety of arable crops when employed in priming media or sprayed on the foliage exogenously (Hussain et al. 2013). The integration of *Moringa* leaves into the soil before sowing is efficient in controlling damping-off disease, according to Yisehak et al. (2011). Some soil-borne fungi in cereals, such as *Rhizoctonia*, *Pythium* and *Fusarium*, were resistant to seed treatment with MLE (Stoll 1988). MLE works on a systemic level to protect seedlings from pathogens. In an *invitro* study, MLE suppressed the growth of several fungal pathogens and reduced the prevalence of *Pythium* damping-off in legumes, vegetables and cereals (Stoll 1988; Adandonon et al. 2003; Hussain et al. 2013). In living systems, the aqueous and ethanolic extracts of *Moringa* leaf increase glutathione (GSH) synthesis and reduce malondialdehyde (MDA) levels, while also having high free radical scavenging abilities in a concentration-dependent manner (Luqman et al. 2012). The mycelium growth of *S. rolfsii* in cowpea (*Vigna unguiculata*) was decreased by seed treatment with concentrated MLE (15 or 20 g leaves 10 mL^{-1} water in the lab and 15 kg leaves 10 L^{-1} water in the field) (Adandonon et al. 2006). To investigate the allelopathic potential of *Moringa* for weed control in field conditions, it is required to optimize MLE lethal doses to a range of weeds in laboratory conditions (Hussain et al. 2013). Under the adverse conditions of the studied soil salinity, common bean seed and plant treatments with salicylic acid (SA; 1 mM) and *Moringa oleifera* leaf extract (MLE; 1 extract: 30 glasses of water by volume), used as seed soaking and foliar spray, respectively, significantly improved plant growth characteristics, plant productivity and anatomy, as well as osmoprotectants and ascorbic acid (AsA).

Essential macronutrients and micronutrients such as calcium, magnesium, potassium, phosphorus, iron, manganese, copper and zinc have been found in MLE. As osmoprotectants, MLE contains antioxidants such as proline and AsA, as well as proline and total soluble sugars. It also contains phytohormones like indole-3-acetic acid (IAA), gibberellins (GAs) and cytokinin zeatin. MLE's diversified composition suggests that they could be used as a plant bio-stimulant (Rady et al. 2013; Rady 2011).

11.7.3 ROLE OF *MORINGA* SEED EXTRACT FOR PROTECTION IN BIOTIC CONDITIONS

Hussain et al. (2013) found that exposing fungal cells to *Moringa* seed crude extract for 24 hours caused the cytoplasmic membrane to rupture and the intracellular components to be damaged (Adandonon et al. 2006). Despite this, *Moringa* extracts combined with other biocontrol agents demonstrated a synergistic effect against infections. In greenhouse and targeted field conditions, seed treatment with *Moringa* extract combined with *Trichoderma* soil drenching resulted in 94% and 70% control of *Sclerotium* damping-off and stem rot of cowpea, respectively, with a considerable improvement in yield (Adandonon et al. 2006). In another study, pearl millet (cv. 7042S) seeds primed with *Moringa* gum biopolymers (1:3 w/v) and 3 g kg⁻¹ metalaxyl 35 SD demonstrated improved germination and a significant reduction in pearl millet downy mildew caused by *Sclerospora graminicola* in both laboratory and greenhouse environments. Furthermore, seed priming with *Moringa* gum biopolymers (1:3 w/v) and 3 g kg⁻¹ metalaxyl 35 SD resulted in increased plant height, fresh and dry weight of seedlings in both laboratory and greenhouse environments, but superior grain yield was only seen in greenhouse conditions (Sudisha et al. 2009). Some studies have highlighted the insecticidal properties of lectins extracted from *Moringa* seed powder, particularly against stored grain pests. At a concentration of 1%, coagulant lectins of *Moringa* (cMoL) increased the pupal mortality rate of the moth flour (*Anagasta kuehniella*) by 27.6%. After 12 hours of incubation, the moth flour midgut proteases were unable to digest cMoL. The insecticidal activity of cMoL is thought to be due to carbohydrate lectin interactions on the surface of the digestive system in the midgut, as well as resistance to enzymatic digestion (De Oliveira et al. 2011; Santos et al. 2012). WSMoL (water-soluble *M. oleifera* lectin) shows dengue virus vector *Aedes aegypti* larvicidal action (Coelho et al. 2009; Santos et al. 2012). Similarly, *Moringa* seed powder integrated at 12.5% and 25.5% (w/w) into 20 g of wheat grains increased the mortality of maize weevil (*Sitophilus zeamais*) adults after 24, 48, 72 and 96 hours of incubation as compared to untreated seeds (Ileke and Oni 2011). Similarly, incorporating *Moringa* powder into wheat grains inhibited the appearance of maize weevil adults and showed no signs of grain weight loss or seed viability even after 3 months of seed treatment, implying that *Moringa*-treated seeds are acceptable for consumption and as planting material (Ileke and Oni 2011). As a result, *Moringa* seed powder can be used as an alternative to pesticides to combat maize weevil (Hussain et al. 2013).

11.8 CONCLUSIONS

The miracle tree, *Moringa oleifera*, offers a wide range of applications in crop science, ranging from improved stand establishment and growth promotion to resistance to biotic and abiotic challenges. *Moringa* leaf and seed extracts can thus be utilized for seed treatment and foliar application. *Moringa* cultivation should consequently be encouraged and new uses and applications explored. In addition, a substantial amount of studies regarding understanding many parts of the *Moringa* genome in respect to biotic stressors must be explored.

In addition, studies on the association between phenotypic and molecular data and genetic maps (both related and physical maps) are needed to identify genes in reproductive competition. Next-generation sequencing (NGS) has the potential to become an accessible tool for discovering genome-wide genetic markers and generating saturated gene maps at a reasonable cost and time.

REFERENCES

Abay, A., Birhane, E., Taddesse, T., & Hadgu, K. M. (2016). *Moringa stenopetala* Tree Species Improved Selected Soil Properties and Socio-economic Benefits in Tigray, Northern Ethiopia. *Science, Technology and Arts Research Journal, 4*(2), 68. https://doi.org/10.4314/star.v4i2.10

Abdul Qados, A.M.S. (2015). Effects of salicylic acid on growth, yield and chemical contents of pepper (*Capsicum annuum* L) plants grown under salt stress conditions. *International Journal of Agriculture and Crop Science, 8*(2), 107–113.

Abu-Taleb, A.M, E.-D. K. and O. A.-O. F. (2011). Assessment of antifungal activity of *Rumex vesicarius* (L.) and Ziziphus spina-christi (L.) wild extracts against two phytopathogenic fungi. *African Journal of Microbiology Research, 5*(9), 1001–1011.

Adandonon, A., Aveling, T.A.S., Labuschagne, N., & Ahohuendo, B.C. (2003). Epidemiology and biological control of the causal agent of damping-off and stem rot of cowpea in the Oueme' valley, Benin. *Annales des Sciences Agonomiques du Benin, 6,* 21–36.

Adandonon, A., Aveling, T.A.S., Labuschagne, N., & Tamo, M. (2006). Biocontrol agents in combination with *Moringa oleifera* extract for integrated control of Sclerotium-caused cowpea damping-off and stem rot. *Eur J Plant Pathol., 115*(4), 409–418.

Aftab, T., Khan, M.M.A., Idrees, M., Naeem, M., & Moinuddin. (2010). Salicylic acid acts as potent enhancer of growth, photosynthesis and artemisinin production in *Artemisia annua* L. *Journal of Crop Science and Biotechnology, 13,* 183–188.

Ahmed, T., Elezz, A. A., & Khalid, M. F. (2021). Hydropriming with moringa leaf extract mitigates salt stress in wheat seedlings. *Agriculture (Switzerland), 11*(12). https://doi.org/10.3390/agriculture11121254

Anosheh, H. P., Emam, Y., Ashraf, M., & Foolad, M. R. (2012). Exogenous Application of Salicylic Acid and Chlormequat Chloride Alleviates Negative Effects of Drought Stress in Wheat. In *Advanced Studies in Biology* (Vol. 4, Issue 11).

Anwar, F., Ashraf, M., & Bhanger, M.I. (2005) Interprovenance variation in the composition of *Moringa oleifera* oilseeds from Pakistan. *J Am Oil Chem Soc., 82,* 45–51.

Anwar, F., Latif, S., Ashraf, M., & Gilani, A. H. (2007). *Moringa oleifera*: A Food Plant with Multiple Medicinal Uses. *Phytother. Res, 21,* 17–25. https://doi.org/10.1002/ptr

Apel, K., & Hirt, H. (2004) Reactive oxygen species: metabolism, oxidative stress, and signal transduction. *Annu Rev Plant Biol., 55,* 373–399.

Ashraf, M., Akram, N. A., Arteca, R. N., & Foolad, M. R. (2010). The Physiological, Biochemical and Molecular Roles of Brassinosteroids and Salicylic Acid in Plant Processes and Salt Tolerance. *Critical Reviews in Plant Sciences, 29*(3), 162–190. https://doi.org/10.1080/07352689.2010.483580

Baker, N. A., Hussein, M. M., & Abou-Baker, N. H. (2014). *Article no. IJPSS.2014.004 SCIENCEDOMAIN international Original Research Article Hussein and Abou-Baker; IJPSS, Article.* www.sciencedomain.org

Barta, C., & Loreto, F. (2006). The relationship between the methyl-erythritol phosphate pathway leads to the emission of volatile isoprenoids and abscisic acid content in leaves. *Plant Physiology, 141*(4), 1676–1683.

Bartel, D. P. (2009). MicroRNAs: target recognition and regulatory functions. *Cell, 136*(2), 215–233.

Beckett, M., Loreto, F., Velikova, V., Brunetti, C., Di Ferdinando, M., Tattini, M., ... & Farrant, J. M. (2012). Photosynthetic limitations and volatile and non-volatile isoprenoids in the poikilochlorophyllous resurrection plant Xerophyta humilis during dehydration and rehydration. *Plant, Cell & Environment, 35*(12), 2061–2074.

Boja, E. S., Kinsinger, C. R., Rodriguez, H., & Srinivas, P. (2014). Integration of omics sciences to advance biology and medicine.

Boumenjel, A., Papadopoulos, A. Ammari, Y. (2021). Growth response of *Moringa oleifera* (Lam) to water stress and to arid bioclimatic conditions. *Agroforestry Systems, 95*(5), 823–833.

Brodribb T.J. (2013). Abscisic acid mediates a divergence in the drought response of two conifers. *Plant Physiol, 162,* 1370–1377.

Brodribb, T. J., & McAdam, S. A. (2013). Abscisic acid mediates a divergence in the drought response of two conifers. *Plant Physiology, 162*(3), 1370–1377.

Brunetti, C., Guidi, L., Sebastiani, F., & Tattini, M. (2015). Isoprenoids and phenylpropanoids are key components of the antioxidant defense system of plants facing severe excess light stress. *Environmental and Experimental Botany, 119,* 54–62.

Brunetti, C., Loreto, F., Ferrini, F., Gori, A., Guidi, L., Remorini, D., Centritto, M., Fini, A., & Tattini, M. (2018). Metabolic plasticity in the hygrophyte *Moringa oleifera* exposed to water stress. *Tree Physiology*, *38*(11), 1640–1654. https://doi.org/10.1093/treephys/tpy089

Cai Z.Y., Yang Y., Liu Y.X., Zhang Y.M., L. J. M. & L. H. Q. (2016). First Report of Colletotrichum chlorophyte causing *Moringa oleifera* Anthracnose in China. *Plant Disease*, *100*(10), 2164–2164.

Cao, X., Wu, Z., Jiang, F., Zhou, R., & Yang, Z. (2014). Identification of chilling stress-responsive tomato microRNAs and their target genes by high-throughput sequencing and degradome analysis. *BMC genomics*, *15*(1), 1–16.

Carbungco, E.S., Pedroche, N.B., Panes, V.A., & De la Cruz, T. E. (2017). Identification and characterization of endophytic fungi associated with the leaves of Moringa oleifera Lam. *Acta Hortic*, *1158*, 373–380. 10.17660/ActaHortic.2017.1158.42

Chang, Y., Liu, H., Liu, M., Liao, X., Sahu, S. K., Fu, Y., Song, B., Cheng, S., Kariba, R., & Muthemba, S. (2019). The draft genomes of five agriculturally important African orphan crops. *GigaScience*, *8*(3), giy152.

Chitiyo, S. T., Ncube, B., Ndhlala, A. R., & Tsvuura, Z. (2021). Biochemical responses of *Moringa oleifera* Lam. plants to graded moisture deficit. *South African Journal of Botany*, *138*, 41–49.

Chittoor, J. M., Leach, J. E., & White, F. F. (1999). Induction of peroxidase during defense against pathogens. *Pathogenesis-Related Proteins in Plants*, 171–193.

Choi, H. W., & Hwang, B. K. (2015). Molecular and cellular control of cell death and defense signalling in pepper. *Planta*, *241*(1), 1–27.

Chuang, P.-H., Lee, C.-W., Chou, J.-Y., Murugan, M., Shieh, B.-J., & Chen, H.-M. (2007) Anti-fungal activity of crude extracts and essential oil of *Moringa oleifera* Lam. *Bioresour Technol.*, *98*, 232–236.

Clark, L. J. G., Gowing, D. J., Lark, R. M., Leeds-Harrison, P. B., Miller, A. J., Wells, D. M., et al. (2005). Sensing the physical and nutritional status of the root environment in the field: a review of progress and opportunities. *J. Agric. Sci.* 143, 347–358. doi: 10.1017/S0021859605005253Coelho, J.S., Santos, N.D.L., Napoleão, T.H., Gomes, F.S., Ferreira, R.S., Zingali, R.B., et al. (2009). Effect of *Moringa oleifera* lectin on development and mortality of *Aedes aegypti* larvae. *Chemosphere 77*, 934–938.

Coupel-Ledru, A., Tyerman, S. D., Masclef, D., Lebon, E., Christophe, A., Edwards, E. J., & Simonneau, T. (2017). Abscisic acid down-regulates hydraulic conductance of grapevine leaves in isohydric genotypes only. *Plant Physiology*, *175*(3), 1121–1134.

Cutler, S. R., Rodriguez, P. L., Finkelstein, R. R., & Abrams, S. R. (2010). Abscisic acid: the emergence of a core signaling network. *Annu. Rev. Plant Biol.* 61, 651–679. doi: 10.1146/annurev-arplant-042809-112122.

Dall'Osto L, Cazzaniga S, & Havaux M, B. R. (2010). Enhanced photo_protection by protein-bound vs free xanthophyll pools: a comparative analysis of chlorophyll b and xanthophyll biosynthesis mutants. *Mol Plant*, *3*, 576–593.

Dall'Osto, L., Cazzaniga, S., Havaux, M., & Bassi, R. (2010). Enhanced photoprotection by protein-bound vs free xanthophyll pools: a comparative analysis of chlorophyll b and xanthophyll biosynthesis mutants. *Molecular plant*, *3*(3), 576–593.

Davison, P. A., Hunter, C. N., & Horton, P. (2002). Overexpression of β-carotene hydroxylase enhances stress tolerance in Arabidopsis. *Nature*, *418*(6894), 203–206.

Delauney, A. J., & Verma, D. P. S. (1993). Proline biosynthesis and osmoregulation in plants. *The Plant Journal*, *4*(2), 215–223.Demoz, B.T., & Korsten, L. (2006) *Bacillus subtilis* attachment, colonization, and survival on avocado Xowers and its mode of action on stem-end rot pathogens. *Biol Control 37*, 68–74.

De Oliveira, C.F.R., Luz, L.A., Paiva, P.M.G., Coelho, L.C.B.B., Marangoni, S., & MacEdo, M.L.R. (2011). Evaluation of seed coagulant *Moringa oleifera* lectin (cMoL) as a bioinsecticidal tool with potential for the control of insects. *Process Biochem* [Internet]. *46*(2), 498–504. Available from: http://dx.doi.org/10.1016/j.procbio.2010.09.025

Dettmer, K., & Hammock, B. D. (2004). Metabolomics--a new exciting field within the" omics" sciences. *Environmental health perspectives*, *112*(7), A396–A397

Do, H. M., Hong, J. K., Jung, H. W., Kim, S. H., Ham, J. H., & Hwang, B. K. (2003). Expression of peroxidase-like genes, H2O2 production, and peroxidase activity during the hypersensitive response to Xanthomonas campestris pv. vesicatoria in Capsicum annuum. *Molecular Plant-Microbe Interactions*, *16*(3), 196–205.

Du, H., Wang, N., Cui, F., Li, X., Xiao, J., & Xiong, L. (2010). Characterization of the β-carotene hydroxylase gene DSM2 conferring drought and oxidative stress resistance by increasing xanthophylls and abscisic acid synthesis in rice. *Plant physiology*, *154*(3), 1304–1318.

Edreva, A., Velikova, V., Tsonev, T., Dagnon, S., Gürel, A., Aktaş, L., & Gesheva, E. (2008). Stress-protective role of secondary metabolites: diversity of functions and mechanisms. *Gen Appl Plant Physiol*, *34*(1–2), 67–78.

Esteban, R., Moran, J. F., Becerril, J. M., & García-Plazaola, J. I. (2015). Versatility of carotenoids: An integrated view on diversity, evolution, functional roles and environmental interactions. *Environmental and Experimental Botany*, *119*, 63–75.

Fahey, J. W. (2005). Moringa oleifera: A review of the medical evidence for its nutritional, therapeutic, and prophylactic properties. *Trees Life J.*, *1*, 1–15.

Falak, S., & Jamil, A. (2013). Expression profiling of bioactive genes from a medicinal plant Nigella sativa L. *Applied Biochemistry and Biotechnology*, *170*(6), 1472–1481.Fariduddin, Q., Hayat, S., and Ahmad, A. (2003). Salicylic acid influences net photosynthetic rate, carboxylation efficiency, nitrate reductase activity, and seed yield in *Brassica juncea*. *Photosynthetica*, *41*, 281–284.

Fariduddin, Q., Khan, T. A., Yusuf, M., Aafaqee, S. T., & Khalil, R. (2018). Ameliorative role of salicylic acid and spermidine in the presence of excess salt in Lycopersicon esculentum. *Photosynthetica*, *56*(3), 750–762.

Farooq, M., Wahid, A., Kobayashi, N., Fujita, D., & Basra, S.M.A. (2009) Plant drought stress: effects, mechanisms and management. *Agron Sustain Dev 29*, 185–212.

Farooq, M., Jabran, K., Cheema, Z.A., Wahid, A., & Siddique, K.H.M. (2011) The role of allelopathy in agricultural pest management. *Pest Manag. Sci.,* *67*, 493–506.

Fini, A., Brunetti, C., Loreto, F., Centritto, M., Ferrini, F., & Tattini, M. (2017). Isoprene responses and functions in plants challenged by environmental pressures associated to climate change. *Frontiers in plant science*, *8*, 1281.

Foidl, N., Makkar, H.P.S. and Becker, K. (2001) The potential of *Moringa oleifera* for agricultural and industrial uses. In: Fuglie, L.J., Ed., The Miracle Tree: The Multiple Attributes of Moringa, CTA, Wageningen, CWS, Dakar, 45–76, 177.

Fuglie, L.J. (1999). The Miracle Tree: Moringa oleifera: Natural Nutrition for the Tropics (Dakar: Church World Service), pp. 68; revised in 2001 and published as The Miracle Tree: The Multiple Attributes of Moringa, pp. 172. http://www.echotech.org/bookstore/advanced_search_result.php?keywords=Miracle+Tree.

Ganguly, S. (2013). Indian ayurvedic and traditional medicinal implications of indigenously available plants, herbs and fruits: A review. *Int. J. Res. Ayurveda Pharm*, *4*, 623–625.

Gepstein, S., & Glick, B. R. (2013). Strategies to ameliorate abiotic stress-induced plant senescence. *Plant Molecular Biology*, *82*(6), 623–633.

Gill, S.S., & Tuteja, N. (2010) Reactive oxygen species and antioxidant machinery in abiotic stress tolerance in crop plants. *Plant Physiol. Biochem.*, *48*, 909–930.

Guizh J., (2008). Fungal diseases of drumstick trees. *Journal of Plant Protection*, *34*(4), 121.

Gull, A., Lone, A. A., & Wani, N. U. I. (2019). Biotic and abiotic stresses in plants. *Abiotic and biotic stress in plants*, 1–19.

Guo, J., Wu, J., Ji, Q., Wang, C., Luo, L., Yuan, Y., Wang, Y., & Wang, J. (2008). Genome-wide analysis of heat shock transcription factor families in rice and Arabidopsis. *Journal of Genetics and Genomics*, *35*(2), 105–118.

Gupta, B. M., & Ahmed, K. K. (2020). *Moringa Oleifera*: A Bibliometric Analysis of International Publications during 1935–2019. *Pharmacognosy Reviews*, *14*(28), 83.

Gupta, O. P., Meena, N. L., Sharma, I., & Sharma, P. (2014). Differential regulation of microRNAs in response to osmotic, salt and cold stresses in wheat. *Molecular Biology Reports*, *41*(7), 4623–4629.

Hamdia, M.A., & Shaddad, M.A.K. (2010) Salt tolerance of crop plants. *J Stress Physiol Biochem.*, *6*, 64–90.

Harris, J. (2015). Abscisic acid: hidden architect of root system structure. *Plants* 4, 548–572. doi: 10.3390/plants4030548

Havaux, M., Dall'Osto, L., & Bassi, R. (2007). Zeaxanthin has enhanced antioxidant capacity with respect to all other xanthophylls in Arabidopsis leaves and functions independent of binding to PSII antennae. *Plant physiology*, *145*(4), 1506–1520.

Hayashi, H., Mustardy, A.L., Deshnium, P., Ida, M., & Murata, N. (1997) Transformation of *Arabidopsis thaliana* with the codA gene for choline oxidase; accumulation of glycinebetaine and enhanced tolerance to salt and cold stress. *The Plant Journal*, *12*, 133–142.

Hayat, Q., Hayat, S., Irfan, M., & Ahmad, A. (2010). Effect of exogenous salicylic acid under changing environment: A review. In *Environmental and Experimental Botany* (Vol. 68, Issue 1, pp. 14–25). https://doi.org/10.1016/j.envexpbot.2009.08.005

Hebers, K., and Sonnewald, V. (1998). Altered gene expression: brought about by inter and pathogen interactions. *Journal of Plant Research*, *111*, 323–328.

Hernández, J.A., Francisco, F.J., Corpas, G.M., Gómez, L.A., & Del Río, F.S. (1993). Salt induced oxidative stresses mediated by activated oxygen species in pea leaves mitochondria. *Plant Physiology*, *89*, 103–110.

Hilaire, E., Young, S. A., Willard, L. H., McGee, J. D., Sweat, T., Chittoor, J. M., Guikema, J. A., & Leach, J. E. (2001). Vascular defense responses in rice: peroxidase accumulation in xylem parenchyma cells and xylem wall thickening. *Molecular Plant-Microbe Interactions*, *14*(12), 1411–1419.

Horváth, E., Szalai, G., & Janda, T. (2007) Induction of abiotic stress tolerance by salicylic acid signaling. *J Plant Growth Regul.*, *26*(3), 290–300.

Hussain, M., Farooq, M., Basra, S. M. A., & Lee, D. J. (2013). Application of moringa allelopathy in crop sciences. In *Allelopathy: Current Trends and Future Applications* (pp. 469–483). Springer Berlin Heidelberg. https://doi.org/10.1007/978-3-642-30595-5_20

Ileke, K.D., & Oni, M.O. (2011). Toxicity of some plant powders to maize weevil, *Sitophilus zeamais* (motschulsky) [Coleoptera: Curculiondae] on stored wheat grains (*Triticum aestivum*). *Afr J Biotechnol.*, *6*, 3043–3048.

Iqbal, et al. (2020) Comparative study of water extracts of Moringa leaves and roots to improve the growth and yield of sunflower. *South Afr. J. Bot. 129*, 221–224.

Jabeen, R., Mustafa, G., UlAbdin, Z., Iqbal, M. J., & Jamil, A. (2014). Expression profiling of bioactive genes from *Moringa oleifera*. *Applied Biochemistry and Biotechnology*, *174*(2), 657–666. https://doi.org/10.1007/s12010-014-1122-9

Jabeen, R., Shahid, M., Jamil, A., & Ashraf, M. (2008). Microscopic evaluation of the antimicrobial activity of seed extracts of *Moringa oleifera*. *Pak J Bot*, *40*(4), 1349–1358.

Jamil, A., Shahid, M., Khan, M. M., & Ashraf, M. (2007). Screening of some medicinal plants for isolation of antifungal proteins and peptides. *Pakistan Journal of Botany (Pakistan)*.

Johnson, M.P., Havaux, M., Triantaphylides, C., Ksas, B., Pascal, A.A., Davison, P.A., and Ruban A.V., (2007). Elevated zeaxanthin bound to oligomeric LHCII enhances the resistance of Arabidopsis to photo_oxidative stress by a lipid-protective, antioxidant mechanism. *J Biol Chem, 282*, 22605–22618.

Johnson, M. P., Havaux, M., Triantaphylides, C., Ksas, B., Pascal, A. A., Robert, B., ... & Horton, P. (2007). Elevated zeaxanthin bound to oligomeric LHCII enhances the resistance of Arabidopsis to photooxidative stress by a lipid-protective, antioxidant mechanism. *Journal of Biological Chemistry*, *282*(31), 22605–22618.

Kant, R., Joshi, R.C., & Faleono, I. (2017). Survey of insect pests on Moringa oleifera Lam. in Samoa. *Acta Hortic, 1158*, 195–200. 10.17660/ActaHortic.2017.1158.23

Kaydan, D., and Okut, M.Y. (2007). Effects of salicylic acid on the growth and some physiological characters in salt stressed wheat (*Triticum aestivum* L.). *Tarim Bİlimleri Dergisi*, *13*(2), 114–119.

Khan, S.-A., Li, M.-Z., Wang, S.-M., & Yin, H.-J. (2018). Revisiting the role of plant transcription factors in the battle against abiotic stress. *International Journal of Molecular Sciences*, *19*(6), 1634.

Kim, T. H., Böhmer, M., Hu, H., Nishimura, N., & Schroeder, J. I. (2010). Guard cell signal transduction network: advances in understanding abscisic acid, CO_2, and Ca^{2+} signaling. *Annu. Rev. Plant Biol.* 61, 561–591. doi: 10.1146/annurev-arplant-042809-112226

Kshirsagar C.R. & D'Souza T.F. (1989). A new Disease of Drumstick. *Journal of Maharashta Agricultural University*, *14*(2), 241–242.

Kumar S., Singh R., & Saini, D.C. (2013). *Moringa oleifera, A new host record of Cercospora apii s. lat. from Uttar Pradesh, India.* (Vol. 3, Issue 1).

Kwaik, Y. A., & Pederson, L. L. (1996). The use of differential display-PCR to isolate and characterize a Legionella pneumophila locus induced during the intracellular infection of macrophages. *Molecular Microbiology*, *21*(3), 543–556.

Lako, J., Trenerry, V.C., Wahlqvist, M., Wattanapenpaiboon, N., Sotheeswaran, S., & Premier, R. (2007) Phytochemical flavonols, carotenoids and the antioxidant properties of a wide selection of Fijian fruit, vegetables and other readily available foods. *Food Chem.*, *101*, 1727–1741.

Lalas, S., & Tsaknis, J. (2002). Extraction and identification of natural antioxidant from the seeds of the *Moringa oleifera* tree variety of Malawi. *Journal of the American Oil Chemists' Society*, 79(7), 677–683.

Lalas, S., & Tsaknis, J. (2002). Extraction and identification of natural antioxidant from the seeds of the *Moringa oleifera* tree variety of Malawi. *Journal of the American Oil Chemists' Society*, 79(7), 677–683. https://doi.org/10.1007/s11746-002-0542-2

Li, M., Xie, F., Li, Y., Gong, L., Luo, Y., Zhang, Y., Chen, Q., Wang, Y., Lin, Y., & Zhang, Y. (2020). Genome-wide analysis of the heat shock transcription factor gene family in Brassica juncea: structure, evolution, and expression profiles. *DNA and Cell Biology*, 39(11), 1990–2004.

Li, P.-S., Yu, T.-F., He, G.-H., Chen, M., Zhou, Y.-B., Chai, S.-C., Xu, Z.-S., & Ma, Y.-Z. (2014). Genome-wide analysis of the Hsf family in soybean and functional identification of GmHsf-34 involvement in drought and heat stresses. *BMC Genomics*, 15(1), 1–16.

Li, Q., Su, X., Ma, H., Du, K., Yang, M., Chen, B., Fu, S., Fu, T., Xiang, C., & Zhao, Q. (2021). Development of genic SSR marker resources from RNA-seq data in Camellia japonica and their application in the genus Camellia. *Scientific Reports*, 11(1), 1–11.

Li, Y., Liu, Z., Zhang, K., Chen, S., Liu, M., & Zhang, Q. (2020). Genome-wide analysis and comparison of the DNA-binding one zinc finger gene family in diploid and tetraploid cotton (Gossypium). *Plos One*, 15(6), e0235317.

Liang, X., Zhang, L., Natarajan, S. K., & Becker, D. F. (2013). Proline mechanisms of stress survival. *Antioxidants & Redox Signaling*, 19(9), 998–1011.

Lommen, W. J. M., Bouwmeester, H. J., Schenk, E., Verstappen, F. W. A., Elzinga, S., & Struik, P. C. (2008). Modelling processes determining and limiting the production of secondary metabolites during crop growth: the example of the antimalarial artemisinin produced in Artemisia annua. *Acta Horticulturae*, 765, 87–94.

Loreto F, F. S. (2015). Reconciling functions and evolution of iso_prene emission in higher plants. *New Phytol*, 206, 578–582.

Loreto, F., & Fineschi, S. (2015). Reconciling functions and evolution of isoprene emission in higher plants. *New Phytologist*, 206(2), 578–582.

Loreto, F., Dicke, M., Schnitzler, J. P., & Turlings, T. C. (2014). Plant volatiles and the environment. *Plant, cell & environment*, 37(8), 1905–1908.

Luqman, S., Srivastava, S., Kumar, R., Maurya, A.K., & Chanda, D. (2012) Experimental assessment of *Moringa oleifera* leaf and fruit for its antistress, antioxidant, and scavenging potential using in vitro and in vivo assays. *Evid Based Complement Altern Med*. doi:10.1155/2012/519084

Lyu, T., Liu, W., Hu, Z., Xiang, X., Liu, T., Xiong, X., & Cao, J. (2020). Molecular characterization and expression analysis reveal the roles of Cys2/His2 zinc-finger transcription factors during flower development of Brassica rapa subsp. chinensis. *Plant Molecular Biology*, 102(1), 123–141.

Macedo, M.L.R., Freire, M.G.M., Silva, M.B.R., & Coelho, L.C.B.B. (2007) Insecticidal action of *Bauhinia monandra* leaf lectin (BmoLL) against *Anagasta kuehniella* (Lepidoptera: Pyralidae), *Zabrotes subfasciatus* and *Callosobruchus maculates* (Coleoptera: Bruchidae). *C. Comp Biochem Physiol 146*, 486–498.

Madsen, M., Schlunt, J., & Omer, E. (1987) *Am. J. Trop. Med. Hyg.*, 90, 101–109.

Mahmood, K., Mugal, T., & Haq, I. U. (2010). Moringa oleifera: A natural gift-A review. *J. Pharm. Sci. Res.*, 2, 775–781.

Makkar, H.P.S., & Becker, K. (1996). Nutritional value and antinutritional components of whole and ethanol extracted *Moringa oleifera* leaves. *Animal Feed Science and Technology*, 63, 211–228.

Mandal, S., Acharya, P., & Kar, I. (2014). Reactive oxygen species signaling in eggplant in response to Ralstonia solanacearum infection. *Journal of Plant Pathology*, 525–534.

Mandokhot, A.M, Fugro, P.A, & Gonkhalekar, S. (1994). *A New Diseases of Moringa oleifera in India, Central Experimental Station, Wakawali 415 711, India*.

Marino G, Brunetti C, Tattini M, Romano A, Biasioli F, Tognetti R, L. F., & Ferrini F, C. M. (2017). Dissecting the role of isoprene and stress-related hormones (ABA and ethylene) in Populus nigra exposed to unequal root zone water stress. *Tree Physiol*, 37, 1637–1647.

Marino, G., Brunetti, C., Tattini, M., Romano, A., Biasioli, F., Tognetti, R., ... & Centritto, M. (2017). Dissecting the role of isoprene and stress-related hormones (ABA and ethylene) in Populus nigra exposed to unequal root zone water stress. *Tree Physiology*, 37(12), 1637–1647.

McAdam S., (2014). Separating active and passive influences on stomatal control of transpiration. *Plant Physiol*, *174*, 1578–1586.

McAdam, S.A.M., & Brodribb, T.J. (2014) Separating active and passive influences on stomatal control of transpiration. *Plant Physiol.*, *164*(4), 1578–86.

Mridha, M. A. U., & Barakah, F. N. (2017). Diseases and pests of moringa: A mini review. *Acta Horticulturae*, *1158*, 117–124. https://doi.org/10.17660/ActaHortic.2017.1158.14

Mridha, M.A.U., & Al-Barakah, F. N. (2015). Report on the association of fungi with the fruits of Moringa oleifera (Manila, Lam. a highly valued medicinal plant. *First International Symposium on Moringa ISHS), Philippines*.

Mutheeswaran, S., Pandikumar, P., Chellappandian, M., & Ignacimuthu, S. (2011). Documentation and quantitative analysis of the local knowledge on medicinal plants among traditional Siddha healers in Virudhunagar district of Tamil Nadu, India. *J. Ethnopharmacol*, *137*, 523–533.

Mutlu, S., Atici, Ö., & Nalbantoglu, B. (2009). Effects of salicylic acid and salinity on apoplastic antioxidant enzymes in two wheat cultivars differing in salt tolerance. In *Biologia Plantarum*, *53*, Issue 2.

Nadeem, M., Azeem, M. W., & Rahman, F. (2015). Assessment of transesterified palm olein and *Moringa oleifera* oil blends as vanaspati substitutes. *Journal of Food Science and Technology*, *52*(4), 2408–2414. https://doi.org/10.1007/s13197-014-1271-4

Nadkarni, K. M. (1976). Indian Materia Medica, Vol. I & II. Popular Prakashan Private Limited, Bombay, India.

Nakabayashi ,R., Yonekura-Sakakibara, K., et al. (2014). Enhancement of oxidative and drought tolerance in Arabidopsis by overaccumulation of antioxidant flavonoids. *Plant J*, *77*, 367–379.

Nakabayashi, R., Yonekura-Sakakibara, K., Urano, K., Suzuki, M., Yamada, Y., Nishizawa, T., ... & Saito, K. (2014). Enhancement of oxidative and drought tolerance in Arabidopsis by overaccumulation of antioxidant flavonoids. *The Plant Journal*, *77*(3), 367–379.

Noctor, G., & Foyer, C.H. (1998) Ascorbate and glutathione: Keeping active oxygen under control. *Annu Rev Plant Physiol Plant Mol Biol.*, *49*, 249–279

Nouman, W., Basra, S.M.A., Siddiqui, M.T., Yasmeen, A., Gull, T., & Alcayde, M. (2014). Potential of Moringa oleifera L. as livestock fodder crop: A review. *Turk J. Agric. For.*, *38*, 1–14.

Olsen, A. (1987) *Water Res.*, *21*, 517–522.

Orwa, C., Mutua, A., Kindt R., Jamnadass, R., & Anthony, S. (2009). *Agroforestree Database: A tree reference and selection guide version 4.0.*

Palada M C, C. L. C. (2003). Suggested cultural practices for Moringa. *International Cooperators Guide, 03*, 545. www.moringa.co.il/article/moringa%02palada.pdf

Palada, M. C. (1996). Moringa (*Moringa oleifera* Lam.): A versatile tree crop with horticultural potential in the subtropical United States. *HortScience*, *31*, 794–797.

Paliwal, R., & Sharma, V. (2011). A review on horse radish tree (*Moringa oleifera*): A multipurpose tree with high economic and commercial importance. *Asian J. Biotechnol*, *3*, 317–328.

Pan, G., Chen, A., Li, J., Huang, S., Tang, H., Chang, L., Zhao, L., & Li, D. (2020). Genome-wide development of simple sequence repeats database for flax (*Linumusitatissimum* L.) and its use for genetic diversity assessment. *Genetic Resources and Crop Evolution*, *67*(4), 865–874.

Pande A. (1998). A new wilt diseases of wild Moringa (*Moringa concanensis*). *Journal of Economic and Taxonomic Botany*, *22*(2), 423–425.

Parrotta, J. A. (1993). *Moringa Oleifera Lam: Resedá, Horseradish Tree, Moringaceae, Horseradish-tree Family*. International Institute of Tropical Forestry, US Department of Agriculture, Forest Service.

Patricio, H.G., & Palada, M. C. (2017). Adaptability and horticultural characterization of different moringa accessions in Central Philippines. *Acta Hortic.*, *1158*, 45–54. 10.17660/ActaHortic.2017.1158.6

Pirrò, S., Matic, I., Guidi, A., Zanella, L., Gismondi, A., Cicconi, R., Bernardini, R., Colizzi, V., Canini, A., & Mattei, M. (2019). Identification of microRNAs and relative target genes in *Moringa oleifera* leaf and callus. *Scientific Reports*, *9*(1), 1–14.

Pollastri, S., Tsonev, T., & Loreto, F. (2014). Isoprene improves photochemical efficiency and enhances heat dissipation in plants at physiological temperatures. *Journal of Experimental Botany*, *65*(6), 1565–1570.

Popoola, J.O., & Obembe, O. O. (2013). Local knowledge, use pattern and geographical distribution of Moringa oleifera Lam. (Moringaceae) in Nigeria. *J. Ethnopharmacol*, *150*, 682–691.

Punja, Z.K. (1985) The biology, ecology and control of *Sclerotium rolfsii*. *Annu Rev Phytopathol.*, *23*, 97–127.

Rady, M. M. (2011). Effect of 24-epibrassinolide on growth, yield, antioxidant system and cadmium content of bean (*Phaseolus vulgaris* L.) plants under salinity and cadmium stress. *Sci. Hort. 129*, 232–237. doi: 10.1016/j.scienta. 2011.03.035

Rady, M. M., & Mohamed, G. F. (2015a). Modulation of salt stress effects on the growth, physio-chemical attributes and yields of Phaseolus vulgaris L. plants by the combined application of salicylic acid and *Moringa oleifera* leaf extract. *Scientia Horticulturae, 193*, 105–113. https://doi.org/10.1016/j.scienta.2015.07.003

Rady, M. M., & Mohamed, G. F. (2015b). Modulation of salt stress effects on the growth, physio-chemical attributes and yields of Phaseolus vulgaris L. plants by the combined application of salicylic acid and *Moringa oleifera* leaf extract. *Scientia Horticulturae, 193*, 105–113. https://doi.org/10.1016/j.scienta.2015.07.003

Rady, M.M., Bhavya Varma, C., and Howladar, S.M. (2013). Common bean (*Phaseolus vulgaris* L.) seedlings overcome NaCl stress as a result of presoaking in Moringa oleifera leaf extract. *Scientia Horticulturae, 162*, 63–70.

Rady, M. M., Mohamed, G. F., Abdalla, A. M., & Ahmed, Y. H. M. (2015). Integrated application of salicylic acid and *Moringa oleifera* leaf extract alleviates the salt-induced adverse effects in common bean plants. *Journal of Agricultural Technology, 11*(7), 1595–1614.

Ragno, S., Estrada-Garcia, I., Butler, R., & Colston, M. J. (1998). Regulation of macrophage gene expression by *Mycobacterium tuberculosis*: down-regulation of mitochondrial cytochrome c oxidase. *Infection and Immunity, 66*(8), 3952–3958.

Ramachandran, C., Peter K.V., & Gopalakrishnan, P. K. (1980). Drumstick (*Moringa oleifera*): A multipurpose Indian Vegetable. *Economic Botany, 34*(3), 276–283.

Ramachandran, C., Peter, K. V., & Gopalakrishnan, P. K. (1980). Drumstick (*Moringa oleifera*): a multipurpose Indian vegetable. *Economic botany*, 276–283.

Ramel F., Birtic S., Cuiné S., Triantaphylidés C., & Ravanat J.-L. (2012). Chemical quenching of singlet oxygen by carotenoids in plants. *Plant Physiol, 158*, 1267–1278.

Ramel, F., Birtic, S., Cuiné, S., Triantaphylidès, C., Ravanat, J. L., & Havaux, M. (2012). Chemical quenching of singlet oxygen by carotenoids in plants. *Plant physiology, 158*(3), 1267–1278.

Randhir, R., Lin, Y.T., & Shetty, K. (2004) Phenolic, their antioxidant and antimicrobial activity in dark germinated fenugreek sprouts in response to peptide and phytochemical elicitors. *Asia Pac J Clin Nutr., 13*, 295–307.

Raskin, I. (1992). Role of salicylic acid in plants. *Annu. Rev. Plant Physiol. Plant Mol. Biol, 43*. www.annual reviews.org

Rivas-San Vicente, M., & Plasencia, J. (2011). Salicylic acid beyond defence: its role in plant growth and development. *Journal of Experimental Botany, 62*(10), 3321–3338.

Roloff, A., Weisgerber, H., Lang, U., & Stimm, B. *Enzyklopädie der Holzgewächse,*. (2009). Wiley-VCH: Weinheim, Germany.

Saint S.A., (2010). Growing and Processing Moringa Leaves. *Moringanews / Moringa Associated of Ghana*, 69.

Santos, A. F. S., Napoleão, T. H., Paiva, P. M. G. and Coelho, L. C. B. B. (2012). *Lectins: Important Tools for Biocontrol of Fusarium Species*. T. F. Rios and E. R. Ortega (Ed.), Nova Science Publishers, Inc., New York, p. 161.

Scharf, K.-D., Berberich, T., Ebersberger, I., & Nover, L. (2012). The plant heat stress transcription factor (Hsf) family: structure, function and evolution. *Biochimica et Biophysica Acta (BBA)-Gene Regulatory Mechanisms, 1819*(2), 104–119.

Schützendübel, A., & Polle, A. (2002) Plant responses to abiotic stresses: heavy metal-induced oxidative stress and protection by mycorrhization. *J Exp Bot., 53*, 1351–1365.

Selmar, D., & Kleinwächter, M. (2013). Stress enhances the synthesis of secondary plant products: the impact of stress-related over-reduction on the accumulation of natural products. *Plant and Cell Physiology, 54*(6), 817–826.

Senaratna, T., Touchell, D., Bunn, E., & Dixon, K. (2000). Acetyl salicylic acid (Aspirin) and salicylic acid induce multiple stress tolerance in bean and tomato plants. *Plant Growth Regul., 30*(2), 157–161.

Shamsul, H., & Aqil, A. (2007) *Salicylic Acid – A Plant Hormone*. Springer Prints; 2007 Available: www.Spring com. Life Sci. Plant Sci.

Sharkey, T. D., Wiberley, A. E., & Donohue, A. R. (2008). Isoprene emission from plants: why and how. *Annals of botany, 101*(1), 5–18.

Sharma, S. S., & Dietz, K.-J. (2009). The relationship between metal toxicity and cellular redox imbalance. *Trends in Plant Science*, *14*(1), 43–50.

Sharma, V., Paliwal, R., Sharma, P., & Sharma, S. (2011). Phytochemical analysis and evaluation of antioxidant activities of hydro-ethanolic extract of Moringa oleifera Lam. pods. *J. Pharm. Res.*, *4*, 554–557.

Shriram, V., Kumar, V., Devarumath, R. M., Khare, T. S., & Wani, S. H. (2016). MicroRNAs as potential targets for abiotic stress tolerance in plants. *Frontiers in Plant Science*, *7*, 817.

Shyamli, P. S., Pradhan, S., Panda, M., & Parida, A. (2021). De novo whole-genome assembly of *Moringa oleifera* helps identify genes regulating drought stress tolerance. *Frontiers in Plant Science*, *12*.

Sreelatha, S., & Padma, P. R. (2009). Antioxidant activity and total phenolic content of *Moringa oleifera* leaves in two stages of maturity. *Plant Foods for Human Nutrition*, *64*(4), 303–311.

Sreelatha, S., & Padma, P. R. (2009). Antioxidant activity and total phenolic content of *Moringa oleifera* leaves in two stages of maturity. *Plant Foods for Human Nutrition*, *64*(4), 303. https://doi.org/10.1007/s11 130-009-0141-0

Stoll, G. (1988). Protection naturelle des Vegetaux en zone tropicale (Natural protection of plants in tropical zone). In: eds Margraf Verlag, CTA, AGRECOL.

Suarez, M., Haenni, M., Canarelli, S., Fisch, F., Chodanowski, P., Servis, C., Michielin, O., Freitag, R., Moreillon, P., & Mermod, N. (2005). Structure-function characterization and optimization of a plant-derived antibacterial peptide. *Antimicrobial Agents and Chemotherapy*, *49*(9), 3847–3857.Szepesi, A., Csiszar, J., Bajkan, S., Gemes, K., Horvath, F., Erdei, L., Deer, A.K., Simon, M.L., & Tari, I. (2005). Role of salicylic acid pre-treatment on the acclimation of tomato plants to salt- and osmotic stress. *Acta Biologica Szegediensis*, *49*, 123–125.

Tang, M., Xu, L., Wang, Y., Cheng, W., Luo, X., Xie, Y., Fan, L., & Liu, L. (2019). Genome-wide characterization and evolutionary analysis of heat shock transcription factors (HSFs) to reveal their potential role under abiotic stresses in radish (Raphanus sativus L.). *BMC Genomics*, *20*(1), 1–13.

Tattini, M., Loreto, F., Fini, A., Guidi, L., Brunetti, C., Velikova, V., & Gori, A. (2015). Isoprenoids and phenylpropanoids are part of the antioxidant defense orchestrated daily by drought-stressed *Platanus* × *acerifolia* plants during Mediterranean summers. *New Phytol*, *207*, 613–626.

Tattini, M., Velikova, V., et al. (2014). Isoprene production in transgenic tobacco alters isoprenoid, non-structural carbohydrate and phenylpropanoid metabolism, and protects photosynthesis from drought stress. *Plant Cell Environ*, *37*, 1950–1964.

Tattini, M., Loreto, F., Fini, A., Guidi, L., Brunetti, C., Velikova, V., ... & Ferrini, F. (2015). Isoprenoids and phenylpropanoids are part of the antioxidant defense orchestrated daily by drought-stressed P latanus× acerifolia plants during Mediterranean summers. *New Phytologist*, *207*(3), 613–626.

Tattini, M., Velikova, V., Vickers, C., Brunetti, C., Di Ferdinando, M., Trivellini, A., ... & Loreto, F. (2014). Isoprene production in transgenic tobacco alters isoprenoid, non-structural carbohydrate and phenylpropanoid metabolism, and protects photosynthesis from drought stress. *Plant, Cell & Environment*, *37*(8), 1950–1964.Tetley, R.M., & Thimann, K.V. (1974). The metabolism of oat leaves during senescence: I. Respiration, carbohydrate metabolism, and the action of cytokinins. *Plant Physiology*, *54*, 294–303.

Thakur, M., & Sharma, A.D. (2005). Salt-stress-induced proline accumulation in germinating embryos: evidence suggesting a role of proline in seed germination. *Journal of Arid Environment*, *62*, 517–523.

Tian, Y., Zeng, Y., Zhang, J., Yang, C., Yan, L., Wang, X., Shi, C., Xie, J., Dai, T., & Peng, L. (2015). High quality reference genome of drumstick tree (*Moringa oleifera* Lam.), a potential perennial crop. *Science China Life Sciences*, *58*(7), 627–638.

TNAU Agrictech (2016) https://agritech.tnau.ac.in/crop_protection/crop_prot_crop_insect-veg_Drumst ick.html

Vailati-Riboni, M., Palombo, V., & Loor, J. J. (2017). What are omics sciences. *In Periparturient Diseases of Dairy Cows* (pp. 1–7). Springer, Cham.

Vanzo, E., Jud, W., Li, Z., Albert, A., Domagalska, M. A., Ghirardo, A., ... & Schnitzler, J. P. (2015). Facing the future: effects of short-term climate extremes on isoprene-emitting and nonemitting poplar. *Plant Physiology*, *169*(1), 560–575.

Vélez-Gavilán, J. (2019). *Moringa oleifera* (horse radish tree). *Forestry Compendium*, (34868).

Velikova, V., Edreva, A., et al. (2004). Endogenous isoprene protects Phragmites australis leaves against singlet oxygen. *Physiol Plant*, *122*, 219–225.

Velikova, V., Brunetti, C., Tattini, M., Doneva, D., Ahrar, M., Tsonev, T., ... & Loreto, F. (2016). Physiological significance of isoprenoids and phenylpropanoids in drought response of Arundinoideae species with contrasting habitats and metabolism. *Plant, Cell & Environment*, *39*(10), 2185–2197.

Velikova, V., Edreva, A., & Loreto, F. (2004). Endogenous isoprene protects Phragmites australis leaves against singlet oxygen. *Physiologia Plantarum*, *122*(2), 219–225.

Velikova, V., Várkonyi, Z., Szabó, M., Maslenkova, L., Nogues, I., Kovács, L., ... & Loreto, F. (2011). Increased thermostability of thylakoid membranes in isoprene-emitting leaves probed with three biophysical techniques. *Plant Physiology*, *157*(2), 905–916.

Verma, S., Nizam, S., & Verma, P. K. (2013). Biotic and abiotic stress signaling in plants. In *Stress Signaling in Plants: Genomics and Proteomics Perspective, Volume 1* (pp. 25–49). Springer, New York, NY.

Vickers, C. E., Gershenzon, J., Lerdau, M. T., & Loreto, F. (2009a). A unified mechanism of action for volatile isoprenoids in plant abiotic stress. *Nature chemical biology*, *5*(5), 283–291.

Vickers, C. E., Possell, M., Cojocariu, C. I., Velikova, V. B., Laothawornkitkul, J., Ryan, A., ... & Nicholas Hewitt, C. (2009b). Isoprene synthesis protects transgenic tobacco plants from oxidative stress. *Plant, Cell & Environment*, *32*(5), 520–531.

Wang, P., Song, H., Li, C., Li, P., Li, A., Guan, H., Hou, L., & Wang, X. (2017). Genome-wide dissection of the heat shock transcription factor family genes in Arachis. *Frontiers in Plant Science*, *8*, 106.

Wani, S. H., Anand, S., Singh, B., Bohra, A., & Joshi, R. (2021). WRKY transcription factors and plant defense responses: latest discoveries and future prospects. *Plant Cell Reports*, *40*(7), 1071–1085.

Wojtaszek, P. (1997). Oxidative burst: an early plant response to pathogen infection. *Biochemical Journal*, *322*(3), 681–692.

Yap, Y. K., El-sherif, F., Habib, E. S., & Khattab, S. (2021). *Moringa oleifera* leaf extract enhanced growth, yield and silybin content while mitigating salt-induced adverse effects on the growth of silybum marianum. *Agronomy*, *11*(12). https://doi.org/10.3390/agronomy11122500

Yasmeen, A., Basra, S.M.A., Ahmad, R., & Wahid, A. (2012). Rendimiento de trigo sembrado tarde en respuesta a la aplicación foliar de extracto de hojas de moringa oleifera lam. *Chil J Agric Res.*, *72*(1), 92–97.

Yisehak, K., Solomon, M., & Tadelle, M. (2011). Contribution of *Moringa* (*Moringa stenopetala*, Bac.), a highly nutritious vegetable tree, for food security in South Ethiopia: a review. *Asian J App Sci.*, *4*, 477–488.

Yoon, Y., Seo, D. H., Shin, H., Kim, H. J., Kim, C. M., & Jang, G. (2020). The role of stress-responsive transcription factors in modulating abiotic stress tolerance in plants. *Agronomy*, *10*(6), 788.

Younis, A., Akhtar, M. S., Riaz, A., Zulfiqar, F., Qasim, M., Farooq, A., Tariq, U., Ahsan, M., & Bhatti, Z. M. (2018). Improved cut flower and corm production by exogenous moringa leaf extract application on gladiolus cultivars. *Acta Sci. Pol. Hortorum Cultus*, *17*(4), 25–38.

Yusuf, M., Fariduddin, Q., Varshney, P., & Ahmad, A. (2012). Salicylic acid minimizes nickel and/or salinity-induced toxicity in Indian mustard (Brassica juncea) through an improved antioxidant system. *Environmental Science and Pollution Research*, *19*(1), 8–18.

Yusuf, M., Hayat, S., Alyemeni, M. N., Fariduddin, Q., & Ahmad, A. (2013). Salicylic acid: physiological roles in plants. In *Salicylic acid* (pp. 15–30). Springer.

Zandalinas, S. I., Mittler, R., Balfagón, D., Arbona, V., & Gómez-Cadenas, A. (2018). Plant adaptations to the combination of drought and high temperatures. *Physiologia plantarum*, *162*(1), 2–12.

Zeinali, N., Altarawneh, M., Li, D., Al-Nu'airat, J., &Dlugogorski, B. Z. (2016). New mechanistic insights: Why do plants produce isoprene. *Acs Omega*, *1*(2), 220–225.

Zhang, Yue, Xujun Zhu, Xuan Chen, Changnian Song, Zhongwei Zou, Yuhua Wang, Mingle Wang, Wanping Fang, and Xinghui Li (2014). "Identification and characterization of cold-responsive microRNAs in tea plant (Camellia sinensis) and their targets using high-throughput sequencing and degradome analysis." *BMC plant biology* 14, no. 1: 1–18.

Zimin, A. v, Marçais, G., Puiu, D., Roberts, M., Salzberg, S. L., & Yorke, J. A. (2013). The MaSuRCA genome assembler. *Bioinformatics*, *29*(21), 2669–2677.

12 Nigella sativa and Stressful Conditions

Amira A. Ibrahim[1*], Elsayed S. Abdel Razik[2],
Sawsan Abd-Ellatif[3] and Samah Ramadan[4]
[1]Botany and Microbiology Department, Faculty of Science,
Arish University, Al-Arish 45511, Egypt
[2]Plant Protection and Biomolecular Diagnosis Department,
Arid Lands Cultivation Research Institute, City of Scientific Research and
Technology Applications, Borg EL-Arab, Alexandria, Egypt
[3]Bioprocess Development Department, Genetic Engineering and
Biotechnology Research Institute, City of Scientific Research and
Technology Applications, Borg EL-Arab, Alexandria, Egypt
[4]Botany Department, Faculty of Science, Mansoura University, Egypt
*Corresponding author: amiranasreldeen@yahoo.com;
amiranasreldeen@sci.aru.edu.eg

CONTENTS

12.1   INTRODUCTION

12.1.1   ORIGIN, HISTORY AND USES

Black seed has been discovered in a flask from a Hittite temple ruin that dates to around 1650 BC in central Turkey. It is not certain, but strongly assumed, that this herb was used in ancient Egypt because it is unclear what its exact name was at that time (Salih et al. 2009). The seeds also were discovered in Tutankhamun's tomb. In the first century AD, the Greek herbalist Dioscorides mentioned a substance he called melanthion in section 79 of his third volume of his *Materia medica* (Zohary and Hopf 2000; Dioscorides and Beck 2005). His description of the plant, which included the fact that it was used to bake bread, which is still the case in some regions of the Middle East, clearly identifies it as *Nigella sativa*. In Asia, Eastern Africa and the Middle East, *Nigella sativa*, also known as black cumin or black seed and Habbat Al-barakah in Arabic, plays a major role in herbal *Materia medica*. On the other hand, Western Europe and the Americas are significantly less familiar with this herb.

For millennia, cultures in the Mediterranean region have employed *Nigella sativa*. The herb with the same name has jet-black seeds that are renowned for having terpenoids, mainly thymoquinone, and fatty acids in them. The traditional use of this herb in treating atopic dermatitis has been supported by clinical studies. Additionally, there is preliminary clinical study evidence that suggests using black seed to lower seizure frequency, enhance lipid profiles and/or help opioid addicts quit using the drug. A native angiosperm from Asia and the Middle East, *Nigella sativa* belongs to the Ranunculaceae family. This miraculous herb was utilized in folk medicine around the world before the development of modern medicine to heal a wide range of illnesses and disorders. In contrast to current allopathic medicine, ancient medical traditions saw a person's health as a synthesis of their mind, body, soul and nature. The written traditional medical texts of important civilizations, including the Ayurveda, Siddha, Unani, Greek-Roman, Malay, Tibb-e-Nabwi and Jewish literature, all refer to this plant. *Nigella* seeds also go by the names black cumin and black seed, and they are widely utilized as medicines throughout all Abrahamic cultures.

Black seed also has anti-diabetic, anti-cancer, anti-microbial, anti-inflammatory, anti-redox, anti-immunomodulator, analgesic, hepatoprotective, nephroprotective, and gastroprotective properties. Studies on animals indicate that this herb might be a helpful addition to acetaminophen and cancer chemotherapy. If consumed in the right amounts, black seed is very safe and is a common part of the diet in the Middle East and India. Although additional study is required, the West should make greater use of this resource. Additionally, it has been discovered that thymoquinone (TQ), one of the main bioactive components of essential oils (Ahmad et al. 2013), together with other valuable components including linoleic acid and dithymoquinone, is responsible for the majority of *N. sativa*'s medicinal qualities (Woo et al. 2011; Butt et al. 2019). It is now well established that TQ, one of the main components of *N. sativa*'s volatile oil, is primarily responsible for the majority of the therapeutic advantages of the substance (Butt et al. 2019).

Elicitation requires improvement of the qualitative and quantitative bioactive secondary metabolites (BSMs) that enhance pronating characteristics for health (Świeca 2016; Sharma 2016). Additionally, it has changed growth and development and could have a beneficial impact on morphological, physiological and biochemical traits that increase biomass crop production and quality (Sharifi-Rad et al. 2016; Ahamed and El-Sayed 2018; Jalali et al. 2019).

Elicitation was particularly important in preventing plant and pathogen attacks that significantly reduced the yield when agrochemicals were used. They improve their resistance to microbial diseases and pest infestations by regulating the expression of genes involved in the production and accumulation of SM phytoalexins (PHS), toxins characterized as broad-spectrum antipesticide and antimicrobial agents, and traditional agriculture, which is frequently implicated in induced systemic resistance (ISR) (Zheng 2005; Gabaston et al. 2018). Elicitation plays a crucial role in how plants respond to biotic and abiotic stimuli and adapt to their environmental changes (Treutter 2005; Jansen et al. 2008; Edreva et al. 2008). Black cumin is a tiny, annual plant that grows to a height of 45 cm. It has a few thin, pinnatisect leaves that are 2–4 cm long and divided into oblong segments. The seeds are triangular and black in hue, with pale, blue blossoms on single, long peduncles.

The black cumin plant has an unusually rigid, erect, branched stem, intensely cut greyishy-green leaves, terminal greyish-blue flowers and unusual, serrated seed vessels. These seed vessels are filled with small, somewhat compressed seeds that are typically three-cornered, with two flat sides and one curved side, black or brown on the outside and white and greasy on the inside (Figure 12.1). Flowers from the *N. sativa* plant are small, fragile and typically have 5–10 petals and are light blue and white. The blooms are naturally bluish, have a variable number of sepals, and are distinguished by the existence of nectars.

The gynoecium is made up of a variable number of multiovule carpels that, following pollination, grow into follicles containing individual fruits that are only loosely linked to form a capsule-like

FIGURE 12.1   Flowers and seeds of black cumin (*Nigella sativa*).

structure. A huge, inflated capsule with three to seven joined follicles that individually carry seeds makes up the fruit. The seeds of *N. sativa* have corrugated integuments and are tiny (1–5 mm). The fruit tastes sharply bitter and smells faintly of strawberries (Varghese 1996; Dwivedi 2003; Tiruppur Venkatachallam et al. 2010).

12.2   RESPONSES TO BIOTIC AND ABIOTIC STRESSES

In the current period of significant climate change, plants face a variety of climate-induced biotic and abiotic difficulties, for example, global warming, unpredictable rainfall and the reduction of arable land and water resources (Atkinson et al. 2013; Narsai et al. 2013; Prasch and Sonnewald 2013; Suzuki et al. 2014; Mahalingam 2015; Pandey et al. 2015; Ramegowda and Senthil-Kumar 2015). Stress is a situation that is harmful to the growth and development of plants and can be brought on by biological, environmental or combined factors.

Under natural conditions, it is more detrimental to global agricultural production when two or more distinct types of stresses occur simultaneously, for instance, salinity and drought or heat stress and drought. Abiotic challenges that occur simultaneously are more harmful to plant metabolism and harvest than abiotic stresses that occur independently at various growth stages. Examples of abiotic stresses that are combined include drought and heat stress in the summer or drought and salinity stress. In order to control pest, disease, bug and weed outbreaks, abiotic stressors are also crucial (Coakley et al. 1999; Scherm and Coakley 2003; McDonald et al. 2009; Ziska et al. 2010; Peters et al. 2014).

By changing the functional and adaptive responses of plants, these pressures also affect the interaction between plants and pests (Scherm and Coakley 2003). Under abiotic stress, weeds outcompete crops as they are more effective water users (Brown and Hovmøller 2002; Ziska et al. 2010; Valerio et al. 2013). Abiotic stress significantly affects plant development and, as a result, causes substantial production losses. The ensuing growth is decreased in the majority of plant species by as much as 50% (Wang et al. 2003).

Apart from numerous combinations of abiotic challenges, plants frequently experience several biotic stresses as a result of contemporaneous or sequential pathogen or herbivore attacks. Biological stress is a new danger that hinders plant production (Strauss and Zangerl 2002; Brown and Hovmoller 2002; Maron and Crone 2006; Maron and Kauffman 2006; Mordecai 2011). One typical instance of coupled biotic stress is when bacterial and fungal diseases attack plants simultaneously. The overall findings of different experiments demonstrated salt stress's inhibitory effects on the germination characteristics of black cumin seeds. Lowering the temperature from the ideal level makes the seeds more vulnerable to saline, and at 15°C, a temperature that is comparable to those experienced during the sowing of black cumin in the Mediterranean regions, seed germination rapidly decreases. In addition to seed germination and mean germination time, other factors like seedling emergence, root length, shoot length and weight may be useful markers in breeding programs to create cultivars that can withstand salt stress.

Additionally, the deleterious impact of salt on seed germination in inadequate temperature circumstances may be mitigated by priming procedures. The germination temperature range for *N. sativa* was between 17°C and 21°C, which is a relatively narrow range. It was discovered that this species is temperature sensitive. According to Baskin and Baskin (1998), temperatures between 20°C and 25°C are acceptable for the majority of species of shrubs in hot semi-deserts and deserts to achieve 60–100% germination.

12.3   RESPONSES OF *NIGELLA SATIVA* TO ABIOTIC STRESSORS

12.3.1   RESPONSES TO SALINITY

A high salt content in soil is the most pervasive problem that restricts plant dispersion and yield (Qin et al. 2010). Managing salt stress to preserve food production is a key issue in many arid and semiarid areas (Kaya et al. 2013). The biological answer to this problem, according to Ashraf et al. (2008), focuses on controlling, using or creating plants that can flourish in salt-damaged soils. Seed priming is one biological technique for raising future crops' tolerance to salt. Prior to full germination, several physiological germination processes can occur in controlled conditions with limited water supply (Sharma et al. 2014).

A physiological defense mechanism known as priming prepares a plant to react to upcoming abiotic stress more swiftly or brutally (Jisha et al. 2013). Many studies have shown that pre-treating seeds with water or inorganic or organic salt solutions before sowing might increase a plant's resilience to salt (Kaya et al. 2006; Patade et al. 2009; Srivastava et al. 2010; Yadav et al. 2011; Jafar et al. 2012).

Around 20% of the world's irrigated soil has reduced fertility due to its high salt level (Flowers and Yeo 1995). Plant transcriptome, metabolome and proteome alterations, as well as multilayer changes in molecular responses, cause plants to be tolerant to salt stress.

By either killing the embryo or decreasing soil potential to the point that water intake is hindered, salt stress has an impact on germination. When seeds of *Ocimum basilicum*, chamomile, *Eruca sativa*, *Petroselinum hortense*, *Thymus maroccanus* and sweet marjoram were planted in saline soil, all suffered delayed germination (Miceli et al. 2003; Ramin 2005; Ali et al. 2007; Belaqziz et al. 2009).

Nigella sativa, Salvia officinalis, Achillea fragratissima, Withania somnifera, Chamomilla recutita, Ocimum basilicum and *Catharanthus roseus* showed growth-inhibiting features when exposed to salt stress (Jaleel et al. 2008a, 2008b; Abd El-Azim and Ahmed 2009; Ben Taarit et al. 2009; Hussain et al. 2009; Ghanavati and Sengul 2010). Aggregation of phenolics with growing saline levels was also seen in *Nigella sativa* and *Mentha pulegium* (Bourgou et al. 2010; Queslati et al. 2010).

Nigella plants allegedly produced more phenolic chemicals like quercetin, apigenin and trans-cinnamic acid when grown in highly saline soil (Bourgou et al. 2010). During halopriming treatments, seeds germinated more successfully in sodium chloride than in pure water. Salinity may cause the seeds to absorb Na^+ and Cl^-. The reason why NaCl performs better at seed germination than distilled water may be due to this absorption, which maintains an open water potential gradient for water absorption during seed germination (Kaya et al. 2006). During osmopriming, seeds are steeped in aerated, low-water-capacity solutions. Less external water potential limits the volume and pace of imbibition and speeds up a number of metabolic processes in seeds that occur before germination (Jisha et al. 2013).

The imbibition process in primed seeds took place faster because of the larger osmotic potential gradients between the distilled water in the germination papers and the osmotic potential on the seed surfaces (Foti et al. 2008). It has been demonstrated that a variety of priming techniques can boost the salt tolerance of seedlings (Mahmoudi et al. 2012).

The usage of NaCl to encourage seed germination was found to have significant positive effects in both saline and non-saline environments. By boosting antioxidant enzyme activity and osmolyte accumulation for osmotic adjustments, Saha et al. (2010) found that pretreating seeds with a sub-lethal dose of NaCl might be a helpful way of mitigating the detrimental impacts of salt stress on

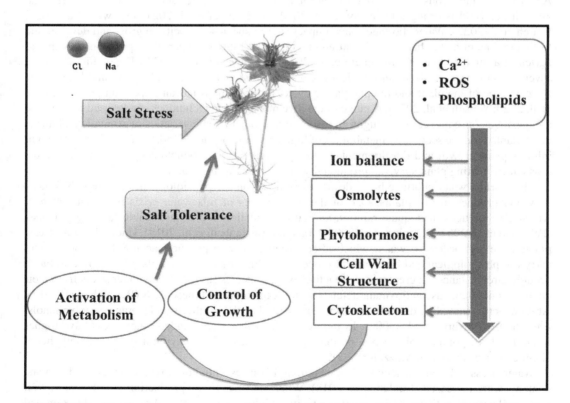

FIGURE 12.2 *Nigella sativa* responses to salinity stress. Ca^{2+}, Calcium ions; ROS, reactive oxygen species.

plants. NaCl pretreatments reduced the quantity of Na that accumulated in the stems and increased K and Ca ion concentrations in the leaves and stems of melon, which lessened the harmful impacts of salinity (Sivritepe et al. 2005). Significant differences were found in the cellular reactions of several *N. sativa* genotypes to *in vitro* NaCl stress. According to the results, under the effect of extreme saline conditions, the total anthocyanin concentrations and the total flavonol content of *N. sativa* calluses can be largely credited for the antioxidant activity of the plant (250 mM of NaCl). This discovery demonstrated the potential for the secondary metabolites to be biosynthesized in *N. sativa* even under the most severe salt stress. The salt tolerance of *N. sativa* is illustrated in Figure 12.2.

Black seed germination was significantly influenced by the priming substances $CaCl_2$ and $ZnSO_4$. $CaCl_2$'s advantageous effects may stem from calcium's role as a second messenger in plant cells, its capacity to protect against detrimental environmental stress or its effects on hormonal balance (Iqbal et al. 2006).

In line with the findings of Basra et al. (2006) for wheat, the performance of black seed germination improved after treatment with a modest $CaCl_2$ concentration. For seedling vigor and plant tolerance to environmental stressors, seeds must contain an adequate content of Zn (Cakmak 2008). Prom-u-thai et al. (2012) observed that rice seedlings grown in Zn-deficient soil had greater Zn concentrations following $ZnSO_4$ seed priming; Zn content is necessary for protein synthesis and gene expression. The production of reactive oxygen species (ROS) occurs often throughout seed germination, and Zn may play a crucial role in the detoxification of these substances in tissues (Cakmak 2000).

12.3.2 RESPONSES TO DROUGHT STRESS

According to numerous studies, one of the most significant abiotic factors restricting worldwide agricultural yield is drought (Mahdavi et al. 2020; Naservafaei et al. 2021; Baldwin et al. 2022; Biareh et al. 2022). Water shortages have impacted crop species' capacity to grow and develop due to factors like climate change, deficient and inconsistent rainfall dispersal, population growth and agricultural malpractice (Karimmojeni et al. 2021; Gharibvandi et al. 2022). Depending upon the severity and length of the water deficiency, environmental elements including salt and heat stress, the genotype of the plant and the plant's life cycle, drought stress has an impact on plants (Kumar et al. 2020; Yousefi et al. 2020). A range of creative management techniques can be used to decrease the negative consequences of drought. In response to drought, some varieties of aromatic and medicinal plants produce secondary metabolites of high economic value (Mahdavie et al. 2020). Drought inhibits plant growth and development, lowering average crop production by up to 50% in most crops and affecting plant survival, productivity and spread (Wang et al. 2003).

The world's water shortage has a destructive impact on the physiology, morphology, biochemistry and productivity of plants. One of the main causes of this water crisis is climate change, which also includes inadequate rainfall, an aging population and unsustainable farming practices (Paknejad et al. 2007; Razmjoo et al. 2008; Ghadyeh-Zarrinabadi et al. 2019). Under drought stress, plants have less access to water, which affects their ability to produce, accumulate biomass and carry out physiological and biochemical processes. Additionally, when a plant is exposed to harsh drought circumstances, it creates more reactive oxygen species (ROS). When under stress from their environment, plants use antioxidant defense systems to counteract these ROS (Bettaieb et al. 2012). Abiotic factors including drought stress in plants stimulate the synthesis and storage of polyphenols, according to Navarro et al. (2006). In light of the fact that plants under water stress have higher tissue levels of polyphenols despite producing less biomass, they may be a source of polyphenol chemicals (De Abreu and Mazzafera 2005).

Water stress reduced black cumin's seed output, but its essential oil, thymoquinone, and carvone contents were increased (Emdad et al. 2018). When it comes to the production of substances with medicinal uses, *N. sativa* is an industrial plant with a high economic value. An earlier study examined how the irrigation schedule affected this plant's ability to produce more fruit (Takruri and Dameh 1998; Ghamarnia and Jalili 2013). Traditional medicine has used black cumin seeds for a variety of ailments for a long time, including asthma, hypertension, diabetes, inflammation, sore throat, bronchitis, headache, dermatitis, fever, dizziness and influenza. They are also used for their vermifuge, lactagogue, diuretic, carminative and other characteristics (Ali and Blunden 2003).

The fatty acid composition can be used to assess whether a seed is used for food, industry or medicinal purposes. Although the two *Nigella* species (*N. sativa* and *N. damascene*) that have been reported have similar oil contents, there are quantitative differences between these species. The most important saturated fatty acids in these two species are palmitic, stearic and myristic, whereas linoleic and oleic acids are the most important unsaturated fatty acids (Telci et al. 2014). Extensive research has been carried out on the advantages of using goods made from the *Nigella* species for industry, medicine and other purposes. However, little research has been done on how drought stress impacts the various fatty acid and oil compositions of *Nigella* species.

According to Bannayan et al. (2008) and Ghamarnia et al. (2010), *N. sativa* tolerated water scarcity situations reasonably well. However, *N. sativa* is vulnerable to water stress, according to other studies (Ghamarnia and Jalili 2013; Shahbazi 2019). These species may have a change in their fatty acid composition when under drought stress, however, this is not yet known. Studies on the impact of the irrigation regime on essential oil levels and yield in several cumin genotypes as well as oil and oil composition have been published. In a pot experiment employing 40%, 60%, 80% and 100% of the water needed by the plant, Ghamarnia and Jalili (2013) discovered that cumin's seed and oil yields decreased as water stress increased. According to research by Haj Seyed Hadi et al. (2012), during the budding and flowering initial stages, under typical irrigation termination, seed output decreased while essential oil, thymoquinone and carvone contents rose. However, in addition to a reduction in oil %, Alaghemand et al. (2018) found that the same irrigation approach caused a fall in the levels of oleic, linoleic and linolenic acids. Black cumin under salt stress produced fewer seeds but had higher levels of essential oils.

Rezaei-Chiyaneh et al. (2018) also pointed out that greater transfer of photo-assimilates from leaves to seed as a consequence of a longer period of photosynthesis as a result of a better remaining green trait under stress may have led to a superior grain output in drought stress circumstances (Khakwani et al., 2011). Gamma-aminobutyric acid (GABA) is thought to have a general role in promoting plant growth and development. GABA's ability to promote growth and remobilize assimilates to the seeds may, however, account for the rise in plant length, follicle number, seed and essential oil output in black cumin as a result of its administration. The majority of studies (Gamble and Burke 1984; Türkan et al. 2005; Wang et al. 2009) indicate that the main impacts of water shortage stress on higher plants include a rise in respiration, a decline in photosynthesis and cell plasma membrane destruction. In addition, a shortage of water causes the production of ROS, which lowers the amounts of carotenoids and chlorophyll (Kiani et al. 2008; Chéour et al. 2014). However, it should be highlighted that the rise in soluble sugars and proline can be more of a sign of existing water stress than a protective measure against future water stress. Therefore, in the lack of any measurements of plant water relations, this investigation cannot assess the significance of soluble sugars and proline to osmoregulation.

Black cumin has been reported to increase proline production, catalase (CAT) and peroxidase (POX) activity when exposed to water stress (Ahmadpour Dehkordi and Balouchi 2012). However, given the inverse relationship between CAT, POX or superoxide dismutase (SOD) activity and chlorophyll content, increasing the antioxidant enzyme activity may be viewed as a strategy to maintain photosynthesis's effectiveness in the face of water stress.

This goal is regulated by the osmotic actions of substances to minimize the detrimental effects of oxidative stress brought on by water shortage stress, according to the favorable correlations between the activity of antioxidant enzymes and the accumulation of soluble sugars or prolines. Black cumin's reaction to water deficiency stress was improved by exogenous GABA therapy, albeit

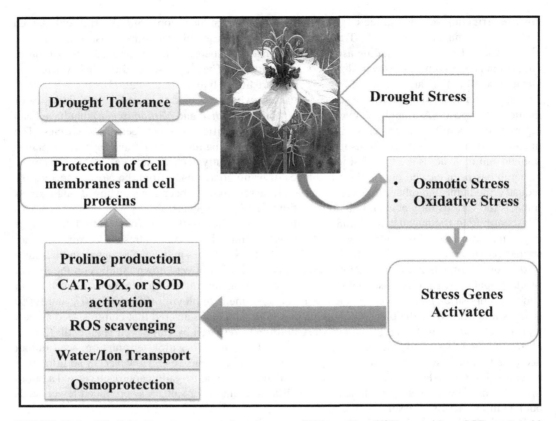

FIGURE 12.3 *Nigella sativa* responses to drought stress. CAT, catalase; POX, peroxidase; SOD, superoxide dismutase; ROS, reactive oxygen species.

it remains unknown how GABA particularly affects this plant's chlorophyll, soluble sugars, proline and CAT activity. According to the study's findings, foliar application of GABA strengthens black cumin's resistance to water deficiency stress by boosting water uptake, chlorophyll content and photosynthetic efficiency. Other crops with different growing conditions have reportedly benefited from GABA therapy on plant sensitivity to water shortage stress for successfully farming black cumin with a limited water source. More research is still needed to fully understand the role of GABA-intermediated control in plant growth and the related biochemical pathways at the molecular and genetic levels. The *Nigella sativa* response to drought stress is described in Figure 12.3.

12.3.3 RESPONSES TO HEAVY METAL STRESS

Heavy metals (HMs) are metal elements with a high mass. These are dangerous even at low concentrations (Diaconu et al. 2020). The atomic density of heavy metals is larger than 4 g/cm^3 (Singh et al. 2016). One group of heavy metals is arsenic (As), and it also includes iron (Fe), cadmium (Cd), copper (Cu), chromium (Cr), mercury (Hg), lead (Pb), platinum (Pt) and silver (Ag). Owing to increased industrialization, improvements in the use of farming synthetics and urbanization, heavy metal pollution is a significant environmental issue that may be deadly due to bioaccumulation through the food chain. Due to the aggregation of heavy metals in the environment and the consumption of food crops tainted with these hazardous substances, the population's health has been negatively impacted (Zukowska et al. 2008). Animals and people are put in danger when heavy metals are assimilated by plants through uptake and accretion throughout the food chain (Jordao et al. 2006; Sprynskyy et al. 2007). The environment contains heavy metals from a range of sources, including natural, agrarian, industrial and other sources. Both manmade and natural activities have the potential to contaminate an area with heavy metals. The planet has been heavily contaminated by mining, refining and agriculture.

With the exception of a few that work as micronutrients at limited levels, plants that are allowed to grow in contaminated areas are typically exposed to high concentrations of heavy metals that are otherwise not needed for their survival. These heavy metals enter the plant species through contaminated soil, water or air, and then they build up in various parts of the plant. At various stages of the plant life cycle, these heavy metals' harmful impacts have been documented (Shamshed et al. 2018). Seed germination, a vital stage in the life of flowering plants, is adversely affected by heavy metals.

Inefficient sugar and protein metabolism, delayed germination, membrane integrity damage from oxidation and nutrient loss are all effects of heavy metal toxicity in seeds that lower yield (Wang et al. 2003; Ahmad and Ashraf 2011). Radicles accumulate more heavy metals like Cd, Ni and Pb than plumules do when seeds are germinated. Hydrolyzing enzymes that slow down the process of seed germination and radicle emergence in a range of plant species include α-amylases, acid invertases, acid phosphatases, proteases and ribonucleases. As a result, the developing embryo receives less reserve food supplies, such as proteins and carbohydrates (Rahoui et al. 2010; Ashraf et al. 2011; Singh et al. 2011). Lead modifies the genetic DNA profile of the seedlings (Mohamed 2011). A buildup of lipid peroxidation products, nutritional loss due to membrane leakage, reduced embryo growth and a lower overall percentage of seed germination are all reportedly caused by cadmium hyperaccumulation in seeds (Ahsan et al. 2007; Sfaxi-Bousbih et al. 2010; Smiri et al. 2011). However, relatively higher concentrations of these heavy metals are associated with a decrease in the total percentage of seed germination, while their relatively lower amounts just slow down the process.

The physiological and biochemical functions of plants are adversely affected by heavy metal poisoning, which hinders their capacity to expand and mature (Sheldon and Menzies 2005; Katare et al. 2015; Pichhode and Nikhil 2015). Even though it is well known that heavy metals impair plant metabolism, several of these metals are essential for keeping certain metabolic reactions in their ideal state since they are parts of several enzymes and proteins (Küpper et al. 2003).

Superoxide radicals ($O_2^{\bullet-}$), hydroxyl radicals ($OH^{\bullet-}$) and hydrogen peroxide (H_2O_2) are just a few of the many ROS that are produced as a result of heavy metal stress and are all capable of harming the lipids, proteins and nucleic acids in cell walls (Cuypers et al. 2011). Plants have an antioxidant resistance system that consists of non-enzymatic antioxidants like ascorbic acid (AsA), glutathione (GSH) and tocophe, as well as enzymes like superoxide dismutase (SOD), catalase (CAT), glutathione reductase (GR), peroxidase (POD), ascorbate peroxidase (APX) and guaiacol peroxidases (GPX) (Sytar et al. 2013).

For maintaining membrane integrity and offering plants resilience to environmental challenges, osmoprotectants like proline and glycine betaine (GB) are crucial. Enzymes are not affected by proline and GB in large concentrations, but they are hydrated, which helps with the regeneration of enzyme activity (Kishor et al. 2005). Specific medicinal plants can be used to aid in the repair of heavy metal-contaminated soil. Microorganisms in the rhizosphere are connected by phytoremediation systems in order to eliminate, reduce or accumulate environmental pollutants (Ouyang 2002).

For dealing with heavy metal phytoremediation contamination, woody plant species with high biomass, such as crop, medicinal and fragrant plants, are indicated (Zheljazkov et al. 2008). It is unacceptable to consume crops that have undergone phytoremediation since there is a chance that the heavy metal will enter the food chain through consumption by both people and animals (Gupta et al. 2014). The production and use of several therapeutic plants in polluted soil require strict monitoring because of their increased ability to collect metals in their edible parts, as advised by WHO (2005).

The defenses against heavy metal stress in plants are simple and well-organized. Strong cuticles, physiologically active parts like trichomes, cell walls and mycorrhizal symbiosis serve as the first line of defense against heavy metal stress. However, when heavy metal ions breach these barriers and accumulate in various plant tissues and cells, plants launch a number of defense mechanisms to reduce the effects of heavy metals (Harada et al. 2010). The production of various biological molecules, such as protein metallochaperones, chelators like organic acids, glutathione (GSH), phytochelatins (PCs) and microtubules (MTs), plant exudates (flavonoid and phenolic compounds), phytohormones and many others, is among the fundamental and most important ways to lessen the effects of heavy metal stress (Dalvi and Bhalerao 2013; Viehweger 2014).

However, oxidative stress conditions emerge from the excessive generation of ROS if the afore-mentioned procedures are unsuccessful in lowering heavy metal stress (Mourato et al. 2012). To reduce ROS levels, plants' antioxidant defense mechanisms—both enzymatic and non-enzymatic antioxidants—are compelled to engage in ROS scavenging activities (Sharma et al. 2012). Through PC synthesis or ROS quenching, GSH is essential in protecting from heavy metal stress. When exposed to heavy metals from GSH, the PCS enzyme, which is helpful for metal remediation and managing cellular homeostasis, creates PCs, or heavy metal-binding moieties (Yadav 2010). As-phytochelatin synthase 1 (AsPCS1) and yeast cadmium factor 1 (YCF1) in *Arabidopsis thaliana* demonstrated improved resistance to Cd and As and increased the probability of metal uptake (Guo et al. 2012). *Elsholtziahai chowensis* metallothionein type 1 (EhMT1) decreased Cu stress and hence reduced oxidative damage in tobacco genotypes by lowering H_2O_2 levels and increasing POD activity (Xia et al. 2012). By reducing ROS levels and raising SOD enzyme activity, the chemical known as the metallothionein gene, or TaMT3, which is introduced to tobacco, increases tolerance to Cd toxicity; nevertheless, it also hampers POD activity (Zhou et al. 2014).

12.3.4 PROTECTIVE EFFECTS OF *NIGELLA SATIVA* AND ITS DERIVATIVES AGAINST HEAVY METAL TOXICITY

Heavy metals are extremely toxic and have a bad impact on both the environment and human health. These ions build up and cause serious dangers to the environment and human health. There are many different ways that toxic metals can be exposed to individuals and the environment, including the air, food, water, waste and industries. There are a variety of strategies to remove heavy metals or reduce their harmful impacts, depending on the type. Physical, biological and chemical techniques are frequently used in the treatment and elimination of organic pollutants. Herbal treatment is one of the best known and efficient strategies to minimize the harmful consequences of organic pollutants like heavy metals in animal and human organs.

In the meantime, medicinal plants like *N. sativa* or black seed are commonly used as an anti-dote and protective agent due to the powerful component thymoquinone (TQ). There is a great deal of data on its biological activities and protective advantages in a variety of organs and tissues, including the brain, genome, liver, kidney and lung. It also serves as an example of one of the first toxins that people became aware of, which is widely dispersed in the atmosphere in a variety of ways. Human activity can cause environmental pollution, as can natural weathering of geological structures, mining, manufacturing, wash-off, burning of fossil fuels, agriculture and incineration (Firdaus et al. 2016).

Weakness, dehydration, pain and diarrhea are among the worst side effects that can result from being exposed to a high dose of arsenic (Vahidnia et al. 2007).

The long-term effects of As exposure also include liver disease, anemia and digestive problems. These conditions are in addition to hyperkeratosis, Bowen's disease, squamous and basal cell carcinoma, diabetes mellitus and Blackfoot disease (Chen et al. 2007; Sattar et al. 2016). ROS-mediated oxidative damage acts as a frequent mechanism for As pathogenesis because of As' involvement in the creation of ROS and its capacity to cause lipid peroxidation (LPO), which in turn leads to the generation of active oxygen and the nitrile molecule (Chen et al. 2007). TQ significantly reduces the toxicities caused by a number of heavy metals (Ma et al. 2010).

As an illustration, Firdaus et al. (2016) showed that pre-treatment with TQ dramatically decreased the As-induced neurotoxicity in brain preparations of male Wistar rats (Firdaus et al. 2016). The comet experiment further showed that pre-treatment with TQ reduced the DNA damage brought on by As (Firdaus et al. 2016). TQ's ability to protect against As-induced kidney damage has been demonstrated by Sener et al. (2016). They concluded that TQ could be used as an adjuvant (Sener et al. 2016). Similarly, Fouad and Jresat (2015) reported that TQ significantly increased serum testosterone and glutathione (GSH) levels and significantly decreased malondialdehyde (MDA) and NO levels caused by As administration in the testicular tissue, in addition to decreasing the expression of inducible nitric oxide (NO) synthase and caspase-3 that As-induced (Fouad et al. 2017). TQ has recently been discovered to reduce the neurotoxic effects of arsenic (As) and, through its antioxidant function, to reduce the oxidative stress that is caused in the nervous system (Kassab and El-Hennamy 2017).

12.4 RESPONSES OF *NIGELLA SATIVA* TO BIOTIC STRESSORS

Fusarium root rot caused by *F. comptoceras*, *F. solani* and *F. lateritium* has been frequently found in fields of black cumin in the Assiut governorate provinces, Egypt. Infected black cumin fields displayed disease symptoms such as drooping and yellowing leaves, browning of underground plant parts, and eventually plant death. One of the most common diseases caused by the vascular wilt pathogen *Fusarium* sp. Koike, which can also be found in soil and seeds, is black cumin root rot. In the infected roots of black cumin plants, eight fungal isolates were discovered. Domsch et al. (1980) described the physical characteristics of mycelia and spores.

The isolates, later identified as *F. comptoceras*, *F. solani* and *F. lateritium*, were highly pathogenic and caused symptoms such as stunted growth, root rot and plant mortality. This research backs up previous findings that *F. solani* and *F. equiseti* are pathogenic to black cumin plants (Mohammadi and Mofrad 2009; Ramchandra and Bhatt 2012). Various *Pseudomonas* strains have been mentioned in numerous reports by numerous experts worldwide as being capable of significantly controlling a variety of bacterial, nematode and fungal diseases in cereals, horticultural crops, oil seeds and other plants. In many cases, bacterial antagonism proved to be more effective at disease control than fungicides. *Pseudomonas* spp. is a well-known bacterium that promotes plant development.

Black seeds treated with the studied bacterial PGPR (plant growth-promoting rhizobacteria) isolates in a greenhouse encouraged plant growth more successfully than the untreated control plant. It has been noted that biological management strategies for crop growth promotion are advantageous for other crop species. It's possible that microorganisms aid in the fixation of nitrogen. In contrast to infected controls, black cumin seeds treated with bioagent as a suspension had a positive effect on fungal populations, according to the results of a study by Mohamed et al. (2020).

12.5 CONCLUSIONS

Consuming *Nigella sativa* seeds is generally safe and may have medical applications. Future research should focus on the mechanisms through which *N. sativa* seeds create their therapeutic effects. With a greater understanding of its mechanism of bioactivity, it may one day be able to incorporate this medicinal herb as a supplement drug into traditional medical research. Plants are physically unable to leave their current location to avoid the impacts of environmental stress, such as abiotic stress. Abiotic variables such as soil salinity, heavy metal contamination and drought stress are all known to dramatically lower agricultural production. Plant metabolism is perturbed as a result of these alterations under abiotic stress, which makes it simpler to reorganize the metabolic network and sustain the active metabolic processes.

The findings suggest that, based on seed and oil yields, water stress led to a rapid drop in all black cumin plant traits, including seed and oil yields, as well as in plant and water usage efficiency. These results suggest that black cumin is a therapeutic plant that is water stress sensitive. Black cumin's crop water demand was suggested as the water stress threshold value, and it was not encouraged to consider how water stress might affect the crop's water needs. Plants have created a sophisticated defense network in order to survive in challenging environments. Drought is a common negative limiting factor that has an impact on plant growth, development, physiology, metabolism, yield and output. One method employed by plants to endure drought is changes in morphology, physiology and biochemistry at different stages of plant growth. In the past 10 years, various metabolic genes have been used to improve drought tolerance in crops via genetic engineering approaches.

REFERENCES

Abd EL-Azim, W.M., and S. T. Ahmed. 2009. Effect of salinity and cutting date on growth and chemical constituents of *Achillea fragratissima* Forssk, under Ras Sudr conditions. *Research Journal of Agriculture and Biological Sciences* 5:1121–1129.

Ahamed, T.E.S., and S.A. El-Sayed. 2018. Verification and Validation of Dandelion (*Taraxacum officinal*) Seeds-Gamma Irradiated under Elicitation with Nano- and Micro-Zinc for Potential Optimization Biomass and Ennghanci Phenolics, Flavonoids and Antioxidant Activity. *International Journal of Innovative Science and Research Technology* 3: 398–403.

Ahmad M.S., and M., Ashraf. 2011. Essential roles and hazardous effects of nickel in plants. *Reviews of environmental contamination and toxicology* 214: 125–167.

Ahmad, A., A., Husain, M. Mujeeb, et al. 2013. A Review on Therapeutic Potential of *Nigella sativa*: A Miracle Herb. *Asian Pacific Journal of Tropical Biomedicine* 5: 337–352. https://doi.org/10.1016/S2221-1691(13)60075-1

Ahmadpour Dehkordi, S., and H.R. Balouchi. 2012. Effect of seed priming on antioxidant enzymes and lipids peroxidation of cell membrane in black cumin (*Nigella sativa*) seedling under salinity and drought stress. Electron. *Journal of Crop Production*5: 63–85.

Ahsan N., S.H. Lee, D.G. Lee, et al. 2007. Physiological and protein profiles alternation of germinating rice seedlings exposed to acute cadmium toxicity. *Comptes Rendus Biologies* 330: 735–746.

Alaghemand, A., S. Khaghani, M.R. Bihamta, et al. 2018. Green synthesis of zinc oxide nanoparticles using black cumin (*Nigella sativa* L.) extract: the effect on the height and number of branches. *Journal of Nanostructure in Chemistry* 8: 82–88.

Ali, R.M., H.M. Abbas, and R.K. Kamal. 2007. The effects of treatment with polyamines on dry matter, oil and flavonoids contents in salinity stressed chamomile and sweet marjoram. *Plant, Soil and Environment* 53:529–543.

Ali, B., and G. Blunden, 2003. Pharmacological and toxicological properties of black cumin (*Nigella sativa*). *Phytotherapy Research* 17: 299–305.

Al-Sman, K.M., A.M.K. Abo-El-yousr, A. Elraky, et al. 2017. Isolation, Identification and Biomanagement of Black Cumin (*Nigella sativa*) using Selected Bacterial Antagonists . *International Journal of Phytopathology* (03): 47–56.

Anosheh, H.P., H. Sadeghi, and Y. Emam. 2011. Chemical priming with urea and KNO3 enhances maize hybrids (*Zea mays* L.) seed viability under abiotic stress. *Journal of Crop Science and Biotechnology* 14: 289–295.

Ashraf, M., H.R. Athar, P.J.C. Harris, et al. 2008. Some prospective strategies for improving crop salt tolerance. *Advances in Agronomy* 97: 45–110.

Ashraf, M.Y., N. Azhar, M. Ashraf, et al. 2011. Influence of lead on growth and nutrient accumulation in canola (*Brassica napus* L.) cultivars. *Journal of Environmental Biology* 32(5): 659–666.

Atkinson, N.J., C.J. Lilley, and P.E. Urwin. 2013. Identification of genes involved in the response to simultaneous biotic and abiotic stress. *Plant Physiology* 162:2028–2041. https://doi.org/10.1104/pp.113.222372

Aycicek, M., O. Kaplan, and M. Yaman. 2008. Effect of cadmium on germination, seedling growth and metal contents of sunflower (*Helianthus annus* L.). *Asian Journal of Chemistry* 20: 2663.

Baldwin, T.C., and A. Mastinu. 2022. Physiological, Biochemical, and Agronomic Trait Responses of *Nigella sativa* Genotypes to Water Stress. Horticulturae 8: 193. https:// doi.org/10.3390/ horticulturae8030193.

Bannayan, M., F. Najafi, M. Azizi, L. et al. 2008. Yield and Seed Quality of *Plantago ovata* and *Nigella sativa* under Different Irrigation Treatments. *Industrial Crops and Products* 27: 11–16.

Baskin, C.C. and Baskin, J.M. 1998., Seeds Ecology, Biogeography, and Evolution of Dormancy and Germination, Academic Press, San Diego.

Basra, S.M.A., M. Farooq, Afzal, I., and Hussain, M. 2006. Influence of Osmopriming on the germination and early seedling growth of coarse and fine rice. *International Journal of Agriculture and Biology* 8: 19–22.

Bayati, P.; H. Karimmojeni, J. Razmjoo, et al. 2009. Salt stress effects on germination, growth and essential oil content of an endemic thyme species in Morocco (*Thymus maroccanus* Ball.). *Journal of Applied Sciences Research* 5: 858–863.

Belaqziz, R., Romane, A., and Abbad, A. 2009. Salt stress effects on germination, growth and essential oil content of an endemic thyme species in Morocco (*Thymus maroccanus* Ball.). *Journal of Applied Sciences Research* 5(7): 858–863.

Ben Taarit M.K., K. Msaada, M. Hosni, et al. 2009. Plant growth, essential oil yield and composition of sage (*Salvia officinalis* L.) fruits cultivated under salt stress conditions. *Industrial Crops and Products* 30: 333–337.

Bettaieb, I., I. Jabri-Karoui, I. Hamrouni-Sellami, et al. 2012. Effect of drought on the biochemical composition and antioxidant activities of cumin (*Cuminum cyminum* L.) seeds. *Industrial Crops and Products* 36: 238–245.

Biareh, V., F. Shekari, S. Sayfzadeh, et al. 2022. Physiological and Qualitative Response of *Cucurbita pepo* L. to Salicylic Acid under Controlled Water Stress Conditions. *Horticulturae* 8: 79.

Bourgou, S., M.E. Kchouk, A. Bellila, et al. 2010. Effect of salinity on phenolic composition and biological activity of *Nigella sativa*. *Acta Horticulturae* 853:57–60.

Brown, J.K., and Hovmøller, M.S. 2002. Aerial dispersal of pathogens on the global and continental scales and its impact on plant disease. Science 26: 297(5581): 537–541. doi: 10.1126/science.1072678.

Brown, J.K.M., L. Chartrain, P. Lasserre-Zuber, et al. 2015. Genetics of resistance to *Zymoseptoria tritici* and applications to wheat breeding. *Fungal Genetics and Biology* 79:33–41.

Butt, A.S., N. Nisar, N. Ghani, et al. 2019. Isolation of Thymoquinone from *Nigella sativa* L. and *Thymus vulgaris* L., and Its Anti-Proliferative Effect on HeLa Cancer Cell Lines. *Tropical Journal of Pharmaceutical Research* 18: 37–42.

Cakmak, I. 2000. Role of zinc in protecting plant cells from reactive oxygen species. *New Phytologist* 146: 185–205.

Cakmak, I. 2008. Enrichment of cereal grains with zinc: Agronomic or genetic biofortification? *Plant Soil* 302: 1–17.

Chen, C.J., S.L. Wang, J.M. Chiou, et al. 2007. Arsenic and diabetes and hypertension in human populations: a review. *Toxicology and Applied Pharmacology* 222(3):298–304. doi: 10.1016/j.taap.2006.12.032

Chéour, F., I. Kaddachi, D. Achouri, et al. 2014. Effects of water stress on relative water: chlorophylls and proline contents in barley (*Hordeum vulgare* L.) leaves. *Journal of Agriculture and Veterinary Sciences* 7:13–16.

Coakley, S.M., H. Scherm, and S. Chakraborty. 1999. Climate change and plant disease management. *Annual Review of Phytopathology* 37:399–426.

Cuypers, A., S. Karen, R. Jos, et al. 2011. The cellular redox state as a modulator in cadmium and copper responses in Arabidopsis thaliana seedlings. *Journal of Plant Physiology* 168: 309–316.

Dalvi, A.A., and S. A. Bhalerao. 2013. Response of plants towards heavy metal toxicity: an overview of avoidance, tolerance and uptake mechanism. *Annals of Plant Sciences* 2: 362–368.

De Abreu, I. N., and P. Mazzafera. 2005. Effect of water and temperature stress on the content of active constituents of *Hypericum brasiliense* Choisy. *Plant Physiology and Biochemistry* 43: 241–248.

Demir I., and K. Mavi. 2004. The effect of priming on seedling emergence of differentially matured watermelon (*Citrulus lanatus* (Thunb) Matsum and Nakai) seeds. *Scientia Horticulturae* 102: 467–473.

Diaconu M., L.V. Pavel, R.M. Hlihor, et al. 2020. Characterization of heavy metal toxicity in some plants and microorganisms—A preliminary approach for environmental bioremediation. *New Biotechnology* 56: 130–139.

Dioscorides P., and L.Y. Beck. 2005. transl. The Materia Medica: Classical Students and Texts, vol. 38 [in German]. Hildesheim, Germany: Olms-Weidmann, 2005.Domsch, K.H., Gams, W., and Anderson, T.H. 1980. Compendium of soil fungi. Academic Press; London, p. 809.

Dwivedi, S.N. 2003. Ethnobotanical studies and conservational strategies of wild and natural resources of Rewa district of Madhya Pradesh. *Journal of Economic and Taxonomic Botany* 27: 233–234.

Edreva, A., V. Velikova, T. Tsonev, et al. 2008. Stress Protective Role of Secondary Metabolites: Diversity of Functions and Mechanisms. *General and Applied Plant Physiology* 341: 67–78.

Emdad, M.R., Tafteh, A., and Ghalebi, S. 2018. Validation of aquacrop model for simulating wheat yield in different irrigation events. *Water and Soil* 32(3): 463–473. doi: 10.22067/jsw.v32i3.70189

Esmaeil, R., S. M. Seyyedi, E. Elnaz, et al. 2018. Exogenous application of gamma-aminobutyric acid (GABA) alleviates the effect of water deficit stress in black cumin (*Nigella sativa* L.). *Industrial Crops and Products* 112:741-–748.

Firdaus, F., M.F. Zafeer, E. Anis, et al. 2016. Antioxidant potential of thymoquinone against arsenic mediated neurotoxicity. *Free Radicals and Antioxidants* 6(1):115-–23. doi: 10.5530/fra.2016.1.14

Flowers, T.J., and A.R. Yeo. 1995. Breeding for salinity resistance in crop plants. Where next? *Australian Journal of Plant Physiology* 22:875–884.

Foti, R., Aburenia, K., Tigerea, A., Gotosab, J., and Gerec, J. 2008. The efficacy of different seed priming osmotica on the establishment of maize (*Zea mays* L.) caryopses. *J Arid Environ* 72: 1127–1130.

Fouad, A., and Jresat, I. 2015. Thymoquinone therapy abrogates toxic effect of cadmium on rat testes. *Andrologia* 47, 417–426. doi: 10.1111/and.12281

Fouad, A.A.,W.H. Albuali, and I. Jresat. 2014. Protective effect of thymoquinone against arsenic-induced testicular toxicity in rats. Kuala Lumpur, Malaysia: International Conference on Pharmacology and Pharmaceutical Medicine (ICPPM); 2014:8.

Gabaston, J., T. EL-Khawand, P. Waffo-Teguo, et al. 2018. Stilbenes from Grapevine Root: A Promising Natural Insecticide against Leptinotarsa decemlineata. *Journal of Pest Science* 91: 897–906. https://doi.org/10.1007/s10340-018-0956-2

Gamble, P.E., and J.J. Burke. 1984. Effect of water stress on the chloroplast antioxidant system I. Alterations in glutathione reductase activity. *Plant Physiology* 76: 615–621.

Ghadyeh-Zarrinabadi, I., J. Razmjoo, A.A. Mashhadi, et al. 2019. Physiological response and productivity of pot marigold (*Calendula officinalis*) genotypes under water deficit. *Industrial Crops and Products* 139: 111–488.

Ghamarnia, H., and Z. Jalili. 2013. Water Stress Effects on Different Black Cumin (*Nigella sativa* L.) Components in a Semi- Arid Region. *International Journal of Agronomy and Plant Production* 4: 753–762.

Ghamarnia, H., H. Khosravy, and S. Sepehri. 2010. Yield and Water Use Efficiency of (*Nigella sativa* L.) under Different Irrigation Treatments in a Semi Arid Region in the West of Iran. *Journal of Medicinal Plants Research* 4: 1612– 1616.

Ghanavati, M., and S. Sengul. 2010. Salinity effect on the germination and some chemical components of *Chamomilla recutita* L. *Asian Journal of Chemistry* 22: 859–866.

Gharibvandi, A., H. Karimmojeni, P. Ehsanzadeh, et al. 2022. Weed management by allelopathic activity of Foeniculum vulgare essential oil. *Plant Biosystems - An International Journal Dealing with all Aspects of Plant Biology* 1–15.

Guo, J., W. Xu, and M. Ma. 2012. The assembly of metals chelation by thiols and vacuolar compartmentalization conferred increased tolerance to and accumulation of cadmium and arsenic in transgenic *Arabidopsis thaliana. Journal of Hazardous Materials* 199: 309–313.

Gupta, S.K., M. Chabukdhara, P. Kumar, et al. 2014. Evaluation of ecological risk of metal contamination in river Gomti, India: a biomonitoring approach. *Ecotoxicology and Environmental Safety* 110: 49–55.

Haj Seyed Hadi, M.R., M.T. Darzi, and Z. Ghandehari. 2012. Effect of irrigation treatment and Azospirillum inoculation on yield and yield component of black cumin (*Nigella sativa* L.). *Journal of Medicinal Plants Research* 6: 4553–4561.

Harada, E., J.A. Kim, A.J. Meyer, et al. 2010. Expression profiling of tobacco leaf trichomes identifies genes for biotic and abiotic stresses. *Plant and Cell Physiology* 51: 1627–1637.

Hussain, K., A. Majeed, K. H. Nawaz, et al. 2009. Effect of different levels of salinity on growth and ion contents of black seeds (*Nigella sativa* L.). *Current Research Journal of Biological Sciences* 1: 135–138.

Idris, E. E. S., R. H. Bochow and R. Borriss. 2004. Use of Bacillus subtilis as biocontrol agent. VI. Phytohormone like action of culture filtrates prepared from plant growth-promoting *Bacillus amyloliquefaciens* FZB24, FZB42, FZB45 and *Bacillus subtilis* FZB37. *Journal of Plant Diseases and Protection* 111: 583–597.

Iqbal, M., M. Ashraf, A. Jamil, et al. 2006. Does seed priming induce changes in the levels of some endogenous plant hormones in hexaploid wheat plants under salt stress? *Journal of Integrative Plant Biology* 48: 181–189.

Jafar, M.Z., M. Farooq, M.A. Cheema, et al. 2012. Improving the performance of wheat by seed priming under saline conditions. *Journal of Agronomy and Crop Science* 198: 38–45.

Jalali, S.M., D.M. Movahhedi, A. Salehi, et al. 2019. Effect of Irrigation Regimes and Nitrogen Sources on Biomass Production, Water and Nitrogen Use Efficiency and Nutrients Uptake in Coneflower (*Echinacea purpurea* L.). *Agricultural Water Management* 213:358–367.

Jaleel, C.A., G.M.A. Lakshmanan, M. Gomathinayagam, et al. 2008a. Triadimefon induced salt stress tolerance in *Withania somnifera* and its relationship to antioxidant defense system. *South African Journal of Botany* 74: 126–132.

Jaleel, C.A., B. Sankar, R. Sridharan, et al. 2008b. Soil salinity alters growth, chlorophyll content, and secondary metabolite accumulation in *Catharanthus roseus*. *Turkish Journal of Biology* 32: 79–83.

Jansen, M.A.K., K. Hectors, N.M. O'Brien, et al. 2008. Plant Stress and Human Health: Do Human Consumers Benefit from UV-B Acclimated Crops. *Plant Science* 175: 449–458. https://doi.org/10.1016/j.plant sci.2008.04.010

Jisha, K.C., K. Vijayakumari, and J.T. Puthur. 2013. Seed priming for abiotic stress tolerance: an overview. *Acta Physiologiae Plantarum* 35: 1381–1396.

Jordao C.P., C.C. Nascentes, P.R. Cecon, et al. 2006. Heavy metal availability in soil amended with composted urban solid wastes. *Environmental Monitoring and Assessment* 112: 309–326.

Karimmojeni, H., H. Rahimian, H. Alizadeh, et al. 2021. Competitive Ability Effects of *Datura stramonium* L. and *Xanthium strumarium* L. on the Development of Maize (*Zea mays*) Seeds. *Plants* 10: 1922.

Kassab, R.B., and R.E. El-Hennamy. 2017. The role of thymoquinone as a potent antioxidant in ameliorating the neurotoxic effect of sodium arsenate in female rat. *Egyptian Journal of Basic and Applied Sciences* 4 (3): 160–167. doi: 10.1016/j.ejbas.2017.07.002

Katare J., M. Pichhode, and K. Nikhil. 2015. Growth of *Terminalia bellirica* [(gaertn.) roxb.] on the malanjkhand copper mine overburden dump spoil material. *International Journal of Research -GRANTHAALAYAH* 3: 14–24.

Kattimani, K.N., Y.N. Reddy, and R.B. Rao. 1999. Effect of presoaking seed treatment on germination, seedling emergence, seedling vigour and root yield of ashwagandha (*Withania somnifera* Daunal.). *Seed Science and Technology* 27: 483–488.

Kaya, C., O. Sönmez, S. Aydemir, et al. 2013. Mitigation effects of glycinebetaine on oxidative stress and some key growth parameters of maize exposed to salt stress. *Turkish Journal of Agriculture and Forestry* 37: 188–194.

Kaya, M.D., G. Okçu, M. Atak, et al. 2006. Seed treatments to overcome salt and drought stress during germination in sunflower (*Helianthus annuus* L.). *European Journal of Agronomy* 24: 291–295.

Khakwani, A.A.,M.D. Dennett, and M. Munir. 2011. Early growth response of six wheat varieties under artificial osmotic stress condition. *Pakistan Journal of Agricultural Sciences* 48:121–126.

Kiani, S.P., P. Maury, A. Sarrafi, et al. 2008. QTL analysis of chlorophyll fluorescence parameters in sunflower (*Helianthus annuus* L.) under well-watered and waterstressed conditions. *Plant Science* 175: 565–573.

Kishor, P.K., S. Sangam, R.N. Amrutha, et al. 2005. Regulation of proline biosynthesis, degradation, uptake and transport in higher plants: its implications in plant growth and abiotic stress tolerance. *Current Science* 88: 424–438.

Kumar, A., M. Memo, and A. Mastinu. 2020. Plant behaviour: An evolutionary response to the environment? *Plant Biology* 22: 961–970.

Kumar, B. S. D., I. Berggren, and A. M. Martensson. 2001. Potential for improving pea production by co-inoculation with fluorescent Pseudomonas and Rhizobium. *Plant and Soil* 229: 25–34.

Küpper H., I. Šetlík, E. Šetliková, et al. 2003. Copper-induced inhibition of photosynthesis: limiting steps of in vivo copper chlorophyll formation in *Scenedesmus quadricauda*. *Functional Plant Biology* 30: 1187–1196.

Ma N., M. Sasoh, S. Kawanishi, et al. 2010. Protection effect of taurine on nitrosative stress in the mice brain with chronic exposure to arsenic. *Journal of Biomedical Science* 17 Suppl 1:S7. doi: 10.1186/1423-0127-17-s1-s7

Mahalingam, R. 2015. Consideration of combined stress: a crucial paradigm for improving multiple stress tolerance in plants. In: Mahalingam R (ed) Combined stresses in plants. Springer International Publishing, Cham. https://doi.org/10.1007/978-3-319-07899-1_1

Mahdavi, A., P. Moradi, and A. Mastinu. 2020. Variation in Terpene Profiles of *Thymus vulgaris* in Water Deficit Stress Response. *Molecules* 25: 1091.

Mahmoudi, H., R. Ben Massoud, O., Baatour, et al. 2012. Influence of different seed priming methods for improving salt stress tolerance in lettuce plants. *Journal of Plant Nutrition* 35: 1910–1922.

Maron, J.L., and E. Crone. 2006. Herbivory: effects on plant abundance, distribution and population growth. *Proceedings of the Royal Society B: Biological Sciences* 273:2575–2584.

Maron, J.L., and M. Kauffman Ms. 2006. Habitat-specific consumer impacts on plant population dynamics. *Ecology* 87:113–124.

McDonald, A., S. Riha, A. DiTommasob, et al. 2009. Climate change and the geography of weed damage: analysis of US maize systems suggests the potential for significant range transformations. *Agriculture, Ecosystems & Environment* 130:131–140. https://doi.org/10.1016/j.agee.2008.12.007

Miceli, A.,A. Moncada, and F. D'Anna. 2003. Effect of water salinity on seeds-germination of *Ocimum basilicum* L., *Eruca sativa* L. and *Petroselinum hortense* Hoffm. *Acta Horticulturae* 609: 365–370.

Mohamed, H.I. 2011. Molecular and biochemical studies on the effect of gamma rays on lead toxicity in cowpea (*Vigna sinensis*) plants. *Biological Trace Element Research* 144: 1205–1218.Mohamed, N., Hassan, E., Khater, R. 2020. Response of black cumin (*Nigella sativa* L.) plants to the addition of natural fertilizers and the inoculation by bacteria mix and seaweed liquid extract. *Archives of Agriculture Sciences Journal* 3(2): 1–15.

Mohammadi, A., and N. N. Mofrad. 2009. Genetic diversity in population of *Fusarium solani* from cumin in Iran. *Journal of Plant Protection Research* 49: 283–286.

Mordecai, E.A. 2011. Pathogen impacts on plant communities: unifying theory, concepts, and empirical work. *Ecological Monographs* 81:429–441.

Mourato, M., R. Reis, and L.L. Martins. 2012. Characterization of plant antioxidative system in response to abiotic stresses: a focus on heavy metal toxicity. In: Montanaro G. (eds.). *Advances in Selected Plant Physiology Aspects* 12: 1–17.

Narsai, R., C. Wang, J. Chen, et al. 2013. Antagonistic, overlapping and distinct responses to biotic stress in rice (*Oryza sativa*) and interactions with abiotic stress. *BMC Genomics* 14:93. https://doi.org/10.1186/1471-2164-14-93

Naservafaei, S., Y. Sohrabi, P. Moradi, et al. 2021. Biological Response of *Lallemantia iberica* to Brassinolide Treatment under Different Watering Conditions. *Plants* 10: 496.

Navarro, J.M., P. Flores, C. Garrido, et al. 2006. Changes in the contents of antioxidant compounds in pepper fruits at different ripening stages, as affected by salinity. *Food Chemistry* 96: 66–73.

Ouyang, Y. 2002. Phytoremediation: modeling plant uptake and contaminant transport in the soil–plant–atmosphere continuum. *Journal of Hydrology* 266 (1–2): 66–82.

Paknejad, F., M. Nasri, H.T. Moghadam, et al. 2007. Effects of drought stress on chlorophyll fluorescence parameters, chlorophyll content and grain yield of wheat cultivars. *Journal of Biological Sciences* 7: 841–847

Pandey, P.,V. Ramegowda, and M. Senthil-Kumar. 2015. Shared and unique responses of plants to multiple individual stresses and stress combinations: physiological and molecular mechanisms. *Frontiers in Plant Science* 6:723.

Patade, V.Y., S. Bhargava, and P., Suprasanna. 2009. Halopriming imparts tolerance to salt and PEG induced drought stress in sugarcane. *Agriculture, Ecosystems & Environment* 134: 24–28.

Peters, K., L., Breitsameter, B., Gerowitt. 2014. Impact of climate change on weeds in agriculture: a review. *Agronomy for Sustainable Development* 34:707–721. https://doi.org/10.1007/s13593-014-0245-2

Pichhode, M., and K. Nikhil. 2015. Effect of copper dust on photosynthesis pigments concentrations in plants species. *International Journal of Engineering and Management* 2: 63–66.

Prasch, C.M., and U. Sonnewald .2013. Simultaneous application of heat, drought, and virus to *Arabidopsis* plants reveals significant shifts in signaling networks. *Plant Physiology* 162(4): 1849–1866. https://doi.org/10.1104/pp.113.221044

Prom-u-thai, C., B. Rerkasem, A. Yazici, et al. 2012. Zinc priming promotes seed germination and seedling vigor of rice. *Journal of Plant Nutrition and Soil Science*. 175: 482–488.

Qin, J.,W.Y., Dong, K.N. He, et al. 2010. NaCl salinity- induced changes in water status, ion contents and photosynthetic properties of *Shepherdia argentea* (Pursh) Nutt. Seedlings. *Plant, Soil and Environment* 56: 325–332.

Queslati, S., N. Karray-Bouraoui, H. Attia, et al. 2010. Physiological and antioxidant responses of *Mentha pulegium* (Pennyroyal) to salt stress. *Acta Physiologiae Plantarum* 32:289–296.

Rahoui S., A. Chaoui, and E.J. El Ferjani. 2010. Membrane damage and solute leakage from germinating pea seed under cadmium stress. *Hazard Material* 178: 1128–1131.

Ramchandra, S., and P. N. Bhatt. 2012. First report of *Fusarium equiseti* causing vascular wilt of cumin in India. Plant Disease 96: 1821.

Ramegowda, V., and M. Senthil-Kumar. 2015. The interactive effects of simultaneous biotic and abiotic stresses on plants: mechanistic understanding from drought and pathogen combination. *Journal of Plant Physiology* 176:47–54.

Ramin, A.A. 2005. Effects of salinity and temperature on germination and seedling establishment of sweet basil (*Ocimum basilicum* L.). *Journal of Herbs, Spices & Medicinal Plants* 11:81–90.

Razmjoo, J., P. Heydarizadeh, and M. Sabzalian. 2008. Effect of salinity and drought stresses on growth parameters and essential oil content of Matricaria chamomile. *International Journal of Agriculture And Biology* 10: 451–454.

Rezaei-Chiyaneh, E., Seyyedi, S.M., Ebrahimian, E. et al. 2018. Exogenous application of gamma-aminobutyric acid (GABA) alleviates the effect of water deficit stress in black cumin (*Nigella sativa* L.). *Industrial Crops and Products* 112: 741–748.

Saha, P., P. Chatterjee, and A.K. Biswas .2010. NaCl pretreatment alleviates salt stress by enhancement of antioxidant defense system and osmolyte accumulation in mungbean (*Vigna radiata* L. Wilczek). *Indian Journal of Experimental Biology* 48: 593–600.

Salih, B., T. Sipahi, and E. Oybak Donmez. 2009. Ancient nigella seeds from Boyali Hoyuk in north-central Turkey. *Journal of Ethnopharmacology* 124:416–420.

Sattar, A., S. Xie, M.A. Hafeez, et al. 2016. Metabolism and toxicity of arsenicals in mammals. *Environmental Toxicology and Pharmacology* 48:214–24. doi: 10.1016/j. etap.2016.10.020

Scherm, H., and S.M. Coakley. 2003. Plant pathogens in a changing world. *Australasian Plant Pathology* 32:157–165. https://doi.org/10.1071/AP03015

Sener, U., R. Uygur, C. Aktas, et al. 2016. Protective effects of thymoquinone against apoptosis and oxidative stress by arsenic in rat kidney. *Renal Failure* 38(1):117–23. doi: 10.3109/0886022x.2015.1103601

Sfaxi-Bousbih, A., A. Chaoui, and E. El Ferjani. 2010. Cadmium impairs mineral and carbohydrate mobilization during the germination of bean seeds. *Ecotoxicology and Environmental Safety* 73: 1123–1129.

Shahbazi, E. 2019. Genotype Selection and Stability Analysis for Seed Yield of *Nigella sativa* Using Parametric and Non-Parametric Statistics. *Scientia Horticulturae* 253: 172–179.

Shamshed S., Shahis M., Ratia M., et al. 2018. Effect of organic amendment on cadmium stress to pea: A multivariate comparison of germination vs young seedlings and younger vs older leaves. *Ecotoxicology and Environmental Safety* 151: 91–97.

Sharifi-Rad, M., J.A. Sharifi-Rad, and T. da Silva .2016. Morphological, Physiological and Biochemical Responses of Crops (*Zea mays* L., *Phaseolus vulgaris* L.), Medicinal Plants (*Hyssopus officinalis* L., *Nigella sativa* L.) and Weeds (*Amaranthus retroflexus* L., *Taraxacum officinale* F.H. Wigg) Exposed to SIO_2 Nanoparticles. *Journal of Agriculture, Science and Technology* 18:1027–1040.

Sharma, A.S., Rathoreb, K., Srinivasana, et al. 2014. Comparison of various seed priming methods for seed germination, seedling vigour and fruit yield in okra (*Abelmoschus esculentus* L. Moench). *Scientia Horticulturae* 165: 75–81.

Sharma, P., Jha, R.S., et al. 2012. Reactive oxygen species, oxidative damage, and antioxidative defense mechanism in plants under stressful conditions. *Journal of Botany* 1–26.

Sharma, R.Z. 2016. Optimization of Methyl Jasmonate and cyclodextrin-enhanced Taraxerol and Taraxasterol in (*Taraxacumofficinale Weber*) Cultures. *Plant Physiology and Biochemistry* 103: 24–30. https://doi.org/10.1016/j.plaphy.2016.02.029

Sheldon, A.R., and N.W., Menzies. 2005. The effect of copper toxicity on the growth and root morphology of Rhodes grass (*Chloris gayana* Knuth.) in resin buffered solution culture. *Plant Soil* 278: 341–349.

Singh R., N. Gautam, A. Mishra, et al. 2011. Heavy metals and living systems: an overview. *Indian Journal of Pharmacology* 43(3): 246–253.

Singh, S., P. Parihar, R. Singh, et al. 2016. Heavy metal tolerance in plants: role of transcriptomics, proteomics, metabolomics, and ionomics. *Frontiers in Plant Science* 6: 1143.

Sivritepe, H.Ö., N. Sivritepe, A. Eris, et al. 2005. The effects of NaCl pre-treatments on salt tolerance of melons grown under long-term salinity. *Scientia Horticulturae* 106: 568–581.

Smiri, M., A. Chaoui, N. Rouhier, et al. 2011. Cadmium affects the glutathione/glutaredoxin system in germinating pea seeds. *Biological Trace Element Research* 142: 93–105.

Sprynskyy, M., P. Kosobucki, T. Kowalkowski, et al. 2007. Influence of clinoptilolite rock on chemical speciation of selected heavy metals in sewage sludge. *Journal of Hazardous Materials* 149: 310–316.

Srivastava, A.K., V.H. Lokhande, V.Y. Patade, et al. 2010. Comparative evaluation of hydro-, chemo-, and hormonal priming methods for imparting salt and PEG stress tolerance in Indian mustard (*Brassica juncea* L.). *Acta Physiologiae Plantarum* 32: 1135–1144.

Strauss, S.Y., and A.R., Zangerl. 2002. Plant–insect interactions in terrestrial ecosystems. In: Herrera CM, Pellmyr O (eds) Plant–animal interactions: an evolutionary approach. Blackwell Science, Oxford, UK, pp 77–106.

Suzuki, N., R.M. Rivero, V. Shulaev, et al. 2014. Abiotic and biotic stress combinations. *New Phytologist* 203:32–43. https://doi.org/10.1111/nph.12797

Świeca, M. 2016. Elicitation and Treatment with Precursors of Phenolics Synthesis Improve Low-Molecular Antioxidants and Antioxidant Capacity of Buckwheat Sprouts. *Acta Scientiarum Polonorum, Technologia Alimentaria* 15: 17–28. https://doi.org/10.17306/J.AFS.2016.1.2

Sytar, O., A. Kumar, D. Latowski, et al. 2013. Heavy metal-induced oxidative damage, defense reactions, and detoxification mechanisms in plants. *Acta Physiologiae Plantarum* 35: 985–999.

Takruri, H.R., and M.A. Dameh. 1998. Study of the nutritional value of black cumin seeds (*Nigella sativa* L). *Journal of the Science of Food and Agriculture* 76: 404–410.

Telci, I., A. Sahin-Yaglioglu, F. Eser, et al. 2014. Comparison of Seed Oil Composition of *Nigella sativa* L. and N. *damascena* L. during Seed Maturation Stages. *Journal of the American Oil Chemists' Society* 91: 1723–1729.

Tiruppur Venkatachallam, S.K., Pattekhan, H., Divakar, S., and Kadimi U.S. 2010. Chemical composition of Nigella sativa L. seed extracts obtained by supercritical carbon dioxide. J Food Sci Technol. 47(6): 598–605. doi: 10.1007/s13197-010-0109-y

Treutter, D. 2005. Significance of Flavonoids in Plant Resistance and Enhancement of Their Biosynthesis. *Plant Biology* 7: 581–591. https://doi.org/10.1055/s-2005-873009

Türkan, I., M. Bor, F. Özdemir, et al. 2005. Differential responses of lipid peroxidation and antioxidants in the leaves of drought-tolerant *P. acutifolius* Gray and drought-sensitive *P. vulgaris* L. subjected to polyethylene glycol mediated water stress. *Plant Science* 168:223–231.

Vahidnia, A., G.B. van der Voet, and F.A. de Wolff. 2007. Arsenic neurotoxicity--a review. *Human & Experimental Toxicology* 26(10):823–832. doi: 10.1177/0960327107084539

Valerio, M., S. Lovelli, M. Perniola, et al. 2013. The role of water availability on weed–crop interactions in processing tomato for southern Italy. *Acta Agriculturae Scandinavica, Section B* 63:62–68.

Varghese, E.S.V. 1996. Applied Ethnobotany: A Case Study Among the Kharias of Central India. Deep Publications, New Delhi, ISBN-13: 978-8185622064, Pages: 307.

Viehweger, K. 2014. How plants cope with heavy metals. *Botanical Studies* 55: 35.

Wang, W., B. Vinocur, and A. Altman. 2003. Plant responses to drought; salinity and extreme temperatures: towards genetic engineering for stress tolerance. *Planta* 218:1–14.

Wang, W., B. Vinocur, and A. Altman. 2003. Plant responses to drought, salinity and extreme temperatures: Towards genetic engineering for stress tolerance. *Planta* 218: 1–14.

Wang, M.C., D. Bohmann, and H. Jasper. 2003. JNK signaling confers tolerance to oxidative stress and extends lifespan in *Drosophila*. *Developmental Cell* 5: 811–816.

Wang, X.D., B.T. Xie, M.W. Tan, et al. 2009. Effects of exogenous GABA on yield, quality, and high temperature tolerance during anthesis stage of winter wheat. *Journal of Triticeae Crops* 4: 623–626.

WHO . 2005. National policy on traditional medicine and regulations of herbal medicines. Geneva.

Woo, C.C., S.Y. Loo, V. Gee, et al. 2011. Anticancer Activity of Thymoquinone in Breast Cancer Cells: Possible Involvement of PPAR-γ Pathway. *Biochemical Pharmacology* 82: 464–475. https://doi.org/10.1016/j.bcp.2011.05.030

Xia, Y., Y. Qi, Y. Yuan, et al. 2012. Overexpression of Elsholtzia haichowensis metallothionein 1 (EhMT1) in tobacco plants enhances copper tolerance and accumulation in root cytoplasm and decreases hydrogen peroxide production. *Journal of Hazardous Materials* 233: 65–71.

Yadav, P.V., M. Kumari, and Z. Ahmed. 2011. Chemical seed priming as a simple technique to impart cold and salt stress tolerance in capsicum. *Journal of Crop Improvement* 25: 497–503.

Yadav, S.K. 2010. Heavy metals toxicity in plants: an overview on the role of glutathione and phytochelatins in heavy metal stress tolerance of plants. *South African Journal of Botany* 76: 167–179.

Yousefi, A.R., S. Rashidi, P. Moradi, et al. 2020. Germination and Seedling Growth Responses of Zygophyllum fabago, Salsola kali L. and Atriplex canescens to PEG-Induced Drought Stress. *Environments* 7: 107.

Zheljazkov, V.D., E.A. Jeliazkova, N. Kovacheva, et al. 2008. Metal uptake by medicinal plant species grown in soils contaminated by a smelter. *Environmental and Experimental Botany* 64(3): 207–216.

Zheng, L., F. S. Hong, S. P. Lu, et al. 2005. Effect of Nano-Tio$_2$ on Strength of Naturally and Growth Aged Seeds of Spinach. *Biological Trace Element Research* 104: 83–91. https://doi.org/10.1385/BTER:104:1:083.

Zhou, B.,W. Yao, S. Wang, et al. 2014. The metallothionein gene, TaMT3, from Tamarix and rossowii confers Cd^{2+} tolerance in tobacco. *International Journal of Molecular Sciences* 15(6): 10398–10409.

Ziska, L.H., M.B. Tomecek, and D.R. Gealy. 2010. Evaluation of competitive ability between cultivated and red weedy rice as a function of recent and projected increases in atmospheric CO$_2$. *Agronomy Journal* 102:118–123. https://doi.org/10.2134/agronj2009.0205

Zohary, D., and M. Hopf. 2000. Domestication of Plants in the Old World: The Origin and Spread of Cultivated Plants in West Asia Europe and the Nile Valley. Oxford, UK: Oxford University Press.

Żukowska, J., and M. Biziuk. 2008. Methodological evaluation of method for dietary heavy metal intake. *Journal of Food Science* 73: R21–R29.

13 Olea europaea and Stressful Conditions

Jalal Kassout[1,2*], Jean-Frédéric Terral[3], Rachid Azenzem[2],
Abdeltif El Ouahrani[1], Abdelouahab Sahli[1],
Mhammad Houssni[1], Soufian Chakkour[1], Khalil Kadaoui[1]
and Mohammed Ater[1]

[1]Laboratory of Applied Botany, BioAgrodiversity Team, Faculty
of Sciences of Tétouan, Abdelmalek Essaâdi University, Tétouan,
Morocco
[2]National Institute of Agronomic Research, INRA-CRRA, Marrakech-Safi,
BP 533, Marrakech, Morocco
[3]ISEM, Université de Montpellier, Equipe DBA, CNRS, IRD, EPHE,
Montpellier, France
*Corresponding author: jalalkassout@gmail.com;
jalal.kassout@inra.ma

CONTENTS

13.1 INTRODUCTION

Plants are continuously exposed to various environmental factors. They are prone to a large spectrum of stressful conditions arising from both biotic and abiotic stressors. Abiotic stressors include drought, low or high temperature, high salinity and heavy metals, while biotic stressors refer to damage caused by fungi, bacteria, insects, nematodes, viruses or herbivores. Both stressors end up limiting plant growth, development and productivity (Cipollini 2004; Zhao et al. 2017; Xu et al. 2019; Lamers et al. 2020). As a result, they cause enormous economic losses

DOI: 10.1201/9781003242963-13

by reducing the yield and/or quality of agricultural products and distressing the functioning of natural ecosystems (van Lierop et al. 2015; Savary et al. 2019; Graziosi et al. 2019). In the current context of global change, the vulnerability of plant species to biotic and abiotic stresses is increasing with the observed and predicted climate changes (Allen et al. 2015; Zhao et al. 2017; Wu et al. 2020), as well as global pathogen and pest outbreaks (Lee et al. 2012; Saponari et al. 2014; Spence et al. 2020). The rapid rate of changing climate is likely to affect the long-term persistence and resilience of plant species (Allen et al. 2010; Vellend et al. 2017; Zhan et al. 2018). For instance, the impact of climate change and anthropogenic pressures in the Mediterranean region is already evident (Polade et al. 2014; Matesanz and Valladares 2014). These changes are likely to pose severe conditions for Mediterranean plant species (Blondel et al. 2010; Lloret et al. 2016), particularly for those with specific biogeographical requirements, such as the emblematic Mediterranean olive tree, which plays a dual role as a forest and agricultural fruit tree.

The olive tree (*Olea europaea* L.) is arguably the most iconic fruit tree in the Mediterranean region (Besnard et al. 2018), embracing extraordinary economic, cultural and ecological interests (Kaniewski et al. 2012). Thus, its agroecosystems' sustainability is significantly worth particular attention from all stockholders in order to build a long-term conservation strategy. Today, six sub-species are recognized (Green 2002; Figure 13.1), but only one of them has been domesticated "*Olea europaea* subsp. *europaea* var. *sylvestris*" also known as oleaster (Besnard et al. 2018). This latter is the progenitor of all cultivated olive varieties grouped nowadays under the "*O. e.* subsp. *europaea* var. *sativa*" denomination. From an archeological perspective, the exploitation of oleaster is traced back to the Upper Paleolithic and early Neolithic, mostly as fodder and firewood sources (Terral et al. 2004; Carrión et al. 2010; Kaniewski et al. 2012). Recent paleobotanical, archeological, historical and genetic studies (e.g., Terral et al. 2004; Carrión et al. 2010; Besnard et al. 2013; Díez et al. 2015) have identified multiple centers of olive domestication from the autochthonous oleaster population. These studies have revealed that the expansion of olive cultivation in the Mediterranean is mostly linked to human activities and favorable climatic conditions (Terral and Mengüal 1999; Figueiral and Terral 2002; Kaniewski et al. 2012). Nowadays, the Mediterranean region remains the main olive cultivation hotspot for producing oil and table olives (FAO 2019). Widespread around the Mediterranean region, oleaster could be found under various climatic conditions, representing *per se* a fundamental element of several woody plant communities (Benabid and Fennane 1994; Kassout et al. 2019; Gianguzzi and Bazan 2019). For instance, the olive tree can withstand a wide spectrum of climate variability through several morpho-physiological and biochemical traits that are involved in maintaining the main physiological processes (Terral et al. 2004; Bacelar et al. 2009; Ahmadipour et al. 2018; Kassout et al. 2019, 2021). However, in the light of rapid global warming causing the expansion of pests and diseases, both wild and cultivated olive trees are under several threats. Within this context, it is crucial to understand the possible response of the olive tree to biotic and abiotic stressors to develop adequate strategies to maintain its genetic resources and to develop suitable breeding programs. In this chapter, we present a review of the main responses of the olive tree to biotic and abiotic stressors.

13.2   BOTANY OF *OLEA EUROPAEA*

The olive tree (*Olea europaea* L. subsp. *europaea*) traces back its origin to several million years and it is linked to the Mediterranean climate (Suc 1984; Palamarev 1989). The genus *Olea* (Oleaceae) contains approximately 40 species and subspecies (Green 2002). However, phylogeny revision of *Olea* genus demonstrated that this genus is polyphyletic and should be restricted to only 17 species or subspecies, which belong to three lineages extant throughout Africa, southern Europe, Asia and Oceania (Besnard et al. 2009). As of today, six subspecies of wild olive, also known as the "olive complex," are recognized (Médail et al. 2001; Green 2002): subsp. *europaea* in the

FIGURE 13.1　Native distribution of wild olive (*Olea europaea* L.) subspecies (according to Rubio de Casas et al. 2006).

Mediterranean basin; subsp. *laperrinei* (Batt. & Trab.) Cif. in Saharan massifs (Hoggar, Aïr, Jebel Marra); subsp. *cuspidata* (Wall. *ex* G.Don) Cif. in southern Africa to southern Egypt and from Arabia to China; subsp. *guanchica* P. Vargas *et al.* in the Canary Islands; subsp. *maroccana* (Greut. & Burd.) P. Vargas *et al.* in southern Morocco; and subsp. *cerasiformis* G. Kunkel & Sunding in Madeira. These subspecies are allogamous (Green 2002) and diploid forms (2n=46), except for subsp. *cerasiformis* and subsp. *maroccana*, which are tetraploid and hexaploid, respectively (Besnard et al. 2008). Among the six recognized subspecies, *O. e.* subsp. *europaea* has a large distribution area over the Mediterranean Basin (Besnard et al. 2018). Two botanical varieties of *O. e.* subsp. *europaea* are distinguished: var. *europaea* for cultivated forms and var. *sylvestris* for wild and spontaneous trees that are usually named "oleaster," which is considered the wild ancestor of all the cultivated varieties (Green 2002; Besnard et al. 2018). Oleaster is widespread around the Mediterranean Basin and represents a characteristic element of its flora (Carrión et al. 2010; Gianguzzi and Bazan 2019). It can be found, often mixed with numerous feral individuals escaped from cultivation, and is currently present in the Levant, Turkey, the Peloponnese, and coasts of mainland Greece, the Maghreb, the southern Iberian peninsula, southern Italy, Cyrenaica (Libya), the Mediterranean islands and, much more sporadically on the northern Mediterranean coast (Zohary et al. 2012). The olive tree remains a keystone species of traditional agro-systems in the Mediterranean (Loumou and Giourga 2003; Besnard et al. 2018; Khadari et al. 2019). This is the case for Moroccan traditional agroecosystems, where the olive tree is tightly associated with ecological, social and economic dynamics (Haouane 2012; Ater et al. 2016; Lauri et al. 2019). For instance, within the traditional agroecosystems in northern Morocco, oleaster trees were preserved and used for grafting with olive varieties, which is considered an ongoing process of domestication (Aumeeruddy-Thomas et al. 2017).

The olive tree is a medium-sized evergreen, sclerophyllous, long-lived and wind-pollinated tree (Lewington and Parker 1999; Rhizopoulou 2007; Arnan et al. 2012). Olive leaves are simple, with opposite leaflets in terminal or axillary inflorescences, and they exhibit a large variation in shape and size (Kassout et al. 2019). Thus, olive leaf anatomy and surface characteristics exhibit drought-resistant features, including thick cuticle, high stomatal density, high thickness and compact mesophyll cells (Bacelar et al. 2004; Kassout et al. 2019). Generally, every two or three years, the leaves are replaced by new ones that appear in spring and/or autumn. The flowers are radially symmetrical, usually hermaphrodites, but sometimes unisexual. The tree produces a large number of flowers, but the fruit amount is usually less than 5% (Martin et al. 1994). It shows high phenotypic diversity in tree architecture (Rallo et al. 2008). Olive fruit is a drupe, composed of three principal tissues, the endocarp, mesocarp and epi- or exocarp, together they form the pericarp, initiated as the ovary wall. Oleaster wood anatomical features are classified as diffuse-porous based on the constitutive elements involved in sap conduction, mainly vessels. Vessels are more or less numerous, isolated or mainly joined in radial lines of from two to five or seven. Rays are uni- to biseriate, the latter being most common and the axial parenchyma is paratracheal. Growth ring boundaries are generally distinguishable in immature wood, such as in young branches and twigs, which are still free of any mechanical constraint. Nevertheless, in mature wood of the trunk and older branches, growth rings are indistinct owing to a heterogeneous cambium functioning on its circumference, and biomechanical constraints linked to orientation and weight of the branch (Terral 2000).

Looking to its ecology, oleaster is indigenous to the thermo-Mediterranean climate from humid to semi-arid, but appears eliminated by the cold and extreme aridity (Carrión et al. 2010; Kassout et al. 2022). However, its distribution area is less than that of cultivated olive, which is very fragmented mainly due to anthropogenic disturbances on natural habitats amplified by climatic changes and aridification (Benabid 1984, 1985; Gianguzzi and Bazan 2019). In the past, oleaster persisted during the Pleniglacial in the Mediterranean region in which several refugia were identified by palaeoecological studies (Figueiral and Terral 2002; Carrión et al. 2010) and inferred by genetic computational models (Besnard et al. 2013). The Holocene climatic warming would then have allowed the rapid expansion of oleaster (Figueiral and Terral 2002; Carrión et al. 2010). Recent genetic studies revealed an east–west differentiation of oleaster populations (Breton et al. 2009; Besnard et al. 2013; Díez et al. 2015), with three distinct plastid DNA lineages characterizing the geographical structure of its genetic diversity, namely E1, from the Peloponnese (Greece) to the Levant; E2 and E3 in the western part of the Mediterranean Basin (Besnard et al. 2013). Moreover, the phylogeographic and population genetic studies have highlighted that most olive cultivars are derived from the eastern-oleaster population (Lumaret et al. 2004; Breton et al. 2009; Díez et al. 2015), and therefore, 90% of olive cultivars are characterized by the E1 lineage (Besnard et al. 2013). Nowadays, more than 10 million hectares are under olive tree cultivation with expansion to Argentina, Australia, Chile, China and the United States (FAO 2019). As a result of selection and breeding programs, more than 1200 olive varieties are selected and cultivated around the Mediterranean (FAO 2019) and are grown to produce high-quality fruit for oil and table consumption (Bartolini et al. 2005).

Olive cultivars could be classified according to their use as table olives, oil olive or both table olives and oil. Thus, they can be classified according to their geographic origins such as *Picual*, *Arbequina* and *Cornicabra* in Spain, *Picholine marocaine*, *Menara* and *Haouzia* in Morocco, *Chemlali* in Tunisia, *Frantoio*, *Moraiolo* and *Leccino* in Italy, and *Ayvalik* in Turkey (Jacoboni and Fontanazza 1981; FAO 2010; Muzzalupo et al. 2014; Belaj et al. 2004; Gemas et al. 2004; Martins-Lopes et al. 2007; Khadari et al. 2008; Fabbri et al. 2009).

13.3 MEDICINAL AND NUTRITIVE USES OF *OLEA EUROPAEA*

The olive tree has long been considered a valuable tree by ancient civilizations around the Mediterranean region due to its oil and fruit's medicinal and nutritive virtues (Kaniewski et al. 2012). During antiquity, the olive was mentioned by Theophrastus [371–287 (BCE)] in *De Causis Plantarum* and *De Historia Plantarum* as the most valuable tree of all times, thus, Dioscorides [40–90 Current Era (CE)] mentioned the olive in *De Materia Medica* (CE 50–70) as a beneficial medicinal tree (Rhizopoulou 2007; Riddle 1985). *O. europaea* has been widely described in traditional medicine recipes in which several parts of the plant are used, such as fruits, leaves, seeds and wood (Table 13.1). Medicinal and nutritive uses of *O. europaea* are recorded throughout the Mediterranean region and the world, where it has been used to treat various ailments. Thus, isolated components have shown various pharmacological activities, attesting that the olive represents one of the most valuable tree species for the Mediterranean region and the world.

Olive leaves and fruits are used in traditional medicine and consumed alone or mixed with herbs. For instance, they have been used to treat diabetes, hypertension, inflammation, diarrhea, respiratory and urinary tract infections, stomach and intestinal diseases, asthma, hemorrhoids and rheumatism, and also as a laxative, mouth cleanser and vasodilator (Bellakhdar 1997). Decoction of dried leaves and fruit is used orally to treat diarrhea, respiratory and urinary tract infections, stomach and

TABLE 13.1
Some Traditional Uses of *Olea europaea* L. Parts

Part used	Preparation used	Medicinal use	References
Leaves and fruits	Infusions and/or macerations	Hypoglycemic, hypotensive	Bouzabata (2013), Bellakhdar (1997)
	Decoction or infusion	Antidiabetic	Ali-Shtayeh et al. (2012)
Fruits	Olive oil + lemon juice	To treat gallstones	Hashmi et al. (2015)
	Olive oil applied on scalp	To prevent hair loss	Zargari (1997)
	Olive oil	Applied over fractured limbs	Ghazanfar and Al-Sabahi (1993)
		Skin cleanser	Fujita et al. (1995)
Seeds	Seeds oil taken orally	Laxative	Al-Khalil (1995)
Leaves	Leaves extract in hot water	Diuretic	Vardanian (1978), Bellakhdar (1997)
	Infusion of leaves and taken orally	Antipyretic, anti-inflammatory, tonic, antibacterial	Ribeiro et al. (1986), Haloui et al. (2010)
	Infusion	Eye infections treatment, to treat gout	Guerin and Reveillere (1985), Flemmig et al. (2011)
	Boiled and taken orally	To treat asthma	Lawrendiadis (1961)
Dried leaves	Boiled and taken orally	To treat hypertension	Ribeiro et al. (1988), Bellakhdar et al. (1991), Ribeiro et al. (1986)
Dried leaves and fruit	Decoctions with oral use	Diarrhea, respiratory, and urinary tract infections	Bellakhdar et al. (1991)

intestinal diseases, and also as mouth cleansers (Bellakhdar et al. 1991) and to treat hypertension and diabetes (Tahraoui et al. 2007). Olive oil is used for hair loss and taken orally to treat hypertension and agitation, as a laxative and vermicide (Bellakhdar 1997). Table 13.1 gives some traditional uses of olive parts.

The use of olive leaves to treat diabetes has been proven by several experimental studies and is mainly linked to the presence of oleuropein and hydroxytyrosol due to their ability to restrain oxidative stress, which is widely associated with pathological complications of diabetes (de Bock et al. 2013; El-Amin et al. 2013; Eidi et al. 2009). Moreover, leaf extract has been demonstrated to have anti-microbial, anti-viral, anti-oxidant and anti-diabetic effects. For instance, olive leaf extract has been reported to be an effective antimicrobial agent against various pathogens, including *Salmonella typhi*, *Vibrio parahaemolyticus* and *Staphylococcus aureus* (including penicillin-resistant strains); and also *Klebsiella pneumonia* and *Escherichia coli*, causal agents of intestinal or respiratory tract infections in humans (Bisignano et al. 1999). Hence, antiviral activity has been observed in leaf extract, reportedly caused by the constituent calcium elenolate, a derivative of elenolic acid (Periera et al. 2007; Adnan et al. 2014). Thus, the anti-oxidant effect has been associated with oleuropein with the ability to decrease the oxidation of low-density lipoprotein (LDL) cholesterol (Visioli et al. 1994). Moreover, phytochemical studies on olive fruits and leaves have highlighted the diversity of their composition in natural compounds that are of great interest in medicine and nutrition (Esti et al. 1998; Bianco and Uccela 2000; Jerman et al. 2010). For instance, flavonoids, flavone glycosides, flavanones, iridoids, iridane glycosides, secoiridoids, secoiridoid glycosides, triterpenes, biophenols, benzoic acid derivatives, xylitol, sterols, isochromans, sugars and a few other types of secondary metabolites have been isolated from different parts of the olive tree (Obied 2013; Jerman et al. 2010). Olive fruits contain a considerable amount of flavonoids, secoiridoids, secoiridoid glycosides and phenolics such as tyrosol, hydroxytyrosol,, and their derivatives (Bianco et al. 1993; Esti et al. 1998; Bianco and Uccela 2000; Owen et al. 2003). Thus, oleuropein represents the most abundant component in olive fruits, and is highly appreciated for its medicinal interest (Cardoso et al. 2011). A large number of flavonoids have been reported from the fruits such as quercetin (Bouaziz et al. 2005), luteolin-7-O-rutinoside (Peralbo-Molina et al. 2012; Bouaziz et al. 2005), apigenin-7-O-rutinoside (Bouaziz et al. 2005; Meirinhos et al. 2005), rutin (Savarese et al. 2007; Bouaziz et al. 2005; Meirinhos et al. 2005), vicenin-2, chrysoeriol, chrysoeriol-7-O-glucoside (Bouaziz et al. 2005), luteolin-7-O-glucoside (Savarese et al. 2007; Bouaziz et al. 2005), quercetin-3-rhamnoside, apigenin (Savarese et al. 2007) and quercitrin (Jerman et al. 2010).

13.4 RESPONSES OF *OLEA EUROPAEA* TO ABIOTIC AND BIOTIC STRESSORS

Unequivocally, abiotic and biotic stressors induce a range of morphological, physiological, biochemical and molecular changes in plant species (Grime 2006; Ghalambor et al. 2007; Osakabe et al. 2014). However, little is known about the specificity of this induction. An understanding of how plants respond to environmental stressors will provide the know-how to help improve plant stress tolerance in breeding and conservation programs. From this perspective, understanding *O. europaea* responses to various environmental stressors will surely provide valuable information on the adaptive strategies of this iconic Mediterranean tree.

13.4.1 RESPONSES OF *OLEA EUROPAEA* TO ABIOTIC STRESSORS

Abiotic stress factors, such as drought, low or high temperature, deficient or excessive water, high salinity, heavy metals and ultraviolet (UV) radiation, among others, largely influence plant growth and development, crop production and ecosystem sustainability. These stresses pose a severe threat to natural and agricultural ecosystems (Wang et al. 2003; Xu et al. 2019). Olive (*O. europaea*) trees are widespread in Mediterranean agroecosystems and are extensively cultivated in many regions around the world (FAO 2019). The rise in the economic value of olive products, especially olive oil, has led to changes in olive cultivation systems from traditional with low-density and low input to high density and high input, and consequently, the emergence of new challenging conditions resulting in abiotic stress factors. The effects of abiotic factors on olive trees might have strong consequences on both yield production and quality. However, the olive has a high capacity to grow under various environmental conditions, especially under the Mediterranean climate variability from mild and wet winters to hot and dry summers (Driouech et al. 2010; Kassout et al. 2019). Olive adaptability has been associated with the changes in its morphological characteristics and physiological mechanisms that occur under various abiotic stress factors (Terral et al. 2004; Connor and Fereres 2005; Fernández 2014; Kassout et al. 2019, 2021). Generally, abiotic stress in the olive tree can lead to changes in the anatomical and morphological traits of roots, stems and leaves, thus on physiological and molecular mechanisms (Terral et al. 2004; Fernández 2014; Kassout et al. 2021). These changes are likely to have negative impacts on olive growth, development and performance (Fernández et al. 1991; Rossi et al. 2013; Kassout et al. 2019, 2021).

13.4.1.1 Responses of *Olea europaea* to Drought

Drought stress is recognized as the most limiting factor affecting plant growth, development, productivity and distribution (Choat et al. 2018; Berdugo et al. 2020; Kassout et al. 2022). Increasing aridity and drought have been identified as the major causes of forest die-off (Allen et al. 2010), the decrease in terrestrial primary production and nutrient cycling (Xu et al. 2019) and the increase in the frequency and intensity of disturbances such as fire and land-cover degradation (Piekle et al. 2011; Dullinger et al. 2012). Moreover, the changes in precipitation regimes and global warming have increased the frequency and intensity of drought (Dai 2013; Spinoni et al. 2019). For instance, in the Mediterranean region, considered a valuable hotspot for cultivated and wild olive, the climate scenarios predict an increase in temperature and a decrease in precipitation (Driouech et al. 2010; Polade et al. 2014). Such changes could affect plant communities and agricultural systems with a prominent threat to growing conditions (Bréda et al. 2006; Choat et al. 2012; Lloret et al. 2016).

Drought stress is considered multidimensional stress leading to changes in the physiological, morphological, ecological, biochemical and molecular traits of plants (Choat et al. 2018; Kassout et al. 2019; Yang et al. 2021). Drought stress significantly reduces crop production and yield performance (Bray et al. 2000; Jaleel et al. 2009; Fang et al. 2010; Cohen et al. 2021). Drought stress could lead to a significant decrease in plant water content and therefore, loss of turgor, which causes a drastic reduction in leaf water potential (Ψ) and cell division and expansion (Farooq et al. 2012). This could result in a significant reduction in leaf size and number or even a fall of leaves, and thus, a reduction in the number of branches and shoots (Farooq et al. 2012), and a limitation in the photosynthetic capacity of the whole canopy (Taiz and Zeiger 2006). In the hydraulic system, xylem water potential decreases under prolonged drought stress, which increases the susceptibility to cavitation and embolism (Terral et al. 2004; Fernández 2014; Choat et al. 2018) and leads to the interruption of whole-plant water transport. However, plant species have developed various and complex morphological, anatomical, physiological and biochemical mechanisms to reduce the impacts of drought stress (Michaletz et al. 2015; Sack et al. 2016; Brodribb et al. 2020; Kassout et al. 2021). For instance, a range of leaf traits of Moroccan populations of wild olive (e.g., dry matter content, surface area and stomata density) was found by Kassout et al. (2019) to be strongly related to ecological parameters associated with aridity. Hence, they showed that mechanistic traits, such as stomatal density and leaf water content, play a crucial role in wild olive responses to the climate variability of the Mediterranean region (Kassout et al. 2019). Thus, the sclerophylly characteristic of the olive leaves plays an important role in maintaining favorable gas exchange and leaf water content and, therefore, allowing trees to persist under water-deficit conditions (Marchi et al. 2008). Under drought conditions, olive leaves are usually thicker with higher leaf tissue (Kassout et al.

2019), which could be considered a mechanism contributing to greater stress tolerance because of the greater resistance to desiccation damage (Mediavilla et al. 2012). Moreover, Kassout et al. (2021) showed that increasing aridity and changes in available water resource, control and modulate the sap-conduction performance of wild olive trees. The authors demonstrated that decreasing vessel lumen area (SVS) was strongly affected by increasing aridity. Moreover, they showed that the individuals growing under lower water resources seem to produce wood with higher hydraulic conductivity early in the branch diameter. Consequently, the inability to produce larger vessels during the development of the growth unit is potentially lethal to the branch of the olive tree and in consequence to the persistence of the population (Kassout et al. 2021). Thus, Terral et al. (2004) reported that the size of the vessel in the olive tree is strongly influenced by seasonal or annual variations in precipitation. Moreover, the anatomical response was even more important when trees are irrigated, as shown in the study by Terral and Durand (2006) for cultivated olives. Consequently, intraspecific variation in *O. europaea* wood anatomy could be considered a dynamic compromise between water transport efficiency by forming large vessels, and safe sap conduction associated both with narrower, numerous and joined vessels to limit the risks of cavitation and embolism and with structural support (Terral et al. 2004; Kassout et al. 2021). Additionally, Torres-Ruiz et al. (2013) demonstrated that olives can withstand a low-stem water potential (Ψ_{stem}) approach to − 5 MPa with minor xylem embolism risks. Moreover, under drought conditions, the olive tree displays a strong capacity to leaf and stem osmotic adjustments, which enables the olive to decrease osmotic potential and to establish a soil–plant gradient to maintain water extraction from the soil even at a low water potential (Dichio et al. 2006; Sofo et al. 2008).

In olive leaves, the stomata occur on the abaxial surface (i.e. hypostomatous species) hidden by numerous trichomes, thus limiting water loss (Pallioti et al. 1994). Stomatal regulation represents an important mechanism to withstand aridity, which is influenced by both hydraulic and hormonal signals (Torres-Ruiz et al. 2014). Stomatal conductance (g_s) responses to water deficit in the olive tree were strongly associated with leaf hydraulic conductance (Hernandez-Santana et al. 2016), but also with the hydraulic failure in roots and stems (Torres-Ruiz et al. 2014). Thus, Bacelar et al. (2007a) showed that under severe drought conditions, olive water use-efficiency decreases significantly. In fact, under drought stress, stomatal conduction (g_s) declines as the water-deficit stress increases (Fernández et al. 1991), and consequently, there is a reduction in the photosynthetic rate of olive trees (Fernández 2014). From another perspective, hormonal regulation (abscisic acid, ABA) has been shown to promote stomatal closure by both direct actions on guard cells and indirect hydraulic action through a decrease in leaf water permeability (Pantin et al. 2013). Thus, a significant correlation was reported between ABA and leaf water potential (Ψ_{leaf}) in olive trees (Kitsaki and Drossopoulos 2005). Also, in the drought-tolerant *chemlali* cultivar, lower levels of ABA were observed compared with the drought-sensitive "Chetoui" cultivar (Guerfal et al. 2009). Furthermore, the regulation of the antioxidant system plays a key role in controlling the oxidative stress caused by reactive oxygen species (ROS). Drought stress facilitates the production of ROS, which acts as a signaling molecule or might cause oxidative stress in the plant (Mattos and Moretti 2015). Thus, drought stress induces a decrease in some antioxidant enzyme activities, such as ascorbate peroxidase, catalase and superoxide dismutase, but also in non-enzymatic antioxidant mechanisms, such as the accumulation of phenolic compounds, tocopherols, carotenoids, ascorbate and glutathione (Bacelar et al. 2006; Petridis et al. 2012; Abdallah et al. 2017). Genotypic differences among olive cultivars were found to reflect some differences in drought tolerance. For instance, Bacelar et al. (2009) showed that cultivars native to dry regions exhibit more plasticity to drought conditions compared to cultivars native to the temperate region. Thus, Bacelar et al. (2004) showed that some olive cultivars, such as *Cobrançosa*, exhibited a high foliar-density tissue with thick leaves and trichome layers, while others, such as *Manzanilla* and *Negrinha*, developed thicker parenchyma tissues and cuticle layers. Hence, the *Cobrançosa* cultivar shows an enhanced water use-efficiency compared to *Madural* and *Verdeal Transmontana* cultivars (Bacelar et al. 2007a).

TABLE 13.2
Olive Main Strategies to Improve Drought Tolerance

Main strategies	Mechanism	References
Water extraction from soil	High root density close to trunk surface	Abd-El-Rahman et al. (1966), Fernández and Moreno (1999), Searles et al. (2009)
	Large root system	Lavee (1996), Gómez et al. (2001)
	High root/canopy ratio	Moreno et al. (1996), Celano et al. (1999), Fernández et al. (1991, 1992)
	Small and dense xylem vessels	Terral et al. (2004), Kassout et al. (2021), Tognetti et al. (2009), Diaz-Espejo et al. (2012)
	Hydraulic redistribution	Fernández and Clothier (2002), Terral et al. (2004), Ferreira et al. (2013) Nadezhdina et al. (2012), Rewald et al. (2011)
	Embolism resistance	Salleo and Gullo (1983), Bongi and Palliotti (1994), Terral et al. (2004), Torres-Ruiz et al. (2013)
	Osmotic adjustments	Tognetti et al. (2009), Diaz-Espejo et al. (2012)
	Decrease water potential	Larcher et al. (1981), Gimenez et al. (1996), Moriana et al. (2012)
Water loss control and restriction	Small and sclerophyllous leaves	Bacelar et al. (2004), Diaz-Espejo et al. (2012), Kassout et al. (2019)
	Paraheliotropism	Schwabe and Lionakis (1996), Natali et al. (1999)
	Small and dense stomata	Kassout et al. (2019), Marchi et al. (2008)
	Dense trichome layer	Pallioti et al. (1994), Diaz-Espejo (2000)
	Hypostomatous leaves	Pallioti et al. (1994), Ennajeh et al. (2010)
	Reduced stomatal conductance	Fernández et al. (1997), Tognetti et al. (2009), Boughalleb and Hajlaoui (2011)
	Leaf mesophyll compactness	Bacelar et al. (2004), Marchi et al. (2008), Ennajeh et al. (2010)
	Aquaporins regulation	Secchi et al. (2007), Perez-Martin et al. (2011)
	Increase in carotenoids and the carotenoids/chlorophylls ratio	Abdallah et al. (2017)
	Increment in enzymatic and non-enzymatic	Bacelar et al. (2006, 2007b), Guerfel et al. (2009a,b), Petridis et al. (2012)

The *Arbequina* cultivar is characterized by smaller leaves, which reduce whole-plant water loss, however, it has thinner trichome layers, thus having less protection against water loss (Bacelar et al. 2004). Table 13.2 highlights the main olive strategies to withstand and tolerate drought stress.

13.4.1.2 Responses of *Olea europaea* to Waterlogging

Olive is the most important cultivated tree species in the Mediterranean region, representing 98% of the cultivated surface for this tree with more than 10.2 million hectares. However, to meet the rising demand for olive products, its cultivation has been expanded to soils of marginal quality, such as those affected by waterlogging. In the current context of global warming, flooding events are expected to be more frequent and intense, and consequently, waterlogging is considered a considerable stress factor for plant species (Bailey-Serres et al. 2012; Zhou et al. 2020).

Waterlogging occurs when excess water saturates the soil pores leading to inhibition of the root's respiration (Sasidharan et al. 2017). During waterlogging, gas exchange between the soil and the atmosphere is interrupted and, as a result, root activity is blocked (Armstrong and Drew 2002). Additionally, the accumulation of toxic substances (i.e. H_2S, N_2, Mn^{+2}, Fe^{+2}) combined with ROS in the root environment under waterlogging conditions, create more stressful conditions for plant growth and development (Tian et al. 2019; Zhou et al. 2020). Moreover, the hypoxic conditions resulting from prolonged waterlogging could ultimately lead to the death of the plant (Fukao et al. 2019). However, the intensity, timing and duration of waterlogging stress could generate different responses within plant species. Various morphological and physiological adaptations occur in the same plant species conferring waterlogging tolerance (Pan et al. 2021). For instance, the induction of adventitious roots in tree species represents a major adaptation mechanism to waterlogging stress (Kozlowski and Pallardy 2002; Qi et al. 2019). Thus, other morphological changes, such as the formation of aeration tissues in the roots and stems, facilitate the soil–atmosphere gas exchange (Kozlowski and Pallardy 2002; Pan et al. 2020).

Olive trees, as for many fruit trees, are very susceptible and sensitive to waterlogging and soil hypoxia, however, it depends on the cultivar (Hassan and Seif 1990). Hassan and Seif (1990) showed that under waterlogging conditions of 30 days the cultivar *Mission* died, while the *Kalamon* cultivar stood for a 60-daysperiod after the application of waterlogging stress. Thus, Aragüés et al. (2010) showed that the combined effect of salinity and waterlogging has a determining effect on the *Arbequina* cultivar's survival and growth. Waterlogging is a complex stress factor acting on several aspects of the whole-plant physiological processes (Pan et al. 2020), which could explain the previously reported negative effect on olive growth and performance.

13.4.1.3   Responses of *Olea europaea* to Salinity

Soil salinity is considered the most limiting stress factor for plant species' growth and development (Zhu 2002). The rising temperatures and the decreasing precipitation rate affect soil proprieties (Gelybó et al. 2018), and consequently, impose more challenging conditions for wild and cultivated species. However, the impact of salt stress is evident for crop species, by an enormous reduction in their productivity and performance (Connor and Fereres 2005; Isayenkov 2012; Zhao et al. 2017). Soil salinity arises from the high concentration of sodium (Na^+) and chlorine (Cl^-) ions, and less from calcium (Ca^{2+}) and sulfate (SO_4^{2-}) ions. Furthermore, salt stress in plants is associated with oxidative stress due to the liberation of ROS (Isayenkov 2012). Moreover, the osmotic potential reduction between root and soil solutions was found to be correlated with the uptake of Na^+ and Cl^- ions (Abbasi et al. 2016).

The olive tree is moderately resistant to salinity (Rugini and Fedeli 1990; Gucci and Tattini 2010). Earlier studies on olive tolerance to salinity have shown that electrical conductivity (EC) higher than 4 to 6 dS m^{-1} induced a 10% reduction in olive production (Bernstein 1965). Thus, Maas and Hoffman (1977) showed that olive production is strongly affected by increasing salinity beyond an ECs of 4 to 6 dS m^{-1}. Hence, a reduction in olive growth and total leaf area was associated with an increasing concentration of NaCl (Vigo et al. 2005), which was interpreted by a strong reduction in the photosynthetic capacity of the olive trees (Vigo et al. 2002). However, salt tolerance appears to depend on olive cultivars (Chartzoulakis et al. 2002, 2006; Vigo et al. 2002). For instance, Chartzoulakis et al. (2002, 2006) showed that under different salt concentration treatments, the cultivar *Kalomon* shows resistance aptitude, while *Megaritiki* and *Kothreiki* were salt-tolerant; however, *Mastoidis*, *Amphissis* and *Koroneiki* were less tolerant. Thus, the *Arbequina* cultivar shows decreasing tolerance to salinity depending on the exposure time and concentration (Aragüés et al. 2005). Vigo et al. (2002, 2005) tested the effect of seawater on some olive cultivars, and showed that increasing the concentration of salt water significantly decreases the total plant mineral content, leaf number, leaf area and the leaf development in *Chondrolia Chalkidikis*, *Kalamon* and *Manzanilla de Sevilla* cultivars. Tattini et al. (1992) showed significant differences in the responses to salinity between *Frantoio* and *Leccino* cultivars. However, is noteworthy to mention that the maximum value of NaCl concentration tolerated by olive trees was estimated at around 137 mM (Rugini and Fedeli 1990), while the production significantly dropped above 30 mM (Gucci and Tattini 2010). Besides productivity reduction and growth performance, salt stress in the olive tree is characterized by apparent foliar chlorosis and necrosis, desiccation of flowers, ovary abortion, and for the most severe conditions, necrosis of the stem tip and at the root level (Gucci et al. 2003). However, even though fruit production is negatively affected by salinity stress; it does not alter the olive oil content (Gucci et al. 2003; Chartzoulakis 2005). Finally, salt tolerance in the olive tree is associated with its ability to exclude and retain Na^+ and Cl^- in the root (Chartzoulakis 2005) and is a complex process that implies physiological and morphological adjustments from organ to whole-plant level.

13.4.1.4 Responses of *Olea europaea* to Heavy Metals

In the last few decades, the intensification of agricultural systems and the anthropogenic pressures on natural ecosystems have resulted in a dramatic increase in heavy metal concentrations in water, air and soil. Nowadays, a high concentration of heavy metals poses a serious problem for multiple life forms (Yang et al. 2005; Miransari 2011). Although heavy metals are naturally present in soil (Nagajyoti et al. 2010), the excessive use of inorganic and organic fertilizers and pesticides in agriculture has increased their toxicity levels (Yanqun et al. 2005; Guan et al. 2018). Heavy metals naturally found in the environment include cadmium (Cd), copper (Cu), iron (Fe), lead (Pb), nickel (Ni), cobalt (Co), zinc (Zn), arsenic (As), chromium (Cr), silver (Ag) and mercury (Hg). Some heavy metals, such as Fe, Cu and Zn, are essential for plants and animals and represent important micronutrients (Reeves and Baker 2000).

Essential heavy metals play a key role in the physiological and biochemical functions of plants and animals. For instance, Fe is an important constituent of several plant proteins and enzymes (Marschner 1995). Zn was found to be associated with the maintenance of metabolic process and reproduction (Prasad 2012), thus being a cofactor for RNA polymerase. Hence, Cu is a highly important cofactor involved in key physiological processes of plants like photosynthesis and respiration (Chatterjee et al. 2006). Even with their crucial and vital role in several physiological processes, elevated concentrations of heavy metals cause toxic symptoms in plants (Blaylock and Huang 2000; DalCorso et al. 2013; Farias et al. 2013). For example, elevated levels of Zn cause chlorosis in leaves, a severe decline in photosynthetic pigments and consequently a reduction in plant biomass and production (Mirshekali et al. 2012). Hence, an excessive amount of Cu has been proven to generate oxidative stress and ROS and negatively affect plant growth (Moreno-Caselles et al. 2000; Lewis et al. 2001; Opdenakker et al. 2012). Pb contamination severely influences the morphological characteristics of plants, and thus, their growth performances (Akhtar et al. 2017). However, it is noteworthy that non-essential heavy metals, such as Pb, Cr and Cd, are very toxic even at very low doses (Hayat et al. 2012; Opdenakker et al. 2012).

The responses of the olive tree to a higher concentration of heavy metals depend on the element's identity and its concentration, the age of the plants and the time of year. For instance, Fernández-Escobar et al. (1999) showed that Cu concentration is generally higher in older leaves, which could be explained by copper being used as a fungicide. Liang et al. (2019) showed that an increase in Cd and Pb concentration first increased soluble protein, superoxide dismutase (SOD) activity and soluble sugar but was followed by a significant decrease after a prolonged period of exposure. They highlighted that the combined stress from Cd and Pb contamination reduced the plant resistance and tolerance, and consequently, their growth. Moreover, Luka et al. (2019) highlighted the high level of heavy metals (i.e. Cu, Cd, Cr, As, Pb and Ni) in Cyprus olive oil, which can be a serious health issue. The authors' findings show that heavy metals could easily be transported from contaminated soil through the roots to the fruits. These observations have been reported also for Spanish table olives (López-López et al. 2008). Further, Al-Habahbeh et al. (2021) showed that the use of treated wastewater in cultivation leads to the contamination of all parts of olive trees, including the leaves, fruits and roots. However, they showed that higher concentrations of Fe, Cu and Ni were found in roots and fruits, and thus, in leaves, high values of Fe, Mn, Cd and Pb were recorded. Bourazanis et al. (2016) studied the effects of treated wastewater on *Koroneiki* olive cultivar and showed a significant decrease in the levels of K, Ca, Mg, Mn, Fe and Zn in the leaves. However, they concluded that wastewater use did not have a significant effect on the oil quality characteristics. Following the above-mentioned results, the olive tree could be considered a metal bioaccumulator (Wilson and Pyatt 2007).

13.4.2 Responses of *Olea europaea* to Biotic Stressors

Plants are continuously subjected to adverse environmental factors arising from living components, which are referred to by biotic stressors. In the same manner as abiotic stressors, they could result in several morphological, physiological and biochemical changes (Saijo et al. 2020). The impacts of biotic stress are evident on plant growth, productivity and performance. Fungal pathogens are the most important biotic stressor for plant species; however, bacteria, viruses, nematodes, weeds and parasitic plants are responsible for much plant damage (Fraire-Velázquez et al. 2011; Muller et al. 2012). Hence, insect pests are similarly important as biotic stressors for plant species, especially for crop species (Cera et al. 2017). Therefore, facing these threats plants have developed various adaptation mechanisms to such stressors (Bhar et al. 2021). Several findings have shown that plants developed complex sensory mechanisms to deal with the impact of biotic stressors, and consequently balance their performances (Rizhsky et al. 2004; Wang et al. 2019; Lamers et al. 2020).

13.4.2.1 Responses of *Olea europaea* to Insect Pests

As for many fruit trees, biotic stressors, such as insect pests, affect *O. europaea* performances with evident and harmful consequences on its fruits and oil quality. Major olive pests are given in Table 13.3. One of the most severe insect pests for the olive tree is the monophagous olive fly, *Bactrocera oleae* (Diptera: Tephritidae) (former name: *Dacus oleae*) (Daane and Johnson 2010). Female olive fruit flies lay their eggs inside ripe and unripe olive fruits (Sharaf 1980), and therefore, the larvae feed inside the fruit pulp through three instar stages until their complete larval development (Daane and

TABLE 13.3
Main Olive Tree Pests and Diseases

Pest/disease	Scientific name	Organs attacked
Major insect pests		
Olive fly	*Bractocera oleae*	Olive fruits
Olive moth	*Prays oleae*	Leaves, terminal buds, flowers and fruits
Olive black scale	*Saissetia olea*	Leaves, twigs, inflorescences
Olive psyllid	*Euphyllura olivina*	Leaves, buds, young shoots, stems, inflorescences and fruiting shoots
Olive weevil	*Otiorhynchus cribricollis*	Leaves
Olive scale	*Parlatoria oleae*	Leaves, woody parts, fruits
Olive borer	*Hylesinus oleiperda*	Trunk and branches
Olive Bark Beetle	*Phloeotribus scarabeoides*	Twigs, flower clusters and above all fruiting clusters, pruning wood, trunk, branches and twigs of trees suffering from dieback
Pyralid moth	*Euzophera pinguis*	Branches and trunk
Jasmine moth	*Palpita vitrealis*	Leaves, terminal buds and fruits
Olive thrips	*Liothrips oleae*	Leaves, young stems, terminal shoots, fruits
Leopard moth	*Zeuzera pyrina*	Leaf petioles, young twigs, twigs, branches, trunk
Oleander scale	*Aspidiotus nerii*	Leaves, fruits
Olive bark midge	*Resseliella oleisuga*	Woody stems, bark
Olive leaf gall midge	*Dasineura oleae*	Leaves, vegetative buds, flower stalks and stems
Oystershell scale	*Lepidosaphes ulmi*	Leaves, twigs, fruits

TABLE 13.3 (Continued)
Main Olive Tree Pests and Diseases

Pest/disease	Scientific name	Organs attacked
Mites		
Olive gall mite	*Aceria oleae*	Leaves, buds, shoots, flower clusters, fruits
Bacteria diseases		
Olive quick decline syndrom	*Xylella fastidiosa*	Xylem vessels
Olive knot/tuberculosis	*Pseudomonas savastanoi pv. savastanoi*	Twigs, branches, trunk, leaves
Mycosis		
Verticillium wilt	*Verticillium dahliae*	Vascular disease causing wilting of attacked parts
Olive leaf spot/peacock spot	*Venturia oleaginea*	Leaves especially and fruits
Cercospora leaf spot	*Pseudocercospora cladosporioides*	Leaves and fruits
Anthracnose	*Colletotrichum acutatum, C. gloeosporioides (=Gloeosporium olivarum)*	Leaves, twigs, flowers and fruits
Sooty mold	*Cladosporium herbarium, Limacinula oleae, Alternaria tenuis, Aureobasidium pullulans, Capnodium elaeophilum*	Leaves, flowers, fruits, twigs and branches
Dalmatian disease	*Botryosphaeria dothidea*	Fruits
Leprosy, cylindrosporiosis	*Phlyctema vagabunda* (tel. *Neofabrea alba*)	Leaves, branches and fruits
Other fruit rots	*Alternaria, Aspergillus, Cladosporium, Diplodia, Geotrichum, Fusarium, Phomopsis, Neofusicoccum* spp.	Fruits
Cankers (wilting)	*Neofusicoccum* spp., *Eutipa lata, Phoma incompta, Diplodia* spp.	Shoots
Emerging tracheomycotic diseases	*Lecythophora lignicola, Pleurostomophora richardsiae, P. cava, Phaeoacremonium* spp.	Vascular diseases
Woody root rots,, crown rots, foot rot	*Armillariella mellea, Rosellinia necatrix, Dactylonectria* spp.	Roots
Fine root rots	*Phytophthota, Cylindrocarpon, Fusarium, Pythium* spp.	Roots
Viral diseases		
Yellowing, malformations	*Nepovirus, Cucumovirus, Oleavirus*	Leaves, buds, flowers
Nematodes		
Root knot	*Meloidogyne spp.*	Roots
Root lession	*Pratylenchus spp.*	Roots

Johnson 2010). The pupation of larvae can take place on the soil or inside the olive fruit (Sharaf 1980). In the Mediterranean, *B. oleae* larvae are observed from the end of August, with a pupation stage in mid-September (Bento et al. 1999). However, its population dynamics largely depend on spatial and climatic conditions (Marchi et al. 2016). Hence, the analysis of the genetic diversity of the Mediterranean *B. oleae* population shows the existence of three separate groups (Augustinos et al. 2005). These groups were most likely associated with olive domestication and expansion from the east to the western Mediterranean (van Asch et al. 2012). Olive cultivars show different susceptibility to *B. oleae* infection. For instance, Malheiro et al. (2015) reviewed the susceptibility of

61 cultivars from different geographic locations around the Mediterranean to *B. oleae*. The authors highlighted the existence of three major stimuli increasing the infestation by olive fly, including the physical, chemical and molecular characteristics of olive fruits. Hence, Rizzo et al. (2007) showed that larger and greener olive fruits with lower skin elasticity are significantly more infected. The chemical composition of olive fruits might play an important role in the infestation or protection of olive fruits. Scarpati et al. (1993, 1996) showed that increasing (E)-2-hexenal amounts could have a repellent action, while the olive leaf volatiles toluene and ethylbenzene, stimulate fly olive oviposition. Iannotta et al. (2002) showed that the *Turdunazza antimosca* cultivar is less susceptible to *B. oleae* compared to the *Tonda Iblea*, *Moresca* and *Verdese* cultivars, which seems to be associated with the repellent action of its fruits' oleuropein concentration. Moreover, Iannotta and Scaliercio (2012) reported that concentrations of oleuropein higher than 30 g/kg of fruits inhibit the early development of *B. oleae* in some cultivars, such as *Carboncella di Pianacce*, *Gentile di Chieti*, *Bardhi i Tirana*, *Kokermadh i Berat* and *Nociara* cultivars. Thus, a higher concentration of volatile compounds in leaves of *Cobrançosa* cultivar was associated with a significant decrease in *B. oleae* infestation (Malheiro et al. 2016). Moreover, the authors showed that the seasonal variation in fruit maturation plays a role in the infestation degrees. These observations were reported in the *Cellina di Nardò* and *Cima di Mola* cultivars, with a low infestation rate during the ripening season; however, the *Ogliarola del Vulturecultivar* cultivar shows a low infestation level until the end of October (Iannotta et al. 2006).

Olive black scale (*Saissetia oleae*, Hemeptera) represents a serious threat to the olive tree, and is a widespread pest around the Mediterranean (Teviotdale et al. 2005). The shape of females is hemispherical, being 2–5mm long and 1–4 mm wide, and characterized by a noticeable "H" shape on their back. *S. oleae* start to lay eggs in May, and in July, mature individuals start to feed by sapsucking on leaves, shoots and sometimes fruits. For instance, in Morocco, Ouguas and Chemseddine (2011) showed that the optimum mobile star of *S. oleae* is in July, thus, they reported its tendency to attack young organs rather than old ones. Hence, Van Steenwyk et al. (2002) described the main symptom of *S. oleae* infestation as a reduction in tree vigor, leaf drop and the dieback of twigs in severe cases.

Another serious damage-causing pest to the olive tree is the olive moth (*Prays oleae*, Lepidoptera: Yponomeutidae), also known as the kernel borer. It is found all over the Mediterranean region and in other olive-cultivated regions (EPPO 2011). *P. oleae* is a monophagous olive pest that attacks the flowers, fruits and leaves (Arambourg et al. 1986; Ramos et al. 1998). *P. oleae* is characterized by three different generations (Nave et al. 2016). The first attack is the flowers (anthopagous) in which the larvae attack the flowers and feed on the anthers and ovaries. In the second generation (carpophagous), larvae attack the young fruits close to the stalk and enter the endocarp, which results in fruit dehydration and drop, and consequently the loss of entire production. Then, in the third generation (phyllophagous), the female lays eggs on the upper leaf surface and the larvae feed on leaves (Arambourg et al. 1986; Ramos et al. 1998; Landa et al. 2019). Since *P. oleae* attack the olive flowers, leaves and fruits, infestation by this pest can cause the entire loss of production and even the growth and development of olive trees in the following years, consequently, it can generate tremendous economic losses (Ramos et al. 1992; Rosales et al. 2008).

Olive psyllid (*Euphyllura olivine*) is a biting-sucking insect that causes considerable damage to olive groves in the Mediterranean region (Seljak 2006). It has also been reported in India, Iran, the United Kingdom, Germany and the United States (California) (Meftah et al. 2014). It is most abundant in spring when olive trees bloom and causes up to 60% yield losses in some parts of the Mediterranean Basin (Percy et al. 2012). The insect has seven life stages: egg, five larval stages and adult (Hodkinson 1974). Infected trees can be easily identified from the pale white wax masses surrounding the pupae (Meftah et al. 2014). The tolerance level is approximately 2.5 to 3 nymphs per 100 flower clusters, which is equivalent to a cluster infestation rate of 50–60% (IOC 2007). Besides the abovementioned three major olive pests, some other pests can negatively affect olive productivity and development, such as the oleander scale (*Aspidiotus nerii*), the olive-bark beetle (*Phloeotribus scarabeoides*), the ash-bark beetle (*Hylesinus toranio*) and the pyralid moth (*Euzophera pinguis*).

13.4.2.2 Responses of *Olea europaea* to Diseases

Olive trees, like other crops, are host to several diseases caused by bacteria, fungi and viruses that can have a direct or indirect impact on both quantitative and qualitative production. Of the many diseases recorded, only a few of them are likely to cause economically significant damage in olive groves (Table 13.3). Rapid decline syndrome of olive trees is caused by *Xylella fastidiosa*, which is a Gram-negative bacterium, transmitted by several insect vectors. It infects an extremely wide host range of over 500 plant species (EFSA 2020). When the bacterium invades the xylem vessels, it blocks the transport of mineral nutrients and water. One of the most common symptoms of infection is the sudden and rapid drying of a portion of the leaves, with the edges necrotizing while the adjacent tissues turn yellow or red. Observed symptoms include leaf scorch, wilting of foliage, defoliation, chlorosis or bronzing along the leaf margins, and dwarfing. Bacterial infections can be so severe that they result in the death of infected plants. Symptoms usually appear on only a few branches, but later spread to cover the entire plant (OEPP 2020).

Tuberculosis disease is caused by the Gram-negative plant-pathogenic bacterium *Pseudomonas savastanoi* pv. *savastanoi*, and is characterized by the formation of proliferations (tumors, galls or nodules) in the aerial part of olive trees, mainly on the stems and branches and occasionally on leaves and fruits (Quesada et al. 2012; Ramos et al. 2012). Although olive tuberculosis disease is widespread in most olive-growing areas, there is no accurate estimate of the losses it causes. This is very difficult to measure because many factors can influence the severity of the symptoms. Severe infections can cause branch death and progressive weakening, leading to a loss of tree vigor and thus harvest (Iacobellis et al. 1993).

Olive leaf spot/peacock spot (*Venturia oleaginea*) is one of the most common diseases of olive (*Olea europaea*) in the world (Obanor et al. 2008). Leaf symptoms consist of brown, circular and often zoned spots, possibly surrounded by a yellow halo (Zarco et al. 2007). Infections occur primarily on leaves, but other organs, including shoots, petioles, fruit stalks, fruits and inflorescences, may also be infected. On susceptible cultivars and over time, leaves become chlorotic and severe defoliation can occur, resulting in the loss of plant vigor, shoot death and a reduction in the number of flower buds and, consequently, fruit production (Scibetta et al. 2019).

Verticillium wilt (*Verticillium dahlia*) is considered one of the most devastating diseases of the olive tree and a major limiting factor for olive oil production. *V. dahliae* is a globally distributed soilborne fungus capable of infecting more than 400 plant species, including annual herbaceous plants and weeds, as well as fruit, landscape and ornamental trees, and shrubs (Pegg and Brady 2002). Verticillium wilt is a vascular disease with two categories of symptoms in olive: "apoplexy" (acute form) and "slow decline" (chronic form) (Blanco-López et al. 1984). Rapid dieback of shoots and branches that eventually results in the loss of the entire tree characterizes the acute variant, while the slow decline consists mainly of flower mummification and necrosis of inflorescences and leaves that develop on individual branches. Losses due to olive verticillium wilt include high rates of tree mortality and reductions in fruit yield, particularly in highly susceptible cultivars (Jiménez-Díaz et al. 2012).

13.5 FUTURE RESEARCH PRIORITIES

Plants are continuously living under adverse environmental cues, such as biotic and abiotic stresses. These stressors are limiting factors to the growth, development and productivity of plant species, whether in natural or agricultural systems. Plants' responses to stressful conditions are complex and greatly depend on the extent and type of stressors. However, plants respond to these stresses by adjusting their morphological, physiological and biochemical processes (Ghalambor et al. 2007; Osakabe et al. 2014; Kassout et al. 2021). Nonetheless, these adjustments could be at the expense of resource reserves and the whole-plant balance under severely stressful conditions (Prentice et al. 2011; Maracahipes et al. 2018). In the case of *O. europaea*, several studies have highlighted its responses to stressful conditions and the underlying mechanisms behind its capacity for adaptation (Terral et al. 2004; Fernández 2014; Kassout et al. 2021). Although the olive tree is well adapted to the stressful environmental conditions in the Mediterranean region that allow for good plasticity and adaptation, the ongoing and projected global changes may alter this capacity (Kassout et al. 2021). For instance, considering the relatively restricted geographic range of *Olea europaea* subsp. *maroccana*, projected climate change scenarios underline a high risk of extinction facing this endemic subspecies and point out the necessity to set urgent conservation strategies to allow for the preservation of its genetic and morphological diversity (Kassout et al. 2022). Moreover, in the context of global warming, the rising demand for olive products, such as oil and table olives, is associated with the adoption of extensive management practices that impose several challenges for most growing areas (Lopez-Bellido et al. 2016; Brito et al. 2019). Increasing demand for irrigation, fertilizers and pesticides and the emergence of pests and diseases are likely to threaten the growth and productivity of the olive tree (Therios 2009; Fernández 2014). Therefore, understanding the main responses of the olive tree to abiotic and biotic stressors is crucial for improving crop management and developing adequate breeding programs. For instance, highlighting the hydraulic system adjustments and functioning in the olive tree under drought (Terral et al. 2004; Kassout et al. 2021) will improve the development of new irrigation methods, especially in areas prone to severe drought events. Moreover, the selection of new cultivars adapted to drought and more resistant to pests and diseases will increase the actual growing areas and maintain future production under the pronounced climate warming. Finally, the adaptation mechanism of the olive tree facing stressful conditions is still poorly understood and needs further investigation, particularly in regard to drought (Fernández 2014; Kassout et al. 2021), pests and diseases (Therios 2009).

REFERENCES

Abbasi, H., M. Jamil, A. Haq, S. Ali, R. Ahmad, and Z. Malik. 2016. Salt stress manifestation on plants, mechanism of salt tolerance and potassium role in alleviating it: a review. *Zemdirbyste-Agricul.* 103: 229–238. doi:10.13080/z-a.2016.103.030

Abdallah, M. B., K. Methenni, I. Nouairi, M. Zarrouk, and N. B. Youssef. 2017. Drought priming improves subsequent more severe drought in a drought-sensitive cultivar of olive cv. Chétoui. *Scientia horticulturae* 221: 43–52.

Abd-El-Rahman, A.A., A.F. Shalaby, M.S. Balegh. 1966. Water economy of olive under desert conditions. *Flora* 156: 202–219.

Adnan, M., R. Bibi, S. Mussarat, A. Tariq, Z. K. Shinwari. 2014. Ethnomedicinal and phytochemical review of Pakistani medicinal plants used as antibacterial agents against Escherichia coli. *Annals of clinical microbiology and antimicrobials*, *13*(1): 1–18. doi:10.1186/s12941-014-0040-6

Ahmadipour, S., I. Arji, A. Ebadi, and V. Abdossi. 2018. Physiological and biochemical responses of some olive cultivars (*Olea europaea* L.) to water stress. *Cellular and Molecular Biology* 64(15): 20–29. doi. org/10.14715/cmb/2017.64.15.4

Akhtar, N., S. Khan, I. Malook, S. U. Rehman, and M. Jamil. 2017. Pb-induced changes in roots of two cultivated rice cultivars grown in lead-contaminated soil mediated by smoke. *Environmental Science and Pollution Research* 24(26): 21298–21310.

Al-Habahbeh, K. A., M. B. Al-Nawaiseh, R. S. Al-Sayaydeh, J. S. Al-Hawadi, R. N. Albdaiwi, H. S. Al-Debei, and J. Y. Ayad. 2021. Long-term irrigation with treated municipal wastewater from the wadi-musa region: Soil heavy metal accumulation, uptake and partitioning in olive trees. *Horticulturae* 7(6): 152. doi.org/10.3390/horticulturae7060152

Ali-Shtayeh, M. S., R. M. Jamous, and R. M. Jamous. 2012. Complementary and alternative medicine use amongst Palestinian diabetic patients. *Complementary Therapies in Clinical Practice* 18(1): 16–21.

Al-Khalil, S. 1995. A survey of plants used in Jordanian traditional medicine. *International Journal of Pharmacognosy* 33(4): 317–323.

Allen, C. D., D. D. Breshears, and N. G. McDowell. 2015. On underestimation of global vulnerability to tree mortality and forest die-off from hotter drought in the Anthropocene. *Ecosphere* 6:art129. doi:10.1890/es15-00203.1

Allen, C. D., A. K. Macalady, H. Chenchouni, D. Bachelet, N. McDowell, M. Vennetier, et al. 2010. A global overview of drought and heat-induced tree mortality reveals emerging climate change risks for forests. *Forest ecology and management* 259(4): 660–684. doi.org/10.1016/J.FORECO.2009.09.001

Alloway, B. J. 2013. Heavy metals and metalloids as micronutrients for plants and animals, in *Heavy Metals in Soils*, Whiteknights: Springer, 195–209.

Aragüés, R., A. Puy, A. Royo, and J.L. Espada. 2005. Three-year field response of young olive trees (*Olea europaea* L., cv. 'Arbequina') to soil salinity: trunk growth and leaf ion accumulation. *Plant and Soil* 271: 265–273.

Aragüés, R., Puy, J., and D. Isidoro. 2004. Vegetative growth response of young olive trees (*Olea europaea* L., cv. Arbequina) to soil salinity and waterlogging. Plant and Soil 258(1): 69–80.

Arambourg, Y., Pralavorio, R. 1986. Hyponomeutidae. *P. oleae*. In *Traité d'Entomologie Oléicole*; Arambourg, Y., Ed.; Conseil Oléicole International: Madrid, Spain, pp. 47–91.

Armstrong, W., and M.C. Drew. 2002. Root growth and metabolism under oxygen deficiency. In: Waisel Y, Eshel A and Kafkafi U, eds. *Plant roots: the hidden half*, 3rd edn. New York: Marcel Dekker, 729–761.

Arnan, X., B. C. López, J. Martínez-Vilalta, M. Estorach, and R. Poyatos. 2012. The age of monumental olive trees (*Olea europaea*) in northeastern Spain. *Dendrochronologia* 30(1): 11–14. doi.org/10.1016/j.dendro.2011.02.002

Ater, M., Barbara, H., & J. Kassout. 2016. Importance des variétés locales, de l'oléastre et des pratiques traditionnelles de l'oléiculture dans la région de Chefchaouen (Nord du Maroc). L'oléiculture au Maroc de la préhistoire à nos jours: pratiques, diversité, adaptation, usages, commerce et politiques. Montpellier: CIHEAM.

Augustinos, A. A., Z. Mamuris, E. E. Stratikopoulos, S. D'amelio, A. Zacharopoulou, and K. D. Mathiopoulos. 2005. Microsatellite analysis of olive fly populations in the Mediterranean indicates a westward expansion of the species. *Genetica* 125(2): 231–241.

Aumeeruddy-Thomas, Y., A. Moukhli, H. Haouane, B. Khadari. 2017. Ongoing domestication and diversification in grafted olive–oleaster agroecosystems in Northern Morocco. *Regional Environmental Change* 17:1315–1328. doi.org/10.1007/s10113-017-1143-3.

Bacelar, E. A., C. M. Correia, J. M. Moutinho-Pereira, B. C. Gonçalves, J. I. Lopes, & J. M. Torres-Pereira. 2004. Sclerophylly and leaf anatomical traits of five field-grown olive cultivars growing under drought conditions. *Tree physiology* 24(2): 233–239. doi:10.1093/treephys/24.2.233

Bacelar, E. A., J. M. Moutinho-Pereira, B. C. Gonçalves, H. F. Ferreira, and C. M. Correia. 2007a. Changes in growth, gas exchange, xylem hydraulic properties and water use efficiency of three olive cultivars under contrasting water availability regimes. *Environmental and Experimental Botany* 60(2): 183–192.

Bacelar, E. A., J. M. Moutinho-Pereira, B. C. Gonçalves, J. I. Lopes, & C. M. Correia. 2009. Physiological responses of different olive genotypes to drought conditions. *Acta Physiologiae Plantarum* 31(3): 611–621. doi.org/10.1007/s11738-009-0272-9.

Bacelar, E. A., D. L. Santos, J. M. Moutinho-Pereira, B. C. Gonçalves, H. F. Ferreira, and C. M. Correia. 2006. Immediate responses and adaptative strategies of three olive cultivars under contrasting water availability regimes: changes on structure and chemical composition of foliage and oxidative damage. *Plant Science* 170(3): 596–605.

Bacelar, E. A., D. L. Santos, J. M. Moutinho-Pereira, J. I. Lopes, B. C. Gonçalves, T. C. Ferreira, & C. M. Correia. 2007b. Physiological behaviour, oxidative damage and antioxidative protection of olive trees grown under different irrigation regimes. *Plant and Soil* 292(1): 1–12.

Bailey-Serres, J., S. C. Lee, and E. Brinton. 2012. Waterproofing crops: effective flooding survival strategies. *Plant Physiology* 160: 1698–1709. doi:10.1146/annurev.arplant.59.032607.092752.

Bartolini, G., G. Prevost, C. Messeri, C. Carignani. 2005. Olive Germplasm: cultivars and world-wide collections. FAO/Plant Production and Protection, Rome. Available at: www.apps3.fao.org/wiews/olive/oliv.jsp.

Belaj, A., I. Trujillo, D. Barranco, L. Rallo. 2004. Characterization and identification of Spanish olive germplasm by means of RAPD markers. *HortScience* 39(2): 346–350. doi.org/10.21273/HORTSCI.39.2.346.

Bellakhdar, J. 1997. *Contribution à l'étude de la pharmacopée traditionnelle au Maroc: la situation actuelle, les produits, les sources du savoir (enquête ethnopharmacologique de terrain réalisée de 1969 à 1992)*, Doctoral dissertation, Université Paul Verlaine-Metz.

Bellakhdar, J., R. Claisse, J. Fleurentin, C. Younos. 1991. Repertory of standard herbal drugs in the Moroccan pharmacopoea. *Journal of ethnopharmacology*, 35(2): 123–143.

Benabid, A. 1984. Etude phytoécologique des peuplements forestiers et préforestiers du Rif centro-occidental (Maroc). Institut scientifique Université Mohammed V, Rabat.

Benabid, A. 1985. Les écosystèmes forestiers préforestiers et présteppiques du Maroc: diversité, répartition biogéographique et problèmes posés par leur aménagement. *Foret Mediterranéenne* VII:53–64.

Benabid, A., and M. Fennane. 1994. Connaissances sur la végétation du Maroc: phytogéographie, phytosociologie et séries de végétation. *Lazaroa* 14: 21–97.

Bento, A., L. Torres, J. Lopes, and R. Sismeiro. 1999. A contribution to the knowledge of *Bactrocera oleae* (GMEL) in Tras-os-Montes region (Northeastern Portugal): Phenology, losses and control. *Acta Horticulture* 474: 541–544.

Berdugo, M., M. Delgado-Baquerizo, S. Soliveres, R. Hernández-Clemente, Y. Zhao, Gaitán, J. J., ... & F. T. Maestre. 2020. Global ecosystem thresholds driven by aridity. *Science* 367(6479): 787–790.

Bernstein, L. 1965. *Salt tolerance of fruit crops* (No. 292). Agriculture Research Service, US Department of Agriculture.

Besnard, G., C. Garcia-Verdugo, R. Rubio de Casas, U. A. Treier, N. Galland, & P. Vargas. 2008. Polyploidy in the olive complex (*Olea europaea*): evidence from flow cytometry and nuclear microsatellite analyses. *Annals of Botany* 101(1): 25–30. doi:10.1093/aob/mcm275.

Besnard, G., B. Khadari, M. Navascués, M. Fernández-Mazuecos, A. El Bakkali, N. Arrigo,... & V. Savolainen 2013. The complex history of the olive tree: from Late Quaternary diversification of Mediterranean lineages to primary domestication in the northern Levant. *Proceedings of the Royal Society B: Biological Sciences* 280(1756): 20122833. doi.org/10.1098/rspb.2012.2833

Besnard, G., R. Rubio de Casas, P. A., Christin, & P. Vargas. 2009. Phylogenetics of Olea (Oleaceae) based on plastid and nuclear ribosomal DNA sequences: tertiary climatic shifts and lineage differentiation times. Annals of Botany 104(1): 143–160. doi:10.1093/aob/mcp105

Besnard, G., J. F. Terral, & A. Cornille. 2018. On the origins and domestication of the olive: a review and perspectives. Annals of botany, 121(3): 385–403. doi.org/10.1093/aob/mcx145

Bhar, A., A. Chakraborty, and A. Roy. 2021. Plant responses to biotic stress: Old memories matter. *Plants* 11(1): 84.

Bianco, A., and N. Uccella. 2000. Biophenolic components of olives. *Food Research International*, 33(6): 475–485.

Bianco, A., R. L. Scalzo, and M. L. Scarpati. 1993. Isolation of cornoside from *Olea europaea* and its transformation into halleridone. *Phytochemistry* 32(2): 455–457. doi.org/10.1016/S0031-9422(00)95015-5.

Bisignano, G., A. Tomaino, R. L. Cascio, G. Crisafi, N. Uccella, A. Saija. 1999. On the in-vitro antimicrobial activity of oleuropein and hydroxytyrosol. *Journal of pharmacy and pharmacology*, 51(8): 971–974.

Blanco-López, M. A., R. M. Jiménez-Díaz, and J. M. Caballero. 1984. Symptomatology, incidence and distribution of *Verticillium wilt* of olive trees in Andalucía. *Phytopathologia Mediterranea* 1–8.

Blaylock, M.J., and J.W. Huang. 2000. Phytoextraction of metals. In: Raskin I, Ensley BD (eds) *Phytoremidation of toxic metals-using plants to clean up the environment*. Wiley, New York, pp 53–70

Blondel, J., J. Aronson, J.Y. Bodiou, G. Boeu. 2010. *The Mediterranean region: biological diversity through time and space*. Oxford University Press, Oxford, pp 376.

Bongi, G., and A., Palliotti. 1994. Olive. In: Schaffer, B., Andersen, P.C. (Eds.), *Handbook of Environmental Physiology of Fruit Crops*. Volume I: Temperate Crops. CRC Press, Inc., Boca Raton, Florida, USA, pp 165–187.

Bouaziz, M., R. J. Grayer, M. S. Simmonds, M. Damak, and S. Sayadi. 2005. Identification and antioxidant potential of flavonoids and low molecular weight phenols in olive cultivar Chemlali growing in Tunisia. *Journal of agricultural and food chemistry* 53(2): 236–241. doi:10.1021/jf048859d

Boughalleb, F., and H., Hajlaoui. 2011. Physiological and anatomical changes induced by drought in two olive cultivars (cv. Zalmati and Chemlali). *Acta Physiologiae Plantarum* 33: 53–65.

Bourazanis, G., P. A. Roussos, I. Argyrokastritis, C. Kosmas, and P. Kerkides. 2016. Evaluation of the use of treated municipal waste water on the yield, oil quality, free fatty acids' profile and nutrient levels in olive trees cv Koroneiki, in Greece. *Agricultural Water Management* 163: 1–8.

Bouzabata, A. 2013. Traditional treatment of high blood pressure and diabetes in Souk Ahras District. *Journal of Pharmacognosy and Phytotherapy* 5(1): 12–20.

Bray, E.A., J. Bailey-Serres, and E. Weretilnyk. 2000. Responses to abiotic stresses. In: Buchanan BB, Gruissem W, Jones RL (eds). *Biochemistry and molecular biology of plants*. American Society of Plant Physiologists, Rockville, pp 1158–1203.

Bréda, N., R. Huc, A. Granier, and E. Dreyer. 2006. Temperate forest trees and stands under severe drought: a review of ecophysiological responses, adaptation processes and long-term consequences. *Annals of Forest Science 63*(6): 625–644. doi.org/10.1051/forest:2006042

Breton, C., J. F. Terral, C. Pinatel, F. Médail, F. Bonhomme, & A. Bervillé. 2009. The origins of the domestication of the olive tree. *Comptes rendus biologies* 332(12): 1059–1064. doi.org/10.1016/j.crvi.2009.08.001

Brito, C., L. T. Dinis, J. Moutinho-Pereira, and C. M. Correia. 2019. Drought stress effects and olive tree acclimation under a changing climate. *Plants* 8(7), 232.

Brodribb, T.J., J. Powers, H. Cochard, and B. Choat. 2020. Hanging by a thread? Forests and drought. *Science* 368:261–266.

Cardoso, S. M., S. I. Falcão, A. M. Peres, & M. R. Domingues. 2011. Oleuropein/ligstroside isomers and their derivatives in Portuguese olive mill wastewaters. *Food Chemistry* 129(2): 291–296. doi.org/10.1016/j.foodchem.2011.04.049

Carrión, Y., M. Ntinou, and E. Badal. 2010. *Olea europaea* L. in the North Mediterranean Basin during the Pleniglacial and the Early-Middle Holocene. *Quaternary Science Reviews* 29(7–8): 952–968. doi.org/10.1016/j.quascirev.2009.12.015

Celano, G., B. Dichio, G. Montanaro, V. Nuzzo, A.M. Palese, C. Xiloyannis. 1999. Distribution of dry matter and amount of mineral elements in irrigated and non-irrigated olive trees. *Acta Horticulturae* 474, 381–384.

Cera, R., J. Avelino, C. Gary, P. Tixier, E. Lechevallier, and C. Allinne. 2017. Primary and secondary yield losses caused by pests and diseases: Assessment and modelling in coffee. *PLoS One* 12(1): 1–17.

Chartzoulakis, K., M. Loupassaki, M., Bertaki, and I. Androulakis. 2002. Effects of NaCl salinity on growth, ion content and C02 assimilation rate of six olive cultivars. *Scientia Horticulturae* 96: 235–247.

Chartzoulakis, K., Psarras G., Vemmos S., Loupassaki M., M. Bertaki. 2006. Response of two olive cultivars to salt stress and potassium supplement. *Journal of Plant Nutrition* 29: 2063–2078.

Chartzoulakis, K. S. 2005. Salinity and olive: growth, salt tolerance, photosynthesis and yield. *Agricultural Water Management* 78(1–2): 108–121.

Chatterjee, C., R. Gopal, & Dube, B. K. 2006. Physiological and biochemical responses of French bean to excess cobalt. *Journal of plant nutrition* 29(1): 127–136.

Choat, B., T. J. Brodribb, C. R., Brodersen, R. A., Duursma, R., López, and B. E. Medlyn. 2018. Triggers of tree mortality under drought. *Nature* 558: 531–539. doi:10.1038/s41586-018-0240-x

Choat, B., S., Jansen, T. J., Brodribb, H., Cochard, S., Delzon, R., Bhaskar, et al. 2012. Global convergence in the vulnerability of forests to drought. *Nature* 491(7426): 752–755. doi:10.1038/nature11688

Cipollini, D. 2004. Stretching the limits of plasticity: can a plant defend against both competitors and herbivores? *Ecology* 85: 28–37. doi:10.1890/02-0615

Cohen, I., S.I. Zandalinas, C. Huck, F.B. Fritschi, and R. Mittler. 2021. Meta-Analysis of Drought and Heat Stress Combination Impact on Crop Yield and Yield Components. *Physiologia Plantarum* 171(1), 66–76. doi:10.1111/ppl.13203

Connor D.J., and E. Fereres. 2005. The physiology of adaptation and yield expression in olive. *Horticultural Review* 31: 155–229.

Daane, K. M., and M. W. Johnson. 2010. Olive fruit fly: Managing an ancient pest in modern times. *Annual Review of Entomology* 55: 151–169. doi.org/10.1146/annurev.ento.54.110807.090553.

Dai, A. 2013. Increasing drought under global warming in observations and models. *Nature Climate Change* 3(1): 52–58. doi.org/10.1038/nclimate1633

DalCorso, G., A., Manara, and A. Furini. 2013. An overview of heavy metal challenge in plants: from roots to shoots. *Metallomics* 5: 1117–1132. doi:10.1039/c3mt00038a

De Bock, M., J. G., Derraik, C. M., Brennan, J. B., Biggs, P. E., Morgan, S. C., Hodgkinson,... & W. S Cutfield. 2013. Olive (*Olea europaea* L.) leaf polyphenols improve insulin sensitivity in middle-aged overweight men: a randomized, placebo-controlled, crossover trial. *PloS one* 8(3): e57622.

Diaz-Espejo, A., B. Hafidi, J. E. Fernández, M. J. Palomo, and H. Sinoquet. 2000. Transpiration and photosynthesis of the olive tree: a model approach. In IV International Symposium on Olive Growing 586, pp. 457–460.

Diaz-Espejo, A., T.N., Buckley, J.S., Sperry, M.V., Cuevas, A., de Cires, S., Elsayed-Farag, M.J., Martin-Palomo, J.L., Muriel, et al. 2012. Steps toward an improvement in process-based models of water use by fruit trees: a case study in olive. *Agricultural Water Management* 114: 37–49.

Dichio, B., Xiloyannis, C., Sofo, A., & G. Montanaro. 2006. Osmotic regulation in leaves and roots of olive trees during a water deficit and rewatering. *Tree physiology* 26(2): 179–185.

Diez, C. M., I., Trujillo, N., Martinez-Urdiroz, D., Barranco, L., Rallo, P., Marfil, & B. S. Gaut. 2015. Olive domestication and diversification in the Mediterranean Basin. *New Phytologist* 206(1): 436–447. doi.org/10.1111/nph.13181

Driouech, F., M., Déqué, & E. Sánchez-Gómez. 2010. Weather regimes—Moroccan precipitation link in a regional climate change simulation. *Global and Planetary Change* 72(1–2): 1–10.

Dullinger, S., A., Gattringer, W., Thuiller, D., Moser, N. E., Zimmermann, A., Guisan,... & Hülber, K. 2012. Extinction debt of high-mountain plants under twenty-first-century climate change. *Nature climate change*, 2(8), 619–622.

EFSA. 2020. Update of the Xylella spp. host plant database- systematic literature search up to 30 June 2019. EFSA Journal. 8(4):6114. doi.org/10.2903/j.efsa.2020.6114.

Eidi, A., M., Eidi, R. Darzi. 2009. Antidiabetic effect of *Olea europaea* L. in normal and diabetic rats. *Phytotherapy Research: An International Journal Devoted to Pharmacological and Toxicological Evaluation of Natural Product Derivatives* 23(3): 347–350.

El-Amin, M., P., Virk, M. A., Elobeid, Z. M., Almarhoon, Z. K., Hassan, S. A., Omer, et al. 2013. Anti-diabetic effect of *Murraya koenigii* (L) and *Olea europaea* (L) leaf extracts on streptozotocin induced diabetic rats. *Pak J Pharm Sci*, 26(2): 359–65.

Ennajeh, M., A. M., Vadel, H., Cochard, & H. Khemira. 2010. Comparative impacts of water stress on the leaf anatomy of a drought-resistant and a drought-sensitive olive cultivar. *The Journal of Horticultural Science and Biotechnology* 85(4): 289–294.

EPPO 2011. EPPO Global Database. Available at: https://gd.eppo.int/taxon/PRAYOL/distribution.

Esti, M., L., Cinquanta, E. La Notte. 1998. Phenolic compounds in different olive varieties. *Journal of Agricultural and Food Chemistry* 46(1): 32–35.

Fabbri, A., M., Lambardi, Y. Ozden-Tokatli. 2009. Olive breeding. In: Mohan Jain S, Priyadarshan PM (eds) *Breeding plantation tree crops: tropical species*, 1st edn. Springer, New York, pp 423–465.

Fang, Y., and L. Xiong. 2015. General mechanisms of drought response and their application in drought resistance improvement in plants. Cellular and molecular life sciences 72(4): 673–689. doi:10.1007/s00018-014-1767-0.

Fang, X., N. C. Turner, G. Yan, F. Li, and K. H. Siddique. 2010. Flower numbers, pod production, pollen viability, and pistil function are reduced and flower and pod abortion increased in chickpea (*Cicer arietinum* L.) under terminal drought. *Journal of Experimental Botany* 61(2): 335–345.

FAO, 2010. *The second report on the state of the World's Plant Genetic Resources for Food and Agriculture*, Commission on Genetic Resources for Food, Rome, Italy.

FAO, 2019. Available online: www.fao.org/faostat/en/#home (accessed on 26 May 2022).

Farias, J. G., F. L. G. Antes, P. A. A. Nunes, S. T. Nunes, G. Schaich, L. V. Rossato, et al. 2013. Effects of excess copper in vineyard soils on the mineral nutrition of potato genotypes. *Food and Energy Security* 2: 49–69. doi: 10.1002/fes3.16

Farooq, M., M. Hussain, A. Wahid, & K. H. M. Siddique. 2012. Drought stress in plants: an overview. *Plant responses to drought stress* 1–33.

Fernández, J. E., F. Moreno, F. Cabrera, J. L. Arrue, & J. Martín-Aranda. 1991. Drip irrigation, soil characteristics and the root distribution and root activity of olive trees. *Plant and soil* 133(2): 239–251.

Fernández, J.E. 2014. Understanding olive adaptation to abiotic stresses as a tool to increase crop performance. *Environmental and Experimental Botany* 103: 158–179. doi.org/10.1016/j.envexpbot.2013.12.003

Fernández, J.E. and F. Moreno, F. 1999. Water use by the olive tree. Journal of crop production 2(2): 101–162.

Fernández, J.E., and B.E. Clothier. 2002. Water uptake by plants. In: *The Encyclopaedia of Life Support Systems (EOLSS)*, Developed under the Auspices of the UNESCO. Eolss Publishers, Oxford, UK. www.eolss.net

Fernández, J.E., F., Moreno, I.F. Girón, O.M. Blázquez. 1997. Stomatal control of water use in olive tree leaves. *Plant and Soil* 190: 179–192.

Fernández, J.E., F. Moreno, J. Martín-Aranda, E. Fereres. 1992. Olive-tree root dynamics under different soil water regimes. *Agricoltura Mediterranea* 122: 225–235.

Fernández-Escobar R., R. Moreno and M. García-Creus. 1999. Seasonal changes of mineral nutrients in olive leaves during the alternate-bearing cycle. *Scientia Horticulturae* 82: 25–45.

Ferreira, M.-I., N. Conceição, T.S. David, and N. Nadezhdina. 2013. Role of lignotuber versus roots in the water supply of rainfed olives. *Acta Horticulturae* 991:181–188.

Figueiral, I., and J.F. Terral. 2002. Late Quaternary refugia of Mediterranean taxa in the Portuguese Estremadura: charcoal based palaeovegetation and climatic reconstruction. *Quaternary Science Reviews* 21: 549–558. doi.org/10.1016/S0277-3791(01)00022-1

Flemmig, J., K. Kuchta, J. Arnhold, & H. W. Rauwald. 2011. *Olea europaea* leaf (Ph. Eur.) extract as well as several of its isolated phenolics inhibit the gout-related enzyme xanthine oxidase. *Phytomedicine* 18(7): 561–566.

Fraire-Velázquez, S., R. Rodríguez-Guerra, and L. Sanchez-Calderon. 2011. Abiotic and biotic stress crosstalk in plants. In A. K. Shanker & B. Venkateswarlu (Eds.), *Abiotic stress response in plants–physiological, biochemical and genetic perspectives* (pp. 3–26). Rijeka: In Tech.

Fujita, T., Sezik, E., M. Tabata, E. Yesilada, G. Honda, Y. Takeda, et al. 1995. Traditional medicine in Turkey VII. Folk medicine in middle and west Black Sea regions. *Economic botany* 49(4): 406–422.

Fukao, T., B. E. Barrera-Figueroa, P. Juntawong, and J. M. Peña-Castro. 2019. Submergence and waterlogging stress in plants: a review highlighting research opportunities and understudied aspects. *Frontier in Plant Science* 10:340. doi: 10.3389/fpls.2019.00340

Gelybó, G., E. Tóth, C. Farkas, Á. Horel, I. Kása, and Z. Bakacsi. 2018. Potential impacts of climate change on soil properties. *AgroKémiaés Talajtan* 67: 121–141. doi:10.1556/0088.2018.67.1.9

Gemas, V. J. V., M. C. Almadanim, R. Tenreiro, A. Martins, P. Fevereiro. 2004. Genetic diversity in the Olive tree (*Olea europaea* L. subsp. *europaea*) cultivated in Portugal revealed by RAPD and ISSR markers. *Genetic Resources and Crop Evolution* 51(5): 501–511. doi.org/10.1023/B:GRES.0000024152.16021.40.

Ghalambor, C., J. McKay, S. Carroll, D. Reznick. 2007. Adaptive versus non-adaptive phenotypic plasticity and the potential for contemporary adaptation in new environments. *Functional Ecology* 21: 394–407. doi.org/10.1111/j.1365-2435.2007.01283.x

Ghazanfar, S. A., & A. M. Al-Al-Sabahi. 1993. Medicinal plants of northern and central Oman (Arabia). *Economic Botany* 47(1): 89–98.

Gianguzzi, L., and G. Bazan, 2019. The *Olea europaea* L. var. *sylvestris* (Mill.) Lehr. forests in the Mediterranean area. *Plant Sociology* 56: 3–4. doi:10.7338/pls2019562/01.

Gimenez, C., E. Fereres, C. Ruz, F. Orgaz. 1996. Water relations and gas exchange of olive trees: diurnal and seasonal patterns of leaf water potential photosynthesis and stomatal conductance. *Acta Horticulturae* 449: 411–415.

Gómez, J.A., J.V. Giráldez, E. Fereres. 2001. Rainfall interception by olive trees in relation to leaf area. *Agricultural Water Management* 49: 65–76

Graziosi, I., M. Tembo, J. Kuate, and A. Muchugi. 2019. Pests and diseases of trees in Africa: a growing continental emergency. *Plants, People, Planet* 2(1): 14–28. doi:10.1002/ppp3.31

Green, P. S. 2002. A revision of Olea (*Oleaceae*). *Kew Bulletin.* 57: 91–140. doi:10.2307/4110824

Grime, J. P. 2006. *Plant strategies, vegetation processes, and ecosystem properties.* John Wiley & Sons.

Guan, Q., F. Wang, C. Xu, N. Pan, J. Lin, R. Zhao, et al. 2018. Source apportionment of heavy metals in agricultural soil based on PMF: A case study in Hexi Corridor, northwest China. *Chemosphere* 193: 189–197.

Gucci R., S. Mancuso, and L. Sebastiani. 2003. Resistenza agli stress ambientali. In: Fiorino P (ed) *Olea—Trattato di Olivicoltura. Edagricole.* Edizioni Agricole de Il Sole 24 ORE Edagricole Srl, pp 91–111.

Gucci, R., and M. Tattini. 2010. Salinity tolerance in olive. *Horticultural Reviews*, Janick, J.,(ed.), John Wiley & Sons, Inc.: Oxford, UK, 21, 177–214.

Guerfel, M., A. Beis, T. Zotos, D. Boujnah, M. Zarrouk, and A. Patakas. 2009a. Differences in abscisic acid concentration in roots and leaves of two young olive (*Olea europaea* L.) cultivars in response to water deficit. Acta Physiol. *Plant. 31 (4)*, 825–831 doi.org/10.1007/s11738-009-0298-z

Guerfel, M., Y. Ouni, D. Boujnah, and M. Zarrouk. 2009b. Photosynthesis parameters and activities of enzymes of oxidative stress in two young 'Chemlali'and 'Chetoui'olive trees under water deficit. *Photosynthetica* 47(3): 340–346.

Guerfel, M., O. Baccouri, D. Boujnah, W. Chaïbi, and M. Zarrouk. 2009. Impacts of water stress on gas exchange, water relations, chlorophyll content and leaf structure in the two main Tunisian olive (*Olea europaea* L.) cultivars. *Scientia Horticulturae* 119(3): 257–263.

Guerin, J. C., & H. P. Reveillere. 1985. Antifungal activity of plant extracts used in therapy. 2. Study of 40 plants extracts against 9 fungi species. *Annales Pharmaceutiques Francaises* (France).

Haloui, E., Z. Marzouk, B. Marzouk, I. Bouftira, A. Bouraoui, & N. Fenina. 2010. Pharmacological activities and chemical composition of the Olea europaea L. leaf essential oils from Tunisia. *Journal of Food, Agriculture and Environment* 8: 204–208.

Hamidov, A., K. Helming, G. Bellocchi, W. Bojar, T. Dalgaard, B. B., Ghaley, et al. 2018. Impacts of climate change adaptation options on soil functions: A review of European case-studies. *Land degradation & development* 29(8): 2378–2389.

Haouane, H. 2012. *Origines, domestication et diversification variétale chez l'olivier (Olea europaea L.) à l'ouest de la Méditerranée*. Ph.D. thesis, Montpellier SupAgro, Université Cadi Ayyad, Marrakech, Morocco.

Hashmi, M. A., A. Khan, M. Hanif, U. Farooq, and S. Perveen. 2015. Traditional uses, phytochemistry, and pharmacology of *Olea europaea* (olive). *Evidence-Based Complementary and Alternative Medicine*. doi.org/10.1155/2015/541591

Hassan, M.M. and S.A. Seif. 1990. Response of seven olive cultivars to water logging. *Gartenbauwissenschaft* 55: 223–225.

Hayat, S., G. Khalique M. Irfan, A. S. Wani, B. N. Tripathi, and A. Ahmad. 2012. Physiological changes induced by chromium stress in plants: an overview. *Protoplasma* 249: 599–611. doi:10.1007/s00709-011-0331-0

Hernandez-Santana, V., C. M. Rodriguez-Dominguez, J. E. Fernández, & A. Diaz-Espejo, 2016. Role of leaf hydraulic conductance in the regulation of stomatal conductance in almond and olive in response to water stress. *Tree Physiology* 36(6): 725–735.

Hodkinson, I. D. 1974. The biology of the Psylloidea (Homoptera): a review. *Bulletin of Entomological Research* 64(02): 325. doi:10.1017/s0007485300031217

Iacobellis, N. S., A. Sisto, and G., Surico. 1993. Occurrence of unusual strains of *Pseudomonas syringae* subsp. *savastanoi* on olive in central Italy. *EPPO Bulletin* 23(3): 429–435. doi:10.1111/j.1365-2338.1993. tb01348.x

Iannotta, N., and S. Scalercio. 2012. Susceptibility of cultivars to Biotic Stresses. In: Muzzalupo I (ed) *Olive germplasm and olive cultivation, table olive and olive oil industry in Italy*. In Tech, Rijeka, pp 81–106.

Iannotta, N., B. Macchione, M. E. Noce, E. Perri, and S. Scalercio. 2006. Olive genotypes susceptibility to the *Bactrocera oleae* (Gmel.) infestation. In *Proceedings of the second international seminar on "biotechnology and quality of olive tree products around the mediterranean basin"*, Olivebioteq, Marsala-Mazara del Vallo (Vol. 2, pp. 5–10).

Iannotta, N., D. Monardo, & L. Perri. 2002. Relazione tra contenuto e localizzazione dell'oleuropeina nella drupa e attacco di *Bactrocera oleae* (Gmel.), *Atti Convegno Internazionale di Olivicoltura. Spoleto, 2002*, pp. 361–366.

International Olive Council (IOC). 2007. *Production techniques in olive growing*. Edition 2007. ISBN: 978-84-931663-8-0. Madrid (Spain): IOC: 346 p.

Isayenkov, S. V. 2012. Physiological and molecular aspects of salt stress in plants. *Cytology and Genetics* 46(5): 302–318. doi:10.3103/S0095452712050040

Jacoboni, N., and G. Fontanazza. 1981. *Cultivar*. Rome: REDA L'Olivo, pp. 7–52.

Jaleel, C. A., P. A. Manivannan, A. Wahid, M. Farooq, H. J. Al-Juburi, R. A. Somasundaram, and R. Panneerselvam. 2009. Drought stress in plants: a review on morphological characteristics and pigments composition. *Int.* J. Agric. *Biol* 11(1): 100–105.

Jerman, T., P. Trebše, and B. M. Vodopivec. 2010. Ultrasound-assisted solid liquid extraction (USLE) of olive fruit (*Olea europaea*) phenolic compounds. *Food Chemistry* 123(1): 175–182. doi.org/10.1016/j.foodchem.2010.04.006

Jiménez-Díaz, R. M., M. Cirulli, G. Bubici, M. del Mar Jiménez-Gasco, P. P. Antoniou, and E. C. Tjamos. 2012. *Verticillium Wilt*, A Major Threat to Olive Production: Current Status and Future Prospects for its Management. *Plant Disease* 96(3): 304–329. doi:10.1094/pdis-06-11-0496

Kaniewski, D., E. Van Campo, T. Boiy, J. F. Terral, B. Khadari, & Besnard, G. 2012. Primary domestication and early uses of the emblematic olive tree: palaeobotanical, historical and molecular evidence from the Middle East. *Biological Reviews* 87(4): 885–899. doi:10.1111/j.1469-185X.2012.00229.x

Kassout, J., M. Ater, S. Ivorra, H. Barbara, B. Limier, J. Ros, V. Girard, L. Paradis, & J. F. Terral. 2021. Resisting aridification: adaptation of sap conduction performance in Moroccan wild olive subspecies distributed over an aridity gradient. *Frontiers in plant science* 12: 1288. doi:10.3389/fpls.2021.663721

Kassout, J., J. F. Terral, A. El Ouahrani, M. Houssni, S. Ivorra, et al. 2022. Species Distribution Based-Modelling Under Climate Change: The Case of Two Native Wild *Olea europaea* Subspecies in Morocco, *O. e.* subsp. *europaea* var. *sylvestris* and *O. e.* subsp. *maroccana*. In *Climate Change in the Mediterranean and Middle Eastern Region* (pp. 21–43). Springer, Cham.

Kassout, J., J.- F. Terral, J. Hodgson, and M. Ater. 2019. Trait-based plant ecology a flawed tool in climate studies? The leaf traits of wild olive that pattern with climate are not those routinely measured. *PLoS ONE*. 14, e0219908. doi:10.1371/journal.pone.0219908

Khadari, B., J. Charafi, A. Moukhli, M. Ater. 2008. Substantial genetic diversity in cultivated Moroccan olive despite a single major cultivar: a paradoxical situation evidenced by the use of SSR loci. *Tree Genetics & Genomes* 4(2): 213–221. doi.org/10.1007/s11295-007-0102-4

Khadari, B., A. El Bakkali, L. Essalouh, C. Tollon, C. Pinatel, G. Besnard. 2019. Cultivated olive diversification at local and regional scale: evidence from the genetic characterization of French genetic ressources. *Frontiers in plant science* 10: 1593. doi.org/10.3389/fpls.2019.01593.

Kitsaki, C. K., and J. B. Drossopoulos. 2005. Environmental effect on ABA concentration and water potential in olive leaves (*Olea europaea* L. cv "Koroneiki") under non-irrigated field conditions. *Environmental and experimental botany* 54(1): 77–89. doi.org/10.1016/j.envexpbot.2004.06.002

Kozlowski, T. T., and S. G. Pallardy. 2002. Acclimation and adaptive responses of woody plants to environmental stresses. *The botanical review* 68(2): 270–334.

Lamers, J., T. Van Der Meer, and C. Testerink. 2020. How plants sense and respond to stressful environments. *Plant Physiology* 182: 1624–1635. doi:10.1104/pp.19.01464

Landa, B. B. 2019. EIP-AGRI Focus Group, Pests and diseases of the olive tree. *Funded by European Commission*, 1–20.

Larcher, W., J.A.P.V. Moraes, H. Bauer. 1981. Adaptative responses of leaf water potential CO2-gas exchange and water used efficiency of *Olea europaea* during drying and rewatering. In: Margaris, N.S., Mooney, H.A. (Eds.), *Components of productivity of Mediterranean-Climate Regions*. Basic and Applied Aspects. The Hague, Boston, London, pp. 77–84.

Lauri, P.E., K. Barkaoui, M. Ater, A. Essaadi, A. Rosati. 2019. Agroforestry for Fruit Trees in Europe and Mediterranean North Africa, in: *Agroforestry for Sustainable Agriculture*. doi.org/10.19103/AS.2018.0041.18

Lavee, S. 1996. Biology and physiology of the olive tree. In: Lavee, S., Barranco, D., Bongi, G., Jardak, T., Loussert, R., Martin, G.C., Trigui, A. (Eds.), *World Olive Encyclopaedia*. *International Olive Council*, Madrid, Spain, pp. 61–110.

Lawrendiadis, G. 1961. Contribution to the knowledge of the medicinal plants of Greece. *Planta medica* 9(02): 164–169.

Lee, I. M., E. J. Shiroma, F. Lobelo, P. Puska, S. N. Blair, P. T. Katzmarzyk, & Lancet Physical Activity Series Working Group. 2012. Effect of physical inactivity on major non-communicable diseases worldwide: An analysis of burden of disease and life expectancy. *Lancet (London, England)*, 380(9838), 219–229. doi.org/10.1016/s0140-6736(12)61031-9

Lewington, A. and E. Parker. 1999. *Ancient trees. Trees that live for a thousand years*. Collins & Brown, London.

Lewis, S., M. E. Donkin, & M. H. Depledge. 2001. Hsp70 expression in Enteromorpha intestinalis (Chlorophyta) exposed to environmental stressors. *Aquatic Toxicology* 51(3): 277–291.

Liang, J., W. Yang, Q. Yan, and N. Cui. 2019. Physiological Effects of Combined Stress of Cd, Pb and Ti on Olive. In *IOP Conference Series: Earth and Environmental Science* (Vol. 384, No. 1, p. 012144). IOP Publishing.

Lloret, F., E. G. de la Riva, I. M. Pérez-Ramos, T. Marañón, S. Saura-Mas, R. Díaz-Delgado, and R. Villar. 2016. Climatic events inducing die-off in Mediterranean shrublands: are species' responses related to their functional traits?. *Oecologia* 180(4): 961–973. doi.org/10.1007/s00442-016-3550-4

Lopez-Bellido, P. J., L. Lopez-Bellido, P. Fernandez-Garcia, V. Muñoz-Romero, and F. J. Lopez-Bellido. 2016. Assessment of carbon sequestration and the carbon footprint in olive groves in Southern Spain. *Carbon Management* 7(3–4): 161–170.

Lopez-Lopez, A., R. Lopez, F. Madrid, and A. Garrido-Fernández. 2008. Heavy metals and mineral elements not included on the nutritional labels in table olives. *Journal of agricultural and food chemistry* 56(20): 9475–9483.

Loumou, A., and C. Giourga. 2003. Olive groves: "The life and identity of the mediterranean". *Agriculture and human values* 20(1): 87–95.

Luka, M. F., and E. Akun. 2019. Investigation of trace metals in different varieties of olive oils from northern Cyprus and their variation in accumulation using ICP-MS and multivariate techniques. *Environmental Earth Sciences* 78(19): 1–10.

Lumaret, R., N. Ouazzani, H. Michaud, G. Vivier, M. F. Deguilloux, & F. Di Giusto. 2004. Allozyme variation of oleaster populations (wild olive tree) (*Olea europaea* L.) in the Mediterranean Basin. *Heredity* 92(4): 343–351. doi.org/10.1038/sj.hdy.6800430

Maas, E. V., & G. J. Hoffman. 1977. Crop salt tolerance—current assessment. *Journal of the irrigation and drainage division* 103(2): 115–134.

Malheiro, R., S. Casal, P. Baptista, and J. A. Pereira. 2015. A review of *Bactrocera oleae* (Rossi) impact in olive products: From the tree to the table. *Trends in Food Science & Technology* 44(2): 226–242.

Malheiro, R., S. Casal, S. C. Cunha, P. Baptista, and J. A. Pereira. 2016. Identification of leaf volatiles from olive (*Olea europaea*) and their possible role in the ovipositional preferences of olive fly, *Bactrocera oleae* (Rossi) (Diptera: Tephritidae). *Phytochemistry* 121: 11–19.

Mansur, L., and C. Luiz. 2015. Oxidative stress in plants under drought conditions and the role of different enzymes. *Enzyme Engineering* 5: 1–6.

Maracahipes, L., M. B. Carlucci, E. Lenza, B. S. Marimon, B. H. Marimon Jr, F. A. Guimaraes, & M. V. Cianciaruso. 2018. How to live in contrasting habitats? Acquisitive and conservative strategies emerge at inter-and intraspecific levels in savanna and forest woody plants. *Perspectives in Plant Ecology, Evolution and Systematics* 34: 17–25.

Marchi, S., D. Guidotti, M. Ricciolini, & R. Petacchi. 2016. Towards understanding temporal and spatial dynamics of *Bactrocera oleae* (Rossi) infestations using decade-long agrometeorological time series. *International Journal of Biometeorology* 60: 1681–1694.

Marchi, S., R. Tognetti, A. Minnocci, M. Borghi, and L. Sebastiani. 2008. Variation in mesophyll anatomy and photosynthetic capacity during leaf development in a deciduous mesophyte fruit tree (*Prunus persica*) and an evergreen sclerophyllous Mediterranean shrub (*Olea europaea*). Trees (Berl.) 22 (4), 559–571. doi.org/10.1007/s00468-008-0216-9

Marschner, H. 1995. Mineral nutrition of higher plants. *Academic Press*, London.

Martin, G. C., L. Ferguson, & V. S. Polito. 1994. Flowering, pollination, fruiting, alternate bearing, and abscission. In: L. Ferguson, G. Steven Sibbett, G. C. Martin (Eds.), *Olive Production Manual* (pp. 51–56). University of California Press, Berkeley, CA.

Martins-Lopes, P., J. Lima-Brito, S. Gomes, J. Meirinhos, L. Santos, H. Guedes-Pinto. 2007. RAPD and ISSR molecular markers in *Olea europaea* L.: Genetic variability and molecular cultivar identification. *Genetic Resources and Crop Evolution* 54(1): 117–128. doi.org/10.1007/s10722-005-2640-7

Matesanz, S., & F. Valladares. 2014. Ecological and evolutionary responses of Mediterranean plants to global change. *Environmental and Experimental botany* 103: 53–67. doi.org/10.1016/J.ENVEXPBOT.2013.09.004.

Mattos, L., & C. Moretti. 2015. Oxidative stress in plants under drought conditions and the role of different enzymes. *Enzyme Engineering* 5: 1–6.

Médail, F., P. Quézel, G. Besnard, and B. Khadari. 2001. Systematics, ecology and phylogeographic significance of *Olea europaea* L. ssp. *maroccana* (Greuter & Burdet) P. Vargas et al., a relictual olive tree in south-west Morocco. *Botanical Journal of the Linnean Society*, 137(3): 249–266. doi:10.1006/bojl.2001.0477.

Mediavilla, S., A. Escudero, H. Heilmeier. 2001. Internal leaf anatomy and photosynthetic resource-use efficiency: interspecific and intraspecific comparisons. *Tree Physiology* 21: 251–259.

Mediavilla, S., V. Gallardo-López, P. González-Zurdo, and A. Escudero. 2012. Patterns of leaf morphology and leaf N content in relation to winter temperatures in three evergreen tree species. *International Journal of Biometeorology* 56(5): 915–926.

Meftah H., A. Boughdad, A. Bouchelta. 2014. Comparison of biological and demographic parameters of Euphyllura olivina Costa (Homoptera, Psyllidae) on four varieties of olive. *Olivae* (120) 48: 3–16.

Meirinhos, J., B. M. Silva, P. ValentÃo, R. M. Seabra, J. A. Pereira, A. Dias. et al. 2005. Analysis and quantification of flavonoidic compounds from Portuguese olive (*Olea europaea* L.) leaf cultivars. *Natural product research 19*(2): 189–195. doi:10.1080/14786410410001704886.

Michaletz, S.T., M.D. Weiser, J. Zhou, M. Kaspari, B.R. Helliker, B.J. Enquist. 2015. Plant Thermoregulation: Energetics, Trait–Environment Interactions, and Carbon Economics. *Trends in ecology & evolution* 30: 714–724. doi.org/10.1016/j.tree.2015.09.006.

Miransari, M. 2011. Hyperaccumulators, arbuscular mycorrhizal fungi and stress of heavy metal. *Biotechnology Advances* 29: 645–653. doi:10.1016/j.biotechadv.2011.04.006.

Mirshekali, H., H. A. S. H. E. M. Hadi, R. Amirnia, & H. Khodaverdiloo. 2012. Effect of zinc toxicity on plant productivity, chlorophyll and Zn contents of sorghum (Sorghum bicolor) and common lambsquarter (Chenopodium album). *International Journal of Agriculture 2*(3): 247.

Moreno, F., J.E. Fernández, B.E. Clothier, and S.R. Green. 1996. Transpiration and root water uptake by olive trees. *Plant and Soil* 184: 85–96.

Moreno-Caselles, J., R. Moral, A. Pérez-Espinosa, & M. D. Pérez-Murcia, 2000. Cadmium accumulation and distribution in cucumber plant. *Journal of Plant Nutrition* 23(2): 243–250.

Moriana, A., D. Pérez-López, M.H. Prieto, M. Ramírez-Santa-Pau, J.M. Pérez-Rodríguez. 2012. Midday stem water potential as a useful tool for estimating irrigation requirements in olive trees. *Agricultural Water Management* 112: 43–54.

Muller, C., and M. Paschke. 2012. Response to biotic factors causing plant stress. In *Plant response to stress* (pp. 110–133). Zurich: Zurich-Basel Plant Science Center PSC. www.plantscinces.ch.

Muzzalupo, I., G. G. Vendramin, A. Chiappetta. 2014. Genetic biodiversity of Italian olives (*Olea europaea*) germplasm analyzed by SSR markers. *The Scientific World Journal*.

Nadezhdina, N., T.S. David, J.S. David, V. Nadezhdin, J. Cermak, R. Gebauer, M.I. Ferreira, N. Conceicao, M. Dohnal, M. Tesar, K. Gartner, R. Ceulemans. 2012. Root function: in situ studies through sap flow research. In: Manusco, S. (Ed.), *Measuring Roots: An Updated Approach*. Springer, Heidelberg/Dordrecht/London/NewYork, pp. 267–290.

Nadezhdina, N., V. Nadezhdin, M.I. Ferreira, A. Pitacco. 2007. Variability with xylem depth in sap flow in trunks and branches of mature olive trees. *Tree Physiology* 27: 105–113.

Nagajyoti, P. C., K. D. Lee, & T. V. M. Sreekanth. 2010. Heavy metals, occurrence and toxicity for plants: a review. *Environmental chemistry letters* 8(3): 199–216.

Natali, S., C. Bignami, C. Cammilli, M. Muganu. 1999. Effect of water stress on leaf movement in olive cultivars. *Acta Horticulturae* 474: 445–448.

Nave, A., F. Gonçalves, A.L. Crespí, M. Campos, and L. Torres. 2016. Evaluation of native plant flower characteristics for conservation biological control of Prays oleae. Bulletin of Entomological Research 106: 249–257.

Obanor, F. O., M. V. Jaspers, E. E. Jones, and M. Walter. 2008. Greenhouse and field evaluation of fungicides for control of olive leaf spot in New Zealand. *Crop Protection* 27(10):1335–1342.

Obied, H. K. 2013. Biography of biophenols: past, present and future. *Functional Foods in Health and Disease* 3(6): 230–241. doi:10.31989/ffhd.v3i6.51.

OEPP. 2020. Inspection of places of production for *Xylella fastidiosa*. Bulletin OEPP/EPPO Bulletin 50 (3): 415–428. ISSN 0250-8052. doi:10.1111/epp.12691.

Opdenakker, K., T. Remans, E. Keunen, J. Vangronsveld, & A. Cuypers. 2012. Exposure of Arabidopsis thaliana to Cd or Cu excess leads to oxidative stress mediated alterations in

MAPKinase transcript levels. *Environmental and Experimental Botany* 83: 53–61. doi:10.1016/j.envexpbot.2012.04.003

Osakabe, Y., K. Osakabe, K. Shinozaki, & L. S. P. Tran. 2014. Response of plants to water stress. *Frontiers in plant science* 5:86. doi.org/10.3389/fpls.2014.00086.

Ouguas, Y., and M. Chemseddine. 2011. Effect of pruning and chemical control on *Saissetia oleae* (Olivier) (Hemiptera, Coccidae) in olives. *Fruits* 66(3): 225–234.

Owen, R. W., R. Haubner, W. Mier, A. Giacosa, W. E. Hull, B. Spiegelhalder, & H. Bartsch. 2003. Isolation, structure elucidation and antioxidant potential of the major phenolic and flavonoid compounds in brined olive drupes. *Food and Chemical Toxicology* 41(5): 703–717. doi: 10.1016/s0278-6915(03)00011-5

Palamarev, E. 1989. *Paleobotanical evidences of the Tertiary history and origin of the Mediterranean sclerophyll dendroflora.* Plant Systematics and Evolution 162: 93–107.

Pallioti, A., G. Bongi, P. Rocchi. 1994. Peltate trichomes effects on photosynthetic gas exchange of *Olea europaea* L. leaves. *Plant Physiology* 13: 35–44.

Pan, J., R. Sharif, X. Xu, and X. Chen. 2021. Mechanisms of waterlogging tolerance in plants: Research progress and prospects. *Frontiers in Plant Science* 11: 627331.

Pantin, F., F. Monnet, D. Jannaud, J. M. Costa, J. Renaud, B. Muller. et al. 2013. The dual effect of abscisic acid on stomata. *New Phytologist* 197(1): 65–72.

Pegg, G. F., and B. L. Brady. 2002. *Verticillium wilts.* CABI.

Peralbo-Molina, A., F. Priego-Capote, and M. D. Luque de Castro. 2012. Tentative identification of phenolic compounds in olive pomace extracts using liquid chromatography–tandem mass spectrometry with a quadrupole–quadrupole-time-of-flight mass detector. *Journal of Agricultural and Food Chemistry*, 60(46): 11542–11550. doi:10.1021/jf302896m

Percy, D. M., A. Rung, M. S. Hoddle. 2012. An annotated checklist of the psyllids of California (Hemiptera: Psylloidea). *Zootaxa* 3193: 1–27, doi:10.11646/zootaxa.3193.1.1

Pereira, A. P., I. C. Ferreira, F. Marcelino, P. Valentão, P. B. Andrade, R. Seabra. et al. 2007. Phenolic compounds and antimicrobial activity of olive (*Olea europaea* L. Cv. Cobrançosa) leaves. *Molecules*, 12(5): 1153–1162.

Perez-Martin, A., C. Michelazzo, J.M. Torres-Ruiz, J. Flexas, J.E. Fernández, L. Sebastiani, A. Diaz-Espejo. 2011. Physiological and genetic response of olive leaves to water stress and recovery: implications of mesophyll conductance and genetic expression of aquaporins and carbonic anhidrase. *Acta Horticulturae* 922: 99–106.

Petridis, A., I. Therios, G. Samouris, S. Koundouras, and A. Giannakoula. 2012. Effect of water deficit on leaf phenolic composition, gas exchange, oxidative damage and antioxidant activity of four Greek olive (*Olea europaea* L.) cultivars. *Plant physiology and biochemistry* 60: 1–11.Pielke Sr, R. A., A. Pitman, D. Niyogi, R. Mahmood, C. McAlpine, F. Hossain, et al. 2011. Land use/land cover changes and climate: modeling analysis and observational evidence. *Wiley Interdisciplinary Reviews: Climate Change* 2(6): 828–850.

Polade, S. D., D. W. Pierce, D. R. Cayan, A. Gershunov, & M. D. Dettinger. 2014. The key role of dry days in changing regional climate and precipitation regimes. *Scientific reports* 4(1): 1–8. doi.org/10.1038/srep04364.

Prasad, A. S. 2012. Discovery of human zinc deficiency: 50 years later. *Journal of trace elements in medicine and biology* 26: 66–69. doi:10.1016/j.jtemb.2012.04.004

Prentice, I. C., T. Meng, H. Wang, S. P. Harrison, J. Ni, and G. Wang. 2011. Evidence of a universal scaling relationship for leaf CO2 drawdown along an aridity gradient. *New Phytologist* 190(1): 169–180.

Qi, X., Q. Li, X. Ma, C. Qian, H. Wang, N. Ren. et al. 2019. Waterlogging induced adventitious root formation in cucumber is regulated by ethylene and auxin through reactive oxygen species signalling. *Plant, Cell & Environment* 42(5): 1458–1470. doi:10.1111/pce.13504

Quesada, J. M., R. Penyalver, and M. M. López. 2012. Epidemiology and control of plant diseases caused by phytopathogenic bacteria: the case of olive knot disease caused by *Pseudomonas savastanoi* pv. *savastanoi* in *Plant Pathology*, ed. C. J. Cumagun (INTECH Open Access Publisher), 299–326.

Rallo, P., R. Jiménez, J. Ordovás, & M. P. Suárez. 2008. Possible early selection of short juvenile period olive plants based on seedling traits. *Australian journal of agricultural research* 59(10): 933–940. doi: 10.1071/AR08013.

Ramos, C., I. M. Matas, L. Bardaji, I. M. Aragón, & J. Murillo. 2012. *Pseudomonas savastanoi* pv. *savastanoi*: some like it knot. *Molecular plant pathology* 13(9): 998–1009. doi:10.1111/j.1364-3703.2012.00816.x

Ramos, P., M. Campos, and J. M. Ramos. 1998. Long-term study on the evaluation of yield and economic losses caused by *Prays oleae* Bern. in the olive crop of Granada (southern Spain). *Crop Protection* 17(8): 645–647.

Reeves, R.D., Baker, A.J.M. 2000. Metal-accumulating plants. In: Raskin I, Ensley BD (eds) *Phytoremediation of toxic metals: using plants to clean up the environment.* Wiley, New York, pp 193–229

Rewald, B., C. Leuschner, Z. Wiesman, J.E. Ephrath. 2011. Influence of salinity on root hydraulic properties of three olive varieties. *Plant Biosystems* 145(1): 12–22.

Rhizopoulou, S. 2007. *Olea europaea* L. A botanical contribution to culture. *American-Eurasian Journal of Agricultural & Environmental Sciences* 2(4): 382–387.

Ribeiro, R. D. A., M. M. R. F. De Melo, F. De Barros, C. Gomes, and G. Trolin. 1986. Acute antihypertensive effect in conscious rats produced by some medicinal plants used in the state of Sao Paulo. *Journal of ethnopharmacology* 15(3): 261–269.Ribeiro, R. D. A., F. de Barros, M. M. R. F. de Melo, C. Muniz, et al. 1988. Acute diuretic effects in conscious rats produced by some medicinal plants used in the state of Sao Paulo, Brasil. *Journal of Ethnopharmacology* 24(1): 19–29.

Riddle, J. M. 1985. *Dioscorides on Pharmacy and Medicine.* University of Texas Press, Austin

Rizhsky, L., H. Liang, J. Shuman, V. Shulaev, S. Davletova, and R. Mittler. 2004. When defense pathways collide. The response of *Arabidopsis* to a combination of drought and heat stress. *Plant Physiology* 134: 1683–1696. doi:10.1104/pp.103.033431

Rizzo, M. C., G. Lo Verde, R. Rizzo, V. Buccellato, and V. Caleca. 2007. Introduzione ed acclimatamento di Closterocerus sp. in Sicilia per il controllo biologico di Ophelimus maskelli Ashmead (Hymenoptera, Eulophidae) galligeno esotico degli eucalipti. In XXI Congresso Nazionale Italiano di Entomologia, 387–387.

Rizzo, R., V. Caleca, and A. Lombardo. 2012. Relation of fruit color, elongation, hardness, and volume to the infestation of olive cultivars by the olive fruit fly, *Bactrocera oleae. Entomologia Experimentalis et Applicata* 145(1): 15–22.

Rosales, R., I. Sabouni, F. Chibi, D. Garrido, & J.M. Ramos. 2008. Comparing the benefits between pesticides and ethylene treatments in reducing olive moth population numbers and damage. International Journal of Pest Management 54(4): 327–331.

Rossi, L., L. Sebastiani, R. Tognetti, R. d'Andria, G. Morelli, & P. Cherubini. 2013. Tree-ring wood anatomy and stable isotopes show structural and functional adjustments in olive trees under different water availability. *Plant and Soil* 372(1): 567–579.

Rubio de Casas R., G. Besnard, P. Schönswetter, L. Balaguer, P. Vargas. 2006. Extensive gene flow blurs phylogeographic but not phylogenetic signal in Olea europaea L. Theoretical and Applied Genetics 113: 575–583.

Rugini, E., & E. Fedeli. 1990. Olive (*Olea europaea* L.) as an oilseed crop. In *Legumes and oilseed crops I* (pp. 593–641). Springer, Berlin, Heidelberg.

Sack, L., M.C. Ball, C. Brodersen, S.D. Davis. et al. 2016. Plant hydraulics as a central hub integrating plant and ecosystem function: meeting report for "Emerging Frontiers in Plant Hydraulics". *Plant Cell Environ* 39:2085–2094.

Saijo, Y., and E. P. I. Loo. 2020. Plant immunity in signal integration between biotic and abiotic stress responses. *New Phytologist* 225: 87–104. doi:10.1111/nph.15989.

Salleo, S., and M.A., Lo Gullo. 1983. Water transport pathways in nodes and internodes of 1-year-old twigs of *Olea europaea* L. *Giornale botanico italiano* 117: 3–74.

Saponari, M., G. Loconsole, D. Cornara. et al. 2014. Infectivity and transmission of *Xylella fastidiosa* by *Philaenus spumarius* (Hemiptera: Aphrophoridae) in Apulia, Italy. *Journal of Economic Entomology* 107(4): 1316–1319. doi.org/10.1603/EC14142

Sasidharan, R., J. Bailey-Serres, M. Ashikari, B. J. Atwell, T. D. Colmer, K. Fagerstedt. et al. 2017. Community recommendations on terminology and procedures used in flooding and low oxygen stress research. *New Phytologist* 214: 1403–1407. doi:10.1111/nph.14519

Savarese, M., E. De Marco, & R. Sacchi. 2007. Characterization of phenolic extracts from olives (*Olea europaea* cv. Pisciottana) by electrospray ionization mass spectrometry. *Food Chemistry* 105(2): 761–770. doi.org/10.1016/j.foodchem.2007.01.037

Savary, S., L. Willocquet, S. J. Pethybridge, P. Esker, N. McRoberts, & A. Nelson. 2019. The global burden of pathogens and pests on major food crops. *Nature ecology & evolution* 3(3): 430–439. https://doi.org/10.1038/s41559-018-0793-y

Scarpati, M. L., R. L. Scalzo, and G. Vita. 1993. *Olea europaea* volatiles attractive and repellent to the olive fruit fly (*Dacus oleae*, Gmelin). *Journal of chemical ecology* 19(4): 881–891.

Scarpati, M. L., R. L. Scalzo, G. Vita, and A. Gambacorta. 1996. Chemiotropic behavior of female olive fly (*Bactrocera oleae* GMEL.) on *Olea europaea* L. Journal of chemical ecology 22(5): 1027–1036.

Schwabe, W.W., and S.M. Lionakis. 1996. Leaf attitude in olive in relation to drought resistance. *Journal of Horticultural Science* 71: 157–166.

Scibetta, S., G. E. Agosteo, A. Abdelfattah, M. G. Li Destri Nicosia, S. O. Cacciola, and L. Schena. 2019. Development and Application of a qPCR Detection Method to Quantify *Venturia oleaginea* (Castagne) Rossman & Crous in Asymptomatic Olive (*Olea europaea* L.) Leaves. Phytopathology. doi:10.1094/phyto-07-19-0227-r

Searles, P.S., D.A. Saravia, M.C. Rousseaux. 2009. Root length density and soil water distribution in drip-irrigated olive orchards in Argentina under arid conditions. *Crop and Pasture Science* 60: 280–288.

Secchi, F., C. Lovisolo, A. Schubert. 2007. Expression of OePIP2.1 aquaporin gene and water relations of Olea europaea twigs during drought stress and recovery. *Annals of Applied Biology* 150 (2): 163–167.

Seljak., G. 2006. An overview of the current knowledge of jumping plant-lice of Slovenia (Hemiptera: Psylloidea). *Acta Entomologica Slovenica* 14: 11–34.

Sharaf, N. S. 1980. Life history of the olive fruit fly, *Dacus oleae* Gmel. (Diptera: Tephritidae), and its damage to olive fruits in Tripolitania. *Journal of Applied Entomology* 89: 390–400.

Singh, R., N. Gautam, A. Mishra, and R. Gupta, 2011. Heavy metals and living systems: an overview. *Indian Journal of Pharmacology* 43: 246. doi:10.4103/0253-7613.81505

Sofo, A., S. Manfreda, M. Fiorentino, B. Dichio, and C. Xiloyannis. 2008. The olive tree: a paradigm for drought tolerance in Mediterranean climates. *Hydrology and Earth System Sciences* 12(1): 293–301.

Spence, N., L. Hill, & J. Morris. 2020. How the global threat of pests and diseases impacts plants, people, and the planet. *Plants, People, Planet* 2(1): 5–13. doi.org/10.1002/ppp3.10088

Spinoni, J., P. Barbosa, A. De Jager, N. McCormick, G. Naumann. et al. 2019. A new global database of meteorological drought events from 1951 to 2016. *Journal of Hydrology: Regional Studies* 22:100593. doi.org/10.1016/j.ejrh.2019.100593.

Suc, J.P. 1984. Origin and evolution of the Mediterranean vegetation and climate in Europe. *Nature* 307: 429–432. doi.org/10.1038/307429a0.

Tahraoui, A., J. El-Hilaly, Z. H. Israili, B. Lyoussi. 2007. Ethnopharmacological survey of plants used in the traditional treatment of hypertension and diabetes in south-eastern Morocco (Errachidia province). *Journal of ethnopharmacology*, *110*(1): 105–117.

Taiz, L., and E. Zeiger. 2006. *Fisiologia vegetal* (Vol. 10). Universitat Jaume I.

Taiz, L., E. Zeiger, I. M. Møller, & A. Murphy.2015. *Plant physiology and development* (No. Ed. 6). Sinauer Associates Incorporated.

Tattini, M., P. Bertoni, & S. Caselli. 1992. Genotipic responses of olive plants to sodium chloride. *Journal of plant nutrition 15*(9): 1467–1485.

Terral, J.F. 2000. Exploitation and management of olive tree during Prehistoric times in Mediterranean France and Spain. *Journal of Archaeological Science* 27: 127–133.

Terral, J. F., and A. Durand. 2006. Bio-archaeological evidence of olive tree (*Olea europaea* L.) irrigation during the Middle Ages in Southern France and North Eastern Spain. *Journal of Archaeological Science* 33(5): 718–724.

Terral, J.F., and X. Mengüal. 1999. Reconstruction of Holocene climate in Southern France and Eastern Spain using quantitative anatomy of olive wood and archaeological charcoal. *Palaeogeography, Palaeoclimatology, Palaeoecology* 153: 71–92

Terral, J.-F., E. Badal, C. Heinz, P. Roiron, S. Thiébault, & I. Figueiral. 2004. A hydraulic conductivity model points to post-Neogene survival of the Mediterranean Olive in riparian habitat. *Ecology*. 85: 3158–3165. doi:10.1890/03-308113

Teviotdale, B.L., L. Ferguson, and P.M. Vossen. 2005. *UC IPM Pest Management Guidelines: Olive*. Diseases, University of California, Berkeley, California, Publication 3452, pp. 107–109.

Therios, L. 2009. *Olives: Crop Production Science in Horticulture 18*. Wallingford, UK: CABI Publishing

Thomine, S., and V. Lanquar. 2011. Iron transport and signaling in plants, in *Transporters and Pumps in Plant Signaling*, eds M. Geisler and K. Bemema, Berlin; Heidelberg: Springer, 99–131.

Tian, L., J. Li, W. Bi, S. Zuo, L. Li, Li, W. et al. 2019. Effects of waterlogging stress at different growth stages on the photosynthetic characteristics and grain yield of spring maize (*Zea mays* L.) Under field conditions. *Agricultural Water Management* 218: 250–258. doi:10.1016/j.agwat.2019.03.054.

Tognetti, R., A. Giovannelli, A. Lavini, G. Morelli, F. Fragnito, R. d'Andria. 2009. Assessing environmental controls over conductances through the soil–plant–atmosphere continuum in an experimental olive tree plantation of southern Italy. *Agricultural and Forest Meteorology* 149: 1229–1243.

Torres-Ruiz, J. M., H. Cochard, S. Mayr, B. Beikircher, A. Diaz--Espejo, et al. 2014. Vulnerability to cavitation in Olea europaea current--year shoots: further evidence of an open-vessel artifact associated with centrifuge and air-injection techniques. *Physiologia Plantarum*, 152(3): 465–474.

Torres-Ruiz, J. M., A. Diaz-Espejo, A. Perez-Martin, & V. Hernandez-Santana. 2015. Role of hydraulic and chemical signals in leaves, stems and roots in the stomatal behaviour of olive trees under water stress and recovery conditions. *Tree Physiology* 35(4): 415–424.

Torres-Ruiz, J.M., A. Diaz-Espejo, A. Perez-Martin, and V. Hernandez-Santana. 2013. Loss of hydraulic functioning at leaf, stem and root level and its role in the stomatal behaviour during drought in olive trees. *Acta Hortic 991*: 333–339. doi.org/10.17660/ActaHortic.2013.991.41.

van Asch, B., I. Pereira-Castro, F. Rei, and L. T. Da Costa. 2012. Mitochondrial haplotypes reveal olive fly (*Bactrocera oleae*) population substructure in the Mediterranean. *Genetica* 140(4): 181–187.

van Lierop, P., E. Lindquist, S. Sathyapala, and G. Franceschini. 2015. Global forest area disturbance from fire, insect pests, diseases and severe weather events. *Forest Ecology and Management.* 352: 78–88. doi:10.1016/j.foreco.2015.06.010.

Van Steenwyk, R.A., L. Ferguson, and F.G. Zalom. 2002. *UC IPM Pest Management Guidelines. Olive (Insects and Mites).* UCANR Publication 3452, University of Calfornia, Berkeley, Calfornia.

Vardanian, S. A. 1978. Phytotherapy of bronchial asthma in medieval Armenian medicine. *Terapevticheskii Arkhiv* 50(4): 133–136.

Vellend, M., L. Baeten, A. Becker-Scarpitta, V. Boucher-Lalonde, J. L. McCune, J. Messier, & D. F. Sax. 2017. Plant biodiversity change across scales during the Anthropocene. *Annual Review of Plant Biology* 68: 563–586. doi.org/10.1146/annurev-arplant-042916-040949.

Vigo, C., Therios, I. and A. Bosabalidis. 2005. Plant growth, nutrient concentration and leaf anatomy in olive plants irrigated with diluted seawater. *Journal of Plant Nutrition* 28: 101–102.

Vigo, C., I. Therios, A. Patakas, A. Karatassou, and A. Nastou. 2002. Changes in photosynthetic parameters and nutrient distribution of olive plants (*Olea europaea* L.) cultivar 'Chondrolia Chalkidikis' under NaCl, Na2SO4 and KCl salinities. *Agrochimica* 46: 33–46.

Visioli, F., and C. Galli. 1994. Oleuropein protects low density lipoprotein from oxidation. *Life sciences*, 55(24): 1965–1971.

Wang, W., B. Vinocur, and A. Altman 2003. Plant responses to drought, salinity and extreme temperatures: towards genetic engineering for stress tolerance. *Planta* 28: 1–14. doi:10.1007/s00425-003-1105-5

Wang, Z., L.-Y. Ma, J. Cao, Y.-L. Li, L.-N. Ding, K.-M. Zhu. et al. 2019. Recent advances in mechanisms of plant defense to *Sclerotinia sclerotiorum*. *Frontier in Plant Science* 10: 1314. doi:10.3389/fpls.2019.01314.

Wilson, B., and F. B. Pyatt. 2007. Heavy metal bioaccumulation by the important food plant, *Olea europaea* L., in an ancient metalliferous polluted area of Cyprus. *Bulletin of environmental contamination and toxicology* 78(5): 390–394.

Wu, X., Z. Hao, Q. Tang, V. P. Singh, X. Zhang, and F. Hao. 2020. Projected increase in compound dry and hot events over global land areas. *International Journal of Climatology* 41(1): 393–403. doi:10.1002/joc.6626

Xu, C., N.G. McDowell, R.A. Fisher, L. Wei, Sevanto, S. et al. 2019. Increasing impacts of extreme droughts on vegetation productivity under climate change. *Nature Climate Change* 9: 948–953. doi.org/10.1038/s41558-019-0630-6.

Yang, X., Y. Feng, Z. He, & P. J. Stoffella. 2005. Molecular mechanisms of heavy metal hyperaccumulation and phytoremediation. *Journal of trace elements in medicine and biology* 18(4): 339–353. doi:10.1016/j.jtemb.2005.02.007.

Yang, Y., L. Kang, J. Zhao, N. Qi, R. Li, J. Wen, J. Kassout, C. Peng, G. Lin, H. Zheng. 2021. Quantifying Leaf Trait Covariations and Their Relationships with Plant Adaptation Strategies along an Aridity Gradient. *Biology* 10(10):1066. doi:10.3390/biology10101066.

Yanqun, Z., L. Yuan, C. Jianjun, C. Haiyan, Q. Li, & C. Schvartz. 2005. Hyperaccumulation of Pb, Zn and Cd in herbaceous grown on lead–zinc mining area in Yunnan, China. *Environment International* 31(5): 755–762.

Zarco, A., J. R. V. Puente, L. F. R. Castillo, and A. T. Casas. 2007. Detección de las infecciones latentes de" Spilocaea oleagina" en hojas de olivo. *Boletín de sanidad vegetal Plagas* 33(2): 235–248.

Zargari, A. 1997. *Iranian Medicinal plants*. Tehran University of Medical Sciences, 3: 392.

Zhan, J., L. Ericson, and J. J. Burdon. 2018. Climate change accelerates local disease extinction rates in a long-term wild host–pathogen association. *Global Change and Biology* 24: 3526–3536. doi:10.1111/gcb.14111.

Zhao, C., B. Liu, S. Piao, X. Wang, D. B. Lobell, Y. Huang. et al. 2017. Temperature increase reduces global yields of major crops in four independent estimates. *Proceedings of the National Academy of Sciences* 114(35): 9326–9331. doi:10.1073/pnas.1701762114.

Zhou, W., F. Chen, Y. Meng, U. Chandrasekaran, X. Luo, W. Yang, and K. Shu. 2020. Plant waterlogging/flooding stress responses: From seed germination to maturation. *Plant Physiology and Biochemistry* 148, 228–236. doi:10.1016/j.plaphy.2020.01.020.

Zhu, J.K. 2002. Salt and drought stress signal transduction in plants. *Annual review of plant biology* 53: 247–273.

Zohary, D., M. Hopf, E. Weiss. 2012. Domestication of plants in the Old World: the origin and spread of cultivated plants in Southwest Asia, Europe, and the Mediterranean Basin. *Oxford: Oxford University Press*.

14 *Origanum* taxa and Stressful Conditions

Mohamed Bakha[1*], Jalal Kassout[2,3], Abdelkarim Khiraoui[4],
Noureddine El Omari[5], Abdelhakim Bouyahya[6],
Noureddine El Mtili[7], Kaoutar Aboukhalid[8] and
Chaouki Al Faiz[9]

[1]Unit of Plant Biotechnology and Sustainable Development of Natural
Resources "B2DRN," Polydisciplinary Faculty of Beni Mellal, Sultan
Moulay Slimane University, Mghila, Beni Mellal, Morocco
[2]Laboratory of Applied Botany, BioAgrodiversity Team, Faculty of
Sciences of Tétouan, Abdelmalek Essaâdi University, Tétouan, Morocco
[3]National Institute of Agronomic Research, INRA-CRRA, Marrakech-Safi,
Marrakech, Morocco
[4]Ecology and Biodiversity Team, Faculty of Sciences and Technology,
University of Sultan Moulay Slimane, Beni-Mellal, Morocco
[5]Laboratory of Histology, Embryology, and Cytogenetic, Faculty of
Medicine and Pharmacy, Mohammed V University in Rabat, Morocco
[6]Laboratory of Human Pathologies Biology, Department of Biology,
Faculty of Sciences and Genomic Center of Human Pathologies,
Mohammed V University, Morocco
[7]Laboratory of Biology and Health, Faculty of Sciences of Tétouan,
Abdelmalek Essaâdi University, Tétouan, Morocco
[8]National Institute of Agronomic Research (INRA), 10 Bd. Mohamed VI,
Oujda, Morocco
[9]National Institute of Agronomic Research, CRRA-Rabat, Rabat, Morocco
*Mohamed Bakha (bakha.mohamad@gmail.com)

CONTENTS

DOI: 10.1201/9781003242963-14

14.1 INTRODUCTION

In nature, flora are often exposed to environmental fluctuations that could result in a wide range of stressful conditions mainly related to extreme weather, water efficiency, nutrient availability, interspecies competition and exposure to diseases (Pugnaire and Luque 2001; Farooq et al. 2009). These factors in association with the adaptive potential of plants could affect their distribution, population density, growth, morphology, physiology and genetic diversity (Smirnoff 1998; Woodruff 2010). Rare species with restricted geographical distribution and small populations appear to be the most threatened by environmental changes (Paschke et al. 2003). In addition, genetic variation plays a key role in plant distribution, adaptability and survival, where plants with high genetic diversity possess generally a high ability to resist environmental fluctuations (Ayanogç Lu et al. 2006; Aboukhalid et al. 2017a).

Ongoing climate change is imposing broad changes in ecosystem functions and services with considerable impacts on plant distribution, growth and development (Parmesan and Yohe 2003; Vellend et al. 2017). This is the case in the Mediterranean region, where the impact of climate change and anthropogenic pressures are already evident (Polade et al. 2014; Matesanz and Valladares 2014). These changes expose the Mediterranean flora to stressful conditions (Blondel et al. 2010; Lloret et al. 2016), especially for species and populations having specific ecological requirements. In this context, native medicinal and aromatic plants of the Mediterranean region will be under severe threat as a result of the sustained alteration of their environmental conditions.

Among the remarkable Mediterranean taxa, species of *Origanum* genus are one of the most important aromatic and medicinal plants, widely used for various therapeutic and culinary purposes (Kokkini 1997; Aboukhalid et al. 2017a; Stefanaki and van Andel 2021). It belongs to the botanical Lamiacae family, comprising 45 species, 9 subspecies, and 22 hybrids, structured into three groups and 10 sections (Ietswaart 1980; Duman et al. 1995, 1998; Martin et al. 2020). The geographical distribution of *Origanum* genus is mainly restricted to the Mediterranean region (Ietswaart 1980). *Origanum* taxa are threatened by environmental stress arising from climate change. In addition to the stress caused by natural factors, the impact of anthropogenic activities on the diversity of *Origanum* species is very serious (Tomou et al. 2022). Oregano is known for its economic value due to its diverse therapeutic and culinary properties. Thus, a very significant part of marketed oregano is collected from natural populations. Consequently, the over-exploitation of oregano is considered among the main factors that cause fragmentation of the natural distribution area of this plant and decrease its genetic diversity (Aboukhalid et al. 2017a; Bouiamrine et al. 2017). On the other hand, it is worth mentioning that many *Origanum* species such as *O. vulgare*, *O. majorana* and *O. onites* are successfully cultivated outside their natural geographic range in different countries all around the world, due to their high commercial value (Makri 2002; Farías et al. 2010).

The morphological and physiological processes underlying *Origanum* taxa responses to different stressful conditions are still poorly understood. Therefore, understanding the behaviors and processes through which *Origanum* taxa adapt to stressful conditions is very important to understand their dynamic and biogeographical distribution. Subsequently, these data are very useful for the development of strategies for the protection and conservation of oregano in their areas of origin, as well as for the improvement of their culture under different edaphoclimatic conditions. In this chapter, we report the main findings linked to the impacts of stressful conditions on *Origanum* taxa either in their natural habitats or under agricultural conditions.

14.2 GEOGRAPHICAL DISTRIBUTION OF *ORIGANUM* TAXA AND ADAPTATION TO STRESS CONDITIONS

Origanum genus has its optimum distribution area around the Mediterranean region, particularly in the eastern regions (Figure 14.1). Turkey represents the main center of *Origanum* diversity with more than 20 taxa, followed by Greece with eight taxa (Kokkini 1997). The majority of *Origanum*

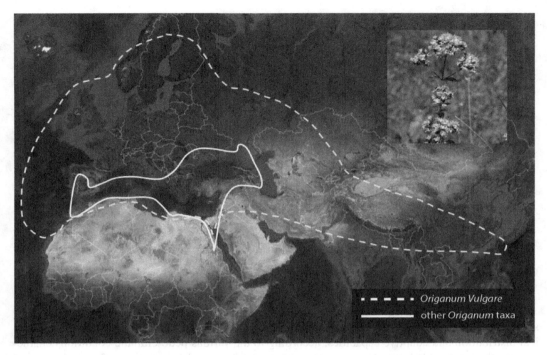

FIGURE 14.1 Geographical distribution of the genus *Origanum* (_ _ _ _*O. vulgare* distribution, _____ the other *Origanum* taxa).

taxa are endemics with a very restricted biogeographical distribution. This limited distribution reflects the specific adaptation of some *Origanum* taxa to the local edaphoclimatic conditions, playing an important role in their growth and persistence. *Origanum* species are mostly perennial herbs and rarely small shrubs, and most species are found in mountainous regions usually in a humid bioclimate (Ietswaart 1980). For instance, *O. dictamus* and *O. microphyllum* are both endemic to Crete (Lianopoulou and Bosabalidis 2014). *O. onites*, widely distributed in Greece and Turkey, is among the *Origanum* species that grow at low altitude levels ranging from sea level to 900 m, mainly on limestone rocky soil (Ietswaart 1980; Vokou et al. 1988). Also, *O. ayliniae* is endemic to western Turkey with a distribution area of less than 10 km² (Dirmenci et al. 2018). *O. sipyleum* is endemic to the western Anatolia region in Turkey. This species grows on calcareous rocks, mainly on sandy-loam, slightly acidic and neutral soils, rich in organic matter, N and P, with pH varying between 6.11 and 6.97 and altitudes between 100 and 1500m (Unal et al. 2013). At this level, stressful conditions induced by environmental and/or anthropic factors appear to be a serious threat to the natural distribution of these endemic taxa that appear to have limited capacities for adaptation.

On the other hand, some *Origanum* taxa have been found able to colonize wide areas as a result of their adaptation capacity to different edaphoclimatic conditions including those considered stressful. Generally, these species are known for their high genetic, morphological and chemical variability (Azizi et al. 2012; Aboukhalid et al. 2017a, 2017b; Alekseeva et al. 2021). At a large scale, *O. vulgare* is the most widespread taxon, growing in the Mediterranean region, and in western, central and eastern Asia. *O. vulgare* is also the most polymorph *Origanum* species with six subspecies described under this species, which reflects its high adaptability (Iestwaart 1980). Furthermore, *O. syriacum* subsp. *sinaicum*, endemic to the Sinai peninsula region in Egypt, is among the rare *Origanum* taxa that are characterized by high growth capacity in arid environments (Zaghloul et al. 2014).

FIGURE 14.2 Hairs of *O. compactum* leaves (photo provided by C. Al Faiz).

At a local scale, five *Origanum* taxa, namely *O. grosii*, *O. elongatum*, *O. compactum*, *O. vulgare* subsp. *virens* and *O. x font-queri* grow naturally in Morocco. These taxa showed different geographical distribution and adaptive capacities, according to the various pedo-climatic conditions characterizing their natural habitats. Our frequent visits to the fields and observations of the behaviors of these taxa concerning some abiotic stresses, such as drought, frost, and high summer temperatures, allow us to have a clear idea about their tolerance (Aboukhalid et al. 2016, 2017a, 2017b, 2019; Bakha et al. 2017, 2019, 2020). Thus. *O. compactum* and *O. elongatum* are considered the most tolerant taxa to the mentioned natural stressful conditions. However, their effects are not responsible for the decrease of the oregano population in comparison with anthropogenic activities.

O. compactum is an endemic species to Morocco and southern Spain (Ietswaart 1980). Hence, it is the most common *Origanum* species in Morocco and shows an important genetic diversity (Aboukhalid et al. 2017a) and chemical polymorphism (Aboukhalid et al. 2016). *O. compactum* is recognized as a potent species in terms of adaptability to stressful conditions. This species possesses a great ability to persist in a semi-arid bioclimate characterized by low mean rainfall, and in altitudes ranging from 126 to 1400 m (Aboukhalid 2019). In addition, it can grow on soils with alkaline pH, silty or sandy texture, with low phosphate levels (Aboukhalid et al. 2017b). The hairiest leaves (Figure 14.2) and the high essential oil (EO) amounts certainly play an important role in the adaptation of *O. compactum* to stressful conditions. Despite its wide distribution range and high adaptability, the anthropogenic activities, related to the over- and irregular exploitation of its natural resources, cause a serious threat to *O. compactum* (Aboukhalid, 2019).

O. elongatum is an endemic species of Morocco that grows in two mountainous regions in the north of the country. This species can persist in different climatic conditions, where some populations could survive in semi-arid conditions in the Rif Mountains, but only on altitudinal limits ranging from 700 to 2000 m (Bakha et al. 2020). *O. elongatum* is distinguished by its developed roots and its capacity to grow on stony soils, sometimes directly on the rocks (Figure 14.3).

O. vulgare subsp. *virens* and *O. grosii* could be considered vulnerable plant species facing both abiotic stresses and over-exploitation. Their growth requires specific edaphoclimatic conditions. *O. grosii* is a rare endemic species of Morocco growing in a very limited area, only under a humid bioclimate, frequently in altitudes ranging from 800 to 2000 m (Bakha et al. 2019). Lack of rainfall seriously affects this species. *O. vulgare* subsp. *virens* also grows in very specific natural conditions,

FIGURE 14.3 Wild *O. elongatum* growing on rocks in the Isaguen region in Morocco (photo provided by K. Aboukhalid).

FIGURE 14.4 Floraison of the glabre *O. vulgare* subsp. *virens* (on the left) and the hairiest *O. compactum* (on the right) (photo provided by M. Bakha).

under humid climates. This species, widely represented in the past, is now considered the most threatened *Origanum* taxa in Morocco (Figure 14.4).

The observed distribution and diversity of oregano species, as for many Mediterranean medicinal and aromatic herbs, is the result of a long history of diversification and evolution within the Mediterranean landscape (Thompson, 2020). Consequently, the observed variation in their form and function could be considered an adaptation strategy to the tremendous heterogeneity found in Mediterranean environmental conditions.

14.3 IMPACTS OF DIFFERENT TYPES OF STRESS ON *ORIGANUM* AND ADAPTATION MECHANISMS

Stressful conditions such as water deficiency, light and temperature stress, and nutrient scarcity induce a range of morphological, physiological, biochemical and molecular changes in plant species (Grime 2006; Ghalambor et al. 2007; Osakabe et al. 2014). The responses of *Origanum* taxa to different types and levels of stress differ from one species to another and depend on the cultivar or variety within the same species. The major investigations concerning the effect of different stresses on some productive parameters in the genus *Origanum* are summarized in Table 14.1. Results related to the effects of stresses on chemical composition are summarized in Table 14.2.

14.3.1 WATER STRESS IMPACT

Water stress, caused by water excess or deficit, is among the main factors influencing the distribution of plants in nature, as well as biomass and EO production. In the wild, *Origanum* taxa persisting in semi-arid and arid conditions are more tolerant to drought and water-deficit conditions, due to their leaf capacity to maintain turgor potentials (Rhizopoulou and Diamantoglou 1991). Dehydration avoidance is the main strategy used by *Origanum* species to overcome stressful conditions. Reduction in the leaf area under water stress conditions helps *Origanum* plants to minimize dehydration (Pereyra et al. 2021). The seasonal dimorphism of leaves is among the adaptive strategies enabling *Origanum* taxa to survive the summer drought by developing different leaf types according to the season (Margaris 1976; Gerami et al. 2016; Lianopoulou and Bosabalidis 2014). For instance, Gerami et al. (2016) showed that the leaf area of Iranian *O. vulgare* decreased significantly when water availability decreased. In addition, salt stress could induce anatomical changes in *O. majorana* by affecting the length of palisade cells and the width of collenchyma (Baâtour et al. 2012).

In terms of productivity, irrigation generally enhances the biomass and EO yields, with a significant reduction of EO content, while water deficit due to limited irrigation or salinity causes a decrease in biomass and EO yields. However, a significant increase in EO contents was observed in some *Origanum* taxa under water stress: *Origanum* sp. (Said-Al Ahl et al. 2009), *O. vulgare* (Azizi et al. 2009; Said-Al Ahl and Hussein 2010; Morshedloo et al. 2017a; Mohammadi et al. 2021), *O. vulgare* subsp. *hirtum* and subsp. *gracile* (Morshedloo et al. 2017b; Ninou et al. 2017; Emrahi et al. 2021), *O. syriacum* (Khraizat 2015), *O. onites* (Hancioglu et al., 2019) and *O. majorana* (Farsi et al. 2019). In nature, water stress linked to the bioclimate conditions displayed a significant effect on the EO content. It has been reported that the EO yield increases significantly from humid to arid zones. For instance, in *O. compactum*, populations growing under a semiarid bioclimate and stressful conditions displayed the highest levels of EOs, whereas populations growing under a subhumid climate showed lower EO content (Aboukhalid et al. 2017a). The transplantation of *O. compactum* and *O. grosii* plants from the wild under natural stressful conditions, to a cultivated plot under more favorable conditions (fertilization and regular irrigation), caused a significant increase in the biomass production and EO amounts and yields, as well as a noticeable change in the chemical composition (Aboukhalid et al. 2019; Bakha et al. 2019). The increase of the EO content under irrigation could be explained by the enhancement of the leaves/stem ratio in these plants. In addition,

TABLE 14.1

Summary of Stress Condition Impacts on Growth, Biomass And Essential Oils (EO) Production of Different *Origanum* Taxa (↑: Increase, ↓: Decrease)

Origanum taxa	Stress type	Growth and biomass	EO yield	References
O. compactum	Transplantation (from nature to experiment field under irrigation)	Transplantation ↑ biomass yield	Transplantation ↑ the amounts and yields of EO	Aboukhalid 2019
O. grosii	Transplantation (from nature to experiment field under irrigation)	Transplantation ↑ biomass yield	Transplantation ↑ the amounts and yields of EO	Bakha et al. 2019
O. majorana	Water stress (soil moisture deficit)	↑ Water stress ↑ dry matter yield	↑ Water stress ↑ EO content, total lipids	Rhizopoulou and Diamantoglou 1991
O. majorana	Water stress (salt treatment: 0, 50, 100 and 150 mM of NaCl)	Growth and water content were affected just at the highest NaCl concentration (150 mM)	EO yield ↓ with ↑ salinity	Baatour et al. 2010
O. majorana Canadian (CV) and Tunisian (TV) varieties	Water stress (salt effects)	Salt treatment led to ↓ in aerial part growth	Salinity ↓ EO yield in TV and ↑ EO yield in CV	Baâtour et al. 2011
O. majorana	Water stress [tolerance to salinity using seed priming with titanium dioxide (TiO_2)]	In no saline soil: TiO_2 did not affect growth and biomass production In saline soils: TiO_2 ↑ growth and physiological characteristics High concentration of TiO_2 (above 80 mg L^{-1}) and salinity (above 75 mM NaCl) ↓ the growth and biomass production	EO yield and content were sensitive to saline conditions above 25 mM In no saline soil: TiO_2 did not affect EO amount In saline soils: TiO_2 ↑ EO yield and content High concentration of TiO_2 (above 80 mg L^{-1}) and salinity (above 75 mM NaCl) ↓ EO production	Jafari et al. 2021
O. majorana	Water stress (growth responses to methyl jasmonic acid under limited irrigation)	Moderate limited irrigation ↓ plant weight Foliar spray of MeJA ↑ growth parameters such as fresh and dry weight just under mild limited irrigation conditions Under moderately limited irrigation, MeJA did not affect biomass production	Moderate limited irrigation ↓ EO yield Both mild and moderate limited irrigations ↑ EO % Under mildly limited irrigation, MeJA ↓ EO content and yield	Farsi et al. 2019

(continued)

TABLE 14.1 (Continued)

Summary of Stress Condition Impacts on Growth, Biomass And Essential Oils (EO) Production of Different *Origanum* Taxa (↑: Increase, ↓: Decrease)

Origanum taxa	Stress type	Growth and biomass	EO yield	References
O. majorana	Water stress (water deficit)	No effect on the plant's dry weight	Did not affect the EO content	Mohammadi et al. 2021
O. majorana	Water stress [chitosan (0, 250 and 500mg/L) under well-watered; chitosan (0, 250 and 500 mg/L) under water deficit stress]	Under well-watered conditions: chitosan at 250 mg/L and 500 mg/L ↑ the dry weight. Under water deficit: chitosan didn't affect the plant dry weight	Under both well-watered and water-deficit conditions: chitosan didn't affect EO content	Mohammadi et al. 2021
O. majorana	Fertilization (N, P and K)	Growth rates, fresh and dry yield ↑ by nutrient supply (N, P and K)	EO yield ↑ by nutrient supply (N, P and K)	Trivino and Johnson 2000
O. onites	Water stress (irrigation water salinity)	The plants growing in 7 and 10 dS/m could not survive; compared to control ↓ in dry yields under salinity levels up to 5 dS/m	↑ Irrigation water salinity increased phenolic, flavonoid and EO content but ↓ decreased EO yield	Hancioglu et al. 2019
O. syriacum	Irrigation and nitrogenous fertilization	Biomass yields ↑ under full crop irrigation and yield ↓ with ↓ water volumes applied. N fertilizer has a minor effect on biomass yield; Biomass yield ↑ when coupled N fertilizer with high irrigation volumes	–	Khraizat 2015
O. syriacum	Irrigation frequency (5, 10 or 15 days)	Irrigation frequency not affect the biomass	–	Abdel Kader et al. 2019
O. syriacum	Irrigation frequency (16 L per plant each: 1, 2, 3 or 4 weeks)	Height and branching rate and biomass significantly ↑ with ↑ of irrigation frequency	EO yields ↑ with ↑ of irrigation frequency.	Atallah et al. 2011
O. syriacum	Photoperiod	Foliage plant ↑ under long photoperiods	EO content ↑ under long photoperiods	Dudai et al. 1992
O. syriacum	Soil type	The soil type significantly affects growth and biomass production. The highest total foliar biomasses were obtained in nursery mixture soils. The lowest biomass for sterilized-inoculated soil	EO yields varied according to the soil type. The highest yield was obtained in sterilized and inoculated soil. The lowest one was registered in 100% manure soil sample	El-Alam et al. 2019

Taxa	Treatment			Reference
O. syriacum	Organic fertilization (pine needle mulch)	The soil coverage with pine needle ↑ the plant dry weight	The soil coverage with pine needle ↑ the plant fresh weight	Abdel Kader et al. 2019
O. syriacum var. *bevanii*	Nitrogen fertilization (0, 40, 60 and 80 kg ha⁻¹) + phonologic period	The highest herb yield was obtained at fall and the end of blooming period with the N rate of 40 kg ha⁻¹	The highest EO content was obtained at full blooming period with the N rate of 60 kg ha⁻¹	Ozguven et al. 2004
O. syriacum var. *syriacum*	Nitrogen fertilization + spraying with foliar CO₂ nano-fertilizer	Nitrogen and foliar spray with CO₂ nano-fertilizer significantly ↑ growth and biomass yield	Nitrogen and foliar spray with CO₂ nano-fertilizer significantly ↑ EO yield	Hamed 2018
O. vulgare	Water stress (no water stress, mild water stress and moderate water stress)	↑ Water stress ↓ the plant growth and biomass yield with ↑ proline. Some populations showed a relative tolerance to water-deficit conditions	—	Morshedloo et al. 2017a
O. vulgare	Irrigation frequency and organic fertilization (cattle manure)	↑ Irrigation intervals ↓ fresh and dry herb yield. Growth and dry herb yield ↑ by ↑ cattle manure levels	The highest EO content and yield were obtained in the highest irrigation intervals and cattle manure levels	Gerami et al. 2016
O. vulgare	Irrigation management (water deficit in vegetative and pre-flowering stages, as well as in whole cycle analysis)	The water restriction during the pre-flowering stage ↑ dry biomass	Water restriction in the soil throughout the entire phenological cycle ↑ EO production	Dos Santos et al. 2020
O. vulgare	Water stress	↑ Water deficit ↓ the plant dry weight	↑ Water deficit ↑ EO contents	Mohammadi et al. 2021
O. vulgare	Chitosan: chitosan (0, 250 and 500mg/ L) under well-watered condition; chitosan at three levels (0, 250 and 500 mg/L) under water-deficit stress	Chitosan under well-watered conditions ↑ the plant's dry weight. Under water deficit: chitosan at 250 mg/L ↓ the plant's dry weight, while 500 mg/L ↑ the dry weight	Chitosan under well-watered condition ↑ EO and phenol contents. Under water deficit: chitosan at 250 mg/L ↑ EO content, while 500 mg/L ↓ EO content	Mohammadi et al. 2021
O. vulgare	Water stress and potassium (K)-humate	Biomass ↓ under low irrigation and saline water irrigation. Biomass ↑ under K-humate application. The maximum of herb fresh was obtained under irrigation with 90% available soil moisture combined with K-humate fertilizer 1.5 g pot⁻¹	EO content and yield ↓ by low irrigation and saline water irrigation. EO production ↑ with K-humate application. The maximum EO amount and yield were obtained from plants irrigated with fresh water combined with K-humate fertilizer	Said-Al Ahl and Hussein 2010

(continued)

TABLE 14.1 (Continued)
Summary of Stress Condition Impacts on Growth, Biomass And Essential Oils (EO) Production of Different *Origanum* Taxa (↑: Increase, ↓: Decrease)

Origanum taxa	Stress type	Growth and biomass	EO yield	References
O. vulgare	Light: (shade-enclosure and open-field conditions) and organic fertilizer: bocashi	The yield of fresh and dry leaf ↑ in shade-enclosure conditions compared to open-field. ↑ doses of bocashi ↑ yield in both environments but more so in shade-enclosure	–	Murillo-Amador et al. 2015
O. vulgare	Abiotic elicitors (proline, sucrose and NaCl)	The addition of each abiotic elicitors alone in culture medium *in vitro* ↓ fresh weight of callus	–	Al-Jibouri et al. 2012
O. vulgare	Nutritional deficiency	Causes a moderate ↑ proline and ↓ biomass	–	Lattanzio et al. 2009
O. vulgare	Heavy metals (Ni, Cu and Zn) Ni: 100, 210 and 500 ppm; Cu: 200, 500 and 1000 ppm; Zn: 720, 1500 and 3000 ppm	Plants accumulated these heavy metals in aerial parts. Zn ↓ the growth. Ni and Cu don't affect the normal growth of plants. All tested metals caused ↓ in photosynthetic Pigments. Cu and Zn caused cellular damage in plants treated	–	Kulbat-Warycha et al. 2020
O. vulgare subsp. *hirtum*	Irrigation	Irrigation ↑ dry matter	Irrigation ↓ EO content but ↑ EO yield	Ninou et al. 2017
O. vulgare (subsp. *hirtum* and subsp. *gracile*)	Water stress (control, mild stress, moderate stress and severe stress)	Water deficit ↓ fresh and dry weight, and relative water and total chlorophyll contents in both subspecies	Mild and moderate water stress ↑ EO content. The EO content ↓ under severe water stress	Emrahi et al. 2021
O. vulgare (var. *creticum*, subsp. *hirtum*, var. *samothrake*)	Soil moisture regimes (optimal, consistent water deficiency and water deficiency from the beginning of flowering) and N supply (0.5 g and 1 g N per pot)	Water deficiency is caused ↓ of dry matter. Higher nitrogen levels ↑ dry matter production	Consistent water deficiency did not affect EO content. Water deficiency in flowering stage ↑ EO content. Higher nitrogen levels ↓ EO content	Azizi et al. 2009

Taxon	Treatment			Reference
O. vulgare subsp. *hirtum*	Nitrogen fertilization (0, 40, 80 and 120 kg ha⁻¹)	N supply ↑ the growth and dry matter yield. The optimum treatment is 80 kg of N ha⁻¹	N supply had no effect on the EO content, but ↑ the EO yield	Sotiropoulou and Karamanos 2010
O. vulgare subsp. *hirtum*	Foliar application of calcium and magnesium (Ca²⁺ and Mg²⁺)	Ca²⁺ and Mg²⁺ ↑ plant height, the number of stems per plant and dry matter yield	Ca²⁺ and Mg²⁺ affected the EO yield, but they did not affect the EO content	Dordas 2009
O. vulgare subsp. *hirtum*	Light (soil full-light treatment; pot full-light treatment; pot 50%-shade treatment)	The pot full-light treatment contained the highest dry matter	The pot full-light treatment contained the highest EO content	Tibaldi et al. 2011
O. vulgare (*O.* subsp. *virens* and *O.* subsp. *gracile*)	Water stress (no stress, mild stress and moderate stress)	Biomass yield is affected by water stress conditions	The mild and moderate water stresses ↑ the EO content of *O. vulgare* subsp. *gracile*, but did not change the EO content of *O. vulgare* subsp. *virens*	Morshedloo et al. 2017b
Origanum hybrids	Carbon dioxide levels (400 and 800 ppm)	–	The ↑ in level of CO₂ could either ↑ or ↓ EO production according to the hybrid	Yasar et al. 2021
Origanum sp.	Water stress and nitrogen fertilization	↑ Irrigation levels ↑ the production of biomass (the optimum irrigation level was 80% available soil moisture). N ↑ biomass yield under well-, moderate-watered, and water deficit conditions. Water level of 80% available soil moisture and 1.2 g N pot⁻¹ gave the best biomass yield	↑ Irrigation levels ↑ EO yield (the optimum irrigation level was 80% available soil moisture). N ↑ EO yield under well-watered, moderate-watered, and water deficit conditions. Supplying plants with a water level of 80% available soil moisture and with 1.2 g N pot⁻¹ gave the best EO yield	Said-Al Ahl et al. 2009

TABLE 14.2

Summary of Stress Condition Impacts on EO Composition of Different *Origanum* Taxa (↑: Increase, ↓: Decrease)

Origanum taxa	Stress type	Chemical composition	References
O. compactum	Transplantation (from nature to experiment field: open field under irrigation)	The transplantation of the same genotypes caused ↑ in thymol and carvacrol with a ↓ in *p*-cyme and γ-terpinene contents	Aboukhalid 2019
O. grosii	Transplantation (from nature to experiment field: open field under irrigation)	The transplantation of the same genotypes caused ↑ in thymol and carvacrol with a ↓ in *p*-cyme and γ-terpinene contents	Bakha et al. 2019
O. majorana	Water stress (salt treatment: 0, 50, 100 and 150 mM of NaCl)	The main components are trans-sabinene hydrate/terpinene-4-ol and the chemical classes of marjoram EO (monoterpene hydrocarbons and oxygenated monoterpenes) ↓ at 100 and 50 mM NaCl treatments	Baatour et al. 2010
O. majorana	Water stress (growth responses to methyl jasmonic acid under limited irrigation)	Limited irrigation did not have a significant effect on the EO composition. MeJA application affected the EO composition under normal and limited irrigation levels. MeJA application ↓ the content of monoterpene hydrocarbons components and ↑ oxygenated monoterpenes	Farsi et al. 2019
O. majorana	Fertilization (N, P and K)	EO composition was little affected by the nutrient regime	Trivino and Johnson 2000
O. majorana	The day length	The day length influenced the composition of the EO: *cis*-sabinene hydrate ↑ with ↑ day length terpinenes ↓ with ↑ day length	Circella et al. 1995
O. onites	Water stress (irrigation water salinity)	Carvacrol, β-cymene and γ-terpinene contents ↓ with ↑ water salinities up to 2.5 dS/m and after this level the amount of these compounds ↑. Linalool content ↑ with ↑ water salinities up to 2.5 dS/m and after this level the amount of linalool ↓	Hancioglu et al. 2019
O. syriacum	Photoperiod and light intensity	↑ Photoperiod and light intensity ↓ *p*-cymene	Dudai et al. 1992
O. syriacum	Irrigation frequency (16 L per plant each: 1, 2, 3 or 4 weeks)	Thymol and carvacrol content are not significantly affected	Atallah et al. 2011
O. syriacum	Soil type	The soil type affects the EO composition. Vegetable compost soils ↑ thymol. Potting mix, professional agriculture mixture, and nursery mixture soils ↑ thymol/carvacrol chemotype	El-Alam et al. 2019

Taxa	Conditions	Findings	Reference
O. syriacum subsp. *syriacum*	Temperature (18, 22 and 26°C)	The major compound (*cis*-sabinene hydrate) was not influenced by temperature	Novak et al. 2010
O. syriacum	Soil inoculation with arbuscular mycorrhizal fungi	Mycorrhizal inoculation ↑ carvacrol and ↓ thymol productions in comparison to non-inoculated conditions	El-Alam et al. 2019
O. vulgare	Water stress and potassium humate	The soil moisture regimes (fresh and saline water irrigation) and potassium humate fertilization affect relatively the EO composition	Said-Al Ahl and Hussein 2010
O. vulgare	Abiotic elicitors (proline, sucrose and NaCl)	Proline and sucrose ↑ the production of thymol. NaCl ↓ the production of thymol	Al-Jibouri et al. 2012
O. vulgare (subsp. *hirtum* and subsp. *gracile*)	Water stress (control, mild stress, moderate stress and severe stress)	Water stress ↑ carvacrol in comparison to control. In subsp. *hirtum*, the highest content of carvacrol was achieved under mild water stress. In the subsp. *gracile*, the highest content of carvacrol was achieved under severe and moderate stresses	Emrahi et al. 2021
O. vulgare var. *creticum*, subsp. *hirtum*, var. *samothrake*)	Soil moisture regimes and nitrogen supply	The EO composition was found independent of cultivation conditions	Azizi et al. 2009
O. vulgare subsp. *hirtum*	Light (soil full-light; pot full-light; pot 50%-shade treatments)	The growing condition significantly affected the γ-terpinene, cis-sabinene hydrate, 4-terpineol, α-terpinene, linalyl acetate and β-bisabolene. Did not affect carvacrol, p-cymene and sabinene	Tibaldi et al. 2011
O. vulgare (*O.* subsp. *virens* and *O.* subsp. *gracile*)	Water stress (no stress, mild stress and moderate stress)	Water stress did not significantly change the main EO constituents. The amount of (E)-b-caryophyllene in *O. vulgare* subsp. *virens* ↑ under water stress conditions	Morshedloo et al. 2017b
O. vulgare (3 subsp. and 1 hybrid).	Light intensity (reduced and normal light intensity conditions)	↓ Light intensity had a minor effect on the EO composition	Shafiee-Hajiabad et al. 2016
Origanum hybrids	Carbon dioxide levels (400 and 800 ppm)	↑ In carbon dioxide level significantly ↑ the thymol and carvacrol components in numerous hybrids. In *O. syriacum* x *O. majorona* ↑ in carbon dioxide level ↓ the thymol and carvacrol contents	Yasar et al. 2021
O. vulgare x *O. majorana*	Temperature (18°C, 22°C and 26°C)	Thymol ↑ with ↓ temperature. Carvacrol ↑ with ↑ temperature	Novak et al. 2010

O. compactum plants were grown under the same cultural and edaphoclimatic conditions and variability in EO content due to environmental parameters was discarded. Therefore, the variability in EO content might be attributed to both environmental and genetic factors, given that *O. compactum* harbored a high level of genetic diversity (Aboukhalid 2019).

Morshedloo et al. (2017a) reported that some *O. vulgare* populations showed a relative tolerance to water-deficit conditions. Also, *O. majorana* plants are more tolerant to water stress compared to other *Origanum* taxa, where the biomass and EO production could be either not or only slightly affected by water deficit (Baatour et al. 2010; Baâtour et al. 2011; Mohammadi et al. 2021; Jafari et al. 2021). Baatour et al. (2010) found that *O. majorana* is tolerant to low and medium salt concentrations, but its growth is negatively affected under the highest NaCl concentrations (Baatour et al. 2010; Baâtour et al. 2011). Moreover, an increase in water stress could increase the dry matter yield and EO content of *O. majorana*, according to the findings of Rhizopoulou and Diamantoglou (1991).

Irrigation management, such as irrigation frequency and water restriction according to the phenological stage, could affect the growth and productivity of oregano. For instance, irrigation frequency showed different results on cultivated *Origanum* plants. Gerami et al. (2016) showed that increasing irrigation frequency negatively affects plant spread, plant height, stem diameter and fresh and dry herb yield, but increased the EO content and yield in *O. syriacum*. In contrast, Atallah et al. (2011) reported that increasing irrigation intervals from 1 to 4 weeks increased both biomass and EO yield in the same species. However, in *O. vulgare*, an irrigation frequency of 5 to 15 days does not appear to affect biomass production (Abdel Kader et al. 2019). The application of water restriction during the pre-flowering stage could increase the dry biomass production, while water restriction in the soil throughout the entire phenological cycle reduces biomass and enhances EO production (Dos Santos et al. 2020). In addition, water deficiency at the flowering stage significantly increases the EO content (Azizi et al. 2009).

In experimentation using combinations of different treatments, such as fertilization and elicitors supply, under water-deficit conditions to induce water-stress tolerance in oregano, the combination of irrigation with k-humate and N fertilizer induced the highest biomass and EO yields compared to plants irrigated without fertilizer supply (Said-Al Ahl et al. 2009; Said-Al Ahl and Hussein 2010). In saline soils, seed priming with titanium dioxide (TiO_2) at concentrations of 20 and 40 mg L^{-1} enhances the tolerance of *O. majorana* to water stress, with the improvement of growth, physiological characteristics as well as EO production (Jafari et al. 2021). However, a high concentration of TiO_2 (80 mg L^{-1}) showed a negative effect on this species (Jafari et al. 2021). In addition, foliar spraying treatment with 100 µM of MeJA enhanced the growth parameters but decreased EO production in *O. majorana* under mild limited irrigation conditions (Farsi et al. 2019). Chitosan is a natural substance that can be used as an elicitor in many plants. In *O. majorana*, chitosan at 250 mg/L and 500 mg/L was found to be able to increase the dry weight only under well-watered conditions. Meanwhile, in *O. vulgare*, chitosan enhanced the plant's dry weight under both well-watered (at 250 mg/L and 500 mg/L) and water-deficit conditions (at 500 mg/L) (Mohammadi et al. 2021).

14.3.2 Nutrient Availability

Nutrient availability, considered a determinant factor for plant species, was also demonstrated to have potent effects on the growth of oregano. Indeed, the soil type significantly affects the growth, as well as the biomass and EO production, in oregâno. For instance, nursery mixture soils have been found to be the most adequate for biomass production in *O. syriacum*. Meanwhile, sterilized soil inoculated with arbuscular mycorrhizal fungi showed the lowest biomass production in the same species (El-Alam et al. 2019).

In addition, growth and biomass production were found to be negatively affected by a nutritional deficiency in oregano, whereas the decrease in biomass was correlated with an increase in proline production (Lattanzio et al. 2009). Proline accumulated by plants in response to biotic and abiotic stresses plays an important role in their adaptation to environmental perturbations *via* the modifications of many physiological pathways. On the other hand, soil fertility, using both organic and inorganic inputs, contributed to improving biomass and EO productivity in many *Origanum* taxa. Soil supply by organic manure, pine needle mulch as well as bocashi, stimulated the growth and enhanced biomass and EO yields in both *O. vulgare* (Murillo-Amador et al. 2015; Gerami et al. 2016) and *O. syriacum* (Abdel Kader et al. 2019). Further, chemical fertilizers were used to improve the quality of soil and consequently increase the productivity of the cultivated crops. In oregano, chemical fertilization using a supply of N, P and K, induced a significant increase in growth rates, fresh and dry yield as well as EO yield (Trivino and Johnson 2000; Ozguven et al. 2004; Khraizat 2015). Nitrogen is considered as the major element for cultivated oregano plant growth, due to its proven positive effects on growth through the increase in number of stems, number of branches, root surface density and chlorophyll concentration per plant (Azizi et al. 2009; Said-Al Ahl et al. 2009; Sotiropoulou and Karamanos 2010). For instance, Sotiropoulou and Karamanos (2010) reported that the optimum treatment for biomass production in *O. vulgare* subsp. *hirtum* was 80 kg of N ha^{-1}. However, in *O. syriacum* var. *bevanii* the highest herb yield was obtained with only 40 kg ha^{-1} of nitrogen, while the highest EO content was obtained with 60 kg of N ha^{-1} (Ozguven et al. 2004). Soil supply of potassium (K)-humate and foliar application of calcium and magnesium (Ca^{2+} and Mg^{2+}) were also found able to improve the growth and biomass yield in oregano (Dordas 2009; Said-Al Ahl and Hussein 2010).

Nevertheless, long-term use and high amounts of chemical fertilizers cause negative impacts on the soil and the environment as well as on human health. At this level, the combination of fertilization with adequate levels of irrigation and CO$_2$ nano-fertilizer enables improving crop production with low amounts of fertilizers (Ozguven et al. 2004; Hamed 2018).

14.3.3 ALTITUDE

Altitude seems to be an important environmental factor influencing the growth and productivity of plants. Altitude is a complex environmental factor that is associated with many other environmental factors such as temperature, precipitation, wind exposure, shortwave solar radiation and air humidity (Aissi et al. 2016). Few studies report the influence of altitudinal gradient on the growth and productivity of *Origanum* taxa. Increasing elevation resulted in a progressive decrease in plant height in *O. vulgare* found in Greece (Kofidis et al. 2003). Thus, they showed that leaf chlorophyll and photosynthetic activity increase significantly in low- and mid-altitude, whereas they decrease in high-altitude oregano populations. These results suggest that *O. vulgare* populations at mid-altitude were better adapted than populations found at high altitudes. In *O. onites*, the increase in the cortical area and vascular tissues, and the decrease in the stem and lower leaf lengths were associated with an increase in altitude (Gönüz and Özörgücü 1999). It has been shown also that adaptation of plants to altitudinal variations could be mediated *via* metabolic changes. In *O. sipyleum*, the plants increase the production of total protein and proline at altitudes of up to 790 m, with a concomitant decrease of photosynthetic pigment. Meanwhile, at altitudes over 1000 m, *O. sipyleum* plants use abscisic acid and antioxidative enzymes for adaptation, with a decrease in total protein and proline (Unal et al. 2013).

14.3.4 TEMPERATURE

In wild *O. dictamnus*, the plants shed large leaves and keep small apical leaves to resist cold stress during winter (Lianopoulou and Bosabalidis 2014). In addition, glandular and non-glandular hairs play a key role in the adaptation of *Origanum* to stressful conditions. This indumentum shows different appearances from one species to another, and also from one season to another (Bosabalidis

FIGURE 14.5 Low temperature impact on the growth of oregano (photo provided by M. Bakha).

and Sawidis 2014; Shafiee-Hajiabad et al. 2014). Glandular and non-glandular hairs are involved in the protection of plants from low-temperature stress (Figure 14.5). Moreover, the EO secreted by the glandular hairs could also avoid the oxidative stress due to low-temperature levels, and consequently help *Origanum* taxa in resisting winter cold stress (Bosabalidis and Sawidis 2014; Lianopoulou and Bosabalidis 2014). For instance, *p*-cymene constitutes the major constituent of *O. dictatmnus* during the winter, while in the summer carvacrol represents the major compound (Lianopoulou and Bosabalidis 2014).

14.3.5 Light Intensity

Light intensity is an important factor in the growth and distribution of plants. The photoperiod has been found able to affect the growth of *O. majorana*, where normal plant growth and flower formation were observed only at a day length of at least 13 h (Circella et al. 1995). In *O. syriacum*, an increase in photoperiods results in an enhancement of foliage plant and EO production (Dudai et al. 1992). Furthermore, light stress induced by reducing light intensity could decrease the size and density of trichomes, as well as the herb production in numerous *O. vulgare* subspecies (Shafiee-Hajiabad et al. 2015). Moreover, *O. vulgare* subsp. *hirtum* cultivated in pots with full-light treatment produced the highest dry matter compared to plants cultivated in full-light soil and in pot 50%-shade treatments (Tibaldi et al. 2011). In contrast, Murillo-Amador et al. (2015) found that *O. vulgare* is characterized by high fresh and dry leaf yields in shade-enclosure conditions compared to open-field. It seems that the effect of light intensity on the growth and productivity of *Origanum* taxa is linked to other conditions such as water and nutrient availability that can improve their growth and performance under both shade-enclosure and open-field conditions.

14.3.6 Heavy Metals

Heavy metals could be present both in the air and the soil. Recent studies have shown that they can interact with plants, especially through the roots. Heavy metals could indeed be absorbed at the root level (Munir et al. 2022). Some of them exist in the air, and could penetrate through the aerial parts, particularly by the leaves. They have a major impact on the physiology, development and growth regulation of plants, including the genus *Origanum*. Recent studies have shown that these heavy metals even can regulate the synthesis of secondary metabolites, in particular EO of different oregano species. However, at high concentrations, these heavy metals could be highly toxic.

Dbaibo et al. (2019) showed that there was a positive correlation between levels of cadmium (Cd), lead (Pb) and nickel (Ni) in soil and the rate of their accumulation in *O. syriacum* tissues. Earlier, Panou-Filotheou et al. (1998, 2004) studied several physiological parameters related to the cultivation of *O. vulgare* subsp. *hirtum* in soil rich in copper (Cu). Their results showed a decrease in the size and number of stems and leaves, and a decrease in the root length with high Cu content detected in roots. Interestingly, at a concentration of 25.5 µM/g of Cu in soil, the photosynthesis rate, as well as the water potential/transpiration ratio, dramatically decreased, while the water potential significantly increased.

To confirm this soil remediation (phytoremediation), Levizou et al. (2019) determined thresholds in oregano responses and evaluated its performance under soil chromium (Cr) stress. In soil containing 150–200 mg/kg of Cr (VI), *O. vulgare* accumulated high Cr concentrations in the roots (up to 1200 mg/kg DM) and aerial parts (4300 mg/kg DM). Oregano responses showed the presence of a threshold defined at 100 mg/kg of Cr(VI) in the soil. However, above this value, negative results were observed. *O. vulgare* phytoremediation potential will then be lower than the Cr contamination levels.

Kulbat-Warycha et al. (2020) studied the involvement of three soil heavy metals, namely zinc (Zn), Cu and Ni, in the mechanisms of allergen production and biochemical defense in *O. vulgare*. Consequently, the growth of plants contaminated with Cu and Ni was normal, however, the phenotypic representation of those contaminated with Zn was characterized by a limited and dose-dependent growth, whereas plants treated with Zn and Cu suffered from the highest cell damage levels. Moreover, these three metallic elements reduced the total content of photosynthetic pigments, in particular chlorophyll. Likewise, Cu (1000 ppm) increased the allergenic protein profiling concentration, while the other two metals (at high doses) decreased the concentration of phenolics and antioxidant activity.

14.3.7 BIOTIC STRESS

Like abiotic stress, biotic stress could result in many changes in plant growth and physiology. In the case of *Origanum* taxa, few investigations are showing the effect of plant pathogens such as fungi, insects, bacteria and viruses, on this plant group. Sulaiman et al. (2021) found that the infestation of *O. vulgare* with whitefly *Trialeurodes vaporariorum*, phloem-feeding insect herbivory, induced a decrease in the photosynthesis rate and an increase in stomatal conductance and terpene emissions, with the inhibition of lipoxygenase pathway volatiles (LOX) and benzenoid emissions. Moreover, the combination of whitefly infestation with heat stress resulted in a resistance enhancement of *O. vulgare* to heat stress.

In the rhizosphere, there is a huge interaction between plants and microorganisms, especially bacteria. These microorganisms can interact with plants at the root level and thus increase plant resistance, elucidation and recycling of mineral and organic matter. Among these phenotypes, there is the ability of microorganisms to emit elicitors that can increase the yield of secondary metabolites responsible for plant growth, their protection and even the synthesis of new molecules belonging to secondary metabolites (Chamkhi et al. 2021). In addition, plants of the genus *Origanum* can interact with these microorganisms.

Mowafy et al. (2016) assessed the ability of certain bacteria (both cultured and wild) to promote the growth of oregano, and to do so, they isolated and identified bacteria associated with *O. syriacum* subsp. *sinaicum*. Characterization of the best microorganisms (endophytes or rhizospheric), plant growth-promoting (PGP), was carried out by testing their nitrogen fixation activity and gibberellic acid synthesis, while their growth-promoting potential for oregano was confirmed by several tests, namely the production of ammonia, phytohormones and hydrogen cyanide and also the determination of phosphate solubilization and 1-aminocyclopropane-1-carboxylate deaminase. Consequently, the authors of this study were able to isolate 25 PGP bacteria, using their power to produce gibberellic acid in large quantities. All these isolates produce varying amounts and different levels of phytohormones, and the same for the production of ammonia and hydrogen cyanide.

Taking into account the excessive food demand following the growth of the global population, the discovery of new bio-fertilizers for crops has been proven necessary. PGP bacteria can play this role and be a promising alternative to replace synthetic fungicides, improve plant health and reduce environmental stresses. Meléndez et al. (2017) estimated the bacterial diversity in a soil sample intended for oregano cultivation, located in the Yaqui Valley (Sonora), by evaluating their tolerance to different abiotic stresses (pesticide resistance, salinity and temperature) as well as their potential as PGP bacteria. The authors isolated 25 bacteria with high tolerance to fungicides, salinity resistance and heat stress. These bacteria also recorded a high production of indoleacetic acid (auxin) and siderophores (iron-chelating compounds). These results showed the involvement of agricultural soil microorganisms in improving the living conditions of *O. vulgare*.

To determine the type of these bacterial interactions on other oregano species (*O. vulgare* L.) and also between the bacteria themselves (molecular communication), a recent study evaluated the impact of three isolates, namely *Moraxella osloensis*, *Bacillus megaterium* and *Bacillus niacini*, on this plant (in consortium or individually), with the exploration of quorum-sensing signaling activity using biofilm assays (Loera-Muro et al. 2021). Positive results were observed on physiological and morphometric parameters in *O. vulgare* inoculated with *B. megaterium*, without any synergistic effect when a bacterial consortium was inoculated. For the genus *Bacillus*, an activation of QS signaling was noted. It should be deduced that in addition to synergistic or positive bacterial interactions, the interactions of PGP bacteria (in the consortium) with oregano and during the formation of biofilms could be antagonistic or neutral.

On the other hand, Hristozkova et al. (2015) showed the impact of mycorrhizal colonization in soil industrially polluted with cadmium (Cd) and lead (Pb) on the accumulation of these heavy metals, and the chemical composition and antioxidant power in *O. majorana*. Subsequently, this colonization modified the content of the major substances of marjoram EO with a decrease in Pb accumulation in the roots and shoots. Thus, due to the presence of high amounts of phenolic compounds, this colonization enhanced the antioxidant potential in *O. majorana* aerial parts.

14.3.8 Seed Germination

Seed germination is a key factor in determining the survival of the species in their environments as well as in the agriculture field (Saatkamp et al. 2014). The germination of *O. compactum* and *O. elongatum* is sensitive to higher temperatures (>25 °C), acid soils (pH<3.5) and high NaCl levels (Laghmouchi et al. 2017; Belmehdi et al. 2018). In addition, it has been revealed that salinity is negatively correlated with seed germination rate in *O. majorana*. These results indicated that changes in abiotic factors related to changes in temperature, soil properties and soil moisture deficit could have negative effects on the natural seed banks of this remarkable species of oregano. To improve seed germination in oregano, seed priming with titanium dioxide (TiO_2) results in enhancement of the germination rate in *O. majorana* in both normal and saline soils (Jafari et al. 2021).

14.4 THE IMPACT OF STRESS ON EO COMPOSITION

Oregano plants are rich in bioactive compounds that belong to different chemical classes such as flavonoids, resins, tannins, sterols, phenolic glucosides and terpenoids (Skoula et al. 1999; Morshedloo et al. 2017b). Generally, the quality of oregano depends on the quantity and chemical composition of their EO, which are known for their diverse and potent biological activities like antibacterial, antifungal, anticancer and antioxidant properties (Bouyahya et al. 2020). Although the chemical composition of EO is genetically controlled (Crocoll et al. 2010; Morshedloo et al. 2017b), several factors could influence their biosynthesis, which results in a wide range of chemical polymorphisms. For instance, the chemical composition of the EO in *Origanum* taxa could differ according to the species (Skoula et al. 1999), organ (Shafaghat, 2011), geographical origin (Bakha et al. 2019), phenological stage (Bouyahya et al. 2017) and edaphoclimatic conditions (Aboukhalid et al. 2017b).

Aboukhalid et al. (2016) studied on a large scale the chemical diversity of 527 *O. compactum* plants covering most of the natural distribution range of the species in Morocco. The influence of environmental factors on the observed chemical polymorphism and the geographical distribution of components were also evaluated (Aboukhalid et al. 2017a). These studies showed that the carvacrol chemotype was the most common in almost all populations growing under semi-arid climates. However, populations growing under humid climates were characterized by an important heterogeneity and represent all the chemotypes reported for this species, with the dominance of thymol and carvacrol. Moreover, the authors showed that environmental factors play a key role in the diversity of secondary metabolites in *O. compactum*. Interestingly, two groups of populations were distinguished according to their chemical composition and edaphoclimatic characteristics. The first group is composed of populations growing in regions with a humid climate and clayey, sandy and alkaline soils. These populations contained high amounts of thymol, α-terpineol and carvacryl methyl oxide. The second group is represented by populations belonging to semi-arid climates, growing at high altitudes with silty soils. These populations showed high carvacrol amounts (Aboukhalid et al. 2017a). In addition, Bakha et al. (2019) studied the chemical composition of *O. grosii* in nature and experimental fields. In nature, the EO of *O. grosii* was dominated by *p*-cymene, thymol and carvacrol. However, the transplantation of the same *O. grosii* plants in an experimental field (under regular irrigation) caused a change in their EO composition, manifested by a significant decrease in *p*-cymene amounts, with an increase of thymol and carvacrol. This change in EO composition could be explained by a positive genotype–environment interaction.

On the other hand, some studies have reported the impact of water stress on EO composition in oregano. Emrahi et al. (2021) found that water stress stimulates the production of carvacrol in *O. vulgare* subsp. *hirtum* and *O. vulgare* subsp. *gracile*. However, Hancioglu et al. (2019) revealed that carvacrol, *p*-cymene and γ-terpinene amounts decrease under water stress conditions in *O. onites*. Meanwhile, Morshedloo et al. (2017b) reported that the major compounds of *O. vulgare* subsp. *virens* and *O. vulgare* subsp. *gracile* are not affected by water stress.

We reported previously that glandular and non-glandular hairs as well as EO protect *Origanum* from low-temperature stress (Bosabalidis and Sawidis, 2014; Shafiee-Hajiabad et al. 2014), where *p*-cymene constituted the major compound during the winter in *O. dictatmnus* (Lianopoulou and Bosabalidis 2014). The production of *p*-cymene decreases with an increase of day length and light intensity in both *O. syriacum* (Dudai et al. 1992) and *O. majrana* (Circella et al. 1995). In addition, Novak et al. (2010) found that the increase in temperature from 18 to 26°C does not affect the chemical composition in *O. syriacum*, but increases the production of carvacrol in the hybrid *O. vulgare* × *O. majorana*. However, EO composition was slightly affected by nitrogen supply (Trivino and Johnson, 2000; Azizi et al. 2009).

Alraey et al. (2019) examined the performance of plant growth-promoting bacteria on the chemical composition of the EO of wild endemic *O. syriacum* subsp. *sinaicum*. They found that the EO content varies in response to different bacterial inoculums with the identification of 15 compounds, where *p*-cymene, γ-terpinene and carvacrol were the major components, with a significant improvement in carvacrol (67%) in response to Serratia SK3. The latter could be a potential bio-factor for improving the growth performance of oregano and the constituents of its EO. In addition, soil inoculation with arbuscular mycorrhizal fungi enhances the production of carvacrol and decreases that of thymol in *O. syriacum* (El-Alam et al. 2019). These findings indicate that the diversity and concentration of chemical compounds in *Origanum* could depend on the type of associated growth-promoting microorganism.

14.5 CONCLUSION

In natural conditions, as in the field, the plant species are generally exposed to more than one stress. Further, the response of plants depends on the kinds and levels of stresses, which explains that the same botanical species could use different strategies under different conditions. In the case of *Origanum* taxa, numerous studies have highlighted how abiotic stresses affect the growth and biomass and EO production. The studied taxa showed various levels of tolerance to stresses, where some species such as *O. compactum* and *O. vulgare* subsp. *hirtum* were found to be more tolerant to drought, high temperature and heavy metals than others. However, a lack of information has been recorded regarding the impact of biotic stress. On the other hand, further studies are necessary to show the physiological mechanisms used by *Origanum* taxa to face stressful conditions, notably the protective role of the EO under abiotic and biotic stresses. Understanding the ways through which *Origanum* taxa interact with the different kinds and levels of stresses will help in the development of techniques to improve biomass and EO production, the stimulation of *Origanum* taxa to product-specific and targeted compounds as well as the conservation of threatened species under natural conditions.

REFERENCES

Abdel Kader, E. H., M. Nakhle, V. Talj, N. Taha, S. Oleik, M. Housein, and H. Rizk. 2019. Effect of Pine Needle Mulch and Irrigation Frequency on the Yield of *Origanum syriacum* under Open Field Condition. *European Journal of Medicinal Plants*, *26*(4): 1–8. https://pesquisa.bvsalud.org/portal/resource/pt/sea-189450

Aboukhalid, K., A. Lamiri, M. Agacka-Mołdoch, T. Doroszewska, A. Douaik, M. Bakha, J. Casanova, F. Tomi, N. Machon, and C. Al Faiz. 2016. Chemical polymorphism of *Origanum compactum* grown in all natural habitats in Morocco. *Chemistry & Biodiversity*, *13*(9): 1126–1139. https://doi.org/10.1002/cbdv.201500511

Aboukhalid, K., N. Machon, J. Lambourdière, J. Abdelkrim, M. Bakha, A. Douaik, G. Korbecka-Glinka, F. Gaboun, F. Tomi, A. Lamiri, and C. Al Faiz. 2017a. Analysis of genetic diversity and population structure of the endangered *Origanum compactum* from Morocco, using SSR markers: Implication for conservation. *Biological Conservation*, *212*: 172–182. https://doi.org/10.1016/j.biocon.2017.05.030

Aboukhalid, K., C. Al Faiz, A. Douaik, M. Bakha, K. Kursa, M. Agacka-Mołdoch, N. Machon, F. Tomi, and A. Lamiri. 2017b. Influence of environmental factors on essential oil variability in *Origanum compactum* Benth. growing wild in Morocco. *Chemistry & biodiversity*, *14*(9): e1700158. https://doi.org/10.1002/cbdv.201700158

Aboukhalid, K. 2019. *Etude de la variabilité chimique et génétique et cartographie d'Origanum compactum du Maroc*. Université Hassan I, Morocco (Doctoral dissertation).

Aissi, O., M. Boussaid, and C. Messaoud. 2016. Essential oil composition in natural populations of *Pistacia lentiscus* L. from Tunisia: Effect of ecological factors and incidence on antioxidant and antiacetylcholinesterase activities. *Industrial Crops and Products*, *91*: 56–65. https://doi.org/10.1016/j.indcrop.2016.06.025

Alekseeva, M., T. Zagorcheva, M. Rusanova, K. Rusanov, and I. Atanassov. 2021. Genetic and Flower Volatile Diversity in Natural Populations of *Origanum vulgare* subsp. *hirtum* (Link) Ietsw. in Bulgaria: Toward the Development of a Core Collection. *Frontiers in Plant Science*, *12*. https://doi.org/10.3389/fpls.2021.679063

Al-Jibouri, A. M. J., A. S. Abd, D. M. Majeed, and E. N. Ismail. 2012. Influence of Abiotic Elicitors on Accumulation of Thymol in callus cultures of *Origanum vulgare* L. *Journal of Life sciences*, *6*(10): 1094.

Alraey, D. A., S. A. Haroun, M. N. Omar, A. M. Abd-ElGawad, A. M. El-Shobaky, and A. M. Mowafy. 2019. Fluctuation of essential oil constituents in *Origanum syriacum* subsp. *sinaicum* in response to plant growth promoting bacteria. *Journal of Essential Oil Bearing Plants*, 22(4): 1022–1033. https://doi.org/10.1080/0972060X.2019.1661794

Atallah, S. S., I. El Saliby, R. Baalbaki, and S. N. Talhouk. 2011. Effects of different irrigation, drying and production scenarios on the productivity, postharvest quality and economic feasibility of *Origanum syriacum*, a species typically over-collected from the wild in Lebanon. *Journal of the Science of Food and Agriculture, 91*(2): 337–343. https://doi.org/10.1002/jsfa.4191 Ayanogç Lu, F., A. Ergül, and M. Arslan. 2006. Assessment of genetic diversity in Turkish oregano (*Origanum onites* L.) germplasm by AFLP analysis, *The Journal of Horticultural Science and Biotechnology*, 81(1): 45–50.

Azizi, A., F. Yan, and B. Honermeier. 2009. Herbage yield, essential oil content and composition of three oregano (*Origanum vulgare* L.) populations as affected by soil moisture regimes and nitrogen supply. *Industrial crops and products, 29*(2–3): 554–561. https://doi.org/10.1016/j.indcrop.2008.11.001

Azizi, A., J. Hadian, M. Gholami, W. Friedt, and B. Honermeier. 2012. Correlations between genetic, morphological, and chemical diversities in a germplasm collection of the medicinal plant *Origanum vulgare* L. *Chemistry & Biodiversity, 9*(12): 2784–2801. https://doi.org/10.1002/cbdv.201200125

Baatour, O., R. Kaddour, W. Aidi Wannes, M. Lachaal, and B. Marzouk. 2010. Salt effects on the growth, mineral nutrition, essential oil yield and composition of marjoram (*Origanum majorana*). *Acta physiologiae plantarum, 32*(1): 45–51. https://doi.org/10.1007/s11738-009-0374-4

Baâtour, O., R. Kaddour, H. Mahmoudi, I. Tarchoun, I. Bettaieb, N. Nasri, S. Mrah, G. Hamdaoui, M. Lachaâl, and B. Marzouk. 2011. Salt effects on *Origanum majorana* fatty acid and essential oil composition. *Journal of the Science of Food and Agriculture, 91*(14): 2613–2620. https://doi.org/10.1002/jsfa.4495

Baâtour, O., M. B. Nasri-Ayachi, H. Mahmoudi, I. Tarchoun, N. Nassri, M. Zaghdoudi, M. Zaghdoudi, W. Abidi, R. Kaddour, S. M'rah, G. Hamdaoui, B. Marzouk, and M. Lachaâl. 2012. Salt effect on physiological, biochemical and anatomical structures of two *Origanum majorana* varieties (Tunisian and Canadian). *African Journal of Biotechnology, 11*(27): 7109–7118. https://doi.org/10.5897/AJB11.3493

Bakha, M., C. Al Faiz, M. Daoud, N. El Mtili, K. Aboukhalid, A. Khiraoui, N. Machon, and S. Siljak-Yakovlev. 2017. Genome size and chromosome number for six taxa of *Origanum* genus from Morocco. *Botany Letters, 164*(4): 361–370. https://doi.org/10.1080/23818107.2017.1395766

Bakha, M., M. Gibernau, F. Tomi, N. Machon, A. Khiraoui, K. Aboukhalid, N. El Mtili, and C. Al Faiz. 2019. Chemical diversity of essential oil of the Moroccan endemic *Origanum grosii* in natural populations and after transplantation. *South African Journal of Botany, 124*: 151–159. https://doi.org/10.1016/j.sajb.2019.05.014

Bakha, M., N. El Mtili, N. Machon, K. Aboukhalid, F. Z. Amchra, A. Khiraoui, M. Gibernau, F. Tomi, and C. Al Faiz. 2020. Intraspecific chemical variability of the essential oils of Moroccan endemic *Origanum elongatum* L. (Lamiaceae) from its whole natural habitats. *Arabian Journal of Chemistry, 13*(1): 3070–3081. https://doi.org/10.1016/j.arabjc.2018.08.015

Belmehdi, O., A. El Harsal, M. Benmoussi, Y. Laghmouchi, N. S. Senhaji, and J. Abrini. 2018. Effect of light, temperature, salt stress and pH on seed germination of medicinal plant *Origanum elongatum* (Bonnet) Emb. & Maire. *Biocatalysis and Agricultural Biotechnology, 16*: 126–131. https://doi.org/10.1016/j.bcab.2018.07.032

Blondel, J., J. Aronson, J-Y. Bodiou, and G. Boeuf. 2010. The Mediterranean region: biological diversity in space and time. Oxford University Press.

Bosabalidis, A. M. and T. Sawidis. 2014. Glandular and non-glandular hairs in the seasonally dimorphic *Origanum dictamnus* L. (Lamiaceae) as a means of adaptation to cold stress. *Acta Agrobotanica, 67*(1). https://doi.org/10.5586/aa.2014.010

Bouiamrine, E. L., L. Bachiri, J. Ibijbijen, and L. Nassiri. 2017. Fresh medicinal plants in middle atlas of Morocco: trade and threats to the sustainable harvesting. *Journal of Medicinal Plants, 5*(2): 123–128.

Bouyahya, A., N. Dakka, A. Talbaoui, A. Et-Touys, H. El-Boury, J. Abrini, and Y. Bakri. 2017. Correlation between phenological changes, chemical composition and biological activities of the essential oil from Moroccan endemic Oregano (*Origanum compactum* Benth). *Industrial crops and products, 108*: 729–737. https://doi.org/10.1016/j.indcrop.2017.07.033

Bouyahya, A., G. Zengin, O. Belmehdi, I. Bourais, I. Chamkhi, D. Taha, T. Benali, N. Dakka, and Y. Bakri. 2020. *Origanum compactum* Benth., from traditional use to biotechnological applications. *Journal of Food Biochemistry, 44*(8): e13251. https://doi.org/10.1111/jfbc.13251

Chamkhi, I., T. Benali, T. Aanniz, N. El Menyiy, F. E. Guaouguaou, N. El Omari, M. El-Shazly, G. Zengin, and A. Bouyahya. 2021. Plant-microbial interaction: The mechanism and the application of microbial elicitor induced secondary metabolites biosynthesis in medicinal plants. *Plant Physiology and Biochemistry*, *167*: 269–295. https://doi.org/10.1016/j.plaphy.2021.08.001

Circella, G., C. Franz, J. Novak, and H. Resch. 1995. Influence of day length and leaf insertion on the composition of marjoram essential oil. *Flavour and Fragrance Journal*, *10*(6): 371–374. https://doi.org/10.1002/ffj.2730100607

Crocoll, C., J. Asbach, J. Novak, J. Gershenzon, and J. Degenhardt. 2010. Terpene synthases of oregano (*Origanum vulgare* L.) and their roles in the pathway and regulation of terpene biosynthesis. *Plant molecular biology*, *73*(6): 587–603. https://doi.org/10.1007/s11103-010-9636-1

Dirmenci, T., T. Yazici, T. Özcan, S. Celenk, and E. Martin. 2018. A new species and a new natural hybrid of *Origanum* L.(Lamiaceae) from the west of Turkey. *Turkish Journal of Botany*, *42*(1): 73–90. https://doi.org/10.3906/bot-1704-35

Dordas, C. 2009. Foliar application of calcium and magnesium improves growth, yield, and essential oil yield of oregano (*Origanum vulgare* ssp. *hirtum*). *Industrial crops and products*, *29*(2–3): 599–608. https://doi.org/10.1016/j.indcrop.2008.11.004

Dos Santos, H. T., R. A. Sermarini, M. A. Moreno-Pizani, and P. A. A. Marques. 2020. Effects of irrigation management and seasonal stages on essential oil content and biomass of *Origanum vulgare* L. *Notulae Scientia Biologicae*, *12*(1): 42–56. https://doi.org/10.15835/nsb12110588

Dudai, N., E. Putievsky, U. Ravid, D. Palevitch, and A. H. Halevy. 1992. Monoterpene content in *Origanum syriacum* as affected by environmental conditions and flowering. *Physiologia Plantarum*, *84*(3), 453–459.

Duman, H., Z. Aytaç, M. Ekici, F. A. Karavelioğulları, A. Dönmez, and A. Duran. 1995. Three new species (Labiatae) from Turkey. *Flora Mediterranea*, *5*: 221–228. www.herbmedit.org/flora/5-2 21.pdf

Duman, H., Başer, K. H. C. and Z. Aytaç. 1998). Two new species and a new hybrid from Anatolia. *Turkish Journal of Botany*, *22*(1): 51–58. https://journals.tubitak.gov.tr/botany/vol22/iss1/8/

El-Alam, I., R. Zgheib, M. Iriti, M. El Beyrouthy, P. Hattouny, A. Verdin, J. Fontaine, R. Chahine, A. Lounès-Hadj Sahraoui, and H. Makhlouf. 2019. *Origanum syriacum* essential oil chemical polymorphism according to soil type. *Foods*, *8*(3): 90. https://doi.org/10.3390/foods8030090

Emrahi, R., M. R. Morshedloo, H. Ahmadi, A. Javanmard, and F. Maggi. 2021. Intraspecific divergence in phytochemical characteristics and drought tolerance of two carvacrol-rich *Origanum vulgare* subspecies: subsp. *hirtum* and subsp. *gracile*. *Industrial Crops and Products*, *168*: 113557. https://doi.org/10.1016/j.indcrop.2021.113557

Farías, G., O. Brutti, R. Grau, P. D. L. Lira, D. Retta, C. van Baren, S. Ventod, and A. L. Bandonid. 2010. Morphological, yielding and quality descriptors of four clones of *Origanum* spp. (Lamiaceae) from the Argentine Littoral region Germplasm bank. *Industrial Crops and Products*, *32*(3): 472–480. https://doi.org/10.1016/j.indcrop.2010.06.019

Farooq, M., A. Wahid, N. S. M. A. Kobayashi, D. B. S. M. A. Fujita, and S. M. A. Basra. 2009. Plant drought stress: effects, mechanisms and management. In *Sustainable agriculture* pp. 153–188. Springer, Dordrecht. https://doi.org/10.1007/978-90-481-2666-8_12

Farsi, M., F. Abdollahi, A. Salehi, and sS. Ghasemi. 2019. Growth responses of *Origanum majorana* L. to methyl jasmonic acid under limited irrigation conditions. *Journal of Essential Oil Bearing Plants*, *22*(2): 455–468. https://doi.org/10.1080/0972060X.2019.1602481

Gerami, F., P. R. Moghaddam, R. Ghorbani, and A. Hassani 2016. Effects of irrigation intervals and organic manure on morphological traits, essential oil content and yield of oregano (*Origanum vulgare* L.). *Anais da Academia Brasileira de Ciências*, *88*: 2375–2385. https://doi.org/10.1590/0001-3765201620160208

Ghalambor, C. K., J. K. McKay, S. P. Carroll, and D. N. Reznick. 2007. Adaptive versus non-adaptive phenotypic plasticity and the potential for contemporary adaptation in new environments. *Functional ecology*, *21*(3): 394–407. https://doi.org/10.1111/j.1365-2435.2007.01283.x

Gönüz, A. and B. Özörgücü. 1999. An investigation on the morphology, anatomy and ecology of *Origanum onites* L. 1. *Turkish journal of botany*, *23*(1): 19–32. https://journals.tubitak.gov.tr/botany/vol23/iss1/3/

Grime, J. P. 2006. *Plant strategies, vegetation processes, and ecosystem properties*. John Wiley & Sons.

Hamed, E. S. 2018. Effect of nitrogenous fertilization and spraying with nano-fertilizer on *Origanum syriacum* L. var. *syriacum* plants under North Sinai conditions. *Journal of Pharmacognosy and Phytochemistry*, 7(4): 2902–2907. www.phytojournal.com/archives/2018/vol7issue4/PartAW/7-4-573-827.pdf

Hancioglu, N. E., A. Kurunc, I. Tontul, and A. Topuz. 2019. Irrigation water salinity effects on oregano (*Origanum onites* L.) water use, yield and quality parameters. *Scientia Horticulturae*, 247: 327–334. https://doi.org/10.1016/j.scienta.2018.12.044

Hristozkova, M., M. Geneva, I. Stancheva, M. Boychinova, and E. Djonova. 2015. Aspects of mycorrhizal colonization in adaptation of sweet marjoram (*Origanum majorana* L.) grown on industrially polluted soil. *Turkish Journal of Biology*, 39(3): 461–468. https://doi.org/10.3906/biy-1408-47

Ietswaart, J. H. 1980. *A taxonomic revision of the genus Origanum (Labiatae)*, 4, p. 158. The Hague: Leiden University Press.

Jafari, L., F. Abdollahi, H. Feizi, and S. Adl. 2021. Improved Marjoram (*Origanum majorana* L.) Tolerance to Salinity with Seed Priming Using Titanium Dioxide (TiO2). *Iranian Journal of Science and Technology, Transactions A: Science*, 1–11. https://doi.org/10.1007/s40995-021-01249-3

Khraizat, Z. M. 2015. *Water productivity of Origanum syriacum under different irrigation and nitrogen treatments* (Doctoral dissertation). https://scholarworks.aub.edu.lb/bitstream/handle/10938/10565/st-6326.pdf?sequence=1

Kofidis, G., A. M. Bosabalidis, and M. Moustakas. 2003. Contemporary seasonal and altitudinal variations of leaf structural features in oregano (*Origanum vulgare* L.). *Annals of botany*, 92(5): 635–645. https://doi.org/10.1093/aob/mcg180

Kokkini, S. 1997. Taxonomy, Diversity and Distribution of *Origanum* Species. In Oregano. Proceedings of the IPGRI International Workshop on oregano, 8-12 May 1996, CIHEAM, Valenzano, Bari, Italy, edited by S. Padulosi, 2–12. Roma (Italy).

Kulbat-Warycha, K., E. C. Georgiadou, D. Mańkowska, B. Smolińska, V. Fotopoulos, and J. Leszczyńska. 2020. Response to stress and allergen production caused by metal ions (Ni, Cu and Zn) in oregano (*Origanum vulgare* L.) plants. *Journal of Biotechnology*, 324: 171–182. https://doi.org/10.1016/j.jbiotec.2020.10.025

Laghmouchi, Y., O. Belmehdi, A. Bouyahya, N. S. Senhaji, and J. Abrini. 2017. Effect of temperature, salt stress and pH on seed germination of medicinal plant *Origanum compactum*. *Biocatalysis and agricultural biotechnology*, 10: 156–160. https://doi.org/10.1016/j.bcab.2017.03.002

Lattanzio, V., A. Cardinali, C. Ruta, I. M. Fortunato, V. M. Lattanzio, V. Linsalata, and N. Cicco. 2009. Relationship of secondary metabolism to growth in oregano (*Origanum vulgare* L.) shoot cultures under nutritional stress. *Environmental and Experimental Botany*, 65(1): 54–62. https://doi.org/10.1016/j.envexpbot.2008.09.002

Levizou, E., A. A. Zanni, and V. Antoniadis. 2019. Varying concentrations of soil chromium (VI) for the exploration of tolerance thresholds and phytoremediation potential of the oregano (*Origanum vulgare*). *Environmental Science and Pollution Research*, 26(1): 14–23. https://doi.org/10.1007/s11356-018-2658-y

Lianopoulou, V. and A. M. Bosabalidis. 2014. Traits of seasonal dimorphism associated with adaptation to cold stress in *Origanum dictamnus* L. (Lamiaceae). *Journal of Biological Research-Thessaloniki*, 21(1): 1–9. https://doi.org/10.1186/2241-5793-21-17

Lloret, F., E. G. de la Riva, I. M. Pérez-Ramos, T. Marañón, S. Saura-Mas, R. Díaz-Delgado, and R. Villar. 2016. Climatic events inducing die-off in Mediterranean shrublands: are species' responses related to their functional traits?. *Oecologia*, 180(4): 961–973. https://doi.org/10.1007/s00442-016-3550-4

Loera-Muro, A., M. G. Caamal-Chan, T. Castellanos, A. Luna-Camargo, T. Aguilar-Díaz, and A. Barraza. 2021. Growth effects in oregano plants (*Origanum vulgare* L.) assessment through inoculation of bacteria isolated from crop fields located on desert soils. *Canadian Journal of Microbiology*, 67(5): 381–395. https://doi.org/10.1139/cjm-2020-035

Makri, O. 2002. Cultivation of oregano. *Oregano, the genera Origanum and Lippia. Taylor and Francis, London and New York*, 153–162.

Margaris, N. S. 1976. Structure and dynamics in a phryganic (East Mediterranean) ecosystem. *Journal of Biogeography*, 249–259. https://doi.org/10.2307/3038015

Martin, E., T. Dirmenci, T. Arabaci, T. Yazici, and T. Özcan. 2020. Karyotype studies on the genus *Origanum* L. (Lamiaceae) species and some hybrids defining homoploidy. *Caryologia*, 73(2): 127–143.

Matesanz, S. and F. Valladares. 2014. Ecological and evolutionary responses of Mediterranean plants to global change. *Environmental and Experimental botany*, *103*: 53–67. https://doi.org/10.1016/j.envexpbot.2013.09.004

Meléndez, M. G., G. Z. Camargo, J. J. M. Contreras, A. H. Sepúlveda, S. de los Santos Villalobos, and F. I. P. Cota. 2017. Abiotic stress tolerance of microorganisms associated with oregano (*Origanum vulgare* L.) in the Yaqui Valley, Sonora. *Open Agriculture*, *2*(1): 260–265. https://doi.org/10.1515/opag-2017-0029

Mohammadi, H., L. A. Dizaj, A. Aghaee, and M. Ghorbanpour. 2021. Chitosan-Mediated Changes in dry Matter, Total Phenol Content and Essential Oil Constituents of two *Origanum* Species under Water Deficit Stress. *Gesunde Pflanzen*, *73*(2): 181–191. https://doi.org/10.1007/s10343-020-00536-0

Morshedloo, M. R., S. A. Salami, V. Nazeri, and L. E. Craker. 2017a. Prolonged water stress on growth and constituency of Iranian of Oregano (*Origanum vulgare* L.). *Journal of Medicinally Active Plants*, *5*(2): 7–19. https://doi.org/10.7275/R5XS5SKW

Morshedloo, M. R., L. E. Craker, A. Salami, V. Nazeri, H. Sang, and F. Maggi. 2017b. Effect of prolonged water stress on essential oil content, compositions and gene expression patterns of mono-and sesquiterpene synthesis in two oregano (*Origanum vulgare* L.) subspecies. *Plant physiology and biochemistry*, *111*: 119–128. https://doi.org/10.1016/j.plaphy.2016.11.023

Mowafy, A. M., D. A. Alraey, M. N. Omar, A. Elshobaky, and S. A. Haroun. 2016. *Origanum syriacum* ssp. *sinaicum* associated growth promoting bacteria. *IOSR J Environ Sci Toxicol Food Technol, 10*: 53–60. https://asset-pdf.scinapse.io/prod/2522694111/2522694111.pdf

Munir, N., M. Jahangeer, A. Bouyahya, N. El Omari, R. Ghchime, A. Balahbib, S. Aboulaghras, Z. Mahmood, M. Akram, S. M. Ali Shah, I. N. Mikolaychik, M. Derkho, M. Rebezov, B. Venkidasamy, M. Thiruvengadam, and M. A. Shariati. 2022. Heavy Metal Contamination of Natural Foods Is a Serious Health Issue: A Review. *Sustainability*, *14*(1): 161. https://doi.org/10.3390/su14010161

Murillo-Amador, B., L. E. Morales-Prado, E. Troyo-Diéguez, M. V. Córdoba-Matson, L. G. Hernández-Montiel, E. O. Rueda-Puente, and A. Nieto-Garibay. 2015. Changing environmental conditions and applying organic fertilizers in *Origanum vulgare* L. *Frontiers in plant science*, *6*: 549. https://doi.org/10.3389/fpls.2015.00549

Ninou, E., K. Paschalidis, and I. Mylonas. 2017. Essential oil responses to water stress in greek oregano populations. *Journal of Essential Oil Bearing Plants*, *20*(1): 12–23. https://doi.org/10.1080/0972060X.2016.1264278

Novak, J., B. Lukas, and C. Franz. 2010. Temperature influences thymol and carvacrol differentially in *Origanum* spp.(Lamiaceae). *Journal of essential oil research*, *22*(5): 412–415. https://doi.org/10.1080/10412905.2010.9700359

Osakabe, Y., K. Osakabe, K. Shinozaki, and L. S. P. Tran. 2014. Response of plants to water stress. *Frontiers in plant science*, *5*: 86. https://doi.org/10.3389/fpls.2014.00086

Ozguven, M., F. Ayanoglu, and A. Ozel. 2004. Effects of Nitrogen Rates and Cutting Times on the Essentia Oil Yield and Components of *Origanum syriacum* L. var. p. *Journal of agronomy*, *5*(1): 101–105.

Panou-Filotheou, E., A. M. Bosabalidis, and S. Karataglis. 1998. Effects Of Cu Excess On Oregano [*Origanum vulgare* Subsp. *hirtum* (Link) Ietswaart]. II Physiological Effects. In *Progress in Botanical Research* (pp. 349–352). Springer, Dordrecht. https://doi.org/10.1007/978-94-011-5274-7_99

Panou-Filotheou, H. and A. M. Bosabalidis. 2004. Root structural aspects associated with copper toxicity in oregano (*Origanum vulgare* subsp. *hirtum*). *Plant science*, *166*(6): 1497–1504. https://doi.org/10.1016/j.plantsci.2004.01.026

Parmesan, C. and G. Yohe. 2003. A globally coherent fingerprint of climate change impacts across natural systems. *Nature*, *421*(6918): 37–42. https://doi.org/10.1038/nature01286

Paschke, M., G. Bernasconi, and B. Schmid. 2003. Population size and identity influence the reactiosn norm of the rare, endemic plant *Cochlearia bavarica* across a gradient of environmental stress. *Evolution*, *57*(3): 496–508. https://doi.org/10.1111/j.0014-3820.2003.tb01541.x

Pereyra, M. S., J. A. Argüello, and P. I. Bima. 2021. Genotype-dependent architectural and physiological responses regulate the strategies of two oregano cultivars to water excess and deficiency regimes. *Industrial Crops and Products*, *161*, 113206.

Polade, S. D., D. W. Pierce, D. R. Cayan, A. Gershunov, and M. D. Dettinger. 2014. The key role of dry days in changing regional climate and precipitation regimes. *Scientific reports*, *4*(1): 1–8. https://doi.org/10.1038/srep04364

Pugnaire, F. I. and M. T. Luque. 2001. Changes in plant interactions along a gradient of environmental stress. *Oikos*, *93*(1): 42–49. https://doi.org/10.1034/j.1600-0706.2001.930104.x

Rhizopoulou, S. and S. Diamantoglou. 1991. Water stress-induced diurnal variations in leaf water relations, stomatal conductance, soluble sugars, lipids and essential oil content of *Origanum majorana* L. *Journal of Horticultural Science*, *66*(1): 119–125. https://doi.org/10.1080/00221589.1991.11516133

Saatkamp, A., P. Poschlod, and D. L.Venable. 2014. The functional role of soil seed banks in natural communities. In: Gallagher RS (ed) Seeds—the ecology of regeneration in plant communities. CABI, Wallingford, pp 263–294.

Said-Al Ahl, H., E. A. Omer, and N. Y. Naguib. 2009. Effect of water stress and nitrogen fertilizer on herb and essential oil of oregano. *International Agrophysics*, *23*(3): 269–275. https://agro.icm.edu.pl/agro/element/bwmeta1.element.agro-article-3fe55777-9a19-4b2a-a07e-432ebaae69b8

Said-Al Ahl, H. A. H. and M. S. Hussein. 2010. Effect of water stress and potassium humate on the productivity of oregano plant using saline and fresh water irrigation. *Ozean Journal of Applied Sciences*, *3*(1): 125–141.

Shafaghat, A. 2011. Antibacterial activity and GC/MS analysis of the essential oils from flower, leaf and stem of *Origanum vulgare* ssp. *viride* growing wild in northwest Iran. *Natural product communications*, *6*(9): 1934578X1100600933. https://doi.org/10.1177/1934578X1100600933

Shafiee-Hajiabad, M., M. Hardt, and B. Honermeier. 2014. Comparative investigation about the trichome morphology of Common oregano (*Origanum vulgare* L. subsp. *vulgare*) and Greek oregano (*Origanum vulgare* L. subsp. *hirtum*). *Journal of Applied Research on Medicinal and Aromatic Plants*, *1*(2): 50–58. https://doi.org/10.1016/j.jarmap.2014.04.001

Shafiee-Hajiabad, M., J. Novak, and B. Honermeier. 2015. Characterization of glandular trichomes in four *Origanum vulgare* L. accessions influenced by light reduction. *Journal of Applied Botany and Food Quality* 88: 300–307. https://core.ac.uk/download/pdf/84118034.pdf

Shafiee-Hajiabad, M., J. Novak, and B. Honermeier. 2016. Content and composition of essential oil of four *Origanum vulgare* L. accessions under reduced and normal light intensity conditions. *Journal of Applied Botany and Food Quality*, 89.

Skoula, M., P. Gotsiou, G. Naxakis, and C. B. Johnson. 1999. A chemosystematic investigation on the mono- and sesquiterpenoids in the genus *Origanum* (Labiatae). *Phytochemistry*, *52*: 649–657. https://doi.org/10.1016/S0031-9422(99)00268-X

Smirnoff, N. 1998. Plant resistance to environmental stress. *Current opinion in Biotechnology*, *9*(2): 214–219. https://doi.org/10.1016/S0958-1669(98)80118-3

Sotiropoulou, D. E. and A. J. Karamanos. 2010. Field studies of nitrogen application on growth and yield of Greek oregano (*Origanum vulgare* ssp. *hirtum* (Link) Ietswaart). *Industrial Crops and Products*, *32*(3): 450–457. https://doi.org/10.1016/j.indcrop.2010.06.014

Stefanaki, A. and T. van Andel. 2021. Mediterranean aromatic herbs and their culinary use. In *Aromatic Herbs in Food* (pp. 93–121). Academic Press. https://doi.org/10.1016/B978-0-12-822716-9.00003-2

Sulaiman, H. Y., B. Liu, E. Kaurilind, and Ü. Niinemets. 2021. Phloem-feeding insect infestation antagonizes volatile organic compound emissions and enhances heat stress recovery of photosynthesis in *Origanum vulgare*. *Environmental and Experimental Botany*, *189*: 104551. https://doi.org/10.1016/j.envexpbot.2021.104551

Thompson, J. D. 2020. *Plant Evolution in the Mediterranean: Insights for conservation*. Oxford University Press, USA.

Tibaldi, G., E. Fontana, and S. Nicola. 2011. Growing conditions and postharvest management can affect the essential oil of *Origanum vulgare* L. ssp. *hirtum* (Link) Ietswaart. *Industrial Crops and Products*, *34*(3): 1516–1522. https://doi.org/10.1016/j.indcrop.2011.05.008

Tomou, E. M., H. Skaltsa, G. Economou, and A. Trichopoulou. 2022. Sustainable diets & medicinal aromatic plants in Greece: Perspectives towards climate change. *Food Chemistry*, *374*: 131767. https://doi.org/10.1016/j.foodchem.2021.131767

Trivino, M. G. and C. B. Johnson. 2000. Season has a major effect on the essential oil yield response to nutrient supply in *Origanum majorana*. *The Journal of Horticultural Science and Biotechnology*, *75*(5): 520–527. https://doi.org/10.1080/14620316.2000.11511278

Unal, B. T., A. Guvensen, A. E. Dereboylu, and M. Ozturk. 2013. Variations in the proline and total protein contents in *Origanum sipyleum* L. from different altitudes of spil mountain Turkey. *Pakistan Journal of Botany*, *45*(S1), 571–576.

Vellend, M., L. Baeten, A. Becker-Scarpitta, V. Boucher-Lalonde, J. L. McCune, J. Messier, I. H. Myers-Smith, and D. F. Sax. 2017. Plant biodiversity change across scales during the Anthropocene. *Annual Review of Plant Biology*, *68*: 563–586. https://doi.org/10.1146/annurev-arplant-042916-040949

Vokou, D., S. Kokkini, and J. M. Bessiere. 1988. *Origanum onites* (Lamiaceae) in Greece: Distribution, volatile oil yield, and composition. *Economic Botany*, *42*(3), 407–412. https://doi.org/10.1007/BF02860163

Woodruff, D. S. 2010. Biogeography and conservation in Southeast Asia: how 2.7 million years of repeated environmental fluctuations affect today's patterns and the future of the remaining refugial-phase biodiversity. *Biodiversity and Conservation*, *19*(4): 919–941. https://doi.org/10.1007/s10531-010-9783-3

Yasar, A., Y. Karaman, I. Gokbulut, A. O. Tursun, N. Tursun, I. Uremis, and M. Arslan. 2021. Chemical Composition and Herbicidal Activities of Essential Oil from Aerial Parts of *Origanum* Hybrids Grown in Different Global Climate Scenarios on Seed Germination of *Amaranthus palmeri*. *Journal of Essential Oil Bearing Plants*, *24*(3): 603–616. https://doi.org/10.1080/0972060X.2021.1951848

Zaghloul, M. S., P. Poschlod, and C. Reisch. 2014. Genetic variation in Sinai's range-restricted plant taxa *Hypericum sinaicum* and *Origanum syriacum* subsp. *sinaicum* and its conservational implications. *Plant Ecology and Evolution*, *147*(2): 187–201. https://doi.org/10.5091/plecevo.2014.838

15 Petrosolinum crispum under Stressful Conditions

Amir Abdullah Khan[1*], Haifa Abdulaziz Sakit AlHaithloul[2],
Suliman Mohammed Alghanem[3], Hoda H. Senousy[6],
Wardah A. Alhoqail[4], Fuchen Shi[1*], Iftikhar Ali[5]
and Mona Hassan Soliman[6,7]
[1]Department of Plant Biology and Ecology, Nankai University
Tianjin, China
[2]Biology Department, College of Science, Jouf University,
Sakaka, Kingdom of Saudi Arabia
[3]Biology Department, Faculty of Science, Tabuk University, Tabuk,
Saudi Arabia s-alghanem@ut.edu.sa
[4]Department of Biology, College of Education, Majmaah University,
Kingdom of Saudi Arabia
[5]Center for Plant science and Biodiversity, University of Swat,
Charbagh, Pakistan
[6]Botany and Microbiology Department, Faculty of Science,
Cairo University, Giza, Egypt
[7]Biology Department, Faculty of Science, Taibah University,
Al-Sharm, Yanbu El-Bahr, Yanbu, Kingdom of Saudi Arabia
*Corresponding author: Amir_nku@hotmail.com;
fcshi@mail.nankai.edu.cn

CONTENTS

DOI: 10.1201/9781003242963-15

15.1 INTRODUCTION

Plants have developed different mechanisms to survive when exposed to severe environmental conditions. Being sessile, plants develop mechanisms that allow them to detect true environmental changes and respond accordingly by decreasing the damage and maintaining the resources for growth and reproduction. When plants are subjected to multiple stresses they activate special and unique mechanisms for countering the negative effects (Rizhsky et al. 2004; Khan et al. 2021). Bearing this in mind, it would be inadequate to impose each stress individually and test the tolerance of plant species (Miller et al. 2010). In the case of biotic and abiotic stresses it's true because the signaling pathways under stress may work antagonistically against each other (Anderson et al. 2004; Asselbergh, De Vleesschauwer and Höfte 2008). Although the results are valuable, however, most of the studies explain the effects of single stress on plants at a time. Environmental conditions are unfavorable and suboptimal to most plant species, restricting plants from achieving the full-scale genetic ability to grow and reproduce (Bray 2000; Rockström and Falkenmark 2000). Unfavorable environmental conditions do not only affect growth but also disturb the yield. For example, the average yield of wheat crop is decreased eight-fold as compared to the yield if plants are grown in completely optimum conditions (Boyer 1982). This difference in production can be better elaborated by unfavorable conditions which ultimately lead the plants into harmful physiological alterations which are known as stresses (Shao et al. 2008). Globally, abiotic stresses such as cold, drought, heat, salinity and nutrient deficiency have shown a huge impact on the agriculture sector and it is estimated that an average reduction of more than 50% yield occurs for most major crops (Wang, Vinocur and Altman 2003). Besides abiotic stress, plants also face a threat from biotic stress. A very large range of organisms such as pests, pathogens, bacteria, viruses, herbivorous insects and nematodes destroy crops (Vranová, Van Breusegem and Dat 2002). In response to stress, each plant elicits a specific mechanism of response that ranges from the cellular to the molecular level, which protects plants from the detrimental effects and ensures survival (Herms and Mattson 1992).

Parsley (*Petroselinum crispum*), a herbaceous medicinal plant, is native to America, Europe and the Middle East. It is a biennial species in its natural habitat; however, it can also grow as a semi-perennial plant under greenhouse conditions (Marthe 2020). It is consumed in fresh or dried form and is also used as a source of essential oil. Its leaves are used as a flavoring and condiment ingredient. Parsley is also a rich source of vitamin C and antioxidants. It's also a rich source of iron (Kmiecik and Lisiewska 1999). Oil extract from its leaf is mainly used for meat flavoring, making sauces, canned food and sometimes in perfumery (Shaath et al. 1988; Sabry, Kandil and Ahmed 2016).

15.2 PARSLEY GLOBAL AVAILABILITY

As compared to other aromatic and medicinal plants globally, the *ex situ* collections of genetic resources of parsley are presented. The biggest sets are 351 in Spain, 242 accessions in the federal *ex situ* gene bank at IPK of Germany, 249 in the USA, 229 in Poland, 163 in Hungary, 66 in the Russian Federation, 163 in Ukraine, 57 accessions in Romania, and 40 accessions in the Czech Republic (Westcott 2010; Kreide, Oppermann and Weise 2019).

15.3 ECONOMIC VALUE OF PARSLEY

The use of the parsley plant dates back 2500 years. Theophrastus, one of the pioneer Greek botanists, mentioned it in 322 BCE (Marthe 2020). Hippocrates (460–370 BCE) used it as a diuretic agent while the Greek people considered it sacred and used it to decorate tombs. Later, the Romans took it to Central Europe. In the Middle Ages, its use as a medicinal plant started. Currently, it is used as an aromatic herb for cooking. At present, in Germany, it is grown in 1800 ha (Plescher, Grohs and Pforte 2014).

15.4 GROWING CONDITIONS AND HARVESTING

The cultivation season for parsley is the beginning of March. After cultivation, the plants make leaves rosette followed by the first harvest after 12 weeks of cultivation. If the temperature and environmental conditions are favorable, new harvests can be obtained after 21–30 days. Plants are ready for the first harvest at the height of 20–25 cm, when an area of 4–5 cm is cut above the ground (Hoppe et al. 2013). The optimum and best conditions for parsley growth are sandy and clay soil, 1–5% humidity and a pH of 6–7 (Marthe 2020). It is a cool-season plant and grows best at 7–16°C, and below 7°C the growth is considerably slowed (Plescher, Grohs and Pforte 2014). If the seeds are sown at low temperature the germination can reach up to 26 days. However, between 12 and 15°C the temperature is best. Usually, the seeds are soaked at 20 ± 2°C for a duration of 18 h, which is followed by re-drying at 30°C for best germination. In the same way, seed dormancy can be broken at 5°C. The plant's leaves can tolerate up to −8°C but the roots can tolerate up to −10°C. It is also important not to expose plants to continuous frost as it can cause the plants to rot. also, to protect the plants from pest attacks there is a strict need for crop rotation every five years (Marthe 2020).

15.5 ECONOMIC VALUATION

It is estimated that in Germany the first dry yield of parsley ranges from 800 to 900 kg/ha, while for the second cut it ranges from 1 to 1.1 t/ha. A total of five cuts are possible in one year which may yield up to 5.3 t/ha (Hoppe et al. 2013). When grown in February the yield reaches 12 kg/ha and a height of between 12–13 cm. The essential oil obtained from parsley is of a lower value as compared to fresh and dry products. It is estimated that the annual production of parsley essential oil from the seed is $0.6 million, while that of the herb essential oil amounts to $0.36 million annually (Verlet 1993).

15.6 EFFECT OF BIOTIC AND ABIOTIC STRESSORS ON PARSLEY

Plants being sessile have to survive in different environmental conditions where they face different stress conditions, as shown in Figure 15.1. However, they have developed specific mechanisms to counter stressful conditions. The adopted mechanisms have helped plants to detect challenges and respond accordingly by minimizing damage while protecting their valuable resources for yield and growth (Rizhsky, Liang and Mittler 2002). Different plants have evolved different mechanisms to overcome stressful conditions. In light of this, it is inappropriate to apply a single stress to each species and identify a single mechanism of response (Seifi et al. 2012). Herein, the effects of biotic and abiotic stress interactions in parsley plants are studied, with an emphasis on explicating the mechanisms involved in parsley to survive in unfavorable conditions.

15.7 BIOTIC STRESS

Cultivation of parsley in greenhouse conditions or fields needs a specific check on the phytopathological aspect of this plant. Globally, parsley yield is significantly reduced by pests such as bacteria, fungi and viruses. That is why the damage in yield quality and quantity persists with or without the application of pesticides. The best way to produce parsley with approved active substances is

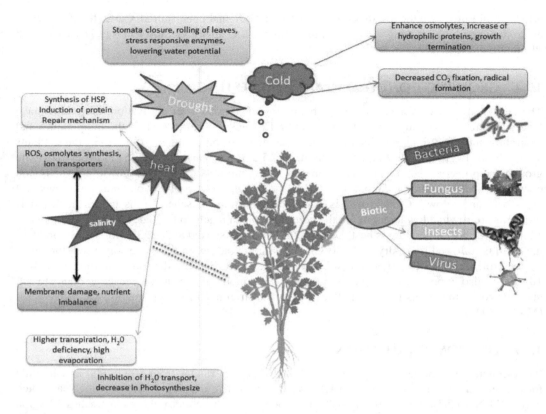

FIGURE 15.1   Stress tolerance mechanisms in *Petrosolinum crispum* under biotic and abiotic stress.

to produce a variety which is resistant to pathogenic attack. In Table 15.1 a list of all the pathogens is given.

15.8   BACTERIAL PATHOGENS

It has previously been reported that the bacterial strain *Pseudomonas syringae* is responsible for causing leaf spot disease and macerations in parsley (Bull et al. 2011; Reintke, Hiddink and Mendez 2016). The common symptoms of this disease are water-soaked lesions on leaves, followed by different colors and shades on leaves from tan to black. The affected spots become dry and papery over time.

15.9   FUNGAL PATHOGENS

Parsley, like other crops, is damaged by fungal pathogens globally. The most common diseases such as *Septoria* blight caused by *Septoria petroselini*, downy mildew (*Plasmopara petroselini*), powdery mildew (*Erysiphe heracle*) and *Alternaria* leaf blight caused by *Alternaria radicina* have been reported globally, with *Septoria* blight being the most common one (Tahvonen 1978). The second most common fungal disease is downy mildew, which was reported many times in Germany during the first half of the 20th century (Brandenburger and Hagedorn 2006). In downy mildew diseases, chlorosis with a yellow faint color occurs on the upper surface of the leaves, while on the lower

TABLE 15.1

Diseases and Pathogenicity of Parsley Plants

Name Of Disease	Pathogen	Nature Of Pathogen	References
Leaf spots	*Pseudomonas* spp.	Bacteria	(Burton et al. 2021)
Septoria blight	*Septoria petroselini* (Lib.)	Fungus	(Lopes and Martins 2008)
Blight	*Septoria blight (S. petroselini)*	Fungus	(Tahvonen 1978)
Downy mildew	*P. petroselini*	Fungus	(Brandenburger and Hagedorn 2006)
Alternaria leaf blight	*Alternaria radicina*	Fungus	(Marthe et al. 2013)
Leaf spots disease	*Alternaria radicina*	Fungus	(Lopes and Martins 2008)
Sclerotinia rot	*Sclerotinia sclerotiorum (Lib.)*	Fungus	(Nowicki 1997)
Fusarium wilt	*Fusarium oxysporum*	Fungus	(Nawrocki 2005)
Pythium root rot	*Pythium* spp.	Fungus	(Tsuchida et al. 2018)
Powdery mildew	*Euleia heraclei*	Insect	(Marthe et al. 2003)
Celery mosaic virus disease (CeMV)	Celery mosaic virus	Virus	
	Apium virus Y (ApVY)	Virus	(Delmas et al. 2019)
Virus complex	Carrot red leaf virus (CtRLV); carrot mottle virus (CMoV); carrot mottle mimic virus (CMoMV); *Apium* virus Y; carrot yellow leaf virus (CYLV)	Virus	(Delmas et al. 2019; Hause et al. 2013; Karlas et al. 2010)

surface of the leaves white or gray mycelium is developed. The lesions increase and finally cover the whole leaf, resulting in rot. The infection spreads very rapidly and can be detected by its typical smell (Amein et al. 2006).

In Germany, in 1990, another fungal disease, namely powdery mildew, was detected which is caused by *E. heraceli* which decreases the yield. This pathogenic fungus is also common in other members of the family Apiaceae and is responsible for decreasing the quality and market value of the product. This fungus has also been introduced in Brazil. The main route of pathogenicity is through affected seeds (Rosa et al. 2008).

Alternaria radicina is another common fungal pathogen that is responsible for *Alternaria* leaf blight disease and seedlings damping off disease (Gärber, Nega and Jahn 2007; Marthe and Scholze 2006). *Alternaria* spp. are usually seed-borne on parsley. Disease spread starts from older leaves and the infestations spread to young leaves which can be identified by lesions on the margins of the leaf that have an irregular shape. Within the lesions, concentric regions of different colors are visible which are followed by gray and whitish conidia formation at the end of each mycelium (Meyer et al. 2010). Another well-known fungal pathogen in parsley is *Sclerotinia sclerotium* which is responsible for causing *Sclerotinia* rot which specifically affects and destroys the macerating stalk (Marthe et al. 2013).

15.10 VIRAL PATHOGENS

Along with bacterial and fungal pathogens, viruses are also responsible for damaging parsley growth and yield globally. The best known viral phytopathogens studied so far are celery mosaic virus (CeMV), *Apium* virus Y (ApVY) and a virus complex, i.e. CemV, from the carrot red leaf virus, the parsley-specific carrot mottle virus (CMoV). Moreover, ApVY and CYLV are also an active part of the virus complex (Hoppe et al. 2013).

15.11 ABIOTIC STRESS TOLERANCE IN PARSLEY

Abiotic stress is a very large issue in the agriculture sector in modern currently. Abiotic stresses such as water, salinity, drought, heat and cold are greatly responsible for extreme crop losses globally. The main effect of abiotic stress is on photosynthesizing in plants. A decrease in photosynthetic ability leads to a low yield in plants, including parsley. That is why mechanisms must be identified through which the global reduction in parsley yield due to abiotic stress is reduced. Here we discuss the impacts of four common abiotic stresses, i.e. water, heat, salinity and drought, on parsley plants and the strategies involved in overcoming these stresses are elaborated on (Khan et al. 2021).

15.12 SALINITY STRESS

Salinity is a vital problem that reduces the growth and yield of crops in areas under salt exposure (Gull, Lone and Wani 2019). It is considered the main issue affecting all crops. The main factors affecting soil salinity are land topography, anthropogenic activities and salt composition (Kotagiri and Kolluru 2017). It has been reported previously that salinity causes ion toxicity and water loss, which result in growth reduction and nutritional value deficiency, which finally lead to the death of the plant, particularly in plants that are sensitive to a saline environment (Mohammadi, Hazrati and Ghorbanpour 2020). Salt stress also causes alterations in the physiochemical levels of plants by lowering the chlorophyll content, which leads to degradation of chlorophyll. This alteration leads to an imbalance in chlorophyll metabolism and incomplete photosynthesis. However, to cope with this issue, some plants have developed different mechanisms which include regulating salt concentration in certain tissues. This mechanism helps to regulate the concentration of Na^+ and Cl^- in the cells and tissues of plants (Khan et al. 2021).

Previously, in a study conducted by Lacramioara et al. (Oprica and Caunic 2013), it was shown that when parsley seedlings were subjected to salinity stress of 50 mM and 100 mM NaCl, they showed a net increase in polyphenols under increasing salt concentration, i.e. one variety showed maximum polyphenol at 50 mM NaCl in the shoot, while the second parsley variety showed maximum polyphenols synthesis at 100 mM NaCl in the root. The production of polyphenol shows the involvement of secondary metabolites. This mechanism of defense has already been discussed as one of the mechanisms adopted by plants to overcome saline conditions. Previously, the same results were described in red pepper (Navarro et al. 2006), mulberry (Agastian, Kingsley and Vivekanandan 2000) and mangrove *Aegiceras corniculatum* (Parida et al. 2004), respectively, when subjected to salt stress.

A study by Desire and Arslan (2021) revealed that when parsley plants were exposed to saline conditions of 2.5 and 4 dSm⁻¹, respectively, they showed a net decrease in fresh weight, height, transpiration, stomatal conductance and chlorophyll content. However, there was a net increase in the total protein, glutathione S-transferase (GST) and peroxidase (POD) activities at the low level of salinity, which helped the plant to overcome the stress conditions. The same study also revealed that exogenous foliar application of salicylic acid enhances all the growth parameters in parsley under a saline regime. These results are also in agreement with those of Aziz et al. (Aziz, Sabry and Ahmed 2013) when they subjected parsley to 0, 1.7, 3.1 and 4.7 dSm⁻¹ salt concentrations and the plants showed a net decrease in all growth parameters, i.e. growth, height and biomass, as compared to the control. However, to cope with stress a net increase in antioxidants, particularly peroxidase (POD) and catalase (CAT), was detected.

15.13 DROUGHT STRESS

The productivity and growth of plants are also limited by another stressor, namely drought (Seleiman et al. 2021). Currently, 70% of water is used for irrigation of crops, which is ever increasing with the increase in population and it may reach up to 83% by 2050 to meet the needs of humans (Facts

2012). The loss of crops is estimated at billions of dollars due to salinity and drought stresses (Shabala, Bose and Hedrich 2014). When plants are exposed to drought stress they face the issue of oxidative stress, which results in a decreased photosynthetic rate. To cope with the negative effects, plants produce phenolic compounds which help in plant protection (Jaafar, Ibrahim and Mohamad Fakri 2012). Petropoulos et al. (2008) exposed three parsley varieties (plain and curly leafed and turnip-rooted) to 35–40% and 45–60% water deficiency stress to evaluate the net impact on growth and essential oil production. The results showed that a net decrease occurred in shoot and root fresh weight, plant growth and leaf number. The reduction in growth parameters was signifi-cant. The drought stress also affected the essential oil and aromatic contents (1,3,8-*p*-menthatriene, myristicin, terpinolene + *p*-cymenene). However the impact on aroma varied from variety to variety. Drought showed a net decrease in essential oil production from roots, which was because the net biomass decreased, but there was no net change in the quality of oil composition. In another study (Najla, Sanoubar and Murshed 2012) two parsley varieties were subjected to 10, 30 and 50% water stress. Parsley plants subjected to stress showed a net decrease in leaf surface area, stem diameter, stem length and number of leaves. There was also a decline in chlorophyll, carotenoids, anthocyanin and vitamin C contents. However, there was not much difference in essential oil contents, which was because of phenol components (Meyer et al. 2010).

15.14 HEAT AND COLD STRESS

Due to anthropogenic activities, the duration, frequency and intensity of extreme weather have been deteriorating very quickly over the last few years, specifically temperature (Ummenhofer and Meehl 2017). Collins et al. ()2013 reported that the change in temperature will be higher in the future, which will affect the climate of cold regions quickly. However, it is also predicted that cold extremes will also occur in warmer temperature climates. Optimum temperature is necessary for the growth, development, quality and productivity of each and every plant species (Hatfield and Prueger 2015). This is why a slight climate change may entail big challenges for crops worldwide (Meena et al. 2018). These estimates emphasize the importance of knowing the effects of extreme temperatures on plant growth, development and production in order to assess potential future crop damage miti-gation techniques.

15.15 GENES EXPRESSION OF *PETROSOLINUM CRISPUM* DURING STRESS TOLERANCE

Complex biological processes such as development, signal transduction and metabolic pathways in plants and animals can be easily studied through gene expression analysis. For the analysis of gene expression, specifically qualitative or quantitative analysis (qPCR analysis) is the most effective tool as it is the most simple, specific and highly sensitive process (Wong and Medrano 2005).

Some important genes such as *ACTIN, UBQ* and *EF-1α* are considered housekeeping candidate genes (Table 15.2) because of their important role in overcoming different conditions and enabling the survival of the cell (Bustin 2002). However, it has been confirmed that the expression of these genes may be different under different environmental conditions of the plant (Suzuki, Higgins and Crawford 2000). In an experiment when carrot was exposed to single abiotic stress conditions, GAPDH showed a maximum level of stability (Tian et al. 2015), whereas in another study GAPDH seems to be the least expressed gene during exposure to stressful conditions (Kang et al. 1998). This is evident from the different studies that genes expressed in one plant species may be different from another species when exposed to the same stress conditions (Gutierrez et al. 2008).

Parsley plants were examined (Table 15.2) for eight candidate genes, i.e. *ACTIN, GAPDH, SAND, UBC, TIP41, eIF-4α, TUB* and *EF-1α* by qPCR analysis under different abiotic stress conditions (cold, drought, heat and salt) and hormone stimuli treatments (such as SA, MeJA, GA and ABA) (Li

TABLE 15.2
Genes Involved in Stress Tolerance of Parsley Plants

Gene	Improved traits	Transgenic species	Function in Parsley	Reference
ACTIN	Tolerance to abiotic stress	*Arabidopsis*	Abiotic stress	(Li et al. 2016)
GAPDH	Sensing stress and tolerance	*Arabidopsis*	Hormone stimuli treatment	(Li et al. 2016)
eIF-4α	Tolerance to heat, cold, salinity and drought	*Arabidopsis*	Hormone stimuli treatment	(Li et al. 2016)
SAND	Tolerance to heat, cold, salinity and drought	*Arabidopsis*	Abiotic stress	(Li et al. 2016)
UBC	Stress protective gene	*Arabidopsis*	Least stable for abiotic stress	(Li et al. 2016)
TIP41	Growth modulation during abiotic stress	*Arabidopsis*		(Li et al. 2016)
EF-1α,	Prevention from apoptosis during stress	*Arabidopsis*	Most stable for abiotic stress	(Li et al. 2016)
TUB	Abiotic stress	*Arabidopsis*	Most stable for abiotic stress/ hormone stimuli treatment	(Li et al. 2016)
Pc16182 (ERF)	Cold, water and abscisic acid response	*Arabidopsis*	Abiotic stress	(Li et al. 2014)
Pc16943 (ERF)	Cold, water and abscisic acid response	*Arabidopsis*	Abiotic stress	(Li et al. 2014)
Pc24931 (ERF)	Cold, water and abscisic acid response	*Arabidopsis*	Abiotic stress	(Li et al. 2014)
Pc32872 (AP2)	Cold, water and salinity response	*Arabidopsis*	Abiotic Stress	(Li et al. 2014)
Pc33331 (AP2)	Cold, water and salinity response	*Arabidopsis*	Abiotic stress	(Li et al. 2014)
Pc41893 (ERF)	Cold, water and salinity response	*Arabidopsis*	Abiotic stress	(Li et al. 2014)
PcWRKY1	Defense against pathogen	*Petrosolinum crispum*/yeast	Biotic stress	(Turck, Zhou and Somssich 2004)

et al. 2016). To understand the stability of candidate gene expression different statistical tools were used, i.e. Norm Finder (Andersen, Jensen, and Ørntoft 2004), Ge Norm (Vandesompele et al.), Best keeper (Pfaffl et al. 2004) and Ref finder (Xie et al. 2012). The results of Li et al. (2016) showed that under abiotic stress conditions the most stable genes were *EF-1α* and *TUB*, while for hormone stimuli treatment the most stable genes were *EF-1α*, *GAPDH* and *TUB*. Moreover, *EF-1α* and *TUB* genes showed the most stability and *UBC* the least one among all the tested samples.

Many other studies have also shown the role of the AP2/ERF family genes in response to abiotic stress (Xu et al. 2011). Recently Li et al. ()2014 studied seven different transcription factors in parsley from the AP2/ERF family genes when exposed to abiotic stress conditions (drought, salinity and temperature). The genes chosen belonged to three subfamilies, i.e. *ERF*, *AP2* and *RAV*. In previous studies of abiotic stress, these genes have already shown sensitivity to cold temperature. *Pc37218* and *Pc41893* genes showed the highest upregulation of 21- and 18-fold, respectively, when plants were exposed to low temperature for 8 hours, while the *Pc16943* gene showed upregulation by nine-fold in heat stress. In the same way, *Pc16182*, *Pc16943*, *Pc24931* and *Pc41893* exhibited sensitivity to salinity stress. When the same parsley plants were exposed to drought stress, genes *Pc24931*, *Pc32872*, *Pc33331* and Pc41893 showed upregulation. When parsley plants were exposed to biotic stress, *WRKY* and *PCPR1* genes expression was upregulated when the qPCR data were analyzed (Turck, Zhou and Somssich 2004). These findings show the involvement of stress-related genes which enable parsley plants to overcome unfavorable environmental conditions.

15.16 CONCLUSIONS AND FUTURE RECOMMENDATIONS

Parsley is globally used as a cultivated spice and medicinal plant as it is rich in antioxidants, flavonoids, vitamins and essential oils. Biotic and abiotic factors are involved in affecting the growth and yield of parsley, and they also affect the physiological functions of parsley. In this chapter, we have collected data from research articles to understand the mechanisms of stress tolerance from the physiological level to the genetic level. Our findings confirm the involvement of flavonoids in stress tolerance and defending the plants from the negative impacts of stress. However, further research is needed to more fully understand the mechanisms of countering stress.

REFERENCES

Agastian, P, SJ Kingsley, and M Vivekanandan. 2000. Effect of salinity on photosynthesis and biochemical characteristics in mulberry genotypes. *Photosynthetica* 38 (2):287–290.

Amein, T, CHB Olsson, M Wikström, R Findus, D Ab, and SAI Wright. 2006. First report in Sweden of downy mildew on parsley caused by Plasmopara petroselini. *Plant disease* 90 (1):111–111.

Andersen, Claus Lindbjerg, Jens Ledet Jensen, and Torben Falck Ørntoft. 2004. Normalization of real-time quantitative reverse transcription-PCR data: a model-based variance estimation approach to identify genes suited for normalization, applied to bladder and colon cancer data sets. *Cancer research* 64 (15):5245–5250.

Anderson, Jonathan P, Ellet Badruzsaufari, Peer M Schenk, et al. 2004. Antagonistic interaction between abscisic acid and jasmonate-ethylene signaling pathways modulates defense gene expression and disease resistance in Arabidopsis. *The Plant Cell* 16 (12):3460–3479.

Asselbergh, Bob, David De Vleesschauwer, and Monica Höfte. 2008. Global switches and fine-tuning—ABA modulates plant pathogen defense. *Molecular Plant-Microbe Interactions* 21 (6):709–719.

Aziz, Eman E, Reham M Sabry, and Salah S Ahmed. 2013. Plant growth and essential oil production of sage (Salvia officinalis L.) and curly-leafed parsley (Petroselinum crispum ssp. crispum L.) cultivated under salt stress conditions. *World Applied Sciences Journal* 28 (6):785–796.

Boyer, John S. 1982. Plant productivity and environment. *Science* 218 (4571):443–448.

Brandenburger, Wolfgang, and Gregor Hagedorn. 2006. *Zur Verbreitung von Erysiphales (Echten Mehltaupilzen) in Deutschland*: Biologische Bundesanst. für Land-und Forstwirtschaft.

———. 2006. *Zur Verbreitung von Peronosporales (inkl. Albugo, ohne Phytophthora) in Deutschland*: Biologische Bundesanstalt für Land-und Forstwirtschaft.

Bray, Elizabeth A. 2000. Response to abiotic stress. *Biochemistry and molecular biology of plants*:1158–1203.

Bull, Carolee T, Christopher R Clarke, Rongman Cai, Boris A Vinatzer, Teresa M Jardini, and Steven T Koike. 2011. Multilocus sequence typing of Pseudomonas syringae sensu lato confirms previously described genomospecies and permits rapid identification of P. syringae pv. coriandricola and P. syringae pv. apii causing bacterial leaf spot on parsley. *Phytopathology* 101 (7):847–858.

Burton, Nicholas O, Alexandra Willis, Kinsey Fisher, et al. 2021. Intergenerational adaptations to stress are evolutionarily conserved, stress-specific, and have deleterious trade-offs. *Elife* 10:e73425.

Bustin, SA. 2002. INVITED REVIEW Quantification of mRNA using real-time reverse transcription PCR (RT-PCR): trends and problems. *Journal of molecular endocrinology* 29:23–39.

Collins, Matthew, Reto Knutti, Julie Arblaster, et al. 2013. Long-term climate change: projections, commitments and irreversibility. In *Climate Change 2013-The Physical Science Basis: Contribution of Working Group I to the Fifth Assessment Report of the Intergovernmental Panel on Climate Change*: Cambridge University Press.

Delmas, Bernard, Houssam Attoui, Souvik Ghosh, Yashpal S Malik, Egbert Mundt, and Vikram N Vakharia. 2019. ICTV virus taxonomy profile: Birnaviridae. *Journal of General Virology* 100 (1):5–6.

Desire, Milostin, and Hakan Arslan. 2021. The effect of salicylic acid on photosynthetic characteristics, growth attributes, and some antioxidant enzymes on parsley (petroselinum crispum L.) under salinity stress. *Gesunde Pflanzen* 73 (4):435–444.

Facts, UNESCO-WWAP. 2012. Figures from the United Nations World Water Development Report 4 (WWDR4). UNESCO: Paris, France.

Gärber, U, E Nega, and M Jahn. 2007. Alternaria radicina an Petersilie (Petroselinum crispum L.)—Schadwirkung und alternative Bekämpfungsmöglichkeiten. *Z Arznei-Gewürzpfla* 12:74–79.

Gull, Audil, Ajaz Ahmad Lone, and Noor Ul Islam Wani. 2019. Biotic and abiotic stresses in plants. *Abiotic and biotic stress in plants*:1–19.

Gutierrez, Laurent, Mélanie Mauriat, Stéphanie Guénin, et al. 2008. The lack of a systematic validation of reference genes: a serious pitfall undervalued in reverse transcription-polymerase chain reaction (RT-PCR) analysis in plants. *Plant biotechnology journal* 6 (6):609–618.

Hatfield, Jerry L, and John H Prueger. 2015. Temperature extremes: Effect on plant growth and development. *Weather and climate extremes* 10:4–10.

Hause, Ben M, Mariette Ducatez, Emily A Collin, et al. 2013. Isolation of a novel swine influenza virus from Oklahoma in 2011 which is distantly related to human influenza C viruses. *PLoS pathogens* 9 (2):e1003176.

Herms, Daniel A, and William J Mattson. 1992. The dilemma of plants: to grow or defend. *The quarterly review of biology* 67 (3):283–335.

Hoppe, B, F Marthe, M Böhme, A Kienast, E Schiele, and E Teuscher. 2013. Petersilie (Petroselinum crispum [Mill.] Nyman ex AW Hill). *Handbuch des Arznei-und Gewürzpflanzenbaus* 5:276–309.

Jaafar, Hawa ZE, Mohd Hafiz Ibrahim, and Nur Farhana Mohamad Fakri. 2012. Impact of soil field water capacity on secondary metabolites, phenylalanine ammonia-lyase (PAL), maliondialdehyde (MDA) and photosynthetic responses of Malaysian Kacip Fatimah (Labisia pumila Benth). *Molecules* 17 (6):7305–7322.

Kang, SK, H Motosugi, K Yonemori, and A Sugiura. 1998. Supercooling characteristics of some deciduous fruit trees as related to water movement within the bud. *The Journal of Horticultural Science and Biotechnology* 73 (2):165–172.

Karlas, Alexander, Nikolaus Machuy, Yujin Shin, et al. 2010. Genome-wide RNAi screen identifies human host factors crucial for influenza virus replication. *Nature* 463 (7282):818–822.

Khan, Amir Abdullah, Tongtong Wang, Tayyaba Hussain, et al. 2021. Halotolerant-Koccuria rhizophila (14asp)-Induced Amendment of Salt Stress in Pea Plants by Limiting Na+ Uptake and Elevating Production of Antioxidants. *Agronomy* 11 (10):1907.

Kmiecik, W, and Z Lisiewska. 1999. Comparison of leaf yields and chemical composition of the Hamburg and leafy types of parsley. I. Leaf yields and their structure. *Folia Horticulturae* 11 (1).

Kotagiri, Divya, and Viswanatha Chaitanya Kolluru. 2017. Effect of salinity stress on the morphology and physiology of five different Coleus species. *Biomedical and Pharmacology Journal* 10 (4):1639–1649.

Kreide, Stefanie, Markus Oppermann, and Stephan Weise. 2019. Advancement of taxonomic searches in the European search catalogue for plant genetic resources. *Plant Genetic Resources* 17 (6):559–561.

Li, Meng-Yao, Xiong Song, Feng Wang, and Ai-Sheng Xiong. 2016. Suitable reference genes for accurate gene expression analysis in parsley (Petroselinum crispum) for abiotic stresses and hormone stimuli. *Frontiers in plant science* 7:1481.

Li, Meng-Yao, Hua-Wei Tan, Feng Wang, et al. 2014. De novo transcriptome sequence assembly and identification of AP2/ERF transcription factor related to abiotic stress in parsley (Petroselinum crispum). *PLoS One* 9 (9):e108977.

Lopes, Maria Cristina, and Victor C Martins. 2008. Fungal plant pathogens in Portugal: Alternaria dauci. *Revista Iberoamericana de Micología* 25 (4):254.

Marthe, F, T Bruchmüller, A Börner, and U Lohwasser. 2013. Variability in parsley (Petroselinum crispum [Mill.] Nyman) for reaction to Septoria petroselini Desm., Plasmopara petroselini Săvul. et O. Săvul. and Erysiphe heraclei DC. ex Saint-Aman causing Septoria blight, downy mildew and powdery mildew. *Genetic resources and crop evolution* 60 (3):1007–1020.

Marthe, F, and P Scholze. 2006. Alternaria-Blattflecken (Alternaria radicina Meier, Drechsler Eddy)—eine bedeutende Erkrankung der Petersilie (Petroselinum crispum [Mill.] Nym.). *Z Arznei-Gewürzpfla* 11:145–149.

Marthe, F, P Scholze, R Kramer, E Proll, K Hammer, and G Wricke. 2003. Evaluation of parsley for resistance to the pathogens Alternaria radicina, Erysiphe heraclei, Fusarium oxysporum, and celery mosaic virus (CeMV). *Plant breeding* 122 (3):248–255.

Marthe, Frank. 2020. Petroselinum crispum (Mill.) Nyman (Parsley). In *Medicinal, Aromatic and Stimulant Plants*: Springer.

Meena, YK, DS Khurana, Nirmaljit Kaur, and Kulbir Singh. 2018. Towards enhanced low temperature stress tolerance in tomato: an approach. *Journal of Environmental Biology* 39 (4):529–535.

Meyer, U, H Blum, U Gärber, M Hommes, R Pude, and J Gabler. 2010. Petersilie (Petroselinum crispum (Mill.) Nyman & AW Hill). *Praxisleitfaden Krankheiten und Schädlinge im Arznei-und Gewürzpflanzenanbau. Deutsche Phytomedizinische Gesellschaft (DPG), Braunschweig*:104–126.

Miller, GAD, Nobuhiro Suzuki, SULTAN Ciftci-Yilmaz, and RON Mittler. 2010. Reactive oxygen species homeostasis and signalling during drought and salinity stresses. *Plant, cell & environment* 33 (4):453–467.

Mohammadi, Hamid, Saeid Hazrati, and Mansour Ghorbanpour. 2020. Tolerance mechanisms of medicinal plants to abiotic stresses. In *Plant Life Under Changing Environment*: Elsevier.

Najla, Safaa, Rabab Sanoubar, and Ramzi Murshed. 2012. Morphological and biochemical changes in two parsley varieties upon water stress. *Physiology and Molecular Biology of Plants* 18 (2):133–139.

Navarro, Josefa M, Pilar Flores, Consuelo Garrido, and Vicente Martinez. 2006. Changes in the contents of antioxidant compounds in pepper fruits at different ripening stages, as affected by salinity. *Food chemistry* 96 (1):66–73.

Nawrocki, Jacek. 2005. Susceptibility of different parsley cultivars to infestation by pathogenic fungi.

Nowicki, Bogdan. 1997. Przyczyny zgorzeli siewek pietruszki korzeniowej [Etiology of root parsley damping-off]. *Acta Agrobotanica* 50 (1–2):35–40.

Oprica, Lacramioara, and Caunic, Mariana. 2013. Variation of flavonoids and total polyphenols contents in two persley (*Petroselinium crispum*) varieties under saline conditions. *Lucrări Ştiinţifice, Universitatea de Ştiinţe Agricole Şi Medicină Veterinară" Ion Ionescu de la Brad" Iaşi, Seria Horticultură* 56 (1):55–59.

Parida, Asish Kumar, Anath Bandhu Das, Yukika Sanada, and Prasanna Mohanty. 2004. Effects of salinity on biochemical components of the mangrove, Aegiceras corniculatum. *Aquatic botany* 80 (2):77–87.

Petropoulos, Spyridon Alexandros, Dimitra Daferera, MG Polissiou, and HC Passam. 2008. The effect of water deficit stress on the growth, yield and composition of essential oils of parsley. *Scientia Horticulturae* 115 (4):393–397.

Pfaffl, Michael W, Ales Tichopad, Christian Prgomet, and Tanja P Neuvians. 2004. Determination of stable housekeeping genes, differentially regulated target genes and sample integrity: BestKeeper–Excel-based tool using pair-wise correlations. *Biotechnology letters* 26 (6):509–515.

Plescher, Andreas, Birgit Grohs, and Lydia Pforte. 2014. Cultivated area and biodiversity of medicinal, spice, dietetic and cosmetic plants in Germany in 2011. AGRIMEDIA GMBH LUCHOWER STR 13A, CLENZE, 29459, GERMANY.

Reintke, J, G Hiddink, and D Sanchez Mendez. 2016. Real-time PCR detection and discrimination of the parsley pathogens Pseudomonas syringae pv. apii and Pseudomonas syringae pv. coriandricola. Paper read at Phytopathology.

Rizhsky, Ludmila, Hongjian Liang, and Ron Mittler. 2002. The combined effect of drought stress and heat shock on gene expression in tobacco. *Plant physiology* 130 (3):1143–1151.

Rizhsky, Ludmila, Hongjian Liang, Joel Shuman, Vladimir Shulaev, Sholpan Davletova, and Ron Mittler. 2004. When defense pathways collide. The response of Arabidopsis to a combination of drought and heat stress. *Plant physiology* 134 (4):1683–1696.

Rockström, Johan, and Malin Falkenmark. 2000. Semiarid crop production from a hydrological perspective: gap between potential and actual yields. *Critical reviews in plant sciences* 19 (4):319–346.

Rosa, DD, CT Ohto, MA Basseto, NL De Souza, and Edson Luiz Furtado. 2008. Brazil, a new location for powdery mildew on parsley and fenchel plants. *Plant Pathology* 57 (2):373–373.

Sabry, Reham M, MAM Kandil, and SS Ahmed. 2016. Growth and quality of sage (Salvia officinalis), parsley (Petroselinum crispum) and nasturtium (Tropaeolum majus) as affected by water deficit. *Middle East J. Agric. Res* 5:286–294.

Seifi, Hamed, Katrien Curvers, Aziz Aziz, and Monica Höfte. 2012. Resistance to Botrytis cinerea in sitiens, an abscisic acid-deficient tomato mutant, involves over-activation of the GABA-shunt to resist pathogen-induced senescence. Paper read at 64th International symposium on Crop Protection.

Seleiman, Mahmoud F, Nasser Al-Suhaibani, Nawab Ali, et al. 2021. Drought stress impacts on plants and different approaches to alleviate its adverse effects. *Plants* 10 (2):259.

Shaath, NA, P Griffin, S Dedeian, and L Paloympis. 1988. chemical composition of Egyptian parsley seed, absolute and herb oil. *Developments in food science*.

Shabala, Sergey, Jayakumar Bose, and Rainer Hedrich. 2014. Salt bladders: do they matter? *Trends in plant science* 19 (11):687–691.

Shao, Hong-Bo, Li-Ye Chu, Cheruth Abdul Jaleel, and Chang-Xing Zhao. 2008. Water-deficit stress-induced anatomical changes in higher plants. *Comptes rendus biologies* 331 (3):215–225.

Suzuki, Toshihide, Paul J Higgins, and Dana R Crawford. 2000. Control selection for RNA quantitation. *Biotechniques* 29 (2):332–337.

Tahvonen, R. 1978. Seed borne fungi on parsley and carrot [Alternaria dauci, Stemphylium radicinum, Septoria petroselini]. *Maataloustieteellinen aikakauskirja*.

Tahvonen, Risto. 1978. Persiljan ja porkkanan siemenlevintäiset sienet Suomessa. *Agricultural and Food Science* 50 (2):91–102.

Tian, Chang, Qian Jiang, Feng Wang, Guang-Long Wang, Zhi-Sheng Xu, and Ai-Sheng Xiong. 2015. Selection of suitable reference genes for qPCR normalization under abiotic stresses and hormone stimuli in carrot leaves. *PLoS one* 10 (2):e0117569.

Tsuchida, CT, SJ Mauzey, R Hatlen, TD Miles, and ST Koike. 2018. First report of Pythium root rot caused by Pythium mastophorum on parsley in the United States. *Plant Disease* 102 (8):1671.

Turck, Franziska, Aifen Zhou, and Imre E Somssich. 2004. Stimulus-dependent, promoter-specific binding of transcription factor WRKY1 to its native promoter and the defense-related gene PcPR1–1 in parsley. *The Plant Cell* 16 (10):2573–2585.

Ummenhofer, Caroline C, and Gerald A Meehl. 2017. Extreme weather and climate events with ecological relevance: a review. *Philosophical Transactions of the Royal Society B: Biological Sciences* 372 (1723):20160135.

Vandesompele, J, K De Preter, F Pattyn, B Poppe, and Roy Van. N., and Speleman, F.(2002). Accurate normalization of real-time quantitative RT-PCR data by geometric averaging of multiple internal control genes. *Genome Biol* 3.

Verlet, N. 1993. Essential oils: supply, demand and price determination. Paper read at International Symposium on Medicinal and Aromatic Plants 344.

Vranová, Eva, Frank Van Breusegem, and J Dat, James. 2002. species in plant signal. *Plant signal transduction* 38:45.

Wang, Wangxia, Basia Vinocur, and Arie Altman. 2003. Plant responses to drought, salinity and extreme temperatures: towards genetic engineering for stress tolerance. *Planta* 218 (1):1–14.

Westcott, Paul. 2010. *USDA agricultural projections to 2019*: DIANE Publishing.

Wong, Marisa L, and Juan F Medrano. 2005. One-step versus two-step real-time PCR. *Biotechniques* 39:75–85.

Xie, Fuliang, Peng Xiao, Dongliang Chen, Lei Xu, and Baohong Zhang. 2012. miRDeepFinder: a miRNA analysis tool for deep sequencing of plant small RNAs. *Plant molecular biology* 80 (1):75–84.

Xu, Zhao-Shi, Ming Chen, Lian-Cheng Li, and You-Zhi Ma. 2011. Functions and application of the AP2/ERF transcription factor family in crop improvement F. *Journal of integrative plant biology* 53 (7):570–s585.

16 *Pheonix dectylifera* and Stressful Conditions

Hamed Hassanzadeh Khankahdani[1]* and
Seyedeh-Somayyeh Shafiei-Masouleh[2]
[1]Horticulture Crops Research Department, Hormozgan Agricultural and Natural Resources Research and Education Center, AREEO, Bandar Abbas, Iran
[2]Department of Genetics and Breeding, Ornamental Plants Research Center (OPRC), Horticultural Sciences Research Institute (HSRI), Agricultural Research, Education and Extension Organization (AREEO), Mahallat, Iran
Corresponding author: hamed51h@gmail.com

CONTENTS

16.1 INTRODUCTION

Phoenix dactylifera L. is an important perennial plant in arid and semi-arid areas, which has economic importance in the Middle East, north and south Africa, and some tropical and sub-tropical regions (Meddich et al. 2018).

Date fruits are good sources of essential nutrients for humans. Some of these important nutrients are carbohydrates, proteins, minerals (calcium, potassium, magnesium, iron, and selenium), and fiber (Meddich et al. 2018). Date fruits are rich in secondary metabolites as well as nutrient compounds, which have medicinal importance for humans. One of the most important compounds in the fruits is phenolic compounds, with the main phenolic compounds in date fruits including kaempferol, malonyl, quercetin, isorhamnetin, 3-methyl-isorhamnetin, luteolin, apigenin, chrysoeriol, and sulfate derivatives. Many medicinal properties such as anti-cancer, diabetic medication, anti-bacterial, and effects on other chronic diseases for the secondary metabolites of date fruits have been reported (for more details see Maqsood et al. 2020; Hussain et al. 2020). Furthermore, a range of traditional medicine use is listed for the date fruit,

DOI: 10.1201/9781003242963-16

including for wound healing, fever, stomach and intestinal disorders, edema, and bronchitis (Taleb et al. 2016; Al-Shwyeh 2019). Secondary metabolites of date fruit are some biochemicals that do not directly affect plant growth processes (i.e. photosynthesis, elements uptake, and all primary metabolisms). The compounds are synthesized in plants against stressful conditions and have protective and antioxidant activities for plants (Hadrami et al. 2011).

Date palm is known as an extremophile (Al-Khateeb et al. 2021), and has moderate tolerance to abiotic stress, however, in extreme stresses, some limitations and changes have been observed in plant growth and development, and metabolism (Meddich et al. 2018). The extreme conditions, including abiotic stress (especially drought and salinity, and heat stresses) and biotic stress (especially Bayoud disease) adversely affect plant growth and development, fruit quality and yield, and even more secondary metabolites and medical features of plants and fruits (Al-Khateeb et al. 2021; Maqsood et al. 2020; Meddich et al. 2018).

16.2 BIOTIC STRESS

Fusarium oxysporum f. sp. Albedinis (Foa) is a soil-borne fungus that infects date palm and causes Bayoud disease. This disease has a rapid rate of spread, and it is an increasing threat to date palm growth and development worldwide (Meddich et al. 2018).

Bouissil and coworkers (2022) assayed a brown algae *Bifurcaria bifurcata*-extracted natural poly-saccharide effects on the Bayoud disease. This extract was alginate and was used to activate the plant defense system because alginate plays a role in phenylpropanoid metabolism through a key enzyme of the metabolism, i.e. phenylalanine ammonia-lyase (PAL). They assayed the alginate as a stimulator in comparison with water (negative control) and laminarin (positive stimulator). They observed the results as alginate-stimulated high activity of PAL in date palm roots (5 times more than water, and 2.5 times more than laminarin). They found overexpression of genes encoding early oxidative enzymes [superoxide dismutase (SOD) and lipoxygenase (LOX)]. About genes encoding β-(1,3)-glucanases and chitinases in response to biotic stress, they found the selected peroxidase (POD) and PR (pathogenesis-related) protein genes do not have a role, and it may be caused by other POD and PR protein genes. They showed that mortality of date palms was decreased by 80% with alginate as compared to control plants (water and laminarin).

Mansoori and Kord (2006) reported the yellowing and death of date palm leaves (fronds) in Iran in 2003. They isolated a fungus, *Fusarium solani*, from the crown (1.5 meters above soil level). They found that fungus infection both by plant inoculation and soil infection could affect the plants. They announced this report as the first and most serious disease of date palm, called yellow death.

Another disease of the date palm is called brittle leaf disease, in which its causal agent is unknown, with no fungi or other pathogens being known to cause it. Mefteh et al. (2017), while exploring and comparing endophytic fungi colonizing internal tissues in healthy and brittle leaf disease of date palm in terms of their composition, antimicrobial activity, and metabolic diversity, isolated 52 endophytic fungi. They found that some of these fungi were shared between two samples, however, some of them were specific to each sample (healthy or infected). They showed that all isolates could have at least two enzymes including amylase, chitinase, cellulase, pectinase, lactase, protease, and lipase. They reported the antibacterial (Gram-positive and -negative bacteria) and antifungal activities in both types of samples. They identified the secondary metabolite of these isolates arsenal for date palm. Among isolates, *Geotrichum candidum* was described as the best endophyte against *Rhizoctonia solani*.

Black scorch is another disease of date palm in which its agent is the soil-borne pathogenic fungus *Thielaviopsis punctulate*. Based on a report, the fungicide Cidely® Top could inhibit its mycelial growth, even *in vitro* (Saeed et al. 2017).

16.3 ABIOTIC STRESS

Besides biotic stress, abiotic stresses, such as water stress, high and low temperatures, and salinity, may limit plant growth and development, and also may cause the spread of photogenic agents, including fungi, bacteria, viruses, insects, and weeds. Abiotic stresses affect plant physiology, nutrient uptake and metabolism, and then plant growth and development, and finally the economic yield of different crops. These detrimental agents of the environment, such as salinity, affect photosynthesis and respiration due to damaging cell organelles. These damages have resulted in nutrient imbalance or secondary stresses such as osmotic or oxidative stresses (Meddich et al. 2018).

Date palms grow in arid and semi-arid regions that are affected by two important abiotic stresses, drought and salinity, stress, with their survival in a wide range of extreme droughts, relatively high levels of salinity, and high temperatures. Both drought and salinity induce secondary stresses in plants, including osmotic and oxidative, and plant metabolic responses against the two main agents are approximately similar (Djibril et al. 2005; Yaish 2015). Therefore, in this chapter, more focus is given to these two stresses. Drought stress leads to osmotic stress, a decrease in CO_2 assimilation and overproduction of free oxygen species (ROS), which are destructive, while salinity leads to toxicity in tissues and causes an accumulation of Na^+ in cells (Yaish 2015).

16.3.1 SALINITY

This species can adapt to gradual salinity stress, however, in prolonged stress, plant growth and development are affected (Al-Khateeb et al. 2021). Different cultivars of date palm may have varied capacities to cope with salinity, with some being more susceptible than others (Al Kharusi et al. 2017, 2019; Yaish and Kumar 2015). Based on a study in 2019, the salinity tolerance characteristics may relate to antioxidation processes and the uptake rates of Na^+ and K^+. In their study, Al Kharusi and co-workers found that a salinity susceptible cultivar, "Zabad," had greater Na^+ than K^+ uptakes and less photosynthetic pigmentation and growth, however, the tolerant cultivar, "Umsila," accumulated less Na^+ and more K^+, and was able to maintain a normal ROS concentration. They found that two methods of antioxidation, non-enzymatic (production of glutathione, phenolic compounds, flavonoids, and proline) and enzymatic [more activities of catalase (CAT), SOD, and ascorbate peroxidase (APX)], are the key mechanisms of tolerant cultivars against salinity. Al Kharusi et al. (2017) found that susceptible cultivars of date palm can be identified based on reducing photosynthesis, the shoot K^+/Na^+ ratio, electrolyte leakage (EL), and relative water content. On the other hand, tolerant cultivars can be determined through Na^+ exclusion from shoots, more photosynthesis, and membrane stability. They introduced cultivars Manoma and Umsila as salinity-tolerant cultivars.

Dynamically, salinity has long-term effects on the plants by increasing relative negative responses to water consumption. Based on these findings, plant growth is affected by either accumulated effects or increasing sensitivity to salinity under long-term stress. Remarkably, plant sensitivity to salinity is steady, and trees in the maturity and productivity period have greater sensitivity than at younger ages (Tripler et al. 2011). These researchers found that trees subjected to low salinity, compared to higher salinity (EC = 1.8 dS m^{-1} vs EC = 4 dS m^{-1}), have twice the amount of growth over 5 years. However, the growth and productivity of date palm under long-term and very high-salinity irrigations (8 and 12 dS m^{-1}) were commercially impossible.

Detrimental effects of salinity on date palms are approximately similar to other crops. It affects plant growth and development, and the survival and yield of the plants are influenced by salt stress-induced oxidative stress, which is due to the over-production of ROS. These occurrences are defense mechanisms against unfavorable environmental changes (Naser et al. 2016).

Furthermore, high salinity may lead to drought stress symptoms, metabolic toxicity and nutritional deficiencies, as well as osmotic and oxidative stresses, because salinity crucially impacts on the biochemical processes of the plants (El Rabey et al. 2015).

16.3.2 WATER STRESS AND WATERLOGGING STRESS

Drought stress can be considered the first abiotic stress that decreases crop productivity worldwide through various processes, including disrupting physiology and retarding plant growth (Harkousse et al. 2021). Al-Khayri and Al-Bahrany (2004) examined the tolerances of two cultivars of date palm (Barhee and Hillali) to drought stress on calli *in vitro* after adding polyethylene glycol (PEG 8000) to liquid Murashige and Skoog medium. They found Barhee more tolerant than Hillali and observed a greater increase in free proline. However, both showed a similar trend against drought, i.e. a reduction in calli growth (fresh weight and relative growth rate).

Date palm cultivation is also performed on coastlines, where plants are exposed to flooding and may be affected by the salt stress of seawater. Du et al. (2021) showed that date palms are tolerant to seawater salinity to some extent and highly tolerant to flooding. They concluded this from the results of reduced CO_2 assimilation, transpiration, and stomatal conductance. They stated that the reduced transpiration upon seawater exposure may have contributed to controlling the movement of toxic ions to leaves, and therefore it may affect plant tolerance to salinity. They found more accumulations of nitrogen compounds in roots, Na and Cl contents in leaves and roots, and less accumulation of sugars and sugar alcohols. The salinity of seawater is due to $MgSO_4$, and they did not find either accumulation of sulfate in roots or stomatal closure, also, higher Na and Cl contents were observed in date palm leaves and roots.

16.3.3 HEAT STRESS

Temperatures more than a threshold level of a plant for the long term are known as heat stress, and may cause serious damage to plant growth and development. This damage depends on the intensity, duration, and rate of increasing temperature. An increase in air temperature increases the soil temperature, and these conditions reduce soil water and lead to secondary stresses, including drought stress (Shafiei Masouleh and Sassine 2020a).

Safronov et al. (2017) in a work on date palm assayed adaptation mechanisms to mild heat, drought and the combined stresses based on transcriptomic and metabolomic data. According to their results, transcriptomic profiling showed a heat response in both heat and combined stress. However, in metabolomic profiling, they found that the plant responses were similar to drought. They observed an increase in soluble carbohydrates (such as fructose and glucose derivatives) in both conditions, and concluded that these three types of stresses affect carbohydrate metabolism and cell wall biogenesis. They stated that transcriptional activation of genes related to reactive oxygen species occurs in all studied stresses, and is comprised of all genes of cytosol, chloroplast, and peroxisome. An interesting finding they found was differential gene expression based on circadian and diurnal rhythm motifs, and showing new strategies of stress avoidance.

Heat stress-induced overproduction of ROS in plants may be the first occurrence that leads to destroying proteins, plant synthesis and enzyme activities, as well as membrane lipid peroxidation and photosynthesis dysfunction. Plants confront heat stress through changing their bodies and the production of molecular signals. These actions play a role in organizing proteins, maintaining cellular structures and turgor, and achieving cellular redox balance. Furthermore, heat-shock protein (HSP) genes will be activated and through cellular proteins and conformational protein functions help plants against heat stress. Another plant strategy under heat stress is the production of abscisic acid (ABA) signaling and accumulation of ABA, which then causes thermotolerance through integration with other hormones and regulatory systems of ROS (Khan et al. 2020b).

16.4 STRATEGIES TO ALLEVIATE STRESS

16.4.1 Plant Strategies

Plants avoid stresses with a series of specific and non-specific strategies, including changing their body structures, synthesis of antioxidants (enzymatic and non-enzymatic), synthesis of hormones, and symbiosis (with bacteria and fungi). It has been reported in numerous studies about date palm plant strategies against biotic and abiotic stress that plants are exposed to based on climatic conditions. Some studies have achieved a general understanding of plant responses as follows.

Youssef and Awad (2008), in their study on date palm, found that salinity had no effects on the carboxylation efficiency of the rubisco enzyme, or the rate of electrons supplied by the electron transport system for regeneration of ribulose 1,5-bisphosphate (RuBP) (Youssef and Awad 2008). Sperling et al. (2014) showed that in date palm, saline water causes a greater accumulation of Na^+ in roots compared to leaves and special tissues of leaves exclude Na^+, and then indirectly affects stomatal conductance as a primary effect on photosynthesis and CO_2 assimilation, but this damage is not permanent. Suhim et al. (2017) investigated the genetic stability of date palms against salinity with NaCl (100, 200, 300, and 400 µM), and made interesting findings, which were the lowest genetic similarities between 400 µM NaCl and the control and 100 µM treatment. According to the analysis of ISSR markers and a dendrogram, they observed three clusters, including 400, the control, 100, 200, and 300 µM. Furthermore, the effects of salinity included an increase in the hydrogen peroxide (H_2O_2) level, malondialdehyde (MDA) concentration and peroxidase activity, and with increasing NaCl level, the effects of salinity were greater on the biochemical responses of plants. Yaish et al. (2016) stated that date palm growing under salinity conditions may relate to microbial communities in soil. They found the communities of endophytic bacteria and fungi through the pyro-sequencing method, and observed that microbial differential abundance in roots under saline stress was slightly altered for microbe diversity, but not for overall microbial community structures. They observed a buffering effect by the host (plant) on the internal environments influencing colonization of microbe communities, which influences date palm tolerance to saline stress. Al-Khateeb et al. (2021) used $CaCl_2$ (0, 5, and 10 mM) for *in vitro* culture of date palm cv. Khalas under NaCl stress (0, 100, and 200 mM). They found that calcium (Ca^{2+}) could alleviate saline stress effects by improving K^+/Na^+ ratios, reducing transcript expression of *NHX1* (sodium/hydrogen exchanger 1) and *HA1* (coding the plasma membrane ATPase) genes, and finally the leaf numbers were increased. The plant strategy for adaptation to salinity could be categorized into four categories, with each of them including some processes and mechanisms that are specific or non-specific (Figure 16.1).

Heat and drought stress are other stresses, followed by salinity, that may affect date palm in special conditions with plants having consistent responses to them through a series of processes. Arab et al. (2016) evaluated the heat and drought tolerance of date palm based on some assumptions, including (1) both heat and drought stresses promoted antioxidant systems, (2) the redox state of date palm will not be affected because of the high tolerance of plants and (3) only heat affects the fatty acid composition and biosynthesis of isoprene, and then the stabilization of membrane integrity. Their results confirmed their assumptions that (1) unaffected redox system indicates unchanged H_2O_2 levels, (2) high isoprene emission shows its possible role as an antioxidant and for stabilization of thylakoid membranes and (3) the response of fatty acids only to drought.

As is well known, amino acid proline in plants acts as a chelator of metals, an active osmolyte, an antioxidative compound and a signaling molecule. Plants usually produce the amino acid against certain abiotic stress conditions, and often breeders use this molecule as a marker-assisted breeding tool for programs for improving drought and salinity tolerance (Yaish 2015). Yaish (2015) investigated whether the proline content increases under drought and high-salinity stresses or other unspecified abiotic stresses, including extreme temperatures, salinity and salinity shock, drought and abscisic acid. They indicated that proline cannot be used as a molecular marker for tolerance improvement of

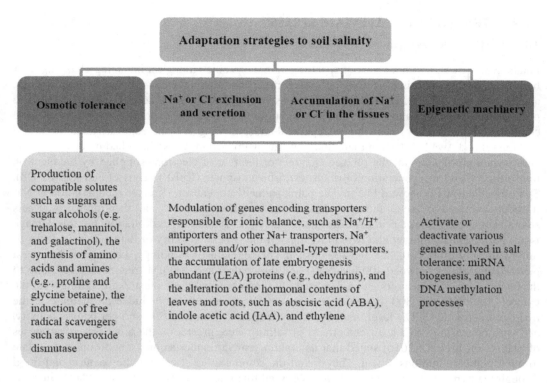

FIGURE 16.1   Major strategies and mechanisms of soil salinity adaptation in plants (Adapted from Yaish and Kumar 2015).

the date palm to drought or salinity, and they indicated that this is consistent with the theory about molecular responses of abiotic stresses often being non-specific.

16.4.2   Agricultural Management

Cultivation regions of date palm include arid and semi-arid regions, and although date palm is a relatively salt- and drought-tolerant plant, insufficient rainfall and improper irrigation practices (for instance, irrigation with saline water), may affect date palm growth and yield, and fruit quality, especially in severe environmental stresses (Patankar et al. 2016). As is well known, rainfall regimes of warmer climates can change on a regional scale, with severe droughts being expected. These changes may have harder impacts on the agro-physiological yield of crops. Therefore, agricultural management is necessary to ensure the best economic performance for products. Understanding the physiological responses of plants to agricultural management may help to provide better decision-making. Today, biofertilizers are of interest to use in combating environmental stresses and therefore enhancing crop quality and yield. Some biofertilizers, which were assayed, include arbuscular mycorrhizal fungi (AMF), plant growth-promoting rhizobacteria (PGPR), and composts. Anli et al. (2020) showed that single or multiple applications of these fertilizers for date palm could provide tolerance to water deficit. They stated that physiological responses to the biofertilizers, including the enhancement of phosphorus in leaves and also in soils, the effects on photosynthetic pigments and efficacy, water status (stomatal conductance), osmolytes (such as sugar content) and defense systems (antioxidant systems) could enhance plant tolerance to the drought. An interesting result that can be seen in their study was a greater beneficial effect of native arbuscular mycorrhizal fungi compared to the exotic ones.

Ait-El-Mokhtar et al. (2020), based on an experiment on date palm, showed that salinity can decrease growth parameters, leaf water potential, AMF colonization, and nitrogen (N), potassium (K^+), phosphorus (P), calcium (Ca^{2+}), and chlorophyll contents through an increase in sodium (Na^+) and chlorine (Cl^-) contents, lipid peroxidation, and H_2O_2 content. They respond to these signals through an increase in proline, soluble sugar. They proposed the use of AMF and/or compost for agricultural management. Based on these operations (separately or combined), it is reported that plant tolerance to salt increased through an increase in mineral uptake, proline content, and water status. Furthermore, they found more antioxidant activities.

One of the agricultural management practices to cope with environmental stress is the exogenous application of stimulants. In a work carried out by Khan et al. (2020a), they used silicon (Si) and gibberellic acid (GA_3) as exogenous applications under heat stress. These operations caused a decrease in heat-induced oxidative stress, superoxide anions and lipid peroxidation in combined treatment compared to the single applications, an increase in catalase, polyphenol oxidase, peroxidase and ascorbate peroxidase activities, as well as upregulation of compounds-related genes expression, viz. CAT, GPX2, SOD, Cyt-Cu/Zn, and glyceraldehyde3-phosphate dehydrogenase gene (GAPDH). Furthermore, the activation of heat shock factor-related genes (especially HsfA3) was observed. They found that signaling of ABA transcript accumulation-related genes (PYL4, PYL8, and PYR1), and then production of ABA was reduced, therefore accumulation of SA antagonism was increased. Finally, heat tolerance was observed in the date palm.

Symbioses of AMF and PGPR as bio-ameliorators of the plant resistance to drought stress have been reported as desirable (Harkousse et al. 2021). Date palm roots naturally have a symbiosis with numerous statins of the fungi and bacteria and enable their colonization. For instance, endophytic bacteria were isolated from date palm seedling roots. Molecular characterization showed that the majority of these strains belonged to the genera *Enterobacter* and *Bacillus* (Yaish et al. 2015a). One of the agricultural management operations is the use of native and/or exotic microorganisms as biofertilizers to cope with numerous stresses.

16.5 MULTI-OMICS APPROACHES TO COPE WITH STRESS

The date palm has global importance in terms of nutritional and medicinal applications for humans, and its breeding to produce new cultivars with better quality and yield of products is important, as conventional breeding is time-consuming, and therefore biotechnological tools and different omics are necessary to achieve targets rapidly (Al-Khayri et al. 2018).

El Rabey et al. (2015) carried out proteomics on 3-month-old seedlings of date palm cv. Sagie that were subjected to salinity and/or drought stress. Based on DIGE (Differential in Gel Electrophoresis) analysis, they found 47 proteins in the leaves of the stressed plants and based on mass spectrometric analysis, they identified 12 proteins. They reported that from later analysis the levels of only three proteins were changed under both stresses, and others were changed only under salinity. These proteins included alpha and beta subunits of ATP synthase, some RubisCO fragments, and an unknown protein. However, significant changes in chlorophyll a-b binding protein, superoxide dismutase, chloroplast light-harvesting chlorophyll a/b-binding protein, light-harvesting complex 1 protein (Lhca1), RubisCO and some RubisCO fragments, RubisCO activase, phosphoglycerate kinase, phosphoribulokinase and transketolase were only observed under salinity.

Different tolerances of date palm cultivars to salinity and drought stresses have been reported. El Rabey and co-workers (2016) assayed severe stress effects (including 48 g L^{-1} NaCl, drought with 82.5 g L^{-1} PEG, and drought with no irrigation) based on leaf proteome analysis. They used a protein 2D electrophoresis method and by mass spectrometry found 55 protein spots. They identified the upregulation of some proteins under three stresses, such as ATP synthase CF1 alpha chains, furthermore, they found some changes in the levels of some spots only for certain stresses. Moreover, they found over-expression or under-expression of proteins under stress. Their interesting findings included the downregulation of all drought (no irrigation) tolerance genes.

As is well known, date palm is not a halophytic species and only can tolerate its particular adverse climatic conditions, and this tolerance varies in its cultivars. Therefore, identifying the genetic nature of different cultivars of date palm may help breeders to improve cultivars, especially in terms of tolerance to three stresses related to its cultivation regions: salt, drought, and heat stress. For example, based on one report, regulatory genes in different categories, including DNA/RNA, protein, membrane, and signaling functional groups play roles in tolerance to salinity. The miRNAs and their targets have been reported which could have critical roles against abiotic stress in date palms (Al-Khayri et al. 2018).

According to proteome analysis of leaves and roots, when molecular signals transfer to roots and leaves, a synergetic responsive network would occur. Although metabolic pathways are similar, these organs receive abiotic stress sooner than other organs by distinct changes. Based on these findings, there is a theory that molecular responses to the environmental stresses are non-specific, and especially, proline cannot be considered as a molecular marker for date palm breeding (El Rabey et al. 2016).

Yaish et al. (2015b) for investigation of the molecular mechanisms for salt tolerance of date palm identified the conserved and novel miRNAs in leaves and roots. They found that uncovered microRNA-mediated post-transcriptional gene regulation plays a role in salt tolerance. Based on deep sequencing, bioinformatics analysis, and semi-quantitative PCR (qPCR), they stated that more than 70% of them are upregulated and cause a tolerance to salinity in date palm. Indeed, miRNA-mediated gene expression occurs in date palm in response to salinity and leads to tolerance. Jana and Yaish (2020) found that the *PdGPX* family has an effective role in date palm against salinity and other abiotic stress. This gene group is related to glutathione peroxidase (GPX) (EC 1.11.1.9), which is an antioxidant enzyme catalyzing the reduction of hydrogen peroxide (H_2O_2 to water and oxygen) and protects cells from ROS. As is well known, tolerance to salinity is a quantitative trait, numerous genes are involved in it, and one of these is a group of putative *PdGPX* genes that were mined in date palms (Jana and Yaish 2020). Patankar et al. (2016), based on quantitative real-time PCR (qPCR), which is introduced as a promising technique for the analysis of stress-induced differential gene expression, and using geNorm, BestKeeper statistical algorithms, NormFinder, and the comparative ΔCT method, found stable reference genes for normalizing gene expression among 12 of the most commonly used reference genes and found the combination of two or three stable reference genes to be suitable for normalization of the target gene expression. They claimed that this facilitates gene expression analysis studies for identifying functional genes related to drought and salinity tolerance in date palm (see Patankar et al. 2016 for more information about the mentioned 12 genes). They concluded that stable reference genes are different for each tissue (leaves and roots). Yaish et al. (2017) showed that date palm differentially responds to salinity, they observed an increase in non-photochemical quenching, a decrease in all gas exchange parameters, and the quantum yield of PSII was not changed. Furthermore, they showed that gene expression profiles in roots and leaves under salt stress are different, and some genes have the same responses in the two tissues. For example, based on gene ontology, leaves had enrichment of transcripts of metabolic pathways, including sucrose and starch metabolism, photosynthesis, and oxidative phosphorylation, but roots were comprised of genes related to phenylpropanoid biosynthesis, membrane transport, casparian strip development, and purine, thiamine and tryptophan metabolism. Moreover, the same genes for both tissues are related to the auxin-responsive gene, GH3, a putative potassium transporter 8 and the vacuolar membrane proton pump. In addition, they reported that different cultivars of date palm respond to salinity differentially, and not all of them are known as relatively salt-tolerant plant species. Patankar et al. (2018), with overexpression of the date palm cDNA library in basic yeast cells (*Saccharomyces cerevisiae*), identified the presence of a group of genes with potential salt-tolerance functions, including aquaporins (*PIP*), serine/threonine protein kinases (*STKs*), ethylene-responsive transcription factor 1 (*ERF1*) and peroxidases (*PRX*). They reported that these genes are expressed in date palm roots and leaves against salt. Al-Harrasi et al. (2020) assayed the function of a salt-inducible vascular highway 1-interacting kinase (*PdVIK*) that is a MAP kinase kinase kinase (MAPKKK) gene from the date palm based on *in vitro* (in yeast) and *in vivo* (in a transgenic *Arabidopsis*) studies against salinity caused by LiCl. They found overexpression of *PdVIK* in yeast and transgenic *Arabidopsis* under salinity conditions, and a tolerance to salt, osmotic and oxidative stress. The date palm has 597 kinases, with *PdVIK* being one of them. The recombinant PdVIK protein with phosphotyrosine activity is reported as having tolerance against salinity in date palm.

Al-Harrasi et al. (2018), based on whole-genome bisulfite sequencing (WGBS) and mRNA sequencing, and enzyme-linked immunosorbent assay (ELISA) in salinity-treated and untreated roots of date palm, found that the level of methylation within the differentially methylated regions (DMRs) was remarkably increased in response to salinity at the mCHG and mCHH sequence contexts [DNA methylation is when one methyl group adds to the fifth carbon of a cytosine (C) and forms 5-methylcytosine (5mC) by DNA methyltransferases (DNMTs)]. DNA methylation may occur in the symmetric CG and CHG contexts and in the asymmetric CHH context ("H" represents A, C or T) (Hsu et al. 2018). Furthermore, based on the integration of high-resolution methylome and transcriptome analyses, a negative correlation between mCG methylation located within the promoters and the gene expression, and a positive correlation was noticed between mCHG/mCHH methylation ratios, and gene expression could be observed especially under stress conditions. Al-Harrasi et al. (2018) stated that DNA methylation can be considered as a mechanism in date palm resistance to salinity, or other abiotic stress. Hsu et al. (2018) expressed that methylation at a promoter region may stop gene expression because it alters the chromatin structure or blocks transcription initiation.

Methylglyoxal (MG) is a by-product of various plant metabolic reactions, especially glycolysis. This compound is a cytotoxic oxygenated short aldehyde, its basal level is low and it acts as an essential signaling molecule to regulate multiple cellular processes. However, under stressful conditions, hyperaccumulation of MG occurs, and this has detrimental effects on multiple developmental processes, including seed germination, photosynthesis, and root growth. Therefore, plants need to have a system for MG detoxification; this system comprises two enzymes, including glyoxalase-I and glyoxalase-II (Jana and Yaish 2020). Jana and Yaish (2020) investigated six putative glyoxalase-I genes (*PdGLX1*) in date palm, and they found that overexpression of the putative *PdGLX1* genes occurs under stressful conditions, and then leads to MG detoxification and removal of ROS.

Garcia-Maquilon et al. (2021) associated the date palm adaptation with drought stress and ABA receptors and molecular mechanisms. They claimed that some crops, and especially date palm, can be adapted to harsh conditions with these receptors, which impact plant–environment interactions. They revealed that Pd27 is the most commonly expressed receptor in date palm, and this is one of the PdPYL8-like receptors. They reported that there are receptor- and ABA-dependent inhibitions of PP2Cs, which trigger activation of the *pRD29B-LUC* reporter in response to ABA in date palm. Furthermore, it is stated that PdPYLs efficiently stop PP2C-mediated repression of ABA signaling. Moreover, date palm, as a Pd27-overexpressing plant, has lower ABA content and therefore has lower transpiration in drought conditions. Therefore, these receptors in date palm efficiently improve ABA signaling in response to abiotic stress, especially drought.

Ghirardo et al. (2021) stated that date palm adapts to severe environmental conditions such as salinity, drought and seasonally high temperatures by adjusting the abundance of proteins related to photosynthesis, abiotic stress, and secondary metabolites. They noticed that heat shock proteins are efficiently expressed proteins as well as the antioxidant systems for the response to heat and water depreciation. Furthermore, proteins related to the secondary metabolism were reported that are downregulated, except the *P. dactylifera* isoprene synthase (PdIspS). This gene is strongly upregulated in response to summer conditions (high temperature and drought).

16.6 EXOGENOUS STIMULANTS

Plants, when confronted with biotic and abiotic stress, may synthesize some compounds called organic solutes, osmolytes, compatible solutes and/or other similar names. These compounds are a part of antioxidant systems in plants which have both osmolyte effects (regulation of cellular water), and also antioxidative effects because they rapidly react with free radical molecules and scavenge them through auto-oxidation changes. These compounds can be categorized into three main groups, including N-containing metabolites, phenols and polyphenols, and carbohydrates. Often these compounds in plants against stresses are non-specific, although some may act specifically (Shafiei Masouleh et al. 2020b; Shafiei-Masouleh, 2021). The application of some of these compounds by inspiration and understanding of nature and their mechanisms could help with agricultural management. Generally, stimulants are natural or synthetic compounds that can be used for the treatment of seeds, plants, and soil. These compounds, by changing the vital and structural processes and mechanisms in plants, increase plant growth and their tolerance to biotic and abiotic stresses (de Vasconcelos and Chaves, 2019). In Table 16.1, some studies of stimulant applications and date palm responses under stressful and non-stressful conditions are listed.

Today, the application of endophytic microbial communities is of interest under both normal and stress conditions to enable more profitability for crops. The endophytic bacteria colonize plant tissues without any disease or damage and cause an improvement in growth and development. Their roles include the production of a range of nutrient products, the facilitation of primary and secondary nutrient uptake through atmospheric nitrogen fixation, the formation of iron siderophores (Wang et al. 1993) and the solubilization of minerals such as phosphate (PO_4^{3-}), potassium (K^+) and zinc (Zn^{2+}) in the rhizosphere. These microorganisms supply growth-promoting phytohormones such as auxin, cytokinin and gibberellin, inhibit ethylene, and also protect against a range of environmental stresses (biotic and abiotic) (Yaish et al. 2015a). AMF are symbiotic microorganisms within the roots of both wild and agricultural plants, and these microorganisms have mutual benefits. They have numerous benefits for plants, including the promotion of mineral nutrition-related plant growth, and tolerance of biotic and abiotic stress. However, the magnitude of the beneficial effects depends on the plant species and the composition of AMF strains. The date palm has a limited root system (low-density root hairs) and this causes more colonization of mycorrhiza and achieves more benefits (Meddich et al. 2018). For instance, in a study by Harkousse et al. (2021), a better response of date palm seedlings against severe water stress was observed under the combination of AMF and PGPR compared to separate applications, and control from the viewpoint of lower activities of SOD, CAT, guaiacol peroxidase (POD) and glutathione S-transferase (GST), and increasing proline content. That said, Baslam et al. (2014) showed that to have more benefits from arbuscular mycorrhizal fungi, the fungal species, and water regime applied for date palm seedlings must be considered. They found that among *Glomus* species of *Glomus intraradices*, *G. mosseae* and *Aoufous complex* (native AMF), the native *Aoufous complex* had better effects in non-stressed conditions, while *G. intraradices* enabled better plant growth despite severe water stress. The interesting point was that the positive effects of *G. intraradices* were not related to the promotion of antioxidant enzyme activity. Instead, it increased the elasticity of leaf cell walls and maintained high water content in leaves without lowering the leaf water potential under stressful conditions. Therefore, the suitable selection of mycorrhiza needs to have more profitability. Benhiba et al. (2013) stated that arbuscular mycorrhizal fungi could be of interest as bio-fertilizers for improving date palm growth and physiology, especially in low water conditions. Although date palm has hardiness and adaptability to arid and semi-arid regions, to meet the highest productivity possible it needs critical amounts of water and nutrients. It has been expressed that symbiosis of arbuscular mycorrhizal fungi with plant roots can have high benefits. These symbiotic associations can enhance the plant availability of water and nutrients, aiding plant growth and productivity. As is well known, salinity is one of the key stresses that may affect date palm growth and productivity, because it disorders Ca/Na and K/Na ratios and causes secondary stress, especially oxidative stress due to membrane lipid peroxidation.

TABLE 16.1

Plant Responses to Exogenous Stimulants under Stressful and Non-stressful Conditions in Date Palm

Exogenous stimulants	Biotic/abiotic stress	Plant responses	References
Putrescine amine (Put), bio–fertilizer (mycorrhizae)	Saline water	Increase in tolerance and adaptation to stress conditions due to: 1. An increase in photosynthetic pigments, activities of antioxidant enzymes (APX, glutathione reductase (GR) and SOD), organic solutes, and/or growth-promoting substances such as GA_3, auxins (IAA), and cytokinin 2. A decrease in lipid peroxidation and inhibitor substances such as ABA 3. A decrease in oxidative damage 4. An increase in diamine oxidase (DAO) and polyamine oxidase (PAO) activities against oxidative damage 5. An increase in H_2O_2 as a molecular signal and a decrease in polyamines against salt-induced oxidative damage	Naser et al. (2016)
Endophytic bacteria isolated from date palm (*Phoenix dactylifera* L.)	Saline stress	1. Some strains increased enzyme 1-aminocyclopropane-1-carboxylic acid (ACC) deaminase and the plant growth regulatory hormone indole-3-acetic acid (IAA) 2. Some strains could chelate ferric iron (Fe^{3+}) and solubilize potassium (K^+), phosphorus (PO_4^{3-}) and zinc (Zn^{2+}), and produce ammonia 3. The strains, including PD-R6 (*Paenibacillus xylanexedens*) and PD-P6 (*Enterobacter cloacae*) could enhance canola root elongation 4. Changes in ethylene and IAA levels, and facilitating the nutrient uptake	Yaish et al. (2015b)
Sulfur, salicylic acid, citric acid, effective microorganism, humic acid, compost enriched with actinomyces, filter mud, Uni-Sal, and Cal-Mor	Soil salinity	1. Alleviating detrimental effects of salinity on the leaf area, yield, and fruit quality 2. Cal-Mor had the best effects 3. Different cultivars responded variously	El-Khawaga (2013)
$Ca(NO_3)_2$	Salt stress	In moderate concentration of bio-stimulant: 1. An increase in chlorophyll content 2. Reduction of Na and Cl in plant tissues	Alturki (2021)
Putrescine amine, plant growth-promoting rhizobacteria and mycorrhiza	Saline water	In date palm zaghloul genotype: 1. A decrease in oxidative damage 2. Promotion of the productivity and quality of fruits 3. An increase in the photosynthetic pigments, organic solutes, growth-promoting substances (GA, IAA, and cytokinins) and activities of oxidant enzymes	

(continued)

TABLE 16.1 (Continued)
Plant Responses to Exogenous Stimulants under Stressful and Non-stressful Conditions in Date Palm

Exogenous stimulants	Biotic/abiotic stress	Plant responses	References
		4. A decrease in the levels of lipid peroxidation and inhibitor substances (ABA)	Helaly and El-Hosieny (2015)
		5. An increase in the activities of APX, GR, and SOD through the combined treatments with Put 2.5 mM and by the decrease in lipid peroxidation	
		6. An increase in DAO and PAO activities for the combined treatments with 5 mM Put and enzymes-induced hydrogen peroxide caused H_2O_2 production, as molecular signaling and inhibited the production of polyamines against salt stress-induced oxidative damage.	
		7. An increase in genotype tolerance to salinity	
Sulfur	Salinity	In Berhi and Sayer cultivars:	Abbas et al. (2015)
		1. Increasing the total chlorophyll, dry weight, RWC, carbohydrates, proline concentration, soluble protein, peroxidase enzyme activities, and endogenous IAA content	
		2. Improvement of tolerance to salinity by maintaining a high ratio of K/Na	
		3. A decrease in catalase enzyme activity and content of ABA	
Copper (CuNPs) and chitosan (CsNPs) nanoparticles	Oxidative browning	*In vitro* conditions:	Mohamed (2019)
		1. Promoting innate immunity of plantlets by modulating the levels of peroxidase, catalase and total phenols seedlings	
		2. An increase in peroxidase, catalase, and a decrease in total phenols under treatment by copper	
		3. An increase in peroxidase, catalase, and total phenols under treatment by chitosan	
		4. Reducing the oxidative browning through the decrease in phenols under copper treatment	
A natural elicitor chitosan, (Ch) 1% alone and in combination with salicylic acid (SA) 2 mM and calcium chloride (Ca) 3%; (Ch, SA, Ca, Ch+Ca, Ch+SA, Ch+SA+Ca)	–	In cultivar "Khasab" during cold storage for 60 days:	Ahmed et al. (2021)
		1. Retarding the senescence/decay of the fruit	
		2. Lowest weight loss, color change, and the least decay after 60 days of storage under Ch+SA more than other treatments	
		3. Lower levels of total soluble solids and highest total phenolic, tannins, and flavonoids contents in some treatments compared to the control	
		4. Higher antioxidant activities	

Jasmonic acid	Bayoud, caused by *Fusarium oxysporum* f. sp. Albedinis (Foa)	5. Extending the shelf life of the fruits In the roots of two cultivars BSTN and JHL: An increase in the activities of peroxidases (POX) and polyphenol oxidase (PPO) in both asymptomatic plants (with limited hypersensitive-reaction like lesions) and JA-treated plants compared to symptomatic plants	Jaiti et al. (2009)
Jasmonic acid (JA)	*Fusarium oxysporum* f. sp. Albedinis	In seedlings of two cultivars: 1. An increase in H_2O_2 content (oxidative burst) and MDA accumulation (indication of lipid peroxidation), and enhancement of peroxidase activity under both conditions, including infected plants, and JA-treated plants 2. Necrotic hypersensitive-reaction-like lesions in the infected plants 3. Lack of these reactions in sensitive plants that show disease symptoms 4. JA acts as a signal for defense reactions	Jaiti et al. (2004)
Jasmonic acid	Salinity	In date palm callus, Shukar cultivar, obtained from culturing the apical and axillary buds in MS media: Enhancement of callus growth indices (rate of biomass, water content, relative growth rate and the number of somatic embryos), decreasing the callus browning and plant tolerance to salinity	Al-Qatrani et al. (2021)
Salicylic acid	*Fusarium oxysporum* f. sp. Albedinis (Foa)	1. In untreated inoculated plants: cell wall degradation and total cytoplasm disorganization 2. In treated plants: plug of intercellular spaces, the deposition of electron-dense materials at the sites of pathogen penetration, and several damages to fungal cells	Dihazi et al. (2011)
Spraying date palm trees with melatonin (Mt) and/or methyl jasmonate (Mj)	–	In date palm (cv. Barhi): 1. In general, Mt is more effective than Mj 2. In combined treatment: more relative chlorophyll and nutrient content of leaves, highest yield of fruits 3. In combined treatment after 28 days of storage at 4°C: the lowest weight loss and fruit decay values, the highest firmness, total soluble solids content, total sugar content, and the lowest total acidity, highest total phenolic content, and activity of peroxidase and polyphenol oxidase enzymes, intact and thick exocarp tissue with a dense layer of epicuticular wax	Fekry et al. (2021)
Effective microorganisms bio-fertilizer (EM), potassium sulfate	Salinity	In "Hayany" date palm as soil treatment (three times a year); increasing the leaf chlorophyll content, fruit set percent, retained fruit percent, yield, fruit quality, and the contents of leaf minerals	Salama et al. (2014)

(*continued*)

TABLE 16.1 (Continued)
Plant Responses to Exogenous Stimulants under Stressful and Non-stressful Conditions in Date Palm

Exogenous stimulants	Biotic/abiotic stress	Plant responses	References
Potassium fertilization (as K_2SO_4 [48% K_2O)] to date palm trees by four methods i.e., control treatment (without potassium), soil application at (1, 2, and 3 kg/palm, foliar application at (1, 2, and 3% K_2SO_4) and injection into the trunk at (1, 2, and 3% K_2SO_4)	–	In injection method: highest vegetative growth, yield, and fruit quality	Elsayd et al. (2018)
0.08% ALA-based (5-aminolevulinic acid-based) functional fertilizer commercially known as Pentakeep-v	Salinity (seawater)	In date palm seedlings: 1. Under salinity: increasing the accumulation of Na^+ in the plant tissues, electrolyte leakage (decreasing membrane integrity), decreasing the assimilation rate due to a decrease in chl a content 2. Under treatment with Pentakeep-v: increasing the photosynthetic assimilation through the increased chl a content, promoting light-harvesting capabilities, decreasing the stomatal limitation to photosynthetic gas exchange, and then improvement of tolerance to salt	Youssef, and Awad, (2008)
NaCl + Ca Cl_2 WW 2:1 (14,000, 16000 and 18,000 mg L^{-1}) and two levels of potassium (2000 and 3000 mg L^{-1})	Salinity (irrigation water)	In 2-year-old *in vitro* produced date palm plantlets cv. Bartomouda: 1. Under salinity: a decrease in plant height, number of leaves and roots, root length, fresh and dry weights of leaves, increased Na, Ca and K contents in leaves with high content of proline 2. Potassium alleviated the negative effects of salt stress	Rasmia and El Banna (2011)
Yeast (*Saccharomyces cerevisiae* 1 × 10^6 cells/ml) and amino acids	Salinity	In date palm cultivar Bartomouda: 1. Amelioration of salt stress effects by balancing the antioxidant enzymes gene expression 2. Date palm plant tolerance to salinity 3. Improvement of plant growth and development, including plant height, number of leaves/plantlet, fresh and dry weights, chlorophyll a and b, and indole contents	Darwesh (2013)

Agent	Stress	Effects	Reference
Twenty-one isolates of microorganisms, including *Bacillus* spp., *Rhizobium* spp., *Ulocladium atrum*, *Candida guilliermondii*, *Pseudomonas* sp., *Rahnella aquatilis* and other bacteria	*Fusarium oxysporum* f.sp. *albedinis* (Foa), the causal agent of Bayoud on date palm	1. Inhibition of Foa mycelium grown 2. Triggering the defense reactions through the accumulation of non-constitutive hydroxycinnamic acid derivatives	El Hassni et al. (2007)
Biological agents (rhizosphere soil of healthy date palm-isolated local actinomycete strains)	*Thielaviopsis punctulata*, the causal agent of black scorch in date palm	*In vitro* and *in vivo* conditions: 1. The strain belonged to *Streptomyces globosus* 2. Antagonistic activity of *Streptomyces globosus* against black scorch agent by inhibition of mycelial growth of the pathogen	Saeed et al. (2017)
Mycorrhizal (colonized with *Rhizophagus intraradices* or *Funneliformis mosseae*)	Long-term drought stress (25% field capacity)	In date palm seedlings: 1. A decrease in shoot height, root length, and shoot and root dry weights under drought 2. Alleviation of detrimental effects of drought 3. A decrease in H_2O_2 and MDA accumulations, and then alleviation of oxidative stress 4. An increase in antioxidant enzymes activities, including catalase, superoxide dismutase, ascorbate peroxidase, and guaiacol peroxidase, and the highest protein and sugar content 5. Better results with the strain *R. intraradices*	Benhiba et al. (2015)
Arbuscular mycorrhizal fungi (AMF), a native AMF consortium (AMF1) and an exotic AMF strain (AMF2); plant growth-promoting rhizobacteria (PGPR); and compost	Salt stress	Improvement of shoot growth and dry weight, an increase in osmolytes and the activity of enzymatic antioxidants, separately or triplicate application of bio-fertilizers	Toubali et al. (2020)
Jasmonic acid	Salt stress under *in vitro* conditions	In date palm callus, Shukar cultivar: 1. A decrease in growth, and an increase in callus browning under high levels of salt 2. Enhancement of callus growth, biomass rate, water content, relative growth rate, somatic embryos number, and a decrease in callus browning 3. Increase in cell and tissue tolerance to saline conditions	Al-Qatrani et al. (2021)
Silicon (Si) (calcium silicate)	Drought stress induced by polyethylene glycol (PEG)- *in vitro* condition	In vitro date palm cv. Barhee: 1. A decrease in growth, protein content, chlorophyll concentration, root induction and length, and an increase in proline accumulation under stress 2. Improvement of shoot growth, root induction, and root length and an increase in leaf wax, protein content, and leaf chlorophyll concentration of stressed plantlets under bio-fertilizer 3. An increase in plant tolerance to drought stress	Al-Mayahi (2016)

Ait-El-Mokhtar et al. (2019) showed that AMF can improve the Ca/Na and K/Na ratios under salinity, and reduce lipid peroxidation and H_2O_2 content, however, the antioxidant activities of enzymes (superoxide dismutase, catalase, peroxidase, and ascorbate peroxidase) increased with biofertilizer inoculation in date palm. The salinity type that was assayed was 240 mM NaCl. It was stated that the biofertilizer effect can mitigate salinity-induced oxidative stress. Meddich et al. (2015) showed that date palm trees had protection against severe drought stress (25% field capacity) under both conditions of mycorrhizal fungi treatment (indigenous or selected) or without treatment when they had higher levels of phenols and the activities of both peroxidase and polyphenol oxidase in their roots, but, under mycorrhizal fungi, the increased levels were greater than without treatment. Furthermore, higher relative water content (RWC), water potential and dry weight of seedlings were observed under mycorrhizal fungi treatment.

Another stress that could affect date palm growth and development is soil pollution, such as cadmium (Cd). Khan et al. (2020a) evaluated the single and combined effects of Cd and salinity stresses on date palms under exogenous silicon. They found some differences between Cd+NaCl, and Cd and salinity. Silicon could alleviate detrimental effects of all types of investigated stress but by different responses of plants. For example, they found downregulation of endogenous salicylic acid, jasmonic acid, and abscisic acid under NaCl stress and combined NaCl+Cd stress, but only lowered accumulation of these phytohormones under Cd stress. Furthermore, under combined stress, they observed a higher ascorbate peroxidase level and cytosolic Cu/Zn superoxide dismutase expression in Si-treated plants and lower metal (Cd) uptake, lipid peroxidation rate, and peroxidase and catalase activities. Moreover, higher transcript accumulations of PROLINE TRANSPORTER 2 and GAPDH and downregulation of ABA RECEPTOR were observed under Si treatment against combined stress. Therefore, silicon may be considered a stress alleviator. Zouari et al. (2016a, b) investigated the impacts of soil cadmium stress on 2-year-old date palms. They observed some physiological responses in plants, including a decrease in growth, macronutrient contents (Ca^{2+}, Mg^{2+} and K^+), membrane stability index (MSI), starch content and POD activity, and an increase in H_2O_2, soluble sugar, APX activity, thiobarbituric acid reactive substances (TBARS), and electronic leakage (EL) as well as total polyphenols. They reported that exogenous proline through irrigation water alleviated detrimental effects of cadmium as well as a decrease in accumulation of Cd in leaves and roots, and finally the growth and nutrient uptake were increased in the stressed plants.

16.7 CONCLUSIONS AND PROSPECTS

Nutritional and medicinal properties of date fruits for humans have been determined and described in numerous researches and reports, and are related to the contents of primary metabolites, including proteins, carbohydrates, amino acids, minerals, and secondary metabolites such as polyamines, phenols, antioxidants, and so on. Date palm is an extremophile plant, which can tolerate moderate biotic and abiotic stresses, however, in extreme conditions, especially salinity, drought, and heat stress, that are encountered in its cultivation area (arid and semi-arid regions) it was observed that plant growth and development suffer some changes to primary and secondary metabolites, and that these impact on the fruit yield and quality (nutritional and medicinal). Although some secondary metabolites with medicinal properties for humans are promoted under some stresses or special stress, general stresses, especially extreme abiotic stress and even slight stress due to biotic stress, could have detrimental effects on fruit quality and yield. Reports have shown that the application of biofertilizers and biostimulants (natural/organic and synthetic) could ameliorate the detrimental effects of stress, and even promote greater fruit quality.

As described in this chapter, numerous studies have been carried out on stress effects on date palm plants but not covering the fruit's medicinal and nutritional properties especially, therefore it is recommended that there be co-participation between medicinal or nutritional experts and horticulturalists.

REFERENCES

Abbas, M.F., A.M. Jasim, and H.J. Shareef. 2015. Role of sulphur in salinity tolerance of date palm (*Phoenix dactylifera* L.) offshoots cvs. Berhi and Sayer. *International Journal of Agricultural and Food Science* 5: 92–97.

Ahmed, Z.F., S.S. Alblooshi, N. Kaur, et al. 2021. Synergistic effect of preharvest spray application of natural elicitors on storage life and bioactive compounds of date palm (*Phoenix dactylifera* L., cv. Khesab). *Horticulturae* 7(6): 145.

Ait-El-Mokhtar, M., M. Baslam, R. Ben-Laouane, et al. 2020. Alleviation of detrimental effects of salt stress on date palm (*Phoenix dactylifera* L.) by the application of arbuscular mycorrhizal fungi and/or compost. *Frontiers in Sustainable Food Systems*, 131.

Ait-El-Mokhtar, M., R.B. Laouane, M. Anli, et al. 2019. Use of mycorrhizal fungi in improving tolerance of the date palm (*Phoenix dactylifera* L.) seedlings to salt stress. *Scientia Horticulturae* 253: 429–438.

Al Kharusi, L., R. Al Yahyai, and M.W. Yaish. 2019. Antioxidant response to salinity in salt-tolerant and salt-susceptible cultivars of date palm. *Agriculture* 9(1): 8.

Al Kharusi, L., D.V. Assaha, R. Al-Yahyai, and M.W. Yaish. 2017. Screening of date palm (*Phoenix dactylifera* L.) cultivars for salinity tolerance. *Forests* 8(4): 136.

Al-Harrasi, I., R. Al-Yahyai, and M.W. Yaish. 2018. Differential DNA methylation and transcription profiles in date palm roots exposed to salinity. *PloS One* 13(1): e0191492.

Al-Harrasi, I., H.V. Patankar, R. Al-Yahyai, et al. 2020. Molecular characterization of a date palm vascular highway 1-interacting kinase (PdVIK) under abiotic stresses. *Genes* 11(5): 568.

Al-Khateeb, S.A., M.N. Sattar, A.A. Al-Khateeb, and A.S. Mohmand. 2021. Calcium supplementation improves in vitro salt tolerance of date palm (*Phoenix dactylifera* L.). *Progress in Nutrition* 23(3).

Al-Khayri, J.M., and A.M. Al-Bahrany. 2004. Growth, water content, and proline accumulation in drought-stressed callus of date palm. *Biologia Plantarum* 48(1): 105–108.

Al-Khayri, J.M., P.M. Naik, S.M. Jain, and D.V. Johnson. 2018. Advances in date palm (*Phoenix dactylifera* L.) breeding. In *Advances in plant breeding strategies: fruits* (pp. 727–771). Springer, Cham.

Al-Mayahi, A.M.W. 2016. Effect of silicon (Si) application on *Phoenix dactylifera* L. growth under drought stress induced by polyethylene glycol (PEG) in vitro. *American Journal of Plant Sciences* 7(13): 1711–1728.

Al-Qatrani, M.K., A.A. Al Khalifa, and N.A. Obaid. 2021. Effect of jasmonic acid on stimulating the growth and development of date palm callus (*Phoenix dactylifera* L.) cultivar shukar in vitro under salt stress conditions. In *IOP Conference Series: Earth and Environmental Science* 923(1): 012017. IOP Publishing.

Al-Shwyeh, H.A. 2019. Date palm (*Phoenix dactylifera* L.) fruit as potential antioxidant and antimicrobial agents. *Journal of Pharmacy and Bioallied Sciences* 11(1): 1.

Alturki, S.M. 2021. The potential use of ca (NO$_3$)$_2$ to improve salinity tolerance in date palm (*Phoenix dactylifera* L.). *Iraqi Journal of Agricultural Sciences* 52(2): 445–453.

Anli, M., M. Baslam, A. Tahiri, et al. 2020. Biofertilizers as strategies to improve photosynthetic apparatus, growth, and drought stress tolerance in the date palm. *Frontiers in Plant Science* 1560.

Arab, L., J. Kreuzwieser, J. Kruse, et al. 2016. Acclimation to heat and drought—lessons to learn from the date palm (*Phoenix dactylifera*). *Environmental and Experimental Botany*, 125: 20–30.

Baslam, M., A. Qaddoury, and N. Goicoechea. 2014. Role of native and exotic mycorrhizal symbiosis to develop morphological, physiological and biochemical responses coping with water drought of date palm, *Phoenix dactylifera*. *Trees* 28(1): 161–172.

Benhiba, L., A. Essahibi, M. Fouad, and A. Qaddoury. 2013. Arbuscular mycorrhizal fungi enhanced date palm tolerance to water deficit. In *The Fifth International Date Palm Conference*. pp. 343–348.

Benhiba, L., M.O. Fouad, A. Essahibi, et al. 2015. Arbuscular mycorrhizal symbiosis enhanced growth and antioxidant metabolism in date palm subjected to long-term drought. *Trees* 29(6): 1725–1733.

Bouissil, S., C. Guérin, J. Roche, et al. 2022. Induction of defense gene expression and the resistance of date palm to *Fusarium oxysporum* f. sp. Albedinis in response to alginate extracted from *Bifurcaria bifurcata*. *Marine Drugs* 20(2): 88.

Darwesh, R.S. 2013. Improving growth of date palm plantlets grown under salt stress with yeast and amino acids applications. *Annals of Agricultural Sciences* 58(2): 247–256.

de Vasconcelos, A.C.F., and L.H. Chaves. 2019. Biostimulants and their role in improving plant growth under abiotic stresses. *Biostimulants in Plant Science*.

Dihazi, A., M.A. Serghini, F. Jaiti, et al. 2011. Structural and biochemical changes in salicylic-acid-treated date palm roots challenged with *Fusarium oxysporum* f. sp. albedinis. *Journal of Pathogens* 2011.

Djibril, S., O.K. Mohamed, D. Diaga, et al. 2005. Growth and development of date palm (*Phoenix dactylifera* L.) seedlings under drought and salinity stresses. *African Journal of Biotechnology* 4(9).

Du, B., Y. Ma, A.M. Yáñez-Serrano, et al. 2021. Physiological responses of date palm (*Phoenix dactylifera*) seedlings to seawater and flooding. *New Phytologist* 229(6): 3318–3329.

El Hassni, M., A. El Hadrami, F. Daayf, et al. 2007. Biological control of bayoud disease in date palm: Selection of microorganisms inhibiting the causal agent and inducing defense reactions. *Environmental and Experimental Botany* 59(2): 224–234.

El Rabey, H.A., A.L. Al-Malki, and K.O. Abulnaja. 2016. Proteome analysis of date palm (*Phoenix dactylifera* L.) under severe drought and salt stress. *International Journal of Genomics* 2016.

El Rabey, H.A., A.L. Al-Malki, K.O. Abulnaja, and W. Rohde. 2015. Proteome analysis for understanding abiotic stress (salinity and drought) tolerance in date palm (*Phoenix dactylifera* L.). *International Journal of Genomics* 2015.

El-Khawaga, A.S. 2013. Effect of Anti-salinity Agents on Grovvth and Fruiting of Different. *Asian Journal of Crop Science* 5(1): 65–80.

Elsayd, I.E.R., S. El-Merghany, and E.M.A. Zaen El–Dean. 2018. Influence of potassium fertilization on Barhee date palms growth, yield and fruit quality under heat stress conditions. *Journal of Plant Production* 9(1): 73–80.

Fekry, W.M., Y.M. Rashad, I.A. Alaraidh, and T. Mehany. 2021. Exogenous application of melatonin and methyl jasmonate as a pre-harvest treatment enhances growth of Barhi date palm trees, prolongs storability, and maintains quality of their fruits under storage conditions. *Plants* 11(1): 96.

Garcia-Maquilon, I., A. Coego, J. Lozano-Juste, et al. 2021. PYL8 ABA receptors of *Phoenix dactylifera* play a crucial role in response to abiotic stress and are stabilized by ABA. *Journal of Experimental Botany* 72(2): 757–774.

Ghirardo, A., T. Nosenko, J. Kreuzwieser, et al. 2021. Protein expression plasticity contributes to heat and drought tolerance of date palm. *Oecologia* 197(4): 903–919.

Hadrami, A.E., F. Daayf, and I.E. Hadrami. 2011. Secondary metabolites of date palm. In *Date Palm Biotechnology*. pp. 653–674. Springer, Dordrecht.

Harkousse, O., A. Slimani, I. Jadrane, et al. 2021. Role of local biofertilizer in enhancing the oxidative stress defense systems of date palm seedling (*Phoenix dactylifera*) against abiotic stress. *Applied and Environmental Soil Science* 2021.

Helaly, M.N., and H.A. El-Hosieny. 2015. Combined Effect of Biofertilizers and Putrescine Amine on Certain Physiological Aspects and Productivity of Date Palm (*Phoenix dactylifera* L.) Grown in Reclaimed-Saline Soil. *Egyptian Journal of Horticulture* 42(1): 721–739.

Hsu, F.M., M. Gohain, P. Chang, et al. 2018. Bioinformatics of Epigenomic Data Generated From Next-Generation Sequencing. In *Epigenetics in Human Disease* (pp. 65–106). Academic Press.

Hussain, M.I., M. Farooq, and Q.A. Syed. 2020. Nutritional and biological characteristics of the date palm fruit (*Phoenix dactylifera* L.)–A review. *Food Bioscience* 34: 100509.

Jaiti, F., A. Dihazi, I. El Hardami, et al. 2004. Effect of exogenous application of jasmonic acid on date palm defense reaction against "*Fusarium oxysporum* "f. sp." albedinis". 1000–1007.

Jaiti, F., J.L. Verdeil, and I. El Hadrami. 2009. Effect of jasmonic acid on the induction of polyphenoloxidase and peroxidase activities in relation to date palm resistance against *Fusarium oxysporum* f. sp. albedinis. *Physiological and Molecular Plant Pathology* 74(1): 84–90.

Jana, G.A., and M.W. Yaish. 2020. Functional characterization of the Glyoxalase-I (PdGLX1) gene family in date palm under abiotic stresses. *Plant Signaling and Behavior* 15(11): 1811527.

Khan, A., S. Bilal, A.L. Khan, et al. 2020a. Silicon-mediated alleviation of combined salinity and cadmium stress in date palm (*Phoenix dactylifera* L.) by regulating physio-hormonal alteration. *Ecotoxicology and Environmental Safety* 188: 109885.

Khan, A., S. Bilal, A.L. Khan, et al. 2020b. Silicon and gibberellins: synergistic function in harnessing ABA signaling and heat stress tolerance in date palm (*Phoenix dactylifera* L.). *Plants* 9(5): 620.

Mansoori, B., and M.H. Kord. 2006. Yellow death: A disease of date palm in Iran caused by *Fusarium solani*. *Journal of Phytopathology* 154(2): 125–127.

Maqsood, S., O. Adiamo, M. Ahmad, and P. Mudgil. 2020. Bioactive compounds from date fruit and seed as potential nutraceutical and functional food ingredients. *Food chemistry* 308: 125522.

Meddich, A., M. Ait El Mokhtar, W. Bourzik, et al. 2018. Optimizing growth and tolerance of date palm (*Phoenix dactylifera* L.) to drought, salinity, and vascular fusarium-induced wilt (*Fusarium oxysporum*) by application of arbuscular mycorrhizal fungi (AMF). In *Root Biology* (pp. 239–258). Springer, Cham.

Meddich, A., F. Jaiti, W. Bourzik, et al. 2015. Use of mycorrhizal fungi as a strategy for improving the drought tolerance in date palm (*Phoenix dactylifera*). *Scientia Horticulturae* 192: 468–474.

Mefteh, F.B., A. Daoud, A. Chenari Bouket, et al. 2017. Fungal root microbiome from healthy and brittle leaf diseased date palm trees (*Phoenix dactylifera* L.) reveals a hidden untapped arsenal of antibacterial and broad spectrum antifungal secondary metabolites. *Frontiers in Microbiology* 8: 307.

Mohamed, E.A. 2019. Copper and chitosan nanoparticles as potential elicitors of innate immune response in date palm: a comparative study. *Archives of Phytopathology and Plant Protection* 52(17–18): 1276–1288.

Naser, H.M., E.H. Hanan, N.I. Elsheery, and H.M. Kalaji. 2016. Effect of biofertilizers and putrescine amine on the physiological features and productivity of date palm (*Phoenix dactylifera* L.) grown on reclaimed-salinized soil. *Trees* 30(4): 1149–1161.

Patankar, H.V., I. Al-Harrasi, R. Al-Yahyai, and M.W. Yaish. 2018. Identification of candidate genes involved in the salt tolerance of date palm (*Phoenix dactylifera* L.) based on a Yeast functional bioassay. *DNA and Cell Biology* 37(6): 524–534.

Patankar, H.V., M. Assaha, D.V. Al-Yahyai, et al. 2016. Identification of reference genes for quantitative real-time PCR in date palm (*Phoenix dactylifera* L.) subjected to drought and salinity. *PloS One* 11(11): e0166216.

Rasmia, D., and A.A. El Banna. 2011. Role of potassium and salinity effects on growth and chemical compositions of date palm plantlets. *Arab Universities Journal of Agricultural Sciences* 19(1): 233–244.

Saeed, E.E., A. Sham, Z. Salmin, et al. 2017. *Streptomyces globosus* UAE1, a potential effective biocontrol agent for black scorch disease in date palm plantations. *Frontiers in Microbiology* 8: 1455.

Safronov, O., J. Kreuzwieser, G. Haberer, et al. 2017. Detecting early signs of heat and drought stress in *Phoenix dactylifera* (date palm). *PLoS One* 12(6): e0177883.

Salama, A.S., O.M. El-Sayed, and O.H. El Gammal. 2014. Effect of effective microorganisms (EM) and potassium sulphate on productivity and fruit quality of "Hayany" date palm grown under salinity stress. *J Agri Vet Sci* 7: 90–99.

Shafiei Masouleh, S.S., and Y.N. Sassine. 2020a. Molecular and biochemical responses of horticultural plants and crops to heat stress. *Ornamental Horticulture* 26: 148–158.

Shafiei Masouleh, S.S., N. Jamal Aldine, and Y.N. Sassine. 2020b. The role of organic solutes in the osmotic adjustment of chilling-stressed plants (vegetable, ornamental and crop plants). *Ornamental Horticulture* 25: 434–442.

Shafiei-Masouleh, S.S. 2021. Contents of Organic Solutes in Horticultural Plants and Crops under Environmental Pollutants. In *Organic Solutes, Oxidative Stress, and Antioxidant Enzymes Under Abiotic Stressors* (pp. 105–122). CRC Press.

Sperling, O., N. Lazarovitch, A. Schwartz, and O. Shapira. 2014. Effects of high salinity irrigation on growth, gas-exchange, and photoprotection in date palms (*Phoenix dactylifera* L., cv. Medjool). *Environmental and Experimental Botany* 99: 100–109.

Suhim, A.A., K.F. Abbas, and K.M. Al-Jabary. 2017. Oxidative responses and genetic stability of date palm *Phoenix dactylifera* L. Barhi cv. under salinity stress. *J. Biology, Agriculture and Healthcare* 7(8): 70–80.

Taleb, H., S.E. Maddocks, R.K. Morris, and A.D. Kanekanian. 2016. Chemical characterisation and the anti-inflammatory, anti-angiogenic and antibacterial properties of date fruit (*Phoenix dactylifera* L.). *Journal of Ethnopharmacology* 194: 457–468.

Toubali, S., A.I. Tahiri, M. Anli, et al. 2020. Physiological and biochemical behaviors of date palm vitro plants treated with microbial consortia and compost in response to salt stress. *Applied Sciences* 10(23): 8665.

Tripler, E., U. Shani, Y. Mualem, and A. Ben-Gal. 2011. Long-term growth, water consumption and yield of date palm as a function of salinity. *Agricultural Water Management* 99(1): 128–134.

Wang, Y., H.N. Brown, D.E. Crowley, and P.J. Szaniszlo. 1993. Evidence for direct utilization of a siderophore, ferrioxamine B, in axenically grown cucumber. *Plant, Cell and Environment* 16(5): 579–585.

Yaish, M.W. 2015. Proline accumulation is a general response to abiotic stress in the date palm tree (*Phoenix dactylifera* L.). *Genet Mol Res* 14(3): 9943–9950.

Yaish, M.W., and P.P Kumar. 2015. Salt tolerance research in date palm tree (*Phoenix dactylifera* L.), past, present, and future perspectives. *Frontiers in Plant Science* 6: 348.

Yaish, M.W., I. Al-Harrasi, A.S. Alansari, et al. 2016. The use of high throughput DNA sequence analysis to assess the endophytic microbiome of date palm roots grown under different levels of salt stress. *Int. Microbiol* 19(3): 143–155.

Yaish, M.W., I. Antony, and B.R. Glick. 2015a. Isolation and characterization of endophytic plant growth-promoting bacteria from date palm tree (*Phoenix dactylifera* L.) and their potential role in salinity tolerance. *Antonie Van Leeuwenhoek* 107(6): 1519–1532.

Yaish, M.W., H.V. Patankar, D.V. Assaha, et al. 2017. Genome-wide expression profiling in leaves and roots of date palm (*Phoenix dactylifera* L.) exposed to salinity. *BMC genomics* 18(1): 1–17.

Yaish, M.W., R. Sunkar, Y. Zheng, et al. 2015b. A genome-wide identification of the miRNAome in response to salinity stress in date palm (*Phoenix dactylifera* L.). *Frontiers in plant science* 6: 946.

Youssef, T., and M.A. Awad. 2008. Mechanisms of enhancing photosynthetic gas exchange in date palm seedlings (*Phoenix dactylifera* L.) under salinity stress by a 5-aminolevulinic acid-based fertilizer. *Journal of Plant Growth Regulation* 27(1): 1–9.

Zouari, M., C.B. Ahmed, W. Zorrig, et al. 2016a. Exogenous proline mediates alleviation of cadmium stress by promoting photosynthetic activity, water status and antioxidative enzymes activities of young date palm (*Phoenix dactylifera* L.). *Ecotoxicology and environmental safety* 128: 100–108.

Zouari, M., N. Elloumi, C.B. N., Ahmed, et al. 2016b. Exogenous proline enhances growth, mineral uptake, antioxidant defense, and reduces cadmium-induced oxidative damage in young date palm (*Phoenix dactylifera* L.). *Ecological Engineering* 86: 202–209.

17 Portulaca oleracea under Salt Stress

Hassan Ahmed Ibraheem Ahmed[1,2],
Peichen Hou[3] and Waqas-ud-Din Khan[4]
[1]Department of Botany, Faculty of Science, Port Said University, Port Said, 42526, Egypt
[2]Tasmanian Institute of Agriculture, University of Tasmania, Hobart, Tasmania, 7005, Australia
[3]Intelligent Equipment Research Center, Beijing Academy of Agriculture and Forestry Sciences, Beijing, China
[4]Sustainable Development Study Centre, Government College University, Lahore, 54000, Pakistan
*Corresponding author: hassan.ahmed@utas.edu.au; hassan.ahmed.sci@gmail.com; houpc@nercita.org.cn; dr.waqasuddin@gcu.edu.pk

CONTENTS

17.1 INTRODUCTION

Portulaca oleracea is a succulent herbaceous plant with an annual life cycle that is known as common purslane (Chauhan and Johnson 2009; Sdouga et al. 2019). The *Portulaca* genus belongs to Portulacaceae (family of Caryophyllales) and includes over 100 morphologically variable species (annuals and/or perennials) that are distributed worldwide, with most species diversity in the tropic and subtropic areas (Ocampo and Columbus 2012). Purslane is of pharmaceutical and nutritional value and has been introduced by researchers as a potential vegetable crop for human consumption (Alam et al. 2015a). It has been considered by the World Health Organization (WHO) as one of the most useful medicinal plants and was named "Global Panacea" (Sultana and Rahman 2013). Its fresh stems and succulent leaves are used widely in medication around the world (Dweck 2001). Moreover, purslane has been described recently as a functional food resource with high antioxidant properties (Petropoulos et al. 2016). The plant contains high amounts of beneficial antioxidant vitamins and minerals (Uddin et al. 2012). It is considered one of the richest vegetable sources of ω-3 and ω-6 fatty acids (especially α-linolenic acid), besides other vital phytochemicals such as polyphenols, α-tocopherol, ascorbic acid, β-carotene and glutathione (Alam et al. 2015a; Sdouga et al. 2019). Petropoulos et al. (2016) introduced a comprehensive review of purslane phytochemicals and its potential biological functions; also, the mineral composition of purslane in relation to the place of origin and plant part was addressed.

Purslane is highly adaptable to various environmental stresses and can dominate in hostile conditions compared to many other cultivated vegetable crops (Yazici et al. 2007). Mulry et al. (2015) proposed the plant as a model system for exploring plant responses to stress. Purslane is introduced with potential for phytoremediation purposes, being a hyperaccumulator of several heavy metals (Petropoulos et al. 2016); removing Cl$^-$ and Na$^+$ ions from moderately saline soils (Kiliç et al. 2008), with an ability to remove ions accounting for 400–500 kg ha^{-1} NaCl (Bekmirzaev et al. 2021); and technological measures, such as co-cultivation with sensitive glycophytic species (Hnilickova et al. 2021). Purslane is deemed to be a very invasive species (Holm et al. 1977), showing a great ability to thrive in arid and saline soils. Hence, it is listed as a halophyte in the HALOPH database (Aronson and Whitehead 1989). The plant shows an ability to produce sufficient biomass yield when treated with moderate salinity levels compared to other vegetable crops (Kafi and Rahimi 2011). Moreover, several environmental factors can affect the final content of bioactive compounds (Petropoulos et al. 2016). Therefore, growing purslane under favorable saline dosages (for growth) could potentially increase its bioactive compound yield and improve their biological potential (Alam et al. 2015a).

Purslane is classified as a C4 plant (Sultana and Rahman 2013), with a distinctive ability to shift to the crassulacean acid metabolism (CAM), for an efficient photosynthesis process under adverse conditions (D'andrea et al. 2014; Ocampo and Columbus 2012). The plant seems to utilize such a kind of adaptive plasticity (Nicotra et al. 2010) to maintain adequate growth and reproduction levels under severe saline conditions (Sdouga et al. 2019). Treating purslane plants with severe salinity was found to inhibit their growth, which is probably ascribed to the inhibition of photosynthesis and the presence of oxidative stress (Xing et al. 2019). He et al. (2021) mentioned that subjecting purslane plants to higher salinity, after being grown under low salinity levels (with enhanced photosynthetic machinery and increased shoot and root biomass accumulations) might improve the production of their phytochemicals. However, the latter notion needs more investigation, and more work is still required for better management of the purslane content of bioactive compounds under saline conditions, and also to add a high nutritional and therapeutic value to its final product, and to recommend the proper cultivation practices.

17.2 GROWTH RESPONSES

P. oleracea is rated as a moderately salt-tolerant halophyte with a salinity threshold of 6.3 dS m^{-1} having been recognized for that plant (Kafi and Rahimi 2011). The available literature shows that purslane exposure to higher/lower salt ranges, as well as short-/long-term applications, comes with remarkable effects on its growth characteristics (Table 17.1). Salinity treatments (NaCl or Na$_2$SO$_4$; 25–200 mM) were found to harm seed germination of purslane, with the light and/or dark combinations (with the stress) being determined for the germination percentage (Naik and Karadge 2017). The overall growth of *P. oleracea* cultivar was greatly reduced with 200 mM NaCl treatment, while the plant showed a satisfactory response under control and/or moderate salt conditions (Zaman et al. 2018). The lowest germination rate of purslane plants was observed under 200 mM NaCl compared to control (Parvaneh et al. 2012). Sdouga et al. (2019) reported that 50 mM NaCl induced *P. oleracea* growth, and plants were still moderately affected up to 100 mM treatment. Similarly, treatment with salinity levels (NaCl) of 5 and 9.8 dS m^{-1} showed no change in the leaf and seed fresh weight of that plant; nevertheless a significant reduction in the total dry weight was recorded in treatments with higher salinity levels of up to 20 dS m^{-1} (Bekmirzaev et al. 2021). Furthermore, treating purslane plants with 140 mM NaCl was found to have a more suppressive effect on plant growth than 70 mM NaCl (Yazici et al. 2007). On the contrary, Hniličková et al. (2019) recorded an evident decrease in *P. oleracea* dry weight by about 32% for plants grown under 50 mM NaCl compared to those of the

TABLE 17.1
Differences in Growth Responses of *Portulaca Oleracea* Plants Treated with Different Salt Ranges

Salt Treatment	Growth responses	Reference
NaCl (0-150 mM)	A 50 mM NaCl induced *P. oleracea* growth, and plants were still moderately affected up to 100 mM treatment	Sdouga et al. (2019)
NaCl (0-32 dS m^{-1})	A considerable variation in growth responses between several purslane accessions was obtained, with a gradual reduction in dry matter contents was recorded with elevated salt dosages and was ranged between 32–63% at 32 dS m^{-1} salinity	Alam et al. (2015a)
NaCl and Na$_2$SO4 (25-200 mM) salinity	Salinity treatments were found to harm seed germination of purslane, with the light and/or dark combinations (with the stress) being determined for the germination percentage	Naik and Karadge (2017)
NaCl (0-32 dS m^{-1})	Elevated salinity caused reduction in several morphological traits such as plant height, number of leaves, number of flowers, as well as fresh and dry weight	Alam et al. (2016)
NaCl (0-300 mmol/L)	A 50 mM NaCl decreased *P. oleracea* dry weight by about 32% compared to those of the control group. No change in dry weight was found with the further increase in salt dosages (100, 200, and 300 mM NaCl)	Hniličková et al. (2019)
NaCl (0-200 mM)	Salt stress induced significant variations in number of leaves, length of the stem, diameter of the stem, and length of roots in two tested *P. oleracea* cultivars with an effect became more obvious at 200 mM NaCl	Zaman et al. (2020)
NaCl (0-100 mM)	The total fresh weight of *P. oleracea* plants was decreased significantly from 213.2 ± 10.2 g under controls down to 138.2 ± 7.1 g in NaCl-treated plants (100 mM NaCl)	Bessrour et al. (2018)
NaCl (0-240 mM)	Extreme salinity treatment reduced growth and yield of purslane plants	Teixeira and Carvalho (2009)
NaCl (0-200 mM)	Both cultivated and wild varieties of purslane reduced their dry matter yield and flowers number under salt stress	Mulry et al. (2015)
NaCl (0-300 mM) hydroponic under LED lightings	Purslane plants grown hydroponically under LED lighting showed greater shoots and roots productivity with 100 mM NaCl compared to no-salt treatment. However, the further increase in NaCl concentration up to 300 mM resulted in reduced shoots and roots	He et al. (2021)
NaCl (1-20 dS m^{-1})	Treatment with NaCl levels (5 and 9.8 dS m^{-1}) showed no change in the leaf and seed fresh weight. A significant reduction in the total dry weight was recorded in treatments with higher salinity levels up to 20 dS m^{-1}	Bekmirzaev et al. (2021)
NaCl (0-200 mM)	The overall growth of *P. oleracea* cultivar was highly declined with 200 mM NaCl treatment, while the plant showed a satisfactory response under control and/or moderate salt conditions	Zaman et al. (2018)
NaCl (0-200 mM)	The lowest germination rate of purslane plants was observed under 200 mM NaCl compared to control	Parvaneh et al. (2012)

(continued)

TABLE 17.1 (Continued)
Differences in Growth Responses of *Portulaca Oleracea* Plants Treated with Different Salt Ranges

Salt Treatment	Growth responses	Reference
NaCl (0-200 mM)	Significant inhibition of purslane growth, in terms of dry and fresh weight of seedlings, shoot and root length as well as leaflet number on the main stem, was obtained under 150 mM and 200 mM NaCl in comparison to zero salt level	Xing et al. (2019)
NaCl (0-140 mM)	Growth of purslane plants was more suppressed under 140 mM NaCl than 70 mM NaCl. The temporal factor of the applied stress sounds important to induce the proper response where differences in several morpho-physiological traits of purslane plants were recorded after 18 as well as 30 days of salt applications onset	Yazici et al. (2007)

control group. However, no change in dry weight was found with the further increase in salt dosages (100, 200 and 300 mM NaCl). Both cultivated and wild varieties of purslane (Figure 17.1) reduced their dry matter yield and flower number under salt stress (NaCl; 0–200 mM) (Mulry et al. 2015). The total fresh weight of *P. oleracea* plants decreased significantly from 213.2 ± 10.2 g for controls down to 138.2 ± 7.1 g in NaCl-treated plants (100 mM NaCl) (Bessrour et al. 2018). Purslane plants grown hydroponically under light-emitting diode (LED) lighting showed greater shoot and root productivity with 100 mM NaCl compared to no-salt treatment. However, the further increase in NaCl concentration up to 300 mM in the same experiment resulted in reduced shoots and roots (He et al. 2021). Significant inhibition of purslane growth, in terms of dry and fresh weight of seedlings, shoot and root length as well as leaflet number on the main stem, was obtained under 150 mM and 200 mM NaCl in comparison to zero salt level (Xing et al. 2019). It seems clear that purslane exposure to salinity above 100 mM NaCl results in ultimate growth retardation. Growth retardation under saline conditions might be attributable to several factors, such as: (1) disturbing the balance between respiration and photosynthesis (Moir-Barnetson et al. 2016; Redondo-Gomez et al. 2010; Rozentsvet et al. 2017); (2) the reduction in cellular osmolality and the accompanied turgor pressure (Clipson et al. 1985; Moir-Barnetson et al. 2016; Rozentsvet et al. 2017); (3) the disruption in enzymatic reactions due to excessive Na^+ and/or Cl^- levels in the cytoplasm (Ayala and O'Leary 1995; Carillo et al. 2011; Moir-Barnetson et al. 2016; Rozentsvet et al. 2017); and (4) the increased energetic costs associated with *de novo* organic solute synthesis (Moir-Barnetson et al. 2016; Munns 2002) and regulation of ion transport (Moir-Barnetson et al. 2016; Yeo 1983).

In addition, salt stress induced significant variations in the number of leaves, length of the stem, diameter of the stem and length of roots in two tested *P. oleracea* cultivars with an effect that become more obvious at 200 mM NaCl (Zaman et al. 2020). Alam et al. (2016) also reported a reduction in several morphological traits such as plant height, number of leaves, number of flowers, as well as fresh and dry weight under elevated salinity (NaCl; 0–32 dS m^{-1}). The application of 200 mM NaCl to the growth medium resulted in a significant decrease in a number of morphological attributes of purslane plants including leaf number, stem diameter, main stem length, root length and the number of roots (Zaman et al. 2018). Besides the salt concentration, the temporal factor of the applied stress appears to be important to induce the proper response of purslane plants. That was manifested by the differences recorded in several morpho-physiological traits of purslane plants after 18 and 30 days of salt applications onset with a range of NaCl (0–140 mM) (Yazici et al. 2007).

A considerable variation in growth responses between several purslane accessions was obtained under the salt range of 0–32 dS m^{-1} NaCl, with a gradual reduction in dry matter contents recorded

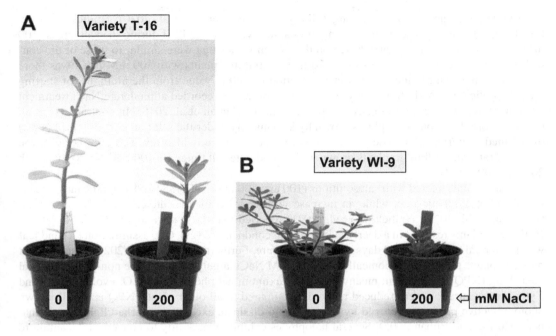

FIGURE 17.1 Variation in growth responses of two isogenic varieties of purslane (cultivated, T-16; and wild, WI-9) under salt treatment. This figure has been slightly modified from Mulry et al. (2015). https://doi.org/10.1371/journal.pone.0138723

with elevated salt dosages which ranged between 32–63% at 32 dS m^{-1} salinity (Alam et al. 2015a). Based on biomass production, Alam et al. (2016) identified the salt tolerance magnitude of 13 purslane accessions, revealing that five of them were moderately susceptible, six accessions were moderately tolerant and the remaining two were found to be salt tolerant. In addition, the latter study also indicated that ornamental purslane accessions showed more salt tolerance than common ones. Therefore, and based on the variation in growth responses of purslane varieties (Figure 17.1), the salt tolerance of purslane accessions seems to be diverse and complex.

17.3 PHOTOSYNTHETIC MACHINERY AND LEAF-GAS EXCHANGE

Treating purslane plants with moderate saline solutions (hydroponically; 0 and 100 mM NaCl) resulted in a significantly higher concentration of total chlorophylls and carotenoids comparable to lower values that have been recorded under severe NaCl concentrations (200 and 300 mM) (He et al. 2021). Both chlorophylls and carotenoids contents were significantly decreased under the treatment with 200 mM salinity stress in two purslane cultivars when compared to plants at zero salt dosage (Zaman et al. 2020). Bessrour et al. (2018) reported a slight reduction (not significant) in both chlorophylls a and b in *P. oleracea* plants when treated with 100 mM NaCl compared to zero-salt conditions. Salt stress induced a marked decrease in the total chlorophyll content of purslane plants at 200 mM NaCl (Zaman et al. 2018). Therefore, the drastic effect of higher salt concentrations on *P. oleracea* plants can be monitored through the enormous reduction of their photosynthetic pigments.

Chlorophyll fluorescence measurement is an important index that demonstrates the changes in photosynthetic pigments in plants' leaves; most important are the photochemical efficiency (Fv/Fm) values. The Fv/Fm ratio was slightly decreased in wild purslane under 200 mM NaCl, while treatment with 0 and/or 100 mM resulted in favorable Fv/Fm values (close to 0.8) in both cultivated

"Tall Green-TG" and wild "Shandong Wild-SD" plants (Zaman et al. 2020). Hnilickova et al. (2021) reported no change in the maximum quantum yield of the PSII during the salt stress (100 mM NaCl) exposure of purslane, based on the Fv/Fm values that were similar to those of the control. However, the latter study also showed that treating the plants with 300 mM NaCl was more effective than 100 mM, and resulted in a reduction in Fv/Fm value from the ninth day of starting the salt application. A slight increase in the Fv/Fm ratio was recorded at moderate NaCl treatment (100 mM) as compared to high stress (up to 200 mM) (Zaman et al. 2018). In contrast, He et al. (2021) reported that purslane plants grown hydroponically under the salt range (0–300 mM NaCl) maintained a Fv/Fm ratio close to 0.8. However, the latter study also revealed a higher electron transport rate and $\Delta F/Fm'$ in plants grown with a moderate salt range (0–100 mM NaCl) than with 200 and 300 mM NaCl.

Purslane plants treated with salt solutions (100 and 200 mM NaCl) showed a significant decrease in their photosynthetic rate, while an increase was recorded for the intercellular CO_2 concentration (Xing et al. 2019). Hnilickova et al. (2021) reported that the application of 300 mM NaCl to *P. oleracea* plants resulted in a decrease in stomatal conductance (gs), CO_2 assimilation (A) and leaf water potential (w) recorded 9 days after salt exposure. Furthermore, He et al. (2021) observed that purslane plants treated hydroponically with 300 mM NaCl largely enhanced the non-photochemical quenching (NPQ) mechanism; meanwhile, the maximum net photosynthetic O_2 evolution rate and Cyt *b6f* concentration were reduced markedly compared to all other plants. NPQ is a well-known photoprotective process employed by halophytes to dissipate excessive absorbed light energy into heat (Horton and Ruban 2005). Several halophytes exhibit great ability to convert violaxanthin to zeaxanthin through the xanthophyll cycle, for an effective NPQ, such as *Sesuvium portulacastrum* and *Tecticornia indica* (Rabhi et al. 2012), and *Salicornia neei* (De Souza et al. 2018). Altogether, these results show that purslane plants grown under salt stress are dealing with a limitation in the availability of water (osmotically) and require stomatal regulation, besides the necessity for maintaining adequate CO_2 supply. Therefore, the transition from the C4 CO_2 fixation mechanism to the crassulacean acid metabolism (CAM) represents the next step for such succulent plants, as was suggested by Hnilickova et al. (2021), where they evidenced a rapid increase in substomatal CO_2 concentration and negative CO_2 assimilation values in purslane plants exposed to 300 mM NaCl for 22 days. Salinity is well-known to promote the C3-CAM transition in the halophyte *M. crystallinum* (Winter and Holtum 2007), and *P. oleracea* shows the ability to inhabit arid and saline environments (Parvaneh et al. 2012).

17.4 IONIC RELATIONS

Elevated NaCl dosages in the range of 0–100 mM resulted in a significant increase in Na^+ content in all organs of the *P. oleracea* plant, while the concentration of K^+ was slightly decreased (Bessrour et al. 2018). Xing et al. (2019) reported that the application of 150 mM and 200 mM NaCl to the culture medium led the Na^+ content to upsurge in purslane leaf, stem and root, while a decrease in K^+ content was recorded, which was probably attributed to membrane depolarization and regulation of ion channels. Treatment with 300 mM NaCl increased the Na^+ content 3.4-fold (71.9 mg/g DW) in *P. oleracea* plants, while K^+ was decreased gradually with increasing NaCl concentrations (Hniličková et al. 2019). Similarly, Sdouga et al. (2019) reported an increase in Na^+ content concomitant with a marked reduction in K^+ of all purslane's organs (leaves, stems and roots); however, the plant maintained a higher K^+ in the aerial part with the highest salt level tested (150 mM NaCl) and it was about 33% in the leaves and 46% in the stems compared to no salt condition. In close relation, Ahmed et al. (2022) reported that the succulent halophyte *Sarcocornia quinqueflora* was able to grow/survive even at the highest NaCl concentration tested (1000 mM), most likely by maintaining a constant Na^+/K^+ ratio in the top regions of the leafy stems, regardless of external salt concentration (0–1000 mM). A high level of K^+ brought about a favorable ratio of K^+/Na^+ even at higher salt concentrations (up to 300 mM) in *P. oleracea* (Hniličková et al. 2019). Bekmirzaev et al. (2021) screened the effect of salinity stress (up to 20 dS m^{-1} NaCl) on micronutrient and macronutrient uptake of purslane leaves and noted: (1) an increase in Na$^+$, Cl$^-$ and Ca^{2+} accumulation with elevated salinity, while K$^+$ content decreased; (2) constant values (mostly) of the macronutrients (P^{3-}, Mg^{2+} and S^{2-}); and (3) very low concentrations of the micronutrients Fe^{2+}, Al^{3+}, Ba^{2+}, Sr^{2+}, Zn^{2+} and Cu^{2+}. Purslane plants grown hydroponically showed the highest concentrations of K$^+$, Ca^{2+} and Mg^{2+} under 0 mM NaCl compared to elevated saline conditions (up to 300 mM) (He et al. 2021). Teixeira and Carvalho (2009) also reported an increase in Na$^+$ and Mg^{2+} levels with a salinity increase (up to 240 mM NaCl) in purslane plants, while K$^+$ and Ca^{2+} decreased. The latter variation in ions accumulation resulted in a marked increase in the Na$^+$/K$^+$, Mg^{2+}/K$^+$, Na$^+$/Ca^{2+} and Mg^{2+}/Ca^{2+} ratios, which was reflected by a reduced growth/yield under extreme salinity treatments. Alam et al. (2016) advised that the salt tolerance of several purslane genotypes was not correlated with shoot Na$^+$ accumulation, but the increment in Na$^+$/Ca^{2+}, Na$^+$/K$^+$ and Mg^{2+}/Ca^{2+} ratios and the discrimination between these ions were key factors and largely ruled salt tolerance among these purslane accessions. The latter findings enabled the authors to deduce the possible existence of a range of salt-tolerant mechanisms involved, taking into account the considerable variation obtained in the salinity tolerance among accessions.

The electrolyte leakage (EL) is mainly related to the efflux of K$^+$, which occurs in plentiful amounts in plant cells (Demidchik et al. 2014). A significant increase in EL was recorded for *P. oleracea* plants grown at 100 mM NaCl in comparison to the control, with the highest EL percentages of 86.7% and 92.4% being recorded for 200 and 300 mM, respectively (Hniličková et al. 2019). The EL was enhanced with elevated salinity (0–200 mM NaCl) in leaves and roots of *P. oleracea* (Zaman et al. 2020; Zaman et al. 2018). Thus, high EL values of purslane may be attributed to a naturally high content of K$^+$ rather than being a sign of stress-induced plasma membrane damage, as also reported by Mansour and Salama (2004). In this regard, very interesting findings were revealed for the K$^+$ content in succulent stems of the extreme halophyte *Sarcocornia quinqueflora*, where the plants maintained a constant K$^+$ concentration of around 50 mM under a wide range of salinity treatments (200–1000 mM NaCl) (Ahmed et al. 2021b), suggesting a key role for K$^+$ homeostasis in maintaining adequate stomatal functioning and enzymatic activity. Flowers and Colmer (2008) mentioned that the metabolism of halophytes in the Chenopodiaceae (belong to the same plant order of *P. oleracea*; Caryophyllales) was mainly evolved to adjust to lower concentrations of cytoplasmic K$^+$.

17.5 WATER RELATIONS AND OSMOTIC ADJUSTMENT

Salinity treatment significantly reduced the osmotic potential of leaves and roots of *P. oleracea* up to 100 mM NaCl, as recorded by Bessrour et al. (2018). Purslane plants grown under the salt range of 0–240 mM NaCl showed no significant differences in their leaves' water content (Teixeira and Carvalho 2009). However, Zaman et al. (2018) recorded the highest relative water content (RWC) percentage at no salt treatment of the salt range 0–200 mM NaCl. RWC of purslane leaves was impacted by stress duration in the study carried out by Yazici et al. (2007), where 18 days of treatment resulted in an increase in leaf RWC, while a decrease was recorded after 30 days, this was correct under both the 70 and 140 mM NaCl treatments. Salinity induced a reduction in the leaf water potential of purslane from the ninth day of salt treatment (0–300 mM NaCl) and was statistically significant relative to the control (Hnilickova et al. 2021). Leaf succulence showed similar values for *P. oleracea* plants grown in a hydroponic system (illuminated by LED lighting) with all salt levels, but plants treated with 300 mM NaCl were exceptional (He et al. 2021).

Osmotic adjustment in a number of purslane accessions was suggested to be linked with the increased Na$^+$ influx under elevated NaCl concentrations (0–32 dS m^{-1}) (Alam et al. 2016). Also, the accumulation of proline under salinity conditions (0–140 mM NaCl) was indicated to have a role in purslane osmoregulation (Yazici et al. 2007). The same attribution was also proposed for the significant increase in proline levels of purslane leaves with elevated salt concentrations (NaCl; 0–150 mM) (Sdouga et al. 2019), however, the authors also recorded a significant decrease in other compatible solutes, soluble sugars and proteins, perhaps ruling out their direct involvement in osmoregulation under these conditions. In this regard, NaCl-treated purslane plants (100 mM) exhibited a marked decrease in their protein content compared to the no-salt condition, with 18.5% and 11.6% reductions recorded in the leaves and roots, respectively (Bessrour et al. 2018). Similarly, Teixeira and Carvalho (2009) reported a reduction in the crude protein content of purslane leaves with elevated salinity (up to 240 mM NaCl), and also an increase in exposure time of salt stress; nonetheless, carbohydrates and mineral residue content showed a considerable increase. Purslane plants grown hydroponically revealed similar levels of total soluble sugars under the entire salt range applied (0–300 mM NaCl) (He et al. 2021). Lipid and protein concentrations were found to decrease with increasing salinity levels (0–200 mM NaCl) (Parvaneh et al. 2012). Furthermore, salinity treatments with 100 mM NaCl induced a three-fold increase in fatty acids relative to the control leaves and 7–8-fold in stems. However, in roots, a marked decrease was observed. Therefore, the osmoprotection of purslane plants under salt stress seems to have an effective reliance on moderate Na^+ uptake and elevated proline concentrations, besides other compatible solutes that might be involved. Also, the increase in the exposure time of salt stress might be interfering with the switching point of the above mechanisms.

17.6 OXIDATIVE STRESS AND ANTIOXIDANTS

Maintaining reactive oxygen species (ROS) homeostasis is a key target of several mechanisms launched by plants experiencing salt stress conditions, for the protection of their cellular structures and the best growth responses (Tanveer and Ahmed 2020). As mentioned above, the photosynthetic machinery of purslane plants is inhibited under salt stress, a factor that predictably provokes ROS levels to unfavorable thresholds, and also causes an upsurge of their scavengers. In some purslane accessions, Alam et al. (2015a) recorded an increase in some antioxidant molecules including total phenolics content (TPC), total flavonoids content (TFC) and ferric reducing antioxidant power (FRAP) activity with 8 dS m^{-1} NaCl treatment. Treatment with NaCl dosages (0–100 mM) came with a negative effect on phenolic compounds in leaves and stems of *P. oleracea* plants, while no change was observed in roots (Bessrour et al. 2018). He et al. (2021) reported that higher salinity treatments of the tested range (0–300 mM NaCl; hydroponically) resulted in lower concentrations of ascorbic acid (Asc) and TPC in purslane, which was attributed to its reduced photosynthetic performance. However, the latter authors recorded an increase in proline content with elevated salinity levels. In this regard, Ahmed et al. (2021b) showed that the succulent halophyte *Sarcocornia quinqueflora* accumulated proline excessively in its shoots and roots above the optimal salt dosage (200 mM NaCl) and up to 1000 mM, with antioxidant activities being suggested. A significant increase in malondialdehyde (MDA) and proline contents was observed in purslane plants treated with 300 mM NaCl (Hnilickova et al. 2021). The same also was found in the study carried out by Yazici et al. (2007), nevertheless, the exposure time (to salt stress; 140 mM NaCl) was shown to be crucial, where 18 days of treatment did not reveal any change in MDA or proline contents, but a significant increase was recorded after 30 days. Both leaf proline and sugar levels of purslane were elevated with NaCl increment (up to 200 mM) (Parvaneh et al. 2012). Mulry et al. (2015) reported an increase in proline and betalains (betacyanins and betaxanthins) under salt stress in two varieties of purslane. However, the authors illustrated that the two varieties perhaps had different strategies in allocating resources to deal with NaCl stress (up to 200 mM); as one variety invested more in betalains, while the other invested more in proline. Betalains are water-soluble pigments that accumulate largely under salt stress, instead of anthocyanins, in plant families belonging to Caryophyllales, such as those reported recently for some succulent halophytes (Ahmed et al. 2021a;

Parida et al. 2018). Betalains and/or anthocyanins were suggested for a potential role in the scavenging of cytotoxic ROS (Parida et al. 2018).

Enzymatic antioxidants appear to play an important role in relieving salinity-induced oxidative stress in purslane, with some concerns about the exposure time and the ROS type. In that respect, Yazici et al. (2007) reported enhancement in catalase (CAT) and glutathione reductase (GR) activities in leaves of purslane after 18 days of salt stress (up to 140 mM), while ascorbate peroxidase (APX) and GR were the prominent members after 30 days of treatment; inhibiting the increase in lipid peroxidation. Treatments of purslane plants with 100–200 mM NaCl resulted in increasing the MDA content and the production rate of O_2^-, inducing activities of CAT, peroxidase (POD) and superoxide dismutase (SOD); nevertheless the prolonged exposure to salinity stress showed a decreasing tendency in their activities (Xing et al. 2019). Yazici et al. (2007) suggested that GR, APX and CAT enzymes are mostly involved in H_2O_2 scavenging in purslane leaves, while non-enzymatic antioxidants like α-tocopherol, glutathione and ascorbate might have been involved in the scavenging of superoxide radicals. The latter suggestion was built upon the fact that SOD activity did not show an increase with elevated salinity.

17.7 GENETIC, TRANSCRIPTOMIC AND METABOLOMIC RESPONSES

Molecular analyses at the genetic level are an important guide for breeding programs targeting high-yielding and stress-tolerant characteristics of purslane agriculture. Genetic diversity and relationships among 45 collected purslane accessions were estimated by Alam et al. (2015b), showing a high level of genetic diversity in the germplasm. Alam et al. (2015a) indicated potential wide genetic diversities among 12 purslane accessions screened under salt stress, where a great variation in their contents of bioactive compounds and their responses to salt treatments was recorded. Ferrari et al. (2020) developed protocols for measuring the mRNA levels of target genes of interest in *P. oleracea* using RT-qPCR, with 12 potential reference genes being considered to be optimal, including CYP and CLAT genes for salinity stress among others involved in several physiological responses such as the variation in leaf development—CLAT and TIP; tissue type and water availability—UBC10 and CYP; day/night sampling time—CLAT and EF1A; subspecies—TIP and HEL; and hormone treatments—TIP and PP2A. The authors added that the transition from C4 to CAM in *P. oleracea* is also promoted under salinity conditions, based on the data gained for PPC-1E1c relative transcript abundance that showed a substantial upregulation in leaves under salt stress. It was found that expression levels of the Pyrroline 5-Carboxylate Synthetase (P5CS) gene in purslane plants are more intense under higher salinity levels (up to 150 mM) compared to zero salt, indicating that proline synthesis is important in plants' adaptation under saline conditions (Sdouga et al. 2019). El-Bakatoushi and Elframawy (2016) reported that *P. oleracea* plants were able to cope with higher saline conditions (up to 200 mM NaCl) by employing the increase in expression levels of PoVHA (*Portulaca oleracea* vacuolar H⁺-ATPase; an energy provider for tonoplast Na^+/H^+ antiporter) for efficient vacuolar sequestration ability of Na^+, concomitant to the decrease in the expression of PoHKT1 (high-affinity potassium transporter; proposed to mediate a substantial Na^+-influx) to hinder Na^+ influx into the cytosol. Salinity stress might inhibit photosynthesis and water uptake in *P. oleracea*, where treating the plant with 200 mM NaCl downregulated genes required for photosynthesis and aquaporin functioning (Xing et al. 2020). However, the latter study also showed an increase in expression levels of aminocyclopropane-1-carboxylic acid synthase, bifunctional L-3-cyanoalanine synthase/cysteine synthase and cyanoalanine synthase, promoting ethylene production.

Transcriptome analysis screened the differentially expressed genes (DEGs) from two ecotypes of purslane (one with the red stem living in an arid environment and one with the green stem living in a humid environment), and most DEGs belonged to the WRKY and NAC families [transcription factors (TFs) families]. The latter were found to play key roles in the regulation of plants' resistance to hostile conditions, including adaptation to water stress (Wu et al. 2021). Transcriptome profiles were compared for purslane leaves in plants grown with 200 mM NaCl for 0, 2, 6, 12 and 24 h (Xing et al. 2019), and the transcriptome sequencing showed hundreds to thousands of DEGs, compared with the control (0 mM NaCl). These results illustrated that salinity stress induced the downregulation of transcription levels of genes involved in photosynthesis, energy metabolism, lignin biosynthesis and signaling transduction. The authors also observed that the number of DEGs increased with exposure time from 2 to 12 h, suggesting a short-term emergency reaction of purslane in response to unfavorable NaCl dosages. The latter findings were confirmed by qPCR analyses of six unigenes, that showed similar changing tendency in RPKM (reads per kilobase of transcript sequence per million mapped reads) values among different treatments, including: (1) the chloroplastic omega-3 fatty acid desaturase (*FAD3*); (2) the endoplasmic reticulum isozyme, omega-6 fatty acid desaturase (*FAD6*); (3) probable plastidic glucose transporter (*PPGT*); (4) sodium/hydrogen exchanger (*SHE*); (5) the ABC transporter G family member (*ABCT-G*); and (6) the chloride channel protein (*CCP*). Overall, enrichment of the KEGG (Kyoto Encyclopedia of Genes and Genomes database) pathway showed that salt stress had significant effects on purslane genes involved in photosynthesis and carbon fixation, oxidative phosphorylation and pentose phosphate, flavone, flavonol and phenylpropanoid biosynthesis, phenylalanine, porphyrin, chlorophyll and glycerolipid metabolism and plant hormone signal transduction (Xing et al. 2019). The study carried out by Du et al. (2021) targeted the full-length transcriptome of *P. oleracea*, revealing that 132,536 transcripts and 78,559 genes were detected. The authors also showed the amounts of DEGs and related KEGG pathways, and the expression profiles of related genes in the biosynthesis of unsaturated fatty acids pathway in leaves and roots of two purslane genotypes were also analyzed. The latter indicated the accumulation of ω-3 fatty acid in different organs and genotypes of purslane.

Zaman et al. (2020) screened the metabolic profiling (using GC-MS analysis) of leaves and roots of two *P. oleracea* cultivars, Tall Green "TG" and Shandong Wild "SD," under salinity stress, proposing a dramatic change when plants were treated with 100 and 200 mM NaCl in comparison to control. Through the latter study, 132 different metabolites were quantified by applying the metabolic pathway analysis in response to salt stress, including 35 organic acids, 26 amino acids, 20 sugars, 14 sugar alcohols, 20 amines, 13 lipids and sterols, and four other acids. The author added that these metabolites are a part of several biochemical pathways, such as the TCA cycle, GS/GOGAT cycle, GABA, glycolysis, proline synthesis, shikimic acid and amino acid metabolic pathway. Metabolomic analyses of purslane plants treated with 200 mM NaCl indicated higher contents of pyrophosphate, D-galacturonic acid and elaidic acid compared to the control, which positively regulate glycolysis, energy supply and the integrity of the cell membrane (Xing et al., 2020).

17.8 STIMULANTS-INDUCED AMENDMENT OF GROWTH/PRODUCTION UNDER SALT STRESS

Several stimulants have been reported for improving purslane plants' performance under saline conditions (Table 17.2). The combination of water salinity (80 mM NaCl) and LED lighting (RB, red and blue; and RB+FR, far red lights) improved the yield and quality of purslane microgreens, reducing the content of anti-nutritional compounds; where oxalate, Na^+ and Cl^- contents were reduced under these combinations and an increase was recorded for total phenolics (Giménez et al. 2021). Although purslane has been widely accepted for commercial agriculture, health concerns were raised considering the potential detrimental impacts of its higher oxalic acid content on human consumption (Palaniswamy et al. 2004; Palaniswamy et al. 2001; Petropoulos et al. 2016). In this regard, Camalle et al. (2020) compared the responses of two purslane ecotypes under elevated NH_4^+ concentrations (33% and 75%) and 50 mM NaCl, and reported that the ecotype ET displayed a significant reduction in oxalic acid content, while total chlorophyll and carotenoid contents, crude protein content, total fatty acid and α-linolenic acid were increased, enhancing its leaf quality in comparison to the ecotype RN.

TABLE 17.2
Stimulants-induced Amendment of Growth/Production of *Portulaca Oleracea* under Salt Stress

Stimulant applied	Effect	References
LED lighting	Combining of water salinity (80 mM NaCl) and LED lighting (RB, red and blue; and RB+FR, far red lights) improved the yield and quality of purslane microgreens, reducing the content of anti-nutritional compounds; where oxalate, Na^+ and Cl^- contents were reduced under these combinations and an increase was recorded for total phenolics	Giménez et al. 2021
Ammonium (NH_4^+)	NH_4^+ concentrations (33% and 75%) with 50 mM NaCl treatment resulted in a significant reduction in oxalic acid content for ecotype ET, while total chlorophyll and carotenoid contents, crude protein content, total fatty acid and α-linolenic acid were increased, enhancing its leaf quality in comparison to ecotype RN	Camalle et al. 2020
Salicylic acid (SA)	SA application alleviated the deleterious effects of salinity stress (up to 80 mM), and 1 mM of SA was more effective in increasing growth and pigments content than 0.5 mM	Fathi et al. 2019
Ascorbic acid (AsA)	The interaction effect of ascorbic acid (0–20 mM) and salinity stress (0–210 mM NaCl) improved all studied characteristics including some morphological traits, cell membrane stability, leaf relative water content and extract yield	Niki et al. 2013
Silicon (Si)	Silicon (sodium silicate; 1 mM) application to purslane grown under a range of NaCl concentrations (0.6, 7, 14 and 21 dS/m) indicated a positive effect on number of leaves per plant, leaf area index and height of main stem	Rahimi et al. 2011

Purslane plants grown under limited water availability showed better growth and biomass yield after a foliar spray with salicylic acid was applied, improving photosynthetic pigments and gas exchange, increasing the accumulation of secondary metabolites and compatible solutes, improving the antioxidant defense system and regulating the fatty acid profile of membrane lipid (Saheri et al. 2020). The latter alterations are comparable to the positive effect of SA in increasing salinity tolerance in several plant species, as reported elsewhere (Li et al. 2014; Nazar et al. 2015). SA application was found to alleviate the deleterious effects of salinity stress (up to 80 mM) on purslane plants, where 1 mM of SA was more effective in increasing growth and pigments content than 0.5 mM (Fathi et al. 2019).

Ascorbate is a strong antioxidant that can improve several plants' responses to salt stress. The interaction effect of salinity stress (0–210 mM NaCl) and ascorbic acid (0–20 mM) on *P. oleracea* was found to improve all the studied characteristics, including some morphological traits, cell membrane stability, leaf relative water content and extract yield (Niki et al. 2013).

Furthermore, the application of silicon (sodium silicate; 1 mM) to purslane grown under a range of NaCl concentrations (0.6, 7, 14, 21 dS/m^{-1}) indicated a positive effect on the number of leaves per plant, leaf area index and height of the main stem (Rahimi et al. 2011).

17.9 CONCLUSION

Purslane shows high adaptability to salt stress compared to many other cultivated vegetable crops. Nevertheless, a considerable variation in salt tolerance among purslane accessions does exist. Available works of literature show that the applied salt ranges in addition to the exposure period drive the proper growth responses of purslane. The plant produces sufficient biomass yield under moderate salinity, while salt dosages above 100 mM NaCl appear to be suppressive. The plant maintains favorable Fv/Fm values, and an efficient non-photochemical quenching (NPQ) mechanism under moderate saline conditions, while extreme salinity induces the transition from the C4 CO_2 fixation mechanism to the CAM. Na^+ may be used as a cheap osmoticum under moderate saline conditions, but rather the accumulation of compatible solutes such as proline and soluble sugars under higher salt concentrations. Maintaining adequate Na^+/K^+, Mg^{2+}/K^+, Na^+/Ca^{2+} and Mg^{2+}/Ca^{2+} ratios is important for the proper purslane responses to saline exposure, while the increase in these ratios results in reduced growth/yield. Severe saline conditions provoke ROS levels to unfavorable thresholds in purslane and also cause an increase in their scavengers (enzymatic and non-enzymatic antioxidants, in addition to betalains). Salt stress affects purslane genes involved in photosynthesis and carbon fixation, oxidative phosphorylation and pentose phosphate, flavone, flavonol and phenylpropanoid biosynthesis, phenylalanine, porphyrin, chlorophyll and glycerol-lipid metabolism and plant hormone signal transduction. Several stimulants are suggested for better plant performances under salt stress, including LED lighting, salicylic acid (SA), ascorbic acid (Asc), NH_4^+ and/or silicon. The above factors may be of importance when recommending the proper practices for cultivating purslane plants under saline conditions, with the ultimate aim of achieving the best pharmaceutical and nutritional value.

REFERENCES

Ahmed, H.A.I., Shabala, L., Shabala, S., 2021a. Tissue-specificity of ROS-induced K^+ and Ca^{2+} fluxes in succulent stems of the perennial halophyte *Sarcocornia quinqueflora* in the context of salinity stress tolerance. Plant Physiology and Biochemistry 166, 1022–1031.

Ahmed, H.A.I., Shabala, L., Shabala, S., 2021b. Understanding the mechanistic basis of adaptation of perennial *Sarcocornia quinqueflora* species to soil salinity. Physiologia Plantarum 172, 1997–2010.

Ahmed, H.A.I., Shabala, S., Goemann, K., Shabala, L., 2022. Development of suberized barrier is critical for ion partitioning between senescent and non-senescent tissues in a succulent halophyte *Sarcocornia quinqueflora*. Environmental and Experimental Botany 194, 104692.

Alam, M., Juraimi, A.S., Rafii, M., Hamid, A.A., Aslani, F., Hakim, M., 2016. Salinity-induced changes in the morphology and major mineral nutrient composition of purslane (*Portulaca oleracea* L.) accessions. Biological research 49, 1–19.

Alam, M.A., Juraimi, A., Rafii, M., Hamid, A., Aslani, F., Alam, M., 2015a. Effects of salinity and salinity-induced augmented bioactive compounds in purslane (*Portulaca oleracea* L.) for possible economical use. Food Chemistry 169, 439–447.

Alam, M.A., Juraimi, A.S., Rafii, M.Y., Hamid, A.A., Arolu, I.W., Latif, M.A., 2015b. Genetic diversity analysis among collected purslane (*Portulaca oleracea* L.) accessions using ISSR markers. Comptes Rendus Biologies 338, 1–11.

Aronson, J.A., Whitehead, E.E., 1989. HALOPH: a data base of salt tolerant plants of the world.

Ayala, F., O'Leary, J.W., 1995. Growth and physiology of *Salicornia bigelovii* Torr. at suboptimal salinity. International Journal of Plant Sciences 156, 197–205.

Bekmirzaev, G., Ouddane, B., Beltrao, J., Khamidov, M., Fujii, Y., Sugiyama, A., 2021. Effects of salinity on the macro-and micronutrient contents of a halophytic plant species (*Portulaca oleracea* L.). Land 10, 481.

Bessrour, M., Chelbi, N., Moreno, D.A., Chibani, F., Abdelly, C., Carvajal, M., 2018. Interaction of salinity and $CaCO_3$ affects the physiology and fatty acid metabolism in *Portulaca oleracea*. Journal of Agricultural and Food Chemistry 66, 6683–6691.

Camalle, M., Standing, D., Jitan, M., Muhaisen, R., Bader, N., Bsoul, M., Ventura, Y., Soltabayeva, A., Sagi, M., 2020. Effect of salinity and nitrogen sources on the leaf quality, biomass, and metabolic responses of two ecotypes of *Portulaca oleracea*. Agronomy 10, 656.

Carillo, P., Annunziata, M.G., Pontecorvo, G., Fuggi, A., Woodrow, P., 2011. Salinity stress and salt tolerance, Abiotic Stress in Plants-Mechanisms and Adaptations. InTech.

Chauhan, B., Johnson, D., 2009. Seed germination ecology of *Portulaca oleracea* L.: an important weed of rice and upland crops. Annals of applied biology 155, 61–69.

Clipson, N.J.W., Tomos, A.D., Flowers, T.J., Jones, R.G.W., 1985. Salt tolerance in the halophyte *Suaeda maritima* L. Dum. Planta 165, 392–396.

D'andrea, R.M., Andreo, C.S., Lara, M.V., 2014. Deciphering the mechanisms involved in *Portulaca oleracea* (C4) response to drought: metabolic changes including crassulacean acid-like metabolism induction and reversal upon re-watering. Physiologia Plantarum 152, 414–430.

De Souza, M.M., Mendes, C.R., Doncato, K.B., Badiale-Furlong, E., Costa, C.S., 2018. Growth, phenolics, photosynthetic pigments, and antioxidant response of two new genotypes of sea asparagus (*Salicornia neei* Lag.) to salinity under greenhouse and field conditions. Agriculture 8, 115.

Demidchik, V., Straltsova, D., Medvedev, S.S., Pozhvanov, G.A., Sokolik, A., Yurin, V., 2014. Stress-induced electrolyte leakage: the role of K^+-permeable channels and involvement in programmed cell death and metabolic adjustment. Journal of Experimental Botany 65, 1259–1270.

Du, H., Zaman, S., Hu, S., Che, S., 2021. Single-molecule long-read sequencing of purslane (*Portulaca oleracea*) and differential gene expression related with biosynthesis of unsaturated fatty acids. Plants 10, 655.

Dweck, A., 2001. Purslane-the global panacea.

El-Bakatoushi, R., Elframawy, A., 2016. Diversity in growth and expression pattern of PoHKT1 and PoVHA transporter genes under NaCl stress in *Portulaca oleracea* taxa. Genetika 48, 233–248.

Fathi, S., Kharazmi, M., Najafian, S., 2019. Effects of salicylic acid foliar application on morpho-physiological traits of purslane (*Portulaca olaracea* L.) under salinity stress conditions. Journal of Plant Physiology and Breeding 9, 1–9.

Ferrari, R.C., Bittencourt, P.P., Nagumo, P.Y., Oliveira, W.S., Rodrigues, M.A., Hartwell, J., Freschi, L., 2020. Developing *Portulaca oleracea* as a model system for functional genomics analysis of C4/CAM photosynthesis. Functional Plant Biology 48, 666–682.

Flowers, T.J., Colmer, T.D., 2008. Salinity tolerance in halophytes. New Phytologist, 945–963.

Giménez, A., Martínez-Ballesta, M.d.C., Egea-Gilabert, C., Gómez, P.A., Artés-Hernández, F., Pennisi, G., Orsini, F., Crepaldi, A., Fernández, J.A., 2021. Combined effect of salinity and LED lights on the yield and quality of purslane (*Portulaca oleracea* L.) microgreens. Horticulturae 7, 180.

He, J., You, X., Qin, L., 2021. High salinity reduces plant growth and photosynthetic performance but enhances certain nutritional quality of C4 halophyte *Portulaca oleracea* L. grown hydroponically under LED lighting. Frontiers in Plant Science 12, 457.

Hniličková, H., Hnilička, F., Orsák, M., Hejnák, V., 2019. Effect of salt stress on growth, electrolyte leakage, Na^+ and K^+ content in selected plant species. Plant, Soil and Environment 65, 90–96.

Hnilickova, H., Kraus, K., Vachova, P., Hnilicka, F., 2021. Salinity stress affects photosynthesis, malondialdehyde formation, and proline content in *Portulaca oleracea* L. Plants 10, 845.

Holm, L., Plucknett, D., Pancho, J., Herberger, J., 1977. *Portulaca oleracea* L. Portulacaceae, purslane family. World's worst weeds: Distribution and biology. University of Hawaii Press, Honolulu, 78–83.

Horton, P., Ruban, A., 2005. Molecular design of the photosystem II light-harvesting antenna: photosynthesis and photoprotection. Journal of Experimental Botany 56, 365–373.

Kafi, M., Rahimi, Z., 2011. Effect of salinity and silicon on root characteristics, growth, water status, proline content and ion accumulation of purslane (*Portulaca oleracea* L.). Soil Science and Plant Nutrition 57, 341–347.

Kiliç, C.C., Kukul, Y.S., Anaç, D., 2008. Performance of purslane (*Portulaca oleracea* L.) as a salt-removing crop. Agricultural Water Management 95, 854–858.

Li, T., Hu, Y., Du, X., Tang, H., Shen, C., Wu, J., 2014. Salicylic acid alleviates the adverse effects of salt stress in *Torreya grandis* cv. Merrillii seedlings by activating photosynthesis and enhancing antioxidant systems. PLOS one 9, e109492.

Mansour, M.M.F., Salama, K.H., 2004. Cellular basis of salinity tolerance in plants. Environmental and Experimental Botany 52, 113–122.

Moir-Barnetson, L., Veneklaas, E.J., Colmer, T.D., 2016. Salinity tolerances of three succulent halophytes (*Tecticornia* spp.) differentially distributed along a salinity gradient. Functional Plant Biology 43, 739.

Mulry, K.R., Hanson, B.A., Dudle, D.A., 2015. Alternative strategies in response to saline stress in two varieties of *Portulaca oleracea* (purslane). PloS One 10, e0138723.

Munns, R., 2002. Comparative physiology of salt and water stress. Plant, Cell & Environment 25, 239–250.

Naik, V.V., Karadge, B.A., 2017. Effect of NaCl and Na₂SO4 salinities and light conditions on seed germination of purslane (*Portulaca oleracea* Linn.). Journal of Plant Stress Physiology 3, 1–4.

Nazar, R., Umar, S., Khan, N.A., 2015. Exogenous salicylic acid improves photosynthesis and growth through increase in ascorbate-glutathione metabolism and S assimilation in mustard under salt stress. Plant Signaling & Behavior 10, e1003751.

Nicotra, A.B., Atkin, O.K., Bonser, S.P., Davidson, A.M., Finnegan, E.J., Mathesius, U., Poot, P., Purugganan, M.D., Richards, C.L., Valladares, F., 2010. Plant phenotypic plasticity in a changing climate. Trends in Plant Science 15, 684–692.

Niki, E.E., Pazoki, A., Rezaei, H., Eradatmande, A.D., Yousefirad, M., 2013. Effects of ascorbate foliar application on morphological traits, relative water content and extract yield of Purslane (*Portulaca oleracea* L.) under salinity stress. Iranian Journal of Plant Physiology 4, 889–898.

Ocampo, G., Columbus, J.T., 2012. Molecular phylogenetics, historical biogeography, and chromosome number evolution of Portulaca (Portulacaceae). Molecular phylogenetics and Evolution 63, 97–112.

Palaniswamy, U.R., Bible, B.B., McAvoy, R.J., 2004. Oxalic acid concentrations in Purslane (*Portulaca oleracea* L.) is altered by the stage of harvest and the nitrate to ammonium ratios in hydroponics. Scientia Horticulturae 102, 267–275.

Palaniswamy, U.R., McAvoy, R.J., Bible, B.B., 2001. Stage of harvest and polyunsaturated essential fatty acid concentrations in purslane (*Portulaca oleracea*e) leaves. Journal of Agricultural and Food Chemistry 49, 3490–3493.

Parida, A., Kumari, A., Panda, A., Rangani, J., Agarwal, P., 2018. Photosynthetic pigments, betalains, proteins, sugars, and minerals during *Salicornia brachiata* senescence. Biologia Plantarum 62, 343–352.

Parvaneh, R., Shahrokh, T., Meysam, H.S., 2012. Studying of salinity stress effect on germination, proline, sugar, protein, lipid and chlorophyll content in purslane (*Portulaca oleracea* L.) leaves. Journal of Stress Physiology & Biochemistry 8, 182–193.

Petropoulos, S., Karkanis, A., Martins, N., Ferreira, I.C., 2016. Phytochemical composition and bioactive compounds of common purslane (*Portulaca oleracea* L.) as affected by crop management practices. Trends in Food Science & Technology 55, 1–10.

Rabhi, M., Castagna, A., Remorini, D., Scattino, C., Smaoui, A., Ranieri, A., Abdelly, C., 2012. Photosynthetic responses to salinity in two obligate halophytes: *Sesuvium portulacastrum* and *Tecticornia indica*. South African Journal of Botany 79, 39–47.

Rahimi, Z., Kafi, M., Nezami, A., Khozaie, H., 2011. Effect of salinity and silicon on some morphophysiologic characters of purslane (*Portulaca oleracea* L.). Iranian Journal of Medicinal and Aromatic Plants Research 27, 359–374.

Redondo-Gomez, S., Mateos-Naranjo, E., Figueroa, M.E., Davy, A.J., 2010. Salt stimulation of growth and photosynthesis in an extreme halophyte, *Arthrocnemum macrostachyum*. Plant Biology (Stuttg) 12, 79–87.

Rozentsvet, O.A., Nesterov, V.N., Bogdanova, E.S., 2017. Structural, physiological, and biochemical aspects of salinity tolerance of halophytes. Russian Journal of Plant Physiology 64, 464–477.

Saheri, F., Barzin, G., Pishkar, L., Boojar, M.M.A., Babaeekhou, L., 2020. Foliar spray of salicylic acid induces physiological and biochemical changes in purslane (*Portulaca oleracea* L.) under drought stress. Biologia 75, 2189–2200.

Sdouga, D., Amor, F.B., Ghribi, S., Kabtni, S., Tebini, M., Branca, F., Trifi-Farah, N., Marghali, S., 2019. An insight from tolerance to salinity stress in halophyte *Portulaca oleracea* L.: Physio-morphological, biochemical and molecular responses. Ecotoxicology and Environmental Safety 172, 45–52.

Sultana, A., Rahman, K., 2013. *Portulaca oleracea* Linn. A global Panacea with ethno-medicinal and pharmacological potential. International Journal of Pharmacy and Pharmaceutical Sciences 5, 33–39.

Tanveer, M., Ahmed, H.A.I., 2020. ROS signalling in modulating salinity stress tolerance in plants, Salt and drought stress tolerance in plants. Springer, pp 299–314.

Teixeira, M., Carvalho, I.S.d., 2009. Effects of salt stress on purslane (*Portulaca oleracea*) nutrition. Annals of Applied Biology 154, 77–86.

Uddin, M., Juraimi, A.S., Ali, M., Ismail, M.R., 2012. Evaluation of antioxidant properties and mineral composition of purslane (*Portulaca oleracea* L.) at different growth stages. International Journal of Molecular Sciences 13, 10257–10267.

Winter, K., Holtum, J.A., 2007. Environment or development? Lifetime net CO_2 exchange and control of the expression of crassulacean acid metabolism in *Mesembryanthemum crystallinum*. Plant Physiology 143, 98–107.

Wu, M., Fu, S., Jin, W., Xiang, W., Zhang, W., Chen, L., 2021. Transcriptome comparison of physiological divergence between two ecotypes of *Portulaca oleracea*. Biologia Plantarum 65, 212–220.

Xing, J.-C., Dong, J., Wang, M.-W., Liu, C., Zhao, B.-Q., Wen, Z.-G., Zhu, X.-M., Ding, H.-R., Zhao, X.-H., Hong, L.-Z., 2019. Effects of NaCl stress on growth of *Portulaca oleracea* and underlying mechanisms. Brazilian Journal of Botany 42, 217–226.

Xing, J., Zhao, B., Dong, J., Liu, C., Wen, Z., Zhu, X., Ding, H., He, T., Yang, H., Wang, M., 2020. Transcriptome and metabolome profiles revealed molecular mechanisms underlying tolerance of *Portulaca oleracea* to saline stress. Russian Journal of Plant Physiology 67, 146–152.

Yazici, I., Türkan, I., Sekmen, A.H., Demiral, T., 2007. Salinity tolerance of purslane (*Portulaca oleracea* L.) is achieved by enhanced antioxidative system, lower level of lipid peroxidation and proline accumulation. Environmental and Experimental Botany 61, 49–57.

Yeo, A., 1983. Salinity resistance: physiologies and prices. Physiologia Plantarum 58, 214–222.

Zaman, S., Bilal, M., Du, H., Che, S., 2020. Morphophysiological and comparative metabolic profiling of purslane genotypes (*Portulaca oleracea* L.) under salt stress. BioMed Research International 2020.

Zaman, S., Shah, S., Jiang, Y., Che, S., 2018. Saline conditions alter morpho-physiological intensification in purslane (*Portulaca oleracea* L.). J. Biol. Regul. Homeost. Agents 32, 635–639.

18 *Portulaca oleracea* under Drought Stress

Peichen Hou[1], Ping Yun[2], Mei Qu[2,3], Aixue Li[1*],*
Hassan Ahmed Ibraheem Ahmed[2,4],
Waqas-ud-Din Khan[5] and Bin Luo[1*]*
[1]Research Center of Intelligent Equipment, Beijing Academy of
Agriculture and Forestry Sciences, Beijing, China
[2]Tasmanian Institute of Agriculture, University of Tasmania, Hobart,
TAS, 7005, Australia
[3]International Research Centre for Environmental Membrane Biology,
Foshan University, Foshan, China
[4]Department of Botany, Faculty of Science, Port Said University,
Port Said, 42526, Egypt
[5]Sustainable Development Study Centre, Government College
University, Lahore, 54000, Pakistan
*Corresponding author: hassan.ahmed@utas.edu.au;
hassan.ahmed.sci@gmail.com; houpc@nercita.org.cn;
luob@nercita.org.cn; liax@nercita.org.cn; dr.waqasuddin@gcu.edu.pk

CONTENTS

18.1 INTRODUCTION

Drought stress is one of the most serious natural disasters worldwide. Plants in arid and semi-arid regions of the world are affected by drought stress (Anjum et al. 2011). The latter issue has been worsening rapidly in recent years, inhibiting plant growth and development, and resulting in crop yield reductions (Munns 2002). Therefore, drought stress is an important constraint on agricultural production (Umezawa et al. 2006). Drought stress causes a limitation in water available to plant roots and indirectly reduces the availability of nutrients in the soil (Baghbani-Arani et al. 2017), and causes changes in plant leaf morphology, root and stem phenotype, chlorophyll content, photosynthesis, nutrient metabolism, phytohormone metabolism and respiration rate (Hura

DOI: 10.1201/9781003242963-18

et al. 2015). Plants are immobile and hence can be stressed by various natural environments, however, after long-term natural selection, they have evolved numerous mechanisms to adapt to such environmental stresses (Basu et al. 2016; Xiong and Yang 2003; Yamaguchi-Shinozaki and Shinozaki 2006; Zhu 2002). In response to drought stress, plants may exhibit a short life cycle, by flowering earlier, to escape drought stress (Shavrukov et al. 2017); changing leaf morphology and structure to adapt to drought conditions (Poorter and Markesteijn 2008); avoiding excessive water loss through stomatal regulation (Qi et al. 2018); hindering excessive water transpiration by changing the structure of the waxy layer (Xue et al. 2017); creating a well-developed root system to use deep groundwater to supply aboveground tissues (Hu and Xiong 2014; Lopes et al. 2011); synthesizing osmotic adjustment (OA) molecules to regulate turgor to maintain cell morphology against drought stress (Xiong et al. 2002); scavenging free radicals through antioxidant action to avoid damage to cell membranes (Farooq et al. 2009a); and reducing energy consumption and maintaining basic life activities through leaf shedding and early senescence (Munne-Bosch and Alegre 2004).

Portulaca oleracea is an annual succulent herbaceous plant that belongs to the genus *Portulaca* in the family Portulacaceae of the Caryophyllales. *Portulaca* is the only genus of the family Portulacaceae, with about 100 varieties distributed around the world, mainly in the tropics and subtropics (Ocampo and Columbus 2012). There are eight *Portulaca* plant populations based on the distribution and classification in the world (Liu et al. 2000). *P. oleracea* is commonly used as a medicinal plant and vegetable (Alam et al. 2014; Dkhil et al. 2011; Duan et al. 2021; Jin et al. 2015; Karkanis and Petropoulos 2017; Petropoulos et al. 2016; Srivastava et al. 2021; Xiang et al. 2005; Zhou et al. 2015) (Figure 18.1a), but also as a horticultural plant (Borsai et al. 2020), as a forage grass in some areas (Siriamornpun and Suttajit 2010), and even as a bioenergy plant (Borland et al. 2009). The commercial prospects of *P. oleracea* are gaining increasing attention (Lara et al. 2011). *P. oleracea* is a special C4/CAM pathway plant (Ferrari and Freschi 2019; Ferrari et al. 2020a; Ferrari et al. 2020b; Koch and Kennedy 1982; Lara et al. 2004), which can be found in well-irrigated soil It is a phenotyped C4 pathway plant, but it exhibits the physiological characteristics of the crassulacean acid metabolism (CAM) pathway under drought stress. Moreover, after re-watering, *P. oleracea* was able to rapidly resume growth and development by the metabolic mechanism of the C4 pathway (D'Andrea et al. 2014; Lara et al. 2003), therefore *P. oleracea* has strong environmental adaptability and plasticity (Ferrari et al. 2020a; Ferrari et al. 2020b) (Figure 18.1b). Another prominent feature of *P. oleracea* is that it can complete its life cycle in 2–4 months (Singh 1973), and so it can escape drought stress. Due to its strong tolerance

(a) (b)

FIGURE 18.1 Field and wild *P. oleracea*: (a) *P. oleracea* is grown as a vegetable, (b) *P. oleracea* is one of the dominant weeds in the field (Photo: Peichen Hou).

to environmental stress, *P. oleracea* has become an important material for the study of plant drought tolerance mechanisms (Ferrari et al. 2020a; Ferrari et al. 2020b; Jin et al. 2015; Parvaneh et al. 2012; Rahdari et al. 2012).

18.2 GROWTH RESPONSES

Drought affects photosynthesis and respiration by reducing the amount of water available around the plant rhizosphere, thereby inhibiting plant growth and development, and reducing grain yield and quality (Anjum et al. 2011; Garcia-Gomez et al. 2000). Drought stress leads to reduced root water uptake, accelerated mature leaf senescence and premature leaf shedding, and it also leads to slower cell elongation and reduced meristem viability, thereby reducing plant growth and yield (Munns 2002). Leaf morphology and structure under drought stress are crucial for water loss and water use efficiency (WUE). Therefore, plants respond to drought stress by adjusting the osmotic pressure to change cell size, showing leaf wilting or curling morphology (Poorter and Markesteijn 2008). Leaves are covered with a waxy layer to reduce water loss (Shepherd and Wynne Griffiths 2006; Xue et al. 2017). Also, regulation of stomatal density and stomatal aperture on leaves are associated with plant responses to drought stress (Qi et al. 2018; Yuan et al. 2019). The plant root is the first tissue to feel the signal of water deficit when drought stress occurs, and some drought-tolerant plants can grow rapidly or increase the root–shoot ratio by using deep groundwater to ensure aboveground water supply (Hu and Xiong 2014; Lopes et al. 2011; Wu and Cosgrove 2000). However, the growth of plant roots is inhibited and the morphology and structure are changed under severe drought stress (Price et al. 2002). Although *P. oleracea* is a drought-tolerant plant, the drought-tolerant ability of different genotypes varies greatly, which is reflected in the genetic variation and morphological characteristics (Karkanis and Petropoulos 2017), physiological indicators (Alam et al. 2015a), agronomic features (Egea-Gilabert et al. 2014) and yield differences (Karkanis and Petropoulos 2017).

At present, the research into the effect of drought stress on the germination of *P. oleracea* seeds mainly includes the following aspects from the areas of medicinal value and edible vegetables (Rahdari et al. 2012; Ren et al. 2011) and weed management (Chauhan and Johnson 2009). The effects of drought stress on the germination of *P. oleracea* seeds may depend on dose and time effects. Studies have shown that under osmotic stress from −0.1 to −1.0 MPa, polyethylene glycol (PEG6000) has little effect on the germination rate of *P. oleracea* seeds (Rahdari et al. 2012). Under moderate water stress of −0.26 Mpa, PEG8000-treated *P. oleracea* seeds, the maximum germination rate is only 50%, so seed germination is inhibited (Chauhan and Johnson 2009). It has also been reported that PEG8000 with a low concentration of less than 10% has little effect on the germination rate of *P. oleracea* seeds, but when the concentration increases to 20–30%, seed germination was significantly inhibited (Ren et al. 2011). Drought stress affects seed germination, but lower osmotic stress has a minor inhibitory effect on *P. oleracea* seed germination, and so *P. oleracea* can complete germination in a mild drought environment.

Drought stress reduces the plant height and yield of *P. oleracea*, and affects the morphological structure and function of leaves and roots. The most significant effects of drought stress on *P. oleracea* were slow plant growth, thinner stems, and reduced dry weight and yield (Ferrari et al. 2020b; Hosseinzadeh et al. 2021; Idrees et al. 2010; Karkanis and Petropoulos 2017; Lara et al. 2003), which may be due to water deficit leading to difficulty in root water absorption and increased root osmotic pressure (Idrees et al. 2010). When the concentration of PEG8000 is between 0–15%, the root weight of most *P. oleracea* cultivars increased, and by comparing root phenotypic differences between drought-tolerant and drought-sensitive cultivars, it was found that the drought-tolerant cultivar *Tokombiya* had longer roots and more branches than the drought-sensitive cultivar *Golden E*, since the drought-tolerant cultivars begin to establish extensive root systems in the early stage of drought stress to ensure water supply (Ren et al. 2011). Therefore, the length of the root system and the number of branches may be used as important indicators for screening *P. oleracea* drought-tolerant varieties. The effects of drought stress on the leaves of *P. oleracea* were manifested in the reduction of leaf number and leaf area, the reduction of stomatal conductance or stomatal aperture, the wilting of the plant and the decrease of relative leaf water content (D'Andrea et al. 2014; D'Andrea et al. 2015; Ferrari et al. 2020b; Idrees et al. 2010; Jin et al. 2015; Karkanis and Petropoulos 2017; Saheri et al. 2020). After 22 days of drought stress, the soil water content was close to 20%, the leaves had a wilting phenotype and the leaf water content began to decrease, but the leaf water content could still be maintained at about 91%, which indicated the survival of *P. oleracea* under severe drought, and the physiological indicators returned to close to control levels after re-watering (Jin et al. 2015). Similar conclusions have also been reported that the relative water content (RWC) of *P. oleracea* leaves decreased under drought stress, but the relative RWC returned to the original level after re-watering (D'Andrea et al. 2015). *P. oleracea* may resist drought stress by maintaining sufficient water reserve and effectively reducing water loss, which may be related to the C4/CAM pathways (Lara et al. 2003). The leaves maintain sufficient water content and high WUE for normal C4/CAM pathways metabolism and resistance to drought stress.

18.3 PHYTOHORMONES REGULATION AND SIGNAL TRANSDUCTION

Phytohormones regulate plant growth and development, and also participate in biotic and abiotic stress processes throughout the life cycle of the plant. Phytohormones are key mediators for the plant to respond to drought stress. After sensing drought stress signals, plants release phytohormones to activate and regulate various plant physiological and developmental processes, including stomatal closure, root growth stimulation and the accumulation of osmotic regulators to alleviate drought stress. The synthesis of phytohormones is induced by drought stress, including abscisic acid (ABA), auxins (IAA), jasmonic acid (JA), salicylic acid (SA), ethylene (ET), gibberellins (GAs), cytokines (CKs) (Ullah et al. 2018) and melatonin (MT) (Fan et al. 2018; Sharma and Zheng 2019). Auxin regulates plant drought stress negatively, and the reduction of IAA content caused by drought stress can promote the expression of the *LEA* gene, thereby improving the drought resistance of rice (Zhang et al. 2009a). BRs are polyhydroxy sterols in the plant. Exogenous application of BRs can improve the drought resistance of *Arabidopsis thaliana* and *Brassica napus* (Kagale et al. 2007), ABA and GA are involved in mitigating drought stress in maize (Cohen et al. 2009) and ethylene aggravates drought stress by causing a reduced maize grain yield (Habben et al. 2014), while overexpression of *cytokinin dehydrogenase* gene improves drought tolerance in barley (Pospisilova et al. 2016) and MT enhances drought tolerance in a few plants (Cui et al. 2017). Foliar spraying of SA improves drought tolerance in wheat (Sedaghat et al. 2020) and JA plays a negative role in drought tolerance in rice (Dhakarey et al. 2017, 2018). The response of the plant to osmotic stress is controlled by ABA-dependent and -independent processes (Zhu 2016). ABA has a sesquiterpene structure and is known as the "stress hormone" of plants, which is closely related to plant stress resistance. ABA is involved in plant drought stress signal transduction. Under normal conditions, the ABA signaling pathway is blocked and its downstream gene expression is inhibited, but drought stress ABA synthesis in the root initiates signal transduction through its transport upwards and binding to receptors (Lin et al. 2021; Ma et al. 2009; Miyakawa and Tanokura 2011; Park et al. 2009), while ABA accumulation is controlled by the expression of the *NCED3* rate-limiting enzyme gene (Fujii et al. 2011). It has been confirmed that *MYB* and *WRKY* transcription factor genes were obtained in the transcriptome of *P. oleracea*, and the expression of these genes was upregulated under drought stress. *MYB* gene may positively regulate ABA-induced stomatal closure of *P. oleracea*, while *WRKY* gene is involved in the synthesis of ABA hormones, so it plays an important role in signal regulation (Wu et al. 2021). However, a single phytohormone often cannot complete the regulation of drought stress, therefore it is essential to regulate the growth and development of plants under drought stress through the cross-talk of different phytohormones (Rowe et al. 2016). However, there is no doubt that the research on phytohormone metabolic pathways and the regulatory mechanism of phytohormones on *P. oleracea* under drought stress is still limited currently.

18.4 LEAF GAS EXCHANGE

Drought stress induces ABA accumulation, stomatal closure and changes in gene expression within hours (Signorelli et al. 2015), and plants close stomata through ABA signaling to reduce water loss under drought conditions (Schroeder et al. 2001). When leaf photosynthesis takes place, the stomata guard cells expand due to water absorption, and the stomata open and absorb CO_2 to release oxygen, but under drought stress, K^+ is pumped out of the guard cells, so that the stomata close to avoid the loss of water in the plant, protecting the plant from drought stress (Cochrane and Cochrane 2009). The stomatal distribution of *P. oleracea* is special compared with other plants. Compared with soybean, the stomata of *P. oleracea* are mainly distributed on the front of the leaf, and the number is significantly less than that of soybean. This distribution may avoid the loss of more water on the back of the leaf under high-temperature stress and drought. When drought stress occurs, *P. oleracea* can quickly close the stomata and maintain the water potential in the plant, so stomatal distribution and stomatal density may be one of the key reasons for the drought tolerance of *P. oleracea* (Ren et al. 2011). Drought stress led to a reduction of stomatal conductance (gs), net photosynthetic rate (Pn), transpiration rate (E) and intercellular CO_2 concentration (Ci) in *P. oleracea* (Saheri et al. 2020). In addition, stomatal closure resulted in changes in CO_2 exchange patterns and a reduction in net CO_2 assimilation, and the positive effects of this physiological mechanism showed a reduction of photodamage and water loss by transpiration (D'Andrea et al. 2014). Drought stress affects the carbon assimilation efficiency of photoreaction and the dark reaction of plants and ultimately affects photosynthetic products (Reddy et al. 2004). Plant assimilate CO_2 mainly through the following pathways: C3, C4 and CAM pathways. In addition, *P. oleracea* is the representative C4/CAM that has both pathways. C4 plants assimilate CO_2 in mesophyll cells and vascular bundle sheath cells, this unique carbon assimilation mechanism makes C4 plants have higher WUE than C3 plants. This characteristic makes C4 plants adapt and survive under drought stress (Chaves et al. 2003). In a well-irrigated environment, *P. oleracea* assimilates CO_2 through the C4 pathway, and assimilates CO_2 through the CAM pathway when drought stress occurs, especially in response to the unfavorable conditions of water deficit; *P. oleracea* has strong drought adaptability (D'Andrea et al. 2014; Lara et al. 2003). The CAM pathway of *P. oleracea* under drought stress involves the mRNAs and photosynthesis-related genes *PPC-1E1c*, *ALMT-12E.1* and *DIC-1.1* either in one genotype (Ferrari et al. 2020a) or in multiple genotypes (Ferrari et al. 2020b); both were significantly upregulated, so the CAM pathway was relatively conserved in all *P. oleracea* varieties (Ferrari et al. 2020b), which was also reported previously, and the CAM pathway is responsible for increased WUE (Winter et al. 2005). However, when analyzing *CCM* gene networks and metabolites, a large amount of overlapping information was found, possibly making C4 and CAM pathways mutually redundant in most lineages, so it seems that plants with both C4 and CAM pathways must regulate the same gene networks in opposite patterns, otherwise, it can lead to ineffective metabolite cycling and inefficient transport (Gilman et al. 2021).

The chlorophyll content affects the photosynthetic efficiency of the plant and affects plant growth, development and yield (Ghorbani et al. 2018). When *P. oleracea* was treated with PEG, it was found that the protein concentration decreased due to stress, and the concentrations of chlorophyll a and b were higher than those of the control, which may be related to the increase of chloroplast number under osmotic stress, which means that *P. oleracea* has a unique drought response mechanism (Rahdari et al. 2012). However, it was found that the chlorophyll content of *P. oleracea* decreased under drought stress, but the chlorophyll content in the leaves began to increase rapidly after re-watering for 3 hours, and the chloroplast function recovered rapidly (Jin et al. 2015). Similarly, the contents of chlorophyll a and b and carotenoids were continuously decreased in the control experiment based on the gradual reduction of field capacity (Saheri et al. 2020). There are differences in the effects of PEG osmotic stress treatment and soil water control treatment on the chlorophyll content of *P. oleracea*, which may be due to the faster response to osmotic stress caused by PEG treatment because *P. oleracea* has high WUE (D'Andrea et al. 2014; Herrera 2009; Lara et al. 2003; Winter et al. 2005), and the PEG treatment time is short, so chlorophyll is not or less degraded in this situation. Soil water control treatment lasted for a long time. When the soil water content was close to 20%, the increase in time and dose effect caused *P. oleracea* to be unable to resist severe drought stress, so the decrease of chlorophyll content of *P. oleracea* was observed, and the decrease of chlorophyll content directly leads to the decline of photosynthesis, therefore affecting the yield. It was found that the biomass and yield of the drought-tolerant variety "Domokos" are positively correlated with the photosynthetic rate (Karkanis and Petropoulos 2017).

18.5 WATER RELATIONS

Drought stress induces osmotic stress in plant cells, disrupts cellular metabolism and inhibits growth and development. Plants can reduce the damage from drought stress by regulating the osmotic balance and maintaining the stability of the cell membrane, achieving high WUE under low water potential, and maintaining the normal physiological metabolism of cells (Tripathy et al. 2000). Studies have shown that CAM pathway plants can maintain a minimal metabolic state for extended periods during severe drought, and rapidly recover during rewatering (Rayder and Ting 1983). *P. oleracea* leaves have succulent characteristics, and the cells have large vacuoles, which are like CAM-type plants. Like other CAM pathway plants, *P. oleracea* can assimilate CO_2 at night and accumulate it as organic acid in vacuoles, and metabolic forms of CAM pathway avoid excessive water loss, increase WUE, avoid respiration-induced carbon loss and prolong cell viability (D'Andrea et al. 2014; Herrera 2009; Lara et al. 2003; Winter et al. 2005). Exogenous regulators can improve the WUE of *P. oleracea*, while application of exogenous SA improved the WUE of *P. oleracea* under severe water stress with a field water-holding capacity of 30% (Saheri et al. 2020).

P. oleracea has high WUE under drought stress, which can maintain the normal cell morphology and function. However, the regulation and maintenance of osmotic pressure guarantee normal cell metabolism (Table 18.1). Plants have a perfect OA mechanism, and can rapidly absorb and accumulate inorganic ions (K^+, Na^+, Cl^-) as osmotic regulators to reduce water potential and resist osmotic stress more cost-effectively (Franco-Navarro et al. 2016; Hou et al. 2021; Munns et al. 2020; Shabala and Lew 2002; Turner 2017). In a salt-induced osmotic stress experiment, the K^+ content in the control materials of the 13 *P. oleracea* varieties maintained the highest level, and the Na^+ content increased significantly after low-salt osmotic stress treatment, and inorganic ions directly participated in OA of *P. oleracea* (Alam et al. 2016).

Another strategy adopted by plants to cope with osmotic stress is to synthesize and accumulate compatible solutes (CSs). These CSs include glucose, sucrose, polyols, amino acids, trehalose and quaternary ammonium compounds, which play a role in promoting cellular OA, scavenging reactive oxygen species (ROS), protecting cell membranes and stabilizing protein structures (Blum 2017; Fang and Xiong 2015; Kaur and Asthir 2015; Per et al. 2017; Shehzadi et al. 2019), however, the above CSs are mediated by substrates synthesized *de novo* and the complex enzymatic reaction steps which consume energy (Rubio et al. 2020; Zhang et al. 2009b), and may regulate plant osmotic balance in a "long-acting" manner. *P. oleracea* showed increased proline and soluble sugar content with increasing PEG concentration, which is beneficial for OA without impairing metabolism. Also, reduced electrolyte leakage suggests that *P. oleracea* has a cell membrane protective mechanism, which is also beneficial for cell OA (Jin et al. 2016; Rahdari et al. 2012). In soil water control experiments, with the decrease of field water-holding capacity, the content of proline and soluble sugar in *P. oleracea* plants increased significantly, which also helped to regulate cell osmotic

TABLE 18.1
Osmotic Regulators Regulate the Turgor Imbalance Caused by the Water Deficit of *P. oleracea*

Osmotic regulators	Description	References
Inorganic ions		
K⁺ and Na⁺	K^+, Na^+ rapidly regulate cell osmotic pressure and maintain cell morphology	Alam et al. (2016)
Compatible solutes		
Proline	Increased proline synthesis by up-regulating *PC5S* expression, Proline participation in cell osmotic adjustment, maintaining cell morphology, and reducing electrolyte leakage	Rahdari et al. (2012) Jin et al. (2016) Sdouga et al. (2019) Alahbakhsh et al. (2019) Saheri et al. (2020)
Soluble sugar	Soluble sugar participation in cell osmotic adjustment, maintaining cell morphology, and reducing electrolyte leakage	Rahdari et al. (2012) (Jin et al. 2016)
Xylitol, sorbitol and pinitol	Xylitol, sorbitol and pinitol protect cell membrane, regulate cell osmotic pressure	D'Andrea et al. (2014)
Inositol and galactitol	Inositol and galactitol improve membrane ester fluidity and stabilize cells	Jin et al. (2016)
Urea	Urea regulate cell osmotic pressure	D'Andrea et al. (2014) Jin et al. (2016)
Glycerol	Glycerol regulates cell osmotic pressure	D'Andrea et al. (2014)

pressure (Saheri et al. 2020). Under salt-induced osmotic stress, the content of proline in *P. oleracea* increased with an increase of salt concentration, which was confirmed to be caused by upregulated expression of *P5CS* gene, which increased proline synthesis, it is speculated that *P5CS* may be an important gene in the OA of *P. oleracea* (Sdouga et al. 2019).

Reasonable fertilization can also promote the resistance of *P. oleracea* to drought stress, which is manifested in a significant increase in the content of osmotic regulators represented by proline (Alahbakhsh et al. 2019). Under drought stress, the polyols in *P. oleracea* including xylitol, sorbitol and pinitol accumulate in the plant during the daytime, while pinitol also accumulates at night, which can protect the stability of the cell membrane and regulate the osmotic balance (D'Andrea et al. 2014), and also can be used for non-enzymatic antioxidant scavenging ROS (Chaves et al. 2003; Smirnoff 1993). Under the combined action of drought and heat stress, the contents of small-molecule metabolites of *P. oleracea* increased, the contents of inositol and galactitol increased, which may be related to the change in the fluidity of membrane esters to stabilize cells, and the increased contents of valine and tyrosine may be related to IAA accumulation. The increase of ornithine as a precursor of arginine biosynthesis may help the synthesis of osmotic regulators to regulate cell osmotic pressure. The increase of tryptophan, tyrosine and phenylalanine are related to cell drought tolerance. In addition, the increase in urea content is thought to be involved in the OA of *P. oleracea* in response to drought (Jin et al. 2016). Drought stress induces the increase of urea in *P. oleracea*, which accumulates in vacuoles to regulate cell osmotic pressure, while polyamine cadaverine can counteract the degradation of intracellular proteins by urea, at which time urea was dominant to regulate osmotic pressure in cells (D'Andrea et al. 2014). In addition, glycerol is considered to be an important osmotic regulator of cells (Shen et al. 1999; Xiong et al. 2002), and the glycerol content in *P. oleracea* increased after re-watering treatment, which could regulate the osmotic pressure of plants (D'Andrea et al. 2014).

18.6 OXIDATIVE STRESS AND ANTIOXIDANTS

Aerobic metabolism provides energy for the plant but is often accompanied by the production of ROS. Drought stress leads to excessive ROS in plant cells, such as superoxide (O_2^-), hydrogen peroxide (H_2O_2), singlet oxygen ($^1O_2$) and hydroxyl radicals (OH·), which are produced in plants. When the rate of ROS generation is greater than the rate of scavenging, it leads to the accumulation of ROS (Cruz de Carvalho 2008; Munne-Bosch and Penuelas 2003), and NADPH oxidase (RBOH) is also a source of ROS generation (Mittler and Blumwald 2010; Ozkur et al. 2009). Excessive ROS damages cell membranes and chloroplasts in plants and induces cell membrane peroxidation to produce toxic metabolites like malondialdehyde (MDA). MDA causes oxidative damage to key macromolecules such as proteins, lipids and deoxyribonucleic acid, resulting in cellular metabolic disorders and even death (Farooq et al. 2009b; Foyer and Fletcher 2001; Gill and Tuteja 2010; Munne-Bosch and Penuelas 2003). However, a moderate amount of ROS acts as a signaling molecule in abiotic stress (Apel and Hirt 2004; Baxter et al. 2014; Dogra et al. 2021; Ma et al. 2012). ROS controls ion homeostasis in the cytosol by regulating the activity of ROS-sensitive ion channels (Demidchik and Shabala 2018; Fu and Huang 2001; Shabala et al. 2016). Under osmotic stress, ROS-induced K^+ and Ca^{2+} stabilization excessively have a significant impact on the growth of *Sarcocornia quinqueflora* under optimal and non-optimal conditions (Ahmed et al. 2021). Excess ROS in plant cells is scavenged by two pathways, one is through enzymatic reaction mechanisms including superoxide dismutase (SOD), catalase (CAT), glutathione reductase (GR) and ascorbate peroxidase (APX), and the other pathway is a non-enzymatic reaction mechanism, including β-carotenoids, α-tocopherol, ascorbic acid, glutathione, anthocyanins and flavonoids (Mittler and Blumwald 2010). Carbohydrates also can act as a non-enzymatic antioxidant (e.g. via the pentose phosphate pathway) to promote ROS scavenging (Proels and Huckelhoven 2014). Studies have shown that *P. oleracea* has high antioxidant enzyme activities (Corrado et al. 2021), with its antioxidant capacity being related to the total phenolic content in the plant (Uddin et al. 2012), while the total phenolic content of *P. oleracea* is also related to the growing season of *P. oleracea* (Lim and Quah 2007).

ROS triggers oxidative stress and induces membrane ester peroxidation, which is considered to be unfavorable for plant growth and development (Chavoushi et al. 2019). Drought stress led to an increase in the H_2O_2 content in *P. oleracea*, but it was accompanied by an increase in SOD, CAT, APX and POD enzymes in seedlings, so *P. oleracea* activated the dynamic regulation mechanism of ROS activity in response to oxidative stress (Saheri et al. 2020). MDA is produced by ROS-induced decomposition of polyunsaturated lipids in the cell membrane system (Pryor and Stanley 1975). Drought stress led to a significant increase in the O_2^- content of *P. oleracea*, and the MDA content increased simultaneously, with electrolyte leakage also being observed during this process. However, it was found that the leakage of electrolytes was not significant, indicating that the membrane system was only slightly damaged, which may be due to the increased activities of SOD and POD contributing to the mitigation of ROS oxidative damage to cell membranes (Jin et al. 2015; Jin et al. 2016). Although drought stress can affect the PSII function of *P. oleracea* and lead to the transition of the photochemical system to the excited state to generate free radicals, the increase in the content of non-enzymatic antioxidants such as flavonoids, betaine and phenolic and the increase in the activity of antioxidant enzymes can regulate the level of ROS and avoid oxidative stress in cells (D'Andrea et al. 2014).

18.7 GENETIC RESPONSES AND KEY GENES EXPRESSED IN RESPONSE TO DROUGHT STRESS

According to the genomic differences of different *P. oleracea* varieties, most *P. oleracea* diploid chromosomes have 2n = 18, 36, 54, and the basal chromosomes are x = 9, 18 and 27 (may be diploid, tetraploid or hexaploid) (Ocampo and Columbus 2012), so *P. oleracea* has abundant genetic diversity. So far, there have been many reports on the analysis of the genetic diversity of *P. oleracea*, based on the AFLP fingerprinting technique to analyze the genetic diversity among *P. oleracea* germplasms, using the average genetic similarity as a cut-off value. Through UPGMA cluster analysis four distinct groups were classified and the *Tokombiya* variety was found to be distinct from all other *P. oleracea* varieties. Taken together, these results indicate that there is a significant genetic variation among *P. oleracea* accessions, and *Tokombiya* is a unique accession with strong drought tolerance, and the AFLP technique is demonstrated to be helpful for the discovery of drought-tolerant *P. oleracea* varieties based on these results (Ren et al. 2011). At the genome level, the DNA genotyping technology inter-simple-sequence-repeat (ISSR) was successfully used to obtain the genetic diversity data of *P. oleracea*, which is helpful for the differentiation of *P. oleracea* germplasm resources and the determination of genetic diversity strains, therefore genetic improvement and hybrid breeding of *P. oleracea* are of great significance (Alam et al. 2015b).

The phylogenetic relationships within *P. oleracea* varieties were also studied. The nuclear ITS, chloroplast ndhF, trnT-psbD intergenic region and ndhA intron DNA of *P. oleracea* were analyzed using maximum likelihood and Bayesian methods, and the conclusion supported that the genus portulaca is monophyletic monophyly with an age of the most recent common ancestor (MRCA) of 23 Myr, and there were two main lineages of *P. oleracea*, OL clade (including anti-foliate varieties) and AL clade (including alternation with sub-anti-foliate varieties), which were distributed in Africa, Asia and Australia. Further available data on the AL clade inferred that the basic chromosome number of *P. oleracea* is x = 9, and many chromosome number change events (polyploidization, demi-polyploidization, gain and loss) occurred in portulaca, especially within the OL clade (Ocampo and Columbus, 2012).

So far, more than 400 plant genomes have been sequenced (Chen et al., 2019). However, no genomic data for *P. oleracea* have been reported, which has lagged far behind the genomics studies of other plants. However, chloroplast genome research has been reported. Based on the medicinal value of *P. oleracea*, the complete chloroplast genome has been sequenced and analyzed. The total length of the *P. oleracea* genome is 156,533 bp, with a ring structure and the same typical tetrad structure as other plants, including 87,436 bp large single copy region (LSC), 18,095 bp small single copy region (SSC) and a pair of 25,501 bp inverted repeat (IR) regions. The genome contains 85 protein-coding genes, 37 transfer RNA (tRNA) genes and eight ribosomal RNA (rRNA) genes. There are 40 repeat units and 111 simple sequence repeats (SSRs) in the genome, which facilitates the analysis of evolutionary patterns of *P. oleracea* and the development of molecular markers for genetic classification and facilitates the study of chloroplast photosynthesis under drought stress (Liu et al., 2018).

The plasticity of both C4/CAM pathways in *P. oleracea* has attracted increasing attention (Holtum et al. 2017; Voznesenskaya et al. 2010; Winter et al. 2019). Under normal water supply, C4 pathway-related genes are upregulated to maintain C4 pathway metabolism levels, however, when drought stress occurs, the C4 pathway is slowly closed, and the corresponding C4 pathway genes *PPC-1E1a'*, *NADME-2E.1* and *ASPAT-1E1* are significantly downregulated, but the plasticity advantage of *P. oleracea* is manifested at this time, the CAM pathway was activated, and the related genes *PPC-1E1c*, *ALMT-12E.1* and *DIC-1.1* were upregulated, which contributed to higher WUE in *P. oleracea* and improved its drought resistance. Therefore, C4/CAM pathway's metabolic pattern may serve as a reference strategy for crop drought tolerance breeding (Ferrari and Freschi 2019; Ferrari et al. 2020a; Ferrari et al. 2020b).

The *FDA* genes in plants indirectly regulate membrane fluidity and are precursors for the synthesis of trienoic acids, such as traumatic acid and JA, a class of signaling molecules (Turner et al. 2002). Based on differential display technology, it was confirmed that the expression of *FAD2* gene in *P. oleracea* was significantly upregulated under drought stress. *FAD2* may be one of the candidate genes for drought tolerance that regulates membrane fluidity and signal transduction in *P. oleracea* (D'Andrea et al. 2015).

Transcription factors can specifically bind to *cis*-acting elements of eukaryotic gene promoters to enhance or inhibit the transcription rate of target genes, therefore regulating gene expression

(Riano-Pachon et al. 2007). Transcription factors and genes, and transcription factors and transcription factors form a gene regulatory network to co-ordinately regulate plant stress tolerance (Shinozaki and Yamaguchi-Shinozaki 2007). In the differential gene screening of red-stem and green-stem *P. oleracea*, *AP2/ERF*, *WRKY*, *bZIP*, *NAC* and *MYB* family transcription factor genes were upregulated, and these transcription factors are involved in OA, ABA signal transduction and cyanine accumulation. These genes also showed the adaptation of *P. oleracea* to drought stress (Wu et al. 2021).

18.8 TRANSCRIPTOMIC AND METABOLOMIC ANALYSES RESPONSES UNDER DROUGHT STRESS

An important technology for studying transcriptomics is RNA sequencing (RNA-seq) technology, which is used to discover new genes and gene regulatory networks (Shu et al. 2018). In the absence of a complete reference genome, RNA-seq technology is more advantageous to perform gene expression data of specific processing conditions or developmental stages (Fan et al. 2013). Based on single-molecule real-time (SMRT) sequencing technology, the transcriptome data of two varieties of *P. oleracea* were obtained, and 94 genes were identified to be involved in the "unsaturated fatty acid biosynthesis" pathway, including the omega-3 fatty acid biosynthesis pathway. *SAD* and *FAD2* are important genes in *P. oleracea* (Du et al. 2021), and it has been reported that the *FAD2* gene may be one of the drought-tolerant candidate genes regulating *P. oleracea* membrane fluidity and signal transduction (Turner et al. 2002). *P. oleracea* contains beneficial omega-3 fatty acids in the form of alpha-linolenic acid (ALA) for humans. Transcriptome and lipid metabolome techniques were used to find that the highest ALA content in RR04, which was mainly related to galactolipids. The lysophosphatidylcholine acyltransferase *PoLPCAT*, a gene involved in lipid metabolism, has been identified (Venkateshwari et al. 2018). However, there are differences in the transcriptomes of various ecotypes of *P. oleracea*, among them, the gene expression profiles of the red-stemmed *P. oleracea* adapted to the arid environment showed large distinction compared with those of the green-stemmed *P. oleracea* adapted to the humid environment. Among the differential DEGs, 1464 were identified. These *TF* genes are divided into 47 TF families, among which the *AP2/ERF*, *WRKY*, *bZIP*, *NAC* and *MYB* families are the most noteworthy. These family genes may play important roles in *P. oleracea* tolerance to drought stress, including increasing soluble sugar, the accumulation of proline regulating the function of osmotic balance, and the increase of anthocyanin content involving the adaptation of *P. oleracea* to the arid environment as a non-enzymatic antioxidant (Wu et al. 2021).

Genes involved in drought adaptation and recovery after *P. oleracea* re-watering at the transcriptional level were identified based on differential display (DD) technology, of which the functions of about half of the differential genes were unknown, but four transcripts encoding ribosomal proteins (RPs) were induced (Seqs 007, 039, 054 and 121), these genes may be related to the translation of proteins involved in the drought stress response or the drought stress-induced CAM pathway. However, the transcripts encoding RPs S5 and L2 may be involved in cellular memory, maintain the expression of specific proteins and allow the plant to respond rapidly when drought stress occurs, the regulatory subunit Seq004 transcript of serine/threonine phosphatase (*PP2A*) expression is induced in both drought stress and re-watering, and there is evidence that *PP2A* enhances drought tolerance in tobacco (Xu et al. 2007). In addition, transcript Seq041 mediates brassinosteroid (BR) receptor kinase signal transduction, which was induced under drought stress, thus there may be cross-talk between drought stress and BR. The gene encoding *FAD2* (Seq086) is induced by drought stress, and *FAD2* is the precursor of traumatic acid and JA, and indirectly regulates membrane fluidity, so FAD2 plays an important role in the adaptation of *P. oleracea* to drought stress (D'Andrea et al. 2015).

Drought-induced changes in transcriptional profiles reveal key components of the C4/CAM metabolic machinery in a single variety of *P. oleracea*, with fewer CAM pathway-specific genes than those in the C4 pathway, and CAM pathway-specific genes involved in nocturnal primary carboxylation reactions and movement of malate across the tonoplast. Analysis of gene transcription abundance regulation and photosynthetic physiology revealed that C4 and CAM metabolic mechanisms coexist in *P. oleracea* under mild drought conditions (Ferrari et al., 2020a). However, the above view was further confirmed from the transcriptome data of multiple *P. oleracea* cultivars, and it was confirmed that under the shutdown of the C4 pathway caused by drought stress, photosynthesis-related genes *PPC-1E1a'*, *NADME-2E.1* and *ASPAT- 1E1* were downregulated, and the CAM pathway was activated, and the related photosynthesis genes *PPC-1E1c*, *ALMT-12E.1* and *DIC-1.1* were upregulated. When the water stress was relieved, the expression of C4 pathway-related photosynthetic genes was restored. Therefore, the *P. oleracea* C4/CAM pathway has strong plasticity (Ferrari et al. 2020b).

Metabolomics is used to understand plant metabolic pathways and analyze the metabolic mechanisms of plant responses to stress (Adamski and Suhre 2013; Saito and Matsuda 2010). In recent years, more studies on the transcriptome and metabolome of *P. oleracea* have also been carried out, and important salt tolerance genes and metabolites have been obtained after NaCl treatment in *P. oleracea* (Xing et al. 2020). The metabolomes of salt- and nitrogen-treated *P. oleracea* were also analyzed (Camalle et al. 2020). Two ecotypes of *P. oleracea* were treated with NaCl, and the metabolic profiles of leaves and roots were obtained, with 132 metabolites being screened (Zaman et al. 2020). *P. oleracea* cultivars from Bulgaria and Greece were analyzed using the UHPLC-Orbitrap-MS method to obtain 14 compounds (Voynikov et al. 2019), and the metabolic profiles of 38 *P. oleracea* cultivars were continued and 50 markers were obtained. A total of 29 compounds were reported for the first time, and a *P. oleracea* metabolic database was established (Balabanova et al. 2020), which is conducive to screening drought-related metabolites and molecular design breeding for specific metabolites. Using transcriptomics and phylogeny to analyze the unique metabolic mechanism of both C4/CAM pathways, using the phylogeny of *PEPC* genes combined with RNA-seq data, the CAM pathway transcriptome data were analyzed, confirming that the varieties of *P. oleracea* shared a CAM pathway-specific PEPC ortholog (*PPC-1E1c*) and C4-specific ortholog (*PPC-1E1a'*) (Christin et al. 2014), and elucidated the evolution of CAM pathway by linking PEPC to starch catabolism through the circadian clock. Although *P. amilis* and *P. oleracea* share some central C4 orthologs, the use of only orthologs from most core gene families highlights the diversity of the C4/CAM-combination system and further demonstrates that C4 evolved largely independently in multiple *P. oleracea* lineages (Gilman et al. 2021). Transcriptome, metabolome, gene function and gene editing research are inseparable from genome sequencing data. It is believed that the future release of *P. oleracea* genome sequencing data will promote the progress of modern molecular design and breeding of *P. oleracea*.

18.9 STIMULANTS-INDUCED AMENDMENT OF GROWTH/PRODUCTION UNDER DROUGHT STRESS

Exogenous plant growth regulators play an important role in regulating plant stress resistance (Sharma and Zheng 2019). Commonly used exogenous plant growth regulators are proline (Ashraf and Harris 2013; Ghafoor et al. 2019; Hayat et al. 2012), SA (Canales et al. 2019), ABA (Finkelstein et al. 2008) and glycine betaine (Shehzadi et al. 2019). Novel phytohormones such as MT also have roles in regulating plant stress tolerance (Alyammahi and Gururani 2020). SA, a commonly used exogenous plant growth regulator, plays an important role in plant antioxidant activity and plant tolerance to biotic and abiotic stresses (Hayat et al. 2008; Zhang et al. 2015). Exogenous SA improves plant drought stress (Abbaszadeh et al. 2020; Sohag et al. 2020), salt stress (Li et al. 2014), heavy metal stress (Belkadhi et al. 2015) and low-temperature stress (Siboza et al. 2014). Different concentrations of SA were used which showed different field water-holding capacities in *P. oleracea*. It was found that SA had a significant effect on improving the drought stress of *P. oleracea*, manifested by increased fresh and dry weight, increased number of leaves and increased contents of chlorophyll a and b, respectively. SA improved the stomatal conductance (gs), net photosynthetic rate (Pn), transpiration rate (E) and intercellular CO_2 concentration (Ci) of *P. oleracea* under drought stress, improved carbon assimilation, and promoted the increase of sugar and proline. SA enhanced the activities of CAT, SOD, POD and APX to reduce the content of H_2O_2 and MDA, and finally confirmed that 1.0 mM SA had a better effect on improving the drought stress of *P. oleracea* (Saheri et al. 2020). Studies have also shown that there were many negative effects of drought stress on *P. oleracea* including the reduction of the number of capsules, the number of seeds in the capsules and the number of grains per plant as well as the biomass, but when exogenous SA and ABA are applied, all indicators were improved, and crude protein production was increased, which has important implications for forage grass yield in water-scarce regions (Panahyankivi et al. 2020). ABA can not only induce stomatal closure (Schroeder et al. 2001; Signorelli et al. 2015), but also promote seed dormancy (Finkelstein et al. 2008). *P. oleracea* seeds showed incomplete germination after being immersed in ABA solution, while after the surface ABA was washed away, the seeds could continue to germinate, so exogenous ABA had an inhibitory effect on the germination of *P. oleracea* seeds (Van Rooden et al. 1970). Therefore, exogenous growth regulators play an important role in the production of *P. oleracea*.

18.10 CONCLUSIONS

Drought stress inhibits the growth and development of *P. oleracea*, but *P. oleracea* has special stomatal distribution and density to reduce water loss under drought stress, *P. oleracea* can resist drought stress through regulating the osmotic balance by organic osmotic regulators and regulating the content of antioxidant enzymes to promote ROS scavenging and avoid excessive oxidative damage. Exogenous plant growth regulators and phytohormones effectively promote *P. oleracea* to resist drought stress. The unique C4/CAM pathways of *P. oleracea* can effectively assimilate CO_2 and maintain a high WUE. When water is sufficient, the C4 pathway is activated to assimilate CO_2, while under drought stress, the CAM pathway is activated to assimilate CO_2, when the water supply is restored, the C4 pathway is activated again. The adaptability and plasticity of *P. oleracea* to drought stress have attracted much attention. *P. oleracea* has become an important drought-tolerant plant research model (Figure 18.2). However, in *P. oleracea*, the mechanism of phytohormone

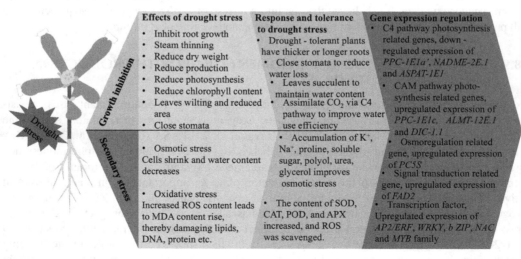

FIGURE 18.2 Mechanism of response and tolerance of *P. oleracea* to drought stress. ROS, reactive oxygen species; MDA, malondialdehyde; SOD, superoxide dismutase; CAT, catalase; POD, peroxidase; APX, ascorbate peroxidase; CAM, crassulacean acid metabolism.

regulation under drought stress has been rarely studied. Although some genes related to drought stress have been identified, gene excavation at the molecular level is very limited. The research on the genome, transcriptome, proteome and metabolome is relatively lagging, which needs to be paid sufficient attention to the unique C4/CAM metabolic pathways as that *P. oleracea* is an important drought-tolerant wild plant research model, and scientists have realized that *de novo* domestication of wild plants with abundant stress resistance traits is the hope for future agricultural production (Fernie and Yan 2019; Yu and Li 2022). To this end, it is an inevitable trend for agricultural production to grow drought-tolerant and water-saving crops like *P. oleracea* for the foreseeable future.

ACKNOWLEDGMENTS

This study was supported by the Beijing Postdoctoral Research Foundation (2018-ZZ-052 and 2018-PC-04). There is no conflict of interest among the authors. We apologize for not citing all of the relevant references due to space limitations.

REFERENCES

Abbaszadeh, B., Layeghhaghighi, M., Azimi, R., and Hadi, N. (2020). Improving water use efficiency through drought stress and using salicylic acid for proper production of Rosmarinus officinalis L. *Industrial Crops and Products* 144. doi: ARTN 11189310.1016/j.indcrop.2019.111893

Adamski, J., and Suhre, K. (2013). Metabolomics plafforms for genome wide association studies-linking the genome to the metabolome. *Current Opinion in Biotechnology* 24(1), 39–47. doi: 10.1016/j.copbio.2012.10.003

Ahmed, H.A.I., Shabala, L., and Shabala, S. (2021). Tissue-specificity of ROS-induced K+ and Ca2+ fluxes in succulent stems of the perennial halophyte Sarcocornia quinqueflora in the context of salinity stress tolerance. *Plant Physiology and Biochemistry* 166, 1022–1031. doi: 10.1016/j.plaphy.2021.07.006

Alahbakhsh, E., Galavi, M., Mosavi Nik, M., Mohkami, Z.J.J.o.W., and Conservation, S. (2019). Effects of irrigation regims and fertilizers on qualitative and quantitative triats of Purslan (Portolaca oleraceae). 26(1), 247–253.

Alam, M.A., Juraimi, A.S., Rafii, M.Y., and Hamid, A.A. (2015a). Effect of Salinity on Biomass Yield and Physiological and Stem-Root Anatomical Characteristics of Purslane (Portulaca oleracea L.) Accessions (Retracted article. See vol. 2019, 2019). *Biomed Research International* 2015. doi: Artn 10569510.1155/2015/105695

Alam, M.A., Juraimi, A.S., Rafii, M.Y., Hamid, A.A., Arolu, I.W., and Latif, M.A. (2015b). Genetic diversity analysis among collected purslane (Portulaca oleracea L.) accessions using ISSR markers. *Comptes Rendus Biologies* 338(1), 1–11. doi: 10.1016/j.crvi.2014.10.007

Alam, M.A., Juraimi, A.S., Rafii, M.Y., Hamid, A.A., Aslani, F., and Hakim, M.A. (2016). Salinity-induced changes in the morphology and major mineral nutrient composition of purslane (Portulaca oleracea L.) accessions. *Biol Res* 49, 24. doi: 10.1186/s40659-016-0084-5

Alam, M.A., Juraimi, A.S., Rafii, M.Y., Hamid, A.A., Uddin, M.K., Alam, M.Z., et al. (2014). Genetic improvement of Purslane (Portulaca oleracea L.) and its future prospects. *Molecular Biology Reports* 41(11), 7395–7411. doi: 10.1007/s11033-014-3628-1

Alyammahi, O., and Gururani, M.A. (2020). Chlorophyll-a Fluorescence Analysis Reveals Differential Response of Photosynthetic Machinery in Melatonin-Treated Oat Plants Exposed to Osmotic Stress. *Agronomy-Basel* 10(10). doi: ARTN 152010.3390/agronomy10101520

Anjum, S.A., Wang, L.C., Farooq, M., Hussain, M., Xue, L.L., and Zou, C.M. (2011). Brassinolide Application Improves the Drought Tolerance in Maize Through Modulation of Enzymatic Antioxidants and Leaf Gas Exchange. *Journal of Agronomy and Crop Science* 197(3), 177–185. doi: 10.1111/j.1439-037X.2010.00459.x

Apel, K., and Hirt, H. (2004). Reactive oxygen species: metabolism, oxidative stress, and signal transduction. *Annu Rev Plant Biol* 55, 373–399. doi: 10.1146/annurev.arplant.55.031903.141701

Ashraf, M., and Harris, P.J.C. (2013). Photosynthesis under stressful environments: An overview. *Photosynthetica* 51(2), 163–190. doi: 10.1007/s11099-013-0021-6

Baghbani-Arani, A., Modarres-Sanavy, S.A.M., Mashhadi-Akbar-Boojar, M., and Mokhtassi-Bidgoli, A. (2017). Towards improving the agronomic performance, chlorophyll fluorescence parameters and pigments in fenugreek using zeolite and vermicompost under deficit water stress. *Industrial Crops and Products* 109, 346–357. doi: 10.1016/j.indcrop.2017.08.049

Balabanova, V., Hristov, I., Zheleva-Dimitrova, D., Sugareva, P., Lozanov, V., and Gevrenova, R. (2020). Bioinformatic Insight into Portulaca Oleracea L. (Purslane) of Bulgarian and Greek Origin. *Acta Biologica Cracoviensia Series Botanica* 62(1), 7–+. doi: 10.24425/abcsb.2020.131662

Basu, S., Ramegowda, V., Kumar, A., and Pereira, A. (2016). Plant adaptation to drought stress. *F1000Res* 5. doi: 10.12688/f1000research.7678.1

Baxter, A., Mittler, R., and Suzuki, N. (2014). ROS as key players in plant stress signalling. *Journal of Experimental Botany* 65(5), 1229–1240. doi: 10.1093/jxb/ert375

Belkadhi, A., De Haro, A., Obregon, S., Chaibi, W., and Djebali, W. (2015). Positive effects of salicylic acid pretreatment on the composition of flax plastidial membrane lipids under cadmium stress. *Environmental Science and Pollution Research* 22(2), 1457–1467. doi: 10.1007/s11356-014-3475-6

Blum, A. (2017). Osmotic adjustment is a prime drought stress adaptive engine in support of plant production. *Plant Cell and Environment* 40(1), 4–10. doi: 10.1111/pce.12800

Borland, A.M., Griffiths, H., Hartwell, J., and Smith, J.A. (2009). Exploiting the potential of plants with crassulacean acid metabolism for bioenergy production on marginal lands. *J Exp Bot* 60(10), 2879–2896. doi: 10.1093/jxb/erp118

Borsai, O., Al Hassan, M., Negrusier, C., Raigon, M.D., Boscaiu, M., Sestras, R.E., et al. (2020). Responses to Salt Stress in Portulaca: Insight into Its Tolerance Mechanisms. *Plants-Basel* 9(12). doi: ARTN 166010.3390/plants9121660

Camalle, M., Standing, D., Jitan, M., Muhaisen, R., Bader, N., Bsoul, M., et al. (2020). Effect of Salinity and Nitrogen Sources on the Leaf Quality, Biomass, and Metabolic Responses of Two Ecotypes of Portulaca oleracea. *Agronomy-Basel* 10(5). doi: ARTN 65610.3390/agronomy10050656

Canales, F.J., Montilla-Bascon, G., Rispail, N., and Prats, E. (2019). Salicylic acid regulates polyamine biosynthesis during drought responses in oat. *Plant Signal Behav* 14(10), e1651183. doi: 10.1080/15592324.2019.1651183

Chauhan, B.S., and Johnson, D.E. (2009). Seed germination ecology of Portulaca oleracea L.: an important weed of rice and upland crops. *Annals of Applied Biology* 155(1), 61–69. doi: 10.1111/j.1744-7348.2009.00320.x

Chaves, M.M., Maroco, J.P., and Pereira, J.S. (2003). Understanding plant responses to drought - from genes to the whole plant. *Funct Plant Biol* 30(3), 239–264. doi: 10.1071/FP02076

Chavoushi, M., Najafi, F., Salimi, A., and Angaji, S.A. (2019). Improvement in drought stress tolerance of safflower during vegetative growth by exogenous application of salicylic acid and sodium nitroprusside. *Industrial Crops and Products* 134, 168–176. doi: 10.1016/j.indcrop.2019.03.071

Chen, F., Song, Y.F., Li, X.J., Chen, J.H., Mo, L., Zhang, X.T., et al. (2019). Genome sequences of horticultural plants: past, present, and future. *Horticulture Research* 6. doi: ARTN 11210.1038/s41438-019-0195-6

Christin, P.A., Arakaki, M., Osborne, C.P., Brautigam, A., Sage, R.F., Hibberd, J.M., et al. (2014). Shared origins of a key enzyme during the evolution of C4 and CAM metabolism. *J Exp Bot* 65(13), 3609–3621. doi: 10.1093/jxb/eru087

Cochrane, T.T., and Cochrane, T.A. (2009). The vital role of potassium in the osmotic mechanism of stomata aperture modulation and its link with potassium deficiency. *Plant Signal Behav* 4(3), 240–243. doi: 10.4161/psb.4.3.7955

Cohen, A.C., Travaglia, C.N., Bottini, R., and Piccoli, P.N. (2009). Participation of abscisic acid and gibberellins produced by endophytic Azospirillum in the alleviation of drought effects in maize. *Botany* 87(5), 455–462. doi: 10.1139/B09-023

Corrado, G., El-Nakhel, C., Graziani, G., Pannico, A., Zarrelli, A., Giannini, P., et al. (2021). Productive and Morphometric Traits, Mineral Composition and Secondary Metabolome Components of Borage and Purslane as Underutilized Species for Microgreens Production. *Horticulturae* 7(8). doi: ARTN 21110.3390/horticulturae7080211

Cruz de Carvalho, M.H. (2008). Drought stress and reactive oxygen species: Production, scavenging and signaling. *Plant Signal Behav* 3(3), 156–165. doi: 10.4161/psb.3.3.5536

Cui, G.B., Zhao, X.X., Liu, S.D., Sun, F.L., Zhang, C., and Xi, Y.J. (2017). Beneficial effects of melatonin in overcoming drought stress in wheat seedlings. *Plant Physiology and Biochemistry* 118, 138–149. doi: 10.1016/j.plaphy.2017.06.014

D'Andrea, R.M., Andreo, C.S., and Lara, M.V. (2014). Deciphering the mechanisms involved in Portulaca oleracea (C4) response to drought: metabolic changes including crassulacean acid-like metabolism induction and reversal upon re-watering. *Physiol Plant* 152(3), 414–430. doi: 10.1111/ppl.12194

D'Andrea, R.M., Triassi, A., Casas, M.I., Andreo, C.S., and Lara, M.V. (2015). Identification of genes involved in the drought adaptation and recovery in Portulaca oleracea by differential display. *Plant Physiol Biochem* 90, 38–49. doi: 10.1016/j.plaphy.2015.02.023

Demidchik, V., and Shabala, S. (2018). Mechanisms of cytosolic calcium elevation in plants: the role of ion channels, calcium extrusion systems and NADPH oxidase-mediated 'ROS-Ca(2+) Hub'. *Funct Plant Biol* 45(2), 9–27. doi: 10.1071/FP16420

Dhakarey, R., Raorane, M.L., Treumann, A., Peethambaran, P.K., Schendel, R.R., Sahi, V.P., et al. (2017). Physiological and Proteomic Analysis of the Rice Mutant cpm2 Suggests a Negative Regulatory Role of Jasmonic Acid in Drought Tolerance. *Front Plant Sci* 8, 1903. doi: 10.3389/fpls.2017.01903

Dhakarey, R., Raorane, M.L., Treumann, A., Peethambaran, P.K., Schendel, R.R., Sahi, V.P., et al. (2018). Corrigendum: Physiological and Proteomic Analysis of the Rice Mutant cpm2 Suggests a Negative Regulatory Role of Jasmonic Acid in Drought Tolerance. *Front Plant Sci* 9, 465. doi: 10.3389/fpls.2018.00465

Dkhil, M.A., Moniem, A.E.A., Al-Quraishy, S., and Saleh, R.A. (2011). Antioxidant effect of purslane (Portulaca oleracea) and its mechanism of action. *Journal of Medicinal Plants Research* 5(9), 1589–1593.

Dogra, V., Singh, R.M., Li, M., Li, M., Singh, S., and Kim, C. (2021). EXECUTER2 modulates the EXECUTER1 signalosome through its singlet oxygen-dependent oxidation. *Mol Plant*. doi: 10.1016/j.molp.2021.12.016

Du, H.M., Zaman, S., Hu, S.Q.Q., and Che, S.Q. (2021). Single-Molecule Long-Read Sequencing of Purslane (Portulaca oleracea) and Differential Gene Expression Related with Biosynthesis of Unsaturated Fatty Acids. *Plants-Basel* 10(4). doi: ARTN 65510.3390/plants10040655

Duan, Y., Ying, Z.M., He, F., Ying, X.X., Jia, L.Q., and Yang, G.L. (2021). A new skeleton flavonoid and a new lignan from Portulaca oleracea L. and their activities. *Fitoterapia* 153. doi: ARTN 10499310.1016/j.fitote.2021.104993

Egea-Gilabert, C., Ruiz-Hernandez, M.V., Parra, M.A., and Fernandez, J.A. (2014). Characterization of purslane (Portulaca oleracea L.) accessions: Suitability as ready-to-eat product. *Scientia Horticulturae* 172, 73–81. doi: 10.1016/j.scienta.2014.03.051

Fan, H., Xiao, Y., Yang, Y., Xia, W., Mason, A.S., Xia, Z., et al. (2013). RNA-Seq analysis of Cocos nucifera: transcriptome sequencing and de novo assembly for subsequent functional genomics approaches. *PLoS One* 8(3), e59997. doi: 10.1371/journal.pone.0059997

Fan, J.B., Xie, Y., Zhang, Z.C., and Chen, L. (2018). Melatonin: A Multifunctional Factor in Plants. *International Journal of Molecular Sciences* 19(5). doi: ARTN 152810.3390/ijms19051528

Fang, Y.J., and Xiong, L.Z. (2015). General mechanisms of drought response and their application in drought resistance improvement in plants. *Cellular and Molecular Life Sciences* 72(4), 673–689. doi: 10.1007/s00018-014-1767-0

Farooq, M., Wahid, A., Kobayashi, N., Fujita, D., and Basra, S. (2009a). "Plant drought stress: effects, mechanisms and management," in *Sustainable agriculture*. Springer), 153–188.

Farooq, M., Wahid, A., Kobayashi, N., Fujita, D., and Basra, S.M.A. (2009b). Plant drought stress: effects, mechanisms and management. *Agronomy for Sustainable Development* 29(1), 185–212. doi: 10.1051/agro:2008021

Fernie, A.R., and Yan, J. (2019). De Novo Domestication: An Alternative Route toward New Crops for the Future. *Mol Plant* 12(5), 615–631. doi: 10.1016/j.molp.2019.03.016

Ferrari, R., and Freschi, L. (2019). "C/CAM as a means to improve plant sustainable productivity under abiotic-stressed conditions: regulatory mechanisms and biotechnological implications. In 'Plant signaling molecules'.(Eds MIR Khan, PS Reddy, A Ferrante, NA Khan) pp. 517–532". Elsevier Inc.: Amsterdam, Netherlands).

Ferrari, R.C., Bittencourt, P.P., Rodrigues, M.A., Moreno-Villena, J.J., Alves, F.R.R., Gastaldi, V.D., et al. (2020a). C4 and crassulacean acid metabolism within a single leaf: deciphering key components behind a rare photosynthetic adaptation. *New Phytol* 225(4), 1699–1714. doi: 10.1111/nph.16265

Ferrari, R.C., Cruz, B.C., Gastaldi, V.D., Storl, T., Ferrari, E.C., Boxall, S.F., et al. (2020b). Exploring C-4-CAM plasticity within the Portulaca oleracea complex. *Scientific Reports* 10(1). doi: ARTN 1423710.1038/ s41598-020-71012-y

Finkelstein, R., Reeves, W., Ariizumi, T., and Steber, C. (2008). Molecular aspects of seed dormancy. *Annu Rev Plant Biol* 59, 387–415. doi: 10.1146/annurev.arplant.59.032607.092740

Foyer, C.H., and Fletcher, J.M. (2001). Plant antioxidants: colour me healthy. *Biologist (London)* 48(3), 115–120.

Franco-Navarro, J.D., Brumos, J., Rosales, M.A., Cubero-Font, P., Talon, M., and Colmenero-Flores, J.M. (2016). Chloride regulates leaf cell size and water relations in tobacco plants. *J Exp Bot* 67(3), 873–891. doi: 10.1093/jxb/erv502

Fu, J., and Huang, B. (2001). Involvement of antioxidants and lipid peroxidation in the adaptation of two cool-season grasses to localized drought stress. *Environ Exp Bot* 45(2), 105–114. doi: 10.1016/ s0098-8472(00)00084-8

Fujii, H., Verslues, P.E., and Zhu, J.K. (2011). Arabidopsis decuple mutant reveals the importance of SnRK2 kinases in osmotic stress responses in vivo. *Proc Natl Acad Sci U S A* 108(4), 1717–1722. doi: 10.1073/ pnas.1018367108

Garcia-Gomez, B.I., Campos, F., Hernandez, M., and Covarrubias, A.A. (2000). Two bean cell wall proteins more abundant during water deficit are high in proline and interact with a plasma membrane protein. *Plant J* 22(4), 277–288. doi: 10.1046/j.1365-313x.2000.00739.x

Ghafoor, R., Akram, N.A., Rashid, M., Ashraf, M., Iqbal, M., and Lixin, Z. (2019). Exogenously applied proline induced changes in key anatomical features and physio-biochemical attributes in water stressed oat (Avena sativa L.) plants. *Physiol Mol Biol Plants* 25(5), 1121–1135. doi: 10.1007/ s12298-019-00683-3

Ghorbani, A., Razavi, S.M., Omran, V.O.G., and Pirdashti, H. (2018). Piriformospora indica inoculation alleviates the adverse effect of NaCl stress on growth, gas exchange and chlorophyll fluorescence in tomato (Solanum lycopersicum L.). *Plant Biology* 20(4), 729–736. doi: 10.1111/plb.12717

Gill, S.S., and Tuteja, N. (2010). Reactive oxygen species and antioxidant machinery in abiotic stress tolerance in crop plants. *Plant Physiology and Biochemistry* 48(12), 909–930. doi: 10.1016/j.plaphy.2010.08.016

Gilman, I.S., Moreno-Villena, J.J., Lewis, Z.R., Goolsby, E.W., and Edwards, E.J.J.b. (2021). Gene co-expression reveals the modularity and integration of C4 and CAM in Portulaca.

Habben, J.E., Bao, X.M., Bate, N.J., DeBruin, J.L., Dolan, D., Hasegawa, D., et al. (2014). Transgenic alteration of ethylene biosynthesis increases grain yield in maize under field drought-stress conditions. *Plant Biotechnology Journal* 12(6), 685–693. doi: 10.1111/pbi.12172

Hayat, S., Hasan, S.A., Fariduddin, Q., and Ahmad, A. (2008). Growth of tomato (Lycopersicon esculentum) in response to salicylic acid under water stress. *Journal of Plant Interactions* 3(4), 297–304. doi: 10.1080/ 17429140802320797

Hayat, S., Hayat, Q., Alyemeni, M.N., Wani, A.S., Pichtel, J., and Ahmad, A. (2012). Role of proline under changing environments: a review. *Plant Signal Behav* 7(11), 1456–1466. doi: 10.4161/psb.21949

Herrera, A. (2009). Crassulacean acid metabolism and fitness under water deficit stress: if not for carbon gain, what is facultative CAM good for? *Annals of Botany* 103(4), 645–653. doi: 10.1093/aob/mcn145

Holtum, J.A.M., Hancock, L.P., Edwards, E.J., and Winter, K. (2017). Facultative CAM photosynthesis (crassulacean acid metabolism) in four species of Calandrinia, ephemeral succulents of arid Australia. *Photosynth Res* 134(1), 17–25. doi: 10.1007/s11120-017-0359-x

Hosseinzadeh, M.H., Ghalavand, A., Boojar, M.M.A., Modarres-Sanavy, S.A.M., and Mokhtassi-Bidgoli, A. (2021). Application of manure and biofertilizer to improve soil properties and increase grain yield, essential oil and omega(3) of purslane (Portulaca oleracea L.) under drought stress. *Soil & Tillage Research* 205. doi: ARTN 10463310.1016/j.still.2020.104633

Hou, P., Wang, F., Luo, B., Li, A., Wang, C., Shabala, L., et al. (2021). Antioxidant Enzymatic Activity and Osmotic Adjustment as Components of the Drought Tolerance Mechanism in Carex duriuscula. *Plants (Basel)* 10(3). doi: 10.3390/plants10030436

Hu, H., and Xiong, L. (2014). Genetic engineering and breeding of drought-resistant crops. *Annu Rev Plant Biol* 65, 715–741. doi: 10.1146/annurev-arplant-050213-040000

Hura, T., Dziurka, M., Hura, K., Ostrowska, A., and Dziurka, K. (2015). Free and Cell Wall-Bound Polyamines under Long-Term Water Stress Applied at Different Growth Stages of xTriticosecale Wittm. *PLoS One* 10(8), e0135002. doi: 10.1371/journal.pone.0135002

Idrees, M., Khan, M.M.A., Aftab, T., Naeem, M., and Hashmi, N. (2010). Salicylic acid-induced physiological and biochemical changes in lemongrass varieties under water stress. *Journal of Plant Interactions* 5(4), 293–303. doi: 10.1080/17429145.2010.508566

Jin, R., Shi, H.T., Han, C.Y., Zhong, B., Wang, Q., and Chan, Z.L. (2015). Physiological changes of purslane (Portulaca oleracea L.) after progressive drought stress and rehydration. *Scientia Horticulturae* 194, 215–221. doi: 10.1016/j.scienta.2015.08.023

Jin, R., Wang, Y.P., Liu, R.J., Gou, J.B., and Chan, Z.L. (2016). Physiological and Metabolic Changes of Purslane (Portulaca oleracea L.) in Response to Drought, Heat, and Combined Stresses. *Frontiers in Plant Science* 6. doi: ARTN 112310.3389/fpls.2015.01123

Kagale, S., Divi, U.K., Krochko, J.E., Keller, W.A., and Krishna, P. (2007). Brassinosteroid confers tolerance in Arabidopsis thaliana and Brassica napus to a range of abiotic stresses. *Planta* 225(2), 353–364. doi: 10.1007/s00425-006-0361-6

Karkanis, A.C., and Petropoulos, S.A. (2017). Physiological and Growth Responses of Several Genotypes of Common Purslane (Portulaca oleracea L.) under Mediterranean Semi-arid Conditions. *Notulae Botanicae Horti Agrobotanici Cluj-Napoca* 45(2), 569–575. doi: 10.15835/nbha45210903

Kaur, G., and Asthir, B. (2015). Proline: a key player in plant abiotic stress tolerance. *Biologia Plantarum* 59(4), 609–619. doi: 10.1007/s10535-015-0549-3

Koch, K.E., and Kennedy, R.A. (1982). Crassulacean Acid Metabolism in the Succulent C(4) Dicot, Portulaca oleracea L Under Natural Environmental Conditions. *Plant Physiol* 69(4), 757–761. doi: 10.1104/pp.69.4.757

Lara, L.J., Egea-Gilabert, C., Ninirola, D., Conesa, E., and Fernandez, J.A. (2011). Effect of aeration of the nutrient solution on the growth and quality of purslane (Portulaca oleracea). *Journal of Horticultural Science & Biotechnology* 86(6), 603–610. doi: Doi 10.1080/14620316.2011.11512810

Lara, M.V., Disante, K.B., Podesta, F.E., Andreo, C.S., and Drincovich, M.F. (2003). Induction of a Crassulacean acid like metabolism in the C(4) succulent plant, Portulaca oleracea L.: physiological and morphological changes are accompanied by specific modifications in phosphoenolpyruvate carboxylase. *Photosynth Res* 77(2–3), 241–254. doi: 10.1023/A:1025834120499

Lara, M.V., Drincovich, M.F., and Andreo, C.S. (2004). Induction of a crassulacean acid-like metabolism in the C(4) succulent plant, Portulaca oleracea L: study of enzymes involved in carbon fixation and carbohydrate metabolism. *Plant Cell Physiol* 45(5), 618–626. doi: 10.1093/pcp/pch073

Li, T.T., Hu, Y.Y., Du, X.H., Tang, H., Shen, C.H., and Wu, J.S. (2014). Salicylic Acid Alleviates the Adverse Effects of Salt Stress in Torreya grandis cv. Merrillii Seedlings by Activating Photosynthesis and Enhancing Antioxidant Systems. *Plos One* 9(10). doi: ARTN e10949210.1371/journal.pone.0109492

Lim, Y.Y., and Quah, E.P.L. (2007). Antioxidant properties of different cultivars of Portulaca oleracea. *Food Chemistry* 103(3), 734–740. doi: 10.1016/j.foodchem.2006.09.025

Lin, Z., Li, Y., Wang, Y., Liu, X., Ma, L., Zhang, Z., et al. (2021). Initiation and amplification of SnRK2 activation in abscisic acid signaling. *Nat Commun* 12(1), 2456. doi: 10.1038/s41467-021-22812-x

Liu, X., Yang, H., Zhao, J., Zhou, B., Li, T., Xiang, B.J.T.J.o.H.S., et al. (2018). The complete chloroplast genome sequence of the folk medicinal and vegetable plant purslane (Portulaca oleracea L.). 93(4), 356–365. doi: https://doi.org/10.1080/14620316.2017.1389308

Liu, L., Howe, P., Zhou, Y. F., Xu, Z. Q., Hocart, C., and Zhang, R. (2000). Fatty acids and β-carotene in Australian purslane (*Portulaca oleracea*) varieties. *Journal of Chromatography A*, 893(1), 207–213.

Lopes, M.S., Araus, J.L., van Heerden, P.D.R., and Foyer, C.H. (2011). Enhancing drought tolerance in C-4 crops. *Journal of Experimental Botany* 62(9), 3135–3153. doi: 10.1093/jxb/err105

Ma, L., Zhang, H., Sun, L., Jiao, Y., Zhang, G., Miao, C., et al. (2012). NADPH oxidase AtrbohD and AtrbohF function in ROS-dependent regulation of Na(+)/K(+)homeostasis in Arabidopsis under salt stress. *J Exp Bot* 63(1), 305–317. doi: 10.1093/jxb/err280

Ma, Y., Szostkiewicz, I., Korte, A., Moes, D., Yang, Y., Christmann, A., et al. (2009). Regulators of PP2C phosphatase activity function as abscisic acid sensors. *Science* 324(5930), 1064–1068. doi: 10.1126/science.1172408

Mittler, R., and Blumwald, E. (2010). Genetic engineering for modern agriculture: challenges and perspectives. *Annu Rev Plant Biol* 61, 443–462. doi: 10.1146/annurev-arplant-042809-112116

Miyakawa, T., and Tanokura, M. (2011). Regulatory mechanism of abscisic acid signaling. *Biophysics (Nagoya-shi)* 7, 123–128. doi: 10.2142/biophysics.7.123

Munne-Bosch, S., and Alegre, L. (2004). Die and let live: leaf senescence contributes to plant survival under drought stress. *Funct Plant Biol* 31(3), 203–216. doi: 10.1071/FP03236

Munne-Bosch, S., and Penuelas, J. (2003). Photo- and antioxidative protection, and a role for salicylic acid during drought and recovery in field-grown Phillyrea angustifolia plants. *Planta* 217(5), 758–766. doi: 10.1007/s00425-003-1037-0

Munns, R. (2002). Comparative physiology of salt and water stress. *Plant Cell Environ* 25(2), 239–250. doi: 10.1046/j.0016-8025.2001.00808.x

Munns, R., Day, D.A., Fricke, W., Watt, M., Arsova, B., Barkla, B.J., et al. (2020). Energy costs of salt tolerance in crop plants. *New Phytol* 225(3), 1072–1090. doi: 10.1111/nph.15864

Ocampo, G., and Columbus, J.T. (2012). Molecular phylogenetics, historical biogeography, and chromosome number evolution of Portulaca (Portulacaceae). *Molecular Phylogenetics and Evolution* 63(1), 97–112. doi: 10.1016/j.ympev.2011.12.017

Ozkur, O., Ozdemir, F., Bor, M., and Turkan, I. (2009). Physiochemical and antioxidant responses of the perennial xerophyte Capparis ovata Desf. to drought. *Environmental and Experimental Botany* 66(3), 487–492. doi: 10.1016/j.envexpbot.2009.04.003

Panahyankivi, M., Alami-Milani, M., and Abbasi, A.J.J.o.P.P.R. (2020). Effect of salicylic acid and abscisic acid on yield and yield components of common purslane (Portulaca oleracea) under water deficit. 27(3), 115–129.

Park, S.Y., Fung, P., Nishimura, N., Jensen, D.R., Fujii, H., Zhao, Y., et al. (2009). Abscisic acid inhibits type 2C protein phosphatases via the PYR/PYL family of START proteins. *Science* 324(5930), 1068–1071. doi: 10.1126/science.1173041

Parvaneh, R., Shahrokh, T., Meysam, H.S.J.J.o.S.P., and Biochemistry (2012). Studying of salinity stress effect on germination, proline, sugar, protein, lipid and chlorophyll content in purslane (Portulaca oleracea L.) leaves. 8(1), 182–193.

Per, T.S., Khan, N.A., Reddy, P.S., Masood, A., Hasanuzzaman, M., Khan, M.I.R., et al. (2017). Approaches in modulating proline metabolism in plants for salt and drought stress tolerance: Phytohormones, mineral nutrients and transgenics. *Plant Physiology and Biochemistry* 115, 126–140. doi: 10.1016/j.plaphy.2017.03.018

Petropoulos, S., Karkanis, A., Martins, N., and Ferreira, I.C.F.R. (2016). Phytochemical composition and bioactive compounds of common purslane (Portulaca oleracea L.) as affected by crop management practices. *Trends in Food Science & Technology* 55, 1–10. doi: 10.1016/j.tifs.2016.06.010

Poorter, L., and Markesteijn, L.J.B. (2008). Seedling traits determine drought tolerance of tropical tree species. 40(3), 321–331.

Pospisilova, H., Jiskrova, E., Vojta, P., Mrizova, K., Kokas, F., Cudejkova, M.M., et al. (2016). Transgenic barley overexpressing a cytokinin dehydrogenase gene shows greater tolerance to drought stress. *New Biotechnology* 33(5), 692–705. doi: 10.1016/j.nbt.2015.12.005

Price, A.H., Cairns, J.E., Horton, P., Jones, H.G., and Griffiths, H. (2002). Linking drought-resistance mechanisms to drought avoidance in upland rice using a QTL approach: progress and new opportunities to integrate stomatal and mesophyll responses. *J Exp Bot* 53(371), 989–1004. doi: 10.1093/jexbot/53.371.989

Proels, R.K., and Huckelhoven, R. (2014). Cell-wall invertases, key enzymes in the modulation of plant metabolism during defence responses. *Mol Plant Pathol* 15(8), 858–864. doi: 10.1111/mpp.12139

Pryor, W.A., and Stanley, J.P. (1975). Letter: A suggested mechanism for the production of malonaldehyde during the autoxidation of polyunsaturated fatty acids. Nonenzymatic production of prostaglandin endoperoxides during autoxidation. *J Org Chem* 40(24), 3615–3617. doi: 10.1021/jo00912a038

Qi, J., Song, C.P., Wang, B., Zhou, J., Kangasjarvi, J., Zhu, J.K., et al. (2018). Reactive oxygen species signaling and stomatal movement in plant responses to drought stress and pathogen attack. *J Integr Plant Biol* 60(9), 805–826. doi: 10.1111/jipb.12654

Rahdari, P., Hosseini, S.M., and Tavakoli, S.J.J.o.M.P.R. (2012). The studying effect of drought stress on germination, proline, sugar, lipid, protein and chlorophyll content in purslane (Portulaca oleracea L.) leaves. 6(9), 1539–1547.

Rayder, L., and Ting, I.P. (1983). CAM-idling in Hoya carnosa (Asclepiadaceae). *Photosynth Res* 4(3), 203–211. doi: 10.1007/BF00052124

Reddy, A.R., Chaitanya, K.V., and Vivekanandan, M. (2004). Drought-induced responses of photosynthesis and antioxidant metabolism in higher plants. *Journal of Plant Physiology* 161(11), 1189–1202. doi: 10.1016/j.jplph.2004.01.013

Ren, S., Weeda, S., Akande, O., Guo, Y., Rutto, L., and Mebrahtu, T.J.J.o.B.R. (2011). Drought tolerance and AFLP-based genetic diversity in purslane (Portulaca oleracea L.). 3, 51.

Riano-Pachon, D.M., Ruzicic, S., Dreyer, I., and Mueller-Roeber, B. (2007). PlnTFDB: an integrative plant transcription factor database. *BMC Bioinformatics* 8, 42. doi: 10.1186/1471-2105-8-42

Rowe, J.H., Topping, J.F., Liu, J.L., and Lindsey, K. (2016). Abscisic acid regulates root growth under osmotic stress conditions via an interacting hormonal network with cytokinin, ethylene and auxin. *New Phytologist* 211(1), 225–239. doi: 10.1111/nph.13882

Rubio, F., Nieves-Cordones, M., Horie, T., and Shabala, S. (2020). Doing 'business as usual' comes with a cost: evaluating energy cost of maintaining plant intracellular K+ homeostasis under saline conditions. *New Phytologist* 225(3), 1097–1104. doi: 10.1111/nph.15852

Saheri, F., Barzin, G., Pishkar, L., Boojar, M.M.A., and Babaeekhou, L. (2020). Foliar spray of salicylic acid induces physiological and biochemical changes in purslane (Portulaca oleraceaL.) under drought stress. *Biologia* 75(12), 2189–2200. doi: 10.2478/s11756-020-00571-2

Saito, K., and Matsuda, F. (2010). Metabolomics for functional genomics, systems biology, and biotechnology. *Annu Rev Plant Biol* 61, 463–489. doi: 10.1146/annurev.arplant.043008.092035

Schroeder, J.I., Kwak, J.M., and Allen, G.J. (2001). Guard cell abscisic acid signalling and engineering drought hardiness in plants. *Nature* 410(6826), 327–330. doi: 10.1038/35066500

Sdouga, D., Ben Amor, F., Ghribi, S., Kabtni, S., Tebini, M., Branca, F., et al. (2019). An insight from tolerance to salinity stress in halophyte Portulaca oleracea L.: Physio-morphological, biochemical and molecular responses. *Ecotoxicol Environ Saf* 172, 45–52. doi: 10.1016/j.ecoenv.2018.12.082

Sedaghat, M., Sarvestani, Z.T., Emam, Y., Bidgoli, A.M., and Sorooshzadeh, A.J.R.J.o.P.P. (2020). Foliar-applied GR24 and salicylic acid enhanced wheat drought tolerance. 67(4), 733–739.

Shabala, S., Bose, J., Fuglsang, A.T., and Pottosin, I. (2016). On a quest for stress tolerance genes: membrane transporters in sensing and adapting to hostile soils. *J Exp Bot* 67(4), 1015–1031. doi: 10.1093/jxb/erv465

Shabala, S.N., and Lew, R.R. (2002). Turgor regulation in osmotically stressed Arabidopsis epidermal root cells. Direct support for the role of inorganic ion uptake as revealed by concurrent flux and cell turgor measurements. *Plant Physiol* 129(1), 290–299. doi: 10.1104/pp.020005

Sharma, A., and Zheng, B.S. (2019). Melatonin Mediated Regulation of Drought Stress: Physiological and Molecular Aspects. *Plants-Basel* 8(7). doi: ARTN 19010.3390/plants8070190

Shavrukov, Y., Kurishbayev, A., Jatayev, S., Shvidchenko, V., Zotova, L., Koekemoer, F., et al. (2017). Early Flowering as a Drought Escape Mechanism in Plants: How Can It Aid Wheat Production? *Front Plant Sci* 8, 1950. doi: 10.3389/fpls.2017.01950

Shehzadi, A., Akram, N.A., Ali, A., and Ashraf, M. (2019). Exogenously applied glycinebetaine induced alteration in some key physio-biochemical attributes and plant anatomical features in water stressed oat (Avena sativa L.) plants. *Journal of Arid Land* 11(2), 292–305. doi: 10.1007/s40333-019-0007-8

Shen, B., Hohmann, S., Jensen, R.G., and Bohnert a, H. (1999). Roles of sugar alcohols in osmotic stress adaptation. Replacement of glycerol by mannitol and sorbitol in yeast. *Plant Physiol* 121(1), 45–52. doi: 10.1104/pp.121.1.45

Shepherd, T., and Wynne Griffiths, D. (2006). The effects of stress on plant cuticular waxes. *New Phytol* 171(3), 469–499. doi: 10.1111/j.1469-8137.2006.01826.x

Shinozaki, K., and Yamaguchi-Shinozaki, K. (2007). Gene networks involved in drought stress response and tolerance. *J Exp Bot* 58(2), 221–227. doi: 10.1093/jxb/erl164

Shu, Y.J., Li, W., Zhao, J.Y., Liu, Y., and Guo, C.H. (2018). Transcriptome sequencing and expression profiling of genes involved in the response to abiotic stress in Medicago ruthenica. *Genetics and Molecular Biology* 41(3), 638–648. doi: 10.1590/1678-4685-Gmb-2017-0284

Siboza, X.I., Bertling, I., and Odindo, A.O. (2014). Salicylic acid and methyl jasmonate improve chilling tolerance in cold-stored lemon fruit (Citrus limon). *Journal of Plant Physiology* 171(18), 1722–1731. doi: 10.1016/j.jplph.2014.05.012

Signorelli, S., Dans, P.D., Coitiño, E.L., Borsani, O., and Monza, J.J.P.O. (2015). Connecting proline and γ-aminobutyric acid in stressed plants through non-enzymatic reactions. 10(3), e0115349.

Singh, K.J.N.P. (1973). Effect of Temperature and Light on Seed Germination of Two Ecotypes of Portulaca oleracea L. 72(2), 289–295.

Siriamornpun, S., and Suttajit, M. (2010). Microchemical Components and Antioxidant Activity of Different Morphological Parts of Thai Wild Purslane (Portulaca oleracea). *Weed Science* 58(3), 182–188. doi: 10.1614/Ws-D-09-00073.1

Smirnoff, N. (1993). The role of active oxygen in the response of plants to water deficit and desiccation. *New Phytol* 125(1), 27–58. doi: 10.1111/j.1469-8137.1993.tb03863.x

Sohag, A.A., Tahjib-Ul-Arif, M., Brestic, M., Afrin, S., Sakil, M.A., Hossain, M.T., et al. (2020). Exogenous salicylic acid and hydrogen peroxide attenuate drought stress in rice. *Plant Soil and Environment* 66(1), 7–13. doi: 10.17221/472/2019-Pse

Srivastava, R., Srivastava, V., and Singh, A. (2021). Multipurpose Benefits of an Underexplored Species Purslane (Portulaca oleracea L.): A Critical Review. *Environ Manage.* doi: 10.1007/s00267-021-01456-z

Tripathy, J., Zhang, J., Robin, S., Nguyen, T.T., Nguyen, H.J.T., and Genetics, A. (2000). QTLs for cell-membrane stability mapped in rice (Oryza sativa L.) under drought stress. 100(8), 1197–1202.

Turner, J.G., Ellis, C., and Devoto, A. (2002). The jasmonate signal pathway. *Plant Cell* 14 Suppl, S153–164. doi: 10.1105/tpc.000679

Turner, N.C. (2017). Turgor maintenance by osmotic adjustment, an adaptive mechanism for coping with plant water deficits. *Plant Cell Environ* 40(1), 1–3. doi: 10.1111/pce.12839

Uddin, M., Juraimi, A.S., Ali, M., and Ismail, M.R.J.I.j.o.m.s. (2012). Evaluation of antioxidant properties and mineral composition of purslane (Portulaca oleracea L.) at different growth stages. 13(8), 10257–10267.

Ullah, A., Manghwar, H., Shaban, M., Khan, A.H., Akbar, A., Ali, U., et al. (2018). Phytohormones enhanced drought tolerance in plants: a coping strategy. *Environ Sci Pollut Res Int* 25(33), 33103–33118. doi: 10.1007/s11356-018-3364-5

Umezawa, T., Fujita, M., Fujita, Y., Yamaguchi-Shinozaki, K., and Shinozaki, K. (2006). Engineering drought tolerance in plants: discovering and tailoring genes to unlock the future. *Curr Opin Biotechnol* 17(2), 113–122. doi: 10.1016/j.copbio.2006.02.002

Van Rooden, J., Akkermans, L., and Van Der Veen, R.J.A.b.n. (1970). A study on photoblastism in seeds of some tropical weeds. 19(2), 257–264.

Venkateshwari, V., Vijayakumar, A., Vijayakumar, A.K., Reddy, L.P.A., Srinivasan, M., and Rajasekharan, R. (2018). Leaf lipidome and transcriptome profiling of Portulaca oleracea: characterization of lysophosphatidylcholine acyltransferase. *Planta* 248(2), 347–367. doi: 10.1007/s00425-018-2908-8

Voynikov, Y., Gevrenova, R., Balabanova, V., Doytchinova, I., Nedialkov, P., and Zheleva-Dimitrova, D. (2019). LC-MS analysis of phenolic compounds and oleraceins in aerial parts of Portulaca oleracea L. *Journal of Applied Botany and Food Quality* 92, 298–312. doi: 10.5073/Jabfq.2019.092.041

Voznesenskaya, E.V., Koteyeva, N.K., Edwards, G.E., and Ocampo, G. (2010). Revealing diversity in structural and biochemical forms of C4 photosynthesis and a C3-C4 intermediate in genus Portulaca L. (Portulacaceae). *J Exp Bot* 61(13), 3647–3662. doi: 10.1093/jxb/erq178

Winter, K., Aranda, J., and Holtum, J.A.M. (2005). Carbon isotope composition and water-use efficiency in plants with crassulacean acid metabolism. *Funct Plant Biol* 32(5), 381–388. doi: 10.1071/FP04123

Winter, K., Sage, R.F., Edwards, E.J., Virgo, A., and Holtum, J.A.M. (2019). Facultative crassulacean acid metabolism in a C3-C4 intermediate. *J Exp Bot* 70(22), 6571–6579. doi: 10.1093/jxb/erz085

Wu, M., Fu, S., Jin, W., Xiang, W., Zhang, W., and Chen, L.J.B.p. (2021). Transcriptome comparison of physiological divergence between two ecotypes of Portulaca oleracea. 65, 212–220.

Wu, Y., and Cosgrove, D.J. (2000). Adaptation of roots to low water potentials by changes in cell wall extensibility and cell wall proteins. *J Exp Bot* 51(350), 1543–1553. doi: 10.1093/jexbot/51.350.1543

Xiang, L., Xing, D., Wang, W., Wang, R., Ding, Y., and Du, L. (2005). Alkaloids from Portulaca oleracea L. *Phytochemistry* 66(21), 2595–2601. doi: 10.1016/j.phytochem.2005.08.011

Xing, J.C., Zhao, B.Q., Dong, J., Liu, C., Wen, Z.G., Zhu, X.M., et al. (2020). Transcriptome and Metabolome Profiles Revealed Molecular Mechanisms Underlying Tolerance of Portulaca oleracea to Saline Stress. *Russian Journal of Plant Physiology* 67(1), 146–152. doi: 10.1134/S1021443720010240

Xiong, L., Schumaker, K.S., and Zhu, J.K. (2002). Cell signaling during cold, drought, and salt stress. *Plant Cell* 14 Suppl, S165–183. doi: 10.1105/tpc.000596

Xiong, L., and Yang, Y. (2003). Disease resistance and abiotic stress tolerance in rice are inversely modulated by an abscisic acid-inducible mitogen-activated protein kinase. *Plant Cell* 15(3), 745–759. doi: 10.1105/tpc.008714

Xu, C., Jing, R., Mao, X., Jia, X., and Chang, X. (2007). A wheat (Triticum aestivum) protein phosphatase 2A catalytic subunit gene provides enhanced drought tolerance in tobacco. *Ann Bot* 99(3), 439–450. doi: 10.1093/aob/mcl285

Xue, D., Zhang, X., Lu, X., Chen, G., and Chen, Z.H. (2017). Molecular and Evolutionary Mechanisms of Cuticular Wax for Plant Drought Tolerance. *Front Plant Sci* 8, 621. doi: 10.3389/fpls.2017.00621

Yamaguchi-Shinozaki, K., and Shinozaki, K. (2006). Transcriptional regulatory networks in cellular responses and tolerance to dehydration and cold stresses. *Annu Rev Plant Biol* 57, 781–803. doi: 10.1146/annurev. arplant.57.032905.105444

Yu, H., and Li, J.J.N.C. (2022). Breeding future crops to feed the world through de novo domestication. 13(1), 1–4.

Yuan, W., Suo, J., Shi, B., Zhou, C., Bai, B., Bian, H., et al. (2019). The barley miR393 has multiple roles in regulation of seedling growth, stomatal density, and drought stress tolerance. *Plant Physiol Biochem* 142, 303–311. doi: 10.1016/j.plaphy.2019.07.021

Zaman, S., Bilal, M., Du, H.M., and Che, S.Q. (2020). Morphophysiological and Comparative Metabolic Profiling of Purslane Genotypes (Portulaca oleracea L.) under Salt Stress. *Biomed Research International* 2020. doi: Artn 482704510.1155/2020/4827045

Zhang, S.W., Li, C.H., Cao, J., Zhang, Y.C., Zhang, S.Q., Xia, Y.F., et al. (2009a). Altered architecture and enhanced drought tolerance in rice via the down-regulation of indole-3-acetic acid by TLD1/OsGH3.13 activation. *Plant Physiol* 151(4), 1889–1901. doi: 10.1104/pp.109.146803

Zhang, Y.H., Primavesi, L.F., Jhurreea, D., Andralojc, P.J., Mitchell, R.A.C., Powers, S.J., et al. (2009b). Inhibition of SNF1-Related Protein Kinase1 Activity and Regulation of Metabolic Pathways by Trehalose-6-Phosphate. *Plant Physiology* 149(4), 1860–1871. doi: 10.1104/pp.108.133934

Zhang, Y.P., Xu, S., Yang, S.J., and Chen, Y.Y. (2015). Salicylic acid alleviates cadmium-induced inhibition of growth and photosynthesis through upregulating antioxidant defense system in two melon cultivars (Cucumis melo L.). *Protoplasma* 252(3), 911–924. doi: 10.1007/s00709-014-0732-y

Zhou, Y.X., Xin, H.L., Rahman, K., Wang, S.J., Peng, C., and Zhang, H. (2015). Portulaca oleracea L.: a review of phytochemistry and pharmacological effects. *Biomed Res Int* 2015, 925631. doi: 10.1155/2015/925631

Zhu, J.K. (2002). Salt and drought stress signal transduction in plants. *Annu Rev Plant Biol* 53, 247–273. doi: 10.1146/annurev.arplant.53.091401.143329

Zhu, J.K. (2016). Abiotic Stress Signaling and Responses in Plants. *Cell* 167(2), 313–324. doi: 10.1016/j.cell.2016.08.029

19 *Ricinus communis* and Stressful Conditions

Maksud Hasan Shah[1]*, Shoumik Saha[2],
Mainak Barman[3], Santanu Kundu[4], Saidul Islam[5],
Kalipada Pramanik[6], Akbar Hossain[7] and
Sk. Md. Ajaharuddin[8]
[1]Department of Agronomy, Bidhan Chandra KrishiViswavidyalaya
(BCKV), Mohanpur, Nadia, West Bengal, India
[2]Department of Genetics and Plant Breeding, Bidhan Chandra
KrishiViswavidyalaya (BCKV), Mohanpur, Nadia, West Bengal, India
[3]Department of Genetics and Plant Breeding, Bidhan Chandra
KrishiViswavidyalaya (BCKV), Mohanpur, Nadia, West Bengal, India
[4]Department of Agronomy, Professor Jayashankar Telangana State
Agricultural University, Hyderabad, Telangana, India
[5]Nadia KrishiVigyan Kendra, Bidhan Chandra KrishiViswavidyalaya
(BCKV), Gayeshpur, Nadia, West Bengal, India
[6]Assistant Professor, Department of Agronomy, PalliSikshaBhavana,
Institute of Agriculture(Visva- Bharati), Sriniketan, Birbhum,
West Bengal, India
[7]Bangladesh Wheat and Maize Research Institute, Dinajpur 5200,
Bangladesh.
[8]Department of Agricultural Entomology, Bidhan Chandra Krishi
Viswavidyalaya (BCKV), Mohanpur, Nadia, West Bengal, India
*Corresponding author: maksudhasanshah@gmail.com

CONTENTS

DOI: 10.1201/9781003242963-19

19.1 INTRODUCTION

Ricinus communis L. is a large oil seed crop, also referred to as castor bean. This plant, which is mostly cultivated in India, Brazil, Mozambique and China, is believed to have a polyphyletic origin and four centers of variability. This crop is cross-pollinated, sexually polymorphic and extremely environment-sensitive. Castor bean seeds, which contain 40–60% oil, are what make the plant profitable. The main triglyceride present in large quantities in castor oil is retinol. The 12-hydroxylic acid (18:1OH) hydroxyl fatty acid is only commercially available from this source and is used in industrial items such as lubricants, coatings, paints, soaps, plastics, cosmetics and medications. Increasing the seed oil content has huge benefits because it is one of the main contributors to the production of oil. Several biotic and abiotic stresses are more prevalent now that large-scale agriculture is being practiced. There is the chance that castor, a non-edible oil seed crop, could be cultivated in poor soil. Therefore, in this circumstance, breeding for abiotic stress tolerance becomes more crucial. A total of 300 defense-response-associated transcription factors, including members of the nucleotide-binding site leucine-rich repeat (NBS-LRR) family, have been identified in the castor bean genome by genomic and transcriptome analysis (Sood and Chauhan, 2018). Additionally, genetic markers linked to disease resistance genes have been gathered, and comparative genomics has revealed general molecular descriptors linked to disease resistance in castor. Pyramiding R genes is regarded as the most environmentally responsible technique to increase resistance tenacity. To accomplish this, researchers must first identify the corresponding genes or markers inside the genomes. However, because of their duplicated nature, frequent paralogue presence and clustered properties, NLRs are difficult to predict using the usual artificial gene annotation processes. Many "omics" studies, including those in metabolomics, proteomics and transcriptomics, have shown how castor can tolerate biotic and abiotic stress in a variety of ways.

19.2 ORIGIN, DISTRIBUTION AND TAXONOMY

19.2.1 ORIGIN

There is debate over the castor bean's place of origin. The crop is grown all across India, although De Candolle (1860), stated that it is indigenous to Africa, whereas Fluckiger and Hanbury, Bentley, and Trimen asserted that it is indigenous to India (Watt 1892). The crop is also believed to have four centers of diversity and a polyphyletic origin. China, the Indian subcontinent, north and southwest Asia, the Arabian Peninsula, and Ethiopian east Africa are among them. The Ethiopian east African region is the most likely location for the castor bean's origin, according to Moshkin (1986).

19.2.2 DISTRIBUTION

Wild castor bean variants thrive in areas close to the Middle East, as well as east and north Africa. It was grown in ancient Egypt as far back as 4000 BCE. It was then introduced to India and other countries, and was first documented in China during the Tang dynasty (618–906 CE). The crop was carried to the New World quickly after Columbus had been there (Purseglove 1968). In Illinois, in 1818, castor bean use was first documented in the United States (Weibel, 1948). The crop is used in many tropical and subtropical nations.

19.3 ECONOMIC BOTANY

Castor oil is of extreme purgative effectiveness. It is a beneficial lotion to treat skin complications like stretch marks, dry skin, fine lines and wrinkles, sunburn, etc. Castor oil is also used to prevent various infections like chronic itching, ringworm, acne, athlete's foot, boils, warts, etc. Castor oil is widely used in mold inhibitors, sweets, flavoring and food additives. It is used as a bio-based polyol in the polyurethane industry as well as in the food industry. This is a major raw ingredient for the nylon and paint industries. It can also be used to stop cereal and pulse crops from rotting. Animals can be fed detoxified castor bean meal (also known as oilcake, pomace or castor cake), along with the remainder after the oil has been extracted along with the fiber from the husks (Kochhar 1981; Salihu et al. 2014). The husks can also be used as manure, but in order to provide it a better-balanced nutrient for plant growth, it essentially needs to be combined with an N-rich organic substance (Salihu et al. 2014). Castor bean meal has excellent utility as manure because of its anti-nematode property, quick mineralization and high N content. Additionally, wheat and castor bean crops benefit from their use. Castor bean is frequently used in public spaces as a decorative plant, primarily as a "dot plant" in standard bedding arrangements.

19.4 BIOTIC STRESS

19.4.1 DISEASES

Several diseases afflict castor, but only three are thought to be economically significant. Gray mold (*Botryotinia ricini* G.H. Godfrey or *Amphobotrys ricini* N.F. Buchw. in its anamorphic), vascular wilt disease (*Fusarium oxysporum* f.sp. *ricini* Nanda and Prasad), as well as charcoal rot (*Macrophomina phaseolina*) are the three main illnesses causing castor disease (Tassi and Goid). Other diseases, including leaf spots disease brought on by the fungi *Alternaria ricini* (Yoshii) Hansf. and *Cercospora ricinella* Saccardo and Berlese and the bacterium *Xanthomona saxonopodis* PV. Ricini Hasse, can also sporadically create significant epidemics depending on the genotype and climatic circumstances. A. ricini, a fungus that may spread through seeds and cause pod rot and seedling blight and result in seed production losses of up to 70%, deserves additional attention among these (Holliday 1980). Gray mold is the castor disease that causes the most deaths worldwide. In the early 20th century, gray mold was widely researched (Godfrey 1923), but only a little amount of study on the condition has been published thereafter (Soares 2012). As a result, there have been some improvements in the treatment of gray mold. Genotypes with modest levels of tolerance have been effectively screened by Arajo et al. (2007) and Anjani (2012). Researchers have developed diagrammatic scales to assess the severity of disease in the field (Sussel et al. 2009; Chagas et al. 2010) and a method to test germplasm resistance in a restricted environment (Soares et al., 2010) (Table 19.1).

Research into the control and management of *B. ricini* using fungicides is still ongoing. Despite the fact that *B. ricini* is a fungus that thrives on seeds, the extremely long time between planting and blooming makes it unlikely that the seed was the first inoculum source for the disease (Soares 2012). In tropical environments, the underlying inoculum is most likely derived from the conidia produced by wild castor plants. Since the growth disease that affects the primary blooms causes a lot of sporulation and is spread by the wind, rain and likely insects, it allows for repeated contamination. In addition, *B. ricini* has a broad range of hosts in the Euphorbiaceae family, including weeds and ornamentals like *Acalypha hispida* Burm., *Caperonia palustris* L., *Euphorbia supine* Raf., *E. milli* Des Moul., *E. pulcherrima* Willd. Ex Klotzsch, *E. heterophylla* L., and *Jatropha podagrica* Hook (Soares 2012). Vascular wilt disease is one of the most significant castor diseases in India (Desai and Dange 2003). The prevalence of this ailment in some regions may have been overestimated since adverse effects might occasionally be mistaken for charcoal rot. The application of varietal resistance, crop rotation and seed treatment are the best management techniques for this disease. India

TABLE 19.1
Various Economically Important Diseases Affecting Castor and Their Causal Organism Along with Identifying Symptoms

Disease	Caused by	Symptoms	References
Cercospora leaf spot	*Cercospora coffeicola*	Black and purple necrotic lesions with light yellow haloes and pale white core are present. In the core of the lesion, there is intense sporulation	(Souza and Maffia 2011)
Phyllosticta leaf spot	*Phyllostictaca pitalensis*	Round dots with gray cores and yellow haloes surrounded by initial leaf symptoms. Finally, the spots spread and fused over time	(Tang et al. 2020a)
Alternaria leaf spot	*Alternaria ricini*	Uneven patches on the leaves coated in what looks to be concentric rings of mold are the symptoms. The leaf can shed early if the disease spot fuses and grows larger	(Gahukar 2018)
Corynespora leaf vein spot	*Corynespora cassiicola*	Small, dark brown dots with yellow haloes along leaf veins or midribs were the initial leaf symptoms. The dots also appeared on the mesophylls. The patches often had grayish-white cores, either in veins or mesophylls	(Tang et al. 2020b)
Cladosporium leaf spot	*Cladosporium tenuissimum*	Early signs included pale brown or gray necrotic patches on the damaged leaves. A mold gradually started to develop on both sides of the blotches. The specks gathered, became larger, and took on an erratic form	(Liu et al. 2019)
Podosphaera powdery mildew	*Podosphaeraxanthii*	White on both sides of the leaves, mycelium, conidiophores and conidia patches could be seen	(Zhao et al. 2018)
Leveillula powdery mildew	*Leveillula taurica*	The underside of older leaves develops big, white spots as a result of the symptoms. Mycelium and conidiophores covered the white spots. Additionally, the leaves' top surfaces have chlorosis and necrosis	(Mirzaee et al. 2011)
Cucumber Mosaic Virus	CMV	Blisters, severe mosaic and leaf distortion are all symptoms. Between 5 and 10% of leaves have clinical symptoms	(Mirhosseini et al. 2017), Raj et al. 2010)
Seed Bud Rot of Castor	*Fusarium solani*	Browning and rotting of hypocotyls and cotyledons are the primary disease signs, and this exacerbates the development tendency. The hypocotyls of the seed buds first developed water-stained blemishes. The lesions then developed gradually and darkened in color. Finally, the seed buds began to decay and the cotyledons turned yellow	(Tang et al. 2020c)

has developed particular breeding lines and commercial hybrids that are resistant to vascular wilt (Anjani et al. 2004; Anjani 2005a, 2005c, 2012; Patel and Pathak 2011). Charcoal rot, often referred to as macrophomina root rot, is a serious disease in the majority of nations where castor is grown (Arajo et al. 2007; Rajani and Parakhia, 2009). Besset et al. (1996) discovered the requirements for establishing castor tolerance to charcoal rot, and numerous resistant genotypes have since been created (Anjani et al. 2004; Anjani 2005b). Cultivar resistance is the main means of preventing charcoal rot, while crop rotation and the usage of organic matter can greatly reduce the disease's severity (Rajani and Parakhia 2009). On castor, the identified plant parasitic nematodes frequently don't produce much damage (Kolte 1995). The most significant one is the reniform nematode (*Rotylenchulus reniformis* Linford and Oliveira), which is well-known for predisposing castor to infection by the fungus *F. oxysporum* (Dange et al. 2005). *Meloidogyne paranaensis, Meloidogyne incognita*, *Meloidogyne javanica* Treub and *Carneiro* M. *ethiopica* Whitehead and all varieties of chit wood are resistant to castor (Arieira et al. 2009; Lima et al. 2009a). To improve castor's tolerance to numerous diseases, research into the inheritance of disease resistance will be important, because choosing tolerant genotypes without sufficient information on their hereditary features may be harmful. Standardizing screening procedures among plant pathologists and expanding global cooperation would help to better understand the underlying linkages between the host, the pathogen and the impact of the climate that are required for the development of disease in castor (Soares, personal communication, 2012).

In India, the major insect pests of castor are castor semi looper (*Achaea janata*), castor shoot borer (*Conogethes punctiferalis*), capsule borer (*Dichocrosis punctiferalis*), tobacco caterpillar (*Spodoptera litura*), red hairy caterpillar (*Amsacta* spp.) and leaf miner insects (*Liriomyza trifolii*) (Basappa 2007; Anjani et al. 2010). In Brazil, the most significant pests are stink bug (*Nezara viridula*); leafhopper (*Empoasca* spp.); defoliators including armyworm (*Spodoptera frugiperda*), *A. janata*, and black cutworm (*Agrotis ipsilon*); and the mites *Tetranychus urticae* and *Tetranychus ludeni* (Soares et al. 2001; Ribeiro and Costa 2008). According to Varón et al. (2010), the cotton lace bug (*Corythucha gossypii*) is a pest of castor plants in Colombia. When *A. janata* attacks castor plants early in the growing season, it has been revealed that the seed production decrease is more pronounced (30 DAE). As the attack continues, the damage gradually decreases (60, 90 or 120 DAE) (Basappa and Lingappa 2001). An integrated pest management program in India using insecticides, crop rotation, bug traps and neem extract increases seed production by 28% (Basappa 2007). Castor cultivars with purple leaves and high anthocyanin content have been reported to be resistant to leafminers. Infection and defoliation by *A. janata* and *S. litura* were reduced in castor leaves with epicuticular wax (blooming) (Sarma et al. 2006).

19.4.2 Studies into Disease Resistance Genes in Castor Bean

The creation of breeding programs has been predicated on the identification of genes for disease resistance in order to generate pathogen-resistant, high-yielding castor crops. The identification of disease resistance genes and an understanding of their mechanisms of action are necessary for rapid breeding operations. Excessive levels of polyploidy and heterozygosity, however, pose obstacles to traditional breeding techniques. Gene discovery and molecular cloning are being conducted extensively using cutting-edge genomics-based methods, and so a difficult process like map-based cloning might be avoided (Joyeux et al. 1999; Dracatos et al. 2009; Arafa et al. 2018). Plant resistance to certain diseases and pests is mostly due to single resistance (R) genes (Chauhan and Sood 2013). In several plant species, there have been shown to be numerous R genes that confer resistance to a variety of bacterial, fungal and viral diseases (DeYoung and Innes 2006; McHale et al. 2006; Seo et al. 2006). Despite the fact that pathogens have quite different infection methods, R genes are remarkably well-preserved (Table 19.2).

TABLE 19.2
Castor Bean Yield Losses Caused by Diseases

Pathogen	Disease caused	Pathogen type	References
Xanthomona sricinicola	Leaf blight	Fungus	Sabet (1959)
Phytophthora palmivora	Seed blight	Bacterium	Uchida and Aragaki (1988)
Leveillula taurica	Powdery mildew	Fungus	Mirzaee et al. (2011)
Botryotinia ricini	Graymold	Fungus	Soares (2012)
Fusarium oxysporum	Fusarium wilt	Fungus	Vahunia et al. (2017)
Cercospora coffeicola	Leaf spot	Fungus	Souza and Maffia (2011)

TABLE 19.3
Different Types of Resistance (R) Genes in Plants

S. No.	Major R gene category	Example	References
1	NBS-LRR-CC	RPM1 and RPS2	Bisgrove et al. (1994), Mackey et al. (2003)
2	NBS-LRR-TIR	L6 and RPP5	Parker et al. (1997)
3	LRR-TrD-Kinase	Xa21	Wang et al. (1998), Gao et al. (2013)
4	LRR-TrD	Cf-2 and Cf-4	Joosten and de Wit (1999)
5	LRR-TrD-PEST-ECS	Ve2	Kawchuk et al. 2001
6	TrD-CC	RPW8	Xiao et al. (2001)
7	Enzymatic R genes	Rpg1	Brueggeman et al. (2002)
8	TIR-NBS-LRR-NLS- WRKY	RRS1R	Deslandes et al. (2002), Zhang et al. (2016)

Note: NBS, nucleotide-binding site; LRR, leucine-rich repeats; CC, coiled-coil; TIR, toll/interleukin-1 receptor; TrD, transmembrane domain; PEST, amino acid domain; ECS, endocytosis cell signaling domain; NLS, nuclear localization signal; WRKY, amino acid domain.

Castor beans also have been implicated with viral infections. In Andhra Pradesh, India, Reddy et al. (2014) discovered the tobacco streak virus. Vein mosaic and necrotic patches on the underside of leaves are signs of the disease. The highest mortality rate in castor crops was observed in Greece and Italy, according to the report of Parrella et al. (2008), which identified olive latent virus 2 as the cause of leaf yellowing, speckling, mottling and arabesque line patterns. Cucumber mosaic virus, exhibiting blistering, leaf distortion and heavy mosaic symptoms, caused significant production losses of castor beans in Lucknow, India, according to Raj et al. (2010). In some regions of Iran, Mirhosseini and Nasrollah-Nejad (2017) reported uniform symptoms in castor for the same virus. The molecular basis of disease resistance in castor crops is poorly understood. The availability of its genomic sequence and transcriptome data has opened up new opportunities for detailed research and characterization of castor bean resistance genes. The eight main groups of R genes are listed in Table 19.3. All of these classes, with the exception of enzymatic R genes, contain nucleotide-binding sites (NBS), similar motifs and/or leucine-rich repeat (LRR) domains.

19.5   CASTOR BEAN DISEASE RESISTANCE GENES

In the castor genome, Sood et al. (2014) discovered 121 potential disease resistance genes in addition to the 47 noble NBS-LRR resistance genes. These genes' alignments to the castor genome sequence show that they had not previously been annotated in this manner. Each of the discovered NBS-LRR resistance genes had more TNLs than CNLs. TIR and CC, two N-terminal domains, are often involved in pathogen detection. In this transcriptome investigation, 318 transcription factors relevant to the defensive response were also reported (Sood et al. 2014). The transcription factor families to which these genes belong include NAM, Homeo-domain, WRKY, bZIP, ERF/AP2-EREBP, SBP and Whirly. Another genome-wide investigation of castor revealed the presence of many AP2/ERF transcription factors that are essential for dicots to tolerate biotic and abiotic stress (Xu et al. 2013). Through the invention of cross-species markers and the application of polymerase chain reaction (PCR) with degenerate primers in various plant species, the conserved area in those gene families may be utilized to perceive distinct putative resistance genes, also known as resistance gene analogs (RGAs). These RGAs typically occur close to important resistance genes or QTLs controlling the resistance characteristic. Therefore, RGAs have been employed as useful markers for breeding disease resistance through marker-assisted selection. Many significant RGAs were identified from castor and other significant Euphorbiaceae family members by Gedile et al. in 2012. In castor, a total of 86 RGAs have been recognized by using sequence homology and degenerate PCR, in order to aid future characterization and confirmation of anticipated R genes, and biotic stress resistance through molecular breeding. Although almost 170 R genes were recognized in castor, it is further possible that a lot of different R genes are as yet unidentified, because many other members of the Euphorbiaceae family have revealed larger numbers of R genes e.g., around 200 in physic nut (Sato et al. 2011), more than 300 in cassava (Lozano et al. 2015; Wolfe et al. 2016; Kayondo et al. 2018), and over 480 in rubber (Lau et al. 2016). Through biotechnological interventions, the identification of additional R genes in castor with the aid of cutting-edge omics techniques would speed up the resistance breeding program. Different plant species have undergone CRISPR/Cas9-based gene editing to modify many aspects of their immune systems, including susceptibility genes (S-genes) to major disease resistance. However, pathogen-specific resistance is offered via single-R gene editing. This restriction can be bypassed and resistance granted against a much wider variety of diseases by creating synthetic immunological receptors with numerous pathogen recognition sites. Additionally, it is possible to introduce several naturally occurring or artificially created resistance genes inside the same cultivar in order to obtain broad disease resistance.

19.6   INSECT RESISTANCE BREEDING

More than 100 insects harm castor bean. In some crops, it is frequently employed as a trap crop. There are certain sources of resistance against some pest insects. For instance, Jayaraj (1966, 1967) reported C3 Pakistan as one of the tolerant sources against leaf hoppers (*Empoasca flavescens*) and identified the accessions RC1096 Cimmerron Coonoor, RC1094, RC1098 Baker and RC1092 Italy, as resistant sources. According to reports, castor plants with double and triple blooms show more resistance to leaf hoppers as compared to those with no bloom or single bloom (Jayaraj 1968; Srinivas Rao et al. 2000; VijayaLakshmi et al. 2005). Lakshminarayana and Anjani (2009) found numerous stable leafhopper-resistance sources in Indian collections (Lakshminarayana 2003; Lakshminarayana and Anjani, 2009). For instance, the Indian accession RG 304 (Liu et al. 1943) is resistant to leaf hopper, nematode and wilt. Some sources of resistance have resistance against multiple insect pests. Five accessions, *viz.*, RG-43, RG-631, RG-1621, RG-3037 and RG-3067, were shown to be resistant to leaf hoppers, according to Anjani et al. (2018). When all the leaves of other plants died, several purple varieties also perished, but two wild genotypes, HY1 and HY2, collected from southern China, maintained resistance to leaf hoppers except for hopper burn (Xuegui Yin, unpublished). According to Ramanathan (2004), EC 103745 is an unusual accession that is resistant to whiteflies. A total of 43 Indian accessions were also noted as possible white fly resistance sources (Lakshminarayana 2003; Anjani and Jain 2004). The most dangerous castor defoliators are the tobacco caterpillar and semi-looper. Five accessions (RG 5, 33, 221, 224 and 449) were able to survive the tobacco semi-looper, according to the study of Thanki et al. (2001), which found that cultivar CO-1 had moderate resistance to the tobacco caterpillar. In confined environments, five Indian accessions (RG 1934, 2546, 2770, 2543 and 2786) showed resistance to the capsule borer (Lakshminarayana 2003; Lakshminarayana and Anjani 2010). Two morphological variants of castor, one with a purple color morphotype and the other with a morphotype resembling a papaya leaf, are reportedly resistant to leaf miners. For instance, the morphotypes of the Indian resistant accessions RG 1930 and RG 2008 were dark purple, whilst those of RG 1766 and RG 1771 were similar to papaya leaves. The leaf miner resistant purple color accession RG 1930, according to Anjani et al. (2007), is tied to maternal inheritance and only exhibited resistance when used as a female parent. Another investigation revealed a link between leaf miner resistance and total phenol content. It was found that the total phenol content of resistant genotypes was high (Prasad and Anjani 2000; Anjani et al. 2010)

19.7 APPLICATION OF MOLECULAR BREEDING FOR DISEASE RESISTANCE IN CASTOR BEAN

Traditional plant breeding techniques are still important and in use, although they are time-consuming and frequently ineffective at controlling disease outbreaks. Molecular markers are frequently used in crop improvement. When a linked marker is employed to select the desired characteristic rather than the gene that generates it, marker-assisted selection (MAS) operates on the theory of marker-trait connections. MAS has gained popularity as a technique for breeding disease resistance over the past few decades. The growing accessibility of genomic sequences is an essential source of information for developing novel markers for fine-mapping important traits like resistance to biotic stresses (Neumann et al. 2011). High-throughput genotyping methods facilitate the tracking of introgressed R genes in crossover programs. R genes have been identified using molecular markers in numerous investigations, mostly in food crops. Only a few of these studies, however, have been carried out for crops important to the business, including castor. For instance, Dhingani et al. (2012) identified the genetic diversity of *Fusarium* wilt resistance genes in castor using simple sequence repeats (SSRs), randomly amplified polymorphic DNA (RAPD) and inter-simple sequence repeats (ISSRs) Markers. Reddy et al., in 2011, found two RAPD markers that are closely linked to the same characteristics. Three RAPD markers connected to the castor wilt-resistant gene were obtained by combining traditional breeding techniques with MAS to map these features (Singh et al. 2011). The investigation of genetic diversity was carried out in castor using molecular markers to find naturally occurring resistance to *Fusarium* wilt (Anjani 2010). Target region amplification polymorphism (TRAP) markers are also created from mRNA sequences in large populations of castor to evaluate disease resistance and the genetic diversity of other features (Simoes et al. 2017). The selection of resistant features utilizing markers and the quantitative study of disease resistance loci will both be sped up by advancements in genome automation. Castor bean disease yield losses are minimized, while the production of superior specialty oils is maximized, thanks to the development of novel molecular markers and the discovery of more disease-resistant genes.

19.8 ABIOTIC STRESS

Castor adapts well to soil dryness, salinity, sodicity and other unfavorable circumstances, and has strong stress resistance (Jiao et al. 2019). It has increased in popularity in recent years in the creation

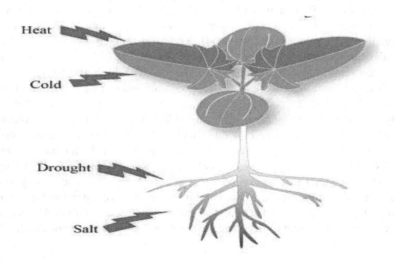

FIGURE 19.1

of ecological environments and soil improvement (Zhou et al. 2010; Wang et al. 2019). The mechanism of abiotic stressors on castor must therefore be studied. Based on this, the study of abiotic stressors on castor has been thoroughly summarized in this chapter.

19.8.1 DROUGHT STRESS

One of the most severe environmental difficulties in the world is the lack of access to water. Several other issues, such as climate change, industrialization, rapid demographic expansion and urbanization, the evolution of economic development, pollution, etc., place significant pressure on the available water reserves in addition to agriculture, which uses 70% of the world's water resources. However, as agriculture is the largest sector affected by water scarcity, it is crucial to discover ways to lower this industry's water usage. One solution is to find crops that can tolerate drought, and castor bean is the top contender. Drought tolerance is the capacity of plants to develop and endure when challenged with a lack of water supplies. Drought is one of the main causes of past mass famines and could now pose a danger to global food security. Due to its robust root structure that can reach deeper into the soil, castor is a drought-tolerant industrial crop. The benefit of castor's ability to survive in drought-prone areas is that it will minimize competition for space with other food crops. The castor crop is extremely vulnerable to water scarcity during its early growth phases. Because of a moisture decline, the cellular level of the castor plant experienced decreased initiation of callus, chlorophyll content and activity of nitrate reductase (Manjula et al. 2003). Water scarcity raises the cuticular wax load and abscisic acid (ABA) level in the sap of phloem (Zhong et al. 1996; Lakshmamma et al. 2009). One strategy for drought tolerance used by castor plants is an osmotic adjustment, which keeps the water balance under osmotic stress. Osmotic adjustment causes an accumulation of many osmotically active compounds and ions, including proline, organic acids, soluble carbohydrates, and Cl^-, Ca and K (Shanker et al. 2014). The cell's osmotic potential is decreased under low water supply situations in order to attract water into the cell and maintain the maintenance of the turgor. Osmotic adjustment accelerates plant growth, assimilation and photosynthetic processes by preserving the normal rate of cytoplasmic activity and organelles (Blum 2017). Osmotic adjustments in castor leaves differed significantly depending on the genotypes' degree of influence. The osmotically reactive chemicals accumulated by the genotypes showed the greatest concentration of soluble sugars (61%), followed by uninterrupted amino acids (17%) and proline (12%), and a very low concentration of potassium (2.8%) (Babita et al. 2010). The genotypes with the maximum osmotic adjustment seed production were greater than those with less osmotic adjustment, at 53% (Babita et al. 2010).

The castor plant's net CO_2 fixation remained high in dry conditions, and stomatal conductance was unaltered. The fast stomatal closure reduced the rate of water loss through transpiration (Sausen and Rosa 2010). Castor plants' photosynthetic components were found to be protected against drought stress, and diffusive resistance was frequently to blame for restrictions on a photosynthetic pigment (Sausen and Rosa 2010). However, after being exposed to this stress for 24 hours, the plant fully recovered its normal photosynthetic activity. Acute water stress drastically reduced the castor's ability to photosynthesize. The plant, however, was quite sensitive to inadequate lighting (Funk and Zachary 2010).

19.8.2 COLD STRESS

Different plant species, especially crop kinds, require a specific temperature range to function normally. In addition to the cultivar itself, these temperatures depend on the cultivar's stage of growth (Sala et al. 2012). A deviation from the optimal temperature range causes temperature stress, which disrupts plant function. There are two types of low-temperature stressors: chilling stress and freezing stress (Kolaksazov et al. 2013). The castor plant is very susceptible to cold stress in the early stages of its development cycle. Because of this, among the key issues that many breeders have thought about, but to no avail, are the seed's late germination, uneven germination and cold-sensitive germination. The plant requires temperatures between 31 and 36°C, which are both suitable. Its germination is enhanced by temperatures between 14 and 15°C (Salihu et al. 2014). Within this range, the rate of emergence and germination both increase as the temperature rises. When seeds are placed in cold soil, germination and seedling emergence are postponed, which leads to an uneven establishment (Moshkin 1986). Castor seedlings may sustain serious damage or even die at temperatures below 1°C since the crop is extremely sensitive to cold. When the temperature falls to 5°C or lower, the plant is impacted by the cold (Moshkin 1986). An average daily temperature of 15°C promotes the growth of seedlings, and between 20 and 28°C is the ideal temperature for the entire growing season. The blossoming and maturation cycle may be shortened by an increase in temperature. The pistil and stamen flowers will blossom if the daily average temperature is higher than 18°C (Yin et al. 2019). Castor can be developed to be more resistant to cold stress, and a variety of physiological factors may be used to choose resistant varieties. Chlorophyll content and specific leaf area are two factors that are genetically linked to cold stress. Using these traits (Škori´c 2016) and castor can be researched to increase resistance to cold stress. Since proteins directly affect how plants react to stress, the identification of protein markers at the proteome level would aid in understanding the physiological signs under cold stress. Studies have shown that the number of cold-regulated/ late embryogenesis-abundant (COR/LEA) proteins increases in response to cold stress. Heat shock proteins (HSPs) can be crucial in preventing protein misfolding while under the influence of cold stress. Recently, a variety of plants have been studied extensively. Using proteomic techniques, a group of DAPS (differentially abundant protein species) associated with cellular responses to cold stress was discovered (Wang et al. 2019).

19.8.3 SALINITY STRESS

One of the main abiotic stresses, salt, affects over 7% of the world's land surface, which further lowers crop productivity (Li et al. 2012). According to a 2008 FAO assessment, salinity is anticipated to affect more than 830 million hectares of land. Due to activities connected with the availability of an excessive amount of sodium, 434 million hectares of land are salinized, and about 403 million hectares of land are salinized as a result of the presence of salt water (FAO 2008). All of these areas have parts that are unsuited for cultivation owing to salt damage and drought. Because most agriculturally productive land is negatively flooded by saline water, the need to sustainably feed an increasing population has become a serious concern. In light of this, it is crucial to growing crops that can survive salt. Castor crops may grow and survive in marginal lands with a wide range of external salinity, as opposed to other types of crops (Oleiwi et al. 2016). Exposure to 160 mol m^{-3} NaCl did not affect the castor plant's ability to grow (Jeschke and Wolf 1988). A large amount of salt (Na) in the soil or irrigation water hampered the castor development and yield (Silva et al. 2008). The most delicate times of plant development are the early stages (Pinheiro et al. 2008). Castor seed emergence was delayed and reduced as salinity increased, however, the genotype CSRN 367 was substantially more resistant to salinity effects than BRS Paraguaçu (Silva et al. 2005). In castor plants, increasing Na salinity proved to be detrimental to the photosynthetic system and proline accumulation (Li et al. 2010). The nutrient intake and accumulation remained unaltered at this extreme level of salt stress, which is higher than the threshold of growth and emergence developed for castor in Na salinity (Li et al. 2010). Castor must meet parameters of 7.1 dS m^{-1} and 13.6 dS m^{-1} in order to grow and emerge. After being exposed to the same salt level for 11 days, the emergence rate was 50% lower and 60% of the seedlings died after 9 days (Zhou et al. 2010). Osmotic stress and ion toxicity are brought on by high salinity stress, which also affects the growth and development of plants. This stress disturbs the ions' equality (Kumar et al. 2013). Salt stress's hyperosmotic and hyperionic actions impair plants' ability to absorb water, and their greatest concentrations inside the plant become dangerous (Kumar et al. 2013). Despite the cationic nature of water, high salt content harmed castor's growth. The plant was very sensitive to Na ions when they were present in the irrigation water, and Ca and Mg were unable to mitigate their detrimental effects throughout the emergence stage and during the initial stages of castor growth (Severino et al. 2014; Lima et al. 2018). Electrical conductivity had a greater influence on the growth and development of castor than did the irrigation water's cationic nature, and the detrimental consequences of the cations were in the following order: Na^+> Na^{++} Ca^{2+}> Ca^{2+}>Na^{++} Ca^{2++} Mg^{2+}>K^+. The times of inflorescence development and flower bud opening were more strongly influenced by the irrigation water's cationic makeup, while plants watered with calcic water showed the effects more prominently (de Lima et al. 2016). High levels of salinity stress in plants can impair important biochemical and physiological processes as well as hormones produced by reactive oxygen species (ROS) (Yan et al. 2017). Gibberellic acid (GA_3) is one of the main hormones that plants use to respond to stress, and it has been discovered that pre-soaking castor seeds in 250 M GA_3 enhanced seedling growth, even under the most severe salinity stress (Jiao et al. 2019). Additionally, plants use various evolutionary techniques including compartmentalization and secretion to counteract salt stress, which can maintain the low cytosolic balance by eliminating sodium (Purty et al. 2008; Zhang et al. 2014b). Therefore, similar techniques can be tested in castor to increase its ability to withstand salt stress.

19.8.4 FLOODING AND SUBMERGENCE STRESS

Although water is necessary for plant growth and development, too much of it can be detrimental to their survival and productivity (Fukao et al. 2019). Floods have grown due to unanticipated events, localizations, regimes and global warming (Wilson et al. 2019). One of the major important limitations on agricultural production is that about 10% of the globe's land area is covered in water (Patel et al. 2015). Considerations for plant existence in floods include the type of plant and soil, growth phase, flooding circumstances and levels of fertility (Fukao et al. 2019). Castor is a crop that does not become waterlogged and is very sensitive to soil anoxia brought on by waterlogging. After castor plants were subjected to waterlogging for a minimum of 2 hours and a maximum of 6 hours, the amount of ABA accumulated by flooded roots, CO_2 intake, leaf proliferation, hydraulic conductivity of roots, the conductance of stomata and transpiration were all decreased (Else et al. 2001). Castor crops exposed to soil floods suffered serious damage and eventually died after three days (Severino et al. 2005). In plants exposed to hypoxia, the amount of starch, α-amylase activity, protein and soluble carbohydrates increased, whereas nitrate reductase activity and seed production decreased (Beltro et al. 2006; Baldwin and Cossar 2009). Although studies have demonstrated that flooding affects castor growth and development, it is still unknown how plants respond to excessive water physiologically. This necessitates more study of these systems.

19.8.5 HEAVY METAL STRESS

Due to increasing industrialization and urbanization, more heavy metals like cadmium, lead and nickel have infiltrated the ecosystem. These heavy metals pose a significant risk to all living things on the planet because of their severe toxicity and lack of biodegradability (Jha et al. 2017). Intake of nutrients, biomass, photosynthetic pigments, protein levels and seedling growth can all be impacted by heavy metal poisoning in the castor plant. By using a number of mechanisms, including the activation of antioxidant enzymes, proline exclusion and accumulation, compartmentalization, organic acid exudation and phytochelatins, castor plants may be moderately resistant to high quantities of heavy metals (Yeboah et al. 2020). The ability of the plant to withstand and accumulate heavy metals in the soil appears to be the most efficient and long-lasting treatment. Castor bean is a topic of extensive research for the remediation of metal soils because of its ability to tolerate heavy metals in contaminated soils, as well as its high biomass content and high potential to accumulate heavy metals (Jha et al. 2017). The roots, stems and leaves of the castor plant are the areas where heavy metal accumulation occurs most frequently (Bauddh and Singh 2012; Celik and Akda 2019). The formation of metal complexes, which prevent the metal's translocation from damaging the plant's physiology and overall growth, may be the source of the high concentration of metals in the roots (Yeboah et al. 2020). Additionally, the larger concentration of metals in plant roots raises the possibility that they may help stabilize metal in soils (Olivares et al. 2013). Castor plants are recommended as potential participants in phytostabilization as a result of their notable root deposition of heavy metal ions. The high rate of biomass formation in the castor plant makes it more resilient to heavy metal stress. However, the metal stress tends to somewhat decrease plant biomass. Castor biomass's susceptibility to metal effects varies depending on the type, intensity and duration of the metal stress (Bauddh and Singh 2012; Olivares et al. 2013; Celik and Akda 2019). In a pot experiment, castor tolerance was tested for 4 months in soils that were contaminated with Cd, Cu and Zn. The results showed that the plant had a maximum biomass level and an average height of 136 cm (Wang et al. 2016). In reaction to elevated concentrations of Pb (96 mg/L) and Cd (16 mg/L), the biomass of castor plants in a hydroponic system decreased and increased, respectively. It was found that the castor crop could be used as a Cd soil indicator and a Pb tolerance indicator in Cd/Pb-contaminated soils (Costa et al. 2012). Castor plants adopt several coping mechanisms to deal with high amounts of heavy metal stress in the growing medium. Some of these mechanisms include phytochelatin production, proline accumulation, compartmentalization, antioxidant production and organic acid secretion (Huang et al. 2016; Nascimento and Marques 2018; Celik and Akdas 2019; Ye et al. 2018; Roychowdhury et al. 2019). Castor plant growth can be accelerated by using these procedures, which also reduce metal stress. To boost castor tolerance, many antioxidant systems are engaged when exposed to heavy metals. Due to the presence of GSH (glutathione), POD (peroxidase) and SOD (superoxide dismutase) activities in the roots and leaves of these plants, the Zibo No. 8 and Zibo No. 5 castor cultivars grew more swiftly under Cd stress. Zibo No. 8's root had significantly ($P = 0.05$) higher GSH activity than Zibo No. 5 due to the increased Cd concentration in the development medium. Zibo No. 8's root overexpression of antioxidants demonstrates that, despite the substantial metal buildup in roots, the roots have a working mechanism that removes the metal (Zhang et al. 2014a). In castor plants intercropped with alfalfa under Cd stress, the activity of the enzymes ascorbate peroxidase (APX), catalase (CAT) and POD was similarly considerably elevated (Xiong et al. 2018). Castor was able to endure metal stress because of proline accumulation. Proline, a stress metabolite, prevents plant cells from drying out and maintains cell membrane integrity by acting as an osmoprotectant and cell wall plasticizer (Singh et al. 2016). The level of stress has a positive link with its expression in castor. Due to the concentration of proline in castor, elevated Pb levels (up to 400 M) have no negative impact on it (Kiran and Prasad 2017). Castor leaves developed more of the osmoprotectant proline when grown in Ni-contaminated soil. It was shown that the amount of proline increases during Ni stress depending on the kind of plant and the severity of metal toxicity (Bauddh and Singh 2015). Without exhibiting any toxic symptoms, the castor plant compartmentalizes as one of its great defensive systems against high metal stress. The plant either expels the metal ions from the cell in order to achieve detoxification or sequesters them in the cell wall or vacuole in order to lessen the influence of the metal on other delicate metabolic processes. Numerous studies have shown metal deposition in the roots, stems and leaves of the castor plant in the growing media (Nascimento and Marques 2018). Metal transporters like the class of active ABC transporter protein increased castor trap metal ions in the subcellular compartment (Pal et al. 2013). The subcellular partitioning, which comprises cellular debris, organelles, heat-stable proteins (HSP), metal-rich granules (MRG) and heat-denatured proteins (HDP), is characterized based on the differential centrifugation of tissues. This partitioning sheds light on plants' metal sensitivity (Zhang et al. 2015). Castor plant exposure to 2 and 5 mg/L Cd had greater impacts on organelles, MRG and soluble fractions (Zhang et al. 2015). The castor plant's limited excretion of organic acids from the roots also helped it grow in heavy metal-contaminated soils. In nickel-contaminated soil, tartaric, malate, low cysteine and oxalic acids developed in the roots, which then stimulated root growth when subjected to 750 mol/L of copper stress (Bauddh and Singh 2015; Huang et al. 2016). Last but not least, castor crops are encouraged to reduce metal stress by phytochelatins and other high-affinity ligands, especially in soil that has been contaminated with nickel (Adhikari and Kumar 2012).

19.9 CONCLUSIONS

Castor has a lot of potential as a biofuel crop and industrial feedstock in the future due to its greater oil content, strong modification of fatty acid content, extremely high oil production, wider range of adaptation and the capacity to grow in marginal areas vulnerable to drought and salty conditions, which are all advantages. As a result, the majority of scientists working on castor anticipate it will become a key crop for the production of plant lipids for both energy and manufacturing purposes. There have been many studies and advancements on the stress breeding of castor. Recent genome sequencing data of castor will give a heavy thrust in molecular breeding as well as genomics, proteomics, transcriptomics and metabolomics studies into the biotic and abiotic stresses of castor. Future research efforts will play a crucial role in the improvement of castor under different stresses. An integrated crop improvement program could lead to further development.

ACKNOWLEDGMENT

There is no conflict of interest among the authors. We apologize for not citing all of the relevant references due to space limitations.

REFERENCES

Adhikari, T. and A. Kumar. 2012. Phytoaccumulation and tolerance of *Riccinus communis* L. to nickel. *International Journal of Phytoremediation* 14:481–492.

Anjani, K. 2010. Pattern of genetic diversity among Fusarium wilt resistant castor germplasm accessions (*Ricinuscommunis* L.). *Electronic Journal of Plant Breeding* 1(2):182–187.

Anjani, K. 2005a. Purple-colored castor (*Ricinuscommunis* L.)-A rare multiple resistant morphotype. *Current Science* 88:215–216.

Anjani, K. 2005b. RG 2722, Castor (*Ricinuscommunis* L.) germplasm with resistance to Macrophomina root rot. *Indian Journal of Plant Genetic Resources* 65:74.

Anjani, K. 2005c. RG1608, Castor (*Ricinuscommunis* L) germplasm with resistance to Fusarium Wilt. *Indian Journal of Plant Genetic Resources* 65:74.

Anjani, K. 2010b. Extra-early maturing germplasm for utilization in cas-tor improvement. *Industrial Crops and Products* 31:139–144. 10.1016/j.ind-

Anjani, K. 2012. Castor genetic resources: A primary gene pool for exploi- tation. *Industrial Crops and Products* 35:1–14.10.1016/j.indcrop.2011.06.011. doi:10.1016/j.indcrop.2011.06.011

Anjani, K. 2010. Pattern of genetic diversity among Fusarium wilt resistant castor germplasm accessions (*Ricinus communis* L.). *Electronic Journal of Plant Breeding* 1:182–187.

Anjani, K., Raoof, M.A., Reddy, A.V.P., and Rao, C.H. 2004 Sources of resistance to major castor (Ricinus communis) diseases. *Plant Genet Resour Newsl* 137:46–48.

Anjani, K., M. Pallavi, and S.N. Sudhakara Babu. 2007. Uniparental inheritance of purple leaf and the associated resistance to leafminer*Liriomyzatrifolii*(Burgess) (Diptera: Agromyzidae) in castor bean (*Ricinus communis* L.). *Plant breeding* 126:515–520.

Anjani, K., M.A. Raoof, M. Santha Lakshmi Prasad, P. Duraimurugan, C. Lucose, P. Yadav, R.D. Prasad, L.J. Jawahar, andc. Sarada. 2018. Trait-specific accessions in global castor (*Ricinus communis* L.) germplasm core set for utilization in castor improvement. *Industrial Crops and Products* 112:766–774.

Arafa, R.A., N.E.K. Soliman, O.M Moussa, S.M. Kamel, and K. Shirasawa. 2018. Characterization of Egyptian Phytophthorainfestans population using simple sequence repeat markers. *Journal of General Plant Pathology* 84:104–107.

Araújo, A.E., N.D. Suassuna, and W.D. Coutinho. 2007. Doenças e seu con-trole. In: D.M.P. Azevedo and N.E.M. Beltrão, editors, O agronegóciodamamona no Brazil. 2nd ed. EmbrapaInformaçãoTecnológica, Brasília.p. 281–303.

Arieira, C.R.D., S.M. Santana, M.L. Silva, C. Furlanetto, R.C.F. Ribeiro, and E.A. Lopes. 2009. Reaction of castor bean (*Ricinuscommunis* L.) and sunflower (*Helianthus annuus* L.) cultivars to *Meloidogynejavanica*, *M.incognita* and *M. paranaensis. NematologiaBrasileira* 33:61–66. (In Portuguese, with English abstract.)

Babita, M., M. Maheswari, L.M. Rao, A.K. Shanker, and D.G. Rao. 2010. Osmotic adjustment, drought tolerance and yield in castor (*Ricinuscommunis* L.) hybrids. *Environmental and Experimental Botany* 69:243–249. doi:10.1016/j. envexpbot.2010.05.006

Baldwin, B.S., and R.D. Cossar. 2009. Castor yield in response to planting date at four locations in the south-central United States. *Industrial Crops and Products* 29:316–319.

Bauddh, K., and R. P. Singh. 2012. Growth, tolerance efficiency and phytoremediation potential of *Ricinuscommunis*(L.) and *Brassica juncea*(L.) in salinity and drought affected cadmium contaminated soil. *Ecotoxicology and Environmental Safety* 85:13–22.

Basappa, H. 2007. Validation of integrated pest management modules for castor (*Ricinuscommunis*) in Andra Pradesh. *Indian Journal of Agricultural Science* 77: 357–362.

Basappa, H., and Lingappa, S. 2001 Damage potential of *Achaea janata* Linn. at different phonological stages of castor. *Indian J Plant Protec* 29: 17–24.

Bauddh, K., and Singh, R.P. 2015. Assessment of metal uptake capacity of castor bean and mustard for phytoremediation of nickel from contaminated soil. *Bioremed J* 19: 124–138.

Beltrão, N., J. Souza, and J. Santos. 2006. Metabolic alterations happened in castor (BRS 149-Nordestina) due to the hydric stress for deficiency and excess in the soil. *Rev Bras De Oleaginosas e Fibrosas* 10: 977–984.

Besset, B.G., N. Lucante, V. Kelechian, R. Dargent, and H. Muller. 1996. Evaluation of castor resistance to sclerotial wilt deases caused by Macro-phominaphaseolina. *Plant Disease* 80: 842–846. doi:10.1094/PD-80-0842

Bisgrove, S.R., M.T. Simonich, N.M., A. Sattler and R.W. Innes. 1994. A disease resistance gene in Arabidopsis with specificity for two different pathogen avirulence genes. *Plant Cell* 6: 927–933.

Blum, A. 2017. Osmotic adjustment is a prime drought stress adaptive engine in support of plant production. *Plant Cell Environment* 40: 4–10.

Brueggeman, R., N. Rostoks, D. Kudrna, A. Kolian, F. Han, J. Chen, A. Druka, B. Steffenson, and A. Kleinhofs. 2002. The barley stem rust-resistance gene Rpg1 is a novel disease-resistance gene with homology to receptor kinases. *Proceedings of the National Academy of Sciences*USA 99: 9328–9333.

Çelik, Ö., and E.Y. Akdas. 2019. Tissue-specific transcriptional regulation of seven heavy metal stressresponsive miRNAs and their putative targets in nickel indicator castor bean (*R. communis* L.) plants. *Ecotoxicology and Environmental Safety* 170: 682–690.

Chagas, H.A., M.A. Basseto, D.D. Rosa, M.D. Zanotto, and E.L. Furtado. 2010. Diagramatic scale to assess gray mold (Amphobotyrsricini) in castor bean(*Ricinuscommunis* L.). (In Portuguese, with English abstract.). *Summa Phytopathologica* 36: 164–167. Doi: 10.1590/S0100- 54052010000200011

Chauhan, R.S., and A. Sood. 2013. Comparative genomics in Euphorbiaceae. In: Bahadur B, Sujatha M, Carels N (Eds) Jatropha: challenges for a new energy crop, vol 2. Springer, New York, pp 351–374.

Cook, D., S. Fowler, O. Fiehn, and M.F. Thomashow. 2004. A prominent role for the CBF cold response pathway in configuring the low-temperature metabolome of Arabidopsis. *Proceedings of the National Academy of Sciences*USA 101(42): 15243–15248. crop.2009.09.016. doi:10.1016/j.indcrop.2009.09.016

Costa Ds, E.T., L.R.G. Guilherme, E.E.C de Melo, B.T. Ribeiro, B.I. Euzelina dos Santos, E. da Costa Severiano, V. Faquin, and B.A. Hale. 2012. Assessing the tolerance of castor bean to Cd and Pb for phytoremediation purposes. *Biological Trace Element Research* 145: 93–100.

Dange, S.R.S., A.G. Desai, and S.J. Patel. 2005. Diseases of castor. In: G.S.Saharan, N. Mehta, and M.S. Sangwan, editors, Diseases of oilseed crops.Indus Publishing Co., New Delhi. p. 211–235.

De Candolle. 1860. Prodromous systematic natural is regni vegetabilis. 17: 1016.

De Lima, G. S., H.R. Gheyi, R.G. Nobre, D.A. Xavie, L.A. dos Anjos Soares, L.F. Cavalcante, and J.B. dos Santos. 2016. Emergence, growth, and flowering of castor beans as a function of the cationic composition of irrigation water. *Semi CiênciasAgrárias* 37: 651–664.

Desai, A.G., and S.R.S. Dange. 2003. Standardization of root dip inoculation technique for screening of resistance to wilt of castor. *Journal of Mycology and Plant Pathology* 33: 73–75.

Deslandes, L., J. Olivier, F. Theulieres, J. Hirsch, D.X. Feng, P. Bittner-Eddy, J. Beynon, and Y. Marco. 2002. Resistance to Ralstoniasolanacearum in Arabidopsis thaliana is conferred by the recessive RRS1-R gene, a member of a novel family of resistance genes. *Proceedings of the National Academy of Sciences USA* 99: 2404–2409.

DeYoung, B.J., and R.W. Innes. 2006. Plant NBS-LRR proteins in pathogen sensing and host defense. *Nature Immunology* 7(12): 1243–1249.

Dhingani, R.M., R.S. Tomar, M.V. Parakhia, S.V. Patel, and B.A. Golakiya. 2012. Analysis of genetic diversity among different Ricinuscommunis genotypes for macrophomina root rot through RAPD and microsatellite markers. *International Journal of Plant Protection* 5: 1–7.

Dracatos, P.M., N.O.I. Cogan, T.I. Saw bridge, A.R. Gendall, K.F. Smith, G.C. Spangenberg, and J.W. Forster. 2009. Molecular characterization and genetic mapping of candidate genes for qualitative disease resistance in perennial ryegrass (*LoliumperenneL.*). *BMC Plant Biology* 9: 62.

Else, M.A., D. Coupland, L. Dutton, and M.B. Jackson. 2001. Decreased root hydraulic conductivity reduces leaf water potential, initiates stomatal closure and slows leaf expansion in flooded plants of castor oil (*Ricinuscommunis*) despite diminished delivery of ABA from the roots to shoots in xylem sap. *Physiologia Plantarum* 111: 46–54. doi:10.1034/j.1399- 3054.2001.1110107.xethiopica control

FAO. 2008. FAO land and plant nutrition management service. Available at: www.fao.org/ ag/agl/agll/spush

Fukao, T., B.E. Barrera-Figueroa, P. Juntawong, and J.M. Peña-Castro. 2019. Submergence and waterlogging stress in plants: a review highlighting research opportunities and understudied aspects. *Frontiers in Plant Science* 10: 340.

Funk, J.L., and V.A. Zachary. 2010. Physiological responses to short-term water and light stress in native and invasive plant species in southern California. *Biological Invasions* 12: 1685–1694.

Gahukar, R.T. 2018. Management of pests and diseases of castor (*Ricinus communis* L.) in India: current status and future perspective. *Archives of Phytopathology and Plant Protection* 51: 956–978.

Gao, L., Y. Cao, Z. Xia, G. Jiang, G. Liu, W. Zhang, and W.Zhai. 2013. Do transgenesis and marker-assisted backcross breeding produce substantially equivalent plants?—a comparative study of transgenic and backcross rice carrying bacterial blight resistant gene Xa21. *BMC Genomics* 14: 738.

Godfrey, J.T.A. 1923. Gray mold of castor. *Journal of Agricultural Research* 23: 679–716.

Guy, C., F. Kaplan, J. Kopka, J. Selbig, and D.K. Hincha. 2008. Metabolomics of temperature stress. *Physiology Plant* 132 (2): 220–235.

Holliday, P. 1980. Fungus diseases of tropical crops. General Publ. Co. Ltda, Toronto, O. N., and J.D. Hooker. 1890. Flora of British India 5. L. Reeve and Co, London.

Huang, G., G. Guo, S. Yao, N. Zhang, and H. Hu. 2016. Organic acids, amino acids compositions in the root exudates and Cu-accumulation in castor (*Ricinus communis* L.) under Cu stress. *International Journal of Phytoremediation* 18: 33–40.

Hutchinson, J. 1959. Families of flowering plants, vol I, II, 2nd edn, Oxford, LondonJayaraj, S. 1968 Studies on plant characters of castor associated with resistance to *Empoasca flavescens* Febr (Homoptera Jassidae) with reference to selection and breeding of varieties. *Indian J Agric Sci* 38: 1–6.

Joosten, M., and P. de Wit. 1999. The tomato-Cladosporiumfulvum interaction: a versatile experimental system to study plant–pathogen interactions. *Annual Review of Phytopathology* 37: 335–367.

Jeschke, W.D., and O. Wolf. 1988. Effect of NaCI salinity on growth, development, ion distribution, and ion translocation in castor bean (*Ricinus communis* L.). *Journal of Plant Physiology* 132: 45–53.

Jha, A.B., A.N. Misr, and P. Sharma. 2017. Phytoremediation of heavy metal-contaminated soil using bioenergy crops. In: Bauddh K, Singh B, Korstad J (eds) Phytoremediation potential of bioenergy plants. Springer, pp 63–96.

Jiao, X., W. Zhi, G. Liu, G. Zhu, G. Feng, N. Eltyb Ahmed Nimir, I. Ahmad, and G. Zhou. 2019. Responses of foreign GA3 application on seedling growth of castor bean (*Ricinus communis* L.) under salinity stress conditions. *Agronomy* 9: 274.

Joyeux, A., M.G. Fortin, R. Mayerhofer, and A.G. Good. 1999. Genetic mapping of plant disease resistance gene homologues using a minimal *Brassica napus* L. population. *Genome* 42: 735–743.

Kaplan, F., J. Kopka, D.W. Haskell, W. Zhao, K.C. Schiller, N. Gatzke, D.Y. Sung, and C.L. Guy. 2004. Exploring the temperature-stress metabolome of Arabidopsis. *Plant Physiology* 136(4): 4159–4168.

Kawchuk, L.M., J. Hachey, D.R. Lynch, F. Kulcsar, G. van Rooijen, D.R. Waterer, A. Robertson, E. Kokko, R. Byers, R.J. Howard, R. Fische, and D. Prufer. 2001. Tomato ve disease-resistance genes encode cell surface-like receptors. *Proceedings of the National Academy of Sciences*USA 98: 6511–6515.

Kayondo S. I., Del Carpio D. P., Lozano R., Ozimati A., Wolfe M., et al. 2018. Genome-wide association mapping and genomic prediction for CBSD resistance in *Manihot esculenta*. *Sci. Rep.* 8: 1–11. 10.1038/s41598-018-19696-1

Kiran, B.R., and M.N.V. Prasad. 2017. Responses of *Ricinus communis* L. (castor bean, phytoremediation crop) seedlings to lead (Pb) toxicity in hydroponics. *Selcuk Journal of Agriculture and Food Sciences* 31: 73–80.

Kochhar, S.L. 1981. Economic botany in the tropics. MacMillan India Ltd, Delhi, India, pp 170–171.

Kolte, S.J. 1995. Castor: Diseases and crop improvement. Shipra Publ., Delhi, India.

Kolaksazov, M., F. Laporte, K. Ananieva, P. Dobrev, M. Herzog, and E. Ananiev. 2013. Effect of chilling and freezing stresses on jasmonate content in Arabis alpina. *Bulgarian Journal of Agricultural Science* 19: 15–17.

Kumar, K., M. Kumar, S. R. Kim, H. Ryu, and Y. G. Cho. 2013. Insights into genomics of salt stress response in rice. *Rice* 6: 1–15.

Lakshmamma, P., P. Lakshimmi, C. Lavanya, and K. Anjani. 2009b. Growth and yield of different castor genotypes varying in drought tolerance. *Annals of Arid Zone* 48: 35–39.

Lakshmamma, P., M. LakshminarayanM, L. Prayaga, K. Alivelu, and C. Lavanya. 2009. Effect of defoliation on seed yield of castor (*Ricinus communis* L.). *Indian Journal of Agricultural Sciences* 79: 630–623.

Lakshminarayana, M. 2003. Host plant resistance in castor against major insectpests. In: Extend summaries: national seminar on stress management in oilseeds for attaining self-reliance in vegetable oils, Hyderabad, India, 28–30 Jan, p 107.

Lakshminarayana, M., and K. Anjani. 2009. Screening for confirmation of reaction of castor genotypes against leafhopper (*Empoascaflavescens*Fabr.) and capsule borer (*Conogethespunctiferalis*Guen.). *Journal of Oilseeds Research* 26 (SplIss): 457–459.

Lakshminarayana, M., and K. Anjani. 2010. Sources of resistance in castor against capsule borer *Conogethes*(*Dichocrocis*) *punctifaralis*Guen *Journal of Oilseeds Research* 27 (SplIss): 273–274.

Lau, N.S., Makita, Y., Kawashima, M., et al. 2016. The rubber tree genome shows expansion of gene family associated with rubber biosynthesis. Sci. Rep. 6, 28594. https://doi.org/10.1038/srep28594

Li, G., S. Wanb, J. Zhoua, Z. Yanga, and P. Qina. 2010. Leaf chlorophyll fluorescence, hyperspectral reflectance, pigments content, malondialdehyde and proline accumulation responses of castor bean (*Ricinuscommunis* L.) seedlings to salt stress levels. *Industrial Crops and Products* 31: 13–19. doi:10.1016/j.indcrop.2009.07.015

Lima, E.A., J.K. Mattos, A.W. Moita, R.G. Carneiro, and R.M.D.G. Carneiro. 2009a. Host status of different crops for Meloidogyneethiopicacontrol. *Tropical Plant Pathology* 34: 152–157. 10.1590/S1982- 56762009000300003

Lima, R.L.S., L.S. Severino, G.B. Ferreira, M.I.L. Silva, R.C. Albuquerque, N.E.M. Beltrão, and L.R. Sampaio. 2007a. Castor bean growth on soil containing high aluminum level on the presence and absence of organic matter. (in Portuguese, with English abstract.). *RevistaBrasileira de Oleaginosas e Fibrosas* 11: 15–21.

Lima, R.L.S., L.S. Severino, R.C. Albuquerque, G.B. Ferreira, L.R. Sampaio, and N.E.M. Beltrão. 2009b. Neutralization of exchangeable aluminum by wood ash and bovine manure and the effect on castor plants growth. (in Portuguese, with English abstract.). *RevistaBrasileira de Oleaginosas e Fibrosas* 13: 9–17.

Li., J.G, L.J. Pu, M. Zhu, R. Zhang. 2012. The present situation and hot issues in the salt-affected soil research. *Acta GeographicaSinica* 67: 1233–1245.

Lima, G.S.D., R.G. Nobre, H.R. Gheyi, L.A.D.A. Soares, C.A. de Azevedo, and V.L. de Lima. 2018. Salinity and cationic nature of irrigation water on castor bean cultivation. *RevistaBrasileira de EngenhariaAgrícola e Ambiental* 22: 267–272.

Liu, Y.L., X.G. Yin, J.N. Lu, Y. Li, and Y.H. Zhou. 2019. First report of castor leaf spot caused by *Cladosporium tenuissimum*in Zhanjiang. China. *Plant Diseases* 103: 375.

Lozano R., Hamblin M. T., Prochnik S., and Jannink J. L. 2015. Identification and distribution of the NBS-LRR gene family in the Cassava genome. *BMC Genomics* 16: 1–14. 10.1186/s12864-015-1554-9

Mackey, D., Belkhadir, Y., Alonso, J.M., Ecker, J.R., and Dangl, J.L. 2003. Arabidopsis RIN4 is a target of the type III virulence effector AvrRpt2 and modulates RPS2-mediated resistance. *Cell* 112, 379–389.

Manjula, K., P. Sarma, R. Thatikunt, and T.N. Rao. 2003. Evaluation of castor (*Ricinus Communis* L.) genotypes for moisture stress. *Indian Journal of Plant Physiology* 8: 319–322.

Manjula, K., P.S. Sarma, T. Ramesh, and T.G.N. Rao. 2003a. Evaluation of castor genotypes for moisture stress. *Indian Journal Plant Physiology* 8: 319–322.

Manjula, K., P.S. Sarma, T. Ramesh, and T.G.N. Rao. 2003b. Screening of castor genotypes for dry lands using PEG induced stress. *Journal of Oilseeds Research* 20: 170–171.

McHale, L., X. Tan, P. Koehl, and R.W. Michelmore. 2006. Plant NBS-LRR proteins: adaptable guards. *Genome Biology* 7: 212.

Mirhosseini, H.A., and S. Nasrollah-Nejad. 2017. First report of cucumber mosaic virus infecting *Ricinuscommunis* in India. *Plant Diseases* 101(12): 2154.

Mirhosseini, H.A., S. Nasrollah-Nejad, V. Babaeizad, and H. Rahimian. 2017. First report of *Cucumber mosaic virus* infecting *Ricinus communis* in Iran. *Plant Diseases* 101: 2154.

Mirzaee, M.R., M.A. Khodaparast, M. Mohseni, S.H.R. Ramazani, and M. Soltani-Najafabadi. 2011. First record of powdery mildew of castor-oil plant (*Ricinus communis*) caused by the anamorphic stage of *Leveillulatauraica*in Iran. *Australasian Plant Disease* 6:.36–38.

Moshkin, V. 1986. Castor. Taylor & Francis: Amerind, New Delhi, India.

Moshkin, V.A. 1986. Castor. Amerind Publishing Co, New Delhi.

Neumann, K., B. Kobiljski, S. Denčić, R.K. Varshney, and A. Börner. 2011. Genome-wide association mapping: a case study in bread wheat (*Triticumaestivum* L.). *Molecular Breeding* 27: 37–58.

Nascimento,C.W.Ad., and M.C. Marques. 2018. Metabolic alterations and X-ray chlorophyll fluorescence for the early detection of lead stress in castor bean (*Ricinus communis*) plants. *Acta Scientiarum - Agronomy* 40: 2–9.

N.E.M. Beltrão, editors, O agronegócio da mamona no Brazil. EmbrapaAlgodão/EmrapaInformaçãoTecnológica, Campina Grande, Brasília, Brazil. p. 213– 227.

Obata, T., and A.R. Fernie. 2012. The use of metabolomics to dissect plant responses to abiotic stresses. *Cellular and Molecular Life Sciences* 69(19): 3225–3243. https://doi.org/10.1007/s00018-012-1091-5

Oleiwi, A.M., N.M. Elsahookie, and L.I. Mohammed. 2016. Performance of castor bean selects in saline sodic soil. *International Journal of Applied Agricultural Sciences* 2: 64.

Olivares, A.R., R. Carrillo-González, Md. C.A. González-Chávez, and R.M.S Hernández. 2013. Potential of castor bean (*Ricinus communis* L.) for phytoremediation of mine tailings and oil production. *Journal of Environmental Management* 114: 316–323.

Palma, F., F. Carvajal, C. Lluch, M. Jamilena, and D. Garrido. 2014. Changes in carbohydrate content in zucchini fruit (*Cucurbitapepo* L.) under low temperature stress. *Plant Science* 217–218: 78–86.

Parker, J.E., M.J. Coleman, V. Szabo, L.N. Frost, R. Schmidt, E.A. van der Biezen, T. Moores, C. Dean, M.J. Daniels, and J.D.G. Jones. 1997. The Arabidopsis downy mildew resistance gene RPP5 shares similarity to the toll and 6 Genomics of Disease Resistance in Castor Bean 111 interleukin-1 receptors with N and L6. *Plant Cell* 9: 879–894.

Parrella, G., A. De Stradis, and C.Vovlas. 2008. First report of olive latent virus 2 in wild castor bean (*Ricinuscommunis* L.) in Italy. *Plant Pathology* 57: 392.

Patel, M.K., M. Joshi, A. Mishra, and B. Jha. 2015. Ectopic expression of SbNHX1 gene in transgenic castor (Ricinuscommunis L.) enhances salt stress by modulating physiological process. *Plant Cell, Tissue and Organ Culture* 122: 477–490.

Patel, P.B., and H.C. Pathak. 2011. Genetics of resistance to wilt in castor caused by *Fusariumoxysporum* f. *spricini. Agricultural Science Digest* 31: 30–34.

Pal, R., A. Banerjee, and R. Kundu. 2013. Responses of castor bean (*Ricinus communis* L.) to lead stress. *National Academy of Sciences, India, Section B: Biological Sciences* 83: 643–650.

Patel, R.R., D.D. Patel, P. Thakor, B. Patel, and V.R. Thakkar. 2015. Alleviation of salt stress in germination of Vigna radiata L. by two halotolerant Bacilli sp. isolated from saline habitats of Gujarat. *Plant Growth Regulator* 76: 51–60.

Pinheiro, H.A., J.V. Silva, L. Endres, V.M. Ferreira,C.C. de Albuquerque, F.F. Cabral, J.F. Oliveira, L.W.T. de Carvalho, J.M. dos Santos, and F.B.G. dos Santos. 2008. Leaf gas exchange, chloroplastic pigments and dry matter accumulation in castor bean (*Ricinuscommunis*L) seedlings subjected to salt stress conditions. *Industrial Crops and Products* 27: 385–392.

Prasad, Y.G., and K. Anjani. 2000. Resistance to serpentine leafminer*Liriomyzatrifolii*(Burgess) in castor (*Ricinus communis* L.). *Indian Journal of Agricultural Sciences* 71: 351–352.

Pinheiro, H.A., J.V. Silva, L. Endres, V.M. Ferreira, C.A. Camara, F.F. Cabral, J.F. Oliveira, L.W.T. Carvalho, J.M. Santos, and B.G. Santos. 2008. Leaf gas exchange, chloroplastic pigments and dry matter accumulation in castor bean (*Ricinuscommunis* L) seedlings subjected to salt stress conditions. *Industrial Crops and Products* 27: 385–392. doi:10.1016/j.indcrop.2007.10.003

Poire, R., H. Schneider, M.R. Thorpe, A.J. Kuhn, U. Schurr, and A. Walter. 2010. Root cooling strongly affects duel leaf growth dynamics, water and carbohydrate relations in *Ricinuscommunis*. *Plant Cell Environment* 33: 408–417. doi:10.1111/j.1365-3040.2009.02090.x www.isaaa.org; www. rahan.co.il

Purseglove, J. W. 1968. Tropical crops: dicotyledons. Vol 1 & 2 combined 1st ELBS edn 1974. ELBS and Longman, UK, pp 180–186 Reprint 1984.

Purty, R.S., G. Kumar, S.L. Singla-Pareek, and A. Pareek. 2008. Towards salinity tolerance in Brassica: an overview. *Physiology and Molecular Biology of Plants* 14: 39–49.

Qi, H., L. Hua, L. Zhao, and L. Zhou. 2011. Carbohydrate metabolism in tomato (*Lycopersiconesculentum* Mill.) seedlings and yield and fruit quality as affected by low night temperature and subsequent recovery. *African Journal of Biotechnology* 10(30): 5743–5749.

Raj, S., S.K. Snehi, K.K. Gautam, and M.S. Khan. 2010. First report of association of *Cucumber mosaic virus* with blister and leaf distortion of castor bean (*Ricinus communis*) in India. *Phytoparasitica* 38: 283.

Rajani, V.V., and A.M. Parakhia. 2009. Management of root rot disease (*Macrophomina phaseolina*) of castor (*Ricinus communis*) with soil amendments and biocontrol agents. *Journal of Mycology and Plant Pathology* 39: 290–293.Ramanathan, T. 2004 Applied genetic of oilseed crops. Dava Publishing House, New Delhi.

Rao, P.V.R., Shankar, V.G., and Reddy, A.V. 2009. Variability studies in castor (*Ricinus communis* L.). *Res Crops* 10: 696–698

Reddy, B.B.V., L.Prasanthi, Y. Sivaprasad, A. Sujitha, and T.G. Krishna. 2014. First report of tobacco streak virus in castor bean. *Journal of Plant Pathology* 96: 431–439.

Reddy, R.N., M. Sujatha, A.V. Reddy, and A.P. Reddy. 2011. Inheritance and molecular mapping of wilt resistance gene (s) in castor (*Ricinuscommunis* L.). *International Journal of Plant Breeding* 5: 84–87.

Ribeiro, L.P., and E.C. Costa. 2008. Occurrence of Erinnyisello and Spodopteramarima in castor bean plantation in Rio Grande do Sul State, Brazil. *Ciencia Rural* 38: 2351–2353. (In Portuguese, with English abstract.) doi:10.1590/S0103-84782008000800040

Rizhsky, L., H.J. Liang, J. Shuman, V. Shulaev, S. Davletova, and R. Mittler. 2004. When defense pathways collide. The response of Arabidopsis to a combination of drought and heat stress. *Plant Physiology* 134(4): 1683–1696.

Roychowdhury, R., M. Roy, S. Zaman, and A. Mitra. 2019. Phytoremediation potential ofcastor oil plant (*Ricinus Communis*) grown on fly ash amended soil towards lead bioaccumulation. *Journal of Emerging Technologies and Innovative Research* (JETIR) 6: 156–160.

Sabet, K.A. 1959. Bacterial leaf blight disease of castor. *Ann Appl Biol* 47: 49–55.

Salihu, B.Z., A.K. Gana, and B.O. Apuyor. 2014. Castor oil plant (*Ricinuscommunis* L.): botany, ecology and uses. *International Journal of Scientific Research* 3(5): 1333–1341.

Sarma, A.K., M.P. Singh, and K.I. Singh. 2006. Resistance of local castor genotypes to Achaea janata Linn. andSpodopteralituraFabr. *Journal of Applied Zoological Research* 17: 179–181.

Salihu, B., A. Gana, and B. Apuyor. 2014. Castor oil plant (*Ricinus communis* L.): botany, ecology and uses. *International Journal of Scientific Research* 3: 1334–1341.

Sato, S., et al. 2011. Sequence analysis of the genome of an oil-bearing tree, *Jatropha curcas* L. *DNA Res.* 18, 65–76.

Sausen, T.L., and L.M.G. Rosa. 2010. Growth and carbon assimilation limitations in *Ricinuscommunis* (Euphorbiaceae) under soil water stress conditions. *ActaBotanicaBrasilica* 24: 648–654. doi:10.1590/S0102-33062010000300008

Seo, Y.S., M.R. Rojas, J.Y. Lee, S.W. Lee, J.S. Jeon, P. Ronald, W.J. Lucas, and R.L. Gilbertson. 2006. A viral resistance gene from common bean functions across plant families and is upregulated in a non- virus-specific manner. *Proceedings of the National Academy of SciencesUSA* 103: 11856–11861.

Severino, L.S., C.L.D. Lima, N.E.M. Beltrão, G.D. Cardoso, and V.A. Far-ias. 2005. Comportmentdamamoneira sob encharcamento do solo. EmbrapaAlgodão, Campina Grande, Brasilia, Brazil.

Severino, L.S., R.L. Lima, N. Castillo, A.M. Lucena, D.L. Auld, and T.K. Udeigwe. 2014. Calcium and magnesium do not alleviate the toxic effect of sodium on the emergence and initial growth of castor, cotton, and safflower. *Industrial Crops and Products* 57: 90–97.

Shanker, A.K., M. Maheswari, S. Yadav, S. Desai, D. Bhanu, N.B. Attal, and B. Venkateswarlu. 2014. Drought stress responses in crops. *Functional & Integrative Genomics* 14:11–22

Silva, S., A. Alves, H. Ghey, N. Beltrão, L. Severino, and F. Soares. 2005. Germination and initial growth of two cultivars of castor bean under saline stress. (Portuguese, with English abstract). *Rev Bras de EngenhariaAgrícola e Ambiental Suppl* 347–352.

Silva, S., A. Alves, H. Ghey, Nd. M. Beltrão, L. Severino, and F. Soares. 2008. Growth and production of two cultivars of castor bean under saline stress. (Portuguese, with English abstract). *Rev Bras de EngenhariaAgrícola e Ambiental* 12: 335–342.

Singh, S., P. Parihar, R. Singh, V.P. Singh, and S.M. Prasad. 2016. Heavy metal tolerance in plants: role of transcriptomics, proteomics, metabolomics, and ionomics. *Frontiers in Plant Science* 6: 1143.

Silva, S.M.S., A.N. Alves, H.R. Ghey, N.E.M. Beltrão, L.S. Severino, and F.A.L. Soares. 2005. Germination and initial growth of two cultivars of castor bean under saline stress. (in Portuguese, with English abstract.). *RevistaBrasileira de EngenhariaAgrícola e Ambiental Supplement*: 347–352.

Silva, S.M.S., A.N. Alves, H.R. Ghey, N.E.M. Beltrão, L.S. Severino, and F.A.L. Soares. 2008. Growth and production of two cultivars of castor bean under saline stress. (in Portuguese, with English abstract.). *RevistaBrasileira de EngenhariaAgrícola e Ambiental* 12: 335–342. (In Portuguese, with English abstract.) doi:10.1590/S1415-43662008000400001

Simões, K.S., S.A. Silva, E.L. Machado, and H.S. Brasileiro. 2017. Development of TRAP primers for *Ricinuscommunis* L. *Genetics and Molecular Research* 16: 1–13.

Singh, M., I. Chaudhuri, S.K. Mandal, and R.K Chaudhuri. 2011. Development of RAPD markers linked to fusarium wilt resistance gene in castor bean (*Ricinuscommunis* L.). *Journal of Genetic Engineering and Biotechnology* 28: 1–9.

Škori´c, D. 2016. Sunflower breeding for resistance to abiotic and biotic stresses. In: Shanker, A.K., Shanker, C. (eds) Abiotic and biotic stress in plants—recent advances and future perspectives. IntechOpen, London, pp 585–635

Soares, D.J. 2012. The gray mold of castor bean: A review. In C.J.R. Cumagun, editor, Plant pathology. InTech Publisher, Rijeka, Croatia.

Soares, D.J. 2012. Thegray mold of castor bean: A review. In C.J.R. Cumagun, editor, Plant pathology. InTech Publisher, Rijeka, Croatia.

Soares, D.J., J.F. Nascimento, and A.E. Araújo. 2010. Componentesmonocícli-cos do mofocinzento (Amphotrysricini) emfrutos de differencesgenótiposdemamoneira. Proc. CongressoBrasileiro de Mamona, 4th, 2010, JoãoPessoa. EmbrapaAlgodão, Campina Grande, Brasilia, Brazil. www.cbmam ona.com.br/pdfs/FIT-01.pdf (accessed 21 Mar. 2012). p. 957–962.

Soares, J.J., L.H.A. Araújo, and F.A.S. Batista. 2001. Pragas e seu Controle. In: Azevedo, D. M.P. de (Org.). O agronegócio da mamona no Brasil. Brasília: Embrapa Informações Tecnológicas,1: 213–227.

Sala, C. A., M. Bulos, E. Altieri, and M.L. Ramos. 2012. Sunflower: improving crop productivity and abiotic stress tolerance. In: Tuteja N, Gill SS, Tiburcio AF, Tuteja R (eds) Improving crop resistance to abiotic stress. Wiley-VCH, pp 1203–1249.

Sood, A., Jaiswal, V., Chanumolu, S. K., Malhotra, N., Pal, T., and Chauhan, R. S.(2014). Mining whole genomes and transcriptomes of jatropha (*Jatropha curcas*) and castor bean (*Ricinus communis*) for NBS-LRR genes and defense response associated transcription factors. *Mol. Biol. Rep.* 41: 7683–7695. doi:10.1007/s11033-014-3661-0

Souza, A.G.C., and L.A. Maffia. 2011. First report of *Cercosporacoffeicola*causing Cercospora leaf spot of castor beans in Brazil. *Plant Diseases* 95: 1479.

Sood R. S. Chauhan. 2018. Genomics of Disease Resistance in Castor Bean C. Kole and P. Rabinowicz (eds.), The Castor Bean Genome, Compendium of Plant Genomes pp 105–114. https://doi.org/10.1007/978-3-319-97280-0

Srinivas Rao, T., Lakshminarayana, M., and Anjani, K. 2000. Studies on the influence of bloom character of castor germplasm accessions on jassids and thrips infestation. In: Extended summaries of national seminar on oilseeds and oils-research and development needs in the millennium, Hyderabad, India, 2–4 Feb 2000, p. 291.

Sussel, A.B., E.A. Pozza, and H.A. Castro. 2009. Elaboration and validation ofdiagrammatic scale to evaluate gray mold severity in castor bean. *Tropical Plant Pathology* 34: 186–191. (In Portuguese, with English abstract.)Doi: 10.1590/S1982-56762009000300010

Tang, J.R., Y.L. Liu, X.G. Yin, J.N. Lu, and Y.H. Zhou. 2020a. First report of castor dark leaf spot caused by *Phyllostictacapitalensis*in Zhanjiang, China. *Plant Diseases* 104: 1856.

Tang, J.R., Y.L. Liu, X.G. Yin, J.N. Lu, and Y.H. Zhou. 2020b. First report of leaf vein spot caused by *Corynesporacassiicola*on castor bean in China. *Plant Diseases* 104: 3056.

Tang, J.R., Y.L. Liu, L.P. Wu, J.N. Lu, and X.G. Yin. 2020c. Identification of pathogen causing castor seed bud rot. *Chinese Journal of Oil Crop Sciences* 42: 1–6.

Thanki, K.V., Patel, G.P., and Patel, J.R. 2001 Varietal resistance in castor to Spodoptera litura Fabricius. *Gujarat Agri Univ Res* J 26: 39–43.

Uchida, J.Y., and M. Aragaki. 1988. Seedling blight of castor bean in Hawaii caused by *Phytophthora palmivora*. *Plant Disease* 72: 1994.

Vahunia, B., P. Singh, N.Y. Patel and Rathava, A. 2017. Management of *Fusarium* wilt of castor (*Ricinus communis* L.) caused by *Fusarium oxysporum* f. sp. ricini with antagonist, botanical extract and pot experiment. *Int. J. Curr. Microbiol. App. Sci.* 6(9): 390–395. doi: https://doi.org/10.20546/ijc mas.2017.609.048

Varón, E.H., M.D. Moreira, and J.P. Corredor. 2010. Effect of Corythucha gossipy on castor oil plant leaves: Sampling criteria and control by insecticides. *RevistaCorpoicaCienciaTecnologiaAgropecuario* 11: 41–47.

Vijaya Lakshmi, P., Satyanarayana, J., Singh, H., and Ratnasudhakar, T. 2005 Incidence of green leafhopper *Empoasca flavescens* Fab. on castor *Ricinus communis* L. in relation to morphological characters and date of sowing. *J Oilseeds Res* 22: 93–99.

Wang, G.L., D.L. Ruan, W.Y. Song, S. Sideris, L.L. Chen, L.Y. Pi, S.P. Zhang, Z. Zhang, C. Fauquet, B.S. Gaut, M.C. Whalen, and P.C. Ronald. 1998. Xa21D encodes a receptor-like molecule with a leucine-rich repeat domain that determines race-specific recognition and is subject to adaptive evolution. *Plant Cell* 10: 765–779.

Wang, X., M. Li, X. Liu, L. Zhang, Q. Duan, and J. Zhang. 2019. Quantitative proteomic analysis of castor (Ricinuscommunis L.) seeds during early imbibitions provided novel insights into cold stress response. *International journal of molecular sciences* 20(2): 355.

Watt, G. 1892. A dictionary of economic products of India. 165p.

Wang, S., Y. Zhao, J. Guo, and L. Zhou. 2016. Effects of Cd, Cu and Zn on *Ricinus communis* L. growth in single element or co-contaminated soils: pot experiments. *Ecological engineering* 90: 347–351.

Wang, S., Y. Zhao, J. Guo, and L. Zhou. 2019. Effects of Cd, Cu and Zn on *Ricinus communis* L. growth in single element or co-contaminated soils: pot experiments. *Ecological engineering* 90: 347–351.

Weibel, R.O. 1948. The castor-oil plant in the United States. *Economic Botany* 2(3): 273–283.

Wilson, C.O., B. Liang, and S.J. Rose. 2019. Projecting future land use/land cover by integrating drivers and plan prescriptions: the case for watershed applications. *GIScience& Remote Sensing* 56: 511–535.

Wolfe M. D., Rabbi I. Y., Egesi C., Hamblin M., Kawuki R., et al. 2016. Genome-wide association and prediction reveals genetic architecture of cassava mosaic disease resistance and prospects for rapid genetic improvement. *Plant Genome* 9: 342–356. 10.3835/plantgenome2015.11.0118

Xiong, P. P., C.Q. He, K. Oh, X. Chen, X. Liang, X. Liu, X. Cheng, C.L.Wu, and Z.C. Shi. 2018. Medicago sativa L. enhances the phytoextraction of cadmium and zinc by *Ricinus communis* L. on contaminated land in situ. *Ecological engineering* 116: 61–66.

Xiao, S., S. Ellwood, O. Calis, E. Patrick, T. Li, M. Coleman, and J.G. Turner. 2001. Broad-spectrum mildew resistance in Arabidopsis thaliana mediated by RPW8. *Science* 291: 118–120.

Xu, W., Li, F., Ling, L. et al. 2013. Genome-wide survey and expression profiles of the AP2/ERF family in castor bean (*Ricinus communis* L.). *BMC Genomics* 14, 785. https://doi.org/10.1186/1471-2164-14-785

Yeboah, A., J. Lu, T.Yang, Y. Shi, H.Amoanimaa-Dede, K.G.A. Boateng, and X. Yin. 2020. Assessment of castor plant (*Ricinuscommunis* L.) tolerance to heavy metal stress-A review. *Phyton* 89(3): 453.

Yan, K., S. Zhao, L. Bian, and X. Chen. 2017. Saline stress enhanced accumulation of leaf phenolics in honeysuckle (*Lonicera japonica* Thunb.) without induction of oxidative stress. *Plant Physiology and Biochemistry* 112: 326–334.

Ye, W., G. Guo, F. Wu, T. Fan, H. Lu, H. Chen, X. Li, and Y. Ma. 2018. Absorption, translocation, and detoxification of Cd in two different castor bean (*Ricinus communis* L.) cultivars. *Environmental Science and Pollution Research* 25: 28899–28906.

Yeboah, A., J. Lu, T. Yang, Y. Shi, H. Amoanimaa-Dede, K.G. Agyenim-Boateng, and X.Yin. 2020. Assessment of castor plant (*Ricinus communis* L.) tolerance to heavy metal stress. *Phyton-International Journal of Experimental Botany* 89: 1–20.

Yin, X., J. Lu, K.G. Agyenim-Boateng, and S. Liu. 2019. Breeding for climate resilience in castor: Current status, challenges, and opportunities. In: Kole C (ed) Genomic designing of climate-smart oilseed crops. Springer, Cham, pp 441–498.

Zhang, W., L. Song, J.A. Teixeira da Silva, and H. Sun. 2013. Effects of temperature, plant growth regulators and substrates and changes in carbohydrate content during bulblet formation by twin scale propagation in Hippeastrumvittatum _Red lion'. *Scientia Horticulturae* 160: 230–237.

Zhang, H., Guo, Q., Yang, J., Chen, T., Zhu, G., Peters, M., Wei, R., Tian, L., Wang, C., Tan, D., Ma, J., Wang, G., and Wan, Y. 2014a Cadmium accumulation and tolerance of two castor cultivars in relation to antioxidant systems. J Environ Sci 26: 2048–2055.

Zhang, X., Lu, G., Long, W., Zou, X., Li, F., and Nishio, T. 2014b Recent progress in drought and salt tolerance studies in Brassica crops. Breed Sci 64: 60–73.

Zhang, H., Q. Guo, J. Yang, J. Shen, T. Chen, G. Zhu, H. Chen, and C. Shao. 2015. Subcellular cadmium distribution and antioxidant enzymatic activities in the leaves of two castor (*Ricinus communis* L.) cultivars exhibit differences in Cd accumulation. *Ecotoxicology and Environmental Safety* 120: 184–192.

Zhang, C., H. Chen, T. Cai, Y. Deng, R. Zhuang, N. Zhang, Y. Zeng, Y. Zheng, R. Tang, R. Pan, and W. Zhuang. 2016. Overexpression of a novel peanut NBS-LRR gene AhRRS5 enhances disease resistance to *Ralstoniasolanacearum* in tobacco. *Plant Biotechnology Journal* 15(1).

Zhao, T.T., S.E. Cho, I.Y.Choi, M.K. Choi, and H.D. Shin. 2018. First report of powdery mildew caused by *Podosphaeraxanthii*on *Ricinus communis* in Korea. *Plant Diseases* 102: 1179.

Zhong, W., Hartung, W., Komor, E, and Schobert, C. 1996. Phloem transport of abscisic acid in *Ricinus communis* L. seedlings. Plant Cell Environ 19: 471–477.

Zhou, G., Ma, B., Li, J., Feng, C., Lu, J., and Qin, P. 2010. Determining salinity threshold level for castor bean emergence and stand establishment. Crop Sci 50: 2030–2036.

Zhong, W., W. Hartung, E. Komor, and C. Schobert. 1996. Phloem transport of abscisic acid in *Ricinuscommunis* L. seedlings. *Plant Cell Environment* 19: 471–477. doi:10.1111/j.1365- 3040.1996.tb00339.x

Zhou, G., B.L. Ma, J. Li, C. Feng, J. Lu, and P. Qin. 2010. Determining salinity threshold level for castor bean emergence and stand establishment. *Crop Science* 50: 2030–2036. doi:10.2135/cropsci2009.09.0535

20 *Salvadora persica* L. and Stressful Conditions

Sulaiman[1]*, Arafat Abdel Hamed Abdel Latef[2],
Karine Pedneault[3], Sikandar Shah[1] and Sheharyar Khan[1]
[1]Department of Botany, University of Peshawar, Khyber Pakhtunkhwa, Pakistan
[2]Department of Botany and Microbiology, Faculty of Science, South Valley University, Qena 83523, Egypt
[3]Centre de Recherche et d'Innovation sur les Végétaux, Département de Phytologie, Université Laval, Québec, QC, Canada
*Corresponding author: sulaiman097@uop.edu.pk

CONTENTS

20.1 INTRODUCTION

Nature has long been regarded as a valuable reservoir of therapeutic substances. Nowadays, around the world, many natural ingredients are utilized in medicine. Novel chemical substances have a broad range of structural variations found in nature and can serve as precursors for synthetic and semi-synthetic analogs (Farnsworth and Soejarto 1991). It can be observed easily how frequently natural items or their analogs are used to cure a variety of ailments. The use of natural products, particularly plant-based, has recently attracted increasing attention. This high interest in medicines derived from plants is due to many factors, including the abundance of chemically and physiologically unscreened

DOI: 10.1201/9781003242963-20

herbs and the long history of traditional medicine, which supports the efficacy and safety of using natural products (Atanasov et al. 2015). The plant kingdom has abundant and distinctive resources that may be used to create novel ideas and medications for a variety of pharmaceutical targets.

20.1.1 SALVADORA PERSICA

Salvadora persica is a frequently used folk plant. People, especially Muslims, use it extensively throughout the world. This has drawn our attention to conduct our biological and phytochemical research on isolated compounds and extracts from *S. persica* as part of our attempts to discover potential medicinal plants in the area (Haque and Alsareii 2015). The family Salvadoraceae consists of 10 species belonging to three genera (Azima, Dobera and Salvadora), which primarily reside in the tropical and subtropical regions of Africa and Asia. Recently, a novel record, *Salvadora alii*, has been identified from the province of Sindh, Pakistan. The most important species, *S. persica*, is commonly known throughout the world by various names such as Arak, Caday in Somalia, Merge, Miswak in Pakistan, Pilaue in India, Siwak in Saudi Arabia, Omungambu in South Africa and toothbrush in English (Mekhemar et al. 2021).

In 1949, Dr. Laurent Garcin, after Juan Salvador Bosca from Barcelona, for the first time proposed the term Salvadora. The term persica is derived from Persia. The utilization of Miswak dates back to the introduction of Islamic culture. In Pakistan, these old trees were usually connected with cemeteries, much like the cypress plant is in English culture (Ahmad and Rajagopal 2014). This plant has been utilized from ancient times in traditional medicine, and its uses range from medicine to food and from cosmetics to fuel. Their fruits and leaves were consumed as a green vegetable and their resin could be used to polish furniture. The young twigs and fresh leaves were useful for animals (camels, cows, goats and sheep) with malnutrition disorders, lactation problems and weight loss (Arshad et al. 2021; Shah et al. 2020). According to some reports, *S. persica* honey has therapeutic properties, and the plant's blossoms are an excellent site for honey bee feeding. Additionally, *S. persica* has a remarkable history of utilization as medicine by ethnic groups, especially in Africa and Asia. All parts of the plants including the root, shoot, stem, leaves, bark and buds have been used to treat a number of neurological, blood, excretory and gastroenterological system disorders. This plant was used in ancient times by the people of Saudi Arabia for teeth whitening and gum bleeding. Moreover, the Holy Prophet Muhammad (SAWW0 in 543 CE recommended the use of Miswak (Khan et al. 2015). The Prophet also demonstrated the value of using Miswak frequently for proper dental hygiene by doing it every time he awoke from sleep and before entering his home for prayers (Owis et al. 2020).

20.1.2 NUTRITIONAL USES

Salvadora persica L. is a therapeutically important salt-tolerant plant with vital ethnoecological significance. It has numerous medicinal uses, including oral hygiene. The leaf extracts of *S. persica* have been shown to have antioxidant and nutraceutical benefits (Abhary and Al-Hazmi 2016). Their fruits are dried and cooked, and have a pleasant aroma and sweet taste. The fruits are stony with persistent calyx and corolla. In the mature stage, the fruit is round, fleshy, pinkish, smooth, 5–10 mm in diameter with one seed. In many countries, a fermented juice that has gastric, lithontriptic, carminative and de-obstruent properties is prepared from their seeds (Shah et al. 2020; Tahir et al. 2015). The fruits are thought to be effective against snake bites and are also used to reduce inflammation and gastrointestinal problems. In addition, they are regarded as liver tonics, diuretics and purgatives.

20.1.3 PHYTOCHEMICAL PROFILE

Phytochemical analysis of *S. persica* stem was performed by Aumeeruddy, resulting in the first documented four benzylamides for the first isolation. A number of compounds were isolated and identified as N1, N-benzylbenzamide (III), succindiamide, N4 N-bis(phenylmethyl)-2(S)-hydroxybutanediamide (I), N-benzyl-2-phenylacetamide (II) and benzylurea (IV) (Aumeeruddy et al. 2018). Further investigation revealed that it was comprised of linolic, oleic and stearic acids, while the other identified compounds included esters of aromatic acids, fatty acids and a few terpenoids. The essential oil extracted from the *S. persica* stem contains 1,8-cineole (46%) which is eucalyptol, α-caryophellene (13.4%), β-pinene (6.3%) and 9-epi-(E)-caryophellene. GC-MS analysis of the extracted volatile oil revealed the presence of β-caryophyllene, eucalyptol, thymol, isothymol, benzyl nitrile and eugenol. Sticks of the plant have been reported to be rich in calcium, phosphorus, fluoride and silica (Albabtain et al. 2017).

The ashes obtained from the burning of Miswak stem reported a significant amount of silica. By applying capillary electrophoresis methods, the antibacterial anionic components of the leaf extracts of the stem and root have also been examined. The root and stem extracts are said to include thiocyanate, nitrate and sulfate chloride. The medicine obtained contains a significant number of alkaloidal constituents, such as trimethylamine and unidentified alkaloids, as well as traces of a number of tannins and saponins. There were also observed higher levels of fluoride, sterols, silica, sulfur, vitamin C and a few flavonoids (Farag et al. 2021). Various types of lignin glycosides and flavonoids (rutin and quercetin) have been identified in the stem of *S. persica*. A compound called salvadourea and benzylisothiocynate has been reported from the root. Similarly, a novel alkaloid (salvadoricine) was reported in the leaves (Al Bratty et al. 2020).

20.2 STRESSFUL CONDITIONS

External factors that negatively impact a plant's growth, maturation or yield are referred to as stresses in plants. A variety of plant reactions are brought on by stress, including changed gene regulation, metabolic activity, modifications to rates of growth, alterations in crop productivity, etc. (Dubey 2018). Abiotic stress in plants typically results from certain abrupt alterations in the environment. However, exposure to a given stress results in acclimatization to that highlighted stress in stress-tolerant plant species in a time-dependent manner (Latef and Chaoxing 2011). Abiotic stress and biotic stress are the two main classifications of plant stress. While biotic stress is caused by a biological entity such as diseases, pests, etc. that agricultural plants are exposed to, abiotic stress is imposed on plants by the environment and can be either physical or chemical. Plants were harmed by several stressors, which cause them to display some metabolic disorders (De Bigault Du Granrut and Cacas 2016). If the stress is minor or short-lived, the damage is transient, and the plants can recuperate. However, extreme stressors cause agricultural plants to die by blocking blooming, seed development and initiating senescence. Such plants will be regarded as being vulnerable to stress. However, some plants, such as halophyte plants (ephemerals), can completely avoid the stress. Living things, in particular viruses, fungi, bacteria, nematodes, insects, arachnids and weeds, induce biotic stress in plants. Plant mortality may result from the agents generating biotic stress, which directly depletes their host of their nutrients. Due to pre- and post-harvest declines, biotic stress may become severe (Dolferus 2014). Plants can withstand biotic stressors despite lacking an innate immune response, however, it regulates the stress through various sophisticated techniques. The genetic code of the plant controls the defensive systems that respond to these challenges. The plant genotype has multiple genes that are resistant to various biotic stresses. Abiotic stress, which agricultural plants experience as a result of changes in environmental variables including salinity, sunshine, precipitation, heat, flood and drought, is distinct from biotic stress. The sort of biotic stress that may be applied to agricultural plants depends on the environment in which they are grown, as does their capacity to withstand that specific form of stress (Lamers et al. 2020). Abiotic stressors that negatively influence the growth, development, productivity and seed parameters include drought (shortage of water), flooding stress (waterlogging condition), salinity (high percentage of salt in soil), varying temperature (cold and heat stress), radiations and heavy mineral harmfulness. Freshwater shortages are expected to worsen in the future, which will ultimately result in more abiotic stresses. Therefore, it is imperative to create crop varieties that are resistant to abiotic stressors to regulate food security and safety in the upcoming years (Van de Peer et al. 2021). The roots of a plant act as its primary line of defense against abiotic and biotic stresses. If the soil in which the plant is growing is nutritious and ecologically varied, the plant will have a good chance of surviving adverse conditions. Distortion of the Na^+/K^+ pump in the cytosol of the plant cell is one of the main reactions to abiotic stress, such as excessive salinity. Abscisic acid (ABA), a phytohormone, is crucial for the plant response to environmental stressors such as excessive salt, dehydration, cold temperatures or mechanical injury.

20.2.1 ABIOTIC STRESSES

Any environmental condition that might restrict plant development and productivity is referred to as abiotic stress. Due to the exponential expansion in human societies, environmental stresses have historically become the emphasis of plant breeders and horticulturalists in the last 50 years (which has exposed international food security) (El-Beltagi et al. 2022). It was inevitable given the rising food demand that factors contributing to decreased agricultural output, which were frequently disregarded, would be looked into. Around this time, researchers began to estimate the amount of crop loss due to environmental restrictions and to look into the causes. Some of these pressures have long-lasting consequences on cultivated areas, which in turn influence crop quality and quantity. The impacts of abiotic stressors on arable land are difficult to assess, but scientists have made a rough estimate relying on FAO statistics that around 96.5% of rural land worldwide is influenced by environmental stresses (Ali and Baek 2020). A description of the proportion to which each stress contributed was also presented. Numerous unfavorable elements, including heat, cold, drought and salt (among others), can have an impact on both farming land and crop output in a constantly changing climate.

20.2.2 *SALVADORA PERSICA* AND ABIOTIC STRESS

20.2.2.1 Seed Germination and Seedling Growth under Abiotic Stress

Germination reactions to fluctuations in environmental circumstances, such as salt, sunlight and temperature regulation, determine where and when seedling recruitment occurs as well as the likelihood that seedlings will survive (Foyer et al. 2016). In general, poor competitors can occupy harsh areas in a habitat that superior rivals would avoid because of their better stress responses during germination and seedling emergence. Therefore, comprehensive knowledge of the ecology of plant germinating seeds has been a key study area in plant sciences. However, the ecology of seed emergence of only 100 out of 1469 species of salt-tolerant plants is known www.sussex.ac.uk/ affiliates/ halophytes/).

20.2.2.2 Salinity and Drought

A xero-halophyte, *Salvadora persica* L. is often widespread in dry, saline soils. It is a perennial shrub. Salinity tolerance in *S. persica* occurs spontaneously during the adult vegetative phase. However, as salinization rises, their seed emergence often declines (Kumari and Parida 2018). The range between 200 and 1700 mM NaCl is the highest limit (10% emergence rate) of salinity tolerance for halophyte seedlings. Many halophyte seedlings that have not yet germinated usually survive excessive soil salinity by going into a state of unconditional or induced dormancy and quickly emerge (recover) after sufficient precipitation (Rangani et al. 2020). Temperature and photoperiod fluctuations are said to have an impact on halophyte seed emergence percentage, recovery from stress and salt tolerance. The overall seedling growth ratio, salt resistance and restoration of emergence are often decreased by variations in sub- and supra-optimal temperatures and the absence of light. Salinity and variations in the duration of light can cause different forms of seed dormancy, which under severe environments may even result in seed demise (Rangani et al. 2018).

S. persica is a naturally occurring halophyte that can grow in a variety of environments, including arid, swampy and waterlogged locations, as well as non-salty to very saline soils. A review of the literature revealed that *S. persica* can withstand salt concentrations that are comparable to those in saltwater. It was cultured in a culture medium containing up to 250 mM NaCl without experiencing any noticeable negative effects on its development (Rangani et al. 2016). One of the initial reactions of plants to various abiotic stressors, such as salt, is a reduction in leaf growth. In the current study, *S. persica* plants flourishing in saline-dry environments (ECe between 19.5 and 25.7 dSm^{-1}) had a mean leaf area (LA) that was 27% lower than that in trees growing in non-saline areas. By minimizing the flux of photons seized by the photosystems' antenna elements and thereby lowering the vibrational energy load, a reduction in leaf area can reduce energy consumption, reduce the production of harmful reactive oxygen species (ROS) and increase the availability of protein for osmoregulation. Additionally, the reduction in (LA minimizes water loss through transpiration, which in turn reduces the movement of harmful ions from the base to the branches (Patel and Parida 2021). It can endure water deficiency stress for 14 days and can withstand soil watered with 1000 mM NaCl for up to 60 days, according to first investigations on salinity and dry spell resistance. The results of the physiological investigation demonstrated that under salt and drought stress, development and productivity drastically decreased without affecting the quality of photosystem II. In *S. persica*, integrated antioxidative enzyme controls scavenging stress-induced ROS to maintain proper levels of ROS (Falasca et al. 2015). In addition to enzymes, other antioxidative substances are also engaged in the sequestration of ROS. As a result, knowledge of the roles played by different metabolites and metabolic pathways in the ability of *S. persica* to tolerate drought is essential for understanding how drought tolerance works. This is because the pathophysiology of the plant and metabolite makeup are closely intertwined.

20.2.2.3 Heavy Metal(loid)s Pollution

The contamination of soil with heavy metal(loid)s as a result of numerous man-made and natural processes poses a serious risk to human health on a worldwide scale. Heavy metal(loid)s limit germination rate, development and eventually yield, which has a variety of effects on plant metabolism. Additionally, they might hinder the absorption of crucial minerals and impact a number of biochemical activities in various cellular components (Patel and Parida 2021). By interfering with the shape and porosity of the plastid membrane, stopping the electron transport chain of the photosystem II (PSII) and decreasing the manufacture of pigment, heavy metal(loid)s restrict the photosynthesis process. Plant cytotoxicity is brought on by changes in mineral status caused by heavy metal contamination. Additionally, heavy metals cause oxidative stress and impair the water status of plants. Arsenic (As), one of the heavy metal(loid)s, is a particularly poisonous metalloid that is derived through both natural and human activities, including mining, smelting and the use of As-containing insecticides, herbicides, paints, colors, detergents and medications (Patel and Parida 2022). As(0), As(V), As(0) and As(-III) are the most common forms of inorganic arsenic to be found in soil. Those that are known to disrupt several plant metabolic pathways include As(V) and As(III), which are primarily carried by plants via different phosphate exchangers, water pores and NIP channels, respectively. Additionally, As(V) is an analog of inorganic phosphate (Pi), and As(V) is known to change Pi digestion in plants as a result of its analog nature. When plants are exposed to arsenic stress, their net photosynthetic rate, chloroplast breakdown, membrane structure breakdown, fresh and dry plant cell mass accretion and yield are reduced, while the formation of ROS and oxidative damage increases (Shi et al. 2022). In addition, arsenic can act directly with catalysts' thiol groups, interfering with several metabolic processes. Arsenic, a phosphate analog, interferes with a number of essential cellular functions, including ATP production and the electron transport chain. Different defense mechanisms have been developed by *persica* to combat As' adverse effects. Making heavy metal-binding, cysteine-rich peptides called phytochelatins is one method for removing arsenic from the body. Phytochelatin synthase converts GSH into phytochelatins (PCs). Since the synthesizing enzyme needs metal as a cofactor for its action, the accumulation of toxic metals in plants is necessary for PC production (Rangani et al. 2016). As(III) interacts with PCs and/or glutathione and aids the vesicles' internalization of As–PCs combination through the binding protein cassette (ABC) transporter since As(V) has no attraction for the sulfhydryl group of PCs. By doing this, the plant decreases both the amount of free arsenic present in the cytoplasm and the toxicity of arsenic.

20.2.2.4 Flooding

Similarly, flooding impairs LA expansion also, mainly due to a reduction in nutrient uptake. Tounekti et al. reported that flooding increased the LA of Miswak plants, suggesting that the roots of this species are fairly tolerant to hypoxia. Specific leaf area (SLA) is highly correlated with plant growth rate (Tounekti et al. 2018). As in many other halophytes, SLA decreased (or LMA increased) while LDMC (or LD) increased for Miswak plants growing either in saline-dry (sabkha) or saline-flooded habitats. The decline of SLA due to flooding seems to make the mesophyll parenchyma less compact, thus facilitating O_2 diffusion into the mesophyll. Besides, lower SLA values contribute to leaf longevity, nutrient retention and protection from dehydration. As in many halophytes, the Lth and LDMC (LD) of *S. persica* increased under saline conditions. This agrees with reports on *Alchornea triplinervia*, in which flooding increased the thickness of palisade and spongy parenchyma by 1.4-fold (Nafees et al. 2019). Furthermore, leaves became more succulent (S increased) when the trees were subjected to both salinity and flooding. Such a response is considered an adaptation to save water and dilute toxic ions in halophytes. Generally, the anatomical changes which occur in response to stress are designed to optimize leaf gas exchange. Our results show that transient flooding of Miswak plants grown in non-saline habitats led to an increase in stomatal density (SD), indicating an increased demand for CO_2. In contrast, in saline-dry and saline-flooded habitats, SD decreased, possibly to reduce water loss by transpiration (Falasca et al. 2015). The detrimental effects of salinity on halophytes and glycophytes are generally due to osmotic effects and/or ionic imbalances as a result of nutrient deficiencies or ion toxicities. For instance, when plants fail to hydrate their tissues properly on saline soils they suffer from physiological drought. To maintain a suitable cell turgor in their leaves, halophytes accumulate high amounts of solutes, mainly Na^+, Cl and K^+, in their vacuoles. *S. persica* was able to adequately lower its water potential (Ψw) and osmotic potential ($\Psi \pi$) in the saline-dry sabkha, similarly to what generally happens in other halophytes. Shoot Ψw reached its lowest values (down to -2.3 MPa) in the sabkha, indicating that the trees were capable of osmotically adjusting under both saline-dry and saline-flooded conditions (Khan et al. 2016). This aided salt-stressed Miswak plants to preserve leaf cell turgor as indicated by positive pressure potential (Ψp) values. Flooding increased slightly the Ψw and $\Psi \pi$ of the plants growing in the sabkha, allowing them to maintain their Ψp. In non-saline-flooded habitats, tree Ψw did not change and small changes in leaf water content (LWC) were seen, which suggests that flooding did not cause a physiological drought in these plants. Flooding caused stomata to close (decline of gs) thus depressing Anet and E under both non-saline and saline conditions. The effect of flooding on gas exchange is independent of the plant's water status (Abdeltawab et al. 2022). Stomata closure under such conditions seems to be the result of a signaling process rather than a general loss of turgor (indicated by the high Ψw and Ψp values). Flooding (hypoxia) causes a change in the hormonal balance that causes stomata to close and decrease the passive absorption of water by roots. Furthermore, the increase in the A/gs ratio under such conditions suggests a stomatal limitation to photosynthesis (gs decreased far more than Anet).

20.2.2.5 Genomic Approaches

A series of molecular pathways regulate how plants adapt to environmental challenges. In this sense, the use of functional genomics has a greater influence on our comprehension of how plants react to abiotic challenges. The ability to change a plant's endurance for abiotic stress to increase yield under stressful circumstances has been made possible by technology, which has shown exceptional success in this area (Fouda et al. 2021). The decisive initiation of a few genes through gene editing has also developed into a tangible and alluring method, in contrast to conventional rearing and marker-assisted shortlisting programs, as a quick way to increase the plant's stress resistance, restore homeostasis, and fix broken proteins and glands. *Arabidopsis thaliana* has become a top model organism for research into the mechanisms underlying abiotic stress tolerance since most agricultural plants are glycophytes. *Mesembryanthemum crystallinum*'s genome information, which, when compared to the genomic sequence, seems to comprise several excerpts that have no apparent function, has made it possible to research unique methods specific to halophytes, such as *S. persica* or stress-tolerant plants, which may be difficult to study with *Arabidopsis*. Following that, several halophytes, including *M. crystallinum*, *Suaeda* species and *Atriplex* species, were used to analyze the molecular underpinnings of the salt-loving plants (Khan et al. 2021). Recently, a species of the family Brassicaceae called *Thellungiella halophila* (salt cress) has gained the attention of scientists for understanding the molecular adaptation of halophytes to environmental stresses. This attention is due to its same homology as the model of glycophyte (*A. thaliana*).

In contrast to its salt-sensitive cousin *A. thaliana* and other agricultural plants, this halophyte can thrive in high salinity that would normally be detrimental to their growth. *T. halophila* stands out due to its small diploid genetic code (240 Mb and 2 n = 14), tight and self-fertile life span, and ease of floral dip-coating method of transformation. These characteristics make *T. halophila* an excellent candidate for molecular studies of its rebuttal to abiotic stress conditions and comparative similarity with *A. thaliana*. and *T. halophila*. The genetic analysis provided substantial and original information on the presence of distinct genes that are responsible for abiotic stress responses in *T. halophila* compared to *A. thaliana*.

Scientists studied a full-length *Arabidopsis* cDNA microarray to analyze the patterns of gene expression and compare the modulation of salinity tolerance in salt cress and *Arabidopsis*. Relative to salt stress in *Arabidopsis*, only a small number of genes were triggered by 250 mM NaCl in the salt cress plant (Monfared et al. 2018). Numerous reported abiotic and biotic stress-inhibitory genes, including Fe-SOD, P5CS, PDF1.2, At-NCED, P-protein, b-glucosidase and SOS1, were highly expressed even in the face of stress. In addition, the study discovered that salt cress is more resistant to osmotic damage than *Arabidopsis*. Changes in the expression of the same basic set of genes engaged in salinity tolerance in these plants may be the cause of the differences in salinity tolerance methods between salt-sensitive glycophytes and salt-tolerant halophytes (Kumari et al. 2017). *Arabidopsis* gene-specific markers were used by physiologists, and both A and B showed comparable real-time PCR amplification efficiency. Both *thaliana* and *T. halophila* cDNA concluded that stressed and undisturbed plants of both species express several salt tolerance orthologs differently.

20.2.3 BIOTIC STRESS

Plants experience a variety of biotic stressors brought on by diverse living things such as nematodes, insects, fungi, viruses and bacteria. These biotic stress factors impair agricultural yield by inflicting numerous diseases, infections and destruction on crop plants. To combat biotic stressors, however, many systems have been created through research methods (Behmann et al. 2015). By researching the genetic mechanisms of the agents producing these stressors, biotic stresses in crops can be mitigated. By creating resilient kinds of agricultural plants, genetically modified plants have been demonstrated to be a tremendous aid in overcoming biotic stressors in plants.

Viral, bacterial and fungal infections, parasitic plants and insect pests are just a few of the biotic agents that are constantly attacking plants, and have a significant negative economic and ecological impact. Medicinal plants and their adversaries are engaged in an evolutionary arms race, and in response to this onslaught, they have developed several defenses. Plant metabolism must weigh conflicting demands for nutrients to support defense against needs for cellular upkeep, growth and fertility once an attack has been detected (Signorelli et al. 2019). When various active compounds, such as pathogen-associated molecular patterns (PAMPs), viral coat peptides or fatty acid conjugates, are introduced into insect saliva aerosol droplets, a huge reprogramming of plant gene expression, hormonal and chemical defense responses begins, and a procedure that can be expensive in terms of growth of plants and fitness. A loss in photosynthetic capability in surviving leaf tissues may indicate a "hidden cost" of defense, in addition to triggering responses to deter pathogen and herbivore assaults by shifting resources from development to defense.

20.2.3.1 Biotic Stress and *Salvadora persica* Defense Responses

Several bugs, worms and diseases are accountable for attacking plants and causing biotic stress. Fungal pathogens can be either necrotrophic (destroy host cells by toxin release) or biotrophic (feed on living host cells). They can cause vascular shrivels, leaf blight and pustules in plants. Nematodes feed on plant components and are mostly responsible for soil-borne diseases that cause nutritional deficiencies, stunted development and withering. Similarly, viruses can cause local and systemic harm, causing chlorosis and retardation (Rangani et al. 2018). On the other hand, pests and lice harm plants by either producing eggs on them or feeding (pricking and sipping) on them. Additionally, the insects may function as carriers of various germs and viruses. To fight such challenges, plants have evolved a sophisticated immune system. *S. persica* uses physical barriers including epidermis, waxes and rhizomes as part of its passive first line of defense to ward off viruses and insects. Additionally, plants can produce chemical substances to protect themselves against diseases. Plants also use two layers of pathogen detection to initiate defense against biotic invaders (Korejo et al. 2017). ETI stimulates hypersensitive responses (HRs) and triggers programmed cell death (PCD) in infected and surrounding cells. The proteins encoded by a majority of *R* genes have a specific domain with the conserved nucleotide-binding site (NBS). The second next important domain is a leucine-rich repeat (LRR). Pathogen effectors are recognized directly (physical association) or indirectly (association of an accessory protein) by NB-LRR receptors. Pattern recognition receptors (PRRs) recognize microbe-associated molecular patterns at the initial level of pathogen identification (PAMPs). This type of plant immunity is known as PAMP-triggered immunity (PTI). Herbivore-associated elicitors (HAEs), herbivore-associated molecular patterns (HAMPs) or PRR herbivore effectors are used by phytophagous pests to respond. The second level of pathogen identification revolves around plant resistance (R) proteins, which detect particular pathogen receptors (Avr proteins). It is regarded as an efficient strategy of pest resistance in plants, involving effector-triggered immunity (ETI).

A higher level of protection known as systemic acquired resistance is occasionally triggered by a plant's R gene-mediated reaction to an invading pathogen (SAR). SAR produces systemic whole-plant resistance against a variety of diseases. In SAR, a local encounter causes intra-plant communication to stimulate resilience to the other plant organs (Parida et al. 2016). Both types of plant immune responses often result in the same reaction, although ETI is thought to be more aggressive in response to pathogen attacks.

When plants are exposed to biotic stress, changes in cytosolic calcium (Ca^{2+}) levels are the first signaling events to take place. Immune signaling mechanisms in plants are centered on Ca^{2+} signals. Ca^{2+} concentration changes that happen quickly and briefly are essential for the gene reprogramming needed to provide an effective response. The Ca^{2+} fingerprints of the plant immune function vary. In contrast to ETI, which entails a protracted rise in cytosolic Ca^{2+} levels lasting for many hours, Ca^{2+} transients during PTI activation revert to basic levels in a matter of minutes (Islam et al. 2016). It has been claimed that the Ca^{2+} channel blocker lanthanum prevents the immunological responses connected to both PTI and ETI. PTI and ETI precisely activate the Ca^{2+} ion channels in response to the biological invasion, raising the cytoplasmic Ca^{2+} levels.

20.2.3.2 Photosynthetic Processes and Reactive Oxygen Species Accumulation in Biotic Stress

Through the electron transport chain (ETC), photosynthesis absorbs light energy to assimilate carbon dioxide and support the repair and expansion of the plant's physical structure. The carbon fixation process, fatty acid biosynthesis, absorption of nitrogen into amino acids and other processes use the essential metabolites that sunlight produces. These light-driven chloroplast pathways may affect momentarily activated plant defensive responses. It is interesting to note that the plastids also participate in the biosynthetic processes for the plant defense hormones ABA, JA and SA (Moustakas 2021). Due to the physiological cross-talk in plant–microbe interactions, this may have an effect on plant defense at night. Additionally, in the sense of stress, chloroplast serves as a location for ROS production. During growth and development, leaves become used to variations in light levels as Calvin cycle enzymes and illumination complexes are tuned to effectively use the available light. However, when carbon fixation is stopped or light variations happen, photosynthetic electron transport generates additional electrons. More electrons are produced as a result, which benefits the electron acceptor NADP$^+$. In these conditions, oxygen receives free electrons from ETC that then transfer to produce ROS (Mayta et al. 2018). Additionally, through photorespiration, which produces H_2O_2 in the peroxisomes, light-dependent processes and pathways in the chloroplast influence both short- and long-term-induced plant defensive responses. Acute light stress conditions can also impede chlorophyll production and disturb chloroplasts, which can result in the buildup of ROS. This could be more effective than the chloroplast's antioxidant system. However, ROS has also been strongly linked to plant defense against pathogens, and any change in the chloroplast's redox balance can affect ROS-regulated plant defense (Rejeb et al. 2014). For example, lipid peroxidation happens when ROS build up after the sensing of biotic stress. Beyond ROS's direct signaling capabilities, the effects of the necessary light/dark variations for chloroplast-derived ROS are extensive.

Regarding the aftermath of a pathogen attack, some of the by-products of lipid peroxidation include carbonyl-based reactive electrophiles. These electrophiles are either by-products of lipoxygenase enzyme activity or an effect of ROS on membrane lipids. Many of these electron acceptors are essential signaling molecules involved in the control of cell death and the production of defense-related genes. Therefore, changes in light/dark affect the creation of ROS-derived electrophiles. These incidents are described as occurring when infections or their elicitors contact plants. According to the reaction of the well-known elicitor cryptogenic, cell death was caused by a buildup of ROS when exposed to light (Foyer 2018; Gull et al. 2019). In contrast, when plants are exposed to darkness, cell death is unrelated to ROS buildup and is associated with a particular lipoxygenase activity. The chloroplast is not the main generator of ROS during the biotic stress response. The respiratory burst oxidase, or NADPH oxidase, is what is found in the plasma membrane. This suggests that pathogen protection may not be aided by ROS produced by chloroplasts in the presence of light. However, since NADPH oxidase does not prevent the formation of ROS generated from chloroplasts, this may or may not remain true. To better understand the fundamental processes governing signaling biotic stress tolerance, lesion mimic mutants with randomly occurring necrotic lesions are studied. Similar to those produced in reaction to HR, these cause necrotic lesions on the leaves. Lesion mimic mutants have increased PR gene expression and improved pathogen assault resistance (Hasanuzzaman et al. 2020).

Based on two observations, these mutants emphasize the connection between the biotic response to stress and chloroplast ROS. First, lesion mimic mutants are light-dependent in terms of lesion development. Second, genes involved in chlorophyll production or degradation are highlighted by these mutants' functional characterization. Additionally, the altered expression patterns of the genes involved in the production of chlorophyll result in phenotypes that resemble light-dependent lesions and subsequently improve disease resistance. This might be a result of the production of ROS brought on by the photosensitizing effects of light on intermediates in chlorophyll. When photosensitizers absorb light energy, the electrons are activated (Zhang et al. 2020). Thus, generated ROS serve as signals for infection resistance reactions. Plant defense signaling may be affected by light-derived ROS from either free photosensitive pigments or photosynthetic light-harvesting complexes.

20.3   CONCLUSION AND PROSPECTS

Worldwide food production is affected to a large extent by environmental extremities, and the sensitivity of crops is a major limitation for achieving higher plant productivity. Consequently, due to its ability to overcome multiple and simultaneous stresses, *S. presica* can be considered an excellent woody perennial species to replant degraded sabkhas and coastal habitats. In addition to its role in folk medicine and as a potential source of bioactive compounds, the plant is also a good source of fodder for livestock and an excellent habitat for wildlife. Functional leaf traits reveal that this species overcomes stresses by shifting from fast growth to slow growth to save water and nutrients. Additionally, this could ease the strain on agricultural plants, which are struggling with production issues as a result of exposure to numerous abiotic stressors. The occurrence of floods and droughts must constantly be borne in mind due to the constant increase in temperature and unequal distribution of rainfall. Salt stress may be further exacerbated by anthropogenic activities such as overuse of fertilizers, improper irrigation and resource exploitation. Plants will likely experience both biotic and abiotic stressors more frequently under these conditions. To maintain the security of the food supply and the prosperity of farmers, plant breeders must create cultivars that can withstand stress. This will take molecular effort at the DNA level to create plant defense systems against various stress situations. The plants will continually be exposed to such pressures unless response mechanisms are created against biotic and abiotic stresses, which will ultimately prove to be a major danger to global agriculture. In light of these uses, effective policies must be created and put into action to safeguard the world's non-renewable resources, with the diversity of halophytes accessible serving as a key source. Translation of this knowledge to other salt-sensitive crops will be aided by research into gene regulation and the harmony of individual stress tolerance mechanisms.

ACKNOWLEDGMENT

There is no conflict of interest among the authors. We apologize for not citing all of the relevant references due to space limitations.

REFERENCES

Abdeltawab, S.S., T.S. Abu Haimed, H.A. Bahammam, W.T. Arab, E.A. Abou Neel, and L.A. Bahammam. 2022. Biocompatibility and Antibacterial Action of Salvadora persica Extract as Intracanal Medication (In Vitro and Ex Vivo Experiment). *Materials* 15(4): 1373.

Abhary, M., and A.-A. Al-Hazmi. 2016. Antibacterial activity of Miswak (Salvadora persica L.) extracts on oral hygiene. *Journal of Taibah University for Science* 10(4): 513–520.

Ahmad, H., and K. Rajagopal. 2014. Salvadora persica L.(Meswak) in dental hygiene. *The Saudi Journal for Dental Research* 5(2): 130–134.

Al Bratty, M., H.A. Makeen, H.A. Alhazmi, S.M. Syame, A.N. Abdalla, H.E. Homeida, S. Sultana, W. Ahsan, and A. Khalid. 2020. Phytochemical, cytotoxic, and antimicrobial evaluation of the fruits of miswak plant, Salvadora persica L. *Journal of Chemistry* 2020.

Albabtain, R., M. Azeem, Z. Wondimu, T. Lindberg, A.K. Borg-Karlson, and A. Gustafsson. 2017. Investigations of a possible chemical effect of Salvadora persica chewing sticks. *Evidence-Based Complementary and Alternative Medicine* 2017.

Ali, M.S., and K.-H. Baek. 2020. Jasmonic acid signaling pathway in response to abiotic stresses in plants. *International Journal of Molecular Sciences* 21(2): 621.

Arshad, H., M.A. Sami, S. Sadaf, and U. Hassan. 2021. Salvadora persica mediated synthesis of silver nanoparticles and their antimicrobial efficacy. *Scientific reports* 11(1): 1–11.

Atanasov, A.G., B. Waltenberger, E.-M. Pferschy-Wenzig, T. Linder, C. Wawrosch, P. Uhrin, V. Temml, L. Wang, S. Schwaiger, and E.H. Heiss. 2015. Discovery and resupply of pharmacologically active plant-derived natural products: A review. *Biotechnology advances* 33(8): 1582–1614.

Aumeeruddy, M.Z., G. Zengin, and M.F. Mahomoodally. 2018. A review of the traditional and modern uses of Salvadora persica L.(Miswak): Toothbrush tree of Prophet Muhammad. *Journal of ethnopharmacology* 213: 409–444.

Behmann, J., A.-K. Mahlein, T. Rumpf, C. Römer, and L. Plümer. 2015. A review of advanced machine learning methods for the detection of biotic stress in precision crop protection. *Precision Agriculture* 16(3): 239–260.

De Bigault Du Granrut, A., and J.-L. Cacas. 2016. How very-long-chain fatty acids could signal stressful conditions in plants? *Frontiers in plant science* 7: 1490.

Dolferus, R. 2014. To grow or not to grow: a stressful decision for plants. *Plant science* 229: 247–261.

Dubey, R.S. (2018). Photosynthesis in plants under stressful conditions. In *Handbook of photosynthesis* (pp. 629–649). CRC Press.

El-Beltagi, H.S., S. Shah, S. Ullah, A.T. Mansour, and T.A. Shalaby. 2022. Impacts of Ascorbic Acid and Alpha-Tocopherol on Chickpea (Cicer arietinum L.) Grown in Water Deficit Regimes for Sustainable Production. *Sustainability* 14(14): 8861.

Falasca, S., S. Pitta-Alvarez, and C.M. del Fresno. 2015. Salvadora persica agro-ecological suitability for oil production in Argentine dryland salinity. *Science of the Total Environment* 538: 844–854.

Farag, M., W.M. Abdel-Mageed, A.A. El Gamal, and O.A. Basudan. 2021. Salvadora persica L.: Toothbrush tree with health benefits and industrial applications–An updated evidence-based review. *Saudi Pharmaceutical Journal* 29(7): 751–763.

Farnsworth, N.R., and D.D. Soejarto. 1991. Global importance of medicinal plants. *The conservation of medicinal plants* 26: 25–51.

Fouda, M.S., M.H. Hendawey, G.A. Hegazi, H.M. Sharada, N.I. El-Arabi, M.E. Attia, and E.R. Soliman. 2021. Nanoparticles induce genetic, biochemical, and ultrastructure variations in Salvadora persica callus. *Journal of Genetic Engineering and Biotechnology* 19(1): 1–12.

Foyer, C.H. 2018. Reactive oxygen species, oxidative signaling and the regulation of photosynthesis. *Environmental and Experimental Botany* 154: 134–142.

Foyer, C.H., B. Rasool, J.W. Davey, and R.D. Hancock. 2016. Cross-tolerance to biotic and abiotic stresses in plants: a focus on resistance to aphid infestation. *Journal of Experimental Botany* 67(7): 2025–2037.

Gull, A., A.A. Lone, and N.U.I. Wani. 2019. Biotic and abiotic stresses in plants. *Abiotic and biotic stress in plants*: 1–19.

Haque, M.M., and S.A. Alsareii. 2015. A review of the therapeutic effects of using miswak (Salvadora Persica) on oral health. *Saudi medical journal* 36(5): 530.

Hasanuzzaman, M., M. Bhuyan, F. Zulfiqar, A. Raza, S.M. Mohsin, J.A. Mahmud, M. Fujita, and V. Fotopoulos. 2020. Reactive oxygen species and antioxidant defense in plants under abiotic stress: Revisiting the crucial role of a universal defense regulator. *Antioxidants* 9(8): 681.

Islam, F., T. Yasmeen, M.S. Arif, S. Ali, B. Ali, S. Hameed, and W. Zhou. 2016. Plant growth promoting bacteria confer salt tolerance in Vigna radiata by up-regulating antioxidant defense and biological soil fertility. *Plant growth regulation* 80(1): 23–36.

Khan, A.L., S. Asaf, A. Al-Rawahi, and A. Al-Harrasi. 2021. Decoding first complete chloroplast genome of toothbrush tree (Salvadora persica L.): insight into genome evolution, sequence divergence and phylogenetic relationship within Brassicales. *BMC genomics* 22(1): 1–16.

Khan, A.U., F. Sharif, and A. Hamza. 2016. Establishing a baseline on the distribution and pattern of occurrence of Salvadora persica L. with meteorological data and assessing its adaptation in the adjacent warmed-up zones. *International journal of biometeorology* 60(12): 1897–1906.

Khan, M., A.H. Al-Marri, M. Khan, M.R. Shaik, N. Mohri, S.F. Adil, M. Kuniyil, H.Z. Alkhathlan, A. Al-Warthan, and W. Tremel. 2015. Green approach for the effective reduction of graphene oxide using Salvadora persica L. root (Miswak) extract. *Nanoscale research letters* 10(1): 1–9.

Korejo, F., R. Noreen, S.A. Ali, F. Humayun, A. Rahman, V. Sultana, and S. Ehteshamul-Haque. 2017. Evaluation of antibacterial and antifungal potential of endophytic fluorescent Pseudomonas associated with Salvadora persica and Salvadora oleoides decne. *Pak. J. Bot* 49(5): 1995–2004.

Kumari, A., and A.K. Parida. 2018. Metabolomics and network analysis reveal the potential metabolites and biological pathways involved in salinity tolerance of the halophyte Salvadora persica. *Environmental and Experimental Botany* 148: 85–99.

Kumari, S., K. Yadav, and N. Singh. 2017. Evaluation of genetic fidelity among micropropagated plants of Salvadora persica L. using DNA-based markers. *Meta Gene* 14: 129–133.

Lamers, J., T. Van Der Meer, and C. Testerink. 2020. How plants sense and respond to stressful environments. *Plant Physiology* 182(4): 1624–1635.

Latef, A.A.H.A., and H. Chaoxing. 2011. Effect of arbuscular mycorrhizal fungi on growth, mineral nutrition, antioxidant enzymes activity and fruit yield of tomato grown under salinity stress. *Scientia Horticulturae* 127(3): 228–233.

Mayta, M.L., A.F. Lodeyro, J.J. Guiamet, V.B. Tognetti, M. Melzer, M.R. Hajirezaei, and N. Carrillo. 2018. Expression of a plastid-targeted flavodoxin decreases chloroplast reactive oxygen species accumulation and delays senescence in aging tobacco leaves. *Frontiers in plant science* 9: 1039.

Mekhemar, M., M. Geib, M. Kumar, Y. Hassan, and C. Dörfer. 2021. Salvadora persica: Nature's gift for periodontal health. *Antioxidants* 10(5): 712.

Monfared, M.A., D. Samsampour, G.R. Sharifi-Sirchi, and F. Sadeghi. 2018. Assessment of genetic diversity in Salvadora persica L. based on inter simple sequence repeat (ISSR) genetic marker. *Journal of Genetic Engineering and Biotechnology* 16(2): 661–667.

Moustakas, M. (2021). Plant photochemistry, reactive oxygen species, and photoprotection. In (Vol. 2, pp. 5–8): MDPI.

Nafees, M., M.A. Bukhari, M.N. Aslam, I. Ahmad, M. Ahsan, and M.A. Anjum. 2019. Present status and future prospects of endangered Salvadora species: A review. *J. Glob. Innov. Agric. Soc. Sci* 7(2): 39–46.

Owis, A.I., M.S. El-Hawary, D. El Amir, O.M. Aly, U.R. Abdelmohsen, and M.S. Kamel. 2020. Molecular docking reveals the potential of Salvadora persica flavonoids to inhibit COVID-19 virus main protease. *RSC advances* 10(33): 19570–19575.

Parida, A.K., S.K. Veerabathini, A. Kumari, and P.K. Agarwal. 2016. Physiological, anatomical and metabolic implications of salt tolerance in the halophyte Salvadora persica under hydroponic culture condition. *Frontiers in plant science* 7: 351.

Patel, M., and A.K. Parida. 2021. Salinity alleviates the arsenic toxicity in the facultative halophyte Salvadora persica L. by the modulations of physiological, biochemical, and ROS scavenging attributes. *Journal of Hazardous Materials* 401: 123368.

Patel, M., and A.K. Parida. 2022. Salinity mediated cross-tolerance of arsenic toxicity in the halophyte Salvadora persica L. through metabolomic dynamics and regulation of stomatal movement and photosynthesis. *Environmental Pollution* 300: 118888.

Rangani, J., A. Panda, and A.K. Parida. 2020. Metabolomic study reveals key metabolic adjustments in the xerohalophyte Salvadora persica L. during adaptation to water deficit and subsequent recovery conditions. *Plant Physiology and Biochemistry* 150: 180–195.

Rangani, J., A. Panda, M. Patel, and A.K. Parida. 2018. Regulation of ROS through proficient modulations of antioxidative defense system maintains the structural and functional integrity of photosynthetic apparatus and confers drought tolerance in the facultative halophyte Salvadora persica L. *Journal of Photochemistry and Photobiology B: Biology* 189: 214–233.

Rangani, J., A.K. Parida, A. Panda, and A. Kumari. 2016. Coordinated changes in antioxidative enzymes protect the photosynthetic machinery from salinity induced oxidative damage and confer salt tolerance in an extreme halophyte Salvadora persica L. *Frontiers in plant science* 7: 50.

Rejeb, K.B., C. Abdelly, and A. Savouré. 2014. How reactive oxygen species and proline face stress together. *Plant Physiology and Biochemistry* 80: 278–284.

Shah, S., M. Amin, B. Gul, and M. Begum. 2020. Ethnoecological, elemental, and phytochemical evaluation of five plant species of lamiaceae in Peshawar, Pakistan. *Scientifica* 2020.

Shah, S., S. Khan, R.W. Bussmann, M. Ali, D. Hussain, and W. Hussain. 2020. Quantitative ethnobotanical study of Indigenous knowledge on medicinal plants used by the tribal communities of Gokand Valley, District Buner, Khyber Pakhtunkhwa, Pakistan. *Plants* 9(8): 1001.

Shi, R., L. Liang, W. Liu, and A. Zeb. 2022. Kochia scoparia L., a newfound candidate halophyte, for phytoremediation of cadmium-contaminated saline soils. *Environmental Science and Pollution Research*: 1–10.

Signorelli, S., Ł.P. Tarkowski, W. Van den Ende, and D.C. Bassham. 2019. Linking autophagy to abiotic and biotic stress responses. *Trends in plant science* 24(5): 413–430.

Tahir, K., S. Nazir, B. Li, A.U. Khan, Z.U.H. Khan, A. Ahmad, and F.U. Khan. 2015. An efficient photo catalytic activity of green synthesized silver nanoparticles using Salvadora persica stem extract. *Separation and Purification Technology* 150: 316–324.

Tounekti, T., M. Mahdhi, T.A. Al-Turki, and H. Khemira. 2018. Physiological responses of the halophyte Salvadora persica to the combined effect of salinity and flooding. *Int. J. Agric. Biol* 20: 2211–2220.

Van de Peer, Y., T.-L. Ashman, P.S. Soltis, and D.E. Soltis. 2021. Polyploidy: an evolutionary and ecological force in stressful times. *The Plant Cell* 33(1): 11–26.

Zhang, Y., Y. Zhou, D. Zhang, X. Tang, Z. Li, C. Shen, X. Han, W. Deng, W. Yin, and X. Xia. 2020. PtrWRKY75 overexpression reduces stomatal aperture and improves drought tolerance by salicylic acid-induced reactive oxygen species accumulation in poplar. *Environmental and Experimental Botany* 176: 104117.

21 Salvia officinalis and Stressful Conditions

Es-sbihi Fatima Zohra[1], Hazzoumi Zakaria[2] and Amrani Joutei Khalid[1]*

[1]Laboratory of Microbial Biotechnology and Bioactive Molecules: Faculty of Science and Technology Fez, Sidi Mohamed Ben Abdellah University, B.P. 2202-Road of Imouzzer, Fez, Morocco

[2]Laboratory of Plant and Microbial Biotechnology, Moroccan Foundation for Advanced Science, Innovation and Research, Rabat Design Center, Madinat Al Irfan, Rabat, Morocco

Corresponding author: Es-sbihi Fatima Zohra: Fatimazohra@gmail.com

CONTENTS

21.1 INTRODUCTION

Salvia officinalis is a pharmacologically active plant that is characterized by specific antioxidant, spasmolytic, antimicrobial, anti-hidrotic, astringent and sensory activities. Several data also indicate antigenotoxic and chemo-preventive activities of different extracts of *S. officinalis* L. species. Its essential oil contains cineole, borneol, thujone, camphor, pinene and camphene (Hazzoumi et al. 2019). Like other Lamiaceae, *S. officinalis* essential oil is located in specialized structures called glandular trichomes (Hazzoumi et al. 2019). According to their mode of secretion, glandular hairs can be classified into two main types, capitate and peltate. The capitate glands release their products to the outside through a single pore shortly after their production. In contrast, the peltate glands release their products progressively into a space formed by the rise of the cuticle and outwards after the cuticle bursts.

However, the environmental conditions strongly influence the physiology and phytochemistry of this plant. Notably, stresses (saline, metallic, etc.), cause serious morphological, physiological and biochemical alterations, namely decreased growth and productivity of plants, resulting in osmotic stress, specific ionic effects and nutritional imbalances. Following dysfunction of the photosystem, there is decreased photosynthetic activity and oxidative explosion is induced. In addition, several studies have shown the positive effect of exogenous growth regulators, in particular salicylic acid (SA), in reducing the harmful effects on various plants. Our data show that the treatment of *S. officinalis* with salicylic acid improves plant growth. This positive effect is attributed to stimulation of absorption and translocation of nutrients, stimulation of photosynthetic pigment synthesis, regulation of water status and the antioxidant system of *S. officinalis*. It also improves the quality and quantity of essential oils. Using these data, we try in this work to provide a scientific overview of the impact of stress conditions on *S. officinalis* adaptation, from the physiologic and biochemical aspects, and how the application of some mediator compounds like salicylic acid could remediate the negative impacts.

21.2 BOTANY OF *SALVIA OFFICINALIS*

Salvia officinalis L. belongs to the mint family Lamiaceae, subfamily Nepetoideae, tribe Mentheae and genus *Salvia*. Approximately 240 genera and 7000 species belong to the Lamiaceae family, which is the largest family of the order Lamiales (Dinç et al. 2009). The genus *Salvia* is the largest genus of the Lamiaceae family, comprising around 1000 species (Walker and Sytsma 2007), which are either herbaceous or shrubby perennials, rarely biennials or annuals, and often strongly aromatic species. *Salvia* species commonly grow all around the world; however, they are most abundantly distributed in Europe, around the Mediterranean.

21.3 MEDICINAL/NUTRITIVE USES OF *SALVIA OFFICINALIS*

Salvia officinalis is antiseptic, stimulating, tonic and stomachic. It also has varying degrees of antispasmodic and emmenagogue properties. In external use (in decoction), its properties are resolving, vulnerary and healing. In herbal tea and aromatics, it facilitates digestion. It is also used in the treatment of diabetes because it lowers blood sugar levels. In traditional Austrian medicine, *S. officinalis* has been used for the treatment of disorders of the respiratory tract, mouth, gastrointestinal tract and skin.

21.4 RESPONSES OF *SALVIA OFFICINALIS* TO ABIOTIC STRESS

21.4.1 RESPONSES OF *SALVIA OFFICINALIS* TO SALINITY

Es-sbihi et al. (2021) showed the NaCl toxicity at 150 mM on the shoots and roots growth of *S. officinalis*. This effect is linked to a large accumulation of Na^+ and Cl^- ions in the leaves and roots, with an especially large accumulation in the roots. Efficient NaCl translocation of ions to leaves damages chlorophyll pigment synthesis and therefore results in reduced growth of *S. officinalis* in the presence of stress. Mei et al. (2015) reported that the decrease in chlorophyll levels may be related to the inhibition of chlorophyll biosynthesis regulatory enzymes such as δ-aminolevulinic acid dehydratase and proto-chlorophyllide reductase, or inhibition of essential mineral absorption which is necessary for the biosynthesis of chlorophyll as Mg^{2+}. Among the aspects of stress, toxicity is the accumulation of reactive oxygen species (ROS). We have shown in a study carried out on *S. officinalis* plants (Figure 21.1) a strong H_2O_2 accumulation. The ROS production induced by stress directly led to lesions of the lipid membranes of the cells [malondialdehyde (MDA) accumulation] (Figure 21.2). Stress-induced ROS accumulation, which is a potential cause of oxidative stress, is often positively correlated with MDA content in plants (Akram et al. 2017). In this sense, these results suggest that *S. officinalis* is a species sensitive to different types of stress. Among the defense mechanisms of plants against stress, we note the phenolic compounds, which intervene in a mechanism of acclimatization of the plants under stress conditions. Phenolic compounds minimize the toxic effects of all types of biotic or abiotic stress on the tissues, which explains the increase in the levels of the phenolic compounds in stressed plants.

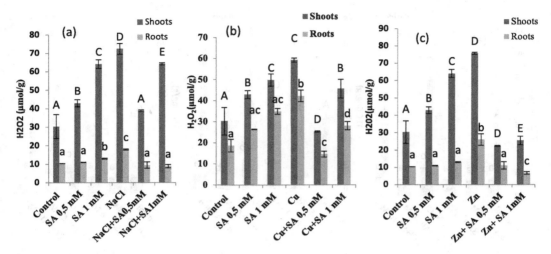

FIGURE 21.1 Changes to H_2O_2 contents in *S. officinalis* exposed to different SA and stress treatments. The values followed by different letters are significantly different ($P \leq 0.05$): salinity (a), copper stress (b), Zn stress (c).

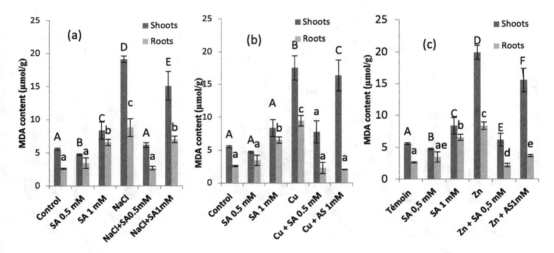

FIGURE 21.2 Changes of MDA contents in S. officinalis exposed to different treatment the SA and stress. The values followed bydifferent letters are significantly different ($P \leq 0.05$) Salinity (a), copper stress (b), Zn stress (c).

Our studies showed an increase in the phenolic compound synthesis (Figure 21.3) and condensed tannins (Figure 21.4) under salt stress. Bettaieb et al. (2011) also reported an increase in phenolic compound concentrations in *S. officinalis*. It appears that stress tolerance in *S. officinalis* is related to the plant's ability to modulate its phenolic compounds in order to cope with oxidative stress caused by stressful conditions. Other molecules contribute strongly to the fight against stress conditions in *Salvia*, namely nitrogen metabolites, such as amino acids, in particular proline, which accumulates in plant tissues under stress.

Increasing the amount of proline in *S. officinalis* plants leads to resistance against water loss, accelerating plant growth and can be considered a powerful non-enzymatic antioxidant, which plants need to reduce the damage of oxidative stress induced by environmental stress. In this sense,

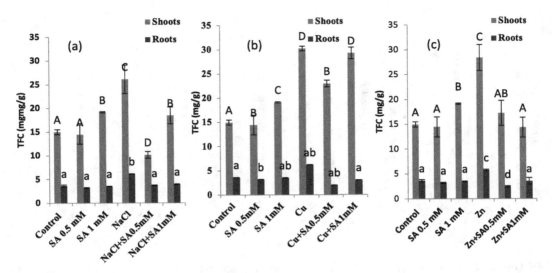

FIGURE 21.3 Changes of TFC contents in *S. officinalis* exposed to different treatment the SA and stress. The values followed by different letters are significantly different ($P \leq 0.05$) Salinity (a), copper stress (b), Zn stress (c).

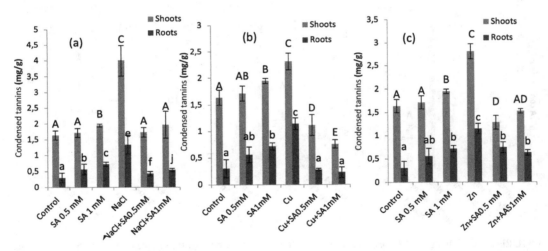

FIGURE 21.4 Changes of condensed tannins contents in *S. officinalis* exposed to different treatment the SA and stress. The values followed by different letters are significantly different ($P \leq 0.05$) Salinity (a), copper stress (b), Zn stress (c).

we have shown in studies carried out on *S. officinalis* that this plant accumulates high concentrations of proline saline stress (NaCl) (Figure 21.5).

21.4.2 RESPONSES OF *SALVIA OFFICINALIS* TO HEAVY METALS

Several works have shown the negative effects of heavy metal on aerial and root growth of *S. officinalis*. Es-sbihi et al. (2020a, 2020b), according to their studies showed how heavy metals, in particular copper (Cu at 40 mM) and zinc (Zn at 40 mM), adversely affect *S. officinalis* productivity.

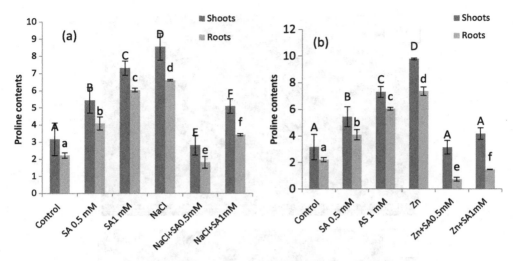

FIGURE 21.5 Changes of proline contents in *S. officinalis* exposed to different treatment the SA and stress. The values followed by different letters are significantly different ($P \leq 0.05$) Salinity (a), Zn stress (b).

This negative effect was attributed to excessive metal absorption by the plant. The latter accumulated much more in the roots than in the leaves of *S. officinalis*. This strong accumulation of metals was accompanied by a decrease in the K^+, P and Ca^{2+} contents. Indeed, the metal toxicity led to a decrease in mineral nutrient accumulation essential to plant growth (El-Metwally et al. 2010; Adrees et al. 2015).

Our studies showed an increase in the phenolic compound synthesis (Figure 21.3) and condensed tannins (Figure 21.4) under the metal stress effect (Cu, Zn). Gerami et al. (2018) showed that cadmium stress decreased essential oil yield in *S. officinalis*. Es-sbihi et al. (2020b) also showed a decrease in the essential oils yield under the effect of Zn (40 mM) in *S. officinalis*. This variation in yield is directly linked to the gland number and maturity.

21.4.3 EFFECT OF STRESS ON GLANDULAR HAIRS AND ESSENTIAL OIL COMPOSITION OF *SALVIA OFFICINALIS* L.

Secondary metabolism, specifically terpenoids, is strongly impacted by stress conditions. In *S. officinalis* this variation is reflected in terms of the quantity synthesized and the quality of the molecules present. In this sense, Taarit et al. (2010) showed that salt stress increased *S. officinalis* yield in concentrations of NaCl between 25 and 75 mM, on the other hand 100 mM decreased the yield of the essential oil. Es-sbihi et al. (2021) also showed a decrease in the essential oils yield under NaCl at 150 mM. Mzabri et al. (2018) showed that 0.8 g/L concentration decreased the essential oils yield of *S. officinalis*. Soltanbeigi et al. (2021) showed that a water deficit also decreased the quantity of the essential oil in *S. officinalis*. Es-sbihi et al. (2020b, 2021) confirmed this decrease in the essential oils yield which was confirmed by a decrease in the glands (peltate glands) density under Zn and NaCl stress.

All observations made by environmental scanning electron microscopy on the leaves of *S. officinalis* showed the presence of peltate glands, capitate glands and non-glandular trichomes (protective hairs). Hazzoumi et al. (2019) also noted in *S. officinalis* the presence of peltate glands with a unicellular stalk and a large round head, and capitate glands with a longer stalk and a smaller head than that of the peltate glands. Es-sbihi et al. (2020a,2020b) also reported the presence of deformed glands under Zn stress (Figures 21.6b and 21.7) and NaCl (Figures 21.8b and 21.9).

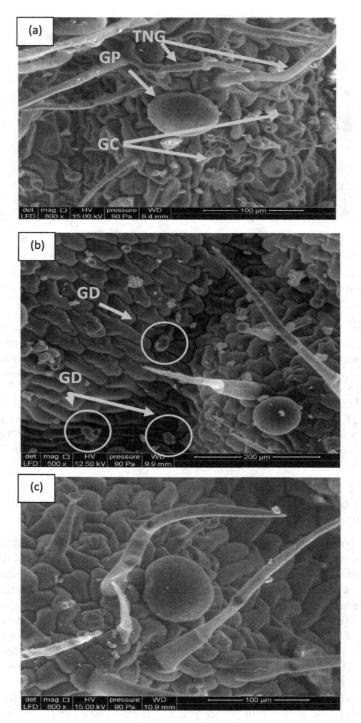

FIGURE 21.6   Observation by environmental scanning electron microscopy in the leaves of Salvia ofcinalis. Control (a), Zn (b) and Zn+0.5mM (Essbihi et al., 2020b). PG: peltate glands; PNG: non-glandular hairs; CG: capitate glands, DG: deformed glands.

Therapeutic advantages of *S. officinalis*, namely antioxidant, antimicrobial, spasmolytic, astringent, anti-hidrotic and specific sensory properties are due to certain molecules and active principles. In terms of composition, the essential oil is mainly composed of 1,8 cineole, thujone and camphor, which are responsible for some of these effects. Russo (2013) showed the importance of the effect of thujone and camphor in the anticancer activity of *S. officinalis*. The essential oils composition varies according to environmental conditions, phenology and the interaction between plants and microorganisms. These modifications result in the appearance or disappearance of certain compounds and modifications of the content of pre-existing compounds. Taarit et al. (2010) showed that salinity at 75 mM stimulated the biosynthesis of all compounds', but mainly thujone. On the other hand, Es-sbihi et al. (2021a) showed in *S. officinalis* that NaCl at 150 mM leads to a reduction in the majority of the compounds, in particular 1,8-cineole, α-thujone and camphor. These same authors noted the appearance of a new major molecule, thujone (content of 15.73%). Stancheva et al. (2014) reported that heavy metals induce a decrease in the content of the major compound. This indicated a deterioration in the essential oil quality under stress.

21.4.4 RESPONSES OF *SALVIA OFFICINALIS* TO DROUGHT

Drought negatively affects the growth and productivity of *salvia officinalis*. Corell et al. (2012) showed that drought caused a decrease in the dry weight (73% of the commercial production of the plant) and essential oil production (69%), and an increase in the nitrogen (N) content of the airborne organs of the plant (15%), together with a decrease in the levels of phosphorus (P), potassium (K) and magnesium (Mg) (21, 25 and 10%, respectively). These results confirm that *S. officinalis* production is limited by the availability of water. Under water-stress conditions the content of the determined compounds or ions increases, performing as osmoregulators (as K), although these osmoregulation processes do not always lead to a variation in the K content (Morgan 1984). After a prolonged period of drought, growth parameters were reduced drastically. Nowak et al. (2010) showed the influence of drought stress on the accumulation of cineole, camphor and α/ß-thujone in sage plants (*Salvia officinalis* L.). Leaves of sage grown under moderate drought stress reveal significantly higher concentrations of monoterpenes (about 33%) than those of plants cultivated under well-watered conditions.

21.5 IMPROVING STRESS TOLERANCE IN *SALVIA OFFICINALIS* BY FOLIAR APPLICATION OF STIMULANT

Several authors have used exogenous SA against biotic (Hayat et al. 2010) and abiotic stresses, such as salt stress (Rouphael et al. 2018), water stress (Damalas 2019), thermal stress (Jahan et al. 2019) and cadmic stress (Szalai et al. 2013). In this sense, we tested in several works the effect of SA spraying at 0.5 and 1 mM on *S. officinalis* subjected to metal and saline stress (Es-sbihi et al., 2020a, 2020b, 2021). In this light, our data showed that SA treatment mainly at 0.5 mM on *S. officinalis* improved the *S. officinalis* growth under stress. In this case, the improvement in growth was confirmed by an improvement in the K, P and Ca contents, which were negatively affected by salinity, accompanied by a decrease in the Na levels (Es-sbihi et al. 2021). Our work on metals also improved *S. officinalis* growth under Cu and Zn stress (Es-sbihi et al., 2020a, 2020b). Growth stimulation induced by SA is related to the stimulation of photosynthetic capacity. The authors showed the positive effect of SA on photosynthesis in the presence of stress by improving the RuBisCo activity, carbonic anhydrase (Ahmed et al. 2018) and stomatal conductance. Our studies also provided evidence that environmental stress-induced oxidative stress is effectively attenuated by spraying *S. officinalis* with SA mainly at a low concentration (0.5 mM) (Figure 21.1). The observed low level of MDA levels (Figure 21.1) in the roots and leaves indicates the protective role of SA in decreasing the peroxidation of cellular organelles in *S. officinalis*. Authors have linked the alleviation of the

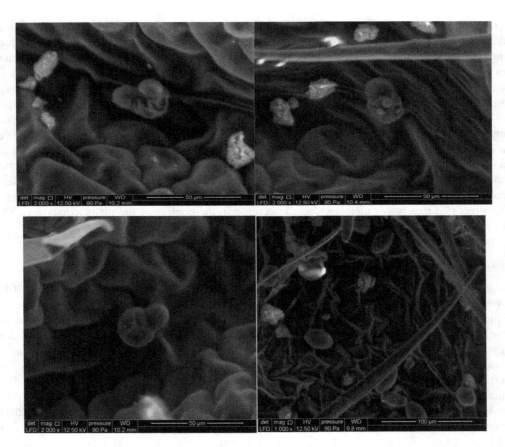

FIGURE 21.7   Observation by environmental scanning electron microscopy in the leaves of Salvia ofcinalis treated with Zn (Essbihi et al., 2020b).

negative effects of stress to an improvement in the antioxidant enzymes activity (Mutlu et al. 2009; Lee et al. 2020; Aftab et al. 2011). SA decreased the concentrations of polyphenols (Figure 21.3), condensed tannins (Figure 21.4) and proline (Figure 21.5) accumulated under the effects of stress. The decrease in these stress markers showed the positive effect of SA on *S. officinalis* in the presence of stress.

The essential oil synthesis increased under the SA effect, and the best result was obtained under the 0.5 mM concentration (Es-sbihi et al., 2020a, 2020b, 2021). This improvement is linked to an improvement in the glands' density under the SA effect, mainly at 0.5 mM. In stress conditions, Khanam and Mohammad (2018) also showed an increase in essential oil content under the SA effect in *Mentha piperita* L. These authors linked this increase in essential oil content under SA during stress conditions to improved growth, photosynthesis, nutrient assimilation and peltate gland density. Others have also demonstrated that SA spraying increased peltate gland density (Idrees et al. 2011; Rowshan et al. 2010). Geram et al. (2018) also showed the positive effect of SA on the growth and productivity of *S. officinalis* cultured under cadmium stress. Spraying SA on plants grown under Zn (Figure 21.6c) and NaCl toxicity (Figure 21.8c) stress showed the absence of deformed glands, which were observed under stresses, showing the remedial effect of SA on *S. officinalis* (Es-sbihi et al. 2020b, 2021). However, jasmonic acid (JA) and its derivatives are widely distributed in plants and affect a variety of processes, including fruit ripening, plant growth and development, and plant responses to injury, infection and insects. Studies have revealed that exogenously supplied

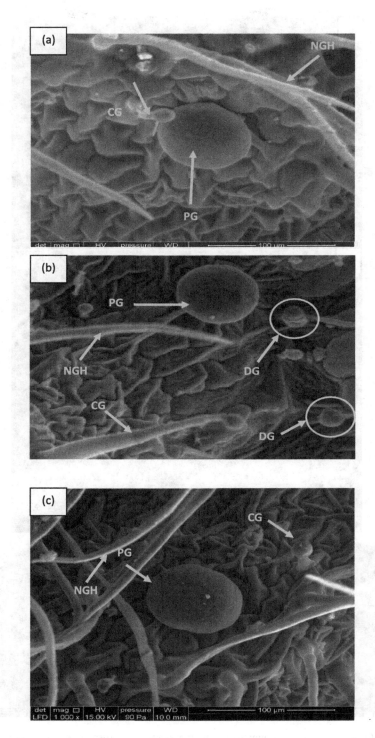

FIGURE 21.8 Observation by environmental scanning electron microscopy of the adaxial of sage leaves (S. officinalis) from control (a), palnts traited with NaCl (b) and plants traited with NaCl + SA (Essbihi et al., 2021) PG: Peltate gland; NGH: Non-glandular hairs; CG: Capitates glands, DG: deformed gland.

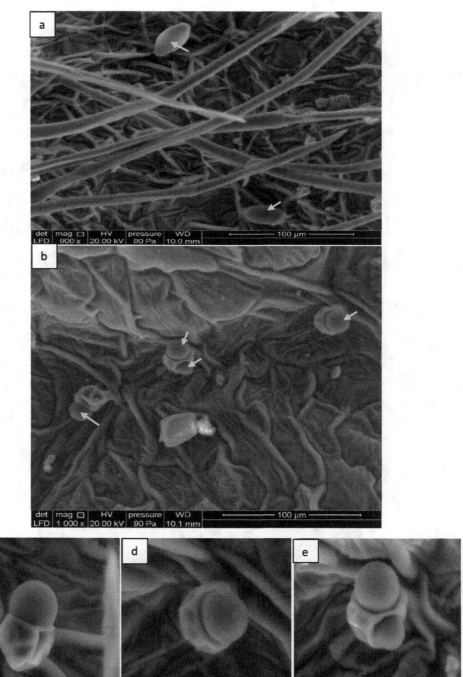

FIGURE 21.9 Observation by environmental scanning electron microscopy of the adaxial of sage leaves (*Salvia officinalis*) from plants treated with NaCl presents deformed glands under the effect of saline stress (Es-sbihi et al. 2021).

JA plays a key role in regulating plant growth, stress responses and development, and can modulate various developmental processes in plants (Per et al. 2018). Karimia et al. (2020) found that exogenous JA along with inoculation of arbuscular mycorrhiza improves the growth characteristics and quality of sage essential oil and improves the tolerance to water deficit. Enhancement of essential oil components and yield by foliar use of JA could be due to increased nutrient uptake, growth cycle or alterations in leaf sebaceous gland populations and biosynthesis of monoterpenes (Ghasemi et al. 2014).

21.6 CONCLUSIONS

Environmental conditions strongly influence *S. officinalis* physiology and phytochemistry. Stress increased free radical levels in *S. officinalis* leading to lipid peroxidation in these plants. Essential oil quality is also affected by stress through the reduction in the contents of the majority compound. Essential oil quantity decreased under different types of stress. These data suggest that *S. officinalis* is a species sensitive to different types of stress.

REFERENCES

Adrees M, Ali S, Rizwan M, Ibrahim M, Abbas F, Farid M, Bharwana SA. 2015. The effect of excess copper on growth and physiology of important food crops: a review. Environ Sci Pollut Res. 22: 8148–8162.

Aftab T, Khan MMA, da Silva JAT, Idrees M, Naeem M. 2011. Role of salicylic acid in promoting salt stress tolerance and enhanced artemisinin production in *Artemisia annua* L. J Plant Growth Regul. 30: 425–35.

Ahmad P, Alyemeni MN, Ahanger MA, Egamberdieva D, Wijaya L, Alam P. 2018. Salicylic Acid (SA) Induced Alterations in Growth, Biochemical Attributes and Antioxidant Enzyme Activity in Faba Bean (*Vicia faba* L.) Seedlings under NaCl Toxicity. Russian J Plant Physiol. 65(1): 104–14.

Akram S, Siddiqui MN, Hussain BMN, Bari MA, Mostofa MG, Hossain MA, Tran LSP. 2017. Exogenous glutathione modulates salinity tolerance of soybean [Glycine max (L.) Merrill] at reproductive stage. J Plant Growth Regul.; 36: 877–88.

Bettaieb, I., Hamrouni-Sellami, I., Bourgou, S., Limam, F., & Marzouk, B. 2011. Drought effects on polyphenol composition and antioxidant activities in aerial parts of *Salvia officinalis* L. Acta Physiologiae Plantarum. 33(4): 1103–1111.

Corell M., Garcia M. C., Contreras J. I., Segura M. L., Cermeño P. 2012. Effect of Water Stress on *Salvia officinalis* L. Bioproductivity and Its Bioelement Concentrations. Communications in Soil Science and Plant Analysis. 37–41

Damalas CA. 2019. Improving drought tolerance in sweet basil (*Ocimum basilicum*) with salicylic acid. Sci Hort. 246: 360–5.

Dinç M., Pinar N. M., Dogu S., Yildirimli S. 2009. Micromorphological studies of Lallemantia l. (Lamiaceae) species growing in Turkey. Acta Biologica Cracoviensia Series Botanica. 51: 45–54.

El-Metwally AE, Abdalla FE, El-Saady AM, Safina SA, EI-Sawy SS. 2010. Response of wheat to magnesium and copper foliar feeding under sandy soil condition. J Am Sci. 6: 818–823.

Es-sbihi FZ, Hazzoumi Z, Joutei KA. 2020a. Effect of salicylic acid foliar application on growth, glandular hairs and essential oil yield in *Salvia officinalis* L. grown under zinc stress. Chem Biol Technol Agric. 7(1): 1–11.

Es-sbihi FZ, Hazzoumi Z, Benhima R, Joutei KA. 2020b. Effects of salicylic acid on growth, mineral nutrition, glandular hairs distribution and essential oil composition in Salvia officinalis L. grown under copper stress. Environ Sustain. 3: 199–208.

Es-sbihi, F. Z., Hazzoumi, Z., Aasfar, A., & Amrani Joutei, K. Improving salinity tolerance in *Salvia officinalis* L. by foliar application of salicylic acid. Chemical and Biological Technologies in Agriculture, 2021;8(1): 1–12.

Gerami, M., Ghorbani, A. and Karimi, S. 2018. Role of salicylic acid pretreatment in alleviating cadmium-induced toxicity in Salvia officinalis L. Iranian Journal of Plant Biology.; 10(1), 81–95.

Ghasemi AP, Rahimmalek M, Elikaei-Nejhad L, Hamedi B. 2014. Essential oil compositions of summer savory under foliar application of jasmonic acid and salicylic acid. J Essential Oil Res. 26(5): 342–347. doi: 10.1080/10412905.2014.922508

Hayat Q, Hayat S, Irfan M, Ahmad A. 2010. Effect of exogenous salicylic acid under changing environment: a review. Environ and experm botany. 68: 14–25.

Hazzoumi Z, Moustakime Y, Joutei KA. 2019. Essential oil and glandular hairs: diversity and roles. In: Essential oils-oils of nature. IntechOpen.

Idrees M, Naeem M, Aftab T, Khan MMA. 2011. Salicylic acid mitigates salinity stress by improving anti-oxidant defense system and enhances vincristine and vinblastine alkaloids production in periwinkle (Catharanthus roseus L. G. Don). Acta Physiol Plant. 33: 987–999.

Jahan MS, Wang Y, Shu S, Zhong M, Chen Z, Wu J, Guo S. 2019. Exogenous salicylic acid increases the heat tolerance in tomato (*Solanum lycopersicum* L) by enhancing photosynthesis efficiency and improving antioxidant defense system through scavenging of reactive oxygen species. Sci Hortic. 247: 421–9.

Khanam D, Mohammad F. 2018. Plant growth regulators ameliorate the ill effect of salt stress through improved growth, photosynthesis, antioxidant system, yield and quality attributes in *Mentha piperita* L. Acta Physiol Plant. 40: 188.

Karimia A., Rahmania F., Ghasemi Pirbaloutib A., and Mohamadid M. 2020. Growth, Physiological and Biochemical Traits of Sage under the Exogenous Stimulating and Stress Factors. Russian Journal of Plant Physiology. 67(5): 933–944.

Lee, C., Choo, K., & Lee, S. J. 2020. Active transposition of insertion sequences by oxidative stress in *Deinococcus geothermalis*. Frontiers in microbiology. 11.

Mei L, Daud MK, Ullah N, Ali S, Khan M, Malik Z, Zhu SJ. 2015. Pretreatment with salicylic acid and ascorbic acid significantly mitigate oxidative stress induced by copper in cotton genotypes. Environ Sci and Pollut Res. 22: 9922–31.

Morgan, J. M. 1984. Osmoregulation and water stress in higher plants. Annual Review of Plant Physiology. 35: 299–319.

Mutlu, S., Atici, Ö., & Nalbantoglu, B. 2009. Effects of salicylic acid and salinity on apoplastic antioxidant enzymes in two wheat cultivars differing in salt tolerance. *Biologia Plantarum.*; *53*(2), 334–338.

Mzabri, M. I., Aamar, A., Boukroute, A., & Chetouani, N. 2018. Effet du stress salin sur la teneur et la composition de l'huile essentielle de la Sauge (*Salvia officinalis*). Annales des sciences de la santé. 1(20): 29–36.

Noreen S, Muhammad A. 2008. Alleviation of adverse effects of salt stress on sunflower (*Helianthus annuus* L.) by exogenous application of salicylic acid: growth and photosynthesis. Pak J Bot.;40(4): 1657–63.

Nowak,. M. 2010. Kleinwaechter M.,Manderscheid R., Selmar D. Drought stress increases the accumulation of monoterpenes in sage (Salvia officinalis), an effect that is compensated by elevated carbon dioxide concentration. Journal of Applied Botany and Food Quality. 83(2):133–136

Per TS, Iqbal M, Khan T, Anjum NA, Masood A, Hussain SJ, Khan NA. 2018. Jasmonates in plants under abiotic stresses: Crosstalk with other phytohormones matters. Environ Experiment Bot. 145: 104–120. https://doi.org/10.1016/j.envexpbot.2017.11.004

Rouphael Y, Raimondi G, Lucini L, Carillo P, Kyriacou MC, Colla G, Cirillo V, Pannico A, El-Nakhel C, Pascale SD. 2018. Physiological and metabolic responses triggered by omeprazole improve tomato plant tolerance to NaCl stress. Front Plant Sci. 9: 249.

Rowshan V, Khoi MK, Javidnia K. 2010. Effects of salicylic acid on quality and quantity of essential oil components in *Salvia macrosiphon*. J Biol Environ Sci. 4:77–82.

Russo A, 2013. Chemical composition and anticancer activity of essential oils of Mediterranean sage (*Salvia officinalis* L.) grown in different environmental conditions. Food Chem Toxicol. 55: 42–7.

Soltanbeigi, A., Yıldız, M., Dıraman, H., Terzi, H., Sakartepe, E., Yıldız, E. 2021. Growth responses and essential oil profile of *Salvia officinalis* L. Influenced by water deficit and various nutrient sources in the greenhouse. Saudi journal of biological sciences. 28(12): 7327–7335.

Stancheva, M., Makedonski, L., & Peycheva, K. 2014. Determination of heavy metal concentrations of most consumed fish species from Bulgarian Black Sea coast. *Bulg Chem Commun.* 46(1), 195–203.

Szalai G, Krantev A, Yordanova R, Popova LP, Janda T. 2013. Influence of salicylic acid on phytochelatin synthesis in Zea mays during Cd stress. Turk J Bot. 37: 708–14.

Taarit MB, Msaada K, Hosni K, Hammami M, Kchouk ME and Marzouk B. 2009. Plant growth, essential oil yield and composition of sage (*Salvia officinalis* L.) fruits cultivated under salt stress conditions. Indust Crops and Prod. 30: 333–337

Taarit Mouna, Kamel Msaada Karim Hosni Brahim Marzouk. 2010. Changes in fatty acid and essential oil composition of sage (*Salvia officinalis* L.) leaves under NaCl stress. J Food Chem. 119(3): 951–956.https://doi.org/10.1016/j.foodchem.2009.07.055

Walker J. B. and Sytsma K. J. 2007. Staminal Evolution in the Genus Salvia (Lamiaceae): Molecular Phylogenetic Evidence for Multiple Origins of the Staminal Lever. Annals of Botany. 100: 375–391.

22 The Role of Secondary Metabolites in *Thymus vulgaris* under Abiotic Stress

Ali Bandehagh[1], Zahra Dehghanian[2], Khashayar Habibi[3] and Arafat Abdel Hamed Abdel Latef[4*]*
[1*]Department of Plant Breeding and Biotechnology, Faculty of Agriculture, University of Tabriz, Tabriz, Iran
[2]Department of Biotechnology, Faculty of Agriculture, Azarbaijan Shahid Madani University, Tabriz, Iran
[3]Department of Biotechnology, College of Agriculture, Isfahan University of Technology, Isfahan, Iran
[4*]Department of Botany and Microbiology, Faculty of Science, South Valley University, Qena 83523, Egypt
*Corresponding authors: Ali Bandehagh, e-mail: bandehhagh@tabrizu. ac.ir, Arafat Abdel Hamed Abdel Latef, e-mail: moawad76@gmail.com

CONTENTS

22.1   INTRODUCTION

Thymus vulgaris L. (thyme) is a *Lamiaceae* family aromatic plant, which is a medicinally significant plant that can be found worldwide, but particularly in the Mediterranean area (Mahdavi, Moradi and Mastinu 2020). Food, cosmetics and pharmaceuticals all use various species of thyme (Gedikoğlu, Sökmen and Çivit 2019). Recently, several manuscripts on the metabolome of thyme plants qualitatively and quantitatively have been published (Assiri et al. 2016). Monoterpenes and sesquiterpenes contain most of these aroma compounds (Moradi et al. 2017). Abiotic stresses alter the volatile compounds of plants significantly (Mahdavi et al. 2020), while water (Moradi, Ford-Lloyd and Pritchard 2017), temperature (Wang et al. 2019), light (Lüpke 2018), oxidative stresses (Hedayati et al. 2020) and salt (Zhang et al. 2022) all increase terpene emission in stressed plants, such as isoprene (Xu et al. 2020), monoterpenes (Mochizuki et al. 2020) and sesquiterpenes, for example (Baggesen et al. 2021). Previous research has shown that environmental factors such as drought affect the volatile composition of thyme (Zahrae et al. 2022). Abiotic stress harms plants by drastically altering cell metabolism and producing an excess of reactive oxygen species (ROS) in the plant. These regulators are important in the biological mechanisms of plants, acting like both toxins and key regulators of growth, hormone signaling, cell cycle, programmed cell death (PCD) and cell reactions (Aftab 2019). In higher plants, primary metabolites are the precursors to secondary metabolites, and the concentrations of various plant secondary products are highly dependent on the growing environment. Secondary metabolites play an important role in nutritive, food additive, flavor and pharmaceutical applications. In most cases, abiotic stresses increase secondary metabolite production in aromatic and medicinally important higher plants, increasing phytomedicine production and promoting essential oil production in aromatic plants (Pradhan et al. 2017). The metabolites of plants are classified into two types: primary metabolites and secondary metabolites (SMs). Plant development and growth are directly influenced by carbohydrates, lipid and protein as primary metabolites. SMs, on the other hand, are functionalized metabolites that affect plant protection and ecologic interactions (Ingle, Padole and Khelurkar 2016). They are also linked to plant color, aroma and scent. They are crucial in plant stress responses. SMs, for example, affect infection termination, whether by living or nonliving agents (Ali et al. 2018). Plant SMs also have functions in mitigating abiotic tensions such as drought, UV light, temperature and salinity stress in addition to biotic stress tolerance (Khare et al. 2020). Plants can decrease morphological characteristics such as leaf area, branch and leaf numbers, height and root volume when subjected to biotic and abiotic tensions (Pradhan et al. 2017). Plants have several protection systems that allow them to deal with stressful situations, they reduce environmental stress at the metabolomic level, and raise SM deposition. Plants' receptors and sensors detect threat signals, allowing them to mount defensive responses to protect themselves from harm. Secondary metabolite accumulation is one of these responses. Plant defense is controlled by transcriptional factors (TFs), which detect stress signals and direct downstream defense gene expression. We provide background information on the medicinal importance of *Thymus vulgaris* and the abiotic stress on plant SMs in this chapter.

22.2 *THYMUS VULGARIS'S* MEDICINAL SIGNIFICANCE

Thymol (2-isopropyl-5-methyl phenol and 5-methyl-2-isopropylphenol) is a crystalline, colorless phenolic monoterpenoid with a unique smell. It is an isomer of carvacrol and the primary active substance in thyme oil. It is a Lamiaceae mint family perennial aromatic herb with an expected lifetime of around 10–15 years, that is woody, very branched and small in stature. It is indigenous to the Mediterranean region and surrounding nations, having originated in northern Africa and parts of Asia, and then spread throughout the world (Figure 22.1) (Kuete 2017). *T. vulgaris* blooms in the spring with bilabiate flowers that are white, yellow or purple. It emits a pleasant and sweet scent when its leaves are touched. Thyme is a relative of the oregano genus *Origanum*, and people have used it for many centuries as a food seasoning and medicinal herb, and its antioxidant and antimicrobial properties have also been studied (Preedy 2015). It was thought to also be an antidote to poison in Ancient Rome; taking thyme beforehand or while eating keeping the person safe from poisonous substances. Because of its popularity, it became one of the emperors' preferred herbs. Later, during the Black Death around 1340, it was used as the main ingredient in medicinal concoctions and ointments applied directly to the blistered skin (Dunn and Coo 2013). Although the pharmacological principle was unknown at the time, Caspar Neumann, a scientist, isolated thymol in 1719. When trying to combat the hookworm pandemic (1879–1880), Italian doctors discovered its anti-hookworm action. Hookworm is an infection caused by an intestinal parasite between (Escobar et al. 2020). Thymol is a highly effective disinfectant that is now commonly found in mouth rinse, hand lotion and skin condition treatments among others (Nagoor Meeran et al. 2017).

FIGURE 22.1   An image of *Thyme vulgaris*.

22.3   BIOSYNTHETIC PATHWAY OF SMS

The Krebs cycle and shikimate pathway are primarily responsible for the production of metabolite precursors. Figure 22.2 depicts the proposed SM biosynthesis pathway. Primary metabolites act as important functions in SM formation. The chemical structure, function and distribution of primary metabolites and SMs in plants can also be identified. Most plants have conserved metabolite biosynthetic pathways. Because of the preservation of this metabolic center, only a few basic metabolic frameworks have emerged. A broad range of structural changes is caused by many hydroxylations, methylation, glycosylation, acylation, prenylation, phosphorylation, oxidation and very few synthetic changes caused by enzyme tailoring. SMs are categorized into three classes based on their biosynthesis mechanisms: synthesized phenolic compounds via the shikimate pathway, terpenes produced through the mevalonic pathway and compounds, including nitrogen, which are synthesized by the tricarboxylic acid cycle pathway (Jamwal, Bhattacharya and Puri 2018). The shikimate pathway produces phenolic compounds, the mevalonic pathway produces terpenes and the tricarboxylic acid cycle pathway produces nitrogen-containing compounds (Datta et al. 2022). The condensed tannins, lignins, flavonoids, lignans and phenylpropanoid/benzenoid volatiles are all produced by phenylalanine; tyrosine also generates quinones (for example, tocochromanols and plastoquinone), betalains and isoquinoline alkaloids; alkaloids, the auxin, indole glucosinolates and phytoalexins are all made up of tryptophan (Xu et al. 2020). The shikimate pathway has seven steps that lead to the end product, chorismate. Erythrose-4-phosphate and phosphoenolpyruvate are combined in the first step to form 3-deoxy-o-arabino-heptulosonate 7-phosphate (DAHP), which is a 2-deoxy-D-glucose-6-phosphate derivative. A shikimate pathway's final steps add side chains and

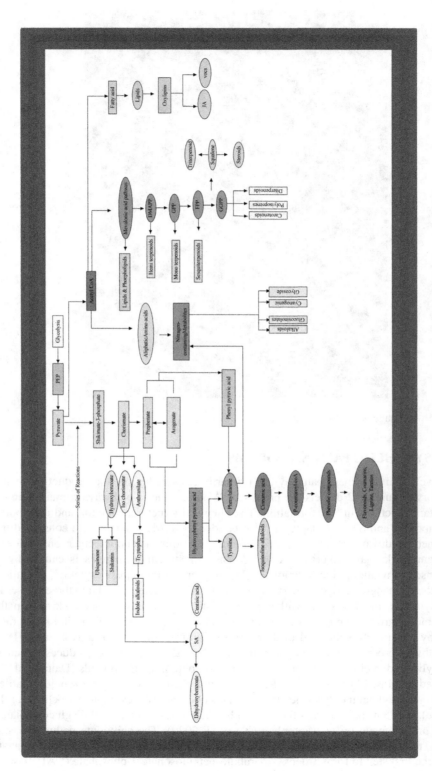

FIGURE 22.2 A schematic representation of the secondary metabolite production pathway (adapted from Jan et al. 2021).

two of the three double bonds required to turn this cyclohexane ring into a benzene ring (the aromatic substance distinguishing feature). Dehydroquinate dehydration to 3-dehydroshikimate, catalyzed by 3-dehydrogenate dehydratase (DHD), is the shikimate pathway's third and fourth responses, and the removable transition of 3-dehydroshikimate to shikimate by shikimate dehydrogenase via NADPH catalysis (SDH). Shikimate 3-phosphate is formed in the fifth stage of the shikimate procedure, shikimate kinase catalyzes the shikimate phosphorylation in the C3 hydroxyl group in this step, which uses ATP as a substrate. 5-Endolpyruvylshikimate 3-phosphate synthase (EPSP) and chorismate synthase catalyze the sixth and seventh steps. EPSP controls a shikimate pathway following the final step, phosphoenolpyruvate's enopyruvyl substituent is converted to shikimate 3-phosphate. Chorismate synthase, the shikimate pathway's final enzyme, catalyzes the reduction of EPSP to chorismate, which is the progenitor of SMs. The second pair connection in the ring is introduced in this final step by 1,4-anti-elimination of the 3-phosphate and C6-pro-R hydrogen from EPSP, resulting in chorismate. Chorismate is a tryptophan, folate, phenylalanine, salicylate, phylloquinone and tyrosine prelude in plant species. Enzymes like amino deoxychorismate synthase iso-chorismate synthase, anthranilate synthase and chorismate mutase control it (Xu et al. 2020).

22.4 SECONDARY METABOLITE GENERATION IN PLANTS AS A SALINITY TOLERANCE STRATEGY

External factors that harm plant growth, development or production are referred to as stress in plants (Verma, Nizam and Verma 2013). Plants respond to stress in various ways, including alterations in the expression of genes, metabolism (cellular), development, production efficiency and so on. Tensions in plants are mainly caused by abrupt alterations in the environment. Therefore, in tension-resistant plant species, exposure to a given tension results in an adaptation to the particular tension in a time-related way (Verma et al. 2013). Plant tension is classified into two types: abiotic and biotic stresses. The environment's abiotic tension on plants can be either chemical or physical, and biotic tension is caused by biological factors such as insects, diseases and so forth (Verma et al. 2013). The abiotic stressors are linked to reduced yield and quality losses in plants (Nivetha et al. 2021). The abiotic stressors reduce crop yield and are acknowledged to be the main restrictions on global food security (Bali and Sidhu 2020). Soil salinity, waterlogging, temperature, drought and heavy metal toxicity are important abiotic stressors (Figure 22.3) (Zhang et al. 2021).

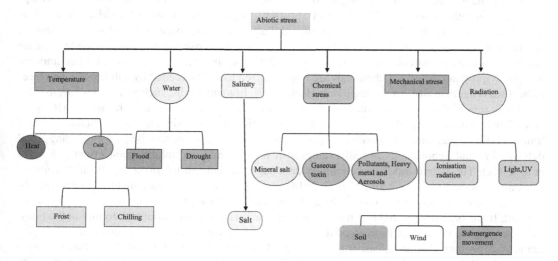

FIGURE 22.3 Plants are influenced by a variety of biotic and abiotic stress signals (Akula and Ravishankar 2011).

Salinity stress is among the most prominent elements which limit the production of plants in many parts of the world, especially in Iran (Khalvandi et al. 2019; Estaji et al. 2018). Anthropogenic activities increase the percentage of the world's salinized landmass, posing a threat to medicinal plant life and also to the availability of the bioactive substances which are generated (Yang et al. 2018). Plants' plasticity (genotypic) allows them to develop a repertoire of SMs required for survival in a shifting and *in vitro* demanding environment and *in vivo* saline development circumstances, which varies by salinity tension amounts, genotype and species. Plants' genotypic plasticity allows them to produce an SM's critical repertoire for living in the physiological disturbance in a modified medium of *in vitro* and *in vivo* saline development circumstances, which varies by salinity stress amounts, genotype and species (Manuka et al. 2019). This becomes more understood as fingerprinting and molecular profiling techniques have been applied to some of the plants and natural products that they generate (Gupta and Huang 2014). The identification of various salinity stress-induced reactions associated with the production of PSMs and the accumulation process has made significant progress (Sytar et al. 2018). Drought, which promotes the accumulation of solutes in greater amounts, due to osmotic regulation, can also generate salinity-induced tension, resulting in the stimulation of SM pathways. The root capability to uptake water is severely diminished during the early phases of salinity-induced stress. This may result in water loss caused by osmotic tension, which is a result of by salt accumulation at greater levels in the plant and soil (Adak et al. 2019). As a result, physiological changes in the functional stability of the membrane, redox equilibrium and nutritional balance affect primary metabolism, which provides the precursors of SM pathways (Xu et al. 2016). These are linked to metabolic alterations related to the biosynthesis of PSM (Iqbal et al. 2020), and include a circadian rhythm reaction. The resistance to the salinity-induced tension refers to a plant's ability to maintain cellular metabolic activities through systemic physiological adjustments (Liu, Zhu and Wang 2015). Physiological modifications such as salt deprivation and sequestration, resistance to accumulated ions, osmotic adjustment, water homeostasis, restricted K^+ loss control development and enlargement modification by biochemical processes are all prevalent in such plants (Van Zelm, Zhang and Testerink 2020). Extremophiles that have adjusted to the development conditions pre-adjust by the SM biosynthesis raising amounts which are driven by salinity stress, and a decrease in their production and salt tension amounts has a deleterious impact on physiological systems (Jha et al. 2019). The molecular effects (and also physiological) of salt-induced tension on crop species' development and essential PSM production have been studied in recent decades. Information about medicinal plants, particularly varying tension and defense reactions connected with the development, is currently limited (El-Hendawy et al. 2019). Significant number of PSMs, classified as terpenoids and steroids, phenolics, flavonoids and alkaloids have been reported to be produced, involved or become activated in cellular stress and defense response function influenced by a saline condition of plant growth physiology (Lucas et al. 2019). Aromatic combinations of higher amounts and phenylpropanoid-derived chemicals are hypothesized to be interceded by salt tension and free-radical scavengers in SM-producing plants (Isah 2019). In many herbs, such as *Thymus vulgaris* L., salinity is one of the most important agents impacting the biochemistry and physiology (Bistgani et al. 2019). The toxicity of ions (Na^+ and Cl^-), oxidative and osmotic tension, restricted CO_2 leaf transport (flux), pigment degradation and photosynthetic inhibition, metabolic pathway change, cell deformation and genotoxicity are all examples of how salinity can inhibit plant growth (Rasouli et al. 2021). The strong link between antioxidant capacity and salt tolerance has been thoroughly recognized in the literature (Valifard et al. 2014). Zrig et al. (2016) reported that higher amounts of NaCl (150 mM) diminish the total leaf dry and fresh mass in *T. vulgaris*, and increase the polyphenol amount. It has been discovered that normal salt tension increased the amount of thymol in the body. Furthermore, the use of NaCl enhanced the synthesis of amino acids, including serine, arginine, glutamic acid, proline, alanine and asparagine. Emami Bistgani et al. (2019) found that saline irrigation (0, 30, 60 and 90 mM) had a significant ($P = 0.05$) effect on T. *vulgaris* physiological properties. By raising the NaCl levels, the dry weight of the plants was dramatically reduced. With the application of NaCl, the Na^+ content in leaves and shoots increased, but the levels of Ca^{2+} and K^+ dropped. In comparison to control plants, the phenolic amount was raised by roughly 20% when 60 mM NaCl was applied. Under salinity stress, phenolics were shown to be enhanced in numerous plant structures and organ systems, according to some publications (Valifard et al., 2014). The typical saline tolerance pathway is hypothesized to be induced by moderate salinity stress by raising total phenolic compound levels (Salem et al. 2014). Genetics and environmental variables both impact the phenolic content (Dykes 2019). Phenolic molecules also act as a critical function in the defense against biotic and abiotic stressors (Akanbong, Senol and Devrim 2021). These metabolites mitigate oxidative tension and scavenge ROS in some plant tissues, acting as antioxidants (Selmar and Kleinwächter 2013). In plants developed under salinity tension, increases in leaf flavonoid amounts of 38.6% and 36.6% were reported after the administration of 60 and 90 mM NaCl, respectively. Phenolic monoterpenes, flavonoids and phenolic acids were abundant in T. *vulgaris*. This species' main component was cinnamic acid. The gallic and rosmarinic acid quantity in T. *Vulgaris* increased considerably at 60 and 90 mM NaCl concentrations (25% and 31.6%, respectively). The vanillic, syringic and caffeic acid quantities in T. *vulgaris*, on the other hand, were unaffected by the NaCl concentration. Surprisingly, a higher level of chlorogenic acid (4.4% increase) was detected in T. *vulgaris* under the control circumstances. Additionally, salt stress had a considerable impact on the plant extracts' antioxidant functions, suggesting that it could be used to boost the medicinal plant's antioxidant level. The reduction in plant development caused by the salt treatment could be due to the development of high salt chloride concentrations in cell membranes and cytoplasm, which has a harmful influence on photosynthesis, carbohydrates and growth regulators (Valifard et al. 2019). Because of hyperosmotic stressors, long-term impacts of salt stress have been observed to limit the plant development rate. Salt content stress during plant growth reduces turgor pressure, which reduces the extension of the wall and lowers the threshold of yield for development, particularly in leaves and stems (Siddique, Kandpal and Kumar 2018). Amor et al. (2019), on the other hand, found that salt influenced the formation of new leaves without affecting cell growth or divisions. According to Ulczycka-Walorska et al. (2020), the plant's essential process diversions under salinity stress are linked to yield reduction owing to biological pathways and carbon metabolism limits. Saeidnejad (2019) found a substantial reduction in mint dry mass after treatment with 50 mM NaCl, which supports the current findings. Because of stomata and reduced net carbon dioxide intake, one of the reasons for the development rate reduction under salinity tension is a fall in the rate of photosynthesis and insufficient ATP supplies (Wu 2018).

22.5 DROUGHT STRESS AND SM PRODUCTION IN THE PLANT

Drought is an abiotic stress that has a significant impact on plant development (Teklić et al. 2021). It is generally related to increased photoinhibition and temperature stresses. It occurs when water availability is reduced to critical levels because of a water shortage and is associated with a high level of solar radioactivity and temperatures (Yuan et al. 2019; Isah, 2019). The increased expression of phenolic content is essential for protecting plants from the negative effects of drought conditions (Naikoo et al. 2019). Drought activated flavonoids and phenolic acid biosynthetic pathways, resulting in their accrual (Gharibi et al. 2019). When the content is increased sufficiently, the potent antioxidants shield plants from the damaging impacts of water demand and supply stress (Rahimi, Taleei and Ranjbar 2018). Once flavonoids accumulate in a plant cell's cytoplasm, they are capable of disinfecting the harmful H_2O_2 molecules that are produced because of a lack of water. Ascorbic acid aids in the reformation of flavonoids into the primary metabolites after they have been completely oxidized (Baskar, Venkatesh and Ramalingam 2018). Drought stress restricts several key gene-encoding enzymes involved in the phenylpropanoid pathway, which leads to the production of phenolic substances (Sharma et al. 2019). When drought conditions existed, the expression of the enzymes which is related to the synthesis of the phenolic compounds, such as CHI, F3H and PAL increased (Hodaei et al. 2018). Drought causes plants to produce more SMs such as terpenes, complex phenols and alkaloids by causing ionic or osmotic stress during the time of *in vivo* and *in vitro* development (Quan et al. 2016). However, in most of the reported cases, the increase was accompanied by decreased biomass production (Selmar and Kleinwächter, 2013). Drought-induced oxidative stress boosted flavonoid biosynthesis (Nakabayashi, Mori and Saito 2014), and has been linked to the protection of plants cultivated in contaminated soils with poisonous metals like aluminum (Bojórquez-Quintal et al. 2017). Drought stress has a significant effect on plant productive output and respiration activities (Zhang et al. 2017). Water deficiency can change secondary metabolites or active compounds in plants (Vosoughi et al. 2018). Drought seems to have a significant impact on crop economic output and plant metabolic activity. Water deficits can alter secondary metabolites or bioactive components in plants (Vosoughi et al. 2018). Drought also can change the yield and composition of metabolites in aromatic and medicinal plants considerably (Fabriki Ourang and Davoodnia 2018). In medicinal plants, spices and herbs under water stress, the density of the aroma profile is one of the most important criterion metrics. Plant cell aroma profiles are frequently altered by a lack of water (Morshedloo et al. 2017). The volatile oils thyme concentration was maximized when plants were grown under extreme environmental stresses, according to Emami Bistgani et al. (2018). The thyme essential oils were significantly increased under significant stress, according to Abdoli and Ghassemi-Golezani (2021). The proportions of carvacrol, α-pinene, β-myrcene and B-caryophyllene essential oils in stressed plants were also significantly greater than in well-watered plants, according to the researchers. However, in underwater deficit irrigation, the percentage of thymol decreased. The α-terpinene, p-cymene, 1,8-cineole, B-caryophyllene, thymol, carvacrol and α–terpinene amounts in the thymol were associated with severe water deficit. Although the results of the aroma profile identified that the effects of treatments for water irrigation differed, adjustments in composition followed nearly identical patterns for the two years of the experiment. Whenever an extreme drought was imposed, thymol, one of the primary ingredients in thyme essential, decreased dramatically (Bistgani et al. 2018). Generally, higher terpene production with a lack of water could be responsible for the increase in the essential oil content. Carbon allocation for the growth of plants has already been limited in this situation, increasing the instability in the plant growth and defensive system capability (Bahreininejad et al. 2013). In addition, under water stress, increased secondary metabolite concentration prevented oxidative stress of the cell surface (Kolupaev et al. 2020).

22.6 MEDICINAL PLANTS' OXIDATIVE STATES AND ANTIOXIDANT REACTIONS TO HEAVY METAL STRESS CONDITIONS

Trace mineral elements are constitutive plant compounds with biological activity as essential or toxic agents in metabolism. The toxic impacts of heavy metals on plants are quite complicated, and they can disrupt pharmacologically active compositions in medicinal plants, compromising the safety, quality and natural plant product performance (Djukic-Cosic et al. 2007). As a result, among the various quality control assessments of medicinal plant raw materials, the identification of metals, particularly toxic metals, is particularly important because industrial contamination of agricultural soils and forests is becoming an important ecological problem in many regions of the world (Zhao et al. 2019).

Heavy metals discharged into the environment can be absorbed by plants via their leaves and/or roots, causing plant metabolism to malfunction, and posing a serious human health risk (Shahid et al. 2017). Several investigations have shown medicinal and aromatic plant pollution, and these plants have a high potential for heavy metal absorption and transfer in edible parts (Solmaz 2021). Heavy metal medicinal plant pollution can occur during the time of cultivation as a result of contamination of the water, soil and/or air, as well as the approaches to processes such as drying, storing and shipping. They are well-known techniques in Chinese and Indian traditional herbal medicine, and are said to provide health advantages (Keshvari et al. 2021). As a result, heavy metal aggregation in medicinal plants' eatable parts must be examined to minimize the heavy metal concentrations in foods (Ozyigit et al. 2018). According to Pirzadah, Malik and Dar (2019), cultivating these plants could modify metal-polluted agricultural land. Plants utilize a variety of techniques to deal with heavy metals that are attracted to root and foliar cells, such as physiological, biochemical, morphological and genetic systems (Shahid and Khalid 2020). the accumulation of heavy metals increases ROS that can be detoxified by antioxidant mechanisms (non-enzymatic and enzymatic) (Rani et al. 2020).

Heavy metal tension produces physiological, morphological and chemical alterations in plant development, synthesis of protein, lipid metabolism, photosynthesis, energy generation and respiration (de Almeida Bezerra et al. 2021; Ranjan et al. 2021). Excessive ROS production, such as singlet oxygen $(1)O_2$, hydrogen peroxide (H_2O_2), hydroxyl $(HO^\bullet)$, superoxide anion $(O_2^\bullet)$, alkoxyl $(RO^\bullet)$ and the radicals of peroxyl $(RO_2^\bullet)$, is the most often seen heavy metal tension early outcome in plant cells (Hussain et al. 2019). While many metabolic activities like photosynthesis and respiration create ROS under normal developmental circumstances, equilibrium takes place between detoxification and ROS generation in the cell. Most abiotic tensions that increase ROS generation, including heavy metals, upset the balance between detoxification and ROS generation, producing oxidative stress in plant cells (Çatav et al. 2020). Table 22.1 depicts the reducing and ROS detoxifying processes by non-enzymatic and enzymatic antioxidants. Furthermore, in Figure 22.4, the oxidative tension pathway generated by several heavy metals is depicted. Metal toxicity can be attributed to three primary causes, as shown in the figure: (1) Because of their affinity for thioyl-, histidine- and carboxyl-groups, proteins can interact directly with them, motivating metals to aim for the cell's structural, catalytic and transfer locations; (2) the ROS increased production, which alters antioxidant defenses and causes oxidative tension; and (3) necessary cations are displaced from certain binding sites, prompting functions to decline (Ranjan et al. 2021). Free redox-active metals, including Fe, Cu and Cr, increase ROS generation by the metal-dependent Haber-Weiss or Fenton responses and could cause oxidative tension in the cells of the plant, unlike the non-redox-active heavy metals (for example, As, Cd, Co, Hg, Mn, Ni, Pb and Zn) (Lone and Gaffar 2021). Lipids (saturated fatty acids peroxidation in the membrane), proteins, nucleic acid and carbohydrates, are the most important cellular constituents that are vulnerable to ROS damage (Voronkova et al. 2018). Lipid peroxidation

TABLE 22.1
Enzymatic Antioxidant Mechanisms for ROS Scavenging and Detoxification

Antioxidant Kind	Catalyzed Reaction	References
Superoxide dismutase	$2O_2^{\bullet-} + 2H^+ \leftrightarrow 2H_2O_2 + O_2$	Nafees et al. 2019
Catalase	$2H_2O_2 \leftrightarrow 2H_2O + O_2$	Nafees et al. 2019
Glutathione peroxidase	$GSH + H_2O_2 \leftrightarrow GSSH + H_2O$	Nafees et al. 2019
Glutathione S-transferase	$RX + GSH \leftrightarrow HX + R\text{-}S\text{-}GSH^a$	Voronkova et al. 2018
Ascorbate peroxidase	$AA + H_2O_2 \leftrightarrow DHA + 2H_2O$	Voronkova et al. 2018
Glutathione reductase	$NADPH + GSSG \leftrightarrow NADP^+ + 2GSH$	Voronkova et al., 2018
Dehydroascorbate reductase	$2GSH + DHA \leftrightarrow GSSG + AA$	Voronkova et al., 2018
Guaiacol peroxidase	$Donor + H_2O_2 \leftrightarrow$ oxidized donor $+ 2H_2O$	Voronkova et al., 2018
Shikimate dehydrogenase	Simple carbohydrates are converted into aromatic amino acids, which serve as a substrate for PAL	Zia ur Rehman et al. 2021
Polyphenol oxidase	Producing melanin-like polymerized phenols	Zia ur Rehman et al. 2021
Cinnamyl alcohol dehydrogenase	Producing lignin biosynthesis precursors	Zia ur Rehman et al. 2021
Phenyl alanine ammonia- lyase	Creating trans-cinnamic acid structures from phenylalanine	Zia ur Rehman et al. 2021

Note: Glutathione (GSH), oxidized form of glutathione (GSSH), ascorbic acid (AA), dehydroascorbate (DHA), phenyl-alanine ammonia lyase (PAL), X may be sulfate, nitrite or halide group.

a R may be aliphatic, aromatic or heterocyclic group.

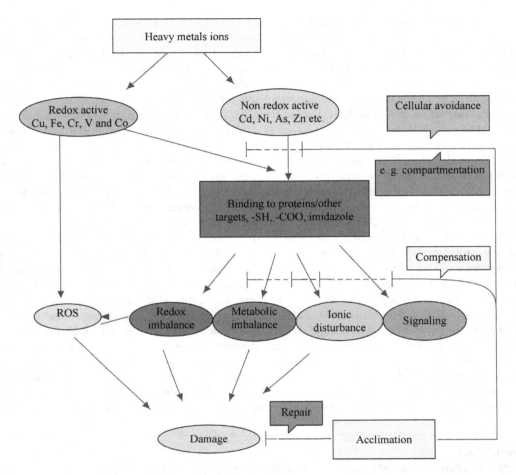

FIGURE 22.4 The main produced oxidative stress mechanisms by various heavy metals (adapted from Sharma and Dietz 2009).

is one of the first signs of oxidative stress in plants, and is measured via malondialdehyde (MDA) amounts (Biswas et al. 2020). MDA is a low-molecular-weight three-carbon aldehyde generated by radical action on polyunsaturated fatty acids (Čamagajevac et al. 2018). Plants use various ways to scavenge excess ROS, including activation of the antioxidant enzymes like catalase, glutathione peroxidase, dehydroascorbate reductase, superoxide dismutase and glutathione S-transferase (Zia ur Rehman et al. 2021), and non-enzymatic antioxidants like ascorbic acid, carotenoids, glutathione and proline (Hussain et al. 2019), and there are several phenolic compositions including anthocyanins, phenolic acid, tannins and lignins (Mohamedshah 2021). As a reaction to physiological stimuli and stress, plants (medicinal) manufacture and aggregate various SMs as potential antioxidants, including cyanogenic glycosides, saponins, cyclic hydroxamic acids, isoflavonoids, sulfur-containing indole derivatives and sesquiterpenes (Lajayer et al. 2017).

22.6.1 Secondary Metabolites and Heavy Metals

Higher plants employ primary metabolites including carbohydrates, amino acids and lipids to generate a variety of SMs (Hatami and Ghorbanpour 2016). Plant SMs are compositions that do not have a direct role in the plant's function; they do, however, play a crucial part in the plant's interaction with its environment for acclimation and resistance (Akula and Ravishankar 2011). Mohanlall (2020) reported nearly 100,000 secondary metabolites with molecular weights of less than 150 kDa and generation with less than 1% of dry matter weight. SMs can be related to the distinctive scents, natural colors, poisons and tastes found in plants (Paramanya, Şekeroğlu and Ali 2021), as well as being the main provenance in food additives, flavors and active compositions pharmaceutically (Kumar and Sharma 2018). SM aggregation is common in plants exposed to many kinds of tensions, signal molecules and elicitors (Hatami and Ghorbanpour 2016) and is influenced by a variety of factors such as genetics, development, growth environment, physiological fluctuations, weather, heavy metals, temperature, light, mineral elements and photoperiod (Dehghan et al. 2018; Choudhary et al. 2021). The efficiency of medicinal plants has been altered by a variety of environmental conditions, although there are limited data on the impact of heavy metals on the amount, quality and SM production in plants (medicinal) (Lajayer et al. 2017). Heavy metal stress has been linked to a considerable rise In secondary metabolites in medicinal plants (Coste et al. 2021). Our recent reports show that after being exposed to heavy metal stress, the generation of secondary metabolites is enhanced (Lajayer, Ghorbanpour and Nikabadi 2017). To put it another way, heavy metal stress increased biosynthesis and aggregated secondary metabolite buildup in pennyroyal cells through a stimulating immunological response (*Mentha pulegium* L.). In the study by Kulbat and Leszczyska (2016) on *Thymus vulgaris* with increasing heavy metal ion levels in the soil, the overall quantity of phenolic compositions, and therefore antioxidant potential, decreased in all stressed plants. Also, plants grown on soils polluted with nickel had a reasonably high level of all antioxidants, even at the minimum dosage. Phenolic compositions are powerful antioxidants and free radical scavengers (Kumar et al. 2020). Moreover, a positive correlation between antioxidant potential and total phenolic content revealed that phenolic compositions were the most abundant antioxidants in the plants studied (Kubalt and Leszczyńska 2016). Phenolic compositions were probably oxidized in plants (*Thymus vulgaris*) growing on soil polluted with the maximum amount of heavy metal ions, and the amount of oxidative harm was too high to be controlled by the cellular antioxidant mechanism (Kubalt and Leszczyńska 2016).

22.7 CONCLUSIONS

In the last decade, there has been a massive increase in the spending and demand for medicinal herbs. The possibilities for medicines and other natural products are expanding because of scientific studies on medicinal plants. Although just 20% of the plant flora has been examined, plants are responsible for roughly 60% of synthetic medicines. On an industrial scale, essential oils, tastes and perfumes of aromatic plants contribute to the economies of developing nations through export revenues and import replacement. Some cancer, diabetes, tumor, thyroid and malarial chemotherapeutic drugs have been isolated from plant origins and synthesized in massive amounts by employing plant hormones that thrive under abiotic stress circumstances (Table 22.2).

TABLE 22.2
Reports of Abiotic Stressors and *Thymus Vulgaris*

Plant	Abiotic stress	References
Thymus vulgaris L.	Drought stress	Llorens-Molina and Vacas 2017; Arpanahi et al. 2020; Bistgani et al. 2018; Ghasemi Pirbalouti et al. 2014
Thymus vulgaris L.	Heavy metals stress	Kubalt and Leszczy-ska 2016; Abu-Darwish and Abu-Dieyeh 2009
Thymus vulgaris L.	Salt stress	Zrig et al. 2016; Bistgani et al. 2019; Valifard et al. 2019; Mahajan, Kuiry and Pal 2020; Noroozisharaf and Rasouli 2021; Bistgani et al. 2019; YAVAS 2021

REFERENCES

Abdoli, Soheila, and Kazem Ghassemi-Golezani. 2021. Salicylic acid: an effective growth regulator for mitigating salt toxicity in plants. *Journal of Plant Physiology and Breeding* 11 (1): 1–15.

Abu-Darwish, Mohammad S, and Ziad HM Abu-Dieyeh. 2009. Essential oil content and heavy metals composition of Thymus vulgaris cultivated in various climatic regions of Jordan. *Int. J. Agric. Biol* 11 (1): 59–63.

Adak, Sanghamitra, Arindam Roy, Priyanka Das, Abhishek Mukherjee, Sonali Sengupta, and Arun Lahiri Majumder. 2019 Soil salinity and mechanical obstruction differentially affects embryonic root architecture in different rice genotypes from West Bengal. *Plant Physiology Reports* 24 (2): 192–209.

Aftab T. 2019. A review of medicinal and aromatic plants and their secondary metabolites status under abiotic stress. *Journal of Medicinal Plants* 7: 99–106.

Akanbong, Elisha Apatewen, Ali Senol, and Alparslan Kadir Devrim. 2021. Phenolic Compounds for Drug Discovery: Potent Candidates for Anti-cancer, Anti-diabetes, Anti-inflammatory and Anti-microbial. *International Journal of Veterinary and Animal Research (IJVAR)* 4 (3): 115–121.

Akula, Ramakrishna, and Gokare Aswathanarayana Ravishankar. 2011. Influence of abiotic stress signals on secondary metabolites in plants. *Plant signaling & behavior* 6 (11): 1720–1731.

Ali S, Ganai BA, Kamili AN, Bhat AA, Mir ZA, Bhat JA, Tyagi A, Islam ST, Mushtaq M, Yadav P. 2018. Pathogenesis-related proteins and peptides as promising tools for engineering plants with multiple stress tolerance. *Microbiological Research* 212: 29–37.

Amor, N, A Jimenez, M Boudabbous, F Sevilla, and C Abdelly. 2019. Implication of peroxisomes and mitochondria in the halophyte Cakile maritima tolerance to salinity stress. *Biol. Plant* 63: 113–121.

Arpanahi, Ali Abdollahi, Mohammad Feizian, Ghazaleh Mehdipourian, and Davood Namdar Khojasteh. 2020. Arbuscular mycorrhizal fungi inoculation improve essential oil and physiological parameters and nutritional values of Thymus daenensis Celak and Thymus vulgaris L. under normal and drought stress conditions. *European Journal of Soil Biology* 100: 103217.

Assiri, Adel MA, Khaled Elbanna, Hussein H Abulreesh, and Mohamed Fawzy Ramadan. 2016. Bioactive compounds of cold-pressed thyme (Thymus vulgaris) oil with antioxidant and antimicrobial properties. *Journal of oleo science* 65 (8): 629–640.

Baggesen, Nanna, Tao Li, Roger Seco, Thomas Holst, Anders Michelsen, and Riikka %J Global change biology Rinnan. 2021. Phenological stage of tundra vegetation controls bidirectional exchange of BVOCs in a climate change experiment on a subarctic heath. 27 (12): 2928–2944.

Bahreininejad, B., Razmjou, J. and Mirza, M. 2013. Influence of water stress on morpho-physiological and phytochemical traits in *Thymus daenensis. International Journal of Plant Production*, 7(1): 151–166. doi: 10.22069/ijpp.2012.927

Bali, Aditi Shreeya, and Gagan Preet Singh Sidhu. 2020. Growth and morphological changes of agronomic crops under abiotic stress. In *Agronomic Crops*: Springer.

Baskar, Venkidasamy, Rajendran Venkatesh, and Sathishkumar Ramalingam. 2018. Flavonoids (antioxidants systems) in higher plants and their response to stresses. In *Antioxidants and antioxidant enzymes in higher plants*: Springer.

Bistgani, Zohreh Emami, Masoud Hashemi, Michelle DaCosta, Lyle Craker, Filippo Maggi, and Mohammad Reza Morshedloo. 2019. Effect of salinity stress on the physiological characteristics, phenolic compounds and antioxidant activity of Thymus vulgaris L. and Thymus daenensis Celak. *Industrial Crops and Products* 135: 311–320.

Bistgani, Zohreh Emami, Seyed Ataollah Siadat, Abdolmehdi Bakhshandeh, et al. 2018. Application of combined fertilizers improves biomass, essential oil yield, aroma profile, and antioxidant properties of Thymus daenensis Celak. *Industrial Crops and Products* 121: 434–440.

Biswas, Koushik, Sinchan Adhikari, Avijit Tarafdar, Roshan Kumar, Soumen Saha, and Parthadeb Ghosh. 2020. Reactive oxygen species and antioxidant defence systems in plants: Role and crosstalk under biotic stress. In *Sustainable agriculture in the era of climate change*: Springer.

Bojórquez-Quintal, Emanuel, Camilo Escalante-Magaña, Ileana Echevarría-Machado, and Manuel Martínez-Estévez. 2017. Aluminum, a friend or foe of higher plants in acid soils. *Frontiers in plant science* 8: 1767.

Čamagajevac Š.I., Ž.T. Pfeiffer, and Š.D. Maronić. 2018. Abiotic stress response in plants: the relevance of tocopherols, Antioxidants and antioxidant enzymes in higher plants. Springer, pp. 233–251.

Çatav, Şükrü Serter, Tuncer Okan Genç, Müjgan Kesik Oktay, and Köksal Küçükakyüz. 2020. Cadmium toxicity in wheat: impacts on element contents, antioxidant enzyme activities, oxidative stress, and genotoxicity. *Bulletin of environmental contamination and toxicology* 104 (1):71–77.

Choudhary, Sadaf, Andleeb Zehra, Mohammad Mukarram. 2021. An insight into the role of plant growth regulators in stimulating abiotic stress tolerance in some medicinally important plants. In *Plant Growth Regulators*: Springer.

Coste, Ana, Carmen Pop, Adela Halmagyi, and Anca Butiuc-Keul. 2021. Secondary metabolites in shoot cultures of Hypericum. *Plant Cell and Tissue Differentiation and Secondary Metabolites: Fundamentals and Applications*: 273–307.

Datta, R., Sharma, A. and Thakur, A. 2022. *Secondary Metabolites from Plants: Role in Plant Diseases and Health Care, Plant Secondary Metabolites*. Springer, pp. 355–369.

de Almeida Bezerra, Laís, Cátia Henriques Callado, Thaís Jorge Vasconcellos. 2021. Chemical and cytotoxical changes in leaves of Eugenia uniflora L., a medicinal plant growing in the fourth largest urban centre of Latin America. *Trees*: 1–14.

Dehghan, Gholamreza, Samaneh Torbati, Reza Mohammadian, Ali Movafeghi, and Amir Hossein Talebpour. 2018. Essential oil composition, total phenol and flavonoid contents and antioxidant activity of Salvia sahendica at different developmental stages. *Journal of Essential Oil Bearing Plants* 21 (4): 1030–1040.

Djukic-Cosic, D, M Curcic, M Cmiljanovic, I Vasovic, and V Matovic. 2007. Heavy metal contents in samples of Hypericum and Thymus spec. collected from different mountain areas in Serbia. *Planta Medica* 73 (09): P_587.

Dunn, J.H. and Koo, J. 2013. Psychological stress and skin aging: a review of possible mechanisms and potential therapies. *Dermatology Online Journal* 19(6).

Dykes, Linda. 2019. Sorghum phytochemicals and their potential impact on human health. *Sorghum*: 121–140.

El-Hendawy, Salah, Nasser Al-Suhaibani, Yaser Hassan Dewir. 2019. Ability of modified spectral reflectance indices for estimating growth and photosynthetic efficiency of wheat under saline field conditions. *Agronomy* 9 (1):35.

Escobar, Angelica, Miriam Perez, Gustavo Romanelli, and Guillermo Blustein. 2020. Thymol bioactivity: A review focusing on practical applications. *Arabian Journal of Chemistry* 13 (12): 9243–9269.

Estaji, Ahmad, Hamid Reza Roosta, Seyed Amin Rezaei, Seyedh Saeedeh Hosseini, and Fatemeh Niknam. 2018. Morphological, physiological and phytochemical response of different Satureja hortensis L. accessions to salinity in a greenhouse experiment. *Journal of Applied Research on Medicinal and Aromatic Plants* 10: 25–33.

Fabriki Ourang, Sedegeh, and Behnam Davoodnia. 2018. Changes in growth characteristics and secondary metabolites in Thymus vulgaris L. under moderate salinity and drought shocks. *Eco-phytochemical Journal of Medicinal Plants* 6 (2): 27–39.

Gedikoğlu, Ayça, Münevver Sökmen, and Ayşe Çivit. 2019. Evaluation of Thymus vulgaris and Thymbra spicata essential oils and plant extracts for chemical composition, antioxidant, and antimicrobial properties. *Food science & nutrition* 7 (5): 1704–1714.

Gharibi, Shima, Badraldin Ebrahim Sayed Tabatabaei, Ghodratolah Saeidi, Majid Talebi, and Adam Matkowski. 2019. The effect of drought stress on polyphenolic compounds and expression of flavonoid biosynthesis related genes in Achillea pachycephala Rech. f. *Phytochemistry* 162: 90–98.

Ghasemi Pirbalouti, A.; Fatahi-Vanani, M.; Craker, L.; Shirmardi, H (2014) Chemical composition and bioactivity of essential oils of Hypericum helianthemoides. Hypericum perforatum and Hypericum scabrum. *Pharm. Biol.* 52: 175–181.

Gupta, Bhaskar, and Bingru Huang. 2014. Mechanism of salinity tolerance in plants: physiological, biochemical, and molecular characterization. *International journal of genomics* 2014.

Hatami, Mehrnaz, and Mansour Ghorbanpour. 2016. Retracted chapter: changes in phytochemicals in response to rhizospheric microorganism infection. *Microbial-mediated induced systemic resistance in plants*: 1–14.

Hedayati, Roghayeh, Davood Bakhshi, Nader Pirmoradian, and Ali Aalami. 2020. On-Tree water spray affects superficial scald severity and fruit quality in'Granny Smith'apples. *Journal of Applied Horticulture* 22 (1).

Hodaei, Mahboobeh, Mehdi Rahimmalek, Ahmad Arzani, and Majid Talebi. 2018. The effect of water stress on phytochemical accumulation, bioactive compounds and expression of key genes involved in flavonoid biosynthesis in Chrysanthemum morifolium L. *Industrial Crops and Products* 120: 295–304.

Hussain, Sajjad, Muhammad Junaid Rao, Muhammad Akbar Anjum. 2019. Oxidative stress and antioxidant defense in plants under drought conditions. In *Plant Abiotic Stress Tolerance*. Springer.

Ingle, K, D Padole, and Khelurkar VC. 2016. Secondary metabolites for plant growth promotion and plant protection. *Adv Life Sci* 5: 10888–10891.

Iqbal, M.S., Iqbal, Z., Hashem, A., Abd_Allah, E.F., Jafri, A., Ansari SA and Ansari, M. I. 2021. Nigella sativa callus treated with sodium azide exhibit augmented antioxidant activity and DNA damage inhibition. Sci Rep. 11(1): 13954. doi: 10.1038/s41598-021-93370-x

Isah, Tasiu. 2019. Stress and defense responses in plant secondary metabolites production. *Biological research* 52.

Jamwal, Komal, Sujata Bhattacharya, and Sunil Puri. 2018. Plant growth regulator mediated consequences of secondary metabolites in medicinal plants. *Journal of applied research on medicinal and aromatic plants* 9: 26–38.

Jan, Rahmatullah, Sajjad Asaf, Muhammad Numan, and Kyung-Min Kim. 2021. Plant secondary metabolite biosynthesis and transcriptional regulation in response to biotic and abiotic stress conditions. *Agronomy* 11 (5): 968.

Jha, Uday Chand, Abhishek Bohra, Rintu Jha, and Swarup Kumar Parida. 2019. Salinity stress response and 'omics' approaches for improving salinity stress tolerance in major grain legumes. *Plant cell reports* 38 (3): 255–277.

Keshvari, Mahtab, Reza Nedaeinia, Mozhdeh Nedaeinia, Gordon A Ferns, Sasan Nedaee Nia, and Sedigheh Asgary. 2021. Assessment of heavy metal contamination in herbal medicinal products consumed in the Iranian market. *Environmental Science and Pollution Research* 28 (25): 33208–33218.

Khalvandi, Masoume, Mohammadreza Amerian, Hematollah Pirdashti, Sara Keramati, and Jaber Hosseini. 2019. Essential oil of peppermint in symbiotic relationship with Piriformospora indica and methyl jasmonate application under saline condition. *Industrial Crops and Products* 127: 195–202.

Khare, Shubhra, NB Singh, Ajey Singh. 2020. Plant secondary metabolites synthesis and their regulations under biotic and abiotic constraints. *Journal of Plant Biology* 63 (3): 203–216.

Kolupaev, Y.E., Horielova, E., Yastreb, T. and Ryabchun, N. 2020. State of antioxidant system in triticale seedlings at cold hardening of varieties of different frost resistance. *Cereal Research Communications* 48(2): 165–171.

Kubalt, Kamila, and Joanna Leszczyńska. 2015. Antioxidants as a defensive shield in thyme (*Thymus vulgaris* L. grown on soil contaminated with heavy metals. Biotechnol Food Sci. 75 (2): 109–117.

Kuete, Victor. 2017. *Medicinal spices and vegetables from Africa: therapeutic potential against metabolic, inflammatory, infectious and systemic diseases*. Academic Press.

Kumar, Indrajeet, and Rajesh Kumar Sharma. 2018. Production of secondary metabolites in plants under abiotic stress: an overview. *Significances of Bioengineering & Biosciences* 2 (4): 196–200.

Kumar, Santosh, Md Abedin, Ashish Kumar Singh, and Saurav Das. 2020. Role of phenolic compounds in plant-defensive mechanisms. In *Plant phenolics in sustainable agriculture*: Springer.

Lajayer, Behnam Asgari, Mansour Ghorbanpour, and Shahab Nikabadi. 2017. Heavy metals in contaminated environment: destiny of secondary metabolite biosynthesis, oxidative status and phytoextraction in medicinal plants. *Ecotoxicology and Environmental Safety* 145: 377–390.

Lajayer, Hemayat Asgari, Gholamreza Savaghebi, Javad Hadian, Mehrnaz Hatami, and Maryam Pezhmanmehr. 2017. Comparison of copper and zinc effects on growth, micro-and macronutrients status and essential oil constituents in pennyroyal (Mentha pulegium L.). *Brazilian Journal of Botany* 40 (2): 379–388.

Liu, Lei, Bin Zhu, and Gao-Xue Wang. 2015. Azoxystrobin-induced excessive reactive oxygen species (ROS) production and inhibition of photosynthesis in the unicellular green algae Chlorella vulgaris. *Environmental Science and Pollution Research* 22 (10):7766–7775.

Llorens-Molina, Juan Antonio, and Sandra Vacas. 2017. Effect of drought stress on essential oil composition of Thymus vulgaris L. (Chemotype 1, 8-cineole) from wild populations of Eastern Iberian Peninsula. *Journal of EssEntial oil rEsEarch* 29 (2): 145–155.

Lone, Irshad A, and Masrat Gaffar. 2021. Phytoremediation Potential of Medicinal and Aromatic Plants. In *Medicinal and Aromatic Plants*: Springer.

Lucas, Marília Souza, Carolina da Silva Carvalho, Giovane Böerner Hypolito, and Marina Corrêa Côrtes. 2019. Optimized protocol to isolate high quality genomic DNA from different tissues of a palm species. *Hoehnea* 46.

Lüpke, Marvin Hermann Friedrich. 2018. Drought effects on isoprenoid emissions of Pinus sylvestris L. and Castanea sativa MILL. assessed by the Tree Drought Emission Monitor, Technische Universität München.

Mahajan, Mitali, Raju Kuiry, and Probir K Pal. 2020. Understanding the consequence of environmental stress for accumulation of secondary metabolites in medicinal and aromatic plants. *Journal of applied research on medicinal and aromatic plants* 18: 100255.

Mahdavi, Atiyeh, Parviz Moradi, and Andrea Mastinu. 2020. Variation in terpene profiles of Thymus vulgaris in water deficit stress response. *Molecules* 25 (5): 1091.

Manuka, R., Karle, S.B. and Kumar, K. 2019. OsWNK9 mitigates salt and drought stress effects through induced antioxidant systems in *Arabidopsis. Plant Physiology Reports* 24(2): 168–181.

Mochizuki, Tomoki, Fumika Ikeda, Koichiro Sawakami, and Akira Tani. 2020. Monoterpene emissions and the geranyl diphosphate content of Acer and Fagaceae species. *Journal of Forest Research* 25 (5): 339–346.

Mohamedshah, Zulfiqar Yusuf. 2021. *Comparative Assessment of Phenolic Bioaccessibility and Bioavailability from 100% Juice and Whole Fruit–a Preclinical Approach*: North Carolina State University.

Mohanlall, Viresh. 2020. Plant cell culture systems for the production of secondary metabolites: a review. *IOSR Journal of Biotechnology and Biochemistry (IOSR-JBB). Vol. 6, Issue 1*.

Moradi, Parviz, Brian Ford-Lloyd, and Jeremy Pritchard. 2017. Metabolomic approach reveals the biochemical mechanisms underlying drought stress tolerance in thyme. *Analytical Biochemistry* 527: 49–62.

Moradi, Parviz, Atiyeh Mahdavi, Maryam Khoshkam, and Marcello Iriti. 2017. Lipidomics unravels the role of leaf lipids in thyme plant response to drought stress. *International journal of molecular sciences* 18 (10): 2067.

Morshedloo, Mohammad Reza, Lyle E Craker, Alireza Salami, Vahideh Nazeri, Hyunkyu Sang, and Filippo Maggi. 2017. Effect of prolonged water stress on essential oil content, compositions and gene expression patterns of mono-and sesquiterpene synthesis in two oregano (Origanum vulgare L.) subspecies. *Plant physiology and biochemistry* 111: 119–128.

Nafees, Muhammad, Shah Fahad, Adnan Noor Shah. 2019. Reactive oxygen species signaling in plants. *Plant Abiotic Stress Tolerance; Springer: Cham, Switzerland*: 259–272.

Nagoor Meeran, Mohamed Fizur, Hayate Javed, Hasan Al Taee, Sheikh Azimullah, and Shreesh K Ojha. 2017. Pharmacological properties and molecular mechanisms of thymol: prospects for its therapeutic potential and pharmaceutical development. *Frontiers in pharmacology* 8: 380.

Naikoo, Mohd Irfan, Mudasir Irfan Dar, Fariha Raghib. 2019. Role and regulation of plants phenolics in abiotic stress tolerance: An overview. *Plant signaling molecules*: 157–168.

Nakabayashi, Ryo, Tetsuya Mori, and Kazuki Saito. 2014. Alternation of flavonoid accumulation under drought stress in Arabidopsis thaliana. *Plant signaling & behavior* 9 (8): e29518.

Nivetha, N, AK Lavanya, KV Vikram. 2021. PGPR-mediated regulation of antioxidants: Prospects for abiotic stress management in plants. In *Antioxidants in Plant-Microbe Interaction*: Springer.

Noroozisharaf, Alireza, and Mehrdad Rasouli. 2021. Effect of brasinosteroid on morphological and physiological traits of garden thyme (Thymusvulgaris) in salinity stress. *Iranian Journal of Seed Sciences and Research* 8 (1):69–92.

Ozyigit, Ibrahim Ilker, Bahattin Yalcin, Senay Turan. 2018. Investigation of heavy metal level and mineral nutrient status in widely used medicinal plants' leaves in Turkey: Insights into health implications. *Biological trace element research* 182 (2):387–406.

Paramanya, Additiya, Nazım Şekeroğlu, and Ahmad Ali. 2021. Secondary Metabolites for Sustainable Plant Growth and Production Under Adverse Environment Conditions. In *Harsh Environment and Plant Resilience*: Springer.

Pirzadah, Tanveer Bilal, Bisma Malik, and Fayaz Ahmad Dar. 2019. Phytoremediation potential of aromatic and medicinal plants: A way forward for green economy. *Journal of Stress Physiology & Biochemistry* 15 (3): 62–75.

Pradhan, J, SK Sahoo, S Lalotra, and RS Sarma. 2017. Positive impact of abiotic stress on medicinal and aromatic plants. *International Journal of Plant Sciences (Muzaffarnagar)* 12 (2): 309–313.

Preedy, Victor R. 2015. *Essential oils in food preservation, flavor and safety*: Academic Press.

Quan, Nguyen Thanh, La Hoang Anh, Do Tan Khang. 2016. Involvement of secondary metabolites in response to drought stress of rice (Oryza sativa L.). *Agriculture* 6 (2): 23.

Rahimi, Yousef, Alireza Taleei, and Mojtaba Ranjbar. 2018. Long-term water deficit modulates antioxidant capacity of peppermint (Mentha piperita L.). *Scientia Horticulturae* 237: 36–43.

Rani, Vibha, Aruj Vats, Aditi Chaudhary Bhavya Bharadwaj, and Shivani Kandpal. 2020. Heavy Metal Toxicity, Mechanism, and Regulation. In *Cellular and Molecular Phytotoxicity of Heavy Metals*: Springer.

Ranjan, Alok, Ragini Sinha, Tilak Raj Sharma, Arunava Pattanayak, and Anil Kumar Singh. 2021. Alleviating aluminum toxicity in plants: Implications of reactive oxygen species signaling and crosstalk with other signaling pathways. *Physiologia Plantarum* 173 (4): 1765–1784.

Rasouli, Fatemeh, Ali Kiani-Pouya, Heng Zhang, and Sergey Shabala. 2021. Mechanisms of Salinity Tolerance in Quinoa. In *Biology and Biotechnology of Quinoa*: Springer.

Saeidnejad, Amir Hossein. 2019. Relationship of Medicinal Plants and Environmental Stresses: Advantages and Disadvantages. In *Handbook of Plant and Crop Stress, Fourth Edition*: CRC Press.

Salem, Nidhal, Kamel Msaada, Wissal Dhifi, Ferid Limam, and Brahim Marzouk. 2014. Effect of salinity on plant growth and biological activities of Carthamus tinctorius L. extracts at two flowering stages. *Acta physiologiae plantarum* 36 (2): 433–445.

Selmar, D. and Kleinwächter, M. 2013. Stress enhances the synthesis of secondary plant products: the impact of stress-related over-reduction on the accumulation of natural products. *Plant and Cell Physiology* 54(6): 817–826.

Shahid, Muhammad, Camille Dumat, Sana Khalid, Eva Schreck, Tiantian Xiong, and Nabeel Khan Niazi. 2017. Foliar heavy metal uptake, toxicity and detoxification in plants: A comparison of foliar and root metal uptake. *Journal of hazardous materials* 325: 36–58.

Shahid, Muhammad, and Sana Khalid. 2020. Foliar application of lead and arsenic solutions to Spinacia oleracea: biophysiochemical analysis and risk assessment. *Environmental Science and Pollution Research* 27 (32): 39763–39773.

Sharma, Anket, Babar Shahzad, Abdul Rehman, Renu Bhardwaj, Marco Landi, and Bingsong Zheng. 2019. Response of phenylpropanoid pathway and the role of polyphenols in plants under abiotic stress. *Molecules* 24 (13): 2452.

Sharma, Shanti S, and Karl-Josef Dietz. 2009. The relationship between metal toxicity and cellular redox imbalance. *Trends in plant science* 14 (1): 43–50.

Siddique, Anaytullah, Geeta Kandpal, and Prasann Kumar. 2018. Proline accumulation and its defensive role under diverse stress condition in plants: An overview. *Journal of Pure and Applied Microbiology* 12 (3): 1655–1659.

Solmaz, Yusuf. 2021. Biofertilizers and their effects on medicinal and aromatic plants. *Sustainable agriculture and livestock for food security under the changing climate*:141.

Sytar, Oksana, Sonia Mbarki, Marek Zivcak, and Marian Brestic. 2018. The involvement of different secondary metabolites in salinity tolerance of crops. In *Salinity Responses and Tolerance in Plants, Volume 2*: Springer.

Teklić, T., Paraðiković, N., Špoljarević, M., Zeljković, S., Lončarić, Z., and Lisjak, M. 2021. Linking abiotic stress, plant metabolites, biostimulants and functional food. *Annals of Applied Biology* 178(2): 169–191.

Ulczycka-Walorska, M., Krzymińska, A., Bandurska, H. and Bocianowski, J. 2020. Response of *Hyacinthus orientalis* L. to salinity caused by increased concentrations of sodium chloride in the soil. *Notulae Botanicae Horti Agrobotanici Cluj-Napoca* 48(1): 398.

Valifard, M, S Mohsenzadeh, B Kholdebarin, and V Rowshan. 2014. Effects of salt stress on volatile compounds, total phenolic content and antioxidant activities of Salvia mirzayanii. *South African Journal of Botany* 93: 92–97.

Valifard, Marzieh, Sasan Mohsenzadeh, Bahman Kholdebarin, Vahid Rowshan, Ali Niazi, and Ali Moghadam. 2019. Effect of salt stress on terpenoid biosynthesis in Salvia mirzayanii: From gene to metabolite. *The Journal of Horticultural Science and Biotechnology* 94 (3):389–399.

Van Zelm, Eva, Yanxia Zhang, and Christa Testerink. 2020. Salt tolerance mechanisms of plants. *Annual review of plant biology* 71:403–433.

Verma, Sandhya, Shadab Nizam, and Praveen K Verma. 2013. Biotic and abiotic stress signaling in plants. In *Stress Signaling in Plants: Genomics and Proteomics Perspective, Volume 1*: Springer.

Voronkova, YS, OS Voronkova, VA Gorban, and KK Holoborodko. 2018. Oxidative stress, reactive oxygen species, antioxidants: a review. *Ecology and noospherology* 29 (1): 52–55.

Vosoughi, Najmeh, Masoud Gomarian, Abdollah Ghasemi Pirbalouti, Shahab Khaghani, and Fatemeh Malekpoor. 2018. Essential oil composition and total phenolic, flavonoid contents, and antioxidant activity of sage (Salvia officinalis L.) extract under chitosan application and irrigation frequencies. *Industrial crops and products* 117: 366–374.

Wang Y, Zeng X, Xu Q, Mei X, Yuan H, Jiabu D, Sang Z, Nyima T. 2019. Metabolite profiling in two contrasting Tibetan hulless barley cultivars revealed the core salt-responsive metabolome and key salt-tolerance biomarkers. AoB Plants 11:plz021.

Wu, Honghong. 2018. Plant salt tolerance and Na+ sensing and transport. *The Crop Journal* 6 (3):215–225.

Xu, Chongzhi, Xiaoli Tang, Hongbo Shao, and Hongyan Wang. 2016. Salinity tolerance mechanism of economic halophytes from physiological to molecular hierarchy for improving food quality. *Current genomics* 17 (3): 207–214.

Xu, Jing-Jing, Xin Fang, Chen-Yi Li, Lei Yang, and Xiao-Ya Chen. 2020. General and specialized tyrosine metabolism pathways in plants. *Abiotech* 1 (2): 97–105.

Yang, Li, Kui-Shan Wen, Xiao Ruan, Ying-Xian Zhao, Feng Wei, and Qiang Wang. 2018. Response of plant secondary metabolites to environmental factors. *Molecules* 23 (4):762.

Yavas, Ilkay. 2021. Effect of salinity on morphological, physiological and biochemical properties of medicinal plants. *Medicinal and Aromatic Plants* 45.

Yuan, Zi-Qiang, Rong Zhang, Bin-Xian Wang, et al. 2019. Film mulch with irrigation and rainfed cultivations improves maize production and water use efficiency in Ethiopia. *Annals of Applied Biology*.175(2): 215–227.

Zahrae RF, Bouhrim M, Mechchate H, Al-zahrani M, Qurtam AA, Aleissa AM, Drioiche A, Handaq N, Zair T. 2022. Phytochemical analysis, antimicrobial and antioxidant properties of Thymus zygis L. and Thymus willdenowii Boiss. essential oils. *Plants* 11.

Zhang, Mingjing, Zhuo Chen, Fang Yuan, Baoshan Wang, and Min Chen. 2022. Integrative transcriptome and proteome analyses provide deep insights into the molecular mechanism of salt tolerance in Limonium bicolor. *Plant Molecular Biology* 108(1): 127–143.

Zhao, Ruirui, Tong Yang, Cong Shi, Meili Zhou, Guoping Chen, and Fuchen Shi. 2019. Effects of urban–rural atmospheric environment on heavy metal accumulation and resistance characteristics of Pinus tabulaeformis in Northern China. *Bulletin of Environmental Contamination and Toxicology* 102 (3): 432–438.

Zia ur Rehman, Muhammad, Muhammad Ashar Ayub, Muhammad Umair, et al. 2021. Synthesis and Regulation of Secondary Metabolites in Plants in Conferring Tolerance Against Pollutant Stresses. In *Approaches to the Remediation of Inorganic Pollutants*: Springer.

Zrig, Ahlem, Taieb Tounekti, Momtaz M Hegab, Sarra Oueled Ali, and Habib Khemira. 2016. Essential oils, amino acids and polyphenols changes in salt-stressed Thymus vulgaris exposed to open–field and shade enclosure. *Industrial Crops and Products* 91: 223–230.

23 Trigonella foenum-graceum and Stressful Conditions

Godwin Anywar[1], Jamilu Edirisa Ssenku[1] and Patience Tugume[1]*
[1]Department of Plant Sciences, Microbiology and Biotechnology, Makerere University, P.O. Box 7062, Kampala, Uganda
*Corresponding author: E-mail: godwinanywar@gmail.com
ganywar@cartafrica.org godwin.anywar@mak.ac.ug

CONTENTS

23.1 INTRODUCTION

Fenugreek (*Trigonella foenum-graecum* L.) is an annual leguminous herb in the Fabaceae family. The plant has triangular leaflets from which the name "*Trigonella*" meaning "little triangle" is derived (Petropoulos 2002). There are several species in the genus *Trigonella*, with different authors putting the number between 50–260 (Kakani and Anwer 2012; Petropoulos 2002). Currently, the Kew database, plants of the world online, lists 287 *Trigonella* species (POWO, 2021). The most cultivated and widely spread species of *Trigonella* genus is *foenum-graecum* (fenugreek), which literally means "Greek hay," derived from its widespread use as forage in the past (Petropoulos 2002). Fenugreek is one of the oldest medicinal plants in use today, with a rich documented history dating back to the first century BCE when used by the Romans (Curry 2017), Egyptians (Rosengarten Jr 1969), in traditional Chinese medicine in 1060 BCE (Yao et al. 2020) and in Ayurveda texts dating as far back as 1000 BCE (Jhajhria and Kumar 2016).

Fenugreek is a fast-growing erect annual herb that reaches maturity in 4–7 months. It grows to a height of 30–60 cm, with a long pink cylindrical stem. It has pinnate, triangular stipules and long-stalked compound leaves. Flowers appear axillary and are 15 cm long, with 2–8 pods, white to yellowish-white color with five petals which are hermaphrodite and insect-pollinated. Seeds are hard, smooth and small in size (5 mm long), with a yellow to golden-yellow color (Ahmad et al. 2016). They grow in curved pods with short hair of 10–18 cm in length and 3.5 × 5 mm in width. The pods are reddish or greenish, turning brown on ripening (Petropoulos, 2002; Wani and Kumar 2018). The plants bear white or yellow flowers, which give rise to long, slender, yellow to brown pods (Yadav and Baquer 2014).

23.2   ORIGINS AND ECOLOGY OF *TRIGONELLA FOENUM-GRACEUM*

Trigonella foenum-graceum is native to eastern Europe and parts of Asia (Petropoulos 2002), although some sources, such as Jiang et al. (2007), have indicated that fenugreek is native to west Africa. Fenugreek is widely distributed in the wild and is currently cultivated on all six continents (Kakani and Anwer 2012; Yadav and Baquer 2014). *Trigonella foenum-graceum* is a fast-growing plant that takes 5–10 days to germinate and 4–7 months to mature. It can also grow in a tropical climate with cool summer or mild winter (Petropoulos 2002; Wani and Kumar, 2018). As different varieties of fenugreek are cultivated in different conditions throughout the world, a wide range of seed yields have been reported by various authors (Petropoulos 2002).

23.3   PHYTOCHEMICAL COMPOSITION OF *TRIGONELLA FOENUM-GRACEUM*

Over 100 different phytochemicals have been identified and isolated from fenugreek seeds (Mandegary et al. 2012; Rayyan et al. 2010). Some of the main compounds are listed in Table 23.1.

TABLE 23.1
Phytochemical Composition of *Trigonella foenum-graceum*

Class of compounds (Phytochemicals)	Examples
Saponins	Disogenin, gitogenin, neogitogenin, homorientin saponaretin, neogigogenin and trigogenin (Yoshikawa et al. 1997; Al-Habori et al. 2001). Furostanol (Petit et al. 1995; Niknam et al. 2021)
Alkaloids	Trigonelline and choline (Yoshikawa et al. 1997; Al-Habori et al. 2001)
Flavonoids	Fenugreekine (Yadav and Baquer 2014), kaempferol-3-O-glucoside, apigenin-7-O-rutinoside, naringenin, quercetin, vitexin and rutin (Niknam et al. 2021)
Amino acids	4-Hydroxyisoleucine
Coumarins	Methyl coumarin, trigocumarin and trimethyl coumarin (Raju et al. 2001)
Minerals	Calcium, zinc, sulfur, phosphorus and iron (Babar et al. 2014; Niknam et al. 2021; Yadav et al. 2011)
Vitamins	A, B_1, C and nicotinic acid (Babar et al. 2014; Yadav et al. 2011)
Other compounds	Fiber, polysaccharides, fixed oils, mucilaginous (Kochhar et al. 2006), polyphenols and coumarin (Niknam et al. 2021)

23.4 MEDICINAL AND OTHER USES AND APPLICATIONS OF FENUGREEK

The seeds of fenugreek have traditionally been used to treat bronchitis, fever, sore throat, wounds, swollen glands, skin irritation, diabetes and ulcers (Yadav and Baquer 2014). Fenugreek has also been used as a carminative, demulcent, expectorant, laxative and stomachic (Yoshikawa et al. 1997).

Fenugreek also has several proven pharmacological properties such as hypoglycemic (Hannan et al. 2007), neuroprotective in painful peripheral neuropathy in diabetes (Morani et al. 2012), hepatoprotective (Meera et al. 2009), significant analgesic and anti-inflammatory activity in the carrageenan-induced rat paw edema (Mandegary et al. 2012), anticancer (Amin et al. 2005; Kawabata et al. 2011; Verma et al. 2010), antibacterial and antifungal (Gangoue et al. 2009; Haouala et al. 2008), antigastric ulcer effect (Pandian et al. 2002) and adaptogenic activity in rodents exposed to anoxia and immobilization stress (Vinod et al. 2012). The fresh leaves of fenugreek are consumed as a vegetable in many societies, while the dried leaves are added to many dishes on the Indian subcontinent and the seeds as a condiment (Yadav and Baquer 2014; Zandi et al. 2015).

Fenugreek contains a hydrocolloid called galactomannan which is used in the pharmaceutical industry as a thickening, emulsifying, stabilizing, gelling and encapsulating agent. It is also used in the food industry to improve the texture and appeal of food (Niknam et al. 2021; Jiang et al. 2007). Fenugreek is generally regarded as safe. Muraki et al. (2011) determined the effective, safe and tolerable dose of fenugreek extract to be around 2.5% (w/w) in experimental rats.

23.5 STRESS IN *TRIGONELLA FOENUM-GRACEUM*

Stress is any external factor that exerts a disadvantageous influence on the plant (Taiz and Zeiger 2003). Stress factors may either be biotic or abiotic. Biotic stress factors include diseases, pests, weeds and insect predation, while abiotic stresses include temperature, moisture and wind (Petropoulos 2002; Taiz and Zeiger 2003). Adaptation and acclimation to environmental stresses result from integrated events occurring at anatomical, morphological, cellular, biochemical and molecular levels. Cellular responses to stress include changes in the cell cycle and cell division, changes in the endomembrane system and vacuolization of cells, and changes in cell wall architecture (Biamonti and Caceres 2009; Taiz and Zeiger 2003). All these changes lead to enhanced stress tolerance of cells. At the biochemical level, plants alter their metabolism in various ways to accommodate environmental stresses, including producing osmoregulatory compounds such as proline and glycine betaine (Taiz and Zeiger 2003). Fenugreek is generally tolerant of most of these biotic and abiotic stresses, especially diseases, insects, drought, high pH, poor soils and salt (Duke 1986).

Stress plays a major role in determining how soil and climate limit the distribution of plant species. Naturally, plants are exposed to various environmental stresses during their growth and development. Whereas the effects of certain factors such as temperature are experienced within a short time frame, others like soil nutrient deficiencies may take a longer time for their stressful effects to be observed. Stress is mainly measured in relation to plant survival, crop yield, growth or the primary assimilation processes of growth. Therefore, understanding the physiological processes and mechanisms underlying stress injury and how plants respond to these insults is of great agricultural significance (Taiz and Zeiger 2003; Mitra et al. 2021).

23.6 RESPONSES OF *TRIGONELLA FOENUM-GRACEUM* TOWARD ABIOTIC STRESS

Abiotic stress refers to the harmful effect of non-living factors such as heavy metals, salinity, water and temperature on the growth, physiology and development of plants (Parwez et al. 2021). Some specific examples of the effects of selected abiotic factors of fenugreek that have been well studied include the following.

23.6.1 Heat and Drought Stress

Plants may wilt in response to drought or water stress. This reduces both water loss from the leaf and exposure to incident light, thereby reducing heat stress on leaves (Taiz and Zeiger 2003). Heat stress affects photosynthesis, respiration and water relations in plants (Wahid et al. 2007), damages plant membrane systems and can lead to loss of electrolytes (Levitt 1980). As a result, plants have developed defense mechanisms against heat stress by capturing reactive oxygen species (ROS), maintenance of cell membrane stability, synthesis of antioxidants, osmoregulation and enhancing the transcription and signal transfer of chaperones (Wahid et al. 2007). Such defense mechanisms lead to changes in genotypic expression (Morimoto 1993; Burke and Orzech 1988) resulting in the synthesis of heat shock proteins (Hsps) or stress proteins (Gupta et al. 2010). Heat shock proteins protect plant cells against deleterious stress effects such as the destruction of cell membranes that would eventually cause cell death. Temperatures in the range 30–40°C induced heat stress in *T. foenum-graceum*, triggering an increase in the synthesis of reactive oxygen species scavenging enzymes (Pant et al. 2013). This suggests that *T. foenum-graceum* is heat tolerant and could be introduced in breeding programs to produce tolerant varieties. Fenugreek tolerates water stress by increasing the endogenous melatonin and trigonelline (Zamani et al. 2020). Water stress causes a reduction in the weight and length of fenugreek shoots, and pigment contents (Zamani et al. 2020). Foliar nourishment of fenugreek under deficit irrigation stress with nano phosphorus enhanced growth and seed yield, coupled with improved physiological, biochemical and anatomical responses. The best results were obtained with the smaller nanoparticles (25 nm) compared to larger particles and conventional fertilizer (Abou-Sreea et al. 2022). Bentonite increases drought tolerance in fenugreek. Bentonite application enhanced relative water content (RWC) and photosynthetic pigments and minimized drought stress indexes such as electrolyte leakage, osmolytes, protein content, hydrogen peroxide, malondialdehyde, antioxidant enzymes, CAT and polyphenol oxidase, total phenol and antioxidant activity under severe drought stress (Mohammadifard et al. 2022).

23.6.2 Salinity Stress

Soil salinization is considered one of the most injurious abiotic stressors in plants. Soil salinity is one of the most significant limiting factors of agricultural productivity and food security because of its detrimental effects (Daliakopoulos et al. 2016). Approximately 20% of the global agricultural land is affected by salt stress, and this acreage is expected to rise to 50% by 2050 (Gupta and Huang 2014; Kumar et al. 2020). Salinity stress is caused by the excess of Na^+ or Cl^- in the soil, impeding plant growth (Munns 2005). In *T. foenum-graecum* germination was not affected up to 140 mM NaCl and the plant survived up to 175 mM of NaCl (Abdelmoumen and El Idrissi 2009). Different plant species and varieties of the same species have different levels of tolerance to salt stress.

The accumulation of high levels of Na^+ limits water conductance, soil porosity and aeration. Soil salinity stress negatively affects the microbial diversity within and around the roots of plants, impedes enzymatic activity, stomatal conductance and photosynthesis rate, causes oxidative stress by enhancing the production of ROS and leads to hypertonic stress (Kumar et al. 2020; Zhang et al. 2018). However, the foliar application of gibberellic acid (10^{-6} M) can significantly mitigate the effects of salinity by maximizing growth and yield in the fenugreek, making it a potential agent for minimizing the effects of salinity-induced stress (Mukarram et al. 2021).

Foliar application of *Moringa oleifera* extract on fenugreek ameliorated the negative impact of salinity against various parameters such as photosynthetic pigments, organic solutes (except proline), total phenols, K^+, Ca^{2+} and Mg^{2+}. *Moringa oleifera* extract could therefore be used to alleviate the negative effect of salinity on fenugreek (Abdel Latef et al. 2017). Also, foliar application of salicylic acid ameliorated salinity stress in fenugreek plants (Abdelhameed et al. 2021).

23.6.3 HEAVY METALS STRESS

Heavy metals are chemical elements that have relatively high density and are toxic when ingested at low concentrations. These include lead, mercury, arsenic, chromium and nickel (Guédon et al. 2008; Shaban et al. 2016). The release of heavy metals into the environment is mainly caused by anthropogenic factors and other natural processes such as erosion, burning of fossil fuels or waste incineration plants, cement plants and leaded petrol (Guédon et al. 2008). Heavy metals are non-biodegradable, persist in soil and bioaccumulate in the food chains of various ecosystems (Fatima et al. 2020).

Heavy metals alter significant physiological processes in fenugreek. Fenugreek has been shown to accumulate heavy metals in root and shoot cells (Kidwai and Dhull 2021). The effects of some specific heavy metals on fenugreek have been studied, with examples given below.

Mercury salts such as H_gCl_2 induced oxidative stress in fenugreek causing anatomical abnormalities in seedlings at different time intervals. Mercury mediated stress parameters such as plant length, dry weight, fresh weight and mitotic index negatively affected the growth and cellular activity of plants having mercury stress (Karmakar 2014).

Lead causes various toxic effects in fenugreek and impairs plant growth. It specifically affects root elongation, seed germination, seedling development, transpiration, photosynthesis and cell division (Kaur 2016). Lead binds to membranes causing separation of the granular material from the plasma membrane. This inhibits polysaccharide chain growth causing the cellulose microfibrils to disorient. High concentrations of lead also led to a reduction in dry matter in fenugreek (Tanwar et al. 2013).

The treatment of fenugreek with arsenic and zinc caused a significant reduction of radicle and cotyledon growth, chlorophyll and carotenoid contents, the accumulation of K, Ca and Cu, soluble proteins and free amino acids in cotyledons and the activities of the hydrolytic enzymes amylase and phosphatase. However, the application of 0.2 mM salicylic acid was able to reverse these harmful effects in fenugreek (Mabrouk et al. 2019).

23.7 RESPONSES OF *TRIGONELLA FOENUM-GRACEUM* TOWARD BIOTIC STRESS

23.7.1 PESTS AND DISEASES

Fenugreek appears very resistant to attacks by insects and animal enemies, since with no serious damage to the plants has been recorded in the literature (Manicas 2002; Duke 1986). The stored seeds of fenugreek are known to last more than 10 years without any treatment or pest attack. Some of the major pests affecting fenugreek include rabbits, hares, leaf miners and game birds, and diseases include collar rot, powdery mildew, leaf spot and *Rhizoctonia* root rot caused by *Rhizoctonia solani* (Manicas 2002).

23.8 GENOMICS OF *TRIGONELLA FOENUM-GRAECUM*

There are limited studies on the metabolomics of fenugreek, with some focusing purely on saponins and some targeting the complete metabolome (Chaudhary et al. 2021). There are several molecular events linking the perception of a stress signal with the genomic responses leading to tolerance (Taiz and Zeiger 2003).

23.8.1 BIOTIC STRESS ELICITATION OF SECONDARY METABOLITES IN *T. FOENUM-GRAECUM*

Generally, plants are continuously exposed to a plethora of biotic stresses such as fungi and bacteria that elicit the production of various secondary metabolites as part of the plant defense mechanisms

(Iriti and Faoro 2009; Prasad et al. 2006). Stress responses of *T. foenum-graceum* have mainly focussed on abiotic stress and thus there is inadequate information specific to its response to biotic stress in literature. However, due to the existence of cross-tolerance in plants to biotic and abiotic stress (Abuqamar et al. 2009; Achuo et al. 2006), biochemical and molecular responses of fenugreek to biotic stress may not differ significantly from those of abiotic stress. According to Mikić (2015), *T. foenum-graceum* may have tolerance to various forms of biotic stresses, that could be mediated through the production of phytoprotectants produced under abiotic stress. These compounds can also be classified as phytoprotectants (Freeman and Beattie 2008; González-Lamothe et al. 2009), among which are phytochemicals that play a key role in promoting human health (González-Chavira et al. 2018; Toscano et al. 2019) by reducing oxidative damage, modulating detoxifying enzymes, stimulating the immune system and showing chemopreventive actions (Giovannetti et al. 2013). These secondary metabolites are involved in altering plant physiology to control both biotic and abiotic stress responses (Jasim et al. 2017). In fenugreek, many bioactive phytochemicals have been reported, including saponins, a class of glycosylated triterpenes that show antimicrobial, antiviral and insecticidal activities (Kharkwal et al. 2012).

23.8.2 STRESS ELICITATION OF TRIGONELLINE PRODUCTION IN FENUGREEK

Trigonella foenum-graceum contains the alkaloid trigonelline which acts as an osmotic agent. The accumulation of trigonelline decreases the cellular osmotic potential, therefore trigonelline plays an important role in protecting the stability of cell membranes when plants are subjected to stress (Tramontano and Jouve 1997). In one study, the trigonelline content increased up to 15-fold under severe stress compared to the control (Jia et al. 2016).

Trigonelline is the major secondary metabolite of fenugreek which is of great importance in the cosmetic and pharmaceutical industries (Beygi et al. 2021) and acts as an osmoregulator, especially in abiotic stress. Thus, there is a vital demand for improving the production of this metabolite in fenugreek to enhance its pharmaceutical value and tolerance to stress conditions in the current climate-changing environment. Trigonelline is regarded as a physiologically active compound in plants that can induce leaf movements, accumulate upon the stress, and act as an osmoprotectant and osmoregulator in response to abiotic stress. In water stress conditions, trigonelline, proline and antioxidant enzyme activity increase in fenugreek increasing resistance to stress conditions (Zamani et al. 2020).

When water stress is induced during and after flowering in fenugreek it may cause an increase in the oil content of the plant (Saxena et al. 2019). Water stress induced by exogenous application of melatonin, methyl jasmonate and salicylic acid has been demonstrated to induce water stress and ultimately trigonelline production (Beygi et al. 2021) and on the contrary a decrease in proline content (Zamani et al. 2020). Induced accumulation of trigonelline and diosgenin by methyl jasmonate in fenugreek has been demonstrated by Irankhah et al. (2020) to be enhanced by simultaneous root symbiotes colonization both under well-watered and water-deficit conditions. The inducement of water stress by applying charcoal enhances the production of trigonelline in different ecotypes of fenugreek under screen house conditions (Bitarafan et al. 2019). However, the enhancement was significantly ecotype dependent.

23.8.3 STRESS ELICITATION OF DIOSGENIN PRODUCTION IN FENUGREEK

Of all the numerous bioactive compounds that are biosynthesized in fenugreek, dioscin, a steroidal saponin, has received a lot of interest due to its antitumor, anticancer and anti-inflammatory activities (Aumsuwan et al. 2016; Tong et al. 2014; Wu et al. 2015). Additionally, the aglycone part of dioscin, called diosgenin, is an important precursor for synthesizing more than 200 steroidal drugs such as contraceptives, testosterone, progesterone and glucocorticoids (Aumsuwan et al. 2016; Patel et al. 2012; Wu et al. 2015). Increasing stress conditions have been reported by Saxena et al. (2017) to enhance the quantity of diosgenin in plants. Optimal concentrations of methyl jasmonate that induce the production of reactive oxygen species (ROS) (Ho et al. 2020) have been reported by De and De (2011) to increase the accumulation of diosgenin in fenugreek by 10.5 times. Similarly, in a study by Ciura et al. (2018) significantly higher concentrations of diosgenin were induced in fenugreek stressed by the exogenous application of methyl jasmonate and treated with diosgenin precursors cholesterol and squalene. However, according to studies by Zhou et al. (2019), elicitation of diosgenin production in fenugreek through exogenous application of methyl jasmonate is dependent on the exposure time and its concentration with significant inducement obtained at 48 h post inducement time.

23.8.4 Effect of Stress on the Secondary Metabolite Genomics of *T. foenum-graecum*

Under normal conditions, the genes and enzymes involved in the biosynthesis of secondary metabolites are kept at low levels (Twaij and Hasan 2022), consequently leading to their low concentration in plant sources. Stress elicitation of their production is usually accompanied by the upregulation transcription of genes and enzymes that control their production to cope with the undesired effects of the stress. Understanding changes in transcriptomes is essential for revealing functional and molecular constituents of cells and tissues during development or during induction by biotic and abiotic factors (Ciura et al. 2018). Some studies aimed at elucidating genomics and transcriptomics of fenugreek during stress elicitation of secondary metabolites production have been conducted. From these studies, it is evident that the catalytic role of the enzymes and the genes can be upregulated by the elicitors among which is biotic stress.

T. foenum-graecum contains numerous nutritionally and pharmaceutically important bioactive constituents, particularly trigonelline and diosgenin (Parwez et al. 2021). Their production is mainly regulated by the transcriptional activities of the gene cassette encoding particular enzymes in their biosynthetic pathway (Twaij and Hasan 2022). Heat stress coupled with 24-epibrassinosteroid treatment of *T. foenum-graecum* has been reported by Sheikhi et al. (2021) to enhance the expression of the enzymes squalene epoxidase, sterol methyltransferase and sterol side chain reductase (SSR) gene involved in the biosynthesis pathway of diosgenin and subsequently to its increase in concentrations in its tissue. Furthermore, treatment of *T. foenum-graecum* with methyl jasmonate (MeJA) has been reported to enhance the diosgenin concentration through upregulation of the expression of 3-hydroxy-3-methylglutaryl-CoA reductase (HMG) and sterol-3-β-glucosyl transferase (STRL), which are pivotal genes of the mevalonate pathway for biosynthesis of diosgenin (Chaudhary et al. 2015).

Transcripts encoding the enzymes involved in the transformation of cholesterol into diosgenin were previously proposed after sequencing differential libraries (Ciura et al. 2017). Direct sequencing confirmed increased levels of steroid 22-α-hydroxylase classified as cytochrome CYP90B1, cytochromes CYP18A1 and CYP734A1 with 26-hydroxylase activity and unspecific monooxygenase (EC:1.14.14.1). Moreover, cytochrome CYP86A2 (fatty acid omega-hydroxylase), CYP72A65 and CYP72A61 were identified.

In the diosgenin biosynthesis pathway, fatty acid ω-hydroxylase (CYP86A2) and steroid 22-alpha-hydroxylase (CYP90B1) genes were annotated in all induced transcriptomes. Moreover, direct sequencing confirmed increased levels of CYP90B1, unspecific monooxygenase and 26-hydroxylase genes in fenugreek plants with an elevated level of diosgenin (Ciura et al. 2018). In the induced fenugreek plants, Ciura et al. (2018) reported eight unigenes annotated as cycloeucalenol cycloisomerase (CPI), one of the key enzymes involved in the phytosterol pathway and showing high FPKM (1393–2088). The CPI genes were expressed at a considerably lower level in the untreated fenugreek.

Expression analysis of fenugreek after stress inducement through the application of methyl jasmonate by Zhou et al. (2019) revealed upregulation of the mevalonate (MVA) pathway and the plastid-targeted 2-C-methyl-D-erythrirtol-4-phosphate (MEP) pathway genes and enzyme that are involved in the formation of 2,3-oxidosqualene. For example, 3-hydroxy-3-methylglutaryl CoA reductase (HMG-CoA reductase) and three HMGR-encoding unigenes (cluster-2140.36411, cluster-2140.78950 and cluster-2140.104061) were found in the fenugreek transcriptome induced by MeJA with their expression levels being improved by 2.65–372.56-fold. In the study, upregulation of the second stage of diosgenin biosynthesis 2,3-oxidosqualene into cholesterol is the backbone of the diosgenin molecule, and in plants, nine enzymes control its formation from cycloartenol (Ciura et al. 2017; Sonawane et al. 2016). Fenugreek transcriptome analysis by Zhou et al. (2019) reported a total of 29 unigenes that code for the nine enzymes, including five for sterol side chain reductase (SSR), eight for sterol 4a-methyloxidase (SMO), two for cycloeucalenol cycloisomerase (CPI), six for CYP51, four for sterol C-14 reductase (C14-R), one for sterol 8,7-isomerase (8,7 SI), two for sterol C-5 (6) desaturase (C5-SD) and one for 7-dihydrocholesterol reductase (7-DR). Most of these cholesterol pathway unigenes are upregulated by the MeJA inducement of drought stress fenugreek. Trigonelline is synthesized by the methylation of nicotinic acid, catalyzed by sadenosyl-L-methionine (SAM)-dependent nicotinate N-methyltransferase (EC 2.1.1.7) (Ashihara 2008; Joshi and Handler 1960).

23.9 CONCLUSIONS

Fenugreek (*Trigonella foenum-graecum* L.) is an important multipurpose commercial crop. It is widely used as a food, spice and medicine. Like other plant species, fenugreek is affected by various stress factors, both abiotic and biotic. However, fenugreek appears to be resilient to most of the stress factors that affect it. Biotic and abiotic factors cause the elicitation of secondary metabolites which are involved in altering the plant physiology to control stress responses. Although fenugreek produces several secondary metabolites, the most significant ones are trigonelline and diosgenin. There are also a limited number of studies on the metabolomics of fenugreek.

LIST OF ABBREVIATIONS

POWO   Plants of the world
NaCl   Sodium chloride
H_gCl_2 Mercuric chloride
MeJA   Methyl jasmonate
HMG   3-Hydroxy-3-methylglutaryl-CoA reductase
STRL   Sterol-3-β-glucosyl transferase
CPI   Cycloeucalenol cycloisomerase
MVA   Mevalonate
MEP   2-C-methyl-D-erythrirtol-4-phosphate
SMO   Sterol 4a-methyloxidase
SSR   Sterol side chain reductase
C5-SD   Sterol C-5 (6) desaturase
7-DR   7-Dihydrocholesterol reductase
SAM   Sadenosyl-L-methionine

REFERENCES

Abdelhameed, R. E., Abdel Latef, A. A. H., and Shehata, R. S. (2021). Physiological responses of salinized fenugreek (*Trigonella foenum-graecum* L.) plants to foliar application of salicylic acid. *Plants, 10*(4), 657.
Abdelmoumen, H., and El Idrissi, M. M. (2009). Germination, growth and nodulation of *Trigonella foenum-graecum*(Fenu Greek) under salt stress. *African Journal of Biotechnology, 8*(11).

Abdel Latef, A. A. H., Abu Alhmad, M., & Ahmad, S. (2017). Foliar application of fresh moringa leaf extract overcomes salt stress in fenugreek (*Trigonella foenum-graecum*) plants. *Egyptian Journal of Botany*, *57*(1), 157–179.

Abou-Sreea, A. I., Kamal, M., El Sowfy, D. M., Rady, M. M., Mohamed, G. F., Al-Dhumri, S. A., Al-Harbi, M.S. and Abdou, N. M. (2022). Small-sized nanophosphorus has a positive impact on the performance of fenugreek plants under soil-water deficit stress: A case study under field conditions. *Biology*, *11*(1), 115.

Abuqamar, S., Luo, H., Laluk, K., Mickelbart, M. V., and Mengiste, T. (2009). Crosstalk between biotic and abiotic stress responses in tomato is mediated by the AIM1 transcription factor. Plant J, 58(2), 347–360. https://doi.org/10.1111/j.1365-313X.2008.03783.x

Achuo, E. A., Prinsen, E., and Höfte, M. (2006). Influence of *drought, salt stress and abscisic acid on the resistance of tomato to Botrytis cinerea and Oidium neolycopersici*. Plant Pathology, 55(2), 178–186. https://doi.org/10.1111/j.1365-3059.2006.01340.x

Ahmad, A., Alghamdi, S. S., Mahmood, K., and Afzal, M. (2016). Fenugreek a multipurpose crop: Potentialities and improvements. *Saudi Journal of Biological Sciences*, *23*(2), 300–310.

Al-Habori, M., Raman, A., Lawrence, M. J., and Skett, P. (2001). In vitro effect of fenugreek extracts on intestinal sodium-dependent glucose uptake and hepatic glycogen phosphorylase A. *International Journal of Experimental Diabetes Research*, *2*(2), 91–99.

Amin A, Alkaabi A, Al-Falasi S, Daoud SA. (2005). Chemopreventive activities of *Trigonella foenumgraecum*(Fenugreek) against breast cancer. Cell Biol Int 29:687–94.

Ashihara, H. (2008). Trigonelline (N-methylnicotinic acid) Biosynthesis and its Biological Role in Plants. Natural Product Communications, 3(9), 1934578X0800300906. https://doi.org/10.1177/1934578X0800300906

Aumsuwan, P., Khan, S. I., Khan, I. A., Ali, Z., Avula, B., Walker, L. A., Shariat-Madar, Z., Helferich, W. G., Katzenellenbogen, B. S., and Dasmahapatra, A. K. (2016). The anticancer potential of steroidal saponin, dioscin, isolated from wild yam (Dioscorea villosa) root extract in invasive human breast cancer cell line MDA-MB-231 in vitro. Archives of Biochemistry and Biophysics, 591, 98–110. https://doi.org/10.1016/j.abb.2015.12.001

Babar, S., Siddiqi, E. H., Hussain, I., Hayat Bhatti, K., and Rasheed, R. (2014). Mitigating the effects of salinity by foliar application of salicylic acid in fenugreek. *Physiology Journal*, *2014*.

Beygi, Z., Nezamzadeh, Z., Rabiei, M., and Mirakhorli, N. (2021). Enhanced accumulation of trigonelline by elicitation and osmotic stresses in fenugreek callus culture. Plant Cell, Tissue and Organ Culture (PCTOC), 147(1), 169–174. https://doi.org/10.1007/s11240-021-02055-w

Biamonti, G., and Caceres, J.F. (2009). Cellular stress and RNA splicing. Trends Biochem. Sci. 34, 146–153.

Bitarafan, Z., Asghari, H. R., Hasanloo, T., Gholami, A., Moradi, F., Khakimov, B., Liu, F., and Andreasen, C. (2019). The effect of charcoal on medicinal compounds of seeds of fenugreek (*Trigonella foenumgraecum* L.) exposed to drought stress. Industrial Crops and Products, 131, 323–329. https://doi.org/10.1016/j.indcrop.2019.02.003

Burke JJ and KA Orzech. (1988). Heat shock response in higher plants: a biochemical model. Plant Cell Environ. 11, 441–444.

Chaudhary, S., Chaudhary, P., and Patel, S. (2021). *Genomics, Transcriptomics, Proteomics and Metabolomics Approaches BT – Fenugreek: Biology and Applications* (M. Naeem, T. Aftab, and M. M. A. Khan (eds.); pp. 355–373). Springer Singapore. https://doi.org/10.1007/978-981-16-1197-1_16

Chaudhary, S., Chikara, S. K., Sharma, M. C., Chaudhary, A., Alam Syed, B., Chaudhary, P. S., Mehta, A., Patel, M., Ghosh, A., and Iriti, M. (2015). Elicitation of Diosgenin Production in *Trigonella foenumgraecum*(Fenugreek) Seedlings by Methyl Jasmonate. International Journal of Molecular Sciences, 16(12), 29889–29899. https://www.mdpi.com/1422-0067/16/12/26208

Ciura, J., Szeliga, M., Grzesik, M., and Tyrka, M. (2017). Next-generation sequencing of representational difference analysis products for identification of genes involved in diosgenin biosynthesis in fenugreek (*Trigonella foenum-graecum*). Planta, 245(5), 977–991. https://doi.org/10.1007/s00425-017-2657-0

Ciura, J., Szeliga, M., Grzesik, M., and Tyrka, M. (2018). Changes in fenugreek transcriptome induced by methyl jasmonate and steroid precursors revealed by RNA-Seq. Genomics, 110(4), 267–276. https://doi.org/10.1016/j.ygeno.2017.10.006

Curry, A. (2017). A 9,000-year love affair. *National Geographic*, *231*(2), 46.

Daliakopoulos, I. N., Tsanis, I. K., Koutroulis, A., Kourgialas, N. N., Varouchakis, A. E., Karatzas, G. P., and Ritsema, C. J. (2016). The threat of soil salinity: A European scale review. *Science of the Total Environment*, *573*, 727–739.

De, D., and De, B. (2011). Elicitation of diosgenin production in *Trigonella foenum-graecum*L. seedlings by heavy metals and signaling molecules. Acta Physiologiae Plantarum, 33(5), 1585–1590. https://doi.org/ 10.1007/s11738-010-0691-7

Duke, J. (1986). *Handbook of Legumes of World Economic Importance. Plenum Press, New York*. Wiley Online Library.

Fatima, A., Farid, M., Alharby, H. F., Bamagoos, A. A., Rizwan, M., and Ali, S. (2020). Efficacy of fenugreek plant for ascorbic acid assisted phytoextraction of copper (Cu); A detailed study of Cu induced morpho-physiological and biochemical alterations. *Chemosphere*, *251*, 126424. https://doi.org/10.1016/j.chem osphere.2020.126424

Freeman, B., and Beattie, G. (2008). An overview of plant defenses against pathogens and herbivores.

Gangoué-Pié boji J, Eze N, Ngongang Djintchui A, et al. (2009). The in vitro antimicrobial activity of some medicinal plants against beta-lactam-resistant bacteria. J Infect Dev Ctries 3:671–80.

Giovannetti, M., Avio, L., and Sbrana, C. (2013). Improvement of nutraceutical value of food by plant symbionts. In Ramawat K. and M. JM. (Eds.), Natural Products. Springer. https://doi.org/10.1007/978-3-642-22144-6_187

González-Chavira, M. M., Herrera-Hernández, M. G., Guzmán-Maldonado, H., and Pons-Hernández, J. L. (2018). Controlled water deficit as abiotic stress factor for enhancing the phytochemical content and adding-value of crops. Scientia Horticulturae, 234, 354–360. https://doi.org/10.1016/j.scie nta.2018.02.049

González-Lamothe, R., Mitchell, G., Gattuso, M., Diarra, M. S., Malouin, F., and Bouarab, K. (2009). Plant antimicrobial agents and their effects on plant and human pathogens. Int J Mol Sci, 10(8), 3400–3419. https://doi.org/10.3390/ijms10083400

Guédon, D., Brum, M., Seigneuret, J.-M., Bizet, D., Bizot, S., Bourny, E., Compagnon, P.-A., Kergosien, H., Quintelas, L. G., and Respaud, J. (2008). Impurities in herbal substances, herbal preparations and herbal medicinal products, IV. Heavy (toxic) metals. *Natural Product Communications*, *3*(12), 1934578X0800301232.

Gupta, B., and Huang, B. (2014). Mechanism of salinity tolerance in plants: physiological, biochemical, and molecular characterization. *International Journal of Genomics*, *2014*.

Gupta, S.C., Sharma, A., Mishra, M., Mishra, R., and Chowdhuri, D.K. (2010). Heat shock proteins in toxicology: how close and how far? Life Sci. 86, 377–384.

Hannan JM, Ali L, Rokeya B, et al. (2007). Soluble dietary fibre fraction of *Trigonella foenum-graecum*(fenugreek) seed improves glucose homeostasis in animal models of type 1 and type 2 diabetes by delaying carbohydrate digestion and absorption, and enhancing insulin action. Br J Nutr 97:514–21.

Haouala R, Hawala S, El-Ayeb A, et al. (2008). Aqueous and organic extracts of *Trigonella foenum-graecum*L. inhibit the mycelia growth of fungi. J Environ Sci (China) 20:1453–7.

Hibasami H, Moteki H, Ishikawa K, et al. (2003). Protodioscin isolated from fenugreek (*Trigonella foenum-graecum*L.) induces cell death and morphological change indicative of apoptosis in leukemic cell line H-60, but not in gastric cancer cell line KATO III. Int J Mol Med 11: 23–6.

Ho, T. T., Murthy, H. N., and Park, S. Y. (2020). Methyl Jasmonate Induced Oxidative Stress and Accumulation of Secondary Metabolites in Plant Cell and Organ Cultures. Int J Mol Sci, 21(3). https://doi.org/10.3390/ ijms21030716

Irankhah, S., Chitarra, W., Nerva, L., Antoniou, C., Lumini, E., Volpe, V., Ganjeali, A., Cheniany, M., Mashreghi, M., Fotopoulos, V., and Balestrini, R. (2020). Impact of an arbuscular mycorrhizal fungal inoculum and exogenous MeJA on fenugreek secondary metabolite production under water deficit. Environmental and Experimental Botany, 176, 104096. https://doi.org/10.1016/j.envexpbot.2020.104096

Iriti, M., and Faoro, F. (2009). Chemical diversity and defence metabolism: how plants cope with pathogens and ozone pollution. Int J Mol Sci, 10(8), 3371–3399. https://doi.org/10.3390/ijms10083371

Jasim, B., Thomas, R., Mathew, J., and Radhakrishnan, E. K. (2017). Plant growth and diosgenin enhancement effect of silver nanoparticles in Fenugreek (*Trigonella foenum-graecum*L.). Saudi Pharmaceutical Journal, 25(3), 443–447. https://doi.org/10.1016/j.jsps.2016.09.012

Jia, X., Sun, C., Zuo, Y., Li, G., Li, G., Ren, L., and Chen, G. (2016). Integrating transcriptomics and metabolomics to characterise the response of Astragalus membranaceus Bge. var. mongolicus

(Bge.) to progressive drought stress. *BMC Genomics*, *17*(1), 188. https://doi.org/10.1186/s12 864-016-2554-0

Jiang, J. X., Zhu, L. W., Zhang, W. M., and Sun, R. C. (2007). Characterization of Galactomannan Gum from Fenugreek (*Trigonella foenum-graecum*) Seeds and Its Rheological Properties. *International Journal of Polymeric Materials and Polymeric Biomaterials*, *56*(12), 1145–1154. https://doi.org/10.1080/009140 30701323745

Jhajhria, A., and Kumar, K. (2016). Fenugreek with its medicinal applications. *Int. J. Pharm. Sci. Rev. Res*, *41*(1), 194–201.

Joshi, J. G., and Handler, P. (1960). Biosynthesis of Trigonelline. Journal of Biological Chemistry, 235(10), 2981–2983. https://doi.org/10.1016/S0021-9258(18)64575-2

Kakani, R. K., and Anwer, M. M. (2012). 16 - Fenugreek. In K. V. B. T.-H. of H. and S. (Second E. Peter (Ed.), *Woodhead Publishing Series in Food Science, Technology and Nutrition* (pp. 286–298). Woodhead Publishing. https://doi.org/10.1533/9780857095671.286

Karmakar N. (2014) Response of fenugreek (*Trigonella foenum-graecumL*) seedlings under moisture and heavy metal stress with special reference to antioxidant system. African Journal of Biotechnology 13(3): 434–440

Kaur, L. (2016). Tolerance and accumulation of lead by fenugreek. *Journal of Agriculture and Ecology*, *1*(1), 22–34. Retrieved from https://journals.saaer.org.in/index.php/jae/article/view/7

Kawabata T, Cui MY, Hasegawa T, et al. (2011). Anti- inflammatory and anti-melanogenic steroidal saponin glycosides from Fenugreek (*Trigonella foenum-graecumL.*). seeds. Planta Med 77:705–10.

Kharkwal, H., Panthari, P., Pant, M. K., Kharkwal, H., Kharkwal, A. C., and Joshi, D. D. (2012). Foaming glycosides: a review. IOSR J Pharm, 2(5), 23–28.

Kidwai, M., and Dhull, S. B. (2021). Heavy Metals Induced Stress and Metabolic Responses in Fenugreek (*Trigonella foenum-graecumL.*) Plants. In *Fenugreek* (pp. 327–353). Springer.

Kochhar A, Nagi M, Sachdeva R. Proximate composition, available carbohydrates, dietary fibre and anti nutritional factors of selected traditional medicinal plants. Journal of Human Ecology. 2006 Mar 1;19(3):195–9.

Kumar, A., Singh, S., Gaurav, A. K., Srivastava, S., and Verma, J. P. (2020). Plant Growth-Promoting Bacteria: Biological Tools for the Mitigation of Salinity Stress in Plants. In *Frontiers in Microbiology* (Vol. 11). www.frontiersin.org/article/10.3389/fmicb.2020.01216

Levitt J. (1980). Responses of plants to environmental stresses, Vol. II. Water, radiation, salt and other stresses, 2nd Edn. New York Academic Press, New York, pp.606.

Mabrouk, B., S. B. Kâab, M. Rezgui, N. Majdoub, J. A. Teixeira da Silva, and L. B. B. Kâab. (2019). Salicylic Acid Alleviates Arsenic and Zinc Toxicity in the Process of Reserve Mobilization in Germinating Fenugreek (*Trigonella foenum-graecum* L.) Seeds." South African Journal of Botany 124: 235–43. doi: https://doi.org/10.1016/j.sajb.2019.05.020.

Mandegary, A., Pournamdari, M., Sharififar, F., Pournourmohammadi, S., Fardiar, R., and Shooli, S. (2012). Alkaloid and flavonoid rich fractions of fenugreek seeds (*Trigonella foenum-graecumL.*) with antinociceptive and anti-inflammatory effects. *Food and Chemical Toxicology*, *50*(7), 2503–2507. https://doi.org/10.1016/j.fct.2012.04.020

Manicas G (2002) Pests and diseases In: *Fenugreek: the genus Trigonella*. Petropoulos, G. A. CRC Press.

Meera R, Devi P, Kameswari B, et al. (2009). Antioxidant and hepatoprotective activities of *Ocimum basilicum* Linn. and *Trigonella foenum-graecumLinn*. against H2O2 and CCl_4 induced hepatotoxicity in goat liver. Indian J Exp Biol 47: 584–90.

Mikić, A. (2015). Brief but alarming reminder about the need for reintroducing 'Greek hay' (*Trigonella foenum-graecumL.*) in Mediterranean agricultures. Genetic Resources and Crop Evolution, 62(6), 951–958. https://doi.org/10.1007/s10722-015-0260-4

Mitra, A. et al. (2021). Plant Stress, Acclimation, and Adaptation: A Review. In: Gupta, D.K., Palma, J.M. (eds) Plant Growth and Stress Physiology. Plant in Challenging Environments, vol 3. Springer, Cham. https://doi.org/10.1007/978-3-030-78420-1_1

Mohammadifard, F., Tarakemeh, A., Moghaddam, M. *et al.* Bentonite Mitigates the Adverse Effects of Drought Stress in Fenugreek (*Trigonella foenum-graecum* L.). J Soil Sci Plant Nutr 22, 1098–1111 (2022). https://doi.org/10.1007/s42729-021-00718-3

Morani AS, Bodhankar SL, Mohan V, Thakurdesai PA. (2012). Ameliorative effects of standardized extract from *Trigonella foenum-graecumL*. seeds on painful peripheral neuropathy in rats. Asian Pac J Trop Med 5:385–90.

Morimoto RI. (1993). Cells in stress: Transcriptional activation of heat shock genes. Science. 259, 1409–1410.

Mukarram, M., Mohammad, F., Naeem, M., and Khan, M. (2021). Exogenous Gibberellic Acid Supplementation Renders Growth and Yield Protection Against Salinity Induced Oxidative Damage Through Upregulating Antioxidant Metabolism in Fenugreek (*Trigonella foenum-graceum* L.). In *Fenugreek* (pp. 99–117). Springer, Singapore.

Munns, R. (2005). Genes and salt tolerance: bringing them together. *New Phytologist*, *167*(3), 645–663.

Munns, R., and Tester, M. (2008). Mechanisms of salinity tolerance. *Annu. Rev. Plant Biol.*, *59*, 651–681.

Muraki, E., Hayashi, Y., Chiba, H., Tsunoda, N., and Kasono, K. (2011). Dose-dependent effects, safety and tolerability of fenugreek in diet-induced metabolic disorders in rats. *Lipids in Health and Disease*, *10*(1), 1–6.

Niknam R., Kiani H., Mousavi Z.E., Mousavi M. (2021) Extraction, Detection, and Characterization of Various Chemical Components of *Trigonella foenum-graecum*L. (Fenugreek) Known as a Valuable Seed in Agriculture. In: Naeem M., Aftab T., Khan M.M.A. (eds) Fenugreek. Springer, Singapore. https://doi.org/10.1007/978-981-16-1197-1_9

Pandian RS, Anuradha CV, Viswanathan P. (2002). Gastroprotective effect of fenugreek seeds (*Trigonella foenum-graecum*) on experi- mental gastric ulcer in rats. J Ethnopharmacol 81:393–7.

Pant, G., Hemalatha, S., Arjunan, S., Malla, S., and Sibi, G. (2013). Effect of heat stress in synthesis of heat shock proteins and antioxidative enzyme response in Trigonella foenum-graceum L. *Journal of Plant Sciences*, *1*(4), 51–56.

Parwez R., Nabi A., Mukarram M., Aftab T., Khan M.M.A., and M., N. (2021). Various Mitigation Approaches Applied to Confer Abiotic Stress Tolerance in Fenugreek (*Trigonella foenum-graecum*L.). In Naeem M., Aftab T., and K. M.M.A. (Eds.), Fenugreek. Springer. https://doi.org/10.1007/978-981-16-1197-1_8

Patel, K., Gadewar, M., Tahilyani, V., and Patel, D. K. (2012). A review on pharmacological and analytical aspects of diosgenin: a concise report. Natural Products and Bioprospecting, 2(2), 46–52. https://doi.org/10.1007/s13659-012-0014-3

Petit PR, Sauvaire YD, Hillaire-Buys DM, Leconte OM, Baissac YG, Ponsin GR, Ribes GR (1995) Steroid saponins from fenugreek seeds: extraction, purification, and pharmacological investigation on feeding behavior and plasma cholesterol. Steroids. 1995 Oct;60(10), 674-80. doi: 10.1016/0039-128x(95)00090-d PMID: 8539775.

Petropoulos, G. A. (2002). *Fenugreek: the genus Trigonella*. CRC Press.

Plants of the world online (POWO, 2021) https://powo.science.kew.org/results?f=andpage.size=480andq=trigonella

Prasad, B. C. N., Gururaj, H. B., Kumar, V., Giridhar, P., and Ravishankar, G. A. (2006). Valine Pathway Is More Crucial than Phenyl Propanoid Pathway in Regulating Capsaicin Biosynthesis in Capsicum frutescens Mill. Journal of Agricultural and Food Chemistry, 54(18), 6660–6666. https://doi.org/10.1021/jf061040a

Raju, J., Gupta, D., Rao, A. R., Yadava, P. K., and Baquer, N. Z. (2001). Trigonella*foenum-graecum*(fenugreek) seed powder improves glucose homeostasis in alloxan diabetic rat tissues by reversing the altered glycolytic, gluconeogenic and lipogenic enzymes. *Molecular and Cellular Biochemistry*, *224*(1–2), 45–51. https://doi.org/10.1023/a:1011974630828

Rayyan, S., Fossen, T., and Andersen, Ø. M. (2010). Flavone C-Glycosides from Seeds of Fenugreek, *Trigonella foenum-graecum*L. *Journal of Agricultural and Food Chemistry*, *58*(12), 7211–7217. https://doi.org/10.1021/jf100848c

Rosengarten Jr, F. (1969). The book of spices. *The Book of Spices*.

Saxena, S., Kakani, R., Sharma, L., Agarwal, D., John, S., and Sharma, Y. (2019). Effect of water stress on morpho-physiological parameters of fenugreek (*Trigonella foenum-graecum*L.) genotypes. Legume Research-An International Journal, 42(1), 60–65. https://doi.org/10.18805/LR-3830

Saxena, S. N., Kakani, R. K., Sharma, L. K., Agarwal, D., John, S., and Sharma, Y. (2017). Genetic variation in seed quality and fatty acid composition of fenugreek (*Trigonella foenum-graecum*L.) genotypes grown under limited moisture conditions. *Acta Physiologiae Plantarum,* 39(10), 218. https://doi.org/10.1007/s11738-017-2522-6

Shaban, N. S., Abdou, K. A., and Hassan, N. E.-H. Y. (2016). Impact of toxic heavy metals and pesticide residues in herbal products. *Beni-Suef University Journal of Basic and Applied Sciences*, 5(1), 102–106. https://doi.org/10.1016/j.bjbas.2015.10.001

Sheikhi, S., Ebrahimi, A., Heidari, P., Amerian, M. R., and Rashidi Monfared, S. (2021). The effect of 24-epi-brassinosteroid on the expression of some genes involved in the diosgenin biosynthetic pathway of fenugreek (*Trigonella foenum-graecum* L.) under high temperature stress. Iranian Journal of Field Crop Science, 52(2). https://doi.org/10.22059/IJFCS.2020.295770.654683

Sonawane, P. D., Pollier, J., Panda, S., Szymanski, J., Massalha, H., Yona, M., Unger, T., Malitsky, S., Arendt, P., Pauwels, L., Almekias-Siegl, E., Rogachev, I., Meir, S., Cárdenas, P. D., Masri, A., Petrikov, M., Schaller, H., Schaffer, A. A., Kamble, A., Giri, A. P., Goossens, A., and Aharoni, A. (2016). Plant cholesterol biosynthetic pathway overlaps with phytosterol metabolism. Nature Plants, 3(1), 16205. https://doi.org/10.1038/nplants.2016.205

Taiz, L., and Zeiger, E. (2003). *Plant physiology* (3rd ed.). Sinauer Associates, Inc. https://doi.org/10.1093/aob/mcg079

Tanwar, A., Aggarwal, A., Charaya, M. U., and Kumar, P. (2013). Enhancement of lead uptake by fenugreek using EDTA and Glomus mosseae. *Communications in soil science and plant analysis*, 44(22), 3431–3443.

Tong, Q., Qing, Y., Wu, Y., Hu, X., Jiang, L., and Wu, X. (2014). Dioscin inhibits colon tumor growth and tumor angiogenesis through regulating VEGFR2 and AKT/MAPK signaling pathways. Toxicology and Applied Pharmacology, 281(2), 166–173. https://doi.org/10.1016/j.taap.2014.07.026

Toscano, S., Trivellini, A., Cocetta, G., Bulgari, R., Francini, A., Romano, D., and Ferrante, A. (2019). Effect of Preharvest Abiotic Stresses on the Accumulation of Bioactive Compounds in Horticultural Produce. Frontiers in Plant Science, 10, 1212–1212. https://doi.org/10.3389/fpls.2019.01212

Twaij, B. M., and Hasan, M. N. (2022). Bioactive Secondary Metabolites from Plant Sources: Types, Synthesis, and Their Therapeutic Uses. International Journal of Plant Biology, 13(1), 4–14. https://doi.org/10.3390/ijpb13010003

Tramontano, W. A., and Jouve, D. (1997). Trigonelline accumulation in salt-stressed legumes and the role of other osmoregulators as cell cycle control agents. *Phytochemistry*, 44(6), 1037–1040. https://doi.org/10.1016/S0031-9422(96)00715-7

Verma SK, Singh SK, Mathur A. (2010). In vitro cytotoxicity of Calotropis procera and *Trigonella foenum-graecum* against human cancer cell lines. J Chem Pharm Res 2:861–5.

Vinod, K. R., Kumar, M. S., Anbazhagan, S., Sandhya, S., Saikumar, P., Rohit, R. T., and Banji, D. (2012). Critical issues related to transfersomes-novel vesicular system. *ACTA Scientiarum Polonorum Technologia Alimentaria*, 11(1), 67–82.

Wahid, A.,, Gelani, S., Ashraf, M., and Foolad, M.R. (2007). Heat tolerance in plants: An overview. Environ. Exp. Bot. 61, 199–223.

Wani, S. A., and Kumar, P. (2018). Fenugreek: A review on its nutraceutical properties and utilization in various food products. *Journal of the Saudi Society of Agricultural Sciences*, 17(2), 97–106.

Wu, S., Xu, H., Peng, J., Wang, C., Jin, Y., Liu, K., Sun, H., and Qin, J. (2015). Potent anti-inflammatory effect of dioscin mediated by suppression of TNF-α-induced VCAM-1, ICAM-1and EL expression via the NF-κB pathway. Biochimie, 110, 62–72. https://doi.org/10.1016/j.biochi.2014.12.022

Yadav, R., Kaushik, R., and Gupta, D. (2011). The health benefits of *Trigonella foenum-graecum*: a review. *Int J Eng Res Appl*, 1(1), 32–35.

Yadav, U. C. S., and Baquer, N. Z. (2014). Pharmacological effects of *Trigonella foenum-graecum* L. in health and disease. *Pharmaceutical Biology*, 52(2), 243–254. https://doi.org/10.3109/13880209.2013.826247

Yao, D., Zhang, B., Zhu, J., Zhang, Q., Hu, Y., Wang, S., Wang, Y., Cao, H., and Xiao, J. (2020). Advances on application of fenugreek seeds as functional foods: Pharmacology, clinical application, products, patents and market. *Critical Reviews in Food Science and Nutrition*, 60(14), 2342–2352.

Yoshikawa M, Murakami T, Komatsu H, et al. (1997). Medicinal Foodstuffs: IV. Fenugreek seeds (1): Structures of trigoneosides Ia, Ib, IIb, IIIa and IIIb new furostanol saponins from the seeds of Indian Trigonella foenum- graecum L. Chem Pharmacol Bull 45:81–7.

Zamani, Z., Amiri, H., and Ismaili, A. (2020). Improving drought stress tolerance in fenugreek (*Trigonella foenum-graecum*) by exogenous melatonin. Plant Biosystems-An International Journal Dealing with all Aspects of Plant Biology, 154(5), 643–655. https://doi.org/10.1080/11263504.2019.1674398

Zandi, P., Basu, S. K., Khatibani, L. B., Balogun, M. O., Aremu, M. O., Sharma, M., Kumar, A., Sengupta, R., Li, X., and Li, Y. (2015). Fenugreek (*Trigonella foenum-graecum* L.) seed: a review of physiological and biochemical properties and their genetic improvement. *Acta Physiologiae Plantarum*, *37*(1), 1–14.

Zhang, S., Fan, C., Wang, Y., Xia, Y., Xiao, W., and Cui, X. (2018). Salt-tolerant and plant-growth-promoting bacteria isolated from high-yield paddy soil. *Canadian Journal of Microbiology*, *64*(12), 968–978.

Zhou, C., Li, X., Zhou, Z., Li, C., and Zhang, Y. (2019). Comparative Transcriptome Analysis Identifies Genes Involved in Diosgenin Biosynthesis in *Trigonella foenum-graecum* L. Molecules, 24(1), 140. https://www.mdpi.com/1420-3049/24/1/140

Index

A

ABC transporter G family member (*ABCT-G*), 340
Abiotic stress, 393
  agricultural/irrigation practices, 105–106
  citrus culture/cold stress, 97–99, 104
  citrus drought stress, 102–103
  citrus lemon, 97
  citrus membrane alterations, to cold, 99–100
  citrus salinity stress/responsive mechanisms, 101
  citrus's molecular reactions to cold stress, 99
  cold/freezing stresses, 7
  Cuminum cyminum, 155
  drought, 5–6
  drought stress resistance mechanisms, 103
  flooding, 6
  genetic approaches, to improve salinity stress, 102
  heat stress, 6–7
  heavy metals stress, 8, 106–108
  ion toxicity interplay, under salinity stress, 102
  osmoprotectors, under cold stress, 100–101
  photosynthesis/photo-inhibition, 100
  physiological response, 99
  salinity/drought stress highlights, 3–4, 101, 104
  water balance, 100
Abiotic stressors, 3, 26, 36, 154, 167, 300
Aboukhalid, K., 291
Abscisic acid (ABA), 26, 65, 109, 350, 394
  ABA deficient 4 (ABA4), 27
  ABA signal transduction, 68
  accumulation, 66
  drought stress-induced synthetic, 70, 351
  phytohormone, 47
  signal transduction, 26
  water scarcity, 377
Acetyl-CoA carboxylation (ACCase) gene, 71
Achillea fragratissima, 227
Aegiceras corniculatum, 304
Aerobic metabolism, 354
Agricultural practices, drought and salinity stress, 105
Ait-El-Mokhtar, M., 317, 326
Alam, M., 337–339
Alchornea triplinervia, 396
Al-Harrasi, I., 319
Al-Khateeb, S.A., 315
Allicin, 22
Alliin, 22
Allium cepa, 2
Allium crops, 28
Allium sativum, 21, 22, 32
Aloe saponaria extract (Ae), 170
1,4-Alpha-glucan-branching enzyme 1 (A0A2H5PA48), 100
Alpha-linolenic acid (ALA), 356
Alraey, D. A., 291
Al-sensitive genotype, 87

Alternaria alternata, 203
Alternaria leaf, 302
Alternaria porri, 8
Alternaria radicina, 303
Aly, A. A., 4
Al-Yassin, A., 102
Amino acid proline, 315
1-Aminocycloprpane-1-carboxylic acid (ACC), 102
Amino deoxychorismate synthase, 423
Amirjani, M. R., 25
Amorim, J. R. d. A., 24
Amplified fragment length polymorphism (AFLP) markers, 109
Analgesic, black seed, 224
Anethum graveolens, 54
  abiotic stress, effect of, 50–52
  agrochemical application effects, 53
  bioactive compounds, pharmacological effects of, 48–49
  biotic/abiotic stress effects, 49, 50
  botanical description, 48
  carbon dioxide enrichment, 54
  fresh/dry leaves of, 48
  global/ecological distribution of, 48
  light quality, 54
  medicinal applications of, 48
  overview of, 47
  pharmacological metabolite production, 52–54
  phytochemical composition, 49
  phytohormonal effects, 53
  plant growth-promoting rhizobacteria/mycorrhizal effects, 52
  salinity stress, 51
  seeds, leaves and inflorescence of, 49
  silicon, exogenous application of, 51
  stressful conditions, 47
  transcriptomics/proteomics of, 54
Anethum graveolens L., 48
Anjani, K., 375
Anthocyanins, 164
  abundance of, 98
  non-enzymatic antioxidants, 69
Anthracnose (*Colletotrichum gloeosporioides*), 174
Anti-cancer, black seed, 224
Anti-diabetic, black seed, 224
Anti-immunomodulator, black seed, 224
Anti-inflammatory, black seed, 224
Anti-microbial, black seed, 224
Antimicrobial peptides, innate immune defenses, 114
Antioxidants, 27
  defenses, 103
  green leaves/green or dry onion, 1
Anti-redox, black seed, 224
Aoufous complex, 320
AP2/ERF family genes, 306
Aphis (Toxoptera) citricidus, 116
Apium virus Y (ApVY), 303

Printed in the United States
by Baker & Taylor Publisher Services